Energy Optimization in
Process Systems and Fuel Cells

Energy Optimization in Process Systems and Fuel Cells

Second Edition

By

Stanisław Sieniutycz

Warsaw University of Technology,
Faculty of Chemical and Process Engineering,
Warsaw, Poland

Jacek Jeżowski[†]

Rzeszów University of Technology,
Department of Chemical and Process Engineering,
Rzeszów, Poland

ELSEVIER

AMSTERDAM • BOSTON • HEIDELBERG • LONDON • NEW YORK • OXFORD
• PARIS • SAN DIEGO • SAN FRANCISCO • SINGAPORE • SYDNEY • TOKYO

Elsevier
Radarweg 29, PO Box 211, 1000 AE Amsterdam, The Netherlands
The Boulevard, Langford Lane, Kidlington, Oxford OX5 1GB, UK
225 Wyman Street, Waltham, MA 02451, USA
525 B Street, Suite 1800, San Diego, CA 92101-4495, USA

Second Edition

British Library Cataloguing in Publication Data
A catalogue record for this book is available from the British Library

Library of Congress Cataloging-in-Publication Data
A catalog record for this book is available from the Library of Congress

ISBN: 978-0-08-098221-2

For information on all Elsevier publications
visit our web site at store.elsevier.com

This book has been manufactured using Print On Demand technology. Each copy is produced to order and is limited to black ink. The online version of this book will show color figures where appropriate.

Working together to grow
libraries in developing countries

www.elsevier.com | www.bookaid.org | www.sabre.org

ELSEVIER   BOOK AID
International   Sabre Foundation

Transferred to Digital Printing in 2013

Contents

Preface

Energy systems are optimized in order to satisfy several primary goals. The first goal (Chapters 3–10 of this book) requires searching for limiting values of some important physical quantities, e.g. limiting power, minimum heat supply, maximum final concentration of a key component, etc. The second goal (Chapters 8 and 11), perhaps the most practical, applies profit or cost analyses to find economically (or exergo-economically) optimal solutions. The third goal (Chapters 12–20) pursues optimal solutions assuring the best system integration.

Optimizations toward energy limits arise in various chemical and mechanical engineering systems (heat and mass exchangers, thermal networks, energy converters, recovery & storage units, solar collectors, separators, chemical reactors, etc.). Associated energy problems are those with conversion, generation, accumulation and transmission of energy. These problems are treated by mathematical methods of optimization such as: nonlinear programming with continuous and mixed variables, dynamic programming, and Pontryagin's maximum principles, in their discrete and continuous versions. The considered processes occur in a finite time and with equipment of finite dimension. Penalties on rate and duration and optimal performance criteria of potential type (obtained within exergy or economic approaches) are effective.

In Chapters 3–10 we define and analyze thermodynamic limits for various traditional and work-assisted processes with finite rates that are important in chemical and mechanical engineering, with a few excursions into ecology and biology. The thermodynamic limits are expressed either as maxima of power or in terms of classical or generalized exergies, where the latter include some rate penalties. We consider processes with heat, work and mass transfer that occur in equipment of finite dimensions and define energy limits of various ranks for these processes. In particular, we show that the problem of energy limits is a purely physical problem that may be stated without any relation to economics. The considered processes include heat-mechanical operations (and are found in heat and mass exchangers), thermal networks, energy converters, energy recovery units, chemical reactors, and separation units. Simple exergo-economics fluidized systems are investigated as those preserving large transfer or reaction area per unit volume. Our analysis is based on the condition that in order to make the results of thermodynamic analyses applicable in industry, it is the thermodynamic limit, not the maximum of thermodynamic efficiency, which must be overcome for prescribed process requirements. Our approach analyzes the physical problem of energy limits as a new direction in nonequilibrium thermodynamics of practical devices in which the optimal control theory is both essential and helpful. Control processes of engine type and heat-pump type are considered, both with pure heat exchange and with simultaneous heat and mass exchange. Links with exergy definition in reversible systems and classical problems of extremum work are pointed out. Practical problems and illustrative examples are selected in order to give an outline of applications. Considerable simplification in analysis of complicated thermal/chemical machines is achieved when some special controls (Carnot variables, T' and μ'—see Chapters 3 and 9) are applied. In particular, the description of classical and work-assisted heat & mass exchangers is unified.

Conclusions may be formulated regarding limits on mechanical energy yield in practical nonlinear systems. It is shown that these limits differ for power generated and power consumed, and that they depend on global working parameters of the system (e.g. total number of heat transfer units, imperfection factor of power generators, average process rates, number of process stages, etc.). New results constitute, between others, limits on multistage production (consumption) of power in nonlinear thermal and chemical devices. They characterize dynamical extrema of power yield (or consumption) for a finite number of stages or a finite time of the resource exploitation. Frequently, these systems are governed by nonlinear kinetics, as in the case of radiation or chemical/electrochemical engines. The generalization of this problem takes into account contribution of transport processes and imperfections of power generators, and includes the effect of drying out the resources of energy and matter. These solutions provide design factors for energy generators that are stronger than the familiar thermostatic bounds (i.e. classical limits for the energy transformation).

Since electrochemical power generation can occur in both thermochemical and biological systems, in this volume fuel cells are treated jointly for chemical and biological systems (Chapter 10). First, the power-efficiency thermodynamics of classical chemical reaction driven fuel cells is developed and power limits are analyzed. Next, two categories of biological fuel cells are considered: enzymatic and microbiological. The former, because of the biocatalyst immobilization, offer high current densities and possibilities of the device miniaturization. The latter, exploiting living organisms, can work for prolonged time and use complex substrates. Bacteria are regarded as self-reproducing catalysts, the property which assures continual production of power. For developing bio-systems, selected evolution examples are briefly considered to determine how system properties change when an evolving organism increases its number of elements (organs or limbs). In some biological systems, evolutionary growth in the number of organs or limbs is accompanied by catastrophes caused by abrupt changes in qualitative properties of the organism or its part. These examples substantiate Williston's law known in the evolution theory, which predicts the evolutionary tendency to reduce number of similar organs or limbs along with a simultaneous modification (specialization) of elements retained by the organism.

This book applies optimization approaches found in second law analysis, finite time thermodynamics, entropy generation minimization, exergo-economics, and system engineering to simulation and optimization of various energy processes. This book promotes systematic thermo-economic methodology and its underlying thermodynamic and economic foundations in various physical and engineering systems. It is a modern approach to energy systems which applies methods of optimization and thermal integration to obtain optimal controls and optimal costs, sometimes in the form of certain potentials depending on the process state, duration and number of stages. The approach, which is common for both discrete and continuous processes, derives optimal solutions by using mathematical models coming from thermophysics, engineering, electrochemistry and economics. It deals with thermodynamic or thermo-economic costs expressed in terms of exergy input, dissipated exergy, or certain extensions of these quantities including time or rate penalties, investment and other economic factors.

When a practical device, apparatus or a machine performs certain engineering tasks (or a 'duty') it is often reasonable to ask about a corresponding lower bound on energy consumption or, if applicable, an upper bound on energy production. The first case occurs in separators, including dryers, the second – in energy generators or engines. Regardless of the economic cost (that may be in some cases quite high or even exceeding an acceptable value), these factors —technical limits—inform an engineer about the system's potential; that of minimum necessary consumption or that of maximum possible yield. Thus, they don't represent economically optimal solutions but rather define limiting extreme possibilities of the system. Technical limits, in particular thermodynamic ones, are important factors in engineering design. In fact, no design is possible that could violate these limits without changes in the system's duty. Classical thermodynamics is capable of providing energy limits in terms of exergy changes. However, they are often too distant from reality; real energy consumption can be much higher than the lower bound and/or real energy yield can be much lower than the upper bound. Yet, by introducing rate-dependent factors, irreversible thermodynamics offer enhanced limits that are closer to reality.

Limits for finite resources are associated with the notion of exergy. They refer either to a sequential relaxation of a resource to the environment (engine mode), or to resources being upgraded in the process going in the inverse direction (heat-pump mode). To deal with these dynamical processes one must first find a general formula for the converter's efficiency and, then, evaluate a limiting work via an optimization. In an irreversible case this limiting work is an extension of the classical work potential. The real work to be optimized is a cumulative effect obtained from a system composed of: a resource fluid at flow (traditional medium or radiation), a set of sequentially arranged engines, and an infinite reservoir.

During the approach to the equilibrium, power is released in sequential engine modes; during the departure—it is supplied in heat-pump modes. In an engine mode a fluid's potential (e.g. temperature T) decreases, to the bath temperature. In a heat-pump mode direction is inverted and the fluid is thermally upgraded. The work (W) delivered in the engine mode is positive by assumption. In the heat-pump mode W is negative, or positive work ($-W$) is supplied to the system. To calculate a generalized exergy, optimization problems are solved, for maximum of work yield [max W] and for minimum of the work supply [min ($-W$)]. The generalized exergy emerges as a function of usual thermal coordinates and a rate or dissipation index, h_σ (in fact, the Hamiltonian value of the extreme process). In some examples we focus on limits evaluated for the work from solar radiation. Limits related analyses answer then the question about a maximum fraction of solar energy that can be converted into mechanical energy. They lead to estimates of maximum work released from a radiation engine and minimum work supplied to a heat pump. Knowing the latter limit, one can calculate lowest supply of solar or microwave energy to a dryer or other separator.

Classical exergy defines bounds on work delivered from (supplied to) slow, reversible processes (Berry et al. 2000). For such bounds the magnitude of the work delivered during a reversible approach to the equilibrium is equal to that of the work supplied when initial and final states are inverted, i.e. when the second process reverses to the first. Yet, bounds predicted by generalized exergies (i.e. those for finite rate processes) are not reversible. In fact, they are different for engine and heat-pump modes. While the reversibility property is lost for a generalized exergy, its bounds are stronger than classical thermostatic bounds.

A remarkable result discussed in this book, is a formal analogy between expressions describing entropy production in operations with thermal machines and in those in traditional heat and mass exchangers, provided that both sorts of operations are described in terms of a suitable control variable. In fact, the analogy emerges when the modeling involves a special control variable T', called Carnot temperature, which represents the joint effect of upper and lower temperatures of the circulating medium, T_1' and T_2'. Since these temperatures are linked by the internal entropy balance (through the power generating part of the machine), there is effectively only one free control, which is just Carnot temperature T' (Chapter 3). When mass transfer is included (Chapter 9), a similar control can be introduced which is Carnot chemical potential μ', a quantity suitable in optimization of diffusion and chemical engines. A detailed formal treatment of these issues is given in a recent publication, (Sieniutycz, 2011f).

This book fills a gap in teaching the process optimization and process integration in energy systems by using scientific information contained in thermodynamics, kinetics, economics and systems theory. Despite numerous works on energy and process integration in real systems (of finite size) appearing regularly in many research journals, no synthesizing treatment linking energy systems optimization with process integration exists so far in the literature. In this book, optimization problems arising in various chemical and mechanical engineering systems (heat and mass exchangers, thermal and water networks, energy converters, recovery units, solar collectors, and chemical separators) are discussed. The corresponding processes run with conversion, generation, accumulation and transmission of energy or substance, and their optimization requires advanced mathematical methods of discrete and continuous optimization and system integration. The methods commonly applied are: nonlinear programming, dynamic programming, variational calculus, Hamilton-Jacobi-Bellman theory, Pontryagin's maximum principles and methods of process integration. Synthesis of thermodynamics, kinetics and economics is achieved through exergo-economic and thermo-kinetic approaches, generalizing classical thermodynamic approaches by taking into account constrained rates, finite sizes of apparatus, environmental constraints, and economic factors.

Heat energy and process water integration within a total site significantly reduces production costs; in particular, costs of utilities commonly applied in process systems such as in the chemical industry and relative branches including waste treatment facilities for environmental protection. However, the presented approaches are also aimed at the total annual cost of subsystems of interest. The integration (Chapters 12–20) requires systematic approaches to design and optimize heat exchange and water networks (HEN and WN). The presentation of these issues, in this book, starts with basic insight-based Pinch Technology for heat recovery to provide problem understanding and, also, short-cut solution techniques. Then systematic, optimization-based, sequential, and simultaneous approaches to design HEN and WN are described. The approaches show how to identify application-specific constraints and requirements and incorporate them into solutions. They also clarify available computational methods. The authors focus on a class of methods that are founded on superstructure concepts. This is the result of their opinion, that such approaches are able to deal efficiently with complex industrial cases. Suitable optimization techniques should be used to achieve the aims. In the case of HEN design problems, special consideration is given to the targeting stage because of its importance at the various levels of the complex process of system design. Also, targets for HEN can be calculated for large scale industrial cases using widely available computer aids. In particular, an advanced simultaneous approach is addressed that generates optimal heat load distribution with regard to total cost. This outcome can be used to devise the final design of HEN in some cases. Selected, advanced methods for HEN synthesis and retrofit are presented. The material here is based on a thorough review of recent literature, with some innovative approaches developed by the authors. In particular a method is given to retrofit a HEN design consisting of standard heat exchangers. The approach employs Genetic Algorithms. In the case of WN design, an innovative approach based on the stochastic optimization method is described. The approach accounts for both grass roots and revamp design scenarios. It is also applicable for calculating targets such as minimum freshwater usage for various raw water sources. Some approaches for HEN and WN design are solved with stochastic/meta-heuristic optimization techniques. The tools are applicable for general nonlinear optimization problems. Hence, in Chapter 1, a separate Section 1.6 contains detailed procedures for these optimization techniques such as Adaptive Random Search, Simulated Annealing, and Genetic Algorithms.

To date, no complete synthesizing treatment of energy systems optimization has been published—in spite of numerous works on energy appearing regularly in many research journals. Yet, a list of some earlier books on optimization or thermal integration can be quoted: (Aris 1961,1964; Beveridge and Schechter 1970; Rosenbrock and Storey 1966; Floudas 1995; Shenoy 1995; El-Halwagi 1997; Biegler, Grossman and Westerberg 1997; Edgar, Himmelblau and Lasdon 2001; Peters, Timmerhaus and West 2003; Smith 2005; Seider, Seader and

Lewin 2004). More recently, several original books appeared (Feidt, 2006; Jaluria 2007; Dincer and Rosen 2007; Mench 2008; Logan 2008), which include, between others, electrochemical systems and fuel cells. While all these books are of considerable value, they do not contain important recent results achieved in the fields of energy optimization and process integration. New results have been obtained for thermal and solar engines, thermal and water networks, and process separators. More recent books amongst those cited above, concentrate on specific topics, such as heat integration (Shenoy, 1995) or mass integration (El-Halwagi 1997, 2005), theory and application of deterministic optimization techniques (Floudas 1995 and Edgar et al. 2001). Some are textbooks for undergraduate students with only basic information on advanced design approaches (e.g. Smith 2005). Some concentrate primarily on simulator application, (e.g. Seider et al. 2004). Though these references are relatively recent, they do not entirely cover new developments on process integration.

Since the important lifecycle problems are omitted in the present book and majority of books cited above, we refer the reader to the excellent paper by MacLean and Lave (2003) on evaluating automobile fuel/propulsion system technologies. The authors of that paper examine the life cycle implications of a wide range of fuels and propulsion systems that could power cars and light trucks in the US and Canada over the next two to three decades, including fossil fuels, hydrogen and electricity, hybrid electric propulsion options, and fuel cells. They review recent studies to evaluate the environmental, performance, and cost characteristics of fuel/propulsion technology combinations that are currently available or will be available in the next few decades.

While nonlinear programming, optimal control, and system integration techniques are basic mathematical tools, this book addresses applied energy problems in the context of the underlying thermodynamics and exergoeconomics. This book can be used as a basic or supplementary text in courses on optimization and variational calculus, engineering thermodynamics, and system integration. As a text for further research, it should attract engineers and scientists working in various branches of applied thermodynamics and applied mathematics, especially those interested in the energy generation, conversion, heat & mass transfer, separations, optimal control, fuel cells, biosystems, etc. Applied mathematicians will welcome a relatively new approach to the theory of discrete processes involving an optimization algorithm with a Hamiltonian constant along the discrete trajectory and its generalization for systems nonlinear in time intervals (which may arise for some special discretizing of underlying continuous models). They should also appreciate numerous commentaries on convergence of discrete dynamic programming algorithms to viscosity solutions of Hamilton–Jacobi–Bellman equations.

This book can be used as a basic or supplementary text in the following courses:

- optimization and variational methods in engineering (undergraduate)
- technical thermodynamics and industrial energetics (undergraduate)

- alternative and unconventional energy sources (undergraduate)
- heat recovery and energy savings (graduate)
- separation operations and systems (graduate)
- thermo-economics of solar energy conversion (graduate)
- thermodynamics of imperfect fuel cell systems (graduate)

The content organization of this book is as follows: in Chapters 1 and 2, an outline of static and dynamic optimization is presented, focusing on methods applied in the examples considered in the book. Chapter 3 treats power limits for steady thermal engines and heat pumps. Chapter 4 develops power optimization theory for dynamic systems modelled as multistage cascades; cascade models are applied to handle the dynamical behavior of engines and heat pumps when the resource reservoir is finite, and the power generation cannot be sustained at a steady rate. Chapters 5–7 analyze various dynamical energy systems characterized by nonlinear models, in particular radiation systems. In Chapter 8, thermally driven and work-assisted drying operations are considered; in particular, the use of an irreversible heat-pump to upgrade a heating medium entering the dryer is described. Chapter 9 treats optimal power yield in chemical reactors, and Chapter 10—efficiencies and power limits in electrochemical and biological fuel cells, and also some related evolutionary systems. Chapter 11 outlines system analyses in thermal and chemical engineering and contains a discussion of the issues at the interface of energy limits, exergo-economics and ecology. Various aspects of the process integration are treated in Chapters 12–20. First, in Chapter 12, introductory remarks are given on heat and water integration in a context of total site integration. A brief literature overview is also supplied. The next chapter addresses the basics of heat Pinch Technology. Chapter 14 gives the foundation for the targeting stage of HEN design. The following chapters address the most important targets in sequence: first maximum heat recovery with systematic tools in Chapter 15; then, in Chapter 16, the minimum number of units and minimum area targets. Approaches for simultaneous targeting are analyzed in Chapter 17; the HEN design problem is dealt with in two chapters: grass roots design in Chapter 18, and HEN retrofit in Chapter 19. Finally, Chapter 20 contains the description of both the insight based and systematic approaches for WN targeting and design.

Acknowledgments

Acknowledgments constitute the last and most pleasant part of this preface. The authors express their gratitude to the Polish Committee of National Research (KBN); under the auspices of which a considerable part of their own research discussed in the book—was performed, in the framework of two grants: grant 3 T09C 063 16 (Thermodynamics of development of energy systems with applications to thermal machines and living organisms) and grant 3 T09C 02426 (Nonequilibrium thermodynamics and optimization of chemical reactions in physical and biological systems). Chapter 9 on chemical reactors was prepared in the framework of the current grant N N208 019434 entitled "Thermodynamics and optimization of chemical and electrochemical energy generators with applications to fuel cells" supported by Polish Ministry of Science. A critical part of writing any book is the process of reviewing, thus the authors are very much obliged to the researchers who patiently helped them read through various chapters and who made valuable suggestions. In preparing this book, the authors received help and guidance from: Viorel Badescu (Polytechnic University of Bucharest, Romania), Miguel J. Bagajewicz, (University of Oklahoma, USA.), R. Steven Berry (University of Chicago, USA.) Roman Bochenek, Alina Jeżowska and Grzegorz Poplewski (Rzeszow University of Technology), Lingen Chen (Naval University of Engineering, Wuhan, China), Guoxing Lin (Physics, Xiamen University, P. R. China), Günter Wozny (TU Berlin, Germany), Vladimir Kazakov (University of Technology, Sydney), Andrzej Kraslawski (Lappeenranta University of Technology, Finnland), Piotr Kuran, Artur Poświata and Zbigniew Szwast (Warsaw University of Technology), Elżbieta Sieniutycz (University of Warsaw), Anatolij M. Tsirlin (Pereslavl-Zalessky, Russia), Andrzej Ziębik (Silesian University of Technology, Gliwice), and Anita Koch (Acquisition Editor) and Anusha Sambamoorthy (Project Manager) from Elsevier. Special thanks are due to Professor A. Ziębik who agreed that the authors exploit, practically in extenso, his 1996 paper System analysis in thermal engineering, published in Archives of Thermodynamics 17: 81–97, which now constitutes an essential part of Chapter 11 of this book. We also acknowledge the consent of the Institute of Fluid Flow Machinery Press in Gdańsk, Poland, the Publisher of Archives of Thermodynamics. Finally, appreciation goes to the book production team in Elsevier for their cooperation, patience, and courtesy.

1 Brief review of static optimization methods

1.1. INTRODUCTION: SIGNIFICANCE OF MATHEMATICAL MODELS

All rational human activity is characterized by continuous striving for progress and development. The tendency to search for the best solution under defined circumstances is called optimization—in the broad sense of the word. In this sense, optimization has always been a property of rational human activity. However, in recent decades, the need for methods which lead to an improvement of the quality of industrial and practical processes has grown stronger, leading to the rapid development of a group of optimum-seeking mathematical methods, which are now collectively called methods of optimization. Clearly, what brought about the rapid development of these methods was progress in computer science, which made numerical solutions of many practical problems possible. In mathematical terms, optimization is seeking the best solution within imposed constraints.

Process engineering is an important area for application of optimization methods. Most technological processes are characterized by flexibility in the choice of some parameters; by changing these parameters it is possible to correct process performance and development. In other words, decisions need to be made which make it possible to control a process actually running. There are also decisions that need to be made in designing a new process or new equipment. Thanks to these decisions (controls) some goals can be reached. For example, it may be possible to achieve a sufficiently high concentration of a valuable product at the end of a tubular reactor at minimum cost; or in another problem, to assure both a relatively low decrease of fuel value and a maximum amount of work delivered from an engine. How to accomplish a particular task is the problem of control in which some constraints are represented by transformations of the system's state and others by boundary conditions of the system. If this problem can be solved, then usually a number of solutions may be found to satisfy process constraints. Therefore, it is possible to go further and require that a defined objective function (process performance index) should be reached in the best way possible, for example, in the shortest time, with the least expenditure of valuable energy, minimum costs, and so on. An optimization problem emerges, related to the optimal choice of process decisions.

In testing a process it is necessary to quantify the related knowledge in mathematical terms; this leads to a mathematical model of optimization which formulates the problem in the language of functions, functionals, equations, inequalities, and so on. The mathematical model should be strongly connected with reality, because it emerges from and finds its application in it. However, the mathematical model often deals with very abstract ideas; thus, finding an optimal solution requires a knowledge of advanced methods. We shall present here only selected methods, suited to the content of this book.

In technology, practically every problem of design, control, and planning can be approached through an analysis leading to the determination of the least (minimum) or the greatest (maximum) value of some particular quantity – physical, technological or economic – which is called the optimization criterion, performance index, objective function or profit function. The choice of decisions (also of a physical, technological or economic nature), which can vary in a defined range, affects the optimization criterion, the criterion being a measure of the effectiveness of the decisions. The task of optimization is to find decisions to assure the minimum or maximum value of the optimization criterion.

The existence of decisions as quantities whose values are not prescribed, but rather chosen freely or within certain limits, makes optimization possible. Optimization – understood as an activity leading to the achievement of the best result under given conditions, always an inevitable part of human activity – only acquired a solid scientific basis when its meanings and methods were described mathematically. Thanks to recent computational techniques and the use of high-speed computing, optimization research has gained economic ground, and the range of problems solved has increased enormously. Apart from the use of digital computers, many optimization problems have been solved by using analog or hybrid computers.

In this book we assume that each optimization problem can be represented by a suitable mathematical model. Clearly, the mathematical model can simulate the behavior of a real system in a more or less exact way. Whenever good agreement is observed between the behavior of a real system and its model, optimization results can be used to improve the performance of the system. However, cases may exist where the process data are not reliable enough and over simplifications may occur in construction of the model; in these cases the results of optimization cannot be accepted without criticism. Clearly, where there is high data inaccuracy or model invalidity, optimization results will not be reliable. However, the models and the data which are now used for optimization are, in fact, the same as those used in design and process control. In many cases these models are well established, so the related optimization is desirable.

The technical implementation of a correct optimization solution may often prove to be difficult. In these cases optimization results can still be useful to expose extremal or limiting possibilities of the system from the viewpoint of an accepted optimization criterion. For example, an obtained limit can be represented by an upper (lower) bound on the amount of electrical or mechanical energy delivered from (supplied to) the system. Real system characteristics, which lie below (above) the limits predicted by the optimal solution, can sometimes be taken into account by considering suboptimal solutions. The latter may be easier to accomplish than the optimal solution.

The mathematical model of optimization is the system of all the equations and inequalities that characterize the process considered, including the optimization criterion. The model makes it possible to determine how the optimization criterion changes with variations in the decisions. In principle, mathematical models can be obtained in two ways. On the basis of physical laws so-called analytical models are formulated. After identification of the system experimental models are determined, often based on regression analysis. Sometimes they are represented by polynomial equations linking outputs and inputs of the system.

In design, analytical models are usually used because only they can make possible the wide extrapolation of data that is necessary when the process scale is changed. In analytical models the number of unknown coefficients to be determined is usually much lower than in empirical models. However, when controlling existing processes, empirical models are still, quite frequently, applied.

If an optimization is associated with planning and doing experiments and its partial purpose is finding the data which help to determine optimal decisions, we are dealing with *experimental optimization*. If a mathematical description is used, which takes into account the process, its environment and a control action, we are dealing with *analytical optimization*. This book deals with analytical optimization. The models we use in most of the chapters are deterministic ones. Yet, since some results in the field of energy limits may be linked with random processes, uncertainty and simulated annealing criteria (Nulton and Salamon, 1988; Harland and Salamon, 1988; Andresen and Gordon, 1994), the final part of this chapter discusses basic techniques of stochastic optimization.

In the working state of a technological process the problem of adaptation of the mathematical model plays an important role. Adaptation should always be made whenever variations are observed in uncontrolled variables of the system, which normally should remain constant. For slowly varying changes, adaptation of the continuous type is possible. Fast varying changes require periodic adaptation for the averaged values of changes which are regarded as noise. Optimization *on line* is carried out simultaneously with adaptation of the model; in this optimization a control action is accomplished directly by the computer. Yet, a computer-aided control involves optimization *off line*.

The optimization criterion (performance index) is an important quantity that appears in the mathematical model in the form of a function or a functional. Usually this is an explicit and analytical form. The choice of optimization criterion in an industrial process must be the subject of very accurate analysis, which often involves both technological and economic terms. This is because the definition of the criterion has an important effect on the problem's solution and expected improvements.

Along with the performance index there appear in the mathematical model some equality and inequality equations (algebraic, differential, integral, etc.) which characterize *constraints* imposed on the process. Both constraints and performance index can contain *decision variables* or *controls*—adjustable variables that an engineer or a researcher can influence.

Some other variables can also appear both in the constraining equations and in the performance index. These are called *uncontrolled variables*. These variables are determined by certain external factors independent of the observer (the composition of a raw material, for example). They cannot be controlled but they can often be measured. For optimization, controlled variables are the most important, as they characterize external action performed on the process and are of utmost importance to the optimization criterion. Important also is sensitivity analysis of the optimization criterion with respect to the decisions, because their number determines to a large extent the difficulty of the optimization problem. Leaving out decisions which affect the optimization criterion in an insignificant way should be treated as a natural procedure which contributes to increased transparency of the results and often facilitates the problem solving.

Imposing constraints on decision variables is typical for all practical problems of optimization. There are, for example, constraints on consumption of some resources, process output, product purity, concentrations of contaminants, and so on. Constraints can also be formulated for thermodynamic parameters of the process in order to specify allowable ranges of temperatures and pressures, intervals of catalyst activities, reaction selectivities, and so on. Constraints which assure reliability and safety of the equipment are important, for example, those imposed on reagent concentrations in combustible mixtures (preventing an explosion), on gas and fluid flows in absorbers (prevention of flood), on the superficial velocity of fluidization (prevention of material blowing out), and so on. Constraints may also be imposed on construction parameters, for example, there are constraints imposed on the size of apparatus in enclosed residential areas or constraints on the lengths of pipes in heat exchangers, which arise from standardization. Although requiring some experience, it is important to leave these constraints out of the optimization formulation because they have a negligible effect on the optimal solution. This enables one to use easier problem-solving techniques yet still maintain precision of the optimization result.

The variety of optimization problems is strongly connected with the variety of constraints (algebraic, differential, integral—each may appear in equality or inequality form). In this chapter only examples of the simplest algebraic constraints are considered; more involved ones may be found in later chapters.

Sets of algebraic constraints often characterize a system at its steady state, hence the name "static optimization".

1.2. UNCONSTRAINED PROBLEMS

Consider an objective function S of several variables

$$S(\mathbf{u}) = S(u_1, u_2, \ldots, u_r) \tag{1.1}$$

In a closed set of independent variables there are several places where an extremum (a minimum or a maximum) may be found. Extreme values (relative minima or maxima of S) may occur at the points where: (a) all partial derivatives $\partial S/\partial u_k$ vanish, (b) $\partial S/\partial u_k$ do not exist, and (c) at a boundary of a closed set. Assume differentiability of S (Hancock, 1960). If we require the first partial derivatives to exist everywhere within the admissible boundary we eliminate the relatively seldom extrema which pertain to case (b). When the existence of the first partial derivatives is presupposed within the admissible region U, the maxima and minima are called ordinary extrema. In most of this book we shall consider the ordinary extrema.

Assume that S has a stationary point $\mathbf{a} = (a_1, a_2, \ldots, a_r)$, that is, the point at which all first derivatives of S vanish. Taylor series expansion of S around point $\mathbf{a}$ gives

$$S(a_1 + \Delta u_1, a_2 + \Delta u_2, \ldots, a_r + \Delta u_r) = S(a_1, a_2, \ldots, a_r)$$

$$+ \sum_{k=1}^{r} \frac{\partial S}{\partial u_k}\bigg|_a \Delta u_k + \frac{1}{2}\sum_{j=1}^{r}\sum_{k=1}^{r} \frac{\partial^2 S}{\partial u_j \partial u_k}\bigg|_a \Delta u_j \, \Delta u_k \tag{1.2}$$

An analysis of the first derivative term shows that since $\Delta u_1, \Delta u_2, \ldots, \Delta u_r$ are independent of each other and may be chosen to be either positive or negative, for an extremum at the point $\mathbf{a} = (a_1, a_2, \ldots, a_r)$ each of the first partial derivatives must be zero at that point, that is

$$\frac{\partial S}{\partial u_k}\bigg|_a = 0, \quad k = 1, 2, \ldots, s \tag{1.3}$$

This equation provides the necessary condition for a minimum of S at point $\mathbf{a}$. For two variables to function, a stationary point is usually either a maximum, minimum or a saddle point.

To determine conditions that are sufficient for a maximum or minimum of S at point $\mathbf{a}$ one needs to examine Equation (1.2) subject to the condition (1.3), that is, to consider the sign of the quadratic form

$$G = \frac{1}{2}\sum_{j=1}^{r}\sum_{k=1}^{r} \frac{\partial^2 S}{\partial u_j \partial u_k}\bigg|_a \Delta u_j \, \Delta u_k \tag{1.4}$$

Clearly, for a minimum of S, the form G of Equation (1.4) is greater than zero for all arbitrary values of Δu_j and Δu_k except $\Delta u_j = \Delta u_k = 0$ for all j and k. Similarly, for a maximum of S, the form G of Equation (1.4) is lower than zero for all arbitrary values of Δu_j and Δu_k except $\Delta u_j = \Delta u_k = 0$ for all j and k. Consequently, at the stationary point, the positive definiteness of the quadratic form G is a sufficient condition for S to be a minimum and the negative definiteness of G is a sufficient condition for S to be a maximum. In matrix notation

$$G = (\Delta u)^T H (\Delta u) \qquad (1.5a)$$

where the element h_{jk} of the matrix H (Hessian matrix) involves the second derivative of objective function S:

$$h_{jk} = \frac{1}{2} \frac{\partial^2 S}{\partial u_j \partial u_k}\bigg|_a \qquad (1.5b)$$

Assume that H is nonsingular. Sylwester's theorem may be used to determine whether the quadratic form is positive definite.

The necessary and sufficient condition for the positive definiteness of G is that each of the principal minors of H be greater than zero, that is, each of the determinants

$$H_1 = h_{11}, \qquad H_2 = \begin{vmatrix} h_{11} & h_{12} \\ h_{21} & h_{22} \end{vmatrix}, \qquad H_s = \begin{vmatrix} h_{11} & h_{12} & \dots & h_{1r} \\ h_{21} & h_{22} & \dots & h_{2r} \\ \cdot & & & \\ \cdot & & & \\ h_{r1} & h_{r2} & \dots & h_{rr} \end{vmatrix} \qquad (1.6)$$

must be greater than zero.

A necessary and sufficient condition for G to be negative definite is that $-G$ be a positive definite or that each of the principal minors of matrix $-H$ be greater than zero according to the conditions specified in Equation (1.6). If this is translated into the direct test for the negative definiteness of G, the necessary and sufficient condition is obtained, which states that the signs of H_i be alternately negative and positive as i goes from 1 to r with H_1 being negative, that is, H_i is negative if i is odd and positive if i is even.

The above conditions may be associated with completing the squares and expressing the quadratic form in its canonical form:

$$G = \sum_{j=1}^{r} \lambda_j (\Delta z_j)^2$$

where λ_j $(j = 1, 2, \dots, r)$ are the eigenvalues of matrix H and increments Δz_j are the distances along the coordinate axes of the new Cartesian system of coordinates. In this new coordinate system the principal axes of the quadratic surface described by G lie along the coordinate axes. Thus the necessary and sufficient condition for the positive definiteness of G can be stated as the requirement that all eigenvalues λ_j be positive and none of λ_j may be zero. Similarly, the necessary and sufficient condition for the negative definiteness of G can be stated as the requirement that all eigenvalues λ_j be negative and none of λ_j may be zero.

Methods for determining eigenvalues of matrices are available in the literature (Amundson, 1966). Cases with nonsingular H are omitted here; they require investigation of the higher order terms in the Taylor series of expansion and testing higher derivatives of S. We recall here only the case of the function of a single variable y where sufficient conditions for ordinary maximum or minimum are as follows. If at a stationary point $y = a$

$$S'(a) = S''(a) = \cdots = S^{(n)}(a) = 0 \tag{1.7}$$

and

$$S^{(n+1)}(a) \neq 0 \tag{1.8}$$

that is, $S^{(n+1)}(a)$ is the first nonvanishing derivative, then $S(y)$ has a point of inflection if n is even and an extremum if n is odd. This extremum is a minimum if $S^{(n+1)}(a) > 0$ and a maximum if $S^{(n+1)}(a) < 0$. The case of odd $n = 1$ with nonvanishing $S^{(n+1)}(a) = S^{(2)}(a)$ is the classical one.

1.3. EQUALITY CONSTRAINTS AND LAGRANGE MULTIPLIERS

Until now we have considered unconstrained optimization problems in which an objective function is expressed entirely in terms of independent variables or decision variables. Classical methods of static optimization also involve the problem of extremum seeking in the presence of equality constraints imposed on certain original variables y_k. In this case only a part of the variables may be treated as independent (or decision) variables, the remaining are the dependent (or state) variables. There are several methods to treat optimization problems of this type. They are usually referred to as the elimination method, the state and decision variable method, the Jacobian method and the method of Lagrange multipliers (Beveridge and Schechter, 1970; Fan et al., 1971).

Consider a general optimization problem with the objective function

$$S = f(y_1, y_2, \ldots, y_p) \tag{1.9}$$

which is subject to s independent equality constraints

$$\begin{aligned}
g_1(y_1, y_2, \ldots, y_p) &= 0 \\
g_2(y_1, y_2, \ldots, y_p) &= 0 \\
&\vdots \\
g_s(y_1, y_2, \ldots, y_p) &= 0
\end{aligned} \tag{1.10}$$

where $s < p$. As p–s variables may be assumed as free, p original variables $y_1, y_2,$
..., y_p may be divided into s dependent or state variables and p–s independent
or decision variables. It is helpful to remember that each constraint contributes
to a new state variable. Let us denote the s state variables as $x_1, x_2, \ldots, x_s$ and
the p–s decision variables as $u_1, u_2, \ldots, u_r$, where $r = p$–s. The number $r = p$–s
is sometimes called the number of degrees of freedom. The division may be
arbitrary, yet its suitable choice should facilitate problem solving. In terms of
state and decision variables, the objective S may be written in the form

$$S = f(x_1, x_2, \ldots, x_s, u_1, u_2, \ldots, u_r) \tag{1.11}$$

and the equality constraints in the form

$$
\begin{aligned}
g_1(x_1, x_2, \ldots, x_s, u_1, u_2, \ldots, u_r) &= 0 \\
g_1(x_1, x_2, \ldots, x_s, u_1, u_2, \ldots, u_r) &= 0 \\
&\vdots \\
g_1(x_1, x_2, \ldots, x_s, u_1, u_2, \ldots, u_r) &= 0
\end{aligned}
\tag{1.12}
$$

The elimination method is conceptually simple: if one can analytically solve the
set of Equation (1.12) for $x_1, x_2, \ldots, x_s$ to obtain these s variables in terms of
r decisions $u_1, u_2, \ldots, u_r$, we can then substitute the solution for $x_1, x_2, \ldots, x_s$
into Equation (1.9) to obtain

$$S = \tilde{f}(u_1, u_2, \ldots, u_r) \tag{1.13}$$

Thus the optimization problem has been reduced to one dealing with a function
of independent variables $u_1, u_2, \ldots, u_r$ (Section 1.2).

The state and decision variable method deals with the set of Equations (1.11)
and (1.12). Once the division of the original variables y_k into x_k and u_j is done,
the first partial derivatives of the objective function f with respect to the decision
variables may be set to zero to yield r necessary conditions of stationarity:

$$\frac{\partial S}{\partial u_j} = \frac{\partial f}{\partial u_j} + \sum_{i=1}^{s} \frac{\partial f}{\partial x_i} \frac{\partial x_i}{\partial u_j} = 0, \quad j = 1, 2, \ldots, r \tag{1.14}$$

To determine the location of the stationary point the set of Equations (1.12)
and (1.14) must simultaneously be solved provided that the expressions for the
derivatives $\partial x_i / \partial u_j$ are obtained by analyzing the differential form of the constraining Equation (1.12). If these partial derivatives can be easily obtained, this
method may be preferred. Yet, this will be true in general whenever each state
variable and each decision variable are present in only a small number of the s
equality constraints. If each state variable and each decision variable appear in
almost all of the equality constraints and if there is a large number of equality

constraints, either the Jacobian method or the method of Lagrange multipliers should be used.

The Jacobian method (Beveridge and Schechter, 1970; Fan et al., 1971) leads to explicit expressions for the partial derivatives $\partial x_i/\partial u_j$ $(i = 1, 2, \ldots, s; j = 1, 2, \ldots, r)$ in terms of partial derivatives of constraining functions g_k with respect to x_i and u_j and, finally, to the partial derivatives $\partial S/\partial u_j$ satisfying Equation (1.14) in terms of the partial derivatives of functions f and g_k with respect to x_i and u_j:

$$\frac{\partial S}{\partial u_j} = \frac{\partial(f, g_1, g_2, \ldots, g_s)/\partial(u_j, x_1, x_2, \ldots, x_s)}{\partial(g_1, g_2, \ldots, g_s)/\partial(x_1, x_2, \ldots, x_s)} = 0, \quad j = 1, 2, \ldots, r \quad (1.15)$$

The set of $s + r$ Equations (1.12) and (1.15) should be solved simultaneously to obtain the values of $x_1, x_2, \ldots, x_s$ and $u_1, u_2, \ldots, u_r$ at the stationary points, and each solution defines the location of the stationary point.

As the elimination of variables is not always simple and obtaining the extremal solution by the Jacobian method may also be complicated, the method of Lagrange multipliers is often used because it places all the original variables on an equal footing.

One may introduce the method of Lagrange multipliers by considering either the original objective function (1.9) subject to constraints (1.10) or the renamed objective function (1.11) subject to constraints (1.12).

We shall focus on Equations (1.11) and (1.12). Let us multiply each equality constraint g_i by an undetermined quantity (Lagrange multiplier) and add the result to the objective function (1.11). We then obtain a modified or Lagrangian objective function

$$\begin{aligned} S_L = f(x_1, x_2, \ldots, x_s, u_1, u_2, \ldots, u_r) \\ + \sum_{i=1}^{s} \lambda_i g_i(x_1, x_2, \ldots, x_s, u_1, u_2, \ldots, u_r) \end{aligned} \quad (1.16)$$

At an extremum point partial derivatives of S_L with respect to the independent (decision) variables must satisfy

$$\begin{aligned} \frac{\partial S_L}{\partial u_j} = \frac{\partial S_L}{\partial x_1}\frac{\partial x_1}{\partial u_j} + \frac{\partial S_L}{\partial x_2}\frac{\partial x_2}{\partial u_j} + \cdots + \frac{\partial S_L}{\partial x_s}\frac{\partial x_s}{\partial u_j} + \frac{\partial f}{\partial u_j} \\ + \lambda_1\frac{\partial g_1}{\partial u_j} + \lambda_2\frac{\partial g_2}{\partial u_j} + \cdots + \lambda_s\frac{\partial g_s}{\partial u_j} = 0 \end{aligned} \quad (1.17)$$

where $j = 1, 2, \ldots, r$, or, equivalently

$$\frac{\partial S_L}{\partial u_j} = \sum_{i=1}^{s} \frac{\partial S_L}{\partial x_i}\frac{\partial x_i}{\partial u_j} + \frac{\partial f}{\partial u_j} + \sum_{i=1}^{s} \lambda_i\frac{\partial g_i}{\partial u_j} = 0 \quad (1.18)$$

where

$$\frac{\partial S_L}{\partial x_i} = \frac{\partial f}{\partial x_i} + \sum_{k=1}^{s} \lambda_k \frac{\partial g_k}{\partial x_i}, \quad i = 1, 2, \ldots, s \qquad (1.19)$$

Assume that the Lagrange multipliers are selected in a way that assures the satisfaction of the stationarity condition of S_L with respect to the state variables x_i:

$$\frac{\partial S_L}{\partial x_i} = \frac{\partial f}{\partial x_i} + \sum_{k=1}^{s} \lambda_k \frac{\partial g_k}{\partial x_i} = 0, \quad i = 1, 2, \ldots, s \qquad (1.20)$$

Equation (1.18) then takes the form

$$\frac{\partial S_L}{\partial u_j} = \frac{\partial f}{\partial u_j} + \sum_{i=1}^{s} \lambda_i \frac{\partial g_i}{\partial u_j} = 0, \quad j = 1, 2, \ldots, r \qquad (1.21)$$

whereas each constraint g_i can be written as the stationarity condition of S_L with respect to λ_i:

$$\frac{\partial S_L}{\partial \lambda_i} = g_i(x_1, x_2, \ldots, x_s, u_1, u_2, \ldots, u_r), \quad i = 1, 2, \ldots, s \qquad (1.22)$$

Thus, in an algorithm applying Lagrange multipliers, the set of Equations (1.20)–(1.22) must be solved for s variables λ_i and x_i and r variables u_j. We observe that the set (1.20)–(1.22) can be obtained directly from the Lagrangian function (1.16) by setting its partial derivatives to zero with respect to each of its variables, x_i, u_j and λ_i, and that all variables are placed on an equal footing, that is, no distinction between them is necessary as in other methods. This means that the Lagrange method applies equally well to the original problem, Equations (1.9) and (1.10).

The Lagrange multipliers have an important property described by

$$\lambda_i = \frac{\partial S}{\partial g_i} \qquad (1.23)$$

which means that each λ_i expresses the change of the objective function that follows from a unit change of the corresponding constraining function. It has been shown (Leonard and Van Long, 1994) that by the very nature of the problem (1.9)–(1.10) or (1.11) and (1.12), the signs of the multipliers of the equality constraints cannot be ascertained.

Sometimes it is convenient first to use the elimination method to eliminate only a part of the state variables, and then to treat the thus-transformed problem by other methods, for example, by the Lagrange multipliers.

Observe that the first-order necessary conditions, presented above, apply equally well to constrained mimimum and to constrained maximum; to distinguish between the two, one must turn to second-order conditions. These conditions involve the second-order derivatives of the Lagrangian S_L; thus, one is required to write down the whole Hessian matrix of S_L as the stage before the formulation of necessity and sufficiency conditions (Leonard and Van Long, 1994). Singular cases and possible failure of the method applying Lagrange multipliers are also discussed in the literature (Beveridge and Schechter, 1970).

1.4. METHODS OF MATHEMATICAL PROGRAMMING

Assume now that an extremum of S is considered subject to some equality and inequality constraints imposed on the variables. When both the constraining equations and the performance index are linear, the optimization problem is a linear programming problem. Linear problems can emerge in both simple units and complex technological networks (Gass, 1958; Barsow, 1961; Charnes and Cooper, 1961; Hadley, 1962, 1964; Llewellyn, 1963; Dantzig, 1968; Beveridge and Schechter, 1970; Kobrinski, 1972; Findeisen, 1974; Findeisen et al., 1974, 1980; Seidler et al., 1980; Jeżowski, 1990b; Jeżowski et al., 2003e; Tan and Cruz, 2004). Linear programming (LP) problems include transportation, distribution from sources to sinks, traveling salesmen, allocation of resources among activities, management decisions, and so on. The simplex method is the most suitable method to solve LP problems (Dantzig, 1968, and the sources cited above). It is not analyzed here because an LP problem is usually too restrictive for applications in energy systems where the objective function is usually nonlinear by nature. Mathematical modeling of energy converters and heat exchange systems (Kanieviec, 1978, 1982) often requires methods of nonlinear programming. The reader is referred to the book by Leonard and Van Long (1994) to see how the linear programming problem can be treated by mathematical programming methods.

When one or more of the constraints, or the objective function, are nonlinear, the static problem is one of nonlinear programming (Luenberger, 1974; Zangwill, 1974; Findeisen et al., 1974; Seidler et al., 1980; Leonard and Van Long, 1994; Banerjee and Ierapetritou, 2003). It can be formulated as follows. Determine values for n variables $\mathbf{y} = (y_1, y_2, \ldots, y_n)$ that optimize the scalar objective function

$$S(\mathbf{y}) = S(y_1, y_2, \ldots, y_n) \tag{1.24}$$

subject to l equality constraints

$$g_i(\mathbf{y}) \equiv g_i(y_1, y_2, \ldots, y_n) = 0, \quad i = 1, \ldots, l \tag{1.25}$$

and to $m-l$ inequality constraints

$$g_i(\mathbf{y}) \equiv g_i(y_1, y_2, \ldots, y_n) \leq 0, \quad i = l+1, \ldots, m \tag{1.26}$$

Nonlinear programming problems often have widely varying properties, and certain limitations on the forms of the functions appearing therein are necessary if the problems are to be solved. Typical algorithms for their solution proceed by absorbing the constraints, Equations (1.25) and (1.26), into an augmented optimization criterion. The Karush–Kuhn–Tucker condition, generalizing the classical Lagrange multipliers to the case involving inequality constraints (Kuhn and Tucker, 1951; Zangwill, 1967; Mangasarian, 1969; Varaiya, 1972; Greig, 1980), is the basic theoretical tool for solving nonlinear programming problems:

$$\nabla S(y) + \sum_{i=1}^{m} \lambda_i \, \nabla g_i(y) = 0 \tag{1.27}$$

$$\lambda_i g_i(y) = 0, \quad i = 1, \ldots, m \tag{1.28}$$

$$\lambda_i \geq 0, \; g_i(y) \leq 0, \; i = l+1, \ldots, m \tag{1.29}$$

We refer the reader to the many books available on the subject (Abadie, 1967; Bracken and McCormick, 1968; Zangwill, 1967; Mangasarian, 1969; Beveridge and Schechter, 1970; Varaiya, 1972; Greig, 1980). We particularly mention an important special case: the equilibrium state of a thermodynamic system is a natural nonlinear programming problem where the free energy of a system is minimized. In the example of a chemical mixture, the free energy is minimized subject to the linear constraints resulting from the conservation of the atoms of the elements; the minimization determines the concentrations at chemical equilibrium (White et al., 1958).

The Kuhn–Tucker method is very general and hence not always the most effective. Sometimes, when only simple equality constraints are present, a number of variables (equal to the number of equality constraints) can be eliminated and the problem can be reduced to that of extremizing an unconstrained function of the remaining variables. For this case (unconstrained optimization of a multivariable function) a variety of iterative nongradient as well as gradient techniques can be used to find the optimum. Most frequently they are based on iterative searches for optima along certain directions in the decision space. These searches start from an arbitrary point and terminate close to the optimum. Many reviews of these methods and their applications are available (Rosenbrock and Storey, 1966; Beveridge and Schechter, 1970; Findeisen et al., 1974; Sieniutycz and Szwast, 1982a) along with associated proofs of convergence (Zangwill, 1967; Mangasarian, 1969).

The constraints can be taken into account not only by the Lagrange multiplier approach, as in the Kuhn–Tucker method, but also by introducing various penalty terms. These add terms to the objective function which penalize violations of the constraints by increasing the objective if any constraint is violated. This forces any search procedure to leave this region quickly (Bracken and McCormick, 1968).

Also, dynamical problems of optimization (Pontryagin et al., 1962; Athans and Falb, 1966; Lee and Marcus, 1967; and Chapter 2 of this book) can be approached by methods of mathematical programming as shown in the book by Canon et al. (1970).

1.5. ITERATIVE SEARCH METHODS

In view of the difficulties in finding extrema in complicated cases, a large number of numerical procedures have been proposed. In most cases, the robustness and efficacy of problem solving depend largely on making the right choice of method. A review of all the numerical procedures that are of potential use is beyond the scope of this book. Here we shall discuss only some of those methods, as there are many sources where these procedures are comprehensively described (Wilde, 1964; Rosenbrock and Storey, 1966; Fiacco and McCormick, 1968; Fan et al., 1971; Findeisen et al., 1974, 1980). A large group of numerical methods consists of those of an iterative nature, each method differing in the way it organizes the directions of optimum seeking.

Before a search method is applied an original optimization problem is broken down to an equivalent unconstrained problem. This may be done either by applying the Lagrangian type objective function of Equation (1.16) or by another constraint-absorbing objective which we shall also denote S_L and call the modified objective function. This term includes, in particular, objectives that contain penalty terms for the violation of constraints (Rosenbrock and Storey, 1966; Bracken and McCormick, 1968; Szymanowski, 1971; Findeisen et al., 1974, 1980; Zangwill, 1974; Sieniutycz, 1978). (However, constrained problems can also be handled by gradient projection methods (Rosen, 1960, 1961).) Next, an iterative procedure is applied.

Typically, an iterative procedure starts at an arbitrary point u^0 and proceeds along a certain direction, say k^1, to assure an increase (maximization) or decrease (minimization) of the modified objective S_L until its extremum in the direction k^1 is reached. To find this "directional extremum" a single variable search for a valley (peak) is applied (Fan et al., 1971). A single variable, l, measures distances covered in various directions. In the first search, the magnitude of the step is selected so as to extremize S_L along the direction k^1. Assuming that the magnitude of the step extremizing S_L is equal to l^1, the corresponding decision is $u^1 = u^0 + k^1 l^1$. The thus obtained u^1 is simultaneously the starting value of u for the second step, and so on.

On the whole, the search for the vicinity of the extremum is a sequential process represented by a number of consecutive steps. For the search towards a minimum the extremizing equation is

$$S_L(u^i) = \min_{l^i} \{S_L(u^{i-1} + k^i l^i)$$
$$\equiv S_L(u_1^{i-1} + k_1^i l^i, u_2^{i-1} + k_2^i l^i, \dots, u_r^{i-1} + k_r^i l^i)\}$$

$$(1.30)$$

and the decision sequence generated in the search process is described by the equation

$$\mathbf{u}^i = \mathbf{u}^{i-1} + \mathbf{k}^i l^i \tag{1.31}$$

where i is the iteration number. The termination condition for each one-step move follows from Equation (1.30) in the form of an orthogonality equation:

$$(\partial S_{\mathrm{L}}/\partial u_1^i)k_1^i + (\partial S_{\mathrm{L}}/\partial u_2^i)k_2^i + \cdots + (\partial S_{\mathrm{L}}/\partial u_r^i)k_r^i = 0 \tag{1.32}$$

Clearly, each ith step move starting at $\mathbf{u}^{i-1}$ should terminate at the point $\mathbf{u}^i$ at which the straight line of the direction $\mathbf{k}^i$ is tangential to the surface of constant S_{L}. In this algorithm all coordinates of the direction vectors are given parameters. The members of the sequence $\mathbf{u}^0, \mathbf{u}^1, \ldots, \mathbf{u}^r$ represent increasingly accurate approximation of the extremum point.

To investigate the nature of the extremum point (maximum or minimum, local or global), the procedure is verified for various starting points $\mathbf{u}^0$. Before any evaluations are made it is appropriate to determine the so-called uncertainty interval in which an extremum is located (Fan et al., 1971). For function $S_{\mathrm{L}}(\mathbf{u}^{i-1} + \mathbf{k}^i l^i)$ in Equation (1.30) a good estimate regarding the extremum location is usually obtained by using the quadratic approximation of S_{L} (Householder, 1953; Rosenbrock and Storey, 1966). Algorithms and computer logic charts applying one-dimensional quadratic approximation and one-dimensional cubic interpolation to minimize S_{L} are available (Fan et al., 1971).

The generation principle for coordinates of vector $\mathbf{k}^i$ is specific for each search method. The direction vector can be defined on the grounds of $S_{\mathrm{L}}(\mathbf{u})$ and its first and second derivatives on the grounds of the function itself. Consequently, the search methods are divided into gradient methods and nongradient methods. The convergence rate of algorithms is important; the criterion is the number of iterations assuring accuracy of the extremum location.

In the simplest, yet moderately effective method of Gauss and Seidel, also called the method of successive variation of independent variables, the successive search directions are parallel to axes, which means that in the r-dimensional space the first direction vector is $\mathbf{k}^1 = (1, 0, 0, \ldots, 0)$ and the rth direction vector is $\mathbf{k}^r = (0, 0, 0, \ldots, 1)$, then the use of the vectors is repeated.

In the method of steepest descent the direction of the gradient vector $\mathbf{g} = \partial S_{\mathrm{L}}/\partial \mathbf{u}$ is evaluated at point $\mathbf{u}^0$, and the direction of the steepest descent is given by the direction of $-\mathbf{g}$. Sometimes the partial derivatives cannot be conveniently obtained analytically, and it is necessary to use their approximate values as respective difference ratios.

A modification of this method called the gradient method eliminates the requirement that the new direction be normal to the old direction. While the direction of steepest descent is evaluated in the gradient method, only a short step of length δ is taken in this direction. Thus, in the gradient method successive approximations are not normal to each other, Equation (1.32) does not apply, and the change in direction at each point is small. Many tests of this method are available (Rosenbrock and Storey, 1966).

From the viewpoint of practical computation the most effective is the group of methods of conjugate gradients, in which the length of the step is again obtained from Equation (1.30). The methods were originated by Davidon (1959) and modified by Fletcher and Powell (1963) and Fletcher and Reeves (1964). They all apply the assumption that in the neighborhood of the minimum, cost function can be approximated by a positive definite quadratic form. Davidon's type of methods are quadratically convergent and only computation of the first partial derivatives of the objective function is necessary.

Two directions, v and w, are said to be conjugate with respect to the positively definite matrix if

$$\mathbf{v}^T \mathbf{A} \mathbf{w} \qquad (1.33)$$

A theorem appropriate to the understanding of conjugate gradients techniques has been proven (Kowalik and Osborne, 1968).

Theorem

If $\xi_1, \xi_2, \ldots, \xi_r$ are a set of vectors mutually conjugate with respect to a positive definite matrix A, then the minimum of the quadratic form

$$F = \mathbf{a} + \mathbf{b}^T \mathbf{u} + \frac{1}{2} \mathbf{u}^T \mathbf{A} \mathbf{u} \qquad (1.34)$$

can be found from an arbitrary starting point $\mathbf{u}^0$ by a finite descent computation in which each of the vectors ξ_i $(i = 1, 2, \ldots, r)$ is applied as the search direction only once. The order in which ξ_i is applied is not essential. It is said that the iterative procedure satisfying this theorem has convergence of the second order (Findeisen et al., 1974).

Thus making r successive one-dimensional searches in the r conjugate directions is sufficient to locate exactly the minimum of a quadratic cost function. All methods using conjugate gradients have convergence of the second order. For an arbitrary function, which is not necessarily quadratic, the convergence is assured provided that the function approaches a quadratic form in the same way that the iterative procedure approaches the minimum.

In conjugate gradient methods, the search direction towards a minimum $\mathbf{k}^i$ is generally different from the direction $-\mathbf{g}^i = -\mathbf{g}(\mathbf{u}^i)$. When the Fletcher and Reeves (1964) method is applied, the vector $\mathbf{k}^i$ is determined as the linear combination of the negative gradient and the direction vector of the previous iteration

$$\mathbf{k}^i = -\mathbf{g}^i + \frac{\mathbf{g}^{iT} \mathbf{g}^i}{\mathbf{g}^{i-1T} \mathbf{g}^{i-1}} \mathbf{k}^{i-1} \qquad (1.35)$$

In Fletcher's and Powell's (1963) modification of Davidon's (1959) method an observation is applied that the gradient near the optimum can be approximated by

$$\mathbf{g}(\mathbf{u}) = \mathbf{H}(\mathbf{u}_{min})(\mathbf{u} - \mathbf{u}_{min}) \qquad (1.36)$$

where $\mathbf{H}(\mathbf{u}_{min})$ is the positively definite, nonsingular matrix of second derivatives. From this equation one could evaluate $\mathbf{u}_{min}$ in the form

$$\mathbf{u}_{min} = \mathbf{u} - \mathbf{H}^{-1}(\mathbf{u}_{min})\mathbf{g}(\mathbf{u}) \qquad (1.37)$$

Yet, since the minimum location is unknown, the matrix $\mathbf{H}^{-1}(\mathbf{u}_{min})$ is not evaluated directly. Instead, a chosen matrix $\mathbf{B}$ is used, which may initially be any positive definite symmetric matrix. This matrix is modified after each iteration by using the information collected in the search.

The search may start with the identity matrix $\mathbf{B}^0 = \mathbf{I}$; in the course of the search a sequence of symmetric matrices is generated which approaches $\mathbf{H}^{-1}$. The direction vector $\mathbf{k}^i$ is determined as

$$\mathbf{k}^{i-1} = -\mathbf{B}^{i-1}\mathbf{g}^{i-1} \qquad (1.38)$$

and the modification of matrix $\mathbf{B}$ proceeds in accordance with the equation

$$\mathbf{B}^i = \mathbf{B}^{i-1} + l^i \frac{\mathbf{k}^{iT}\mathbf{k}^i}{\mathbf{g}^{iT}\mathbf{B}^{i-1}\mathbf{g}^i} - \frac{\mathbf{B}^{i-1}\mathbf{z}^i\mathbf{z}^{iT}\mathbf{B}^{i-1}}{\mathbf{z}^{iT}\mathbf{B}^{i-1}\mathbf{z}^i} \qquad (1.39)$$

where $\mathbf{z}^i = \mathbf{g}^i - \mathbf{g}^{i-1}$ and l^i is the magnitude of the ith step. The procedure is terminated when each of the components of the vectors $l^i\mathbf{k}^i$ and $\mathbf{k}^i$ are less than their prescribed accuracies. Three Pearson methods (Pearson, 1969) are similar. Corresponding algorithms and computer logic charts are available (Findeisen et al., 1974, 1980).

Of the methods using the second partial derivatives of objective function, the Newton–Raphson method applies the search direction in accordance with the principle

$$\mathbf{k}^i = -\mathbf{H}^{-1i}\mathbf{g}^i \qquad (1.40)$$

(compare Equation (1.37)). Yet, the first r iterations are performed using Equations (1.38) and (1.39).

The brief information about the search methods presented here favors those associated with a well-defined sequential process governed by Equations (1.30) and (1.31), and is by no means exhaustive. Consideration of these equations will help the reader to pass smoothly to the dynamic optimization methods outlined in Chapter 2. Many methods which minimize a function without calculating derivatives, that is, do not apply Equations (1.30) and (1.31), are described in the literature (Hooke and Jeeves, 1961; Wilde, 1962; Powell, 1964; Fletcher, 1965, 1969; Nelder and Mead, 1965; Zangwill, 1967; Wilde and Beightler, 1967; Kowalik and Osborne, 1968).

Excellent comparisons of various deterministic search methods are available (Box, 1965a, 1965b; Fletcher, 1965, 1969; Fiacco and McCormick, 1968; Szymanowski, 1971; Szymanowski and Brzostek, 1971; Szymanowski and Jastrzębski, 1971; Findeisen et al., 1974, 1980).

Applications of deterministic search methods in the realm of chemical engineering are covered in the textbook by Edgar et al. (2001) and in the Floudas (1995) monograph. The latter also addresses advanced techniques for nonlinear programming and mixed-integer programming problems. These techniques are needed to solve complex tasks of structural and parameter optimization.

1.6. ON SOME STOCHASTIC OPTIMIZATION TECHNIQUES

1.6.1. Introduction

Most of the sources on optimization techniques, including the book by Floudas (1995), address "classical" deterministic approaches. Not all that many sources deal with modern stochastic (meta-heuristic) approaches and their application to engineering problems. For instance, the book by Edgar et al. (2001) includes only a few short sections on those techniques. Recently, meta-heuristic methods have been gaining increased application in chemical and process engineering. Of the variety of techniques in existence we will consider in the following sections: adaptive random search (ARS), genetic algorithms (GA) and simulated annealing (SA). Observe that other, more recent ones, such as swarm particle optimization, tabu (taboo) search, tunneling algorithm, ant colony approach and differential evolution, have also been applied in process and process system optimization, for example, Rajesh et al. (2000), Mathur et al. (2002), Linke and Kokossis (2003), Lin and Miller (2004), Cavin et al. (2004, 2005), Srinivas and Rangaiah (2006), Babu and Angira (2006). A tendency to try hybrid strategies by implementing two or more different stochastic methods in a unified framework is noticeable. Also, a stochastic approach often serves as a solver to finding a good starting point for a deterministic NLP or MINLP procedure. Usually, the latter refines the global optimum in a small number of (goal) function evaluations (NFE).

Meta-heuristic methods provide only a general framework, thus giving rise to various more detailed algorithms. It is usually difficult to assess their robustness and efficacy. First, there is no proof of convergence for practically meaningful calculation load. Second, insufficient tests have often been performed. Furthermore, even numerous examples do not provide sufficient evidence that a method will be able to solve the problem at hand. The crucial point in successful application of a stochastic method is making a "good" choice of control parameter settings. For some techniques several trials are usually necessary.

In spite of all the drawbacks, meta-heuristic or stochastic approaches have significant advantages when applied to "real life" problems, since they easily account for discontinuous functions, black box models, discrete decisions and logical conditions. The methods addressed in the following have been developed and tested by the research group of J. Jeżowski. We will start with an adaptive random search technique, then genetic algorithms will follow and, finally, simulated annealing. All the methods will first be explained for problems with inequality constraints only. Ways of dealing with equalities will be discussed in the final section of this chapter.

1.6.2. *Adaptive Random Search Optimization*

The adaptive random search/random search (ARS/RS) technique is one of the oldest of the stochastic (meta-heuristic, heuristic) optimization approaches. It had its peak of popularity in chemical and process engineering applications in the 1960s and 1970s. The method is principally aimed at NLP problems, though some of its versions have also been tried to tackle MINLP problems; for example, Salcedo (1992) and Bochenek et al. (1999). At present, there is an opinion that ARS is unable to cope with larger problems. However, this opinion appears to be valid only for MINLP tasks. Zabinsky (1998) stated that for certain ARS algorithms computation load does not increase sharply with the size of the problem. Similar conclusions can be found in Hendrix et al. (2001). The works by Luus (1996, 2002) and Lee et al. (1999), as well as the solutions of the water network optimization problems addressed in Chapter 20, support this conclusion. It is, however, of importance that an appropriate ARS version should be applied and adapted to the problem at hand. Several authors point out the usefulness of the ARS technique for finding good initialization to more sophisticated solvers. The ARS technique can also be applied in hybrid approaches to speed up computation—see, for example, Novak and Kravanja (1999) or Lima et al. (2006).

Last but not least, information in the literature and the experience of the authors allows the claim that a very efficient technique exists for small and medium-size NLP problems featuring multiple optima. Note that, for instance, Liao and Luus (2005) recently reported its superiority over genetic algorithms for some benchmark global optimization tasks.

There exist several versions of the adaptive random search (ARS) optimization method. Here we limit ourselves to those that have been presented in the literature on chemical and process engineering, usually together with their application. See for instance the works of Luus (1973, 1974, 1975, 1993), Luus and Jaakola (1973), Jaakola and Luus (1974), Gaines and Gaddy (1976), Heuckroth et al. (1976), Campbell and Gaddy (1976), Wang and Luus (1978, 1997), Martin and Gaddy (1982), Rangaiah (1985), Mihail and Maria (1986), Luus and Brenek (1989), Salcedo et al. (1990), Salcedo (1992), Banga and Seider (1996), Banga et al. (1998), Bochenek et al. (1999), Michinev et al. (2000), Li and Rhinehart (1998), Bochenek and Jeżowski (2000), Jeżowski and Bochenek (2000,

2002), Luus et al. (2002), Jeżowski and Jeżowska (2003), Jeżowski et al. (2005a), Ziomek et al. (2005).

Basically, almost all their proposals aimed at solving general nonlinear problems with continuous variables ($x_i \in X$), with/without inequality constraints, where we wish to find the extreme of goal function (FC)—here the minimum

$$\min FC(X) \tag{1.41}$$

subject to the constraints

$$g_k(\mathbf{X}) \geq 0, \quad k = 1, \ldots, K \tag{1.42}$$

$$x_i^1 \leq x_i \leq x_i^u, \quad i = 1, \ldots, p \tag{1.43}$$

A generalized algorithm of ARS optimization can be formulated as follows:

1. Choose an initial (starting) point.
2. Calculate values of decision variables in kth iteration from Equations (1.44), (1.44a)

$$x_i^k = x_i^* + \Delta x_i^k, \quad i = 1, \ldots, p \tag{1.44}$$

$$\Delta x_i^k = f(r_i^k, \delta_i^k, \mu_i^k), \quad i = 1, \ldots, p \tag{1.44a}$$

where μ_i is the maximum of density probability distribution for random number r_i, r_i is the random number from certain probability distribution, δ_i is the size of the current search region of variable x_i, x_i^* is currently the best value of variable x_i.
3. Check if constraints (1.42) and (1.43) are met, and if they are, calculate the value of the goal function $FC(X)$, in the opposite case go to point 4.
4. Compare $FC(X)$ with the current best solution $FC(X^*)$; if $FC(X)$ is better (higher for maximization or lower for minimization) set: $X^* \Leftarrow X$ (this is called a success).
5. Check a stopping criterion; if met, stop calculations and accept X^* as the solution.
6. Increment k by 1, update parameters δ_i^k, μ_i^k and go back to Step 2.

Notice that the algorithm employs a so-called "death penalty" in regard to inequality constraints (1.42), that is, infeasible solutions are simply rejected. There is the possibility of using an augmented goal function with penalty terms for inequalities (1.42) but it is commonly considered an inefficient way of dealing with inequality constraints in frames of ARS.

Various probability distributions have been employed for randomly generating the variables. However, the rule of concentrating the generation around currently the best point X^* is always kept. This is achieved by appropriate updating values of parameters μ_i, δ_i in (1.44a). Ideally, the updating should be performed on the basis of information on optimization "history", mainly on history of successes—see point 3 of the general algorithm. A proper approach for updating parameters μ and δ is the key point of ARS algorithms since it directly affects their robustness in regard to probability of locating the optimum, that is, robustness and efficiency related to the number of goal function evaluations or CPU time.

To calculate Δx_i from the general formula (1.44a) we need random parameters r_i and, also, parameters μ_i, δ_i. Parameter r_i is generated for each variable x_i from a specified method and an algorithm specific type of probability distribution. It is important to note that generated parameters are normalized into the range $(-0.5; 0.5)$. The uniform or Gaussian probability distribution (1.45) is most often employed.

$$f(r) = \frac{1}{d\sqrt{2\pi}} \exp\left[-\frac{(r-m)^2}{2d^2}\right] \qquad (1.45)$$

where m and d are the distribution parameters.

It can be concluded from literature results that the type of probability distribution does not greatly influence the optimization. However, a way of calculating values of Δx_i seems to have a larger effect. In most algorithms they are calculated according to Equation (1.46):

$$\Delta x_i^k = \delta_i^k \cdot (r_i)^{\mu_i^k}, \quad i = 1, \dots, p \qquad (1.46)$$

where δ_i^k is the size of the current search region of variable x_i^k—notice that in the first iteration the region is equal to the given initial region size of the variable according to Equation (1.43).

In most algorithms the parameter μ_i^k is kept identical for each variable, that is

$$\mu_i^k = \mu^k, \quad i = 1, \dots, p \qquad (1.47)$$

Luus et al. in their procedures, (called L type in the following) applied μ equal to 1.0 in all iterations and for each variable. Gaddy and co-workers applied odd integer numbers in ascending order, for instance 1, 3, 5, 7, ..., that is, values of μ increase during the course of optimization. Notice that random numbers r_i are from the uniform distribution variables and x_i are from nonuniform distribution. Such algorithms are here called G type. Salcedo and co-authors also followed this nomenclature in algorithms called SGA and MSGA (MSGA is a version of SGA for mixed-integer problems). Algorithm MMA in Mihail and Maria (1986) is also of G type, as well as the ICRS algorithm of Banga and Seider (1996) and Banga et al. (1998). However, they applied Gaussian distribution for r_i (with $m = 0$ and $d = 1$ in Equation (1.45)). A version of the ICRS approach sets μ_i, δ_i at 1 but updates parameters m and d in Gaussian distribution. It is important to note that in all the above-mentioned algorithms the distribution of variables is symmetrical around currently best X^*.

The scheme for updating parameters μ_i, δ_i varies. The common idea is to decrease the sizes of search regions in successive iterations, though they can also be expanded temporarily to provide the means for escaping from local optimum. The sizes are reduced after each success (see point 3 of the general algorithm) in G type and MSGA algorithms. A fixed constant reduction rate independent of successes is used in L type algorithms. Similarly, parameters μ_i vary in G algorithms while in L procedures they are fixed at 1 for each iteration. An increase of μ increases the concentration of generated x_i around currently best x_i^* by changing the profile of the distribution. Hence, in G type methods there are two means to control generation of variables: search region size decrease and distribution profile change, though both are not always employed.

It is difficult to assess the efficacy of the various versions of ARS algorithms, mainly because of lack of sufficient tests. Algorithm L in basic version is the simplest one and has been tested in many works by Luus and co-workers. Other authors have also found it efficient and robust, for instance Rangaiah (1985), Lee et al. (1999), Michinev et al. (2000), Jeżowski and Bochenek (2002), Teh and Rangaiah (2002). Also of importance is that it requires relatively few control parameters. Additionally, in contrast to other approaches it ensures a dense search of regions in iterations, which diminish region sizes after each success. This feature of L algorithms is to some extent similar to the simulated annealing approach and is expected to increase the robustness of optimization.

The original version of the optimization procedure, called the LJ algorithm from Luus and Jaakola, is given in the following. Random numbers r_i are from the uniform distribution.

Given: initial point X^0, initial search sizes δ_i^0 $(i = 1, \ldots, p)$, number of external loops (NEL), number of internal loops (NIL) and size contraction coefficient β in Equation (1.49), usually from range (0.9; 0.99):

1. Set external loop counter k at 1.
2. Calculate X^k from Equation (1.48) NIL times (with $X^* = X^0$, $\delta_i^k = \delta_i^0$ for $k = 1$) and choose the best solution X^* from among the feasible solutions found in NIL loops:

$$x_i^k = x_i^* + r_i \delta_i^{k-1}, \quad i = 1, \ldots, p \qquad (1.48)$$

where r_i is from the range $(-0.5; 0.5)$.
3. Update δ_i^k according to Equation (1.49) and set the current best point at X^*:

$$\delta_i^k = \beta \delta_i^{k-1} \qquad (1.49)$$

4. Increase counter k by 1 up to NEL value and go back to 2.

The LJ algorithm requires three control parameters: β—contraction coefficient of search region size, NEL—number of external loops, NIL—number of trials in external loops (number of internal loops). It does not need any additional choices.

According to Equation (1.49) the sizes of search regions in the LJ procedure are diminished at the same rate for each variable. It seems logical to apply a pace of reduction dependent on a variable. The idea of the modification by Jeżowski and Bochenek (2002), referred to as the LJ–FR algorithm (LJ with final search regions), was to use the final size of the search region for a variable in the data instead of parameter β. For the given final sizes δ_i^f the values of contraction parameter β for all variables are calculated from:

$$\beta_i = \left(\frac{\delta_i^f}{\delta_i^0}\right)^{1/NEL} \tag{1.50}$$

Alternatively, the search region size can be updated according to Equation (1.52) instead of Equation (1.49) and, thus, contraction parameter β_i^k for variable i after k external iterations parameter is calculated from Equation (1.51) instead of Equation (1.50):

$$\beta_i^k = \left(\frac{\delta_i^f}{\delta_i^0}\right)^{1/NEL} \tag{1.51}$$

$$\delta_i^k = \beta_i^k \, \delta_i^0 \tag{1.52}$$

One reason for applying the final region sizes is that they are often easy to assess from "physical" interpretation of the optimization problem and knowledge of the initial sizes of regions. Next, their application has the effect of scaling the variables. Initial hyper-box search space is gradually diminished to a very small final hyper-box with various rates depending on the variable. Jeżowski and Bochenek (2002) showed that good results are also obtained when using variable independent final sizes:

$$\delta_i^f = \delta^f, \quad i = 1, \dots, p \tag{1.53}$$

with δ^f from the range $\langle 10^{-1}; 10^{-4} \rangle$.

It is worth noting that, in spite of the use of final size δ^f identical for all variables, in the majority of problems the modification yields different values of contraction parameters β_i because of different initial sizes. In effect the use of Equation (1.53) gives variable-dependent β parameters. As shown by Jeżowski and Bochenek (2002) the application of final sizes as control parameters in the LJ-FR version instead of parameter β in the LJ algorithm increases the reliability of locating the global optimum by 10–20% for both unconstrained and constrained optimization problems.

Jeżowski et al. (2005a) also developed other version of the algorithm, called LJ-MM, aimed mainly at highly multimodal problems. The basic change in the LJ-MM method in comparison with the original LJ or LJ-FR version relies on a change of profile of search region size reduction rate. In LJ/LJ-FR the profile of reduction rate in terms of external loop number ($k = 1, \dots, NEL$) is very steep in the initial phase of optimization and almost constant in the final stage in which

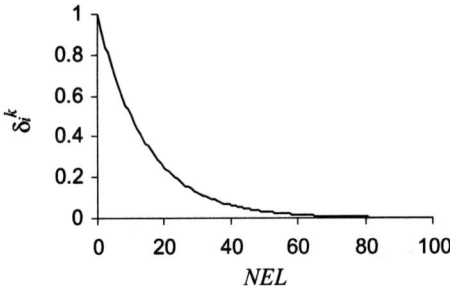

Figure 1.1 The profile of search region reduction rate in the LJ/LJ-FR method.

many iterations are performed. This is illustrated in Figure 1.1 for LJ/LJ-FR algorithm with $\delta_i^0 = 1; \delta_i^f = 10^{-3}$.

Due to a sharp reduction at the beginning of calculations the global optimum region can be cut off and the LJ or LJ-FR algorithm can be trapped into a local optimum. This can be expected for highly multimodal functions with local optima of similar values. Also, a "bad" starting point can cause a similar effect even for more regular functions.

To eliminate this potential pitfall Jeżowski et al. (2005a) applied a size reduction rate that features as close a resemblance to Gaussian distribution as possible, though reduced to the "right" side of the distribution as in Figure 1.2. Hence, size contraction coefficients β_i^k should conform to Formula (1.54).

$$\beta_i = e^{-[z_i/\sigma_i]^2} \tag{1.54}$$

Additionally, they must be kept in the range (0; 1) and the reduction scheme should follow the main idea of the LJ-FR version, that is, the size of the search region of the last major iteration (NEL) has to be equal to the given final size. Jeżowski et al. (2005a) derived the following formula for calculating

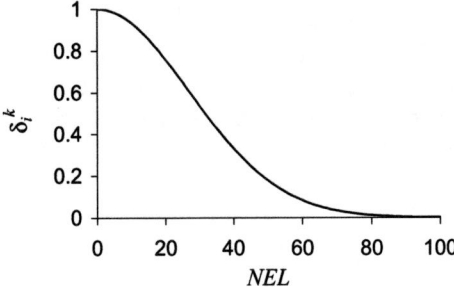

Figure 1.2 The profile of search region reduction rate in the LJ-MM method.

parameters β_i^k:

$$\beta_i^k = \exp\left\{\left(\frac{k}{NEL}\right)^2 \cdot \ln\left(\frac{\delta_i^f}{\delta_i^0}\right)\right\} \tag{1.55}$$

Notice that this requires updating of region sizes with Equation (1.52).

In order to compare profiles of reduction in the LJ-FR and LJ-MM algorithm, Figure 1.2 shows the profile from the latter calculated for the same parameters as those for the profile for the LJ-FR algorithm shown in Figure 1.1. The rate of search region size reduction is slow in the initial phase of optimization of the LJ-MM procedure and, hence, provides good reliability in locating the global optimum. It is, however, achieved by diminishing the number of iterations in the final steps. Hence, the precise location of global optimum can be different, particularly for flat goal function. This potential drawback does not seem to be serious in engineering and industrial applications.

It is interesting that, after a simple manipulation, Equation (1.55) can be changed into:

$$\beta_i^k = \left(\frac{\delta_i^f}{\delta_i^0}\right)^{[k/NEL]^2} \tag{1.56}$$

This equation bears a close resemblance to Formula (1.50) applied in the LJ-FR version. Hence, Jeżowski et al. (2005a) proposed the use of the following semi-empirical equation for the region size reduction parameter:

$$\beta_i^k = \left(\frac{\delta_i^f}{\delta_i^0}\right)^{[k/NEL]^\alpha} \tag{1.57}$$

Notice that the LJ-MM algorithm with Equation (1.57) embeds the LJ-FR algorithm. Setting parameter α at 1.0 in the LJ-MM procedure gives the LJ-FR algorithm. Numerical experiments performed for parameter α higher than 2.5 did not show improvement of the results. The experiments indicated that α in the range (1.5; 2.5) is the best setting.

Application of ARS algorithms as well as other stochastic/meta-heuristic approaches requires a special scheme of dealing with equality constraints. This will be addressed in the following for all stochastic approaches.

The other important issues of ARS method application are choices of:

1. a starting point,
2. initial sizes of search region,
3. values of control parameters.

There exists an opinion in the literature that ARS does not require a feasible initial solution. This is, however, valid only for relatively small problems or those that feature a relatively large space of feasible solutions in comparison with the initial box-like search space defined by Equation (1.43). Generally, both spaces should not differ very much in size. Hence, choice of an initial search space should be performed with great care to find the smallest possible one but without cutting off an optimal solution. An example will be given in Chapter 20.

The LJ-MM algorithm does not require numerous control parameters. As to final sizes a good choice is accepting the scheme defined by Equation (1.53) if the user has no deeper insights into a problem. Values of NEL and NIL are largely problem-size-dependent. According to Jeżowski and Jeżowska (2003) the product of NEL and NIL should be higher than 10^5–10^6 if the number of degrees of freedom exceeds about 12–15. They also advised applying values of NEL smaller than NIL, recommending a fraction NEL to NIL of about 0.5. Nevertheless, the final choice should be made after some trials.

The ARS algorithms by Jeżowski and co-workers, the LJ-MM algorithm in particular, have been tested against an ample set of standard general global optimization benchmark problems taken from various sources such as Floudas and Pardalos (1990), Michalewicz (1996), Michalewicz and Fogel (2002), Price (1978), Ali et al. (1997), Andre et al. (2001), Garcia-Palomares and Rodriguez (2002), Visweswaran and Floudas (1990), Ryoo and Sahinidis (1995), Wang and Luus (1978, 1997), Gouvea and Odloak (1998), Rajesh et al. (2000), and Mathur et al. (2002). Both unconstrained and constrained models have been used. The number of variables varied from 2 to 19. The number of goal function evaluations did not exceed approximately 150 000 for the most difficult tasks. However, it should be noted that the ARS algorithm, similarly to other stochastic techniques, needs several runs. Generally, ARS algorithms require longer CPU than the simulated annealing procedure addressed in Section 1.4. However, CPU time is shorter than for the genetic algorithm from Section 1.6.3. The obvious advantage of ARS is the ability to locate very good solutions regardless of the initial starting point.

Additionally, some chemical and process engineering NLP and (small) MINLP problems have been successfully solved by Jeżowski and co-workers, such as:

- Cross-flow extraction train optimization from Luus (1975), Salcedo et al. (1990), and Cardoso et al. (1996).
- Reactor selection problems from Kocis and Grossmann (1989a), Adjiman et al. (2000), Smith and Pantelides (1997), and Cardoso et al. (1997).
- Structural and parameter optimization of a simple flow sheet for mixture separation from Kocis and Grossmann (1989b).
- Optimization of a multiproduct batch plant formulated first in Sparrow et al. (1975) and addressed also in Grossmann and Sargent (1979), Kocis and Grossmann (1988), Salcedo (1992), and Cardoso et al. (1997).

Only the last problem has not been solved to the global optimum with a sufficient success rate. However, due to very tight constraints, it is a very difficult task for all the kinds of meta-heuristic methods addressed in this chapter. A deeper discussion of this problem is given in the next section. The application of ARS to water network problems will be addressed in Chapter 20.

1.6.3. Genetic Algorithms

Genetic algorithms (GA) are modern stochastic optimization strategies, particularly in comparison with ARS. They belong to a wider class of evolutionary algorithms (EA). Michalewicz and Fogel (2002) explained evolutionary mechanism as follows:

> A population of candidate solutions to the task at hand is evolved over successive iterations of random variation and selection. Random variation provides the mechanism for discovering new solutions. Selection determines which solutions to maintain as a basis for further exploration.

Hence, they differ from many other approaches in that they act on a population of solutions applying competition and selection tools. Genetic algorithms can be seen as procedures that mimic a simplistic mechanism of biological evolution. Even the terms often used, for example, chromosome, stem from biology.

Here we present a general algorithm of GA following Edgar et al. (2001).

1.   Choose an initial population and evaluate the fitness of each member.
2.   **DO** until termination conditions
     - **IF** crossover condition met **THEN** select parents, choose crossover parameters, perform crossover
     - **IF** mutation condition satisfied **THEN** select member, choose mutation parameters, perform mutation
     - Evaluate fitness of offspring.
     - Update population.

In contrast to ARS, evolutionary techniques have found wide application to complex process system designs, particularly to batch operation scheduling. Examples of the latter are contributions by: Azzaro-Pantel et al. (1998), Bernal-Haro et al. (1998), Jung et al. (1998), Wang et al. (1996, 2000, 2004). There are numerous applications for designing heat exchanger networks, mass exchanger networks and other subsystems. Here we list some of those contributions: Androulakis and Venkatasubramanian (1991), Fraga and Matias (1996), Garrard and Fraga (1998), Castell et al. (1998), Lewin (1998), Lewin et al. (1998), Xue et al. (2000), Yu et al. (2000), Tsai and Chang (2001), Björk and Petersson (2003), Björk and Nordmann (2005), Bochenek (2003), Shafiei et al. (2004), Prakotpol and Srinophakun (2004), Ravagnani et al. (2005), Jung et al. (2005), Lavric et al. (2005), Bochenek and Jeżowski (2006), Jeżowski et al. (2007).

It is, however, necessary to note that the GA framework is often associated with another optimization approach, usually NLP or LP solver, to calculate optimal values of continuous variables. It results from the fact that the GA approach is not considered an efficient technique for NLP problems but for combinatorial ones. In Chapter 19 we will address a two-level approach for HEN retrofit that employs only GA as the optimizer.

Because GA is a useful technique for combinatorial problems it has been applied not only to batch scheduling but also to molecular design, for instance in Venkatasubramanian et al. (1994) and Wang and Achenie (2002). GA also appears to be an efficient technique for optimizing processes and apparatus that feature discrete-valued parameters. Here we list some examples of GA application for designing various processes and apparatus, also: Chaudhuri and Diwekar (1997a, 1997b), Upreti and Deb (1997), Tayal and Fu (1999), Zhao et al. (2000), Ravi et al. (2000, 2002), Kasat et al. (2002), Stephan and Chase (2003), Diwekar and Xu (2005), Xu and Diwekar (2005), Selbaş et al. (2006), Mohanty (2006). Finally, the reader is referred to some papers on GA applied to modeling and other issues of process engineering: Edwards et al. (1998), Friese et al. (1998), Nouges (2002), Katare et al. (2004), Singh et al. (2005), Yan and Zhao (2006). It is worth noting that several works address GA application to multiobjective optimization problems.

Similar to other stochastic optimization methods there exist numerous versions of GA. For deeper insight the reader is referred to books by Goldberg (1989), Michalewicz (1996), Corne et al. (1999), Michalewicz and Fogel (2002), and the chapter in Spall (2003). Contributions by Yan and Ma (2001), Costa and Oliveira (2005), Chew et al. (2002), Summanwar et al. (2002), Shopova and Vaklieva-Bancheva (2006), Hwang and He (2006), to mention a few, address some recent versions of the GA procedures. The latter contributed with a hybrid of GA and simulated annealing.

In fact, only a few common rules are applied in the majority of GA optimization techniques published in the literature, such as:

- fixed number of members in generated populations (though there are also other stopping criteria)
- crossover is the basic operator while mutation is treated as an auxiliary one.

Some procedures developed in the literature are quite complex while others are simple. It is difficult to assess the effectiveness of versions suggested in the literature since often insufficient tests have been carried out. It is clear that more sophisticated versions usually ensure higher probability of locating global optimum. Such approaches need, however, more control parameters, which have to be fine-tuned by the user for the problem at hand. This requires a lot of trials and, also, user experience.

In this chapter we explain a GA-based optimization method, called GEN-COM, developed by Bochenek—published in short versions in Bochenek et al. (2001, 2005). It is designed for the general nonlinear mixed-integer optimization problem (MINLP) defined by Equations (1.58)–(1.62) and has been used to solve, amongst others, heat exchanger design problems, as we will show in Chapter 19.

$$\min FX\,(\mathbf{X}, \mathbf{Y}) \tag{1.58}$$

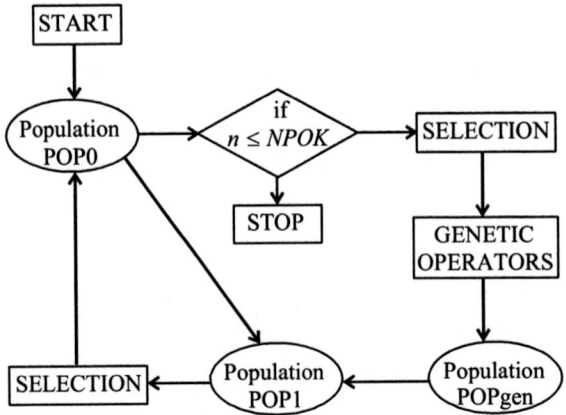

Figure 1.3 Flow sheet of GEN-COM algorithm by Bochenek.

$$h_l(\mathbf{X}, \mathbf{Y}) = 0, \quad l = 1, ..., L \tag{1.59}$$

$$g_k(\mathbf{X}, \mathbf{Y}) \geq 0, \quad k = 1, ..., K \tag{1.60}$$

$$x_i^l \leq x_i \leq x_i^u, \quad i = 1, ..., p \tag{1.61}$$

$$y_j \in D, \quad j = 1, ..., J \tag{1.62}$$

where $\mathbf{X}$ is the vector of continuous variables from $\mathbf{R}^p$ and $\mathbf{Y}$ is the vector of discrete variables from $\mathbf{D}$ space.

In contrast to early GA methods the algorithm does not use a traditional binary chromosome code but employs a vector of decision parameters or other, relevant to the problem, parameter structures. In order to increase the chance of survival of the best-fitted individuals and to decrease the effect of premature degeneration (i.e. local optimum trap) Bochenek applied the so-called "genetically modified subpopulation". Application of genetic operators to selected members of the parent population creates members of this subpopulation. Then, an intermediate population is created that is a superset of the parent population and genetically modified subpopulation. A selection mechanism chooses individuals from the parent population of the next generation. The subpopulation has a smaller number of members than the parent population and the number of its members is determined by a control parameter called modification ratio u_mod. The algorithm of GEN-COM, illustrated additionally by Figure 1.3, is as follows.

1. Input values of basic control parameters: population size *NPOP*, total number of populations generated in solver run *NPOK*, number of genetic operators *NOPER*,

modification ratio u_mod for the subpopulation, probabilities of using the genetic operators p_gen_i $(i = 1, \ldots, NOPER)$.

2. Calculation of parameters for fitness function.
Fitness function FP with a sum of penalty terms CK (if any) is defined by:

$$\text{for minimization}: FP = C^{\max} - FC - CK \qquad (1.63)$$

$$\text{for maximization}: FP = C^{\min} + FC - CK \qquad (1.64)$$

Parameters $C^{\max}$, $C^{\min}$ are calculated as follows. A set of many individuals (e.g. 10 000) is randomly generated and parameters $C^{\max}$, $C^{\min}$ are determined from:

$$\text{for minimization}: C^{\max} = \text{abs}(\max\{FC - CK\}) \qquad (1.65)$$

$$\text{for maximization}: C^{\min} = \text{abs}(\min\{FC - CK\}) \qquad (1.66)$$

where abs $(\cdot)$ is the operator transforming negative into positive value.

3. Calculation of the number of members in genetically modified subpopulation

$$NPOPgen = \text{int}(u\_mod * NPOP) \qquad (1.67)$$

where int $(\cdot)$ is the operator that transforms real number into integer one.

4. Creation of initial population $POP0$.
5. Calculation of fitness function FP for members of population $POP0$ from Equation (1.63) for minimization or from Equation (1.64) for maximization.
6. Calculation of probabilities $p0_i$ $(i = 1, \ldots, NPOP)$ for selecting individuals from population $POP0$.
7. Creation of genetically modified subpopulation:
    (a) Selection of genetic operator i with probability p_gen_i.
    (b) Selection of individual/individuals for genetic modifications according to probability $p0_i$.
    (c) Creation of modified individual/individuals by selected operator.
    (d) Points (a) and (b) are performed until given size of subpopulation $NPOPgen$ is reached.

8. Calculation of fitness function for members of subpopulation $POPgen$.
9. Calculation of probabilities $p1_i$ $(i = 1, \ldots, NPOP + NPOPgen)$ for selecting individuals from the superset consisting of subpopulation and parent population (the superset is intermediate population).
10. Creation of offspring population having $NPOP$ members by choosing members from the superset according to the selection mechanism.
11. The population from the previous step is copied into the parent population of the next generation.
12. Points 5–11 are performed until the generation number is greater than $NPOK$.

Selection for both reproduction (point 7a) and offspring population (point 10) is performed with the roulette mechanism. Some additional rules have to be imposed. They are used to control a compromise between an "elitist" strategy, that is, one that preserves the best-fitted members only, and an alternative purely random mechanism, which does not consider the fitness of members at all. To define the overall selection mechanism it is necessary to choose a probability distribution for members and rules for number of selections. Each member of any population has assigned a probability ($p0_i$ in point 6 and $p1_i$ in point 9 of the algorithm). Three different probability distributions can be chosen in program GEN-COM:

(a) Distribution proportional to fitness function of a member. The probability for member i is determined from:

$$p_i = \frac{FP_i}{\sum_{i=1}^{N} FP_i} \tag{1.68}$$

where N denotes the total number of individuals in the population/subpopulation

(b) Deterministic distribution with probabilities p_i calculated from Equation (1.69). Notice that they have to satisfy Condition (1.70).

$$p_i = \frac{Q(1-Q)^{i-1}}{\sum_{i=1}^{N} Q(1-Q)^{i-1}} \tag{1.69}$$

$$\sum_{i=1}^{N} p_i = 1 \tag{1.70}$$

Parameter Q is the additional control parameter given in the data. Parameter N in Equations (1.68)–(1.70) states the number of members in the population/subpopulation.

(c) Uniform distribution is such that each member is given identical probability p_i independent of its fitness function value. Notice that this distribution is not applied for selection of the new population since effects in a purely stochastic mechanism.

Proportional distribution is most often used in other GA procedures. However, it can result in premature degeneration effects such that members would feature almost identical fitness and the algorithm would fail to locate an optimum. For very small values of parameter Q the deterministic distribution (1.69) becomes similar to the uniform one. In the opposite case, with a high value of Q, the distribution provides a high preference for best-fitted members. Thus, it becomes similar to a pure elitist scheme. The setting of Q has to be done by the user within the range (0.0; 1.0).

Additionally, certain conditions can be imposed on selection mechanisms. They are used to control how many times a member can be chosen for reproduction or a new population.

Three rules are available in program GEN-COM:

(a) Multiple choice of a member—no constraints are imposed on the number of choices for a member.

(b) Single choice—each member can be selected only once.

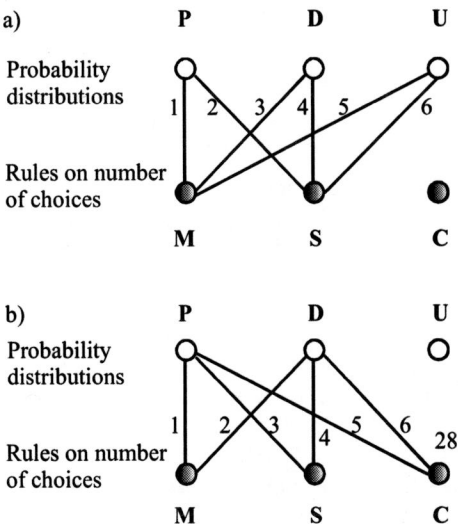

Figure 1.4 Options for choices in reproduction and population updating in GEN-COM: (a) selection for reproduction; (b) selection for population updating. P, proportional; D, deterministic; U, uniform; M, multiple choice; S, single choice; C, constrained choice.

(c) Constrained choice—a member can be selected *eve* times, where parameter *eve*, called expected value, has to be fixed by the user. After each selection the value of *eve* is reduced by 1 if the member has been chosen for crossover or 0.5 if chosen for mutation operation. The member with parameter *eve* equals zero cannot be selected any further.

Condition (b) is not applied at the stage of selection for reproduction in point 7a of the algorithm since it can limit reproduction of best-fitted members.

The overall selection mechanism requires fixing of both choices: the probability distribution type and the rules on number of choices. Six different combinations are possible for selection to reproduction in point 7a and for updating the population. They are shown schematically in Figure 1.4.

To deal with both inequality and equality constraints, two types of penalty mechanisms are available: death penalty and penalty terms in the augmented goal functions with user specified penalty coefficients.

Genetic operators – crossovers and mutations – are often aimed at processing continuous or discrete variables. However, those operators that process discrete valued chromosome can be directly changed into those that can be applied for continuous variables. It suffices to change only the way of generating random numbers, which are used in the operators. The user of GEN-COM has access to various popular operators. To make the explanation clear and compact we assume in the following that an operator processes either a continuous variable or a discrete one. If it can act on both types of variable the proper information will be added. To describe operators we will apply the following notation for chromosome of population member M:

$$\{m_1, m_2, \ldots, m_k, \ldots, m_n\}$$

The following genetic operators are available in GEN-COM.

1. Simple crossover—acts for continuous and discrete variables. Two parents, M1 and M2, exchange information by interchanging segments of their chromosomes. For selected parents and randomly chosen cutting position "k" in both chromosomes the operation is defined by:

$$\begin{aligned} \{m1_1, m1_2, \ldots, m1_k, \ldots, m1_n\} + \{m2_1, m2_2, \ldots, m2_k, \ldots, m2_n\} \\ \{m1_1, m2_2, \ldots, m1_k, m2_{k+1}, \ldots, m2_n\} \\ + \{m2, m2_2, \ldots, m2_k, m1_{k+1}, \ldots, m1_n\} \end{aligned} \tag{1.71}$$

2. Arithmetic crossover—processes only continuous variables. This operates from two parents (M1, M2), and gives only one offspring—the linear combination of the parents is as follows:

$$\begin{aligned} \{m1_1, m1_2, \ldots, m1_k, \ldots, m1_n\} + \{m2_1, m2_2, \ldots, m2_k, \ldots, m2_n\} \\ \{r \cdot m1_1 + (1-r)m2_1, \ldots, r \cdot m1_n + (1-r)m2_n\} \end{aligned} \tag{1.72}$$

Parameter r is a random number from uniform distribution bounded by 0.0 and 1.0.

3. Heuristic crossover—similar to the arithmetic crossover, one offspring P is produced from two parents M1 and M2—formula (1.72). However, the offspring's chromosome depends on the fitness functions of parents according to Equations (1.73a,b):

$$\{m_1, m1_2, \ldots, m1_k, \ldots, m1_n\} + \{m2_1, m2_2, \ldots, m2_k, \ldots, m2_n\} \rightarrow P \tag{1.73}$$

if $FP(M1) > FP(M2)$

$$\begin{aligned} P = \{m1_1 + r(m1_1 - m2_1), m1_2 + r(m1_2 - m2_2), \ldots, m1_n \\ + r(m1_n - m2_n)\} \end{aligned} \tag{1.73a}$$

else

$$\begin{aligned} P = \{m2_1 + r(m2_1 - m1_1), m2_2 + r(m2_2 - m1_2), \ldots, m2_n \\ + r(m2_n - m1_n)\} \end{aligned} \tag{1.73b}$$

Parameter r is a random number from uniform distribution bounded by 0.0 and 1.0

4. Mixed crossover—operates on both types of variable. It requires two parents (M1 and M2) and is similar to simple crossover. The difference is that instead of changing segments of chromosomes only a single element is changed in sequence between parents. The algorithm works as follows:
    - Two pseudo-random numbers A and a_1 are chosen from uniform distribution bounded by 0.0 and 1.0
    - IF $a_1 < A$ THEN m_1 from M1 is inserted into position 1 of offspring chromosome

- **ELSE** m_2 from M2 is put into position 1 of offspring chromosome
- Pseudo-random numbers $a_2, a_3, \ldots$ until the total number of parameters chosen and the procedure for changing elements is repeated.

5. Uniform mutation—processes continuous and discrete variables. One parent M is chosen and, then, randomly chosen position "k" of chromosome is mutated in such a way that its value is changed by the value chosen from the entire range—see Equation (1.61). Selection of the position in chromosome and choice of the value is performed using uniform distribution.

6. Nonuniform mutation—operates on continuous variables. This operator differs from the previous one since the value of parameter m_k of parent M is chosen from the range given by:

$$(m_k - \Delta; m_k + \Delta) \tag{1.74}$$

The + or − sign is chosen with identical probability of 0.5.
Parameter Δ is defined by the formula:

$$\Delta = (m^u - m^1) * (1 - r^{(1 - n/NPOK) * B}) \tag{1.74a}$$

where n is the current number of generation, r is the random number from uniform distribution bounded by 0.0 and 1.0, NPOK is the current number of generations, and B is mutation parameter (default value 2.0).
It is important to note that parameter Δ is the function of the number of populations generated to the moment and it increases with an increase of this number. In consequence, the range of values for parameter "k" diminishes during the course of optimization.

7. Local mutation—randomly selected position k from also randomly chosen individual M is mutated according to:

$$m_k + m_k \pm \Delta \tag{1.75}$$

where $\Delta = (m^u - m^l)/C$, C is the operator's parameter (default value 2.0).
The + or − sign chosen with identical probability of 0.5.

8. Range limit mutation—acts on both continuous and discrete variables. Randomly chosen position k of an individual's M chromosome is given upper (m^u) or lower (m^l) limiting value with uniform distribution. The operator is useful in the case where an optimum value of a variable is at its upper or lower bound.
To date GEN-COM has been applied to solving HEN retrofit problems (see Chapter 19), and also to the following process engineering problems:

- Optimization of multiproduct batch plant formulated first in Sparrow et al. (1975) and considered also in Grossmann and Sargent (1979), Kocis and Grossmann (1988), Salcedo (1992), Cardoso et al. (1997).
- Gibbs free energy minimization for chemical reaction in White et al. (1958), Luus and Jaakola (1973), Rangaiah (1985), Michalewicz (1996).

The latter task caused no problem for GEN-COM. However, the MINLP problem of batch plant optimization is difficult due to the fact that it is ill-defined with regard to constraints at the global optimum. The analysis showed that changes of decision variables of the order of 0.01% around optimum values (in both directions) causes the solution to become infeasible. This causes serious obstacles for stochastic approaches. For the larger data set with five products and six stages (for multiproduct batch plant), solutions far from the optimum are reached, similar to the ARS algorithm in Section 1.2. Also, advanced deterministic approaches failed to solve the case without linearization—see for example Kocis and Grossmann (1988). However, the reformulation proposed by Bochenek et al. (2001, 2005) allows solution of the NLP problem instead of the original MINLP one. GEN-COM was able to solve the problem using 560 000 goal function evaluations.

The solver has been tested for some other test problems in Michalewicz (1996). Due to numerous available options and control parameters it seems almost impossible to find a general best parameter tuning. The algorithm is a rather general GA optimizing framework, giving the user the opportunity to select options.

1.6.4. Simulating Annealing

Simulating annealing (SA) is based on an analogy with annealing ideal crystals in thermodynamics. It is an adaptation for optimization purposes of Metropolis et al.'s (1953) algorithm for simulating thermal moves of molecules at fixed temperature T. Hence, some terms in SA stem from thermodynamics, as for instance "temperature T" applied to name the crucial control parameter. This temperature is reduced systematically in SA procedures. Rapid cooling produces irregularities in crystal structures that do not reach a minimum of energy level. In contrast, a very slow cooling scheme provides ideal crystals with minimum attainable energy but may require a prohibitive calculation time. Movements of molecules, particularly at high temperatures, are chaotic and, hence, there is a possibility of not reaching the minimum energy for these temperatures. In SA optimization, mechanisms mimic this by allowing the acceptance of a worse solution. The probability of such acceptance depends on temperature and should approach zero for low temperatures. The chance of accepting a temporarily worse solution is an important feature of SA strategy. This mechanism should allow the SA algorithm to escape from a local optimum. SA was the first optimization approach that employed such a mechanism; this was followed by tabu search.

A general SA algorithm is as follows.

1. Fix initial temperature (T^0).
2. Generate starting point X^0 (this is the best point X^* at present).
3. Generate randomly point X^s (neighboring point).
4. Accept X^s as X^* (currently best solution) if an acceptance criterion is met. This has to be such a condition that the probability of accepting a worse point is greater than zero, particularly at higher temperatures.
5. If an equilibrium condition is satisfied go to 6 otherwise jump back to 3.

6. If termination conditions are not met, decrease temperature according to a cooling scheme and jump back to 3. If termination conditions are satisfied stop calculations, accepting current best value X^* as final ("optimal") solution.

In general, application of SA in chemical and process engineering is similar to that of GA, though SA is an older technique. Various versions of the basic SA method have been suggested to date. Foundations of SA are addressed in a seminal work by Kirkpatrick et al. (1983) and the book by Aarst and Korst (1989). Various algorithms, often with solutions of some test problems, have been presented among others by: Vanderbilt and Louie (1984), Dueck and Scheuer (1990), Press and Teukolsky (1991), Press et al. (1992), Cardoso et al. (1994, 1996, 1997), Maier and Whiting (1998), Sahin and Ciric (1998), Romeijn et al. (1999), Locatelli (2000), Chen and Su (2002), Chapter 8 in Spall (2003), Jeżowski et al. (2003a, 2003b, 2003c), Őzcelik and Őzcelik (2004). The following works reported application of SA to optimizing process systems: Dolan et al. (1989, 1990), Das et al. (1990), Patel et al. (1991), Ku and Karimi (1991), Floquet et al. (1994), Maia et al. (1995), Maia and Qassim (1997), Murakami et al. (1997), Athier et al. (1997, 1998), Tantimurata et al. (2000, 2001), Poplewski et al. (2002), Jeżowski et al. (2003a, 2003b), Linke and Kokossis (2003), Poplewski (2004), Poplewski and Jeżowski (2003, 2005). Some of these papers dealt with batch process scheduling. Hanke and Li (2000) and Li et al. (2000) applied SA to optimize processes and apparatus in dynamic operation mode. Ourique and Telles (1998), Marcoulaki and Kokossis (2000a, 2000b) and Marcoulaki et al. (2000) employed the technique to attack product design problems. Huber (1994), Zhu et al. (2000) and Teh and Rangaiah (2002) calculated phase equilibrium with SA. Various other applications were reported in Narayan and Diwekar (1996) and Salcedo and Lima (1999). Chaudhuri and Diwekar (1997a, 1997b) used SA for synthesis under uncertainty while Sahin and Ciric (1998) employed it to bilevel optimization. Choi et al. (1999) proposed applying SA with an interval technique to achieve global optimum.

One of the key decisions in SA is a scheme of neighboring point choice. We will concentrate on a method first suggested by Press and Teukolsky (1991). It consists in embedding a well-known simplex optimization approach (Nelder–Mead procedure in particular) into the SA framework. Thus, the method will be referred to as the SA/S algorithm (simulated annealing with simplex). The simplex algorithm is a quite efficient, simple direct optimizer. Hence, its application allows retention of the basic feature of stochastic (meta-heuristic) approaches—ease of application to "ill-behaved" problems. It is of importance that by including the simplex procedure we replace a heuristic, stochastic scheme of generating neighboring points, for instance that from Vanderbilt and Louie (1984), with an efficient, robust local optimizer. However, embedding simplex also has a negative effect since it is limited to continuous variables and does not account directly for constraints.

In the following we will address an algorithm of Jeżowski and Poplewski (Jeżowski et al., 2001b, 2003a, 2003b, 2003c), called the SA/S-1 procedure. It is limited to NLP problems but accounts for constraints. The procedure is similar to the SIMPSA algorithm by Cardoso et al. (1996). Notice that the extension of SIMPSA from Cardoso et al. (1997), called M-SIMPSA, is aimed at MINLP problems. The following explanation will focus on the NLP problem defined by Equations (1.41)–(1.43).

Since the simplex algorithm is deterministic it was necessary to add certain stochastic mechanisms. Press and Teukolsky (1991) suggested randomly disturbing values of the goal function of "deterministic" simplex corners (FC) according to Equation (1.76) for minimization:

$$FC'_i = FC_i - T \log (r_i) \qquad (1.76)$$

where index i denotes the simplex corner, and r is the random number from the uniform distribution normalized to $(-0.5, 0.5)$.

Note that disturbed values of FC' are "worse" than the original ones. Next, the goal function value of new point n calculated for reflected simplex is improved according to Equation (1.77). Similar to disturbances from Equation (1.76) the improvements depend on temperature:

$$FC'_n = FC_n + T \log (r_n) \qquad (1.77)$$

Such simplex movement will be called downhill reflection (for minimization). In order to embed simplex in the SA algorithm it is sufficient to change point 3 of the general algorithm as follows:

3. Disturb simplex corners according to Equation (1.76), perform reflection, disturb reflected point of the simplex according to Equation (1.77) and accept it as the neighboring point.

Jeżowski and Poplewski also included the possibility of uphill movement following suggestions by Press et al. (1992). The choice of movement type is determined by parameter FPR that is random number RND from the uniform distribution and from range $(0; 1)$. FPR is the control parameter of the optimization procedure.

The movement type criterion is:

$$\text{IF} \begin{cases} FPR > RND \text{ THEN uphill movement} \\ FPR \leq RND \text{ THEN downhill movement} \end{cases} \qquad (1.78)$$

The aim of including the uphill movement in the SA/S algorithm was to provide the additional possibility of escaping from a local optimum. One can expect that this mechanism should lead to improvements of optimization robustness, particularly at low temperatures where the SA/S method becomes the practically deterministic Nelder and Mead (1965) procedure.

In the following we will address basic steps of the SA/S-1 method starting from the acceptance criterion. A brief review of some other proposals is also given.

Acceptance criterion

The most common acceptance criterion is that of Metropolis:

$$
P = \begin{cases} 1 & \text{if } \Delta FC < 0 \\ \exp\left(-\dfrac{\Delta FC}{T}\right) & \text{if } \Delta FC \geq 0 \end{cases} \tag{1.79}
$$

The probability of accepting a worse solution drops with a decrease of T and at $T \to 0$ parameter P approaches zero, too.

Some authors advocate criterion (1.80) of Glauber: a solution is accepted with probability P defined by Equation (1.80) independent of whether it is better or worse.

$$
P = \frac{e^{-\Delta FC/T}}{1 + e^{-\Delta FC/T}} \tag{1.80}
$$

For high values of T the probability P from Equation (1.80) approaches zero while for small T parameter P decreases to zero for poor solutions and increases to 1 for better solutions. Also, a threshold condition suggested by Dueck and Scheuer (1990) is sometimes employed, such that the probability of acceptance is:

$$
P = \begin{cases} 1 & \text{if } \Delta FC < 0 \\ \alpha_i & \text{if } \Delta FC \geq 0 \end{cases} \tag{1.81}
$$

Values of parameters α_i are given in data as sequences of real numbers from range (0; 1) in descending order.

Glauber's scheme is considered less efficient by many authors and criterion (1.81) was tested on a limited number of purely combinatorial problems. Metropolis criterion (1.79) is most often used and was also employed in SA/S-1.

Initial simplex generation

SA/S-1 applies the procedure from Cardoso et al. (1996, 1997). Coordinates $(x_i^s = 1, \ldots, N)$ of the first vertex of a simplex are generated from the uniform distribution according to:

$$
x_i^s = x_i^l + r_i(x_i^u - x_i^l), \quad i = 1, \ldots, p \tag{1.82}
$$

Coordinates of the rest of the vertices are calculated on the basis of x_i^s from:

$$
x_i = x_i^s + ((0.5 - r_i) \cdot (x_i^u - x_i^l)), \quad i = 1, \ldots, p \tag{1.83}
$$

Notice that the initial simplex has to be feasible.

Determination of initial temperature

An iterative method from Cardoso et al. (1996, 1997) was applied that requires a number of simplex reflections at a high value of T. Jeżowski and Poplewski applied $T = 10^6 FC^d$, where FC^d denotes the goal function value of the initial vertex. The following parameters are calculated:

m_1 the number of reflections in which improvements of the goal function were observed
m_2 the number of reflections in which there was no improvement of the goal function

The average increase of goal function value (Δf^+) is calculated from:

$$\Delta f^+ = \frac{\sum_{k=1}^{m_2} (FC_k - FC_{k-1})}{m_2} \tag{1.84}$$

where FC_k, FC_{k-1} are values of the goal function of two successive feasible solutions for such simplex reflection which yielded an increase of the goal function.
Acceptance probability (Px) of a solution at temperature T is defined by:

$$Px = \frac{m_1 + m_2 \cdot \exp(-\Delta f^+ / T)}{m_1 + m_2} \tag{1.85}$$

For fixed Px value we can determine initial temperature T^0, such that the acceptance probability of the first point is not less than Px, from:

$$T^0 = \frac{-\Delta f^+}{\ln((Px \cdot (m_1 + m_2) - m_1)/m_2)} \tag{1.86}$$

In the SA/S-1 algorithm, parameter Px was fixed at 0.95.

Temperature decrease—cooling scheme

Often a simple exponential cooling scheme is used:

$$T^{i+1} = T^i \cdot \Delta \tag{1.87}$$

where $\Delta < 1$ (suggested Δ value is in the range 0.9–0.99).
However, several researchers reported that Equation (1.88), called the adaptive cooling scheme or Aarts and van Laarhoven equation, produces more reliable results with a similar number of goal function evaluations:

$$T^{i+1} = \frac{T^i}{1 + ((T^i \cdot \ln(1 + \delta))/3\sigma^i)} \tag{1.88}$$

where σ^i is the standard deviation of the goal function at T^i (Equation (1.88a)):

$$\sigma^i = \sqrt{\frac{\sum_{k=1}^{ip} (FC_k - FC_a)^2}{ip}} \qquad (1.88a)$$

where ip is the number of feasible solutions found in temperature T^i, FC_k is the value of the goal function of solution k at temperature T^i, and FC_a is the arithmetic mean value of the goal function for all feasible solutions calculated at temperature T^i.

Parameter δ in Equation (1.88) is the control parameter of the adaptive cooling scheme. It can take values greater than zero. For small values, less than 1.0, cooling is slow and, in consequence, optimization robustness increases but so does calculation load. Equation (1.88) with (1.88a) has adaptive features, that is, it accounts for "history" of the optimization by correlating a rate of cooling in terms of a distance from equilibrium state—the higher the standard deviation (the system is far away from the equilibrium), the higher the rate of cooling. Note that the exponential cooling scheme does not have such a feature. Because the literature did not give a clear conclusion as to the choice of cooling scheme in the SA/S method, Jeżowski and Poplewski included both schemes in the solver as options. The options were tested and the results proved the superiority of the adaptive cooling scheme. It required less CPU time to yield a similar optimization of performance in regard to the "quality" of the results as the exponential cooling scheme.

Equilibrium condition—point 6 of the general algorithm

In optimization, the equilibrium condition is, most often, simply the number of points generated at temperature level T. Usually the number is fixed as the control parameter and does not depend on the value of T. The condition in SA/S-1 is:

$$\text{No. of generated points} = K \cdot NDF \qquad (1.89)$$

where NDF stands for the number of degrees of freedom.

Stopping (convergence) criterion

The simplest and an often-applied criterion is to stop calculations if the temperature value is smaller than a given number $T^{\min}$. The SA/S-1 procedure also employs this condition:

$$T \leq T^{\min} \qquad (1.90)$$

Several authors applied conditions that are similar to convergence criteria in deterministic methods but account for a stochastic mechanism. For instance, the following conditions can be used:

- Stop calculations if for a certain number of iterations there is only very small improvement of the goal function
- Stop calculations if for a certain number of iterations there is only very small improvement of the average value of the goal function.

One of the above conditions can be applied together with Equation (1.90) or as a single criterion. Jeżowski et al. (2003c) suggested that it would be a good solution; however, it requires an additional control parameter—a number of iterations for which additional conditions should be checked.

Mechanism of dealing with inequality constraints

The problem of dealing with inequality constraints is of importance in SA/S due to the use of the simplex method. There are two types of inequality constraints: limits on variables, often called *explicit* constraints such as Equation (1.43), and typical constraints (Equation (1.42)) called *implicit* constraints. If in a solution generated by simplex movement variable x_i is beyond the upper or the lower bound it is simply set at the lower or the upper limit. In the case of *implicit* inequality constraints there are several techniques available, including penalty terms in augmented goal function, death penalty or repairing option. In SA/S-1 the procedure taken from Cardoso et al. (1996, 1997) was adopted where violated constraints are "maintained" by randomly generated values of variables—a technique similar to repairing solutions in genetic algorithms. Randomly generating points until the violated constraints are met performs the reparation. Notice that an alternative mechanism similar to that in complex algorithms is also possible.

Figure 1.5 shows the flow diagram of the optimization procedure SA/S-1 for problems with inequality constraints.

Control parameters settings

The SA/S-1 method requires four control parameters for the adaptive cooling scheme: control parameter δ in the adaptive cooling scheme (Equation (1.88)), parameter K in the equilibrium criterion (Equation (1.89)), parameter INV controlling inverse movements (Equation (1.79)) and final temperature T^{min} in the convergence criterion (Equation (1.90)). Jeżowski et al. (2003b, 2003c) concluded that there is no universal best parameter setting, unlike Maier and Whiting (1998), who claimed that for the SA method the setting is independent of the problem. However, the conclusions of Jeżowski et al. are in accordance with observations gained from other meta-heuristic approaches. More importantly, Jeżowski et al. (2003b, 2003c) have found that there are limiting values of parameters beyond which optimization robustness improves very slightly. It is of importance that these limiting values of parameters of the SA/S approach are practically problem-independent. The specific conclusions of Jeżowski et al. (2003b, 2003c) are listed below.

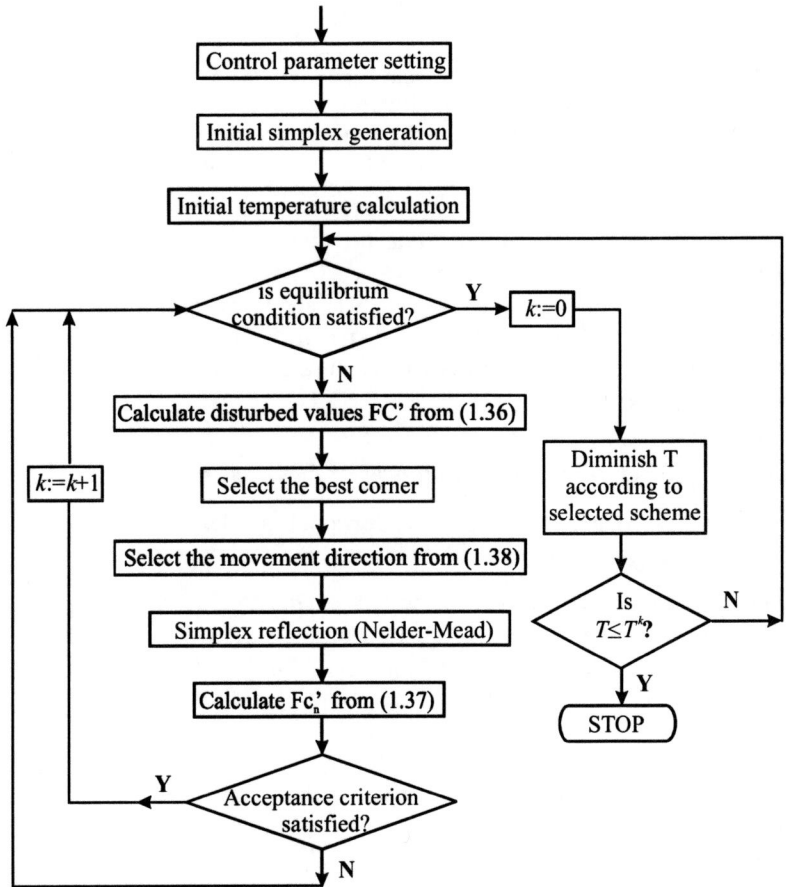

Figure 1.5 Flow sheet of SA/S-1 algorithm.

1. Parameter *INV* controlling the inverse movement of simplex increases the performance of the optimization but should not be higher than about $0.6 \div 0.7$.
2. Parameter *K* – coefficient in equilibrium criterion – should not be higher than 70. Smaller values are sufficient for small and medium-size problems.
3. Parameter δ in the adaptive cooling scheme should not be less than 0.2. Values of the order of 1.0 or higher can be applied for easy optimization problems. Because of the almost regular influence of δ on optimization performance one can find good values of δ in a small number of trials.
4. Since the T^{min} value in the termination criterion does not influence CPU time, small values less than 1.0 can usually be used. The good values of T^{min} can be roughly estimated on the basis of optimization problem dimensionality—the more variables, the less value of T^{min} should be applied.

The SA/S-1 algorithm has been tested on some benchmark global optimization problems with constraints taken from Michalewicz (1996). Also, the following chemical engineering models have been solved:

- Gibbs free energy minimization for chemical reaction from White et al. (1958), Luus and Jaakola (1973), Rangaiah (1985), Michalewicz (1996)
- Optimization of alkylation plant from Amarger et al. (1992), Ryoo and Sahinidis (1995), Zamorra and Grossmann (1998b)
- Optimization of two reactors in the series from Manousiouthakis and Sourlas (1992), Ryoo and Sahinidis (1995), de Gouvea and Odloak (1998), Maranas and Floudas (1997).

The last problem appeared demanding for the SA/S-1 method due to simplex degeneration. To circumvent this a proper choice of decision variables for dealing with equalities has to be applied. Usually, the mechanism of repairing solutions should be modified. Generally, the SA/S-1 algorithm is fast, even an order of magnitude faster than ARS and GA procedures. However, it is limited to NLP problems at present.

1.6.5. Equality constraints handling in ARS, GA, and SA

Generally in all stochastic optimization approaches at least three ways of dealing with equality constraints are available (linearization is not counted here):

1. Application of penalty terms in augmented goal function.
2. Replacement of each equality (1.91) by a pair of inequalities (1.91a,b)

$$h_m(X) = 0, \quad m = 1, ..., M \tag{1.91}$$

$$h_m(X) + \beta_m \geq 0, \quad m = 1, ..., M \tag{1.91a}$$

$$h_m(X) - \beta_m \leq 0, \quad m = 1, ..., M \tag{1.91b}$$

3. Direct solution of equations.

The first two mechanisms are general and easy to use but usually inefficient (in the authors' opinion) in all the techniques considered in this chapter. For augmented goal function we have to choose the form of penalty terms, penalty coefficients and scheme of their changes (if any). Despite numerous suggestions no efficient and general solutions have been found to date—see for example Michalewicz and Fogel (2002) for a discussion of penalty terms in evolutionary approaches. Application of inequalities requires selection of appropriate parameters β. They should be very small numbers, which result in a small feasible solutions space and, in effect, make stochastic optimization inefficient. Direct solution is the most efficient solution to equality constraints, assuming that one is able to make the equality constraints linear with respect to decision variables.

Let variables x_i $(i = 1, ..., p)$ be divided into two groups: decision variables and dependent (state) ones. The values of the former are generated by a stochastic optimization approach while the latter are calculated from equations that are linear with respect to them (for fixed decision variables). Let us consider an illustrative example of constraints (Equation (1.92)):

$$\sum_{i=1}^{p} x_{i,m} y_{i,m} = 0, \quad m = 1, ..., M \tag{1.92}$$

By choosing x_i or y_i as the decision variables we transform constraints (1.92) into the set of linear equations with respect to y_i or x_i, respectively. Notice that constraints (1.92) are sums of bilinear terms $(x_i \, y_i)$. Such terms are common in process engineering problems since they occur in mass and heat balances as products of flow rate and concentration or flow rate and temperature. Hence, balances of processes can be easily dealt with in this manner. Additional possibilities for direct solution of equality constraints result from the sequential procedure of solving constraints that can be applied in stochastic optimization approaches. To show this we employ the next illustrative example:

$$\min (x_1 x_2 x_3) \tag{1.93}$$

s.t.

$$x_1 x_2 = 0 \tag{1.94}$$

$$x_3 \ln(x_2) = 0 \tag{1.95}$$

Let x_1 be the decision variable, and the stochastic optimization algorithm generates its value. Then x_2 can be calculated from Equation (1.94). Finally, the value of x_3 is determined from Equation (1.95) which became linear with regard to x_3.

The example contains three variables and two equations and, thus, has one degree of freedom. It was possible to reduce the number of decision variables to the number of degrees of freedom. In Chapters 19 and 20 we will show that for certain complex process engineering optimization problems such as water network design and heat exchanger network retrofit we can also achieve such an outcome. A substantial reduction in the number of decision variables allows the solution of even large-scale optimization problems with stochastic/meta-heuristic approaches in reasonable CPU time. The reader is also referred to Luus et al. (2002) for additional, more complex examples.

2 Dynamic optimization problems

2.1. DISCRETE REPRESENTATIONS AND DYNAMIC PROGRAMMING ALGORITHMS

In optimization, a process is regarded as dynamic when it can be described as a well-defined sequence of steps in time or space. Dynamic processes can be either discrete or continuous. Cascades (Figure 2.1), which are systems characterized by sequential arrangement of stages, are examples of dynamic discrete processes. The stages can be of finite size, in which case the process is "inherently discrete", or may be infinitesimally small. The latter case refers to a limiting situation where the concept of very many steps serves to approximate the development of a continuous process. In general, optimization theories of discrete and continuous processes differ in their assumptions, formal descriptions and strength of optimality conditions; thus they usually constitute two different fields. Here, however, for brevity, we present a heuristic derivation of optimization conditions focusing on those which in many respects are common for both discrete and continuous processes. Consequently we shall formulate first a basic discrete algorithm for a general model of a discrete cascade process, and then consider its limiting properties when the number of infinitesimal discrete steps tends towards infinity.

The method of dynamic programming (DP; Bellman, 1957; Aris, 1964; Findeisen et al., 1980) constitutes a suitable tool to handle optimality conditions for inherently discrete processes. Yet, the method only enables an easy passage to

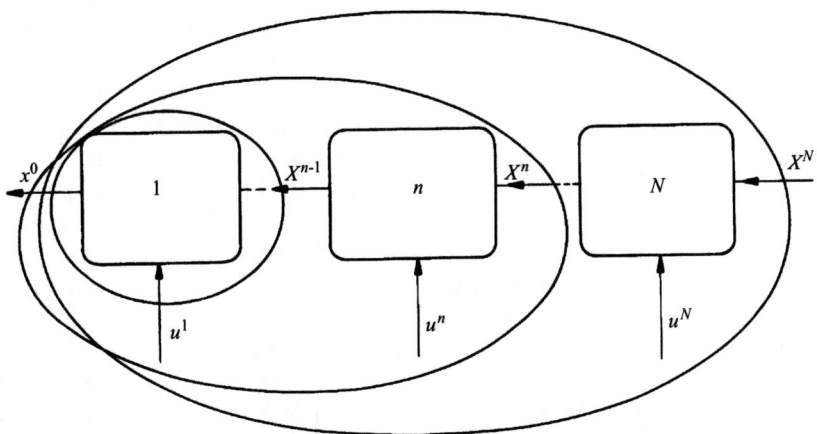

Figure 2.1 Backward optimization algorithm and typical mode of stage numbering in the dynamic programming method. The results are generated in terms of the initial states $\mathbf{x}^n$.

Energy Optimization in Process Systems
©2013 Elsevier Ltd. All rights reserved.

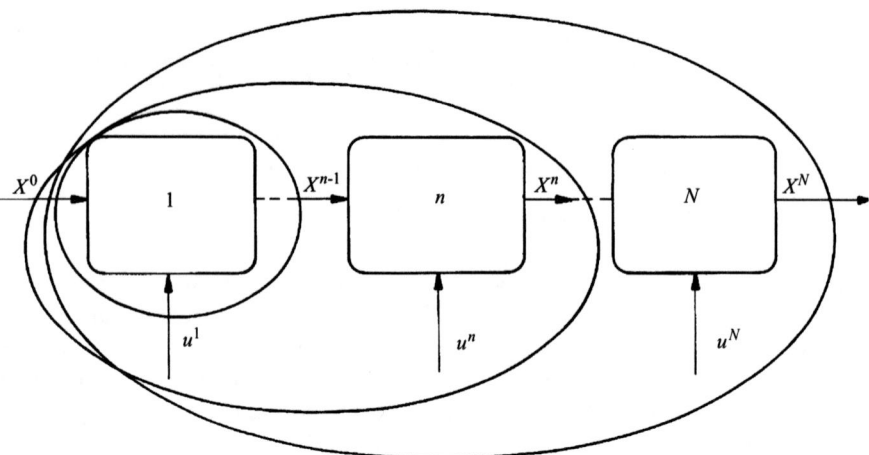

Figure 2.2 Forward optimization algorithm; the results are generated in terms of the final states x^n.

its limiting form for continuous systems under the differentiability assumption. Application of the method is straightforward when it is applied in optimization of control systems without feedback. Dynamic programming is crucial for the existence of the optimal performance potentials discussed in this book, and for the derivation of pertinent equations which describe these potentials. The DP method is based on Bellman's principle of optimality, which makes it possible to replace the simultaneous evaluation of all optimal controls by sequences of local evaluations at sequentially included stages, for evolving subprocesses (Figures 2.1 and 2.2).

Let us focus first on Figure 2.1, where the optimal performance function is generated in terms of the initial states and initial time. The principle of optimality may then be stated as follows: *In a continuous or discrete process which is described by an additive performance criterion, the optimal strategy and optimal profit are functions of the initial state, initial time and (in a discrete process) total number of stages.* A consequence of this property is that each final segment of an optimal path (continuous or discrete) is optimal with respect to its own initial state, initial time and (in a discrete process) the corresponding number of stages. An easy proof of this formulation by contradiction uses the additivity property of the performance criterion (Aris, 1964).

The above formulation of the optimality principle refers to the so-called backward algorithm of the dynamic programming method (Figure 2.1). In this mode, the recursive procedure for applying a governing functional equation begins at the final process state and terminates at its initial state. Consequently, local optimizations take place in the direction opposite to the direction of physical time or the direction of flow of matter. (The process to which this can be applied may be arbitrary: it may be discrete by nature or may be obtained by the discretization of an originally continuous process.) The state transformations possess in the backward algorithm their most natural form, as they describe output states in terms of input states and controls at a stage. The optimization at a stage and optimal functions recursively involve the information generated in earlier subprocesses.

However, one may also generate the optimal profit function in terms of the final states and final time. The optimality principle then has a dual form: *In a continuous or discrete process, which is described by an additive performance criterion, the optimal strategy and optimal profit are functions of the final state, final time and (in a discrete process) total number of stages.* A basic consequence of this property is that each initial segment of the optimal path (continuous or discrete) is optimal with respect to its final state, final time and (in a discrete process) the corresponding number of stages. This formulation refers to the so-called forward algorithm of the dynamic programming method. In this algorithm the recursive optimization procedure for solving the governing functional equation begins from the initial process state and terminates at its final state. With the forward DP algorithm, one makes local optimizations in the direction of real time.

It is the dual (forward) formulation of the optimality principle and the associated forward algorithm that we commonly apply to multistage processes considered later in this chapter. The state transformations used in this case have the form which describes input states in terms of output states and controls at a process stage. Transformations of this sort are directly obtained for multistage processes with an ideal mixing at the stage, otherwise the inverse transformations (applicable to the backward algorithm) might be difficult to obtain in an explicit form. Again, as in the case of the original form of the optimality principle, its dual form makes it possible to replace the simultaneous evaluation of all optimal controls by successive evaluations for evolving optimal subprocesses.

In the continuous case under the differentiability assumption the method of dynamic programming leads to a basic equation of optimal continuous processes called the Hamilton–Jacobi–Bellman equation which constitutes a control counterpart of the well-known Hamilton–Jacobi equation of classical mechanics (Rund, 1966; Landau and Lifshitz, 1971). Moreover, as we shall see later, a similar equation can be derived for special discrete processes: those with unconstrained time intervals θ^n.

2.2. RECURRENCE EQUATIONS

Suitable representation of the optimality principle is contained in recurrence equations. To introduce them, we need first some definitions. We assume that state $\mathbf{x}^n$ and controls $\mathbf{u}^n$ at arbitrary stage n are respectively represented by s- and r-dimensional column vectors $\mathbf{x}^n = [x_1^n, x_2^n, \ldots, x_s^n]^T$ and $\mathbf{u}^n = [u_1^n, u_2^n, \ldots, u_r^n]^T$. Controls are usually constrained to lie in an admissible manifold $\mathbf{u}^n \in U$. A sequence of (optimal) decisions is called (optimal) "strategy" or (optimal) "policy." One may also note that in differential equations of classical variational calculus and analytical mechanics, controls are usually identical to process rates, and the dimensionality of the control vector is that of the spatial state.

We shall focus on Figure 2.1, which depicts the backward optimization algorithm. At each process stage inlet state $\mathbf{x}^n$ is transformed into an outlet state $\mathbf{x}^{n-1}$ depending on an external control $\mathbf{u}^n$. In the vector notation, the state transformation has the form

$$\mathbf{x}^{n-1} = \mathbf{T}^n(\mathbf{x}^n, \mathbf{u}^n) \tag{2.1}$$

Assume that an optimization problem requires us to find a set of controls $\mathbf{u}^n = u_1^n, u_2^n, \ldots, u_r^n$ which satisfy $\mathbf{u}^n \in \mathbf{U}$ and minimize an additive performance criterion

$$S(\mathbf{x}^N, \mathbf{u}^N, \ldots, \mathbf{u}^1) \equiv \sum_{n=1}^{N} T_0^n(\mathbf{x}^n, \mathbf{u}^n), \quad n = 1, 2, \ldots, N \tag{2.2}$$

The discrete functional S is a generalized cost which has to attain a minimum in an optimal process (in an alternative formulation the negative of S, which has the meaning of a profit, can be minimized). Local process constraints (2.1) link the state vector, $\mathbf{x}$, with the control vector $\mathbf{u}$; only their explicit form is specified here.

In the dynamic programming method we define the so-called optimal value function or optimal performance function which describes the extremum value of the performance index $S(\mathbf{x}^n, \mathbf{u}^n, \ldots, \mathbf{u}^1)$. For the backward algorithm depicted in Figure 2.1,

$$f^n(\mathbf{x}^n) \equiv \min_{\mathbf{u}^1, \ldots, \mathbf{u}^n} \sum_{i=1}^{n} T_0^i(\mathbf{x}^i, \mathbf{u}^i), \quad i = 1, 2, \ldots, n \tag{2.3}$$

that is, the function refers to the final subprocess composed of the stages $i = 1, 2, \ldots, n$ for which the inlet state is $\mathbf{x}^n$. Since the number of controls at each stage equal r, then, for $n = N$, Equation (2.3) describes an $N \times r$-dimensional optimization problem. The dynamic programming method transforms this problem into the successive solution of N optimization problems, each partial problem being r-dimensional.

For the one-stage subprocess, Equation (2.3) takes the form

$$f^1(\mathbf{x}^1) = \min_{\mathbf{u}^1} T_0^1(\mathbf{x}^1, \mathbf{u}^1) \tag{2.4}$$

This optimization should take into account constraints imposed on the terminal state $\mathbf{x}^0$. Equation (2.4) describes the first optimization problem among the sequence of those for $i = 1, 2, \ldots, n$. The next components of the sequence can be determined by writing Equation (2.3) in the form

$$f^n(\mathbf{x}^n) = \min_{\mathbf{u}^n} \min_{\mathbf{u}^{n-1}}, \ldots, \min_{\mathbf{u}^1} \{T_0^n(\mathbf{x}^n, \mathbf{u}^n) + \sum_{i=1}^{n-1} T_0^i(\mathbf{x}^i, \mathbf{u}^i)\} \qquad (2.5)$$

Since the decisions undertaken after stage n do not influence the process at stage n (no feedback), Equation (2.5) can be written as follows:

$$f^n(\mathbf{x}^n) = \min_{\mathbf{u}^n} \{T_0^n(\mathbf{x}^n, \mathbf{u}^n) + \min_{\mathbf{u}^{n-1}} \min_{\mathbf{u}^{n-2}}, \ldots, \min_{\mathbf{u}^1} \sum_{i=1}^{n-1} T_0^i(\mathbf{x}^i, \mathbf{u}^i)\} \qquad (2.6)$$

or

$$f^n(\mathbf{x}^n) = \min_{\mathbf{u}^n} \{T_0^n(\mathbf{x}^n, \mathbf{u}^n) + \min_{\mathbf{u}^{n-1}, \ldots, \mathbf{u}^1} \sum_{i=1}^{n-1} T_0^i(\mathbf{x}^i, \mathbf{u}^i)\} \qquad (2.7)$$

Using in this expression the definition of the optimal performance function, Equation (2.3), leads to the equation

$$f^n(\mathbf{x}^n) \equiv \min_{\mathbf{u}^n} \{T_0^n(\mathbf{x}^n, \mathbf{u}^n) + f^{n-1}(\mathbf{x}^{n-1})\} \qquad (2.8)$$

in which output state $\mathbf{x}^{n-1}$ should be evaluated in terms of input $\mathbf{x}^n$ and control $\mathbf{u}^n$. Applying in Equation (2.8) transformation (2.1) leads to the basic equation of dynamic programming:

$$f^n(\mathbf{x}^n) = \min_{\mathbf{u}^n} \{T_0^n(\mathbf{x}^n, \mathbf{u}^n) + f^{n-1}[T^n(\mathbf{x}^n, \mathbf{u}^n)]\} \qquad (2.9)$$

An example is the recurrence equation of a multistage process in which the fluid with specific heat c is heated in a cascade of heat-pumps (Figure 4.3 in Chapter 4) (Figure 2.3):

$$R^n(T^n, t^n) = \min_{\mathbf{u}^n} \left[c \left(1 - \frac{T^e}{T^n + \chi u^n} \right) u^n \theta^n + R^{n-1}(T^n - u^n \theta^n, t^n - \theta^n) \right] \qquad (2.10)$$

where T^n is the temperature at stage n, R^n is the optimal cost function of n-stage subprocess and the c term is specific work consumed at stage n. Because of the work consumption the model involves optimal cost function R rather than an optimal profit function $V = -R$. The control variable u^n is the rate of the temperature change T^n with respect to holdup time t^n; the proportionality coefficient χ is the reciprocal of a time constant and T^e is the temperature of the

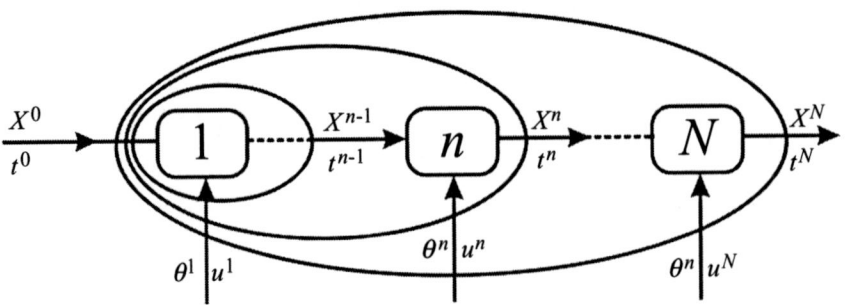

Figure 2.3 A scheme of a multistage control with distinguished time interval, described by the forward algorithm of the dynamic programming method. Ellipse-shaped balance areas pertain to sequential subprocesses which grow by inclusion of proceeding units.

environment. θ^n is the interval of time t^n at stage n, and the minimization should also include θ^n in models where θ^n is a controlled quantity.

The solving procedure for Equation (2.9) runs successively for $n = 1, 2, \ldots, N$, starting with $f^0 = 0$. As a rule a computer is used to solve the problem numerically. At each stage n the previously obtained data of $f^{n-1}(x^{n-1})$ serve to search for optimal controls $\hat{u}^n$ in terms of the coordinates of the inlet states x^n, that is, the optimal vector function $\hat{u}^n(x^n)$ is determined. The optimal sum of T_0^n and f^{n-1} determined for various x^n constitutes the function $f^n(x^n)$. This function is next used to calculate data at stage $n+1$, and so on. When the Nth stage is achieved the initial condition is applied, that is, an optimal process trajectory is found for a definite value $x^n = \alpha$. Simultaneously the corresponding sequence of optimal controls $\hat{u}^n$ is determined.

The virtue of the method is its relative simplicity in handling constraints imposed on controls and/or state. A major flaw of the method is the "curse of the dimensionality," which causes the number of computations to increase dramatically with the dimensionality of state x^n. This serious flaw causes the computational DP algorithms to work efficiently only in problems with low dimensionality of the state vector. However, the dynamic programming method constitutes a general and insightful mathematical tool which allows one to derive and illustrate the structure of basic methods of dynamic optimization, such as Pontryagin's Maximum Principle, variational calculus, and some others, often for both discrete and continuous processes. Some algorithms related to these methods are outlined below.

Analogous algorithms and computational procedures can be designed in the context of the forward DP algorithm (Sieniutycz, 1991). For the forward–backward duality, described above, the performance index is the same in the original and dual problem, yet the numerically generated functions of optimal performance, which describe generalized or time-dependent potentials, are different. Regardless of different, appropriate boundary conditions for f^n (initial in the forward algorithm and final in the backward algorithm), optimal functions of the forward algorithm are potentials in the space of final states whereas those of the backward algorithm are potentials in the space of initial states (remember that the state vector defined here includes the time coordinate, and in discrete problems, total number of stages, N, is the additional variable). Thus, in general, two numerical DP approaches generate different functions which characterize potentials. Only in analytical approaches to the case in which both potential functions include coordinates of fixed-end point as parameters do analytical expressions for forward and backward potentials coincide. Examples of such approaches are shown by Sieniutycz and Berry (2000); with suitable equations obtained in this case, a special problem with free boundary conditions can be treated by considering variations of boundary points.

2.3. DISCRETE PROCESSES LINEAR WITH RESPECT TO THE TIME INTERVAL

In the discrete theory developed in this section a class of multistage control processes is considered which are linear with respect to the residence time interval, or a state variable interval. Applying discrete dynamic programming, the necessary optimality conditions can be determined in a form which contains a discrete equation with a delayed time argument which is of the Hamilton–Jacobi type (see Equation (2.27) below)). To avoid major difficulties associated with solving models of this sort, these conditions are transformed to a form described by a discrete Hamiltonian. It then follows that in multistage autonomous systems with free intervals of time, a Pontryagin-like Hamiltonian emerges which is constant along the optimal discrete trajectory. From a physical standpoint, this constant Hamiltonian condition is a generalization of the energy conservation condition as applied to optimal discrete systems with free intervals of time. A canonical formalism analogous to those in analytical mechanics and the optimal control theory of continuous systems is introduced and analyzed. Benefits and limitations of such a theory are also discussed.

Below, Bellman's method of dynamic programming (DP) is applied to derive necessary optimality conditions for both discrete and continuous processes in terms of canonical equations. A general approach to multistage processes, developed in the form of the so-called stage criterion (Boltyanski, 1969), which applies the DP algorithm in the spirit of Caratheodory's (1935) geometrization of variational calculus, allows the transfer of DP results to a discrete algorithm with a constant Hamiltonian. This discrete algorithm is quite suitable as a computational tool (Berry et al., 2000; Szwast, 1990, 1994; Poświata and Szwast, 2000; Poświata, 2005). See also some earlier, associated works, for example, Denn and Aris (1965) and Jordan and Polak (1964a, 1964b).

Whenever the mathematical model of a cascade is linear with respect to holdup time θ^n or some measure of this time (the case of our multistage models used here), the algorithm of cascade optimization exhibits properties close to those of classical Hamiltonian systems. As these properties are essential in both theoretical and numerical aspects of optimization, they are analyzed below. Discrete models nonlinear with respect to the intervals θ^n or those with locally constrained holdup times θ^n are omitted here (see, however, Chapter 7); they can be optimized with the help of quasi-Hamiltonian structures (Poświata, 2005; Sieniutycz, 2006b).

Assume that the variable t^n describes a cumulative holdup time after the first n stages of the cascade. The variable θ^n is the interval of the time t at the stage n; this must be positive and satisfy the following equation (for $n = 1, \ldots, N$):

$$\theta^n = t^n - t^{n-1} \tag{2.11}$$

However, in the present form, the variable θ^n is not treated jointly with other controls. They are components of the control vector $\mathbf{u}^n$, whereas the decision variable θ^n is treated separately because of its linearity in the model. The mathematical model of cascade considered here is linear with respect to the variable θ^n, but it may be nonlinear with respect to the control vector $\mathbf{u}^n$.

The process dynamics are described by discrete equations of state; they are represented by Equations (2.11) and (2.12), whereas the process performance index S is described by Equation (2.13). All these equations are linear with respect to θ^n and arbitrary with respect to every component of the control vector $\mathbf{u}^n$:

$$x_v^n - x_v^{n-1} = f_v^n(\mathbf{x}^n, t^n, \mathbf{u}^n)\theta^n \tag{2.12}$$

$$S^N = \sum_{n=1}^{N} f_0^n(\mathbf{x}^n, t^n, \mathbf{u}^n)\theta^n + G(\mathbf{x}^N, t^N) - G(\mathbf{x}^0, t^0) \tag{2.13}$$

where $n = 1, \ldots, N$ and $v = 1, \ldots, s$. Note that the definition of the state vector x differs here from that used in some texts as it excludes the performance variable and time τ from the state coordinates. The performance index of Equation (2.13) is generally in Bolza form, which involves a function G of end states. This form is useful in optimization of quite involved problems. In particular, the form is suitable to apply economic optimization criteria.

The performance criterion (2.13) has to be maximized subject to the difference constraints (2.11) and (2.12) operative at each stage n. The constraints link the changes of the state vector, $\mathbf{x}^n$, with the control vector, $\mathbf{u}^n$, and the time interval at the stage, θ^n. Restricting ourselves to the situation where the constraining equation is of the form $\mathbf{u}^n = (\mathbf{x}^n - \mathbf{x}^{n-1})/\theta^n$, we may, in particular, treat the discrete counterpart of the continuous constraint $\mathbf{u} = \dot{\mathbf{x}}$, which is the case in variational calculus. This case implies discrete state transformations in the special form $\mathbf{x}^{n-1} = \mathbf{x}^n - \mathbf{u}^n\theta^n$. The function f_0^n describes the discrete rate of profit generation at stage n. Whenever the mathematical model (including f_0^n) does not explicitly contain time t^n, which is the case for an autonomous process, instead of the original Equation (2.13), one can maximize a certain modified criterion of the "net profit" type, which involves the local performance function $f_0'^n \equiv f_0^n - h$, where h is a constant value of the Hamiltonian H^{n-1} and at the same time the numerical value of the related Lagrange multiplier (see example in Chapter 7).

We shall formulate the optimal cascade problem in terms of a Maximum Principle. Related Hamilton's canonical sets are developed that serve to solve the work optimization problem considered in the main text. Convexity properties for rate functions and constraining sets are assumed, which are stronger conditions than those in Pontryagin's theorem of continuous systems.

Equations (2.11)–(2.13) constitute the mathematical model for the discrete processes linear with respect to the unconstrained decision θ^n. Optimization theory for this class of discrete processes, which, in contrast to traditional discrete systems, is similar to Pontryagin's continuous theory, was developed for multistage separation processes and chemical reactors (Sieniutycz, 1973a, 1974, 1978, 1984; Szwast, 1979, 1994; Sieniutycz and Szwast, 1982a, 1982b). Next, applications of the theory to energy systems were initiated (Sieniutycz, 1997a, 1997b, 1997c, 1999b, 2000c; Berry et al., 2000).

Now we consider necessary optimality conditions and a Maximum Principle. We write the difference constraints (2.11) and (2.12) in the form of explicit transformations of state:

$$t^{n-1} = t^n - \theta^n \tag{2.14}$$

$$x_\nu^{n-1} = x_\nu^n - f_\nu(\mathbf{x}^n, t^n, \mathbf{u}^n)\theta^n \tag{2.15}$$

Defining the optimal performance function $V^n(\mathbf{x}^n, t^n)$ for the n-stage subprocess

$$V^n(\mathbf{x}^n, t^n) = \max_{\mathbf{u}^k, \theta^k} \left\{ \sum_{k=1}^{n} f_0^k(\mathbf{x}^k, t^k, \mathbf{u}^k)\theta^k + G^n(\mathbf{x}^n, t^n) - G^0(\mathbf{x}^0, t^0) \right\} \tag{2.16}$$

and applying Bellman's optimality principle (Bellman, 1957, 1961, 1967; Bellman and Dreyfus, 1962), one obtains the recurrence equation

$$V^n(\mathbf{x}^n, t^n) = \max_{\mathbf{u}^n, \theta^n} \{ f_0^n(\mathbf{x}^n, t^n, \mathbf{u}^n)\theta^n + G^n(\mathbf{x}^n, t^n) - G^{n-1}(\mathbf{x}^{n-1}, t^{n-1}) + V^{n-1}(\mathbf{x}^{n-1}, t^{n-1}) \} \tag{2.17}$$

which is valid subject to the transformation Equations (2.14) and (2.15). Its alternative form

$$
\max_{\mathbf{u}^n, \theta^n} \{ f_0^n(\mathbf{x}^n, t^n, \mathbf{u}^n)\theta^n + [V^{n-1}(\mathbf{x}^{n-1}, t^{n-1}) - G^{n-1}(\mathbf{x}^{n-1}, t^{n-1})]
$$

$$
-[V^n(\mathbf{x}^n, t^n) - G^n(\mathbf{x}^n, t^n)] \} = 0 \tag{2.18}
$$

shows that after we define the function

$$
P^n(\mathbf{x}^n, t^n) \equiv V^n(\mathbf{x}^n, t^n) - G^n(\mathbf{x}^n, t^n) \tag{2.19}
$$

Equation (2.17) can be simplified to the form

$$
\max_{\mathbf{u}^n, \theta^n} \{ f_0^n(\mathbf{x}^n, t^n, \mathbf{u}^n)\theta^n + P^{n-1}(\mathbf{x}^{n-1}, t^{n-1})P^n(\mathbf{x}^n, t^n) \} = 0 \tag{2.20}
$$

With state transformation Equations (2.14) and (2.15) we thus obtain

$$
\max_{\mathbf{u}^n, \theta^n} \{ f_0^n(\mathbf{x}^n, t^n, \mathbf{u}^n)\theta^n + P^{n-1}(\mathbf{x}^n - f^n(\mathbf{x}^n, t^n, \mathbf{u}^n)\theta^n, t^n - \theta^n)
$$

$$
-P^n(\mathbf{x}^n, t^n) \} = 0 \tag{2.21}
$$

To take advantage of Equation (2.21) for arbitrary $\mathbf{x}^n$ and t^n, a function is introduced to abbreviate the expression maximized in that equation:

$$
A^n(\mathbf{x}^n, t^n, \mathbf{u}^n)\theta^n \equiv f_0^n(\mathbf{x}^n, t^n, \mathbf{u}^n)\theta^n
$$

$$
+P^{n-1}(\mathbf{x}^n - f^n(\mathbf{x}^n, t^n, \mathbf{u}^n)\theta^n, t^n - \theta^n) - P^n(\mathbf{x}^n, t^n) \tag{2.22}
$$

This is related to the forward algorithm of the dynamic programming, which we use here. Since Equation (2.22) contains only the inputs and outputs at the (nth) stage, we sometimes call the function $A^n(\mathbf{x}^n, \mathbf{u}^n, t^n, \theta^n)$ the stage criterion. It has to satisfy the relationship

$$
A^n(\mathbf{x}^n, t^n, \mathbf{u}^n, \theta^n) \leq 0 \tag{2.23}
$$

where the equality sign refers to optimal processes. Attaining by the function $A^n(\mathbf{x}^n, \mathbf{u}^n, t^n, \theta^n)$ the maximum for every $n = 1, \ldots, N$, provides the necessary condition for the maximum performance index (2.13). Criterion $A^n(\mathbf{x}^n, \mathbf{u}^n, t^n, \theta^n)$ also frequently provides locally sufficient conditions of optimality. It is assumed that $A^n(\mathbf{x}^n, \mathbf{u}^n, t^n, \theta^n)$ can be differentiated with respect to every component of the state vector $\mathbf{x}^n$ and the control vector $\mathbf{u}^n$, as well as with respect to the time variable t^n and its increment θ^n. With this assumption, the necessary optimality conditions describing stationary extremum states maximizing A^n, Equation

(2.13), are

$$\frac{\partial A^n}{\partial x_i^n} = \left\{ \frac{\partial f_0^n}{\partial x_i^n} - \sum_{i=1}^{s} \frac{\partial P^{n-1}}{\partial x_i^{n-1}} \frac{\partial f_i^n}{\partial x_i^n} \right\} \theta^n + \frac{\partial P^{n-1}}{\partial x_i^{n-1}} - \frac{\partial P^n}{\partial x_i^n} = 0 \qquad (2.24)$$

and

$$\frac{\partial A^n}{\partial t^n} = \left\{ \frac{\partial f_0^n}{\partial t^n} - \sum_{i=1}^{s} \frac{\partial P^{n-1}}{\partial x_i^{n-1}} \frac{\partial f_i^n}{\partial t^n} \right\} \theta^n + \frac{\partial P^{n-1}}{\partial t^{n-1}} - \frac{\partial P^n}{\partial t^n} = 0 \qquad (2.25)$$

They are complemented by the stationarity conditions for the controls θ^n

$$\frac{\partial A^n}{\partial \theta^n} = f_0^n - \sum_{j=1}^{s} \frac{\partial P^{n-1}}{\partial x_j^{n-1}} f_j^n - \frac{\partial P^{n-1}}{\partial t^{n-1}} = 0 \qquad (2.26)$$

and the maximum condition for A^n with respect to $\mathbf{u}^n$ expressed in a special form

$$-\frac{\partial P^{n-1}}{\partial t^{n-1}} + \max_{\mathbf{u}^n} \left\{ f_0^n - \sum_{i=1}^{s} \frac{\partial P^{n-1}}{\partial x_j^{n-1}} f_i^n \right\} = 0 \qquad (2.27)$$

which is a discrete counterpart of the Hamilton–Jacobi–Bellman equation for continuous processes. The maximum property holds under convexity assumptions for functions and constraining sets. Equation (2.27) generalizes the stationary extremum condition $\partial A^{n-1}/\partial u_l^n = 0$ or the expression

$$\frac{\partial A^n}{\partial u_l^n} (\theta^n)^{-1} = \frac{\partial f_0^n}{\partial u_l^n} - \sum_{j=1}^{s} \frac{\partial P^{n-1}}{\partial x_j^{n-1}} \frac{\partial f_j^n}{\partial u_l^n} = 0 \qquad (2.28)$$

to the case of the global extremum of A^n in the allowable region U for the controls u_l^n. Equation (2.27) yields the condition (2.28) in the case of the stationary extremum of $\mathbf{u}^n$.

2.4. DISCRETE ALGORITHM OF PONTRYAGIN'S TYPE FOR PROCESSES LINEAR IN θ^N

With the adjoint variables defined by the equations

$$p_j^{n-1} = -\frac{\partial P^{n-1}}{\partial x_j^{n-1}} \qquad (2.29)$$

and

$$p_t^{n-1} = -\frac{\partial P^{n-1}}{\partial t^{n-1}} \tag{2.30}$$

and the discrete Hamiltonian defined by the equation

$$H^{n-1}(\mathbf{x}^n, \mathbf{p}^{n-1}, \mathbf{u}^n, t^n) = f_0^n(\mathbf{x}^n, \mathbf{u}^n, t^n) + \sum_{\nu=1}^{s} p_\nu^{n-1} f_\nu^n(\mathbf{x}^n, \mathbf{u}^n, t^n) \tag{2.31}$$

it is easy to transform Equations (2.24)–(2.26) to the following canonical form:

$$\frac{x_\nu^n - x_\nu^{n-1}}{\theta^n} = \frac{\partial H^{n-1}}{\partial p_\nu^{n-1}} \tag{2.32}$$

$$\frac{p_\nu^n - p_\nu^{n-1}}{\theta^n} = -\frac{\partial H^{n-1}}{\partial x_\nu^{n-1}} \tag{2.33}$$

$$\frac{H^n - H^{n-1}}{\theta^n} = \frac{\partial H^{n-1}}{\partial t^n} \tag{2.34}$$

$$p_t^{n-1} + \max_{\mathbf{u}^n} H^{n-1}(\mathbf{x}^n, \mathbf{p}^{n-1}, t^n, \mathbf{u}^n) = 0 \tag{2.35}$$

So, if the optimal control $\mathbf{u}^n$ lies in the interior of the admissible region

$$\frac{\partial H^{n-1}\left(\mathbf{x}^n, \mathbf{p}^{n-1}, t^n, \mathbf{u}^n\right)}{\partial u_j^n} = 0 \tag{2.35'}$$

These results hold for $n = 1, \ldots, N$; $\nu\{\tau\sigma\} = 1, \ldots, s$ and $l = 1, \ldots, r$. In the case of the boundary extremum, Equation (2.35') is replaced by a maximum of H^{n-1} with respect to $\mathbf{u}^n$, as implied by Equation (2.35). Thus, Equation (2.35) includes the necessary condition for the stationary optimality of the decision vector $\mathbf{u}^n$, Equation (2.35'), if its optimal solution $\mathbf{u}^n$ falls in the interior of the allowable range U. Equation (2.32) constitutes the Hamiltonian form of the state Equation (2.12); Equation (2.33) is its adjoint equation. Equation (2.34) describes the Hamiltonian change at the stage n, whereas Equation (2.35) states that the enlarged Hamiltonian $\tilde{H}^{n-i} \equiv H^{n-1} + p_t^{n-1}$ of an extremal process is always constant and equal to zero.

When a more generally discrete model is applied in which rates f_v^n and f_0^n in Equations (2.12) and (2.13) also contain an explicit time interval θ^n, Hamiltonians H^{n-1} and $\tilde{H}^{n-1}$ are θ^n dependent. In this case canonical Equations (2.32) and (2.33) are still valid but the distinguishing property of control θ^n in the model is lost and the interior extremum condition (2.35') should be referred to as Halkin's Hamiltonian $h^{n-1} \equiv \theta^n \tilde{H}^{n-1}$ rather than to $\tilde{H}^{n-1}$, that is, $\partial h^{n-1}/\partial\theta^n = 0$ (Halkin, 1966). The equality $\partial h^{n-1}/\partial\theta^n = 0$ means that in terms of the enlarged Hamiltonian $\tilde{H}^{n-1}$

$$\theta^n \frac{\partial \tilde{H}^{n-1}(\mathbf{x}^n, \mathbf{p}^{n-1}, p_t{}^{n-1}, t^n, \mathbf{u}^n, \theta^n)}{\partial\theta^n} + \tilde{H}^{n-1}(\mathbf{x}^n, \mathbf{p}^{n-1}, p_t{}^{n-1}, t^n, \mathbf{u}^n, \theta^n) = 0$$

$$(2.36)$$

(Poświata, 2005; Sieniutycz, 2006b). Essentially, when the convexity assumption for functions and constraining sets is relaxed the weak maximum property attributed to Equation (2.28) may be lost and this equation may refer to minimum or saddle point; see, however, Szwast (1988) for a stronger formulation.

The boundary conditions are determined with respect to the end state coordinates and end times as the vanishing stationarity conditions for the extremum of S, Equation (2.13), in terms of the original function V, Equation (2.16). Partial differentiations of Equation (2.19) and setting to zero partial derivatives of V^n along with use of the definition of adjoint variables and Hamiltonian (2.29)–(2.31) and Equation (2.35) yield

$$p_v^N = \frac{\partial G(\mathbf{x}^N, t^N)}{\partial x_v^N}, \quad v \neq a, b \tag{2.37}$$

$$p_v^0 = \frac{\partial G(\mathbf{x}^0, t^0)}{\partial x_v^0}, \quad v \neq c, d \tag{2.38}$$

$$H^N = -\frac{\partial G(\mathbf{x}^N, t^N)}{\partial t^N}, \quad \text{if } t^N \text{ is free} \tag{2.39}$$

$$H^0 = -\frac{\partial G(\mathbf{x}^0, t^0)}{\partial t^0}, \quad \text{if } t^0 \text{ is free} \tag{2.40}$$

As pointed out above, these equations do not apply to fixed coordinates, such as x_a^N and x_b^N in Equation (2.37). If any component of the boundary state vector, x_v^N or x_v^0, is fixed, the respective component of the adjoint vector, p_v^N or p_v^0, is undetermined. Analogously, if the boundary time, t^N or t^0, is fixed, the respective Hamiltonian, H^N or H^0, is undetermined. Equations (2.37) and (2.38), respectively, describe the boundary conditions for the adjoint vector $\mathbf{p}^n$ at the end and at the beginning of the process, and Equations (2.39) and (2.40), respectively, state boundary conditions for the Hamiltonian H^n at the end and at the beginning of the process.

A number of approximation techniques have been proposed in recent years so that the numerical analysis of dynamic programming algorithms and related Hamilton–Jacobi–Bellman (HJB) equations has experienced a fast growth. In fact, DP algorithms are used to obtain the numerical solutions of related HJB equations. The convergence is not an easy problem, however (see later and Chapter 7). The first challenge posed by HJB and Hamilton–Jacobi equations is the low regularity of solutions, which requires new techniques to design cost-effective approximation schemes. Since the first attempts with monotone finite difference schemes, this has rapidly become an important field of application of the most recent trends in the numerical analysis of PDEs, and in particular of adaptive and high-order methods. The second challenge is posed by the wide range of applications covered by this theory. Real-life problems may require higher dimensions, discontinuous or nonunique solutions, and in general pose difficulties which go far beyond typically academic problems.

2.5. HAMILTON–JACOBI–BELLMAN EQUATIONS FOR CONTINUOUS SYSTEMS

2.5.1. Continuous Optimization Problem

Continuous processes have their own theory which was initially formed under the assumption of differentiable profit function $V(\mathbf{x}, t)$. This early form is described in this section, while in the final part of the chapter nonclassical cases and viscosity solutions are outlined.

We consider a general problem of continuous optimal control associated with the Bolza form of the performance criterion

$$S = \int_{t_i}^{t_f} f_0(\mathbf{x}, \mathbf{u}, t)\mathrm{d}t + G(\mathbf{x}^f(t^f), t^f) - G(\mathbf{x}^i(t^i), t^i) \tag{2.41}$$

where f_0 is the profit intensity and $G(\mathbf{x}, t)$ is a gauging function that depends on the state $\mathbf{x}$ and the time t. For Equation (2.41), a maximum of the criterion S is sought with respect to a suitable choice of the vector functions $\mathbf{x}(t)$ and $\mathbf{u}(t)$. The gauging function G influences an optimal solution only if some of the end coordinates of the state vector or time are not specified.

We may now recall backward and forward DP algorithms considered at the beginning of the present chapter. Dealing with all initial coordinates $x_k^i (k = 1, \ldots, s)$ and initial time t^i as independent variables, we can generate a function describing the maximum value of S in terms of $\mathbf{x}^i$ and t^i. This is called the original optimization problem. However, one can also consider the maximum of S as a function generated in terms of the final coordinates x_k^f and the final time t^f. This is called the dual optimization problem. It is insightful to confront properties of the original problem with those of the dual problem. In terms of coordinates of both end points (initial and final) an optimal performance function $V(x^i, t^i, x^f, t^f)$, such as that in Equation (2.68) below, describes the maximum of S.

The optimization in Equation (2.42) is subject to the differential constraints resulting from a given set of differential equations

$$\frac{dx_k}{dt} = f_k(x, t, \mathbf{u}) \tag{2.42}$$

where $\mathbf{x} = (x_1, x_2, \ldots, x_i, \ldots, x_s)$ is the s-dimensional state vector and $\mathbf{f} = (f_1, f_2, \ldots, f_i, \ldots, f_s)$ is the vector of rates. The r-dimensional vector of adjustable parameters $\mathbf{u} = (u_1, u_2, \ldots, u_r)$ is the control vector. An admissible control usually satisfies certain local constraints, the most typical being

$$\mathbf{u}(t) \in \mathbf{U} \tag{2.43}$$

where $\mathbf{U}$ is an admissible set in the control space. Additional constraints may also exist which link coordinates of the state vector, $\mathbf{x}$, and the control vector, $\mathbf{u}$. They are usually of the type $\mathbf{g}(\mathbf{x}, \mathbf{u}, t) = 0$ or $\mathbf{g}(\mathbf{x}, \mathbf{u}, t) \leq 0$. However, they may be included in the above model by using Lagrange multipliers, special sorts of controls which reside linearly in the model. The presence of these multipliers will increase the dimensionality of $\mathbf{u}$ without changing the general structure of the above model. Thus, the model (Equations (2.41)–(2.43)) is sufficient for quite general considerations. To include the time coordinate in the state vector we can use the enlarged $s + 1$-dimensional vector of state $\tilde{x} = (x_1, x_2, \ldots x_i, \ldots, x_s, x_{s+1})$ where $x_{s+1} \equiv t$ and $s + 1$-dimensional vector of rates $\tilde{\mathbf{f}} = (f_1, f_2, \ldots f_i \ldots f_s, f_{s+1})$ with $f_{s+1}^n \equiv 1$.

We can also take into consideration the performance coordinate, x_0. To this end we define the performance equation

$$\frac{dx_0}{dt} = f_0(x, t, \mathbf{u}) \tag{2.44}$$

where x_0 has an initial value $x_0(t^i) = x_0^i$ and a final value $x_0(t^f) = x_0^f$. One of these values must be free. With Equation (2.44) we find S of Equation (2.41) as a function of both end states and end times:

$$S = g(x_0^f, x^f, t^f, x_0^i, x^i, t^i) \equiv x_0^f - x_0^i + G(x^f, t^f) - G(x^i, t^i) \tag{2.45}$$

Thus, with the help of the performance variable x_0, the original optimization problem becomes that of maximum for $x_0(t^f) + G^f$, subject to differential constraints (2.42) and (2.44) and local constraints (2.43). The dual problem becomes that of minimum for $x_0(t^i) + G^i$, subject to the same set of constraints. Extreme values of these criteria are described by two functions Q, defined below. The notion of the complete state $(x_0, \mathbf{x}, t)$ is useful.

2.5.2. Optimal Performance Functions and Related HJB Equations

The performance variable must be free at the end at which S is extremized. Working first with the backward DP algorithm, we consider an optimal function Q^i of the complete initial state $(x_0^i, \mathbf{x}^i, t^i)$ and of the partially fixed final state $(\mathbf{x}^f, t^f)$. Q^i constitutes an example of an optimal performance function in the complete state space. For simplicity we shall neglect the superscript i in Q. The function is defined as

$$Q(x_0^i, \mathbf{x}^i, t^i, \mathbf{x}^f, t^f) \equiv \max\{x_0^f + G(\mathbf{x}^f, t^f)\} \tag{2.46}$$

To apply Bellman's optimality principle in the context of the backward DP algorithm, we maximize S along a special trajectory that starts at $(x_0^i, \mathbf{x}^i, t^i)$ and possesses a short nonoptimal part as well as a long, remaining optimal part. The long optimal part begins at the point $(x_0^i + \Delta x_0^i, \mathbf{x}^i + \Delta \mathbf{x}^i, t^i + \Delta t^i)$. In order to get Q at $(x_0^i, \mathbf{x}^i, t^i)$, Bellman's optimality principle requires maximizing Q evaluated at the beginning of the long optimal part:

$$Q(x_0^i, \mathbf{x}^i, t^i, \mathbf{x}^f, t^f) \equiv \max_{\mathbf{u}^i} \{Q(x_0^i + \Delta x_0^i, \mathbf{x}^i + \Delta \mathbf{x}^i, t^i + \Delta t^i)\} \tag{2.47}$$

Taylor's expansion of the right-hand side of the above equation yields

$$\begin{aligned}
Q(x_0^i, \mathbf{x}^i, t^i, \mathbf{x}^f, t^f) \equiv \max_{\mathbf{u}^i} \{ & Q(x_0^i, \mathbf{x}^i, t^i, \mathbf{x}^f, t^f) \\
& + \frac{\partial Q}{\partial x_0^i} \Delta x_0^i + \frac{\partial Q}{\partial x^i} \Delta x^i, \frac{\partial Q}{\partial t^i} \Delta t^i)\}
\end{aligned} \tag{2.48}$$

Hence, as Q is independent of $\mathbf{u}$, state equations hold, and the time interval Δt^i is finite:

$$\max_{\mathbf{u}^i} \left\{ \frac{\partial Q}{\partial x_0^i} f_0^i + \sum_{k=1}^{s} \frac{\partial Q}{\partial x_k^i} f_k^i + \frac{\partial Q}{\partial t^i} \right\} = 0 \tag{2.49}$$

This is the Hamilton–Jacobi–Bellman equation of the problem (HJB equation). It can be written in a symbolic form that shows a maximum rate of increase of Q:

$$\max_{\mathbf{u}^i} \left\{ \frac{dQ(x_0^i, \mathbf{x}^i, t^i, \mathbf{x}^f, t^f)}{dt^i} \right\} = 0 \tag{2.50}$$

From Equation (2.45) and the definition of Q, Equation (2.46),

$$\max S = Q(x_0^i, \mathbf{x}^i, t^i, \mathbf{x}^f, t^f) - x_0^i - G(\mathbf{x}^i, t^i) \tag{2.51}$$

Thus, we can evaluate the maximum performance function V, which describes $\max S$, as

$$V(\mathbf{x}^i, t^i, \mathbf{x}^f, t^f) \equiv \max S = Q(x_0^i, \mathbf{x}^i, t^i, \mathbf{x}^f, t^f) - x_0^i - G(\mathbf{x}^i, t^i) \tag{2.52}$$

The function $V(\mathbf{x}^i, t^i, \mathbf{x}^f, t^f) \equiv V(\tilde{\mathbf{x}}^i, \tilde{\mathbf{x}}^f)$ describes extremum values of the criterion S in terms of the generalized end states $\tilde{\mathbf{x}}^i$ and $\tilde{\mathbf{x}}^f$ when the total duration is $T = t^f - t^i$. Since V is a potential and S is not, only S depends on how the controls change during the process. When the process is autonomous, that is, its model does not explicitly contain time t, the form of V is such that $V(t^i, \mathbf{x}^i, t^f, \mathbf{x}^f)$ depends on end times t^i and t^f through the duration $T = t^f - t^i$ only, meaning that the initial time instant does not influence the value of V whenever the duration is fixed. This is associated with the invariance of the Hamiltonian function H along an extremal path. H is then the first integral of the process.

From Equation (2.52),

$$Q(x_0^i, \mathbf{x}^i, t^i, \mathbf{x}^f, t^f) = x_0^i + G(\mathbf{x}^i, t^i) + V(\mathbf{x}^i, t^i, \mathbf{x}^f, t^f) \tag{2.53}$$

$$\frac{\partial Q(x_0^i, \mathbf{x}^i, t^i, \mathbf{x}^f, t^f)}{\partial x_0^i} = 1 \tag{2.54}$$

$$\max_{\mathbf{u}^i} \left\{ f_0^i + \sum_{k=1}^{s} \frac{\partial Q}{\partial x_k^i} f_k^i + \frac{\partial Q}{\partial t^i} \right\} = 0 \tag{2.55}$$

Since the differentiations are with respect to the initial coordinates only, one may work in terms of an effective or overall gauging function, P. While its definition is based on an earlier work (Sieniutycz, 2000c), the present definition is different from the original one:

$$P(\mathbf{x}^i, t^i, \mathbf{x}^f, t^f) \equiv G(\mathbf{x}^i, t^i) + V(\mathbf{x}^i, t^i, \mathbf{x}^f, t^f) \tag{2.56}$$

Alternatively, in view of Equation (2.52),

$$P(\mathbf{x}^i, t^i, \mathbf{x}^f, t^f) \equiv Q(x_0^i, \mathbf{x}^i, t^i, \mathbf{x}^f, t^f) - x_0^i \tag{2.56'}$$

The different sign of gauge G in Equations (2.56) and (2.19) is due to the present use of the backward algorithm. In terms of function P the HJB equation preserves the form derived above:

$$\max_{\mathbf{u}^i} \left\{ f_0^i + \sum_{k=1}^s \frac{\partial P}{\partial x_k^i} f_k^i + \frac{\partial P}{\partial t^i} \right\} = 0 \tag{2.57}$$

Expressing the HJB equation in terms of P is convenient, because P eliminates the effect of profit coordinate x_0^i from function Q and introduces the gauging effect incorporated in $G(\mathbf{x}, t)$.

2.5.3. Optimal Performance in Terms of the Forward DP Algorithm

Now, working with the forward DP algorithm, we consider another function Q^f, which depends on the complete final state $(x_0^f, \mathbf{x}^f, t^f)$, and a partially fixed initial state $(\mathbf{x}^i, t^i)$. Again, for simplicity we shall neglect superscript f in Q. The function is defined as

$$Q(\mathbf{x}^i, t^i, x_0^f, \mathbf{x}^f, t^f) \equiv \min\{x_0^i + G(\mathbf{x}^i, t^i)\} \tag{2.58}$$

In this case a trajectory is considered that terminates at $(x_0^f, \mathbf{x}^f, t^f)$ and that possesses a short nonoptimal final part as well as a long optimal initial part. For the optimal trajectory terminating at $(x_0^f, \mathbf{x}^f, t^f)$ the optimality principle and Taylor expansion yield

$$Q(\mathbf{x}^i, t^i, x_0^f, \mathbf{x}^f, t^f) \equiv \min_{\mathbf{u}^i} \{Q(x_0^f - \Delta x_0^f, \mathbf{x}^f - \Delta \mathbf{x}^f, t^f - \Delta t^f)\} \tag{2.59}$$

$$Q(\mathbf{x}^i, t^i, x_0^f, \mathbf{x}^f, t^f) \equiv \min_{\mathbf{u}^i} \left\{ Q(\mathbf{x}^i, t^i, x_0^f, \mathbf{x}^f, t^f) - \frac{\partial Q}{\partial x_0^f} \Delta x_0^f - \frac{\partial Q}{\partial \mathbf{x}^f} \Delta \mathbf{x}^f - \frac{\partial Q}{\partial t^f} \Delta t^f \right\} \tag{2.60}$$

Again, as Q is independent of $\mathbf{u}$, state equations hold, and the time interval Δt^i is finite:

$$\max_{\mathbf{u}^i} \left\{ \frac{\partial Q}{\partial x_0^f} f_0^f + \sum_{k=1}^s \frac{\partial Q}{\partial x_k^f} f_k^f + \frac{\partial Q}{\partial t^f} \right\} = 0 \tag{2.61}$$

and

$$\max_{\mathbf{u}^i} \left\{ \frac{dQ(\mathbf{x}^i, t^i, x_0^f, \mathbf{x}^f, t^f)}{dt^f} \right\} = 0 \tag{2.62}$$

However, from Equation (2.45):

$$S = g(x_0^f, x^f, t^f, x_0^i, x^i, t^i) \equiv x_0^f - x_0^i + G(x^f, t^f) - G(x^i, t^i)$$

and the present definition of Q, Equation (2.58),

$$\max S = x_0^f + G(x^f, t^f) - Q(x^i, t^i, x_0^f, x^f, t^f) \equiv V(x^i, t^i, x^f, t^f) \quad (2.63)$$

or

$$Q(x^i, t^i, x_0^f, x^f, t^f) = x_0^f + G(x^f, t^f) - V(x^i, t^i, x^f, t^f) \quad (2.64)$$

Whence

$$\frac{\partial Q(x_0^i, x^i, t^i, x^f, t^f)}{\partial x_0^f} = 1 \quad (2.65)$$

and

$$\max_{\mathbf{u}^i} \left\{ f_0^f + \sum_{k=1}^{s} \frac{\partial Q}{\partial x_k^f} f_k^f + \frac{\partial Q}{\partial t^f} \right\} = 0 \quad (2.66)$$

Since the differentiations are with respect to the final coordinates only, we can use the gauging function P similar to that introduced earlier in Equations (2.56) and (2.56′). A suitable definition is now contained in the equation

$$P(x^i, t^i, x^f, t^f) \equiv V(x^i, t^i, x^f, t^f) - G(x^f, t^f) = x_0^f - Q(x_0^i, x^i, t^i, x^f, t^f) \quad (2.67)$$

In terms of this function the HJB equation (2.66) has the form

$$\max_{\mathbf{u}^i} \left\{ f_0^f - \sum_{k=1}^{s} \frac{\partial P}{\partial x_k^f} f_k^f - \frac{\partial P}{\partial t^f} \right\} = 0 \quad (2.68)$$

2.5.4. *Link with Gauged Integrals of Performance*

We shall now interpret the meaning of extremum operations in the HJB equations. Let us start with the dual problem for which the final state is operative and Equation (2.68) holds. As it follows from the previous derivations based on Bellman's optimality principle in terms of the complete final state, the optimal performance function defined by the equation

$$V(\mathbf{x}^i, t^i, \mathbf{x}^f, t^f) \equiv \max S = \max \left\{ \int_{t^i}^{t^f} f_0(\mathbf{x}, \mathbf{u}, t) dt + G(\mathbf{x}^f, t^f) - G(\mathbf{x}^i, t^i) \right\}$$

(2.69)

satisfies in the dual problem (fixed x_0^f and free x_0^i) the condition

$$\max_{\mathbf{u}^i} \{x_0^f - x_0^i - P(\mathbf{x}^i, t^i, \mathbf{x}^f, t^f)\} = 0$$

(2.70)

or the equivalent integral condition

$$\max_{u(t)} \left\{ \int_{t^i}^{t^f} f_0(\mathbf{x}, \mathbf{u}, t) dt - P(\mathbf{x}^i, t^i, \mathbf{x}^f, t^f) \right\} = 0$$

(2.71)

where the function P satisfies Equation (2.56)

$$P(\mathbf{x}^i, t^i, \mathbf{x}^f, t^f) \equiv G(\mathbf{x}^i, t^i) + V(\mathbf{x}^i, t^i, \mathbf{x}^f, t^f)$$

(2.56)

and is the effective or overall gauging function. In fact, Equation (2.71) describes the disappearance of a gauged integral of performance where gauging is performed by P. It follows from the above formulae for V and P that Equations (2.70) and (2.71) are valid for a fixed final boundary point $(x_0^f, \mathbf{x}^f, t^f)$, and the freely varied initial coordinate of profit, x_0^i. The differentiation of Equation (2.71) with respect to the final time t^f proves that the total time derivative of the potential function P satisfies the equation

$$\max_{\mathbf{u}^f} \left\{ f_0(\mathbf{x}^f, t^f, \mathbf{u}^f) - \frac{dP(\mathbf{x}^i, t^i, \mathbf{x}^f, t^f)}{dt^f} \right\} = 0$$

(2.72)

which is just a compact form of the HJB equation (2.68) which deals with the derivatives of the final state. This, in turn, proves a mnemonic rule which states that the differentiation is allowed at the end at which the complete state is operative, and that this differentiation yields a suitable HJB equation.

Otherwise, in the original problem, in which the initial point $(x_0^i, \mathbf{x}^i, t^i)$ is fixed and a maximum of S at a final time t^f is considered, the following analog of Equation (2.72) holds:

$$\max_{\mathbf{u}^i} \left\{ f_0(\mathbf{x}^i, t^i, \mathbf{u}^i) + \frac{dP(\mathbf{x}^i, t^i, \mathbf{x}^f, t^f)}{dt^f} \right\} = 0$$

(2.73)

This result is consistent with Equation (2.57). It might be interpreted as a requirement that, in order to get Equation (2.73), the differentiation with respect to the initial time should be made for the minimum counterpart of Equation (2.70), which is

$$\max_{\mathbf{u}^i} \{x_0^f - x_0^i - P(\mathbf{x}^i, t^i, \mathbf{x}^f, t^f)\} = 0 \qquad (2.74)$$

or

$$\min_{\mathbf{u}(t)} \left\{ \int_{t^i}^{t^f} f_0(\mathbf{x}, \mathbf{u}, t)\mathrm{d}t - P(\mathbf{x}^i, t^i, \mathbf{x}^f, t^f) \right\} = 0 \qquad (2.75)$$

Indeed, the differentiation of the above equations with respect to the initial time t^i leads after changing signs (associated with change of the extremum operation) to Equation (2.73) that agrees with the backward HJB equation (2.57) dealing with derivatives of the initial state.

It is also worth knowing that the optimal performance function V that is related directly to the original criterion S by Equation (2.69) can be used to yield a HJB equation. Indeed, we find from Equations (2.64) and (2.66) or Equations (2.53) and (2.55):

$$\max_{\mathbf{u}^f} \left\{ \tilde{f}_0^f - \sum_{k=1}^s \frac{\partial V}{\partial x_k^f} f_k^f - \frac{\partial V}{\partial t^f} \right\} = 0 \qquad (2.76)$$

and

$$\max_{\mathbf{u}^i} \left\{ \tilde{f}_0^i + \sum_{k=1}^s \frac{\partial V}{\partial x_k^i} f_k^i + \frac{\partial V}{\partial t^i} \right\} = 0 \qquad (2.77)$$

where in either case a gauged profit intensity

$$\tilde{f}_0(x, t, u) = f(x, t, u) + \sum_{k=1}^s \left\{ \frac{\partial G}{\partial x_k} f_k + \frac{\partial G}{\partial t} \right\} \qquad (2.78)$$

appears in a HJB equation describing V.

2.5.5. Diversity of Equivalent Formulations

The Hamilton–Jacobi–Bellman equation is the quasi-linear partial differential equation of the control theory (with the extremum sign) which governs the optimal performance functions P and V (value functions) and the suitable choice of optimal control $\mathbf{u}$. The definition of the performance potential V in Equation (2.69) is the most suitable for processes producing a profit, in which case V is positive. For processes described in terms of a consumed cost, the most suitable definition assuring a positive potential involves an optimal function R which is the negative of V. Indeed, for an arbitrary functional S and the same change of the end states and times,

$$R \equiv \min(-S_{[x^i,x^f]}) = -\max S_{[x^i,x^f]} = -V \qquad (2.79)$$

However, the single extremal function $V(t^i, x^i, t^f, x^f)$ is sufficient to adequately describe the extremum of the functional S.

Contemporary approaches involve a systematic search for the properties and implications of HJB equations in physics, chemistry, thermodynamics and economics. In analytical mechanics such equations are usually derived by the method of variational calculus (Elsgolc, 1960). The control theory approach allows for more general derivations that take into account local constraints imposed on control variables, Equation (2.43).

The HJB relationships, Equations (2.72) and (2.73), are local conditions that describe vanishing maxima of the profit intensity f_0 gauged by the total derivative of the optimal performance function P. Similarly, Equations (2.76) and (2.77) incorporate analogous gauging in terms of the total derivative of function V.

In the latter equations, changing the signs of extremized expressions when change of the extremum operation takes place in S yields, respectively

$$\min_{u^f} \left\{ \frac{\partial V}{\partial t^f} + \sum_{k=1}^{s} \frac{\partial V}{\partial x_k^f} f_k^f - \tilde{f}_0^f(x^f, t^f, x^f) \right\} = 0 \qquad (2.80)$$

and

$$\max_{u^i} \left\{ \frac{\partial V}{\partial t^i} + \sum_{k=1}^{s} \frac{\partial V}{\partial x_k^i} f_k^i + \tilde{f}_0^i(x^i, t^i, u^i) \right\} = 0 \qquad (2.81)$$

The partial derivative of the optimal profit V with respect to time can be taken out of the bracket and the indices f and i omitted in equations of this sort for variable end states. Hence finally, in terms of final states and times,

$$
\begin{aligned}
&\frac{\partial V}{\partial t} + \min_{u} \left\{ \frac{\partial V}{\partial x} . f(x, t, u) - \tilde{f}_0(x, t, u) \right\} \\
&= \frac{\partial R}{\partial t} + \max_{u} \left\{ \frac{\partial R}{\partial x} . f(x, t, u) + \tilde{f}_0(x, t, u) \right\} = 0
\end{aligned}
\qquad (2.82)
$$

whereas in terms of initial states and initial times,

$$\frac{\partial V}{\partial t} + \max_{\mathbf{u}} \left\{ \frac{\partial V}{\partial \mathbf{x}}.\mathbf{f}(\mathbf{x}, t, \mathbf{u}) + \tilde{f}_0(\mathbf{x}, t, \mathbf{u}) \right\}$$

$$= \frac{\partial R}{\partial t} + \min_{\mathbf{u}} \left\{ \frac{\partial R}{\partial \mathbf{x}}.\mathbf{f}(\mathbf{x}, t, \mathbf{u}) - \tilde{f}_0(\mathbf{x}, t, \mathbf{u}) \right\} = 0 \tag{2.83}$$

See Landau and Lifshitz (1971) for analogous properties in classical mechanics. Note that the cost-type functions R appearing in these equations represent extremum actions of mechanics provided that $\tilde{l}_0(\mathbf{x}, t, \mathbf{u}) \equiv -\tilde{f}_0(\mathbf{x}, t, \mathbf{u})$ is a mechanical Lagrangian.

For autonomous systems a mixed description is also possible. In this case initial coordinates of state are operative whereas the varied time coordinate is the process duration $T \equiv t^f - t^i$. The condition of the fixed final time then yields $dT = -dt^i$, thus, from Equation (2.81) or (2.83),

$$-\frac{\partial V}{\partial T} + \max_{\mathbf{u}} \left\{ \frac{\partial V}{\partial \mathbf{x}}.\mathbf{f} + \tilde{f}_0 \right\} = -\frac{\partial R}{\partial T} + \min_{\mathbf{u}} \left\{ \frac{\partial R}{\partial \mathbf{x}}.\mathbf{f} - \tilde{f}_0 \right\} = 0 \tag{2.84}$$

These are still usable HJB equations for optimal functions V or R of the problem. In all equations of this sort, extremized members are certain Hamiltonian expressions. In fact, they refer to Pontryagin-type, "nonextremal" Hamiltonians. An optimal control $\mathbf{u}$ which solves the optimization problem must extremize a Hamiltonian at each point of the extremal path, which means extremizing a wave-front velocity in the HJB equation.

Methods for solving HJB equations can be both analytical and numerical. For some special models, for example, those of thermal machines with heat transfer, solutions can be obtained analytically (Chapter 6), although in many situations classical analytical solutions are not possible or simply do not exist and one has to search for viscosity solutions. Examples in the energy field include problems with free boundary conditions, or with non-Newtonian and radiative transfer and possible constraints imposed on the state and rates. A numerical procedure which works with Bellman's recurrence equation (Chapter 7), is the most common tool to solve an HJB equation in the case of low dimensionality of the state vector (Bellman, 1957; Aris, 1964; Findeisen et al., 1980; Capuzzo Dolcetta and Ishii, 1984; Capuzzo Dolcetta and Falcone, 1989; Falcone and Makridakis, 2001; Falcone and Ferretti, 2006, and Chapter 7).

2.5.6. Passage to the Hamilton–Jacobi Equation

We now consider passage to the Hamilton–Jacobi equation for Equations (2.82) and (2.83), which are treated simultaneously. For a definite physical problem, common formulae describe the effect of variation of final states and final time and of initial states and initial time. We define an adjoint vector, $\mathbf{p}$, as the negative partial derivative $-\partial V/\partial x$ when the varied x is the final state x^f and the partial derivative $\partial V/\partial x$ when the varied x is the initial state x^i. (Note that these adjoints do not coincide with others based on gradients of the function P, which can also be used.) Equation (2.85) below shows an important property: the derivatives $\partial V/\partial x^i$ and $-\partial V/\partial x^f$ (or $\partial P\partial x^i$ and $-\partial P/\partial x^f$) are equal, that is, $\partial V/\partial x^i = -\partial V/\partial x^f$ when two end states x^i and x^f coincide. This substantiates a supposition that p is a physical quantity. In fact, in models of analytical mechanics, the components of p are generalized momenta. In terms of end states and times

$$p \equiv \pm \frac{\partial R}{\partial x} = -\pm \frac{\partial V}{\partial x} \qquad (2.85)$$

where the upper signs refer to final states, Equation (2.41), and lower ones to initial states, Equation (2.42). For the Lagrangian type functions, that is, the integrands $\tilde{l}_0(x, t, u) \equiv -\tilde{f}_0(x, t, u)$, the extremum condition of Hamiltonian expression in Equations (2.82)–(2.84) determines the link between the derivatives of $\tilde{l}_0$ or $-\tilde{f}_0$ with respect to controls u and the state adjoints $p = -\pm\partial V/\partial x$. Consistently, in the following formulae upper signs refer to the final-state variations in Equation (2.82) and lower signs refer to the initial-state variations in Equation (2.83). The extremizing operation is with respect to controls u (in fact, either u^f or u^i), and leads to two equations. The first one links the optimal control u with x, t, and $\pm\partial V/\partial x$; for an unconstrained vector u,

$$\pm \frac{\partial V}{\partial x} \cdot \frac{\partial f(x, t, u)}{\partial u} = \frac{\partial \tilde{f}_0(x, t, u)}{\partial u} \qquad (2.86)$$

The second links the original Equations (2.82) or (2.83) without the extremizing sign:

$$\pm \frac{\partial V}{\partial t} + \pm \frac{\partial V}{\partial x} \cdot f(x, t, u) - \tilde{f}_0(x, t, u) = 0 \qquad (2.87)$$

Upper signs pertain to final states in Equation (2.82) and lower ones to initial states in Equation (2.83).

Now assume that we are able to solve Equation (2.86) in terms of u to obtain the optimal control $u(p, t, x)$ in terms of the state x, time t and state adjoint $p \equiv \mp\partial V/\partial x = \pm\partial R/\partial x$. For models of mechanics, where $f(x, t, u) \equiv u$, the solving u satisfies the familiar relation $p = \partial \tilde{l}_0(x, t, u)/\partial u$ that links momenta and velocities. In other, possibly more general, cases stationary solutions $u(p, t, x)$ follow from Equations (2.85) and (2.86) in the form

$$p \cdot \frac{\partial f(x, t, u)}{\partial u} = -\frac{\partial \tilde{f}_0(x, t, u)}{\partial u} \equiv \frac{\partial \tilde{l}_0(x, t, u)}{\partial u} \qquad (2.88)$$

When we substitute the solution $u(p, t, x)$ of Equation (2.88) into Equation (2.87), an energy-type Hamiltonian of *extremum* process emerges:

$$H(\mathbf{x}, t, \mp \partial V/\partial \mathbf{x}) \equiv \mathbf{p}(\mathbf{x}, t).\mathbf{f}(\mathbf{x}, t, \mathbf{p}(\mathbf{x}, t)) + \tilde{f}_0(\mathbf{x}, t, p(\mathbf{x}, t))$$
$$= \mathbf{p}(\mathbf{x}, t).\mathbf{f}(\mathbf{x}, t, \mathbf{p}(\mathbf{x}, t)) - \tilde{l}_0(\mathbf{x}, t, p(\mathbf{x}, t)) \tag{2.89}$$

where $\mathbf{p} \equiv \mp \partial V/\partial \mathbf{x}$ or $\pm \partial R/\partial \mathbf{x}$ are functions of $\mathbf{x}$ and t. A different font (H instead of H) is used since function (2.89) resides in the state space $(\mathbf{x}, t)$ rather than in the phase space $(\mathbf{p}, \mathbf{x}, t)$. The structure (2.89) is valid for function H even if the controls u_j are constrained, that is, when $\mathbf{p}$ and $\mathbf{u}$ are not linked by Equations (2.86) or (2.88). (For the same range of $\mathbf{x}$, numerical values of $\mathbf{p}$ and H may be different when the control process evolves from state $\mathbf{x}^i$ to state $\mathbf{x}^f$ and then goes back, from $\mathbf{x}^f$ to $\mathbf{x}^i$, in the same amount of time.) In terms of $\partial V/\partial \mathbf{x}$ and $\partial V/\partial t$ Equation (2.87) becomes the Hamilton–Jacobi equation for the characteristic function V:

$$\mp \partial V/\partial t + H(\mathbf{x}, t, \mp \partial V/\partial \mathbf{x}) = 0 \tag{2.90}$$

where the upper sign refers to final states and the lower one to initial states. This form deals with the optimal function of profit type, V; in analytical mechanics functions of optimal cost type, $R = -V$ are used.

Hamilton–Jacobi equations differ from HJB equations insofar as they refer only to extremal paths and H is the *extremum* Hamiltonian. Equation (2.90) should be solved subject to the boundary condition

$$\lim_{f \to i} V(\mathbf{x}^i, t^i, \mathbf{x}^f, t^f) = 0 \tag{2.91}$$

that applies when the final states $(\mathbf{x}^f, t^f)$ approach the initial point $(\mathbf{x}^i, t^i)$.

Associated with the canonical set below, free boundary conditions for H and $\mathbf{p}$ follow from the definition of $V = \max S$ in the form $-H = p_t = \mp \partial V/\partial t = 0$ and $\mathbf{p} = \mp \partial V/\partial \mathbf{x} = 0$. However, in terms of P, Equation (2.57), free boundary conditions are different; the condition $\partial V/\partial t = 0$ implies $\partial P/\partial t = \partial G/\partial t$ at the initial point and $\partial P/\partial t = -\partial G/\partial t$ at the final point. Thus, we find in the common notation

$$\partial P/\partial t = \mp \partial G/\partial t \tag{2.92}$$

and similarly

$$\partial P/\partial \mathbf{x} = \mp \partial G/\partial \mathbf{x} \tag{2.93}$$

where the upper sign refers to final states.

Assuming that the vector of initial values $\mathbf{x}(t^i) = \mathbf{x}^i$ is given, function $V(\mathbf{x}^i, t^i, \mathbf{x}, t)$ is determined as the solution to Equation (2.90) in terms of the final states. The optimal trajectory, or the vector $\mathbf{x}(t)$, is then found from the initial relationship $\mathbf{p}^i = \partial V/\partial \mathbf{x}^i$ which yields $\mathbf{x} = \mathbf{x}(t, t^i, \mathbf{x}^i, \mathbf{p}^i)$. Otherwise, the adjoint vector follows from the final relationship, $\mathbf{p}^f = -\partial V/\partial \mathbf{x}^f$, applied to current states in the form $\mathbf{p} = -\partial V/\partial \mathbf{x}$; this yields $\mathbf{p} = \mathbf{p}(t^i, \mathbf{x}^i, t, \mathbf{x})$. Substituting in this expression the trajectory $\mathbf{x} = \mathbf{x}(t, t^i, \mathbf{x}^i, \mathbf{p}^i)$ we find the evolution of $\mathbf{p}$ in time; $\mathbf{p} = \mathbf{p}(t, t^i, \mathbf{x}^i, \mathbf{p}^i)$. Thus, we obtain both the optimal trajectory and the adjoint vector as functions of the current time t and the initial values $(t^i, \mathbf{x}^i, \mathbf{p}^i)$. In this book we apply the HJB theory to some examples describing continuous systems in Chapter 6, and to discrete systems in Chapter 7.

2.6. CONTINUOUS MAXIMUM PRINCIPLE

Let us now concentrate on the case of minimum cost described by the second equality in Equation (2.82) and limit ourselves to variations of final points. Define the nonoptimal Hamiltonian of Equation (2.82) in the state space $(\mathbf{x}, t)$ by an expression

$$B(\mathbf{x}, t, \mathbf{u}) \equiv \frac{\partial R(\mathbf{x}, t)}{\partial t} + \frac{\partial R(\mathbf{x}, t)}{\partial \mathbf{x}}.f(\mathbf{x}, t, \mathbf{u}) + \tilde{f}_0(\mathbf{x}, t, \mathbf{u}) \qquad (2.94)$$

It then follows from Equation (2.82) that B satisfies the condition

$$\max_{\mathbf{u}} B(\mathbf{x}, t, \mathbf{u}) \equiv \max_{\mathbf{u}} \left\{ \frac{\partial R}{\partial t} + \frac{\partial R}{\partial \mathbf{x}}.f(\mathbf{x}, t, \mathbf{u}) + \tilde{f}_0(\mathbf{x}, t, \mathbf{u}) \right\} = 0 \qquad (2.95)$$

(Boltyanski, 1969). From Equation (2.95) it follows that for an arbitrary point $\tilde{\mathbf{x}} = (\mathbf{x}, t)$ in the (extended) state space a point $\mathbf{u}(\mathbf{x}, t)$ can be found in the control space such that $B(\mathbf{x}, t, \mathbf{u}) = 0$. Assume that that function $R(\mathbf{x}, t)$ possesses second partial derivatives, rate functions f_i have continuous first partial derivatives, and fix a $\mathbf{u}$ in the admissible regime U. It then follows from Equations (2.94) and (2.95) that function $B(\mathbf{x}, t, \mathbf{u})$ also attains a maximum with respect to $\mathbf{x}$ and t, or that the following partial derivatives of $B(\mathbf{x}, t, \mathbf{u})$ must vanish:

$$\frac{\partial B}{\partial x_j} = \sum_{i=1}^{s} \frac{\partial^2 R}{\partial x_i \partial x_j} f_i + \sum_{i=1}^{s} \frac{\partial R}{\partial x_i} \frac{\partial f_i}{\partial x_j} + \frac{\partial^2 R}{\partial t \partial x_j} + \frac{\partial \tilde{f}_0}{\partial x_j} = 0 \qquad (2.96)$$

$$\frac{\partial B}{\partial t} = \sum_{i=1}^{s} \frac{\partial^2 R}{\partial x_i \partial t} f_i + \sum_{i=1}^{s} \frac{\partial R}{\partial x_i} \frac{\partial f_i}{\partial t} + \frac{\partial^2 R}{\partial t^2} + \frac{\partial \tilde{f}_0}{\partial t} = 0 \qquad (2.97)$$

In order to transform these equations into total time derivatives of $\partial R/\partial x_i$ and $\partial R/\partial t$. Since $R(\mathbf{x}, t)$ does not contain controls we obtain

$$\frac{d}{dt} \left(\frac{\partial R}{\partial x_j} \right) = \sum_{i=1}^{s} \frac{\partial^2 R}{\partial x_j \partial x_i} f_i + \frac{\partial^2 R}{\partial x_j \partial t} = \sum_{i=1}^{s} \frac{\partial^2 R}{\partial x_i \partial x_j} f_i + \frac{\partial^2 R}{\partial t \partial x_j} \qquad (2.98)$$

Similarly

$$\frac{d}{dt}\left(\frac{\partial R}{\partial t}\right) = \sum_{i=1}^{s} \frac{\partial^2 R}{\partial x_i \partial t} f_i + \frac{\partial^2 R}{\partial t^2} \tag{2.99}$$

Taking Equations (2.96) and (2.97) into account in Equations (2.98) and (2.99) yields

$$\frac{d}{dt}\left(\frac{\partial R}{\partial x_j}\right) = -\sum_{i=1}^{s} \frac{\partial R}{\partial x_i} \frac{\partial f_i}{\partial x_j} - \frac{\partial \tilde{f}_0}{\partial x_j} \tag{2.100}$$

$$\frac{d}{dt}\left(\frac{\partial R}{\partial t}\right) = -\sum_{i=1}^{s} \frac{\partial R}{\partial x_i} \frac{\partial f_i}{\partial t} - \frac{\partial \tilde{f}_0}{\partial t} \tag{2.101}$$

Equations (2.95), (2.100), and (2.101) can be written in a more transparent form. For this purpose we define the following adjoint variables:

$$p_i = \frac{\partial R(\mathbf{x}, t)}{\partial x_i} \tag{2.102}$$

$$p_t = \frac{\partial R(\mathbf{x}, t)}{\partial t} \tag{2.103}$$

and the phase-space nonoptimal Hamiltonian (Pontryagin et al., 1962):

$$\begin{aligned} H(\mathbf{x}, t, \mathbf{p}, \mathbf{u}) &\equiv \mathbf{p}.\mathbf{f}(\mathbf{x}, t, \mathbf{u}) + \tilde{f}_0(\mathbf{x}, t, \mathbf{u}) \\ &= \mathbf{p}.\mathbf{f}(\mathbf{x}, t, \mathbf{u}) - \tilde{l}_0(\mathbf{x}, t, \mathbf{u}) \end{aligned} \tag{2.104}$$

Equations (2.102) and (2.103) in terms of H, p_i and p_t take the form

$$\frac{dx_\nu}{dt} = \frac{\partial H}{\partial p_\nu} \tag{2.105}$$

$$\frac{dp_\nu}{dt} = -\frac{\partial H}{\partial x_\nu} \tag{2.106}$$

$$\left(\frac{dH}{dt} = \frac{\partial H}{\partial t}\right) \tag{2.107}$$

and

$$p_t + \max_{\mathbf{u}} H(\mathbf{x}, \mathbf{p}, t, \mathbf{u}) = 0 \tag{2.108}$$

Thus an optimal control must maximize H at each instant of time in the interval $t^i \leq t \leq t^f$. This is the famous Maximum Principle (Pontryagin et al., 1962; Athans and Falb, 1966). Equation (2.107) is in brackets since it follows from the canonical equations.

Regardless of some differences in notation, the corresponding boundary conditions for the adjoining vector and Hamiltonian are continuous processes and the same as those given earlier for discrete processes, Equations (2.37) to (2.40). Assuming that some final and initial coordinates, say x_a^f and x_b^f and x_c^i and x_d^i, are fixed, the boundary conditions for free coordinates are

$$p_\nu^f = \frac{\partial G(\mathbf{x}^f, t^f)}{\partial x_\nu^f}, \nu \neq a, b \tag{2.109}$$

$$p_\nu^i = \frac{\partial G(\mathbf{x}^i, t^i)}{\partial x_\nu^i}, \nu \neq c, d \tag{2.110}$$

$$H^f = -\frac{\partial G(\mathbf{x}^f, t^f)}{\partial t^f}, \text{ if } t^N \text{ is free} \tag{2.111}$$

$$H^i = -\frac{\partial G(\mathbf{x}^i, t^i)}{\partial t^i}, \text{ if } t^0 \text{ is free} \tag{2.112}$$

Associated with free values of boundary coordinates, the above equations describe the boundary conditions for the adjoint vector $\mathbf{p}$ and Hamiltonian H at the end and at the beginning of the process. If a coordinate of boundary state, say x_ν^f or x_ν^i, is fixed, the respective component of adjoint vector, p_ν^f or p_ν^i, is undetermined. Analogously, if the boundary time, t^f or t^i, is fixed, the respective Hamiltonian, H^f or H^i, is undetermined.

Methods of averaged optimization in association with the Maximum Principle provide a powerful tool for analyses of dynamical thermal machines, separators and chemical reactors (Doroghov et al., 1973; Tsirlin, 1974, 1997a, 1997b; Tsirlin et al., 1998; Mironova et al., 2000; Berry et al., 2000; Tsirlin and Leskov, 2007).

For the relation between Pontryagin's Maximum Principle and the viscosity solutions for nondifferentiable optimal performance functions (Section 2.8) we refer to the papers by Barron and Jensen (1986), Subbotina (1989), Zhou (1990,

1991), Cannarsa and Frankowska (1991), Tessitore (1995), and the books by Bardi and Capuzzo-Dolcetta (1997) and Vinter (2000).

A number of approximation techniques have been proposed to solve the equations of Maximum Principle (Athans and Falb, 1966; Rosenbrock and Storey, 1966; Sieniutycz and Szwast, 1982a, 1982b, 1983, 1999; Szwast, 1990; Noton, 1972; Findeisen et al., 1974, 1980; Chiang, 1992; Kamien and Schwartz, 1992; Leonard and Van Long, 1994).

Two-point boundary value problems appearing in the Maximum Principle algorithms often cause serious analytical difficulties, so various numerical methods are applied. In a numerical procedure which applies the *gradient method in the control space*, a trial solution $\mathbf{u}(t)$ is assumed which is next amended in the direction of the Hamiltonian increase. First, the state equation (2.42) (equivalent Equation (2.105)) is solved in the natural direction of time, subject to an initial condition, say $\mathbf{x}(t^i) = \alpha$. (Note that knowledge of the adjoint vector $\mathbf{p}(t)$ is not required for this purpose.) Next, the adjoint equation (2.106) is integrated; the profile of vector function $\mathbf{p}(t)$ is determined by the backward integration of Equation (2.106) subject to the final condition (2.109). Simultaneously with the calculation of $\mathbf{p}(t)$, amended control $\mathbf{u}'(t)$ is determined following the rule $\mathbf{u}'(t) = \mathbf{u}(t) + l\delta H/\delta \mathbf{u}$, where l is the step coefficient adjusted to extremize the performance criterion S (e.g. Rosenbrock and Storey, 1966). The iterative procedure is stopped when changes of S and $\mathbf{u}$ become negligible. Block schemes of the method are available (Findeisen et al., 1980; Sieniutycz and Szwast, 1982a, p. 46). Discussion of other methods such as the conjugated gradient method, Newton's method and the method of second variation is also available in the literature (Lasdon et al., 1967; Findeisen et al., 1980).

2.7. CALCULUS OF VARIATIONS

With regard to the optimal control theory the calculus of variations represents its special case associated with the classical method of maximum seeking for functionals

$$S = \int_{t^i}^{t^f} L(\mathbf{x}, \mathbf{u}, t)\mathrm{d}t \tag{2.113}$$

when the state equations have the special form

$$\frac{\mathrm{d}\mathbf{x}}{\mathrm{d}t} = \mathbf{u} \tag{2.114}$$

and satisfy an initial condition. The simplest initial condition fixes the initial state of the arc:

$$\mathbf{x}(t^i) = \alpha \tag{2.115}$$

For functional equations (2.113) subject (2.114) controls **u** are identified with the rates of state changes. The number of controls is equal to the number of state coordinates of vector **x**. Since the control vector in Equation (2.113) can be replaced by the derivative of the state vector, the typical problem of variational calculus is to determine an extremum of the integral:

$$S = \int_{t^i}^{t^f} L(\mathbf{x}, \dot{\mathbf{x}}, t)dt \tag{2.116}$$

subject to the initial condition (2.115) and the final condition (2.121). This is called the fixed end problem. Equation (2.116) constitutes the classical form of the variational problem associated with Equations (2.113)–(2.115). Vector function $\mathbf{x}(t)$ is the argument of functional (2.116).

In variational analysis the meaning of variation is of fundamental importance. The variation of the argument $\mathbf{x}(t)$ is the difference between two functions:

$$\delta\mathbf{x} = \mathbf{x}(t) - \hat{\mathbf{x}}(t) \tag{2.117}$$

The functional $\Delta S = S(\hat{\mathbf{x}} + \delta\mathbf{x}) - S(\hat{\mathbf{x}})$ is called the accretion of the functional S. The main, that is, linear, part of accretion ΔS is called the variation of the functional. For functional (2.116) the variation δS is

$$\delta S = \int_i^{t^f} \sum_{i=1}^{s} \left\{ \frac{\partial L}{\partial x_i}\delta x_i + \frac{\partial L}{\partial \dot{x}_i}\delta\dot{x}_i \right\} dt. \tag{2.118}$$

The fundamental theorem of variational calculus states that the vanishing of variation δS along the extremal curve $\hat{\mathbf{x}}(t)$ is the necessary condition for the extremum of functional (2.116). Integration of Equation (2.118) by parts yields an expression

$$\delta S = \int_t^{t^f} \sum_{i=1}^{s} \left\{ \frac{\partial L}{\partial x_i}\delta x_i + \frac{d}{dt}\left(\frac{\partial L}{\partial \dot{x}_i}\delta x_i \right) - \delta x_i \frac{d}{dt}\left(\frac{\partial L}{\partial \dot{x}_i} \right) \right\} dt \tag{2.119}$$

where the middle term does not contribute to the extremum of S when the end points of the extremal are fixed. This yields the necessary extremum condition satisfied along the extremal curve $\hat{\mathbf{x}}(t)$ in the form of s differential equations:

$$\frac{\partial L}{\partial x_i} - \frac{d}{dt}\left(\frac{\partial L}{\partial \dot{x}_i} \right) = 0, i = 1, 2, \ldots, s \tag{2.120}$$

They are called the Euler–Lagrange equations of variational calculus. The extremum condition (2.120) describes disappearance of the so-called variational derivative $\delta L/\delta x_i$ defined as the left-hand side of Equation (2.120).

The Euler–Lagrange equation (2.120) is equally valid when the right end of the trajectory is fixed or when it moves along a surface or a line. In the first case the final condition for the trajectory has the form

$$\mathbf{x}(t^f) = \beta \qquad (2.121)$$

In the second case, the form of the final condition depends on the manifold along which the end of the trajectory moves. For the motion along a curve described by equations

$$x_i = \varphi_i(t^f), i = 1, 2 \ldots s \qquad (2.122)$$

s conditions must be satisfied in the following form:

$$\sum_{i=1}^{s} \left\{ \frac{\partial L^f}{\partial \dot{x}_i^f} (\varphi(t^f) - \dot{x}_i^f) \right\} + L^f = 0 \qquad (2.123)$$

They are called the transversality conditions.

Solving the set of Euler–Lagrange equation (2.120) with, say, an initial condition, Equation (2.115) and with a final condition, Equation (2.121), or Equation (2.123), leads to the determination of the extremal trajectory which may be an optimal trajectory. The nature of the functional extremum (minimum, maximum or a saddle point) is determined by the investigation of sufficient conditions (Elsgolc, 1960; Moiseiwitsch, 1966; Santilli, 1977). Consideration of the physical properties of the vicinity of extremal and nonoptimal virtual trajectories is also helpful. Information about classical variational calculus can be found in many sources (Lanczos, 1949; Gelfand and Fomin, 1963; Moiseiwitsch, 1966; Rund, 1966; Yourgrau and Mandelstam, 1968; Smidth, 1974; and others).

Important generalizations of the Euler–Lagrange equations (2.120) to cases with many independent variables (e.g. physical fields) and for functionals containing derivatives of higher order can only be mentioned briefly here. Further information can be gained from many books (Rund, 1966; Schechter, 1967; Finlayson, 1972; Logan, 1977; Vujanovic and Jones, 1988; Kupershmidt, 1992; Sieniutycz, 1994).

The inverse problem of variational calculus requires one to find a functional whose extremum conditions are given differential equations. The problem is particularly important in the realm of partial differential equations, where variational principles can be formulated and solutions of important PD equations may be found by direct variations methods (Herivel, 1955; Vainberg, 1964; Stephens, 1967; Yourgrau and Mandelstam, 1968; Penfield and Haus, 1968; Seliger and Whitham, 1968; Finlayson, 1972; Atherton and Homsy, 1975; Santilli, 1977; Berdicevskii, 1983; Glazunov, 1983; Mikhailov and Glazunov, 1985; Vujanovic and Jones, 1988; Kupershmidt, 1992; Sieniutycz, 1994; Sieniutycz and Farkas, 2005). Symmetry principles and conservation laws resulting from the fundamental Nöther's theorem (Nöther, 1918, 1971; Caviglia and Morro, 1987; Caviglia, 1988) are most important results which have contributed significantly to our deeper understanding of the meaning of known physical laws.

Comparing the continuous and discrete results of dynamic optimization, one may formally regard the continuous algorithm of Maximum Principle, Equations (2.105)–(2.108), as a limit of the discrete result, Equations (2.32)–(2.35), when the number of stages tends to infinity. Yet it should be realized that the discrete Maximum Principle requires stronger assumptions than the continuous one. Briefly, the discrete case assumptions must include convexity of rate functions and constraining sets. On the contrary, the sketch of the derivation of the Maximum Principle from the HJB equation (2.82) has been obtained here under severe differentiability assumptions for the HJB solution $R(\mathbf{x}, t)$. In fact, in many examples $R(\mathbf{x}, t)$ is not differentiable and the HJB equation does not admit classical (smooth) solutions at all. These cases are outlined in the following section.

2.8.   VISCOSITY SOLUTIONS AND NONSMOOTH ANALYSES

Hamilton–Jacobi equations are nonlinear equations that arise in many fields of application, including mechanics, geometrical optics, control theory, front propagation, nonequilibrium thermodynamics, image processing, and differential games. Because of their nonlinearity, the related Cauchy problems usually have nonclassical solutions due to the crossing of characteristic curves. Classical analysis under boundary and initial conditions by the method of characteristics is thus limited to local considerations. Global analysis of these problems has been hindered by the absence of an appropriate notion of a solution which has the desired properties of existence and uniqueness. In the Crandall and Lions (1983) paper a notion of solution has been proposed which allows, for example, solutions not to be differentiable but for which strong uniqueness theorems, stability theorems and general existence theorems are all valid.

For scalar equations of the conservation law type, there is a theory regarding the existence and uniqueness of a weak solution called the entropy solution, using the special integral structure of the equations (Lax, 1973). Advanced numerical methods have been developed and widely used to compute approximations that converge to the correct entropy solutions. Nevertheless, this notion of weak solution cannot be applied to many fully nonlinear equations, for example, the eikonal equation, $u_t + |\nabla u| = 0$.

Crandall and Lions (1983) first introduced the notion of the viscosity solution, based on a Maximum Principle and an order-preserving property of parabolic equations. In general, for a given Hamilton–Jacobi equation (Du is limited here to components of the spatial gradient):

$$u_t + H(x, t, u, Du) = 0 \qquad (2.124)$$

where H is a continuous function and nondecreasing in u and Ω, and is an open subset of R^n; there exists a unique uniformly continuous viscosity solution if the initial data are bounded. The continuity of the solution can be understood intuitively from the 1 D case in which "HJ equations are the conservation laws integrated once". The viscosity solution is sometimes understood as the limit of the solutions to the equation with vanishing viscosity (Barles and Perthame, 1988). Accordingly, Crandall and Lions (1984) proved the convergence of two approximations to the viscosity solution of equations whose Hamiltonians only depend on Du. This was generalized by Souganidis (1985a, 1985b), who developed approximation schemes for viscosity solutions of Hamilton–Jacobi equations with variable coefficients. Many sophisticated numerical methods have since been developed.

However, there are problems in control theory and differential games which demand discontinuous solutions. The original viscosity theory does not apply to discontinuous initial data, yet its extensions offer this opportunity. The notion of a semi-continuous viscosity solution was first introduced by Ishii (1985, 1987), who used an extension of Perron's method in his investigation of Hamilton–Jacobi equations with discontinuous Hamiltonians. Because of the nonuniqueness in Ishii's result, other notions of semi-continuous solutions were proposed by various authors (Barron and Jensen, 1990; Barles, 1993) with different kinds of additional properties imposed on the Hamiltonian. Some of these notions need serious restrictions on the Hamiltonians and others are implicit in the sense that the processes of taking supremum and infimum are involved. As a consequence, one cannot develop numerical methods to construct approximations. For an overview of the viscosity theory and applications, see books by Barles (1994) and Bardi and Capuzzo-Dolcetta (1997).

Finally, for the class of equations with Hamiltonians nondecreasing in u, Giga and Sato (2001) introduced a new notion for semi-continuous solution. This notion of solution is defined by the evolution of the zero level curve of the auxiliary level set equation which embeds the original HJ equation. It is thus called the L-solution. Tsai et al. (2001) considered the level set approach for computing discontinuous solutions of a class of Hamilton–Jacobi equations and designed a Lax–Friedrichs-type scheme to compute approximation of the L-solution in its original formulation (i.e. level set). They have shown that, with suitable (CFL) conditions, their schemes keep the discrete version of an important property of this class of HJ equations. Mass-preserving finite-element implementations of the level set method are available (Di Pietro et al., 2006).

In optimal control theory one deals with HJB equations which are quasi-linear partial differential equations and contain an extremizing operation with respect to controls. When an expression describing feedback optimal control is substituted into an HBJ equation, a nonlinear Hamilton–Jacobi equation is obtained. Fleming and Rishel (1975), Findeisen et al. (1977), Lions (1983a, 1983b), and Krylov (1980) are mathematical sources for Hamilton–Jacobi–Bellman equations. In general such equations do not admit global classical solutions (Xiao and Basar, 1997, 1999). Furthermore, the value functions of optimal control problems (and differential games as well) are not differentiable in general, and they are not even continuous for some classes of problems (e.g. Soravia, 1996; Xiao and Basar, 1997, 1999). The theory of viscosity solutions provides a convenient framework to study solutions of HJB equations. The questions of existence and uniqueness for viscosity solutions of HJB equations have been studied by a number of authors, and in particular by Crandall and Lions (1983, 1986), Lions (1982), Crandall (1997), Evans (1994), Ishii (1984), Crandall et al. (1984), Fleming and Soner (1993), Soravia (1996), and Evans and Souganidis (1984).

The notion of viscosity solutions of scalar fully nonlinear partial differential equations of second order provides a framework in which startling comparison and uniqueness theorems, existence theorems, and theorems about continuous dependence may now be proved by efficient and striking arguments. The range of important applications of these results is enormous. The user's guide by Crandall et al. (1992) is a self-contained exposition of the basic theory of viscosity solutions. It covers topics such as partial differential equations, fully nonlinear equations, elliptic equations, parabolic equations, Hamilton–Jacobi equations, dynamic programming, nonlinear boundary value problems, generalized solutions, Maximum Principles, comparison theorems, and Perron's method. The theory of viscosity solutions applies to certain partial differential equations of the form $F(x, u, Du, D^2u) = 0$, where $F: R^N \times R \times R^N \times S(N) \to R$ and $S(N)$ is the set of symmetric $N \times N$ matrices. As stated by Crandall et al. (1992), the primary virtues of this theory are *that it allows merely continuous functions to be solutions of fully nonlinear equations of second order, that it provides very general existence and uniqueness theorems and that it yields precise formulations of general boundary conditions.* Moreover, these features are accompanied by a great flexibility in passing to limits in various settings and relatively simple proofs. In the expression $F(x, u, Du, D^2u)$, u is a real-valued function defined on some subset O of R^N, Du corresponds to the gradient of u and D^2u corresponds to the matrix of second derivatives of u. However, as explained below, Du and D^2u do not have classical meanings and, in fact, examples prove that the theory encompasses classes of equations that have no solutions that are differentiable in the classical sense.

The main point is that a first-order operator $F(x, u, Du)$ is always degenerate elliptic and thus it is proper if and only if $F(x, r, p)$ is nondecreasing in $r \in R$. Proper equations of the form $F(x, u, Du) = 0$ play a fundamental role in the classical calculus of variations and in optimal control theory of ordinary differential equations; in this context they are often called Hamilton–Jacobi and Bellman equations, and then $F(x, r, p)$ is convex in (r, p). These equations, in the full generality of nearly arbitrary proper functions F, are also crucial in differential games theory where they are known as Isaac's equations. The equation

$$-\nu\Delta u + f(x, u, Du) = 0 \qquad (2.125)$$

with $v > 0$ and f nondecreasing in u, may be regarded as a first-order Hamilton–Jacobi equation perturbed by an additional "viscosity" term. Such equations arise in optimal stochastic control.

Hamilton–Jacobi–Bellman and Isaac's equations are, respectively, the fundamental partial differential equations for stochastic control and stochastic differential games. The natural setting involves a collection of elliptic operators of second order, depending either on one parameter a (in the Hamilton–Jacobi–Bellman case) or two parameters a, b (in the case of Isaacs's equations). Their ingredients are expressions of the form

$$\mathcal{L}^\alpha u = -\sum_{i,j=1}^{N} a_{i,j}^\alpha(x)\frac{\partial^2 u}{\partial x_i \partial x_j} + \sum_{i=1}^{N} b_i^\alpha(x)\frac{\partial u}{\partial x_i} + c^\alpha(x)u(x) - f^\alpha(x) \quad (2.126)$$

$$\mathcal{L}^{\alpha,\beta} u = -\sum_{i,j=1}^{N} a_{i,j}^{\alpha,\beta}\frac{\partial^2 u}{\partial x_i \partial x_j} + \sum_{i=1}^{N} b_i^{\alpha,\beta}(x)\frac{\partial u}{\partial x_i} + c^{\alpha,\beta}(x)u(x) - f^{\alpha,\beta}(x)$$

$$(2.127)$$

where all the coefficients are bounded with respect to parameters α and β. Hamilton–Jacobi–Bellman equations include those of the form

$$\sup_\alpha \{\mathcal{L}^\alpha u\} = 0 \quad (2.128a)$$

while a typical Isaac's equation of games theory is of the form

$$\sup_\alpha \inf_\beta \{\mathcal{L}^{\alpha,\beta} u\} = 0 \quad (2.128b)$$

We observe that if $(x, r, p, X) \to (t, x, r, p, X)$ is proper for fixed $t \in [0,T]$, then so is the associated "parabolic" problem

$$u_t + F(t, x, u, Du, D^2 u) = 0 \quad (2.129)$$

when considered as an equation in the $N + 1$ independent variables (t, x). Of the many examples, one describes the evolution of a surface (given by a level set of the initial condition) with a motion along its normal with a speed proportional to the mean curvature.

Rund (1966), Benton (1977), and Lions (1982) discuss (first-order) Hamilton–Jacobi equations. Bryson and Levy (2006) consider related computational issues. Fleming and Rishel (1975), Krylov (1980), Lions (1983a, 1983b), Capuzzo Dolcetta and Ishii (1984), Capuzzo Dolcetta and Falcone (1989), and Bauer et al. (2006) are mathematical sources for Hamilton–Jacobi–Bellman equations and corresponding discrete methods (see more information below).

The notion of viscosity solutions

It is assumed that F satisfies (0.1) (i.e. F is proper) and, unless otherwise stated, is continuous. To motivate the notions, one begins by supposing that u is C^2 (i.e. twice continuously differentiable) on R^N and

$$F(x, u(x), Du(x), D^2u(x)) \leq 0 \tag{2.130}$$

holds for all x (that is, u is a classical subsolution of $F = 0$ or, equivalently, a classical solution of $F \leq 0$ in R^N). Suppose that φ is also C^2 and $\hat{x}$ is a local maximum of $u - \varphi$. Then calculus implies

$$Du(\hat{x}) = D\varphi(\hat{x}) \text{ and } D^2u(\hat{x}) \leq D^2\varphi(\hat{x}) \tag{2.131}$$

and so, by degenerate ellipticity,

$$F(\hat{x}, u(\hat{x}), D\varphi(\hat{x}), D^2\varphi(\hat{x})) \leq F(\hat{x}, u(\hat{x}), Du(\hat{x}), D^2u(\hat{x})) \leq 0 \tag{2.132}$$

The extremes of this inequality do not depend on the derivatives of u and so we may consider defining an arbitrary function u to be (some kind of generalized) subsolution of $F = 0$ if

$$F(\hat{x})u(\hat{x}), D\varphi(\hat{x}), D^2\varphi(\hat{x})) \leq 0 \tag{2.133}$$

whenever φ is C^2 and $\hat{x}$ is a local maximum of $u - \varphi$. Before making formal definitions (Crandall et al., 1992), also note that

$$u(x) \leq u(\hat{x}) - \varphi(\hat{x}) + \varphi(x) \text{ for } x \text{ near } \hat{x}, \varphi \in C^2 \tag{2.134}$$

and the Taylor approximation implies

$$u(x) \leq u(\hat{x}) + \langle p, x - \hat{x} \rangle + \frac{1}{2} \langle X(x - \hat{x}), x - \hat{x} \rangle + o(|x - \hat{x}|^2) \text{ as } x \to \hat{x}$$

$$\tag{2.135}$$

where

$$p = D\varphi(\hat{x}) \text{ and } X = D^2\varphi(\hat{x}) \tag{2.136}$$

One can now define the notions of viscosity subsolutions, supersolutions, and solutions. It is useful to have the notations (Crandall et al., 1992)

$$\text{USC}(\mathcal{O}) = \{\text{upper semicontinuous functions } u : \mathcal{O} \to \mathbb{R}\}$$

$$\text{LSC}(\mathcal{O}) = \{\text{lower semicontinuous functions } u : \mathcal{O} \to \mathbb{R}\} \tag{2.137}$$

Definition

Let F satisfy (0.1) and $O \subset R^N$. A *viscosity subsolution* of $F = 0$ (equivalently, a viscosity solution of $F \leq 0$) on O is a function $u \subset \text{USC}(O)$ such that

$$F(x, u(x), p, X) \leq 0 \text{ for all } \quad x \in \mathcal{O} \quad \text{and} \quad (p, X) \in J_{\mathcal{O}}^{2,+} u(x) \qquad (2.138)$$

Similarly, a viscosity supersolution of $F = 0$ on O is a function $u \in \text{LSC}(O)$ such that

$$F(x, u(x), p, X) \geq 0 \quad \text{for all } x \in \mathcal{O} \text{ and } (p, X) \in J_{\mathcal{O}}^{2,-} u(x) \qquad (2.139)$$

Finally, u is a *viscosity solution* of $F = 0$ in O if it is both a viscosity subsolution and a viscosity supersolution of $F = 0$ in O.

Since these "viscosity notions" are the primary ones, some authors drop the term "viscosity" and simply refer to subsolutions, supersolutions, and solutions. This is right, as the term "viscosity" is an artifact of the origin of this theory in the study of first-order equations and the name was motivated by the consistency of the notion with the method of "vanishing viscosity," which is irrelevant for many second-order equations.

It follows from the discussion preceding the definition that, for example, if u is a solution of $F \leq 0$, φ is C^2 in a neighborhood of O, and $u - \varphi$ has a local maximum (relative to O) at $\hat{x} \in O$, then the inequality holds:

$$F(x, u(x), Du(x), D^2 u(x)) \leq 0 \qquad (2.140)$$

Analogous remarks hold for supersolutions. These remarks motivate the requirement that a subsolution be upper semi-continuous, etc., in the sense that producing maxima of upper semi-continuous functions is straightforward. Solutions, being both upper semi-continuous and lower semi-continuous, are continuous. One might ask if the validity of the discussed inequality for all $\varphi \in C^2$ (with the maxima relative to O) for an upper semi-continuous function u is equivalent to u being a subsolution. This is indeed so. In fact, if $\hat{x} \in O$ then

$$J_{\mathcal{O}}^{2,+} u(\hat{x}) = \{(D\varphi(\hat{x}), D^2\varphi(\hat{x}) : \varphi \text{ is } C^2 \text{ and } u - \varphi \text{ has a } local \text{ maximum at } \hat{x}\}$$

$$(2.141)$$

(Crandall et al., 1992; Crandall, 1997).

Summing up, an upper (respectively, lower) semi-continuous function $u: \Omega \rightarrow R$ is a viscosity subsolution (respectively, viscosity supersolution) of $F = 0$ in Ω if for every $\phi \in C^2(R^n)$ and local maximum (respectively, minimum) point z of $u - \phi$ in Ω one has $F(z, u(z), D\phi(z), D^2\phi(z)) \leq 0$ (respectively, $F(z, u(z), D\phi(z), D^2\phi(z)) \geq 0$). A continuous function $u: \Omega \rightarrow R$ is a viscosity solution of $F = 0$ in Ω if it is both a viscosity subsolution and a viscosity supersolution of $F = 0$ in Ω. The inequalities defining viscosity subsolutions and supersolutions are a consequence of the structure condition and the necessary conditions for extremals if u is a classical solution of $F \leq 0$ or $F \geq 0$ in an open set, a fact which shows a connection between the notion of viscosity solutions and the classical Maximum Principle for second-order elliptic equations.

The importance of this notion lies in the fact that very general uniqueness and existence theorems are valid for viscosity solutions. A typical example is the existence and uniqueness of a unique bounded and uniformly continuous function $u(x)$, $x = (t, y) \in [0,T] \times \mathbf{R}^m$, which is a viscosity solution of $u_t + G(D_y u, D_y^2 u) = 0$ on $[0,T] \times \mathbf{R}^m$ and satisfies $u(0, y) = \psi(y)$ for $y \in \mathbf{R}^m$ whenever $T > 0$, $G(q, Z)$ is continuous in $(q, Z) \in \mathbf{R}^m \times S^m$ and anti-monotone in Z, and ψ is bounded and uniformly continuous on $\mathbf{R}^m$. (S^n is the space of real symmetric ($n \times n$) matrices.) Existence is essentially an effect of the proof of uniqueness, which also establishes monotone and continuous dependence of the solution with respect to ψ and may be proved by an adaptation of the Perron method (Ishii 1985, 1987).

The original presentation of the viscosity solutions introduced by Crandall and Lions (1981, 1983) emphasized the first-order case, because uniqueness results were only obtainable for this case at the time. It was also pointed out that there were several equivalent ways to formulate the notion of viscosity solutions. In these developments, directions were indicated by aspects of nonlinear functional analysis and nonlinear semi-group theory. It became apparent that working with these alternative formulations was superior to the earlier (1981) approach, and perhaps the paper by Crandall et al. (1984) is the first readable account of the early theory. (Note that the definition most emphasized in that paper is not the one put foremost here.) Analogous operations have long been employed by "nonsmooth analysts" (Clarke, 1997).

The first uniqueness proofs for viscosity solutions were given for first-order equations in Crandall and Lions (1983) and then Crandall et al. (1984). The second-order case remained open for quite a period, during which the only evidence that a general theory could be developed was in the results for Hamilton–Jacobi–Bellman equations obtained by Lions (1982, 1983a, 1983b, 1985). A breakthrough was achieved in the second-order theory by Jensen (1988) with the introduction of several key arguments; some of these were simplified in works by Souganidis (1985a, 1985b), Soner (1986), Falcone (1987), Lions and Souganidis (1988), and Barles and Souganidis (1991). The theory of discontinuous viscosity solutions of first-order Hamilton–Jacobi equations has been worked out (Barles, 1993). Barles (1994) synthesizes many of these results.

Motivated by various Hamilton–Jacobi–Bellman equations arising in deterministic optimal control, Barron and Jensen (1990) modified the concept of viscosity solution introduced by Crandall and Lions for convex (or concave) Hamiltonians and semi-continuous solutions. They showed that one can dispense with the Crandall–Lions requirement that we touch the solution by test functions from both above and below, and need only touch from one side.

Which side depends on whether the solution is upper or lower semi-continuous and whether the Hamiltonian is concave. The advantage of testing from only one side is that semi-continuous solutions can only be touched from one side. Barron and Jensen (1990) have shown that this is sufficient to characterize the solution.

A dynamic programming context for the viscosity solutions was worked out by Capuzzo Dolcetta, Ishii and Falcone (Capuzzo Dolcetta, 1983; Capuzzo Dolcetta and Ishii, 1984; Capuzzo Dolcetta and Falcone, 1989; Capuzzo Dolcetta et al., 1985). These papers provided techniques for approximating the viscosity solution of the Bellman equation in deterministic control problems based on discrete dynamic programming. Monotonically convergent schemes, associated with computational algorithms, were obtained and error estimates proved. The approximate solutions converge uniformly to the viscosity solution of the original problem. Discrete time schemes for viscosity solutions of Hamilton–Jacobi–Bellman equations are effective (Falcone and Ferretti, 1994). Stochastic control problems are treated using the dynamic programming approach (Fleming and Soner, 1993). For control Markov diffusion HJB equations are nonlinear PD equations of second order. Viscosity solutions provide a suitable framework to study these HJB equations. Summaries of these approaches are presented in several books (Fleming and Soner, 1993; Bardi and Capuzzo-Dolcetta, 1997; Falcone and Makridakis, 2001). The last book addresses numerical methods for viscosity solutions. Bauer et al. (2006) present an adaptive spline technique to solve HJB equations.

Application-related contributions deal with various algorithms, their convergence, evaluation of their performance, and technical issues related to their implementation. One important issue is that of optimal discretization resolution (Falcone and Ferretti, 1994, 2006; Romanchuk, 1995; Tsitsiklis, 1995; Falcone and Makridakis, 2001; Munos and Moore, 2001), where the viscosity solution approaches serve to rearrange some earlier formulations (Jordan and Polak, 1964a, 1964b; Larson, 1967; Canon et al., 1970; Kaplinskiy and Propoy, 1970; Cullum, 1971; Maikov and Tsirlin, 1973). Other applications contain solutions of concrete problems in mechanics, geometrical optics, control theory, front propagation, differential games, nonequilibrium thermodynamics, and image processing. See the review by Kimia et al. (1994) on optimal control methods in computer vision and image processing. Particularly vast are recent applications to the "shape from shading" problem (Prados and Faugeras, 2003, 2006). Shape from shading is the process of computing the three-dimensional shape of a surface from one image of that surface. Contrary to most of the other three-dimensional reconstruction problems (for example, stereo and photometric stereo), in the Shape From Shading problem, data are minimal as a single image is used. Therefore, this inverse problem is intrinsically difficult. Related publications (Prados and Faugeras, 2003; Prados et al., 2004; Camilli and Prados, 2006) describe the main difficulties, most recent theoretical results, and examples of modeling by rigorous numerical methods.

2.9. STOCHASTIC CONTROL AND STOCHASTIC MAXIMUM PRINCIPLE

There is extensive literature on the Maximum Principle for stochastic *classical* control. See, for example, Kushner, 1965; Kaplinskiy and Propoy, 1970; Bismut, 1978; Bensoussan, 1981; Haussmann, 1986; Elliott, 1990; Peng, 1990; Zhou, 1991; Elliott and Kohlmann, 1994; Cadenillas and Karatzas, 1995; Yong and Zhou, 1999, and references therein; this list of references is by no means exhaustive.

In recent years, singular stochastic control has received considerable attention. The connection between the singular control problem and the optimal stopping problem has been investigated by a number of authors (Karatzas, 1983, 1985; Karatzas and Shreve, 1984, 1985, 1986; Boetius and Kohlmann, 1988; El Karoui and Karatzas, 1988, 1991; Chow et al., 1995; Alvarez, 1999, 2001; Boetius, 2001a, 2001b; Dufour and Miller, 2004). A stochastic Maximum Principle may be proved for optimal singular controls (Cadenillas and Haussmann, 1994; Dufour and Miller, 2006). A step in the proof may require that the original singular control problem is converted into a classical control problem by using a special time transformation (Dufour and Miller, 2006). The optimality conditions can be obtained even for nonlinear systems.

Stochastic control problems may be examined in the context of dynamic programming and HJB equations (Fleming and Soner, 1993). For control Markov diffusion processes HJB equations are nonlinear partial differential equations of second order. Frequently their value functions are not smooth enough to satisfy the HJB equation in a classical sense. The theory of viscosity solutions helps in the study of these HJB equations and to prove continuous dependence of solutions on problem data (Fleming and Soner, 1993).

3 Energy limits for thermal engines and heat pumps at steady states

3.1. INTRODUCTION: ROLE OF OPTIMIZATION IN DETERMINING THERMODYNAMIC LIMITS

In this chapter we define and analyze thermodynamic limits for various traditional and work-assisted processes with finite rates, important in engineering, physics, and biology. The analysis is based on the condition that in order to make thermodynamic analyses usable in a technology, it must be ensured whether a thermodynamic limit (e.g., a lower bound for consumption or work or heat or an upper bound for work or heat production) is an appropriate quantifying criterion for this technology. We limit ourselves here to "static limits," that is, those in steady systems. This chapter also introduces the reader to certain controls called "Carnot variables."

The practical and industrial systems of interest include thermal and chemical generators of mechanical energy (engines) and the apparatus or devices in which this energy is consumed (refrigerators, heat pumps, and separators). In principle, irreversible thermodynamics is a general field suitable to evaluate such limits for processes occurring in finite time and in systems of a finite size. However, theoretical irreversible thermodynamics seldom attack systems with explicit work flux (power) (see De Groot and Mazur (1984) for a basic description). For the purpose of energy systems analyses, irreversible thermodynamics is most often applied in a discrete rather than a continuum form, which means that thermal fields are seldom attacked. Yet the continua are not excluded in treatments of energy problems (Orlov and Berry, 1990, 1991a, 1991b, 1992). A typical irreversible thermodynamics analysis of an energy system refers to a topological structure that belongs in the thermal networks (diverse units connected by appropriate links). Such structures could in principle be treated by network thermodynamics (NT; Peusner, 1986), a general field that transfers meanings and tools of electrical circuit theory to macrosystems described by discrete models.

Yet in view of the strongly theoretical nature and considerable formal complexity of NT at the present time, its two "practical derivatives" are effectively applied. The first is the method of finite time thermodynamics (FTT) considered in the research groups of R.S. Berry in the United States and A. Tsirlin in Russia, and the second is A. Bejan's method of entropy generation minimization (EGM). At this point the reader is referred to some basic sources on both methods for FTT: Andresen et al. (1977, 1984), Andresen (1983), de Vos (1992), and Berry et al. (2000); and for EGM: Bejan and Smith (1974), Bejan and Paynter (1976), Bejan (1982, 1987a, 1987b, 1988a, 1988b, 1996a, 1996b, 1996c, 1996d), and

Energy Optimization in Process Systems

Bejan et al. (1996). For a history of both methods, see Bejan (1996e). While both FTT and EGM were established independently and their original aims and tools were slightly different at the beginning, they principally constitute the same method of treatment of practical irreversible systems. The method is based on the investigation of a governing entity (most often mechanical power in FTT, although other criteria are not excluded, and entropy production in EGM), which is a function or a functional of physicochemical and geometrical characteristics of the system (fluxes, efficiencies, material properties, shapes, dimensions, residence times, etc.). To construct that function or functional, the researcher must build into the system's model the information that comes not only from thermodynamics but also from the theory of transport phenomena (transfer of heat, mass, electricity, and momentum). By extremizing the governing criterion, optimal controls and (in dynamic problems) optimal trajectories are obtained that constitute the solution to the optimization problem from which limiting losses of mechanical energy are found. Estimates of energy limits, such as minimum energy consumption or maximum energy production in the system, are obtained in the form of either numbers (static problems) or potential functions (dynamic problems).

Both FTT and EGM methods present two practical approaches to energy systems, and both of them constitute practical and efficient tools of contemporary industrial energetics. As such, they may be linked with earlier approaches to that field, and even reconstituted on their basis. Those earlier approaches should be given due credit because they frequently contain most valuable results of applied equilibrium thermodynamics (Bosniakovic, 1965; Brodyanskii, 1973; Ciborowski, 1965, 1976; Le Goff, 1979, 1980, and many others). This is said despite the fact that they underestimate, in general, the role of kinetic components in physiochemical models. Their thermostatic origin refers particularly to the research conducted at the University Henri Poincaré of Nancy, France, where the earlier thermodynamic approaches (Le Goff, 1979, 1980, and others) evolved into the present thermokinetic treatments (Tondeur and Kvaalen, 1987; Feidt, 2006). In this book, we often refer to practical achievements of the Le Goff research group in the field of thermal and chemical plants, separation processes (especially distillation), sorption heat pumps, and heat transformers for upgrading waste heat (Le Goff et al., 1986, 1988, 1990a, 1990b, 1990c, 1990d).

The Feidt group, on the other hand, pursues the more recent research, basically in the spirit of FTT and EGM, and it has developed numerous original analyses of various energy systems. They refer to the following problems: thermoeconomics and finite size thermodynamics for optimization of heat pumps, energy cogeneration systems (Benelmir and Feidt, 1997, 1998; Brunin et al., 1997), gas turbines and combined cycles (Cenusa et al., 2003), exergy optimization of heat recovery steam generators (Cenusa et al., 2004), optimal distribution of the heat conductance (Costea and Feidt, 1999), dynamic optimization of heat exchangers (Shaal and Feidt, 1999), Stirling engine optimization (Costea et al., 1999), optimization of the Brayton cycle, reverse cycle machines and some imperfect cycles (Feidt, 1996, 1997, 1999; Popescu et al., 1996), efficiency studies (Feidt, 2001), considerations of greenhouse effect in the context of the design of inverse cycles (Feidt, 2002, 2003), applications to fuel cells (Feidt and Lottin, 2004), and, especially, experimental and theoretical analyses of the operation of natural gas cogeneration systems using polymer exchange membrane fuel cells (Radulescu et al., 2006a, 2006b). This research group has also developed exergy analyses of an adsorption refrigeration unit (Luo and Feidt, 1992; Radcenco et al., 2001; Vasilescu et al., 2007), and formulated problems of technoeconomic optimization of thermal machines and regeneration systems, including some inverse problems (Grosu et al., 1999; Feidt et al., 2002, 2007; Morosuk et al., 2004).

All classical measures of thermodynamic perfection, such as the Carnot efficiency, energy efficiencies, or dissipated energy, have one characteristic in common: they all take the reversible process as their reference (Berry et al., 2000; Sieniutycz, 1998c). Therefore, one can ask whether any real process operates close enough to the reversible limit for the traditional measures to be useful or even relevant. If not, is it possible to extend the concepts of reversible thermodynamics to provide limits for the performance of processes constrained to operate in finite time or at nonvanishing rates? If yes, what are the optimal paths related to the finite time transitions? These questions arose from the need to evaluate how effectively real devices and real industrial processes use energy, from the viewpoint of their own limits rather than the reversible process limits. They stimulated efforts to find bounds described by certain potentials corresponding to a given process performed in a finite time (Salamon et al., 1977; Andresen et al., 1984), in particular the thermodynamic bounds defined by the minimal values of the total energy dissipated or the total energy driving a finite time process (Sieniutycz, 1973b). The choice of optimization criteria can, of course, affect the optimal performance and the corresponding bounds.

Choice of optimization criterion is especially clear in FTT and EGM. In the first method one explicitly searches for conditions of extremism of mechanical power or equivalent condition of maximum work from a steady flow; in the second one an operation with a minimum irreversibility as the suitable limiting solution for the process with a prescribed mean rate is sought. But, by the Gouy–Stodola theorem (Gouy, 1989; Stodola, 1905; Bejan, 1982), the minimum irreversibility ensures a maximum of the work produced by the generator (minimum of work consumed by a separator or a heat pump) that is just the optimization criterion of FTT. Consistently, mechanical energy limits for processes with finite rates can be evaluated either directly or with the help of the minimization of irreversibility (entropy production). Yet this optimization principle refers only to the energy limits and not to economically optimal solutions. These limits are characterized by an upper bound for the mechanical energy production and a lower bound for its consumption.

When an economic optimum is considered, the system's performance criterion should be reformulated in order to describe the sum of the investment costs and production costs. In this case the direct link between optimal power and entropy production is lost. In this book it is not our intention to consider economic optimization problems within their own realm. Our goal is to select and discuss basic evergreen problems formulated in the realm of energy limits and at the interface of energy limits, (thermo)economics, and ecology. Inclusion of thermoeconomic and ecological interfaces makes sense since the most interesting and important phenomena often take place at interfaces of different disciplines. Moreover, the discussion on FTT as it developed in time inherently involves some economic issues. With optimization we may not only consider economic criteria but also focus on physically determined limiting possibilities of practical systems. For the processes considered it is assumed that they must reach the desired extent of completion within a finite time, be realizable with equipment of a finite size, and must produce at least a minimum amount of a required product.

In the fields of separation units and heat and mass exchangers, a great volume of literature on energy limits of separation and exchange operations may be found. This literature often deals with very diverse, single-stage and multistage separation units and a variety of many complex systems composed of such units; optimizations usually run with either technical or quasi-economic criteria with a visible tendency toward complex process economies. Mathematics seems to be a standard tool in this context. What is really novel therein is a unifying, integrating concept of the dynamic (finite time) limits for energy production or consumption. The integrating nature of these limits is important. In fact, traditional chemical engineering approaches to exchange and separation processes dissect the field on the basis of what is specific in processes rather than integrate individual processes. An engineer becomes perplexed whenever a design shifts from one to another operation, and when his or her past experience is of restricted use. Integrating approaches to energy limits are helpful since they are common to various operations within the same generic class.

Diverse approaches to applied thermodynamics (including FTT, EGM, etc.) have many names. Each name often constitutes only one aspect of that approach. In essence, the FTT research procedure includes some typical stages, for example (Tsirlin and Kazakov, 2002a, 2002b):

1. evaluation of minimum of entropy production caused by irreversible processes subject to constraints such as the given average rate of a flux (load), size of equipment, known transfer coefficients, given kinetic formulae, and so on
2. construction of an inequality determining the realizable area that defines thermodynamically feasible processes with the help of thermodynamic balances of energy, mass, and entropy, subject to the condition that entropy production in entropy balance can only be larger than the minimum possible, evaluated in 1
3. finding a limiting value of some objective (rate, efficiency, power, etc.) subject to parameters of the process that are located within the realizable area

The above three stages are encountered in most research where the FTT approach is applied to obtain limiting possibilities of irreversible processes. Fixing the duration of the process is only one of the ways used to define a realistic constraint. There are other characteristic features that define the methodology. Also, the minimization of entropy production (EGM) is merely one of the three stages described above and does not include all remaining stages used in FTT.

Likewise, to assume local equilibrium and to state that the entropy production occurs only within the reservoirs (endoreversibility) is only one of many possible ways to take the total entropy production into account. For example, entropy production can be calculated within the energy generators, at surfaces, and so on.

Consequently, the objective of the FTT approach is not only to take into account the effect of duration of a thermodynamic process but also to obtain efficiency estimates for systems with given rates, bounded sizes, and given mass and heat transfer laws. These estimates could be very loose but they still are generally more realistic than the corresponding reversible estimates. Sometimes an estimate based on reversible processes tends to zero but the corresponding irreversible estimate tends to some finite limit. An example is a process of separating a binary mixture when the key component's concentration is very low, tending toward zero (Berry et al., 2000). Others include the limiting power and the limiting efficiency, for given power, of a heat engine with given (finite) heat transfer coefficients; the corresponding results for heat pumps and refrigerating cycles; the optimal cycles (time paths of pistons) of internal combustion engines; the limiting possibilities of absorption–desorption cycles with given rate; thermodynamic processes distributed over time or length that minimize average entropy production; the minimal work of separation with given average rate. Some of these results and the general methodology used to derive them are described in a recent book by Berry et al. (2000).

The dynamic nature of limits and a constraint for a finite time of operation or a finite size of the system are original features assured in FTT and EGM approaches that are not encountered in classical works on energy problems. Real engineering processes must run to a reasonable degree of completion within a finite amount of time; if a process runs with a vanishing velocity, that is, quasistatically, or if it must relax to equilibrium before it can be considered completed, which takes an infinite amount of time, the process is practically useless. The finite-rate requirement leads to finite production of entropy and losses of exergy; as a process becomes faster the losses increase. One of the major aims of analyses based on energy limits is to work out such conceptual approaches that could lead an engineer to certain generic rather than specific limits or bounds on practical or industrial processes. They describe largely dynamic limits that exhibit a significant degree of universality. The dynamic limits are not only more general than static, the consequence of finite durations and finite rates, but may also be more useful in design. Our interest is in revealing and systematizing such dynamic limits.

For the classical works that describe energy limits following from classical thermodynamics, see the following references: Keenan (1941), Gibbs (1948), Rant (1956), Denbigh (1956), Szargut and Petela (1965), Bosniakovic (1965), Le Goff (1979, 1980), Bockris (1980), Kotas (1985), Strumillo and Kudra (1987), Brodyanskii et al. (1988), Forland et al. (1988), Szargut et al. (1988), and many others. In particular, problems of limits in chemical and separation systems are discussed in Denbigh (1956), Everdell (1965), Forland et al. (1988), Fratcher and Bayer (1981), Malih (1987), and Tsirlin and Kazakov (2002a, 2002b). They generally consider classical (kinetics neglecting) limits of separation processes, except the last one (Tsirlin and Kazakov, 2002a, 2002b), which analyzes the significance of classical limits (Van't Hoff reversible work) for limits predicted by irreversible thermodynamics. All limits (classical or not) usually refer to well-known practical quantities. They may, for example, determine the lower bound for the amount of energy supply, exploitation costs, amount of a key substance, investment, equipment extent, and so on. They describe limitations that an engineer encounters during the design of exchange or separation units running with a prescribed intensity (local or average). Some bounds are so natural and simple (as, e.g., that on the lowest heat consumption) that one may confuse finite-rate limits with the well-known thermostatic limitations. Still, the bounds we are looking for are not classical bounds encountered in textbooks on classical thermodynamics and separation science. Those classical limits pertain to reversible and infinitely slow processes. They are often unrealistically low and hence, quite often, useless.

Critics of FTT often identify this field with merely one paper: an early publication of Curzon and Ahlborn (1975) that has been restricted to an ideal Carnot engine interacting with nonideal, resistance-containing reservoirs. This identification leads to a narrowed and distorted picture of the contemporary FTT whose tools and range have been enriched through the years. Contemporary FTT theory systematically treats the internal entropy generation in thermal machines (Berry et al., 2000) and, as such, it is capable of covering a much broader range of problems than the "endoreversible" approach of Curzon and Ahlborn (1975). This fact seems to be ignored by a number of adversaries of FTT. Criticisms of FTT are explicit in several papers (Gyftopoulos, 1999, 2002, 2003; Moran, 1998; Lior, 2002), always pointing out restrictions following from the acceptance of the endoreversibility. On occasion the criticism is also applied to EGM due to the inherent link between the two fields (Szargut, 1998, 2001b). The view of both fields is too literary and oversimplified; moreover, optima of energy limits are mixed with optima of economic costs. Below we present some (quotations of) typical statements and a related argument.

One of the objections is contained in the incorrect statement: "Real processes are always irreversible and consideration of that irreversibility would imply consideration of its influence on the investment costs and operational costs, which is not possible in the FTT method" (Szargut, 2001b). In fact, irreversibility is a purely physical phenomenon that is completely characterized in terms of physical quantities, as known from thermodynamics, especially irreversible thermodynamics (De Groot and Mazur, 1984) and statistical physics (Landau and Lifshitz, 1974). Consideration of the influence of irreversibility on the investment costs and operational costs not only is possible in FTT but can also be more effective than in other methods (in particular than in classical exergy analysis) due to options in the choice of the optimization criterion in FTT ("control thermodynamics"; Salamon et al., 2001; Chapter 11).

3.2. CLASSICAL PROBLEM OF THERMAL ENGINE DRIVEN BY HEAT FLUX

3.2.1. Maximum Power in Thermal Engines

Curzon and Ahlborn (1975) showed that a heat engine with heat resistances between the reservoirs and the working fluid can operate between two extreme zero-power limits (power being defined as work per cycle time). One extreme is the reversible, Carnot limit at maximum (Carnot) efficiency. This is the limit where the engine cycles extremely slowly (quasistatically); the efficiency is high, but the engine performs too few cycles to deliver much power. If the engine cycles too fast, it operates inefficiently because it does not get hot enough or cold enough to ensure that the ratio of the two extremal temperatures of the working fluid, $T_{1'}/T_{2'}$, is much less than 1. Hence, the other extreme is the "thermal short circuit" limit of zero power at zero efficiency. Heat engines are usually designed to operate somewhere between the limit of the reversible maximum of efficiency and that of maximum power.

In the thermal system described by Newton's law of heat transfer Curzon–Ahlborn efficiency at maximum power is $\eta_{mp} = 1 - (T_2/T_1)^{1/2}$, regardless of the heat transfer coefficients and other material properties. The original concept of Curzon and Ahlborn has been extended to a richer spectrum of problems in FTT by taking the main sources of dissipation (e.g., finite heat conductances) into consideration. This leads to even more realistic bounds on performance. Curzon and Ahlborn analyzed Newtonian heat exchange; models with various functional forms of the temperature difference have also been designed. Of practical importance is radiative heat transfer, where the heat flux is proportional to temperature to the fourth power. Another case is derived from linear irreversible thermodynamics and considers the reciprocal temperature difference. These investigations show that the Curzon and Ahlborn formula is not always simply a function of the reservoir temperatures, T_2 and T_1. It depends, in general, on other variables such as the reservoir heat capacity (for finite reservoirs, see Ondrechen et al., 1981) or the specific heats of the working fluid. This efficiency is not a fundamental upper limit for engines working at maximum power; it depends on the functional temperature dependence of the heat exchange between the working fluid and reservoirs. Moreover, real heat engines with friction and heat leaks exhibit fundamentally different power–efficiency curves than those in which finite-rate heat transfer is the only irreversibility. The presence of friction leads to higher efficiencies when the machine operates more slowly, and heat leaks lead to higher efficiencies when the machine operates at faster rates.

It follows that optimization automatically sorts heat engines into two distinct classes: those dominated by heat leaks and those dominated by friction. (See the power–efficiency diagrams by Gordon (1991).) The power–efficiency curves exhibit a maximum power point and a maximum efficiency point, the latter usually being well below the Carnot efficiency. For the optimal solution

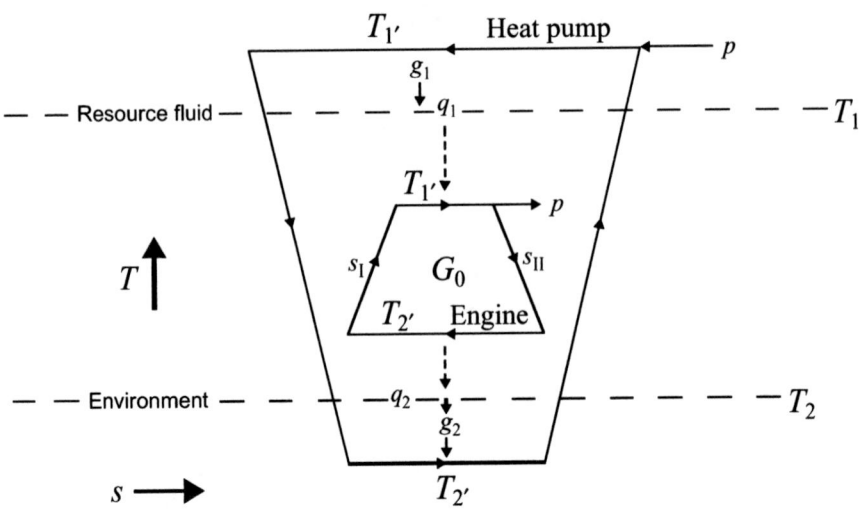

Figure 3.1 Basic designations and comparison of two basic modes with internal and external dissipation: power production in an engine and power consumption in a heat pump.

obtained, effects of friction as well as time-dependent driving functions have been determined (Gordon and Huleihil, 1991).These functions maximize power when heat input and heat rejection are nonisothermal. The optimal performance of an engine that operates between a finite high-temperature source and an infinite heat reservoir and obeys the heat transfer law of Onsager thermodynamics $(q = K\Delta T - 1)$ has been compared with the performance predicted for Newton's law of heat transfer (Yan and Chen, 1990). The analysis shows that the time of the absorbing heat process must be longer than that of the releasing one in the Onsagerian case, but that the times of the two heat exchange processes should be the same for Newton's law of heat transfer. Three heat-source cycles using low-level heat sources such as solar energy, geothermal energy, and waste heat have been analyzed in a unifying treatment (Chen and Yan, 1989). Lagrange multipliers were applied to handle the entropy constraint (Sieniutycz and Kubiak, 2002). This approach is outlined in Chapter 5, Chapter 5. Imperfect systems with internal dissipation in thermal machines were analyzed as engine cycles (Chen et al., 1996; Cheng et al., 1999; Salah El-Din, 2001a, 2001b) or as refrigeration cycles (El Haj Assad, 1999; Chen et al., 2000; Salah El-Din, 2001a, 2001b; Sieniutycz and Szwast, 2003). A common scheme depicting both a nonideal engine and a heat pump is given in Figure 3.1.

Carnot temperatures satisfying Carnot expression for given actual efficiency η were introduced as suitable controls (Sieniutycz, 1999b). Further development of this idea is given in Sections 5.4 and 5.5. Figure 3.2 illustrates fundamentals of the description of irreversible thermal machines in terms of these controls. With Carnot temperatures, the endoreversibility assumption is abandoned to analyze imperfect systems with internal dissipation (Sieniutycz, 1999b).

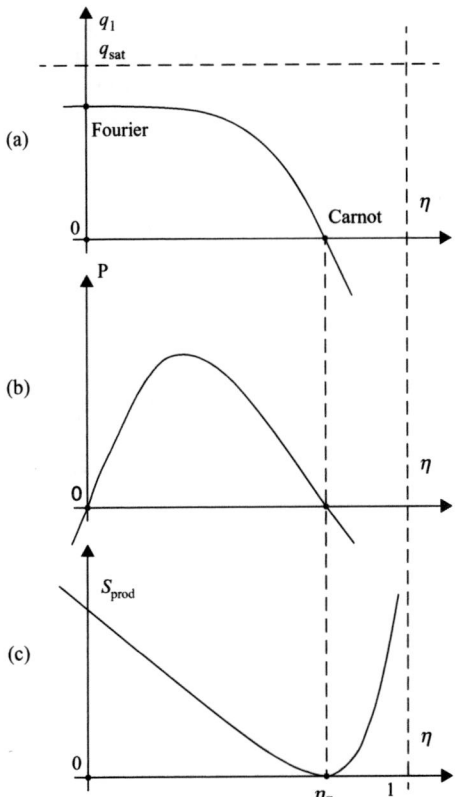

Figure 3.2 Conventional characteristics of Chambadal–Novikov–Curzon–Ahlborn (CNCA) operation when efficiency η is the control variable: (a) driving heat q_1, (b) power yield p, and (c) entropy production S_{prod} or σ_s.

3.2.2. Lagrange Multipliers and Endoreversible System

Here we shall focus on analytical aspects of limits on mechanical energy production in systems composed of two thermal reservoirs and an energy generator. Consider, for example, a single-stage engine process in the standard Chambadal–Novikov–Curzon–Ahlborn operation (CNCA engine) in which c is a resource's specific heat, and g_1 and g_2 are thermal conductances (Novikov, 1958; Curzon and Ahlborn, 1975). Figure 3.1 illustrates engine and refrigerator (heat pump) modes of the process in the general case of internal irreversibilities (factor Φ different than unity). In this section we shall, however, restrict ourselves to the simpler (early) case when $\Phi = 1$ and the loops in Figure 3.1 are rectangular.

This is the so-called "endoreversible" case as there are no internal irreversibilities by assumption. We shall abandon this assumption in the next section.

We shall focus here on a formulation in which the power of the engine is maximized with respect to both temperatures of the circulating fluid, $T_{1'}$ and $T_{2'}$, which are constrained controls. The constraint is the entropy balance across the engine's loop; it is handled by the Lagrange multipliers. In this formulation the engine's power is expressed by an equation

$$p = g_1(T_1 - T_{1'}) \left(1 - \frac{T_{2'}}{T_{1'}}\right) \tag{3.1}$$

It describes the product of the driving heat and endoreversible efficiency. Equation (3.1) can be obtained from the original criterion describing power as difference of heat fluxes, $p = q_1 - q_2$, after the second flux, q_2, is eliminated with the help of the entropy balance constraint

$$\frac{g_1(T_1 - T_{1'})}{T_{1'}} = \frac{g_2(T_{2'} - T_2)}{T_{2'}} \tag{3.2}$$

The constraint is in the form of the continuity of entropy flux. The modified optimization criterion that adjoins constraint (3.2) by the Lagrange multiplier λ has the form

$$p' = p + \lambda C = g_1(T_1 - T_{1'}) \left(1 - \frac{T_{2'}}{T_{1'}}\right)$$
$$+ \lambda \left(\frac{g_1(T_1 - T_{1'})}{T_{1'}} - \frac{g_2(T_{2'} - T_2)}{T_{2'}}\right) \tag{3.3}$$

Now, for the modified criterion p', coordinates of the stationary point with respect to $T_{1'}$, $T_{2'}$, and λ are found. This requires setting to zero respective partial derivatives of p':

$$(p')_{T_{1'}} = 0; \qquad (p')_{T_{2'}} = 0; \qquad (p')_{\lambda} = 0 \tag{3.4}$$

Explicitly, the following set of equations should be solved:

$$(p')_{T_{1'}} = -g_1 + g_1 \frac{T_{2'}T_1}{(T_{1'})^2} - \lambda g_1 \frac{T_1}{(T_{1'})^2}$$
$$= -g_1 \left(1 + \lambda \frac{T_1}{(T_{1'})^2}\right) + g_1 \frac{T_{2'}T_1}{(T_{1'})^2} = 0 \tag{3.5}$$

$$(p')_{T_{2'}} = -\frac{g_1(T_1 - T_{1'})}{T_{1'}} - \lambda g_2 \frac{T_2}{(T_{2'})^2}$$
$$= -\frac{g_2(T_{2'} - T_2)}{T_{2'}} - \lambda g_2 \frac{T_2}{(T_{2'})^2} = 0 \tag{3.6}$$

$$(p')_\lambda = \frac{g_1(T_1 - T_{1'})}{T_{1'}} - \frac{g_2(T_{2'} - T_2)}{T_{2'}} = 0 \tag{3.7}$$

Of course, the last equation is the recovered entropy constraint as the extremum condition of p' with respect to η. This constraint was used in Equation (3.6) to get this last equality. The solution of Equations (3.5) and (3.6) with respect to λ leads, respectively, to the relations

$$\lambda = T_{2'} - \frac{(T_{1'})^2}{T_1} \tag{3.8}$$

$$\lambda = -T_{2'}\frac{T_{2'} - T_2}{T_2} = T_{2'} - \frac{(T_{2'})^2}{T_2} \tag{3.9}$$

Hence, the temperatures of the circulating fluid are linked by an equation

$$\frac{(T_{1'})^2}{T_1} = \frac{(T_{2'})^2}{T_2} \tag{3.10}$$

This is the formula connecting the optimal temperatures $T_{1'}$ and $T_{2'}$ in terms of the temperatures of heat sources. Equation (3.10) leads to a simple relation between temperatures of the circulating fluid

$$T_{1'} = T_{2'}\sqrt{\frac{T_1}{T_2}} \tag{3.11}$$

Substituting the so-expressed temperature $T_{1'}$ into the equation of the entropy balance, we obtain

$$g_1\left(T_1 - \sqrt{\frac{T_1}{T_2}}T_{2'}\right)\left(\sqrt{\frac{T_1}{T_2}}T_{2'}\right)^{-1} = \frac{g_2(T_{2'} - T_2)}{T_{2'}} \tag{3.12}$$

hence, after rearrangements, temperature $T_{2'}$ follows as

$$T_{2'} = \frac{g_1\sqrt{T_1 T_2} + g_2 T_2}{g_1 + g_2} \tag{3.13}$$

Next, with Equation (3.11), temperature $T_{1'}$ is obtained as follows:

$$T_{1'} = \frac{g_1\sqrt{T_1 T_2} + g_2 T_2}{g_1 + g_2}\sqrt{\frac{T_1}{T_2}} = \frac{g_1 T_1 + g_2\sqrt{T_1 T_2}}{g_1 + g_2} \tag{3.14}$$

These are optimal controls, or temperatures of the circulating fluid in the engine at maximum power conditions. One can now calculate optimal heat fluxes

q_1 and q_2:

$$q_1 = g_1 \left(T_1 - \frac{g_1 T_1 + g_2 \sqrt{T_1 T_2}}{g_1 + g_2} \right) = g_1 \left(\frac{g_2 T_1 - g_2 \sqrt{T_1 T_2}}{g_1 + g_2} \right)$$

$$= g\sqrt{T_1} \left(\sqrt{T_1} - \sqrt{T_2} \right)$$

$$q_2 = g_2 \left(\frac{g_1 \sqrt{T_1 T_2} + g_2 T_2}{g_1 + g_2} - T_2 \right) = g_2 \left(\frac{g_1 \sqrt{T_1 T_2} - g_1 T_2}{g_1 + g_2} \right)$$

$$= g\sqrt{T_2} \left(\sqrt{T_1} - \sqrt{T_2} \right)$$

(3.15)

In these equations the overall conductance g has been defined as the harmonic mean

$$g = \frac{g_1 g_2}{g_1 + g_2} \tag{3.16}$$

The maximum power limit of the engine system is

$$p_m = g\sqrt{T_1} \left(\sqrt{T_1} - \sqrt{T_2} \right) - g\sqrt{T_2} \left(\sqrt{T_1} - \sqrt{T_2} \right)$$

$$= g \left(\sqrt{T_1} - \sqrt{T_2} \right) \left(\sqrt{T_1} - \sqrt{T_2} \right) = g \left(\sqrt{T_1} - \sqrt{T_2} \right)^2 \tag{3.17}$$

The optimal efficiency of the energy yield equals $\eta = 1 - q_2/q_1$; thus, from Equations (3.14) and (3.15) we get

$$\eta = 1 - \sqrt{\frac{T_2}{T_1}} \tag{3.18}$$

While this result is well known, it was usually obtained by a method based on the elimination of variables (Curzon and Ahlborn, 1975). Its original derivations are longer. Clearly, the approach using the Lagrange multipliers leads to a solution in a short time and applies a simple methodical procedure. Process controls are here treated on an equal footing. The optimal solution includes controls in terms of states of two fluids participating in *the operation*, (T_1, T_2). Moreover, the optimal decisions identically satisfy the process constraints, in our case the entropy balance of the energy generator. The approach is applicable even when there is an internal entropy source in the thermal machine (Curzon and Ahlborn, 1975). Thus, in spite of the example's simplicity, its consequences are of certain value.

However, if the original criterion, $P = q_1 - q_2$, is treated, in which each of the heat fluxes is expressed in terms of the state variables, T_1 and T_2, and controls, $T_{1'}$ and $T_{2'}$, then extremal Lagrange multipliers are different from those described by Equations (3.8) and (3.9). This difference is the effect of different objective functions in each case while preserving the same constraint. The example proves that trials to attribute to Lagrange multipliers a physical significance should be made carefully. Even if an interesting interpretation is found for one system of constraints, such interpretation has no absolute character. In fact, it will change if transformed constraints are applied, even if they are equivalent constraints that are linear combination of original ones. In other words, the physical significance of λ is not objective with respect to transformation of constraints. Examples of this sort are also known in the variational hydrodynamics of the perfect fluid (Sieniutycz, 1994).

Equation (3.18) describes efficiency at maximum power in the case when internal irreversibilities are absent ($\Phi = 1$). It is often attributed to Curzon and Ahlborn (1975), although the same formula and similar modeling have appeared in a number of engineering papers and books since 1957. They were published by Novikov and his coworkers (Novikov, 1958, 1984; Vukalovich and Novikov, 1972; Novikov and Vosskresenskii, 1977) and promoted in English textbooks by El-Wakil (1962, 1971) and in French by Chambadal (1957, 1963). Bejan has derived Equation (3.18) in his 1982 book on the problem of the optimal allocation of heat transfer area between the two exchangers of a power plant (Bejan, 1982, pp. 46, 146). Bejan has also shown in his 1996 book that when a complete expression for entropy generation is applied, the criteria of maximum power and the minimum of entropy generation lead to the same results (Bejan, 1996a, Bejan, 1996a). These criteria yield Equation (3.18) in the tested case $\Phi = 1$. Yet, various controls can be applied to ensure the limit predicted by Equation (3.18). For example, one can control efficiency η (de Vos, 1992), driving heat flux (Denton, 2002), heat rejected to the environment (Sieniutycz, 2000a, 2000b, 2000c, 2000d, 2000e), entropy flux (Sieniutycz, 1998a, 1998b, 1998c, 1998d, 1998e), Carnot temperature (Denton, 2002), and other quantities.

Equation (3.18) should be properly understood. It refers to the efficiency at maximum power or the efficiency of the thermodynamic limit, and neither to maximum efficiency of irreversible heat engines nor to the efficiency optimizing economic costs. As pointed out (Chen et al., 2001), a frequent interpretative error is that this equation is understood as the upper bound on the engine efficiency, whereas it describes rather the lower bound for economic efficiencies of those thermal engines that are reasonably well described by the model. This correct viewpoint was also presented in de Vos's book (de Vos, 1992). Confusion may also follow from an incorrect assumption that the power plant of a given thermal efficiency operates at the maximum power point. When comparing Equation (3.18) with experimental findings, one must first check if the real operating conditions and the modeling objectives are consistent. This is not often the case. Equation (3.18) refers to the efficiency at maximum power, and, for a free resource, it can also refer to conditions of maximum work that can be generated by the resource fluid at flow. The objective function associated with Equation (3.18) was the maximum power (thermodynamic limit), whereas working power plants have other objectives, usually economic ones.

Chen et al. (2001) also consider the link between the process speed and irreversibility. According to their FTT analysis, when the thermal resistance (equivalent to the internal resistance of a battery) is dominant, a slow process is closer to reversible. But when the heat leak loss (corresponding to the internal discharge of the battery) is dominant, a fast process is closer to reversible. In conclusion, in systems with internal irreversibilities, a fast discharge process may be closer to reversible than a slow one. This is in agreement with experimental data for batteries and with theoretical analysis of an electric battery model with an internal discharge (Bejan and Dan, 1996, 1997).

Generalizations of Equation (3.18) are known that take into account such factors as finite compression ratio, heat leak loss, finite heat source in the working fluid, and fluid flow irreversibilities. Quantitative discussion of maximum power efficiencies in terms of these factors is available, along with discussion of the relevant literature (Chen et al., 2001). The maximum power efficiencies that take these factors into account are in general lower than the efficiency of Equation (3.18).

In a series of experimental works, Ng and Gordon and their coworkers have shown that the thermodynamic behavior of chillers (generalized air conditioning and refrigeration systems) is severely sensitive to internal dissipation and that ignoring its effect can lead to substantial errors in chiller diagnostics and in the prediction of chiller performance characteristics (Ng et al., 1997, 1998; Gordon and Ng, 2000). This substantiates modeling that goes beyond endoreversibility; see next section. However, we shall not repeat there a corresponding analysis with Lagrange multipliers; rather, we shall apply as an independent control the efficiency variable η and then the corresponding temperature $T' = T_2/(1 - \eta)$, which satisfies the Carnot formula by assumption and which we briefly call the Carnot temperature of the problem. In the following we shall show advantages resulting from the process description in terms of the efficiency and Carnot temperature controls.

3.2.3.   Analysis of Imperfect Units in Terms of Efficiency Control

The choice of the control variable is in principle arbitrary. For example, the controls can be: each of heat fluxes q_1 or q_2, each of related entropy fluxes, the efficiency, η, and others. Extensive work by de Vos and coworkers has shown the benefits of using efficiency η as the control variable in many endoreversible operations in thermal, chemical, solar, photothermal, photovoltaic, economic, and hybrid systems (de Vos, 1991, 1992, 1993a, 1993b, 1995, 1997; de Vos and Pauwels, 1981; de Vos and van der Wei, 1992, 1993; de Vos et al., 1993). Figure 3.2 shows typical diagrams illustrating the behavior of the CNCA operation in terms of the efficiency control variable.

We shall now abandon the endoreversibility assumption and focus on imperfect systems with internal dissipation within thermal machines. We shall use the efficiency η as the control variable. The entropy balance of each imperfect thermal machine in Figure 3.1 is

$$\Phi \frac{q_1}{T_{1'}} = \frac{q_2}{T_{2'}} \tag{3.19}$$

where $\Phi = \Delta S_{2'}/\Delta S_{1'} = J_{s_{2'}}/J_{s_{1'}}$ is the parameter defining the internal irreversibility of the thermal machine measured in terms of the entropy changes of the circulating fluid along the two isotherms $T_{1'}$ and $T_{2'}$ (Figure 3.1). In some cases we assume constancy of Φ. Note, however, that whereas one can always introduce internal dissipation in the form represented by Equation (3.19), one may treat Φ as a constant only as the rough approximation. In general, Φ may be a complicated function of the machine's operating variables. Thus, mean values of Φ averaged within operative boundary of system parameters are necessary. Since the theory with an averaged constant Φ usually provides a sufficient insight, it was used in a large number of papers (Chen et al., 1996, 2000; Cheng et al., 1999; Salah El-Din, 2001a, 2001b; El Haj Assad, 1999; Sieniutycz and Szwast, 2003).

Using the entropy balance (3.19) combined with the definition of the first-law efficiency:

$$q_2 = q_1 - p = q_1(1 - \eta) \tag{3.20}$$

yields the stage efficiency given by the pseudo-Carnot formula

$$\eta = 1 - \frac{q_2}{q_1} = 1 - \Phi\frac{T_{2'}}{T_{1'}} \tag{3.21}$$

For an engine stage, where $\Phi > 1$, this efficiency is lower than that of Carnot engine between the temperatures $T_{1'}$ and $T_{2'}$, that is, $\eta < \eta_{C'}$. For heat pump mode $\Phi < 1$ and $\eta > \eta_{C'}$.

By assumption, Newtonian law of cooling holds with conductances g_1 and g_2

$$q_1 = g_1(T_1 - T_{1'}), \qquad q_2 = g_2(T_{2'} - T_2) \tag{3.22}$$

Quantities $T_{1'}$ and $T_{2'}$ are unknown, but they may be expressed in terms of the temperatures T_1 and T_2 and a single control variable at the stage.

The reasoning that uses the notion of efficiency is as follows: as Equations (3.19) and (3.22) hold, temperatures $T_{1'}$ and $T_{2'}$ are not independent but connected by the constraint

$$\frac{g_2(T_{2'} - T_2)}{T_{2'}} - \Phi\frac{g_1(T_1 - T_{1'})}{T_{1'}} = 0 \tag{3.23}$$

which means that the system has only one degree of freedom (one independent control). We use Equation (3.21) to substitute $T_{2'} = (1 - \eta)\Phi^{-1}T_{1'}$ into Equation (3.23). We then obtain an equation for $T_{1'}$ in the form

$$T_{1'} = \frac{g_1 T_1}{g_1 + g_2\Phi^{-1}} + \frac{1}{1 - \eta}\frac{g_2 T_2}{g_1 + g_2\Phi^{-1}} \tag{3.24}$$

and the corresponding equation for $T_{2'} = (1 - \eta)\Phi^{-1}T_{1'}$

$$T_{2'} = \frac{g_1 T_1}{g_1 + g_2\Phi^{-1}}\Phi^{-1}(1 - \eta) + \frac{g_2 T_2}{g_1 + g_2\Phi^{-1}}\Phi^{-1} \tag{3.25}$$

The fluxes of heat in terms of the efficiency η are

$$q_1 = g_1(T_1 - T_{1'})$$

$$= \frac{g_1 g_2 T_1 \Phi^{-1}}{g_1 + g_2 \Phi^{-1}} - \frac{1}{1-\eta} \frac{g_1 g_2 T_2}{g_1 + g_2 \Phi^{-1}} \equiv g' \left(T_1 - \frac{\Phi T_2}{1-\eta} \right) \quad (3.26)$$

and

$$q_2 = \frac{g_2 g_1 T_1}{g_1 + g_2 \Phi^{-1}} \Phi^{-1}(1-\eta) - \frac{g_2 T_2 g_1}{g_1 + g_2 \Phi^{-1}} \quad (3.27)$$

In Equation (3.26) an operational overall conductance has been defined as follows:

$$g' \equiv \frac{g_2 g_1}{\Phi g_1 + g_2} = \left(\frac{\Phi g_1 + g_2}{g_2 g_1} \right)^{-1} = \left(\frac{1}{g_1} + \Phi \frac{1}{g_2} \right)^{-1} \quad (3.28)$$

This is, in fact, the overall conductance of an inactive heat transfer, which is suitably modified due to the presence of coefficient of internal dissipation.

The work flux follows consistently in the form

$$p = \eta q_1 = \eta \left(\frac{g_1 g_2 T_1}{\Phi g_1 + g_2} - \frac{\Phi}{1-\eta} \frac{g_1 g_2 T_2}{\Phi g_1 + g_2} \right) \equiv g'(\Phi) \eta \left(T_1 - \frac{\Phi T_2}{1-\eta} \right)$$

$$(3.29)$$

For $\Phi = 1$ (Carnot cycle) expression for endoreversible power is recovered from Equation (3.29)

$$p = \eta q_1 = \eta g \left(\frac{T_1 - T_2}{1-\eta} \right) \quad (3.30)$$

Maximum of power (3.29) versus η implies $p' = 0$ or $(1-\eta)^2 = \Phi T_2/T_1$. The extremum efficiency is

$$\eta_{mp} = 1 - \sqrt{\frac{\Phi T_2}{T_1}} \quad (3.31)$$

This result is known; it was obtained a decade ago by several researchers; for quotations and modifications see Chen et al. (2001). The result describes the decrease of efficiency of engine due to the internal irreversibilities. Yet for our purposes here not only the result as itself but also the solution property that leads to "Carnot temperatures" is important.

3.2.4. Introducing Carnot Temperature Controls

Below we shall demonstrate the suitability of the so-called Carnot or driving temperature, T', as the control variable. This concept has its roots in the already recognized structure of equations of linear CNCA process. Equations (3.26) and (3.29) lead us to the first encounter with the Carnot temperature, namely, as shown by these equations, the result (3.31) and its associates can also be obtained when the control is the following operative temperature T':

$$T' \equiv \frac{\Phi T_2}{1 - \eta} \tag{3.32}$$

Note that the Carnot formula is obtained by inversion of (3.32) in the endoreversible case of $\Phi = 1$; in the process with internal losses pseudo-Carnot structure follows

$$\eta = 1 - \Phi \frac{T_2}{T'} \tag{3.33}$$

The resulting expressions of Equations (3.26) and (3.29) show that the Newtonian model of the heat and power kinetics assumes in terms of T' a simple form $q_1 = g'(T_1 - T')$ and $p = \eta g'(T_1 - T')$. However, these formulae are burdened by severe assumption of Newtonian mechanism of heat transfer and, therefore, they are insufficient basis for the definition of Carnot temperature T'. Still these equations provide a useful hint.

The important issue of definition of Carnot temperature T' is solved on the basis of thermodynamic rather than kinetic considerations. The definition of T' should be independent of the presence of Φ and specificity of heat transfer law. The thermodynamic definition requires that Equation (3.33) is consistent with the two expressions in Equation (3.21) ($\eta = 1 - q_2/q_1$ and $1 - \Phi T_{2'}/T_{1'}$) that are always valid. Comparison shows that the Carnot temperature of the system satisfies the thermodynamic formula

$$T' \equiv T_2 \frac{T_{1'}}{T_{2'}} \tag{3.34}$$

This equation constitutes the general thermodynamic definition of T'; it is independent of any kinetic mechanism and internal losses Φ (Sieniutycz, 2001a, 2001b). It may also be derived from the invariance of the entropy production with respect to variable transformation, from constrained controls $T_{1'}$ and $T_{2'}$ to unconstrained control T'. Working in the context of radiation engines Kuran (2006) shows the simplification of optimization calculations when passing from two constrained controls $T_{1'}$ and $T_{2'}$ to single unconstrained control T'.

In summary, irreversible efficiency (3.21) can always be written in terms of a single controlling temperature T'. For Newtonian heat transfer T' satisfies the usual equation of overall heat exchange

$$q_1 = g'(T_1 - T') \tag{3.35}$$

where the operative conductance g' satisfies Equation (3.28). The corresponding work flux is

$$p = \eta q_1 = g'\left(1 - \frac{\Phi T_2}{T'}\right)(T_1 - T') \tag{3.36}$$

From the above findings, in particular Equation (3.35), we conclude that Carnot temperature T' is the absolute temperature of an external fluid whose cooling or heating effect exerted on the fluid 1 replaces the joint effect of the Carnot machine and the second fluid under condition of the same operational conductance g'. It is easy to verify that one can derive the maximum power efficiency (3.31) by maximizing (3.36) with respect to T'. Certain properties of physical type, for example, positiveness, can be attributed to T' (Sieniutycz, 1999b). With these properties and the role it plays in the second law, T' determines the direction of heat transfer between the fluid 1 and the thermal machine. In Chapter 4. we show how with using T' one can accomplish optimization of complex multistage processes.

The main virtue of temperature T' is its free-control property in processes of power production. Substituting thermodynamic relation $T_{2'} = T_1 \cdot T_2 / T'$ from Equation (3.34) into entropy balance (3.23) yields temperatures of circulating fluid $T_{1'}$ and $T_{2'}$ in terms of T':

$$T_{1'} = \frac{\Phi g_1 T_1 + g_2 T'}{\Phi g_1 + g_2} \tag{3.37}$$

and

$$T_{2'} = \frac{T_2}{T'} \left(\frac{\Phi g_1 T_1 + g_2 T'}{\Phi g_1 + g_2} \right) \tag{3.38}$$

Calculation of the first heat flux $q_1 = g_1(T_1 - T_{1'})$ yields the already known result $q_1 = g'(T_1 - T')$ (Equation (3.35)). The power $p = \eta(T')q_1$ accordingly follows (Equation (3.36)). Also, with Equations (3.28) and (3.33), we obtain the second heat flux

$$q_2 = g' \Phi T_2 \left(\frac{T_1}{T'} - 1 \right) = g' \frac{\Phi T_2}{T'} (T_1 - T') = q_1 \frac{\Phi T_2}{T'} \tag{3.39}$$

This agrees with Equations (3.19) and (3.34). Next chapters show dynamic applications of T'.

Carnot temperatures (sometimes called driving temperatures) were used implicitly or explicitly in several recent works to effectively describe various work-assisted operations (Sieniutycz, 1999a, 1999b, 1999c, 1999d, 1999e; Berry et al., 2000; Sieniutycz and Kubiak, 2002; Sieniutycz and Szwast, 2003; Kuran, 2006). In terms of these quantities, which are structural properties of active heat exchange in systems with two boundary layers separated by an engine, the actual efficiency is always given by the pseudo-Carnot formula.

Figures 3.3–3.6 illustrate fundamental characteristics of irreversible thermodynamics of thermal machines. With Carnot control T' the classical thermodynamic relations, formulas, and diagrams that describe the entropy production and overall heat transfer in processes without work are extended to irreversible processes with work production or consumption.

It is nice to learn that traditionally defined overall heat conductance, g, does appear in equations of linear thermal systems with power yield (de Vos, 1992;

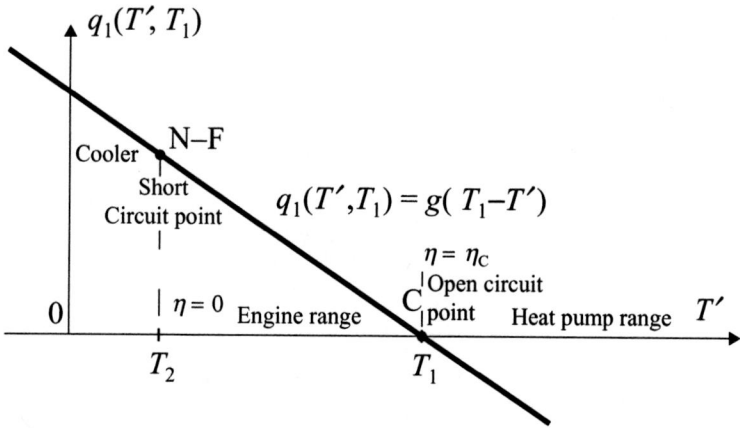

Figure 3.3 Thermal characteristics of linear CNCA operation when Carnot temperature T' is the control variable.

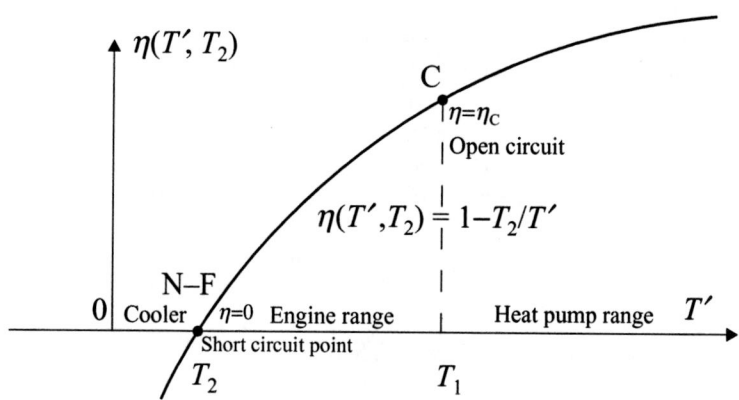

Figure 3.4 Efficiency of CNCA operation when Carnot temperature T' is the control variable.

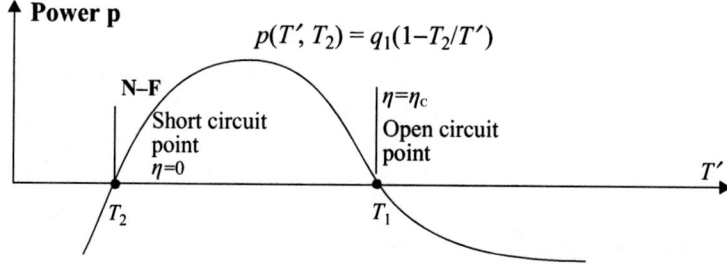

Figure 3.5 Power production in CNCA operation when Carnot temperature T' is the control variable.

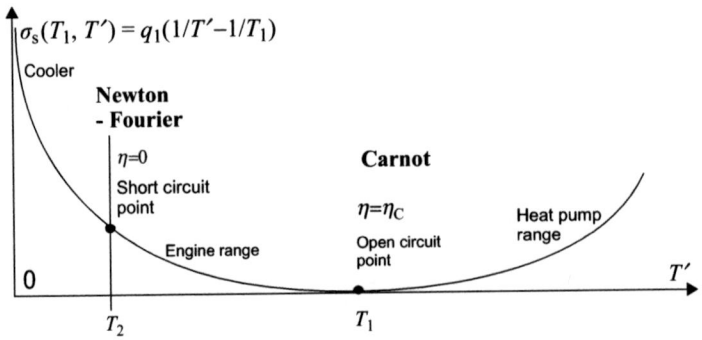

Figure 3.6 Common picture of entropy production for irreversible processes without work and with work production or consumption.

Berry et al., 2000). In imperfect systems an effective conductance g' (generalizing traditional g) emerges if some specific control variables are applied. These controls are Carnot temperatures, T' and T'', where T' is Carnot temperature of Equation (3.34), also called first Carnot temperature, whereas T'' is the second Carnot temperature that is defined below. Both these temperatures ensure Carnot structure of efficiency equations in endoreversible operations.

While the notion of second Carnot temperature is unnecessary to calculate power and driving heat, T'' needs to be introduced if one wants to obtain a Newtonian formula for the heat transfer in the *second* fluid, analogous to Equation (3.35). T'' replaces temperature T_2 and ensures the efficiency expression in the form

$$\eta = 1 - \Phi \frac{T''}{T_1} \tag{3.40}$$

The following equation should be satisfied:

$$1 - \Phi \frac{T_{2'}}{T_{1'}} = 1 - \Phi \frac{T''}{T_1} \tag{3.41}$$

We calculate from this equation $T_{1'}$:

$$T_{1'} = \frac{T_1 T_{2'}}{T''} \tag{3.42}$$

and substitute it to the entropy balance (3.23) expressed in terms of temperatures of circulating medium, $T_{1'}$ and $T_{2'}$. We obtain

$$\Phi g_1 \left(\frac{T''}{T_{2'}} - 1 \right) = g_2 \left(1 - \frac{T_2}{T_{2'}} \right) \tag{3.43}$$

whence an equation follows that describes $T_{2'}$ in the form

$$T_{2'} = \frac{\Phi g_1 T'' + g_2 T_2}{\Phi g_1 + g_2} \tag{3.44}$$

and the heat exchanged with the second reservoir is

$$q_2 \equiv g_2(T_{2'} - T_2) = g_2 \left(\frac{\Phi g_1 T'' + g_2 T_2}{\Phi g_1 + g_2} - T_2 \right)$$

$$= \frac{\Phi g_1 g_2 (T'' - T_2)}{\Phi g_1 + g_2} = g''(T'' - T_2)$$

(3.45)

where $g'' = \Phi g'$ and g' is defined by Equation (3.28). Comparison of Equations (3.34) and (3.41) yields the relations that link both Carnot temperatures

$$\frac{T_{2'}}{T_{1'}} = \frac{T_2}{T'} = \frac{T''}{T_1}$$

(3.46)

or

$$T_1 T_2 = T' T''$$

(3.47)

These equations are of thermodynamic nature, that is, they are valid regardless of the mechanism of heat exchange. In particular, equations are valid for engines driven by energy of solar radiation. Yet, forms of kinetic equations (3.25) and (3.31) are constrained to fluids with Newtonian heat exchange.

Summing up, first and second Carnot temperatures are linked by Equation (3.47); this connection ensures alternative pseudo-Carnot formula for the engine efficiency in the form of Equation (3.40). (This efficiency still refers to the heat q_1, not q_2.) The constraint (3.47) should always be used when *both* Carnot temperatures appear in a mathematical model.

3.2.5. Maximum Power in Terms of Both Carnot Temperatures

We shall briefly discuss power optimization with the use of both Carnot temperatures as decision variables. The constraint represented by the entropy balance must now be written down in terms of Carnot variables, T' and T''. The heat fluxes and power P can be expressed in terms of the first and the second Carnot temperature

$$q_1 = g'(T_1 - T')$$

(3.35)

$$q_2 = \Phi g'(T'' - T_2)$$

(3.48)

The power expression is then

$$P = q_1 - q_2 = g'(T_1 - T') - \Phi g'(T'' - T_2)$$

(3.49)

In the special case of an endoreversible operation, where $g' = g'' = g$:

$$P = g(T_1 + T_2 - T' - T'')$$

(3.49′)

Provided that each flux is expressed in terms of its own Carnot temperature, the role of T' and T'' follows from the observation that in power systems each expression for heat flux (q_1 or q_2) preserves the form of the traditional (Newtonian) heat exchange (Andresen et al., 1977; Andresen, 1983; de Vos, 1992). The connection between the first and the second Carnot temperature may be interpreted as a special form of the continuity equation for the entropy flux. In the method of Lagrange multipliers, the connection is an implicit constraint (the one that prevents elimination of one control). In the subsequent text we shall exploit only those properties of T' and T'' that are essential in analysis of the maximum power, p.

We can now describe the optimization procedure involving both Carnot controls. The modified optimization criterion has the form

$$p' \equiv p + \lambda c = g'(T_1 + \Phi T_2 - T' - \Phi T'') + \lambda(T_1 T_2 - T' T'') \qquad (3.50)$$

We calculate the partial derivatives of function p' with respect to the Carnot temperatures:

$$(p')_{T'} = -(g' + \lambda T'') \qquad (3.51)$$

$$(p')_{T''} = -(g'\Phi + \lambda T') \qquad (3.52)$$

The extremum condition

$$(p')_{T'} = (p')_{T''} = (p')_{\lambda} = 0 \qquad (3.53)$$

yields the following system of three equations with unknowns λ, T', and T'':

$$g' + \lambda T'' = 0 \qquad (3.54)$$

$$\Phi g' + \lambda T' = 0 \qquad (3.55)$$

$$T_1 T_2 = T' T'' \qquad (3.47)$$

From the first and the second equation, the maximum power condition follows in the form

$$T' = \Phi T'' \qquad (3.56)$$

Thus, at the maximum power point of an endoreversible CNCA engine both Carnot temperatures are equal. Substituting this result into the entropy constraint one can evaluate optimal values of T' and T''. Equation of the first Carnot temperature results in the form

$$(T')^2 = \frac{T_1 T_2}{\Phi} \qquad (3.57)$$

Only the positive root of this equation has physical sense; thus

$$T' = \sqrt{\frac{T_1 T_2}{\Phi}} \qquad (3.58)$$

The second Carnot temperature is then

$$T'' = \frac{T'}{\Phi} = \sqrt{\frac{T_1 T_2}{\Phi^3}} \tag{3.59}$$

Hence, the optimal efficiency is $\eta = 1 - T_2/T' = 1 - T_2/\sqrt{T_1 T_2/\Phi}$ or Equation (3.31) is satisfied again. The maximum power may be described by the formula

$$p = g' \left(T_1 + \Phi T_2 - 2\sqrt{\frac{T_1 T_2}{\Phi}} \right) \tag{3.60}$$

which, for $\Phi = 1$, simplifies to the earlier endoreversible result

$$p = g\left(\sqrt{T_1} - \sqrt{T_2}\right)^2 \tag{3.61}$$

The approach based on the Carnot temperatures has an essential virtue that makes it superior with respect to other ones, namely, it ensures a common analytical formalism for processes in traditional exchangers (without work) and in work-assisted exchangers. The traditional exchangers are characterized by the following properties: $p = \eta = 0$, $T' = T_2$, $T'' = T_1$. These properties do not refer, however, to the point that maximizes power P, but to the so-called "short circuit" point. For this point both temperatures of circulating fluid are equal, $T_{1'} = T_{2'}$, engine disappears, and the same heat flux flows through the two resistances, $q_1 = q_2$. In terms of Carnot variables the theory of traditional exchangers is a particular case of the theory of work-assisted operations. Also common for both sorts of operations are analytical expressions describing losses of maximum work and entropy production.

3.2.6. Entropy Production and Flux-Dependent Efficiencies

Throughout the remaining text of this book it will be sufficient to use only the first Carnot temperature. Consider the entropy production in the unit in terms of T'. We shall show that this entropy production satisfies a simple formula with its endoreversible part characteristic for heat exchange between two bodies. Indeed, by evaluating *total* entropy production σ_s as the difference between the outlet and inlet entropy fluxes and with $q_2 = q_1 \Phi T_2/T'$ we find

$$\begin{aligned}
\sigma_s &= \frac{q_2}{T_2} - \frac{q_1}{T_1} = \frac{\Phi q_1}{T'} - \frac{q_1}{T_1} = q_1 \frac{\Phi - 1}{T'} + q_1 \left(\frac{1}{T'} - \frac{1}{T_1} \right) \\
&= q_1 \frac{\Phi - 1}{T'} + \frac{g'(T_1 - T')^2}{T_1 T'}
\end{aligned} \tag{3.62}$$

where Equation (3.39) was used in the transformations. The entropy production, illustrated in Figure 3.6, is here split into sum of two nonnegative terms. The first term, that is, the product $q_1(\Phi - 1)$, is always nonnegative as the signs of q_1 and $(\Phi - 1)$ are the same (positive for engine and negative for heat pump). The second term is obviously positive. The property $T' > 0$ follows for $g' > 0$.

The positiveness of T' and its thermodynamic ramification (3.34), as well as the simple heat formula, Equation (3.35), make quantity T' useful. Moreover, because of the pseudo-Carnot structure of efficiency, Equation (3.33) can be written as follows:

$$
\begin{aligned}
\eta &= 1 - \Phi\frac{T_2}{T'} = 1 - \Phi\frac{T_2}{T_1} - \Phi T_2\left(\frac{T_1 - T'}{T_1 T'}\right) \\
&= \eta_C - (\Phi - 1)\frac{T_2}{T_1} - \Phi T_2\frac{q_1}{g' T_1 T'}
\end{aligned}
\tag{3.63}
$$

This is a relatively simple and suitable expression showing how a finite flux q_1 and internal dissipation influence the process efficiency. Equation (3.63) generalizes earlier endoreversible result (Andresen, 1983; de Vos, 1992). Whenever T' is less than T, meaning that a positive heat q_1 flows to an engine, the first law efficiency of the total system, η, drops below that of the Carnot engine, else in heat pump mode, it raises above η_C.

Finally, a remark communicated to the authors by Prof. A. Tsirlin (Pereslavl Zalessky, Russia) should be added. Heat engine cycle, which corresponds to the maximal power, consists of two isotherms and two adiabats for any law of heat transfer. But the heat engine cycle with given power and maximal efficiency can, in the general case, include three adiabats and three isotherms (such heat transfer laws do exist, although that is a rare case). Thus, a heat engine cycle cannot always be optimized by choosing "Carnot temperatures."

3.3. TOWARD WORK LIMITS IN SEQUENTIAL SYSTEMS

For finite resources or reservoirs intense potentials decay in time; evaluation of energy limits requires a sequential operation (a cascade). A topological scheme of the multistage power yield, leading to generalized exergies, is illustrated in Chapter 4 (Figures 4.1 and 4.2) and that of the steady multistage power consumption in Figure 4.3. These graphs may be referred either to a steady cascade process or to the dynamic time development of a single unsteady engine (refrigerator) that interacts with two sources. The temperature of the first source changes when heat is removed/supplied from/to it and the temperature of the second source is constant. This is a case of the irreversible Lorentz cycle where the heat capacity of one of the sources tends to infinity. Under a piston-like flow assumption equations and results are the same when changes are in time and when changes are in space. Thus, the figures can apply for both steady cascades and dynamically working machines.

Only limits of static processes (infinite resources, infinite reservoirs) are characterized by functions and associated numbers. Changes of driving potentials in machines with finite resources or finite reservoirs lead to variational problems, both continuous and discrete. Essential is construction of a potential function.

For an arbitrary continuous or discrete optimization problem such function can be found by Bellman's method of dynamic programming (Aris, 1964; Bellman, 1967). Both continuous and discrete processes with a single independent variable (time or length) can be treated in the frame of the common discrete formalism of dynamic programming (Aris, 1964; Bellman, 1967; Sieniutycz, 2000c).

In the next chapters we shall quantify limits on production or consumption of mechanical energy in some complex operations with heat and mass exchange. We also compare entropy or exergy criteria for work-assisted operations with those for conventional operations (without work). Our method will deal with arbitrary participating fluids, in particular the radiation fluid. For that fluid endoreversible limits incorporate, as a sole effect, a minimum entropy production caused by simultaneous emission and absorption of radiation. Thus, solar-assisted operations can also be treated by the thermodynamic method. Its dynamic version involves optimization of power and/or related work in an endoreversible sequence of thermal machines, thus generalizing the familiar method of evaluation of the classical exergy in reversible sequences with an evolving resource (Szargut et al., 1988). Treatments of dynamic limits via sequential operations such as multistage heat pumps yield limiting work functions in terms of end states, duration, and (in discrete process) number of stages (Sieniutycz, 1998e, 1999e; Sieniutycz and Berry, 2000).

Modeling a general work-assisted operation for the purpose of limit evaluation is a difficult task as it involves "endoreversible" or higher rank models (Andresen et al., 1983) in place of reversible models of work production (consumption). These extensions are more capable of producing results related to real operations than their reversible counterparts (Sieniutycz and Szwast, 1982a, 1982b). Formal analogies exist between entropy production expressions in work-assisted and in conventional operations that are helpful to develop suitable criteria and models. Endoreversible and irreversible limits are found by optimization of sequential work-assisted and solar-assisted operations, such as heating, evaporation, and drying.

Analogies between entropy production formulae for work-assisted operations and those without work help formulate optimization models of the former. An example is the heat transfer–driven power generation or consumption in an "endoreversible" thermal machine, an engine or heat pump, which interacts with a high-T fluid (e.g., drying gas) at flow (Sieniutycz, 1999b; Berry et al., 2000; Sieniutycz and Berry, 2000). Multistage production (consumption) of power requires the use of the sequence of stages (Chapter 4; Figures 4.1–4.3). To get physical rather than economic limits all stages are those with CNCA process (Ondrechen et al., 1981, 1983; Andresen et al., 1983; Sieniutycz, 1998e; Sieniutycz and Berry, 2000); the limits are those for the mechanical or electrical energy. In an endoreversible engine a resource fluid drives an engine from which the work is taken out. In an endoreversible heat pump a fluid (e.g., drying agent) is driven in the condenser of the heat pump to which work is supplied; in both cases the second fluid is an infinite reservoir. The fluids are of finite thermal conductivity; hence, there are finite thermal resistances in the system.

In multistage heating fluid's temperature increases at each stage; the whole operation is described by the sequence T^0, T^1, ..., T^N.

The popular "engine convention" is used: work generated in a multistage engine, W, is positive, and work generated in a heat pump is negative; this means that a positive work $(-W)$ is consumed in the heat pump. The sign of the optimal work function $V^N = \max W^N$ defines the working mode for an optimal sequential process as a whole. In engine modes, $W > 0$ and $V > 0$. In heat pump modes, $W < 0$ and $V < 0$; therefore, working with a function $R^N = -V^N = \min(-W^N)$ is more convenient. Of special attention are two processes: the one that starts with the state $T^0 = T^e$ and terminates at an arbitrary $T^N = T$ and the one that starts at an arbitrary $T^0 = T$ and terminates at T^e. In these cases the functions V^N are generalization of the classical exergy in discrete processes with finite durations.

Let us focus now on a steady cascade process. The process state space and its influence on the state changes through stages are determined by both the states of the finite-resource fluid flowing through stages and the properties of the thermal bath reservoir. For any finite bath, the state space of the overall (bath-containing) system would necessarily involve variable bath coordinates, and the system changes would then be influenced by a dynamics describing the bath history in terms of these variable coordinates. However, in the case of an infinite bath its intensive parameters, for example, its temperature T^e and chemical potentials μ_i^e, do not change with the cascade process, and this is why they reside in the formal process model only as constant parameters. Thus, it is the condition of an infinite bath reservoir that enables us to treat power functions as reservoir-history independent. The system's history is then expressed exclusively in terms of the fluid coordinates and their time derivatives. The potential function of extremum work obtained via optimization of a work integral is then an exergy-type function that, as usual exergy, contains intensive coordinates of the bath as parameters accompanying state coordinates of the finite-resource fluid.

Use of extremum seeking methods to determine limiting possibilities of sequential systems with power production or consumption could, perhaps, make an impression that the goal here is an economic optimization of thermal machines and/or their diverse topological arrangements, which, in general, are called thermal networks. Should such optimization be the case (Chapters 11–20), a thermal system could be optimized via a customary approach that would require: a detailed network modeling, inclusion of economic accounting, and occasional imbedding of the optimized system into a broader environment to include interacting chains. In fact, the range of optimizing in the present chapter is restricted to thermodynamic limits and a generalized quantity of the exergy type attributed to a definite single stream of substance or energy. Thus, we search for the extremum work for a finite time production (consumption) process, with a single resource stream when the environment is used to dispose the waste heat. An exergy-like quantity appears; it is a generalized potential of extremal work that depends on the end states of the stream and its holdup time (duration). Two sorts of optimization ("limits-oriented" and "economic") are different and any link between them, if at all exists, is indirect at most. In a definite real system, a weak link may be observed when making the exergy balance for participating streams, which is the usual procedure made in the so-called extended second law analysis or thermoeconomic optimization.

In classical, reversible analyses the resource and environment reservoirs are insensitive to the effect of dissipators (boundary layers, resistances, etc.) because the reversible situation forces thermal homogeneity of each reservoir in the space. In an irreversible analysis that involves dissipative phenomena in reservoirs and engines inhomogeneities of transport potentials play a nontrivial role. In the "endoreversible" cascade perfect efficiencies of the Carnot (work-producing) engines along the cascade are essential; in more general cases, for which the CNCA efficiency formula is generalized, internal irreversibilities are included. Work limits follow in terms of the time, state, and properties of boundary layers or other dissipators. Endoreversible models are the simplest (Rubin, 1979a, 1979b; Rubin and Andresen, 1979; Salamon et al., 1980), yet they are of a restricted use in predicting actual characteristics of real thermal machines. As the restriction to external irreversibilities is unnecessary, FTT models go beyond "endoreversible limits," and treat internal irreversibilities as well (Chen et al., 1996, 2000; Cheng et al., 1999; Salah El-Din, 2001a, 2001b; El Haj Assad, 1999; Berry et al., 2000). It is most essential that in either of two methodological versions of FTT, of which the first gives up internal irreversibilities, whereas the second one estimates these from a model, the FTT limits on energy consumption or production are stronger than those predicted by the classical exergy (Szargut et al., 1988); this results from the "rate penalties" taken into account in FTT.

Chapter 4 presents a detailed mathematical analysis of power yield in cascade systems.

3.4. ENERGY UTILIZATION AND HEAT PUMPS

Awareness of limited energy supplies (Berry, 1989) has prompted a significant effort in developing energy recovery processes. The heat pump is in principle the only device that would allow exploitation of the low-exergy sources commonly available in nature and industry. Heat pumps utilize energy by adding low-quality energy flux taken from a low-exergy source to work flux to obtain energy of high quality economically. It is difficult to satisfy restrictions resulting from both thermodynamic and economic balances simultaneously (Keller, 1982). There are heat pumps of various types, including those based on absorption, combined absorption and compression, adsorption, and chemical reactions (Ranger et al., 1990). Sorption heat pumps are most reliable (Le Goff et al., 1990a, 1990b, 1990c, 1990d). Acoustic heat pumps and thermoacoustic engines that achieve a substantial fraction of the Carnot efficiency are also designed (Swift, 1988). Second law analyses of the diverse applications of heat pumps (for heating of houses, upgrading of recycle streams, waste heat recovery for technological and municipal purposes, etc.) are available (Reay and Mac Michael, 1990; Brodowicz and Dyakowski, 1990). Heat pumps working with low-temperature heat sources such as a solar energy collector have been used for these purposes (Duminil, 1976; Kaushik et al., 1985; Kaushik and Kumar, 1987; Tu, 1987). Systems composed of a heat engine and a compression heat pump have been designed (Brodowicz and Dyakowski, 1990; Le Goff et al., 1988). Progress in heat pumps technology has led to ca. threefold increase of their coefficients of performance $Q_1/W = \eta^{-1}$ that attain today values of the order of 6 (Lewandowski, 2001). The compression stage plays an essential role in the performance and classification of heat pumps. It can be driven by work or heat. In the latter case a heat pump actually contains two cycles: the refrigeration cycle itself and the heat engine cycle, which compresses circulating vapor. In a sorption heat pump a working arrangement with medium and sorbent (absorber and desorber) is used in the engine cycle. The net heat supply to the cycle is the difference between the heat supplied to the desorber and that extracted from it by the absorption process. It leads to an increase in the enthalpy of the working vapor between the evaporator and condenser; in compression heat pumps this effect is achieved by a compressor.

Le Goff et al. (1988, 1990a, 1990b, 1990c, 1990d) have developed a reader-friendly schematizing of these systems. Any sorption heat pump or heat transformer is a double energy converter. It first converts low-value thermal exergy into chemical exergy in the separator, which is composed of the desorber and the condenser; reverse conversion of chemical energy into high-value thermal exergy follows in the mixer, composed of the evaporator and the absorber. The "heat pump effect," that is, the upgrading of low-level heat from a lower to a higher temperature, is due primarily to the entropic contribution to the (chemical) exergy of mixing, $\Delta b_m = \Delta h_m - T_0 \Delta s_m$. In absorption and sorption heat pumps the enthalpic contribution is often much smaller than that due to entropy. In chemical heat pumps the enthalpic contribution dominates. It is even possible to design an efficient heat transformer with an ideal mixture, where $\Delta h_m = 0$ (Le Goff et al., 1990a, 1990b, 1990c, 1990d). A general objective of sorption heat pumps is to maximize the "temperature lift" (the temperature difference between the evaporator and absorber) for a given "thermal driving force" (the temperature difference between the desorber and the condenser). These temperature differences are expressed most conveniently in terms of the related nondimensional temperatures $\theta = 1 - T_0/T$. The exergy balance in terms of θ (Le Goff et al., 1990a, 1990b, 1990c, 1990d) shows that the difference between the net consumption (cost) of the exergy in the separation part and the net production of exergy in the mixing part (gain) is just the exergy destroyed by irreversibilities.

Four main types of heat pumps may be distinguished: the classical heat pump, the cold pump or refrigerator, the pump for simultaneous heating and cooling, and the heat transformer. Energy and exergy balances determine the (first law) performance coefficients C, the (second law) overall exergy effectiveness and efficiencies, and the relations between these quantities for the various types of sorption heat pumps (Le Goff et al., 1990a, 1990b, 1990c, 1990d). An exergy analysis shows the reduction of exergy costs K caused by an improvement of the exergy effectiveness. However, the maximum value of C is never attained by its economically optimum value. The effectiveness of commercial sorption heat pumps is low, between 0.1 and 0.3 of the Carnot efficiency. In contrast, thermomechanical machines have an exergy effectiveness on the order of 0.5–0.6. Yet, recent sorption heat pumps use only two stages, whereas in advanced fractionalization adsorption–desorption units (e.g., chromatographic columns) the separation is carried out by an equivalent set of many (several dozen to several hundred) theoretical stages. Le Goff et al. (1988, 1990a, 1990b, 1990c) have investigated some multistage and multiple-effect systems. They have estimated that the effectiveness of sorption heat pumps can be doubled by applying techniques and concepts already known in other sectors of the chemical industry; for example, the countercurrent contacting is only beginning to be applied to heat pumps.

Recent developments in the heat transformers from simple to multieffect arrangements may help achieve the increased effectiveness of sorption heat pumps predicted above. The general purpose of the heat transformer is to produce upgraded heat from a cheap heat source (e.g., of moderate temperature) by rejecting waste heat into the environment. A heat transformer is an arrangement where "the exergy or entropy flux can be transferred from a cold source to a heat sink, thus in a direction inverse to that of natural heat transfer" (Le Goff et al., 1988). It is well known that it is possible to achieve separation in a rectification column; it has now been shown that mixing can be performed in a heat transformer with the reverse operation called "reverse rectification" (Le Goff et al., 1988, 1990d; Ranger et al., 1990). In this process the most volatile fraction enters into the bottom and the poor phase into the top of the column; this allows the upgrading of a low-T heat supplied at the bottom to a higher T heat received at the top of the column. Mass and energy balances in reverse distillation can be set graphically, by chemical engineering methods, from an enthalpy chart by the Ponchon–Savarit method or from a concentration diagram by the McCabe–Thiele method (Ranger et al., 1990); methods of exergy balancing (Bosniakovic, 1965) are possible. In Chapter 8 sequential work-assisted drying using heat pumps is considered; in particular, Figure 8.11 depicts the effect of the utilizing operation in a drying arrangement with a heat pump.

The countercurrent exchange of heat and matter between the separator and mixer combined to form cascades of theoretical stages is essential for multistage transformers. According to Le Goff et al. (1988) number of stages N in the separator and M in the mixer play the roles of the number of turns N in the primary and M in the secondary coil of an electrical transformer. The entropy flux in the former is seen as the electrical current in the latter. However, the analogy of heat transformers to electrical systems is probably not as complete as might be inferred from the above. Entropy is not a conserved quantity in the sense that electric current is; in particular, a steady-state entropy flux is in general not constant along the path (Le Goff et al., 1988, 1990d; Ranger et al., 1990).

Chemical heat transformers use the heat of endothermal and exothermal chemical reactions to obtain high-valued energy. Brodowicz and Dyakowski (1990) have described a chemical heat transformer composed of a rectification column and an exothermal reactor. An example consists of the endothermal dehydrogenation of 2-propanol to acetone and hydrogen, and the exothermal synthesis of 2-propanol by hydrogenation of acetone. These reactions proceed catalytically, with two different catalysts, according to the scheme:

Endothermal, catalyst 1:

$$(CH_3)_2CHOH_{(L)} \rightarrow (CH_3)_2CO_{(g)} + H_2, \qquad \Delta h = 100.4 \, kJ/mol$$

Exothermal, catalyst 2:

$$(CH_3)_2CO_{(g)} + H_{(g)} \rightarrow (CH_3)_2CHOH_{(g)}, \qquad \Delta h = -55 \, kJ/mol$$

Work is obtained from the energy delivered to the boiler of the endothermal reactor (the rectification column). Acetone is more volatile than 2-propanol; hence, gaseous acetone is received with hydrogen at the top of the column and then directed (through a regenerative heat exchanger) to the exothermal reactor. The hydrogenation reaction of acetone to 2-propanol proceeds exothermally on the catalyst bed with a concomitant evolution of heat. As a result, high exergy heat is produced. Diagrams of process paths in multidimensional spaces with some second law analyses are available (Brodowicz and Dyakowski, 1990). Exploitation of evolved heat of chemical reactions in heat pumps is attractive due to its intensity. Several detailed analyses of chemical reactions that might serve in heat pumps have been carried out with the help of second law analyses. Optimization of heat pumps in thermodynamic and economic terms continues in the context of commercial conversions of energy and material resources (Brodowicz and Dyakowski, 1990; Tu, 1987; Wall, 1986).

Energy storage involves a spectrum of problems where the methods of energy limits and thermal efficiency enter in a natural way. Various methods of energy storage (sensible heat, phase changes, compressed gases, sorbents, chemical reactions, batteries, synthetic fuels, electrical capacitors, electromagnets), in particular that of solar energy, have been treated with an indirect reference to their work limits and thermodynamic efficiencies (Domanski, 1990; Garg et al., 1985). A number of systems working in the unsteady regime have been analyzed, and Bejan (1982, 1996a) has explained that the mission of storage devices is to store exergy, not energy. He has obtained limits associated with the optimum duration, corresponding to the minimum of total entropy generated, of storage processes. Extending Krane's (1987) analysis where the energy storage was followed by its removal, Bejan has found the optimum flow rate history of a cold fluid heated while the energy stored in a liquid pool is recovered. The optimal flow of coolant follows constant in time only when its specific heat and the heat transfer coefficient are temperature independent; otherwise this flow should be decreased during relatively poor exchange conditions.

Packed and fluidized beds can store exergy. Wagialla et al. (1991) have modeled energy storage in fluidized beds using the heterogeneous two-phase model of fluidization where the bed is composed of a bubble phase and an emulsion phase. For second law analyses this system is interesting as one with large nonidealities. Only first law efficiencies have been obtained; more appropriate second law efficiencies have not been treated. They can be estimated roughly from the authors' data and lie between 0.18 and 0.03. These low values, which question the use of fluidized beds for storage, are not surprising since large imperfections occur in fluidized beds due to gas bubbling, mixing, and channeling.

Various chemical storage systems and sorption heat pumps based on the separation of chemical components have been designed and classified using second law analyses. A system for heating buildings has been proposed that takes exergy from the weather when it is "bad" in reference to the average climate (Le Goff et al., 1986).

Estimating energy limits from radiation requires analysis of radiative engines. Chapters 5–7 analyze steady and dynamic radiation engines, also called the Stefan–Boltzmann engines (de Vos, 1992), in which both the "upper" and the "lower" heat exchange undergo in accordance with the laws of radiation.

3.5. THERMAL SEPARATION PROCESSES

Energy limits for mechanical separation processes are analyzed in Chapter 8. Here we restrict to brief outline of steady operations of thermal separation, where problems of minimum consumption of heat and limiting productivity are essential (Figure 3.7).

Heat is the form of energy flow that is frequently used in separation operations. In these operations heat is received from a reservoir or source of high temperature T_+ and rejected into a low-T reservoir, at temperature T_-. The estimates for irreversible separation work or power allow us to estimate the lower bound for

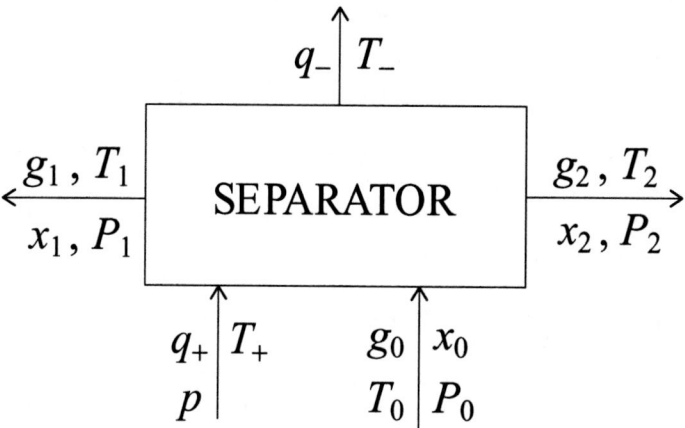

Figure 3.7 Input and output fluxes for computation of separation operation.

heat used in thermal separation. Whenever a minimal power p_{min} is lower than the maximum power p_{max} of a heat engine working between temperatures T_+ and T_-, an irreversible separation process can be accomplished for a prescribed productivity g and fixed concentrations in input and output streams (Equation (8.40) in Chapter 8). Expressions for maximum power p_{max} are known from the theory of CNCA engines and their generalizations. Equation (8.14) of Chapter 8 can then be applied to find estimate for heat consumption after substitution into it the relevant minimum entropy production. As a general formula for σ_s in terms of power p is valid

$$p = \eta q_1 = \eta_C q_1 - T_2 \sigma_s \qquad (3.64)$$

the entropy production can be related to $\eta(p)$ as follows:

$$\sigma_s = -\frac{p}{T_2} + \frac{\eta_C q_1}{T_2} = -\frac{p}{T_2} + \frac{\eta_C p}{T_2 \eta} = \frac{p}{T_2}\left(\frac{\eta_C}{\eta(p)} - 1\right) \qquad (3.65)$$

Using the irreversible estimate of minimum heat consumption (Tsirlin and Kazakov, 2002a, 2002b)

$$q_+ \geq q_+^{rev} + \frac{T_- \sigma_{min}^T}{\eta_C} \qquad (3.66)$$

one finds

$$q_+ = q_+^{rev} + \frac{T_- \sigma}{\eta_C} = q_+^{rev} + \frac{p}{\eta_C}\left(\frac{\eta_C}{\eta(p)} - 1\right) \qquad (3.67)$$

Thus, the heat necessary to accomplish the separation operation is equal to the reversible heat only when the engine exhibits the Carnot efficiency. But this refers to very low, in fact, vanishing, power. For a finite power p Equation (3.67) predicts heats that may be much higher than the reversible heat. To apply Equation (3.67) an expression for the limiting efficiency of transformation of heat into mechanical work at given power can be used in the form

$$\eta(p) = \frac{2\delta k}{\delta k + 1 - \sqrt{(1-k)(1-k\delta^2)}} \qquad (3.68)$$

where $\delta = (\sqrt{T_+} - \sqrt{T_-})/(\sqrt{T_+} + \sqrt{T_-})$ and $k = p/p_{max}$. For $p \to p_{max}$ this efficiency tends to the CNCA expression, $\eta = 1 - \sqrt{T_-}/\sqrt{T_+}$. On the other hand, if $p \to 0$ (the reversible case), then $\eta(p)$ tends to the limiting Carnot efficiency (Berry et al., 2000).

The basic difference between mechanical and thermal separation is that the productivity of the latter is limited, whereas the productivity of the former is not. This property follows from the power constraint $p \leq p_{max}$ and the monotonic increase of productivity g with power p. The limiting productivity of thermal separation in gases can be found by solving a power equation with respect to g and then by substituting a known formula for p_{max} to the result obtained. With this approach charts can be obtained that depict the dependence of maximum productivity on the concentration of a key component in the initial mixture and on the degree of separation (Tsirlin and Kazakov, 2002a, 2002b).

3.6. STEADY CHEMICAL, ELECTROCHEMICAL, AND OTHER SYSTEMS

Early approaches to energy limits in steady chemical systems are known since Denbigh (1956) and Bosniakovic (1965). Future development of electrochemistry in the aspect of future energy issues is the subject of book of Bockris (1980). Bockris and coworkers discuss solar hydrogen energy alternative (Bockris, 1984), conversion of light and water to hydrogen and electric power (Bockris and Kainthla, 1986), and the electrochemistry of waste removal (Bockris et al., 1992). System modeling and analysis perspectives for electrochemical systems and fuel cell development are investigated with the idea of using thermodynamic approaches (Warner and Berry, 1985; Watowich and Berry, 1986; Von Spakovsky and Olsommer, 2002). A proposal to apply thermokinetic methodology to evaluate electrical energy limits in fuel cells has been outlined (Sieniutycz, 2002c).

Ondrechen et al. (1980a, 1980b) and Escher et al. (1985) have studied mechanical heat engines with the heat supplied from exothermic chemical reactions. The first group considered a model with a flow reactor coupled to an engine by a heat exchanger. Maximum fuel efficiency was always attained in these systems at the uninteresting limit of zero flow rate. The criterion of maximum power leads to the requirement that the region of most rapid reaction, "the combustion zone," be located at the downstream end of the reactor. The "combustion zone" becomes narrower as the activation energy of the driving chemical reaction increases, and the maximum power obtainable with a finite, nonzero flow rate is a sensitive function of the activation energy of the reaction. For a chemically driven thermal engine Escher et al. (1985) have shown that dissipation may be decreased and power output and efficiency increased by the variation of external constraints coupled to nonlinearities. Resonances appear when the system approaches Hopf bifurcation instability.

A unified thermodynamic approach to the dynamics of energy and substance convertors in systems with chemical–nonchemical coupling has been constructed using a reformulation of chemical convertors (Shiner, 1992). Chemical convertors are central to chemical engineering but also to biology; they play a key role in almost all biological phenomena. However, chemical and electrochemical convertors did not seem to fit into the dynamic formalism common to other convertors; the dynamic equations of chemical reactions are usually written in a form quite different from that of the equations one has for mechanical or electrical systems. This has made treatment of mechanochemical and electrochemical converters more difficult than electromechanical converters. Thermodynamic research (Shiner, 1984, 1992; Sieniutycz, 1987) has shown that the dynamics of chemical convertors are not as different from those of other sorts of converters as has been thought. Far from equilibrium the chemical flux can be evaluated as the ratio of the affinity to a nonlinear resistance; the apparent differences arise since the chemical resistances are dependent on composition, whereas for ideal electromechanical convertors the resistances are constant. The concept of a chemical resistance has led to general minimum entropy and minimum power formulations of the dynamics of chemical convertors that are useful to nonlinear approaches (Shiner, 1992), replacing more conventional convertors near equilibrium where the chemical resistances can be considered constant. Overall chemical resistances and overall affinities have been applied for the prediction of the total conversions in complex chemical convertors. These formulations are now being applied to analyses of power yield in mechanochemical and electrochemical convertors. Research in electrochemistry investigates some concrete examples of such systems. An electrochemical cell (Hjelmfelt et al., 1991; Shi et al., 2002; Shiner et al., 1996) where chemical reactions are the main source of dissipation has been treated; the ratio of the power output to the free energy change of the reaction is an efficiency of the power yield. For infinite resistances between the electrodes the process efficiency is maximal and equal to unity, but the power output vanishes. To produce a nonvanishing power at a maximum level a finite resistance between electrodes is required, related to nonvanishing chemical rates. Upper limit on the electrical energy production has been determined in a typical FTT analysis (Shi et al., 2002).

By using Pontryagin's principle Watowich and Berry (1986) have determined optimum current paths with maximum work output (and associated criteria) during a finite time interval. They found the constant current strategies for the well-stirred cell and some nonmonotonic trajectories in the diffusive cells. Ratkje and Moller-Holst (1993) have studied the availability rather than the exergy efficiency in (nonisothermal) solid oxide fuel cells. They applied the method of electric work and derived an equation for the cell power that takes into account temperature gradients. The cell emf has been determined from the Gibbs energy of reaction complemented by a term with transported entropies. This term has a negligible influence on cell efficiency but has a large impact on the localization of heat sources and sinks. Economies are affected by both temperature gradients and other factors such as overpotentials, ohmic resistance, and incomplete reactions. Electrochemical cells are power-producing systems with high efficiencies. Since the emission of pollutants is reduced in this case as compared to combustion engines (Band et al., 1982), it is expected that the role of electrochemical conversion will grow in the future.

The growth of fuel cost has stimulated searches for upgrading and modifying fuels. It has been found that the largest imperfections are often in basic chemical steps. The yield of a natural gas substitute from bituminous coal has been analyzed (Kotas, 1986). It has been shown that three chemical steps, pretreatment, gasification, and combustion, dissipate most of the exergy. A pilot plant for coal hydrogasification has also been analyzed, and recommendations for the improvement of plant performance have been made (Tsatsaronis, 1993). A comparison between theoretical minimum and actual irreversibility in catalytic reforming for the purpose of upgrading low octane petroleum has been carried out with a substantial potential for the improvement of plant performance at several definite points (Kotas, 1985, 1986). Exergy analyses of gasification processes have been carried out (Kramarz and Wyczesany, 1989a; Shinnar, 1988, and references therein). The exergy losses of these processes have been compared. Results based on second law analyses have been contrasted with those derived from the first law. Three coal hydrogenation plants differing in the technology of the hydrogen production have been assessed (Kramarz and Wyczesany, 1989b); the three technologies used are based on coke gas, natural gas, and semicoke. The choice of the hydrogen source has been justified by a second law analysis.

Synthetic fuels are under study in many countries. In a series of papers Kramarz and Wyczesany have studied the exergy efficiency of various processes for the conversion of coal to liquid fuel (Kramarz and Wyczesany, 1992, and references therein). In particular, the exergy efficiency of Fischer–Tropsch complexes has been studied using a second law analysis instead of the conventional first law analyses. In this process the synthesis of synthetic liquid fuels from a mixture of hydrogen and carbon monoxide is a catalytic one. The main sources of exergy losses are oxidation reactions in the gasification unit and hydrogenation reactions in the synthesis loop. Due to a sophisticated heat recovery system the exergy efficiencies of the modern Fischer–Tropsch process are quite high, ca. 0.6–0.7.

Exergy analyses of the indirect liquefaction of coal to gasoline have also been performed. The process consists of Lurgi coal gasification, methanol synthesis, and methanol to gasoline conversion. A second law analysis shows that the primary underlying cause of the poor thermal efficiencies in indirect coal liquefaction is the highly irreversible coal gasification reaction. Kramarz and Wyczesany (1992) have also performed exergy analyses of the production of liquid fuels from coal using the so-called high conversion H-coal process, an innovative approach involving direct coal liquefaction followed by deep catalytic hydrogenation and desulfurization. This process shows a high level of engineering and exergy efficiencies of ca. 0.7. The exergy losses in the coal liquefaction process are relatively low because of the direct liquefaction. Kramarz and Wyczesany have further analyzed the so-called consol synthetic fuel process, which is based on the dissolution of fine-ground coal in a process-derived hydrocarbon solvent and the following extraction. Its exergetic efficiencies (ca. 0.75) are superior to those of most processes considered earlier, and there is still potential for improvement. The comparison of the competing processes of coal liquefaction is difficult because they differ in feed, product distribution, and technology; exergy is perhaps the only reasonable criterion (Kramarz and Wyczesany, 1992). Exergy efficiencies, stoichiometry-based and product yield–based criteria, and technological advantages and limitations allow a reasonable comparison. The exergy method is expected to be quite useful in dealing with system designs that involve choices in reaction pathways (Shinnar, 1988). However, for the synthesis of a definite product there are simultaneously many objectives;

therefore, multiobjective optimization techniques have been applied in complex industrial chemical and petrochemical systems. Two possible objectives are the maximization of the exergy change in production and the minimization of the total driving exergy. These multicriterion approaches extend in a sense the basic theory of thermodynamic limits where a single physical criterion is applied.

A large group of research papers deals with minimal dissipation and the optimization of affinity in chemical reactors, a lively line of research (Kjelstrup et al., 1999, 2000; Sauar et al., 1999; Demirel, 2002). While the problem in itself – the development of minimum dissipation strategies in chemical reactors – is quite inspiring, the results obtained to date are not rigorous. The typical problem treated within this group is an extension of the so-called principle of the equipartition of entropy production to nonlinear processes. While the original principle (Tondeur and Kvaalen, 1987) is true, its nonlinear generalizations are based on subjective assumptions, some of which are doubtful approximations and others are simply wrong. Among the invalid ones are the assumption that in any thermodynamic region the reaction rate J can uniquely be linked with the chemical affinity A and the assumption of constancy of Lagrange multiplier λ that some researchers uncritically repeat (Kjelstrup et al., 1999, 2000; Demirel, 2002). In fact, the assumption of a constant λ (and hence of constant thermodynamic forces A_j and ΔT^{-1}) ceases to be valid in nonlinear regimes of chemical reactions where rates are state dependent. For solving this problem notion of a chemical resistance $R(C)$ is important as a state-dependent quantity (Shiner, 1992). Yet, this resistance is a function of all concentrations and temperature, which contradicts the assumption that in arbitrary thermodynamic region the rate of a reaction J_j can uniquely be linked with its affinity A_j. Such rough solutions result in wrong determining of the feasibility regions of thermodynamic data; thus, the literature sources on these problems should be treated with care.

Actually most approaches based on classical second law analyses are global; they frequently rely on the input and output streams of a definite system. By elementary examples and calculations David (1990) has shown that the globality is a dangerous feature, which may lead to basic errors in design. The positiveness of the global entropy production, $\sigma_s V$, for given inputs and outputs does not prove that the process is physically possible. For example, for some positive global values of $\sigma_s V$ the temperature profiles in a countercurrent condenser may show a temperature cross and the required heat transfer will actually not occur (David, 1990). Similar effects are also known for highly nonisothermal sorption, drying processes and for chemical reactors (Ciborowski, 1965, 1976).

Therefore, only local approaches that link the differential balances with kinetics are fully reliable. But such approaches are rare in most classical second law analyses (Szargut and Petela, 1965; Szargut et al., 1988). Because of this limitation such global analyses are more useful for identifying improper processes than for proving that an actual design will work. Yet second law analyses have already proven their usefulness by discovering inefficiencies in existing large plants composed of simple subsystems (objects) and in these subsystems themselves. Along with FTT, Bejan's method of the EGM (Bejan, 1982, 1996a, 1996b, 1996c, 1996d, 1996e) has contributed crucially to the development of analyses that include the simultaneous (local) treatment of balances and kinetics.

In the case of simple kinetics or when equilibrium stage concepts can be applied, classical second law analyses are capable of providing several valuable results for new operations (Le Goff et al., 1988, 1990a, 1990b, 1990c, 1990d). However, in spite of a very large number of works and a few spectacular applications, little theoretical progress has been achieved to date in second law analyses (Szargut, 1998, 2001a). Design based on classical exergy analyses is only justified when the properties of equipment and systems configurations do not vary significantly. The globality property of classical exergy analyses forces them to use models that are not far from black box models (Moser and Schnitzer, 1985). It is doubtful whether global approaches can improve our understanding of practical processes. The treatment of storage units and energy avoidance systems involving global second law analyses is too often out of date, and none of the methods has even approached the rigor of the exact analysis of such systems that extends the classical mathematical theory of regenerators originated by Hausen(Schmidt and Willmott, 1981).

While sufficient in thermal engine theory, the use of constant, diagonal resistances (linear theory without couplings) is insufficient for complex energy convertors in which strong nonlinearities and/or coupling effects are essential. The lost work is decreased in systems with coupling; the magnitude of the effect depends on the relative magnitude of the kinetic cross-coefficients as compared to diagonal ones. Sometimes the effect can be negligible, as in Soret mass transfer. However, it is substantial for transport problems with phase transitions, for example, frost heave (Forland et al., 1988), and for many processes of electrochemical and biochemical energy conversion (Caplan and Essig, 1983). The maximum efficiency depends then on the degree of coupling; only "completely coupled" systems can approach the second law efficiency of unity as a limit. Coupling can sometimes be enhanced by an optimization, change in the type of contacting, etc.

In order to utilize in full the power of thermodynamic analyses in engineering one must often go beyond conventional second law analyses (Shinnar, 1988; David, 1990; Tsatsaronis, 1993; Ziębik, 1996; Valero et al., 1994a, 1994b). In chemical process design the real limits on the reactor yield derive from sources other and more restrictive than the second law. Often they appear as constraints and inefficiencies of available technology; these, however, may have consequences that can be treated by thermodynamic analyses. For example, low overall thermal efficiencies can be a thermodynamic consequence of the lack of a suitable catalyst. The limits related to usual catalysts can, however, be overcome by the use of highly selective catalysts (e.g., shape-selective zeolite catalysts). They integrate selective transport with catalysis in a way that is not only energy efficient but also cheap (Shinnar, 1988). Second law analyses placed in the context of such problems can play both a more subtle and a more useful role in chemical engineering than they do now (Levenspiel, 1988).

Also, according to Sciubba and Wall (2007) it is desirable that the application of exergy methods be expanded into the realm of thermofluid dynamics applications. The local entropy generation in continua can be computed by most present CFD codes; thus, the related exergy destruction can properly be assessed even at local level: this may lead to a group of effective design methods for fins, compact heat exchangers, ailerons, injectors, surfaces, etc. This is an area in which not much has been published to date (Szargut and Petela, 1965; Sieniutycz, 1978; Bejan, 1982; Carrington and Sun, 1992; Fewell et al., 1981; Natalini and Sciubba, 1994, 1999; Poulikakos and Bejan, 1982; Sciubba, 2004), but which might open new perspectives as the exergy efficiency is an important idea.

3.7. LIMITS IN LIVING SYSTEMS

The above considerations and analyses can be extended to systems with living organisms. Here we focus on energy limits associated with the idea of an extremal development governed by the entropy-related criteria of evolution, for example, "complexity" (Shiner et al., 1999). The idea refers to some recent information-theoretic models of multistage living systems characterized by sequentially evolving states (Szwast et al., 2002). The information concept is not only appropriate to complex systems but also well quantitatively defined (Shannon and Weaver, 1969). In living systems diverse models can serve to evaluate energy limits quantitatively (Ahlborn, 1999). Here we restrict to a few basic, qualitative issues. Dynamic aspects of limits in biosystems are considered in Chapter 10.

In living systems a nonequilibrium entropy has to be applied for a sufficiently exact description of the system; in fact, the researcher deals with inherently nonequilibrium systems. Some special paths can terminate at an equilibrium, but generally a transition between two nonequilibrium states is of interest. This state of affair implies an analogy with nonequilibrium statistical physics. There the equilibrium conditions along with Boltzmann distribution are derived from the entropy maximum condition in a closed system, and the extremizing method uses entropy in an arbitrary macroscopic state, not in the equilibrium state (Landau and Lifshitz, 1974). Generally, however, the idea of maximum of a potential function that may even be more general than nonequilibrium entropy is used, subject to given constraints (Szwast et al., 2002).

Complex living systems have developed various strategies to manipulate their self-organization in order to satisfy the principle of minimum complexity increase; ultimately, however, the physical laws set limits to their size, functioning, and rate of development (Ahlborn and Blake, 1999). Species that survive in ecosystems are those that funnel energy into their own production and reproduction and contribute to autocatalytic processes. Poorly developed ecosystems degrade the incoming solar energy less effectively than more mature ecosystems (Schneider and Kay, 1994). Living systems as stable structures increase the degradation of the incoming solar energy, while surviving in a changing environment. One feature is common: all these features increase the systems' ability to dissipate the applied gradient in accordance with the so-reformulated second law of thermodynamics (Schneider and Kay, 1994). In all these situations the second law imposes constraints that are necessary but not sufficient cause for life itself. In fact, thermodynamics proves that the second law underlines and determines the direction of processes observed in the development of living systems. As an ecosystem develops it becomes more effective in removing the exergy part in the energy it captures, and this exergy is utilized to build and support organization and structure. Time and its derivative cycling play a key role in systems' evolution. Evolution itself is a time-dependent process and the understanding of cycling is of great methodological and cognitive importance. Optimal strategies of streets tree networks and urban growth can mimic development of living systems (Bejan, 1997a, 1997b, 1997c, 1997d; Bejan and Ledezma, 1998). Bifurcation phenomena play a role in development and evolution in living systems (Szwast, 1997; Chapter 10). Multiorgan living organisms are treated therein as multistage systems by a complexity criterion based on the statistical entropy. Classical thermodynamics does not appear in this approach, yet the statistical model is governed by an extremum principle, which, as in thermodynamics, implies extremum properties for a potential. Discrete and nonlinear evolutions constitute dynamics in metric spaces that may be curvilinear. Some features of evolving organisms can be predicted from variational principles for shortest paths along with suitable transversality conditions. Criteria of entropy production may also be extremized and Onsager-like symmetries appear in discrete models (Chapter 10; Szwast et al., 2002).

Limits for consumption of substance in biochemical reactions are of importance. A classical-type analysis involves the classical stoichiometry. Yet, when a finite time is required, information must be added on chemical kinetics in the system (Szewczyk, 2002).

3.8.   FINAL REMARKS

In this chapter we have reviewed problem of steady energy limits on the mechanical energy production or consumption in the extended (rate dependent) sense, that is, as the representative for FTT and EGM methods (Hoffmann, 1990). In the first part we have presented verbal discussion of main methodological issues, and in the second an outline of quantitative mathematical formalism for steady CNCA problem. Not only have we reviewed and classified all main methods and results obtained in the field, but we have also considered common objections caused by misunderstanding of these methods and their results. Minimization of entropy production (EGM) that serves to determine the realizability area in FTT (Tsirlin and Kazakov, 2002a, 2002b) may often be only one of three stages and does not include all methods used in FTT. Still we need to articulate the role of so-called "processes of minimal dissipation." They play in control thermodynamics a similar role as equilibrium processes play in equilibrium thermodynamics. The notion of thermodynamic distance (Salamon and Berry, 1983; Mrugala et al., 1990; Hoffmann et al., 1989; Hoffmann, 1990) is related to this concept. The processes of minimal dissipation set the feasibility boundary (realizability area) for real processes (Tsirlin and Kazakov, 2002a, 2002b). The minimal dissipation solutions are available in a number of publications. In particular, they can be found for heat and mass transfer, chemical transformations, and multidimensional kinetics (Tsirlin et al., 1998).

Sometimes a skepticism is expressed whether the principles based on FTT and EGM can be useful to optimize thermal systems with nonequilibrium processes, especially systems of complex topology such as thermal networks (Szargut, 1998; Moran, 1998, Gyftopoulos, 1999, 2002, 2003). This skepticism is expressed by researchers who underestimate the significance of kinetic phenomena in treating practical processes and who forget that only recent approaches (such as FTT, EGM, and control thermodynamics) include explicitly and systematically all information about the "kinetic closure" into optimization models, the inclusion that is generally abandoned in classical second law analyses. While a large portion of objections can be overcome (Chen et al., 2001), we shall not enter here into this debate, as it is enough to recall that an economic problem of a system optimization and the physical problem of work limits for a resource (considered here) are two different problems. The real work may sometimes be dozen times larger than the mechanical energy (exergy) limit associated with the production of a key substance; this is a well-known fact from the theory of Linde operation, for example (Bosniakovic, 1965). In the realm of energy limits, the trade-off between the exploitation and investment costs and the problem of investment reduction by admission of exergy losses are most often irrelevant issues. Yet, at the interface of thermodynamics with economics all issues related to the trade-off are essential (Clark, 1986; El-Sayed, 1999; Chapter 11).

The FTT method yields endoreversible and higher-order generalizations of the energy limits. At the zeroth rank all processes are reversible; then the method yields the classical thermostatic limits. In general, the method serves to evaluate enhanced limits in nonequilibrium processes driven by transfer and rate phenomena. We stress hierarchical nature of the FTT limits, where endoreversible limits are one step better than classical. Of the basic fields compared, exergy analysis, EGM, and FTT, only the latter two can systematically include various concepts of contemporary irreversible thermodynamics. Working with limits of lowest ranks FTT cuts their hierarchy at the level of the endoreversibility.

The simplicity of these models and an aggregated information on which FTT rests are frequently the source of misunderstanding. Adversaries of FTT and EGM commonly ignore the fact that the highly useful notion of the classical exergy is associated with models even simpler than endoreversible: the reversible ones. However, the limits of higher ranks correspond with more realistic models, and the potential of FTT for incorporating results from dissipative fluid mechanics or nonequilibrium field thermodynamics has been shown (Orlov and Berry, 1990).

In this chapter particular results of the energy limits have been obtained for steady engine-type and heat pump–type systems. Considering power maximization in an imperfect CNCA unit we have displayed various representations of the problem differing by a control variable and the optimization approach. We have used efficiencies and Carnot temperatures as independent controls, but also discussed the approach using the Lagrange multipliers where the original controls (temperatures of circulating fluid) are retained. Preparations have been made to pass to analysis of sequential CNCA systems.

In Chapter 4 we present in detail power optimization theory of dynamic systems treated as multistage cascades. In multistage sequential operations one-stage characteristics are repeated in dynamic sequences (cascades) for diverse inputs. Cascade models serve to analyze the dynamic behavior of engines and heat pumps when the resource reservoir is finite and the power yield cannot be sustained at a steady state. The dynamic situations are illustrated in Figures 4.1–4.3 and their computational block scheme in Figure 4.4. To show suitable applications we imbed into the cascade description properties of a single CNCA engine with the resource degrading or a single heat pump with resource's utilization.

4 Hamiltonian optimization of imperfect cascades

4.1. BASIC PROPERTIES OF IRREVERSIBLE CASCADE OPERATIONS WITH A WORK FLUX

This chapter analyzes limits on multistage production or consumption of mechanical energy (work) in sequential heat-mechanical operations characterized by finite rates. The benchmark system, for which these limits are evaluated, is a cascade of imperfect stages through which a resource fluid flows at a finite rate. Each stage consists of a fluid at flow, an imperfect work generator or consumer, and the environment. The problem investigated is that of limiting yield or consumption of power by the fluid that interacts sequentially with the environment in a finite time. A discrete, finite-rate model subsumes irreducible losses of work potential caused by thermal resistances. Dynamical limits on work are obtained which bound one-stage or multistage energy converters with production or consumption of power. These limits are expressed in terms of classical exergy and a residual minimum of entropy generation. A *discrete* generalization of classical exergy is found for systems with a finite number of imperfect stages and finite holdup times. For this generalized exergy a hysteretic property is valid, meaning a difference between the maximum work delivered from engine mode and the minimum work added to the corresponding heat pump mode of the system.

Cascade processes are inevitable whenever resources are finite and bounds on mechanical energy are evaluated. In this case the use of cascade schemes is necessary to define exergy functions and thermodynamic limits in both classical and extended cases. Bejan (1982) posed the basic problem of extracting the most exergy from a continuous hot stream within the reversible limit, whereas its *irreversible* generalizations were solved a decade and half later (Sieniutycz, 1997a, 1997b, 1997c; Bejan and Errera, 1998; Sieniutycz, 1999c, 1999d). This chapter deals explicitly with cascades as genuine discrete systems with a certain number of stages (described by difference equations). When the number of stages N becomes very large, that is, limit $N \to \infty$ is considered for fixed boundary states and process duration, a limiting continuous system is obtained; the particular data for that system generalized to the case of two finite reservoirs are discussed in Section 4.6. To date, despite a considerable number of recent works that treat imperfect thermal machines, either with engine cycles (Andresen et al., 1983; Gordon and Huleihil, 1992; Chen et al., 1996, 1999; Yan and Chen, 1999; Cheng et al., 1999; Salah El-Din, 2001a) or with refrigeration cycles

(Roco et al., 1997; Chen et al., 1996, 1999, 2000; Hoffmann et al., 1997; El Haj Assad, 1999; Kodal, 1999; Salah El-Din, 2001b), mostly endoreversible cascades have been investigated (Sieniutycz, 1999b, 1999d; Sieniutycz and Berry, 2000; Sieniutycz and Szwast, 1999). Assuming both internal and external dissipation, the present chapter transfers the findings from the realm of single-stage engines or refrigerators into the realm of cascades and related sequential operations. A nonideal counterpart of the classical Chambadal–Novikov–Curzon–Ahlborn process (CNCA process; Novikov, 1958; Curzon and Ahlborn, 1975; Chen et al., 2001) is the starting point to derive and then investigate more complex cascade operations. In comparison with the earlier research related to the subject associated with endoreversible cases at most (Bejan, 1982; Bejan and Errera, 1998; Sieniutycz, 1999d), the present analysis enunciates the role of internal irreversibilities as an essential point when treating power production systems. This analysis also involves the implementation of Carnot temperature T', introduced in Chapter 3, as an unconstrained control whose effect replaces that of both (upper and lower) temperatures of the circulating fluid. The existence of T' may be regarded as a direct consequence of the invariance of entropy production with respect to the variable transformation. While the use of T' as a control variable is not necessary (occasionally, e.g., in Section 4.6, another control $u = T' - T$ is used), T' not only facilitates the comparison of work-assisted and conventional exchangers but also allows tracing the effect of internal dissipation in a transparent way. Interesting results include enhanced bounds on the work production (consumption) that go beyond the endoreversibility, and stress the role of internal entropy production in thermal machines.

The role of internal irreversibilities is quantified in experimental verifications of system performance based on elaborated methodology of a single imperfect unit (Edera and Kojima, 2002; Chua et al., 1996; Ng et al., 1997, 1998), a methodology that may be extended to cascades or other complex systems. Because science flows both ways, between basic and applied, a logical organization of experiments in complex systems (of which important representatives are cascades) is difficult or even impossible without a self-consistent formal theory of irreversible cascades. The consistent organization of experiments in single energy objects (chillers, engines, etc.) was achieved no sooner than two decades after the theory of these objects was established. Now experiments are performed for many separate units (Edera and Kojima, 2002; Chua et al., 1996; Ng et al., 1997, 1998) so that verifications and further improvements of their models are possible. For example, the paper by Edera and Kojima (2002) provides numerical cycle simulations and experiments for cooling and heating that illustrate the effect of cooling or heating loads on cooling and heating performance parameters: input of (waste) heat, gas consumption, inlet temperature of waste heat source, and so on. In several works of Gordon and Ng and their co-workers, diverse performance parameters are identified and experimentally verified for reciprocating chillers (Chua et al., 1996; Ng et al., 1997, 1998). The parameters are total ther-

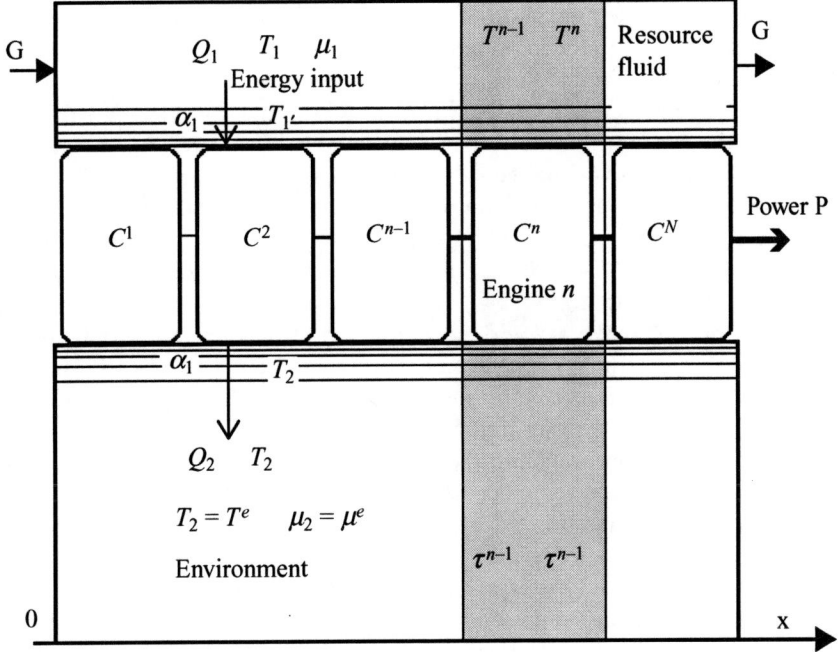

Figure 4.1 Basic scheme and designations for a multistage cascade of engines.

mal resistance, total internal entropy production and an equivalent heat leak (Ng et al., 1997). Yet in all these works both the experiments and numerical tests of internal dissipation remain restricted to single-stage systems. Although a cascade can be regarded as a dynamical development of a single-unit system, there is no paper published that would deal with diagnostics, parameter identification and optimization of imperfect cascades. It is therefore hoped that the establishment of a formal theory of cascades will stimulate experimental studies of complex energy systems. In this regard it is interesting that some features obtained earlier for optimal *endoreversible* cascades remain valid for those with finite internal irreversibilities.

Figure 4.1 depicts a basic cascade system investigated in this chapter and provides a list of basic designations. The system is of engine type and involves two fluids. The first fluid is the driving (downgraded) fluid; it flows through stages of the cascade from which the work is delivered. The second fluid is an infinite reservoir. The fluids are of finite thermal conductivity, hence there are finite thermal resistances representing the dissipative effect of their thermal boundary layers. A finite holdup time is assumed for the driving fluid. The multistage process is, in fact, a sequence of imperfect engines. Engine convention, or positiveness of total power $(P > 0)$, is assumed. The total work delivery, $W = P/G$, has to be maximized. When the final state of the resource fluid is at equilibrium with the environment ($T^N = T^e$), the maximal specific work represents a finite-time exergy of the engine mode, for the fluid at flow.

As there are internal irreversible losses at these stages, models of irreversible cycles use the concept of an internal irreversibility parameter Φ. At stage n the quantity Φ^n satisfies inequality $\Phi^n > 1$ for engine mode and $\Phi^n < 1$ for heat pump mode of the stage. The definitions involve simple local representations of those introduced in the previous chapter for a single unit (stage). At each stage n, by definition $\Phi^n = (\Delta S_{2'}/\Delta S_{1'})^n$, where $\Delta S_{1'}$ and $\Delta S_{2'}$ are respectively entropy changes of the circulating fluid along the "upper" and "lower" isotherm, $T_{1'}^n$ and $T_{2'}^n$. While Φ^n may vary with stage index n, it may temporarily be ignored in the above expressions, yet it is important when the whole cascade is treated. Figure 3.1 of the previous chapter recalls that quantity Φ^n equals the ratio of the entropy fluxes across the thermal machine, $\Phi^n = (J_{s2'}/J_{s1'})^n$. By the second law inequality, inequalities are valid: $(J_{s2'}/J_{s1'})^n > 1$ for engines and $(J_{s2'}/J_{s1'})^n < 1$ for heat pumps. In fact, sequence Φ^1, Φ^2,... Φ^n, provides a synthetic measure of the cascade's imperfection.

Transport quantities are treated in a similar way. Two finite thermal conductances, $g_1^n = \alpha_1^n a_1^n$ and $g_2^n = \alpha_2^n a_2^n$, where a_1^n and a_2^n are the exchange surface areas at the nth stage, represent combined effects of boundary layers and wall conductivities, at the "upper" and "lower" part of the stage. Each part is characterized by its own heat transfer coefficient, α_1^n and α_2^n. The conductances are located between the "upper" or "lower" fluid (environment) and the fluid circulating in the stage; the resistances are, of course, reciprocals of the corresponding conductances. When the fluid is well mixed at each stage, in engine mode the hot fluid releases at stage n heat at a high temperature T_1^n, which reaches the engine part at $T_{1'}^n$. The low-grade heat is released by the engine fluid at the low temperature $T_{2'}^n$ to the environmental fluid, and reaches the bath at a low temperature $T_2 = T^e$. Clearly, for a finite bath, a sequence T_2^n must be considered in place of constant $T_2 = T^e$. When mass transfer is an accompanying driving process (Chapter 9), chemical potentials of active component (μ_1 and μ^e, say) also play a role, as marked in Figure 4.1. Yet, because in this chapter, we assume the absence of such mass transfer, then chemical potentials can be ignored and the efficiency of power production at stage n is simply $\eta^n = p^n/q_1^n$. In engine mode the efficiency η^n is lower than the efficiency η_C^n of the Carnot cycle operating between T_1^n and T_2. In heat pump mode, $T_{1'}^n > T_1^n$, and the flowing fluid is heated by the circulating fluid. This mode causes the fluid's utilization or energy upgrading in the fluid at flow.

Our analysis is aimed at *physical* limits on multistage yield or consumption of mechanical energy (work) in sequential heat-mechanical operations with fluid at flow. To evaluate these limits quantitatively, a benchmark process should be defined. The benchmark process is cascade yield or consumption of power by a resource fluid interacting with the environment in an optimal way, so that extremum work is assured in a given time.

Note that total extremum power from a cascade per unit flow of the driving fluid is the quantity having the units of specific work; this quantity is, in fact, the unit work obtained from the resource fluid at flow. Therefore, whenever the resource flux is constant along the path, it can inform both about power optimization and work optimization. Yet for flows at variable resource fluxes only power optimization is a proper formulation.

At the extremum limit, a proper formal approach is the discrete sequential optimization of total work output at constraints describing dynamics of heat transport and work production or consumption. The suitable optimization algorithm is a discrete maximum principle endowed in the Pontryagin's structure, familiar from optimal control of continuous systems. The optimization shows a link between the process duration, internal irreversibility and the optimal value of the work produced or consumed. Extremum performance functions for this optimum work incorporate a minimum residual entropy production. These functions are determined in terms of number of stages and a holdup time, or the number of transfer units. A *discrete* generalization of classical exergy follows for systems with a finite number of stages and finite holdup time in the cascade. A hysteretic property is valid for this generalized exergy that means a difference between the maximum work delivered from an engine mode and the minimum work added to a corresponding heat pump mode of the system. This conforms with a general physical property of irreversible systems: when only a finite amount of time is allowed, any creative action requires more mechanical energy than the inverted destructive action could provide in the inverse process with the same end thermodynamic states.

Dynamic programming algorithms, applied to *endoreversible* cascades composed of engines or heat pumps, are available (Sieniutycz and Berry, 2000; Sieniutycz, 1999b). In this chapter we focus on the maximum principle rather than the dynamic programming method, and the cascade problem is generalized to be able to treat internal irreversibilities.

A model explaining the role of internal dissipation in thermal cascades that quantifies their limiting work fluxes as well as leading to "imperfect exergies" is the main issue to be worked out in this chapter (Sieniutycz and Berry, 2000; Sieniutycz, 1999b). For a multistage optimization, knowledge of the so-called "equations of state" or "state transformations" is essential. They are derived from analysis of an individual stage, where Carnot temperature T' appears as a suitable control variable. The analysis of a single stage, Section 4.2, is the necessary step that precedes the cascade optimization. The conventional form of state equations describes state outputs in terms of state inputs and controls at stages. For the cascades analyzed here, however, a modified form of these relations is suitable, where state inputs are explicit in terms of state outputs and controls, Section 4.3. This is associated with the so-called forward algorithm of sequential optimization (Sieniutycz and Berry, 2000; Berry et al., 2000). The left-hand sides of Figures 4.1 and 4.2 (the regions surrounding the second Carnot engine) show the principle of designations used for the description of a single stage.

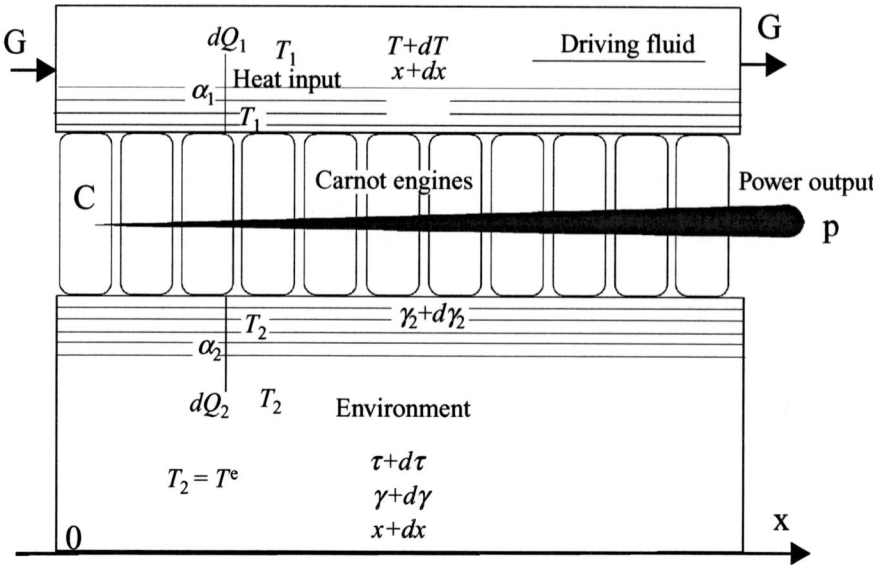

Figure 4.2 Model of power production in an infinite sequence of infinitesimal engines.

4.2. DESCRIPTION OF IMPERFECT UNITS IN TERMS OF CARNOT TEMPERATURE CONTROL

For an arbitrary process stage (omitted superscript n), the entropy balance of each imperfect thermal machine in Figures 4.1–4.3 is

$$\Phi \frac{q_1}{T_{1'}} = \frac{q_2}{T_{2'}} \tag{4.1}$$

where $\Phi = \Delta S_{2'}/\Delta S_{1'} = J_{s2'}/J_{s1'}$ is the parameter defining the internal irreversibility of the thermal machine measured in terms of the entropy changes of the circulating fluid along two isotherms $T_{1'}$ and $T_{2'}$ (Figure 3.1 of the previous chapter). Note, however, that whereas one can always introduce the internal dissipation in the form described by Equation (4.1), one may treat Φ as a constant coefficient only in the first approximation. In general, Φ may be a complicated function of the machine's operating variables. Referring to the variability of Φ we propose an analysis that allows us to exploit data of internal entropy production σ_s^{int} to apply an averaged value of coefficient Φ. This value of Φ is next used in analysis within the boundaries of operative parameters of interest. An alternative analysis using σ_s^{int} may directly lead to a more exact model at the expense of complicated formulae for the stage characteristics. Yet, in this latter case, the results universality is lost and one is usually forced to use numerical approaches instead of analytical ones.

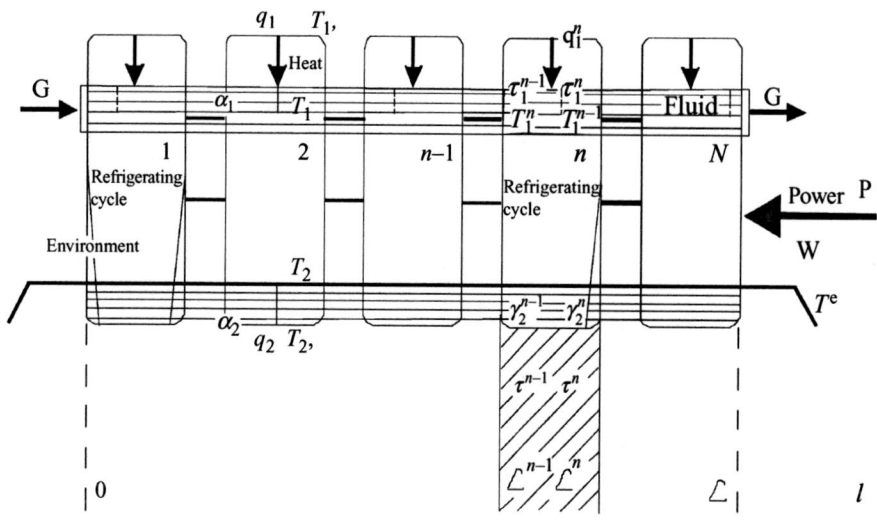

Figure 4.3 Model of power consumption in a sequence of refrigerators or heat pumps.

Cascades of multistage processes are systems difficult to optimize because their efficient optimization requires explicit analytical expression for power produced or consumed at a single stage. In fact, constant irreversibility coefficient Φ assures such an explicit expression. Comparison of Equation (4.1) with the appropriate formula for σ_s^{int} resulting from the entropy balance over the engine part of the system

$$\frac{q_2}{T_{2'}} - \frac{q_1}{T_{1'}} = \sigma_s^{\text{int}} \tag{4.2}$$

leads to the following expression:

$$\Phi = 1 + \frac{T_{1'}\sigma_s^{\text{int}}}{q_1} \tag{4.3}$$

which links the irreversibility coefficient Φ with the internal entropy generation σ_s^{int}. Clearly, σ_s^{int} is not total entropy production in the system but only its internal part referred to the thermal machine. The unknown temperature $T_{1'}$ can be eliminated on account of the fluid's bulk temperature T_1 with the help of heat flux expression, $q_1 = g_1(T_1 - T_{1'})$. The resulting formula can next be averaged within the boundaries of operative parameters of interest. The average value of Φ

$$\Phi = \left\langle 1 + \left(\frac{T_1}{q_1} - \frac{1}{g_1} \right) \sigma_s^{\text{int}} \right\rangle \tag{4.4}$$

is the quantity used in applications. The numerical value $\Phi = 1$ is assured by Equation (4.4) for endoreversible engines or heat pumps, where $\sigma_s^{int} = 0$. Experimental data of σ_s^{int} generated for engines and chillers (Edera and Kojima, 2002; Chua et al., 1996; Ng et al., 1997, 1998; Gordon and Ng, 2000) are helpful to estimate average values of Φ. Also, an effect of heat leak on cascaded heat engines can be taken into account following a recent analysis (Nuwayhid and Moukalled, 2002). Typical numerical values for Φ do not exceed 1.6 for engines and are not less than 0.4 for heat pumps.

Using the energy balance combined with the definition of the first-law efficiency, $q_2 = q_1 - p = q_1(1 - \eta)$, yields the stage efficiency described by the pseudo-Carnot formula

$$\eta = 1 - \frac{q_2}{q_1} = 1 - \Phi\frac{T_{2'}}{T_{1'}} \tag{4.5}$$

For an engine stage, where $\Phi > 1$, this efficiency is lower than that of the Carnot engine $\eta_{C'}$ based on temperatures $T_{1'}$ and $T_{2'}$. For heat pump mode $\Phi < 1$ and $\eta > \eta_{C'}$. Assuming Newtonian law of cooling for the heat exchange process described in terms of conductances g_1 and g_2 we can evaluate power yield as the difference of two heat fluxes:

$$q_1 = g_1(T_1 - T_{1'}) \tag{4.6}$$

$$q_2 = g_2(T_{2'} - T_2) \tag{4.7}$$

The quantities $T_{1'}$ and $T_{2'}$ are unknown, but they may be expressed in terms of temperatures T_1 and T_2 and a single control variable at the stage. The choice of the control variable is, in principle, arbitrary. A possible candidate can be heat flux q_1 or q_2, one of related entropy fluxes, efficiency η, and so on. Here we shall demonstrate the suitability of Carnot temperature, T', to describe optimally controlled (one-stage and multistage) thermodynamic processes.

The reasoning leading to the Carnot temperature may be developed as follows. As Equations (4.1), (4.6), and (4.7) hold, temperatures $T_{1'}$ and $T_{2'}$ are not independent but are connected by constraint

$$\frac{g_2(T_{2'} - T_2)}{T_{2'}} - \Phi\frac{g_1(T_1 - T_{1'})}{T_{1'}} = 0 \tag{4.8}$$

which means that the system has only a single independent control (one degree of freedom). Considering the imperfect efficiency derived in the previous chapter:

$$\eta = 1 - \frac{q_2}{q_1} = 1 - \Phi\frac{T_{2'}}{T_{1'}} \tag{3.21}$$

we note that the single independent control T' that may be introduced therein satisfies the thermodynamic relation

$$T' = T_2 \frac{T_{1'}}{T_{2'}} \tag{4.9}$$

Indeed, substituting temperature (4.9) into Equation (3.21) leads to the Carnot-like efficiency formula

$$\eta = 1 - \Phi \frac{T_2}{T'} \tag{4.10}$$

which holds in the irrreversible case for a given Φ. The superiority of Equation (4.10) in comparison with Equation (3.21) is that the former contains a free independent variable whereas the latter operates with two mutually connected controls, $T_{1'}$ and $T_{2'}$.

We call Equation (4.9) the thermodynamic definition of Carnot temperature as it may be derived from the invariance of efficiency or the entropy production with respect to variables transformation when passing from constrained, traditional controls $T_{1'}$ and $T_{2'}$ to the unconstrained control T'. The temperature T' of Equation (4.9) is independent of any specific mechanisms of heat transfer and coefficient of internal losses Φ.

Summing up, efficiency η may always be written in terms of a single controlling temperature (4.9). Certain properties of physical type, for example, positiveness, can be attributed to T' (Sieniutycz, 1999b). With these properties and the role it plays in the second law, T' determines the direction of heat transfer between the fluid 1 and the thermal machine.

We shall now show how to apply temperature T' to accomplish the power optimization of the single stage. Next we shall show its role in optimization of complex multistage processes (Sections 4.3 and 4.4).

Substituting Equation (4.9) in the form $T_{2'} = T_{1'} \cdot T_2/T'$ into entropy balance (4.8) leads to an equation expressing temperature $T_{1'}$ in terms of T':

$$T_{1'} = \frac{\Phi g_1 T_1 + g_2 T'}{\Phi g_1 + g_2} \tag{4.11}$$

Whereas the use of this result in Equation (4.9) yields $T_{1'}$ in terms of T':

$$T_{2'} = \frac{T_2}{T'} \left(\frac{\Phi g_1 T_1 + g_2 T'}{\Phi g_1 + g_2} \right) \tag{4.12}$$

This means that we have found the representation of the temperatures of circulating fluid in terms of Carnot control T'. With Equation (4.11) also driving heat flux, $q_1 = g_1(T_1 - T_{1'})$ can be expressed in terms of Carnot temperature;

the result is

$$q_1 = g'\left(T_1 - T'\right) \tag{4.13}$$

where g' is an effective overall conductance defined as

$$g' \equiv \frac{g_2 g_1}{\Phi g_1 + g_2} = \left(\frac{\Phi g_1 + g_2}{g_2 g_1}\right)^{-1} = \left(\frac{1}{g_1} + \Phi \frac{1}{g_2}\right)^{-1} \tag{4.14}$$

Also, with Equations (4.1), (4.9), and (4.13) we obtain the second heat flux in the form

$$q_2 = q_1 \frac{\Phi T_2}{T'} = g' \frac{\Phi T_2}{T'}(T_1 - T') = g' \Phi T_2 \left(\frac{T_1}{T'} - 1\right) \tag{4.15}$$

and the corresponding power

$$p = \eta q_1 = g'\left(1 - \frac{\Phi T_2}{T'}\right)(T_1 - T') \tag{4.16}$$

This is consistent with the results of the previous chapter when other controls were used.

Let us consider the *total* entropy production in the unit, σ_s. In terms of T' this entropy production satisfies a simple formula with the endoreversible part characteristic for heat exchange between two bodies. Indeed, by evaluating σ_s as the difference between the outlet and inlet entropy fluxes and using Equation (4.15) to substitute $q_2 = q_1 \Phi T_2/T'$ we find

$$\sigma_s = \frac{q_2}{T_2} - \frac{q_1}{T_1} = \frac{\Phi q_1}{T'} - \frac{q_1}{T_1} = q_1 \frac{(\Phi - 1)}{T'} + q_1 \left(\frac{1}{T'} - \frac{1}{T_1}\right) \tag{4.17}$$

For the linear system where Equation (4.13) holds,

$$\sigma_s = q_1 \frac{(\Phi - 1)}{T'} + \frac{g'(T_1 - T')^2}{T_1 T'} \tag{4.18}$$

In Equations (4.17) and (4.18) the entropy production is split into the sum of two nonnegative terms. The first term containing the product $q_1(\Phi - 1)$ is always nonnegative as the signs of q_1 and $(\Phi - 1)$ are the same (positive for engine and negative for heat pump). The second term is obviously positive. The property $T' > 0$ follows for $g' > 0$. The positiveness of T' as well as the simple heat flux formula, Equation (4.13), make quantity T' quite useful. Moreover, because of the pseudo-Carnot structure of efficiency, Equation (4.10) can be written as follows:

$$\eta = 1 - \Phi\frac{T_2}{T'} = 1 - \Phi\frac{T_2}{T_1} - \Phi T_2 \left(\frac{T_1 - T'}{T_1 T'}\right)$$

$$= \eta_C - (\Phi - 1)\frac{T_2}{T_1} - \Phi T_2 \frac{q_1}{g' T_1 T'}$$

(4.19)

This is a set of relatively simple and suitable expressions showing how finite flux q_1 influences the process efficiency. Whenever T' is less than T, meaning that a positive amount of heat q_1 flows to the engine, the first law efficiency of the whole system, η, drops below that of the Carnot engine; otherwise, in heat pump mode, it raises above η_C.

To stress the role of the efficiency formulae in the entropy production expression we may calculate the latter quantity in an alternative manner with respect to Equation (4.17):

$$\sigma_s = \frac{q_2}{T_2} - \frac{q_1}{T_1} = \frac{(1 - \eta)q_1}{T_2} - \frac{q_1}{T_1} = q_1\left(\frac{1}{T_2} - \frac{1}{T_1}\right) - \frac{\eta q_1}{T_2}$$

$$= \frac{q_1}{T_2}\left\{\left(1 - \frac{T_2}{T_1}\right) - \left(1 - \Phi\frac{T_{2'}}{T_{1'}}\right)\right\}$$

$$= \frac{q_1}{T_2}\left(\Phi\frac{T_{2'}}{T_{1'}} - \frac{T_2}{T_1}\right) = \frac{q_1}{T_2}\left(\Phi\frac{T_2}{T'} - \frac{T_2}{T_1}\right)$$

$$= q_1\left(\Phi\frac{1}{T'} - \frac{1}{T_1}\right) = q_1\left(\frac{\Phi - 1}{T'}\right) + q_1\left(\frac{1}{T'} - \frac{1}{T_1}\right)$$

(4.20)

Clearly, the result is equivalent to Equation (4.17).

Figures 4.1–4.3 present schemes suitable for the evaluation of extremum power and a generalized exergy in cascade processes with possibly imperfect stages. Figure 4.2 may be regarded as a limiting case of Figure 4.1 for a system of the same spatial extent and the same boundary states of the driving fluid, in which the number of discrete units is sufficiently large to attain a good discrete approximation of the continuous system (for which $N \to \infty$). Note that the limiting case of an infinite N corresponds with infinitesimal engines. This special case is analyzed with details in Section 4.6.

Figure 4.3 depicts another cascade system investigated in this chapter. It represents a scheme of multistage power consumption in a sequence of imperfect heat pumps. The first fluid here is utilized (upgraded) fluid; it flows through stages of the cascade to which the work is supplied. The second fluid is an infinite reservoir. As in the previous schemes there are finite thermal resistances in the system representing the dissipative effect of their thermal boundary layers. A finite holdup time is assumed at each process stage for the utilized fluid. The multistage process is, in fact, a sequence of imperfect heat pumps. As the engine convention is still applied, negative sign of the consumed flux of power ($P < 0$) is assumed in heat pump modes. Consequently, in cascades of heat pumps the minimized quantity is the negative of total power per unit mass flux of the driving fluid, $W = (-P/G)$. This quantity is positive and has units of work per unit of mass, so it describes a specific work in a steady process of power consumption. When the initial state of the fluid is at equilibrium with the environment $(T^0 = T^e)$, the minimal work to achieve T^N represents a finite-time exergy of the heat pump mode. The abstract scheme of multistage heat pump system in Figure 4.3 serves to model the fluid's exergy when upgrading thermal energy of fluid at flow. Thin oblique lines accompanying verticals at stages 1 and n show symbolically the possibility of internal dissipation in the refrigerating machines of these stages.

4.3.   SINGLE-STAGE FORMULAE IN A MODEL OF CASCADE OPERATION

In sequential multistage systems one should sum power expressions such as Equation (4.16) over stages. Casting the description in the format of a discrete variational problem we introduce several suitable definitions. One is the heat flux variable of the temperature dimension:

$$u^n = \frac{-q_1^n}{g'^n} \qquad (4.21)$$

Another useful variable is the interval of a nondimensional time at stage n:

$$\theta^n = \tau^n - \tau^{n-1} \qquad (4.22)$$

We apply the nondimensional time τ^n, which is equivalent to the total number of heat transfer units, comprising stages 1, 2, ..., n. This quantity can be linked with the length coordinate, x, or the fluid's residence time t, by an equation

$$\tau \equiv \frac{\alpha' a_v F}{Gc} x = \frac{\alpha' a_v}{\rho c} t = \frac{t}{\chi} \qquad (4.23)$$

where α' is the overall heat transfer coefficient, a_v is the specific area, F is the cross-sectional area for the fluid's flow and $\chi = \rho c / \alpha' a_v$ plays the role of a time constant for the system.

In the multistage cascade description, superscript n is added to all symbols, to point out that the quantities refer to the stage n. Since in the case of the infinite second reservoir the constant $T_2 = T^e$ is only a parameter and not a state variable, we can simplify our designations for the state variable T_1^n and the heat q_1^n by rejecting the unnecessary subscript 1; thus we will use the symbols T^n and q^n for the resource fluid temperature and driving heat flux at the stage n. We shall also use the constant temperature T^e instead of T_2^n. With these modifications Equations (4.10) and (4.13) yield

$$q^n = g'^n \left(T^n - \frac{\Phi^n T^e}{1 - \eta^n} \right) = g'^n (T^n - T'^n) \qquad (4.24)$$

This is the equation of thermal characteristics of the nth stage. The formula inverse with respect to the above,

$$\eta^n = 1 - \Phi^n \frac{T^e}{T^n - q^n/g'^n} = 1 - \Phi^n \frac{T^e}{T'^n} \qquad (4.25)$$

is also useful. Equations (4.24) and (4.25) are equivalent. Each proves that the efficiency at an engine stage decreases with the intensity of the heat flux and the internal irreversibility. However, when furnishing the cascade description, each of these is not equally suitable. Within the discrete optimization formalism it is generally easier to link Equation (4.24) with the energy balance. This happens because, in the cascade formalism, some extensive quantities play essential roles, which are characteristic of the cascade as a whole. They are usually evaluated as cumulative quantities over stages, and may play a role of state coordinates in the cascade optimization. In general, a cumulative quantity may be a state variable whenever there are limits to its consumption or production.

Multiplying Equation (4.24) by the stage efficiency η^n, a state equation for the cumulative work is obtained in terms of η^n. It refers to the cumulative power per unit mass flow, $W^n = p^n/G$, the quantity whose units are those of the specific work

$$W^n - W^{n-1} = c\theta^n \eta^n \left(T^n - \frac{\Phi^n T^e}{1 - \eta^n} \right) \qquad (4.26)$$

where $\theta^n = g'^n/Gc$ is a nondimensional quantity that represents the number of the heat transfer units for the stage n. With Equation (4.25) it is easy to write Equation (4.26) in terms of the Carnot temperature control, T'. The result is

$$W^n - W^{n-1} = -c\theta^n \left(1 - \Phi^n \frac{T^e}{T'^n} \right) (T'^n - T^n)$$

$$= -c \left(1 - \frac{T^e}{T^n} \right) (T^n - T^{n-1}) + cT^e \frac{(\Phi^n - 1)}{T'^n} (T^n - T^{n-1})$$

$$- cT^e \frac{(T'^n - T^n)^2}{T'^n T^n} \theta^n$$

$$(4.27)$$

where the last two terms in the second line are negative and represent the lost work associated, respectively, with irreversible internal losses in engines and external losses in heat conductors. This is clear considering that Equation (4.25) serves as the imperfect-efficiency expression. The negativeness of the middle term in the second line of Equation (4.27) is due to the fact that, in engine mode, $\Phi > 1$ and $T^n < T^{n-1}$ as the temperature of finite resource must decrease when work is produced. In the heat pump mode $\Phi < 1$ and $T^n > T^{n-1}$, thus the term discussed is still negative. However, in that mode, work generation at the stage is negative, which means that more positive work should be added to a heat pump due to internal irreversibilities. In fact, the second line expression of Equation (4.27) shows the split of the work production rate into one classical-type reversible term and two dissipative terms. The lost work appears as a product of T^e and the sum of two entropy production terms, in accordance with the Gouya–Stodola law. The total work is the sum of expressions (4.27) over stages.

Since the optimization problem is that of extremum W^n for $n = N$, it is convenient to treat the variable W as the zeroth state coordinate; then Equations (4.26) and (4.27) are rate formulae describing this coordinate. One may note that a nondimensional variable of the time interval type, $\theta^n \equiv g'^n/(Gc)$, appears linearly in Equations (4.26) and (4.27). Each of these equations is the state equation for the variable W^n in the sense used in optimization. Yet we still need to transform heat equation (4.24) into the second discrete equation of state.

The fluid's temperature changes are related to the second cumulative variable, which is the cumulative driving heat Q^n defined for the first n stages of the cascade: $Q^n = \Sigma q^k$, where $k = 1, 2, \ldots, n$. This quantity satisfies the equality $Q^n - Q^{n-1} = q^n$, where q^n is given by Equation (4.24). The sequence of heat q^n describes allocations of cumulative heat Q^n between the stages, $i = 1, 2, \ldots, n$. But, in terms of the fluid's temperature change ΔT^n at the stage n, each heat q^n equals $-Gc(T^n - T^{n-1})$. Consequently,

$$T^n - T^{n-1} = -\theta^n \left(T^n - \frac{\Phi^n T^e}{1 - \eta^n} \right) = -\theta^n (T^n - T'^n) \qquad (4.28)$$

where $T'^n - T^n = -q^n/g'^n$ is a measure of the heat received by the fluid which is utilized in the nth heat pump. Equation (4.28) is the second discrete equation of state for the cascade process. It is given here either in efficiency or in Carnot-temperature representation. We have assumed here that the fluid specific heat c is constant. In the algorithm of optimal control, the temperature control T' satisfying Equation (4.28) is used in the standardized form

$$\frac{T^n - T^{n-1}}{\theta^n} = T'^n - T^n \qquad (4.29)$$

Equation (4.29) is the discrete counterpart of the usual linear differential equation of heating or cooling in which T^n is the state variable of the problem.

To close the model it suffices to regard $\theta^n = g'^n/(Gc)$ as the increment of the independent variable τ^n. The latter is a cumulative number of the heat transfer units over stages; $\tau^n = \Sigma \theta^k$ for $k = 1, 2, \ldots, n$. As g'^n is influenced by internal irreversibility Φ^n, the quantity τ^n is an operative variable. By definition, variable θ^n is the controlled interval of cumulative number of heat transfer units; thus the third equation of state is

$$\frac{\tau^n - \tau^{n-1}}{\theta^n} = 1 \qquad (4.30)$$

This is, of course, equivalent with Equation (4.22). Time variable τ^n, defined by Equation (4.23), is an independent nondimensional variable equivalent in this model with the overall number of transfer units. It is also a cumulative variable which is a measure of the total residence time of resource fluid in the cascade. Equation (4.30) expresses the essential limitation in all problems with finite exchange area which deals with *constrained* sums of θ^n. In accordance with the terminology used in optimization, Equations (4.27), (4.29) and (4.30) constitute the set of discrete equations of state for the cascade. These equations contain on their right-hand sides the state variables and controls. The purpose of optimization is to determine optimal controls T'^n and θ^n and optimal state coordinates T^n and τ^n that satisfy equations of state (4.27), (4.29), and (4.30) and render the final value of the work function W^N an extremum.

4.4. WORK OPTIMIZATION IN CASCADE BY DISCRETE MAXIMUM PRINCIPLE

Now we shall use the discrete optimization theory to extremize the N-stage sequence of imperfect thermal machines in Figures 4.1–4.3. Summing up works at subsequent stages defined by Equation (4.27) and performing extremization leads to two extremum functions. The first is the function of maximum work released from the cascade system (engine mode):

$$V^N \equiv \max W^N = \max_{\{T'^n \theta^n\}} \sum_{n=1}^{N} \left\{ -c \left(1 - \Phi^n \frac{T^e}{T'^n} \right) (T'^n - T^n)\theta^n \right\} \qquad (4.31)$$

whereas the second is the function of minimum work supplied to the cascade system (heat pump mode)

$$R^N \equiv \min(-W^N) = \min_{\{T'^n \theta^n\}} \sum_{n=1}^{N} \left\{ c \left(1 - \Phi^n \frac{T^e}{T'^n} \right) (T'^n - T^n)\theta^n \right\} \qquad (4.32)$$

Note that function V^N has the meaning of a profit and function R^N has the meaning of a cost and

$$V^N \equiv \max W^N \neq -\min(-W^N) \equiv -R^N$$

Both work functions are nonnegative in the regimes they work (V^N in engine regime and R^N in heat pump regime). They are obtained by extremizing appropriate sums in which T'^n is the Carnot temperature control. In terms of T'^n the efficiency of an imperfect stage is described by the pseudo-Carnot formula (4.10). While the symbol θ^n describes the interval of the nondimensional time τ^n, the time itself (the sum of θ^n) is the operative number of heat transfer units. We are solving, in fact, a "finite size problem" in which intervals of the number of transfer units θ^n are measures of exchange areas at the stages. In the engine modes the inequality $T > T^e$ is associated with $T'^n - T^n < 0$, whereas the inequality $T < T^e$ with $T'^n - T^n > 0$; in this mode the work is released from the system, and by definition $W \geq 0$. In the work pumping mode, the inequality $T > T^e$ is associated with $T'^n - T^n > 0$, whereas the inequality $T < T^e$ with $T'^n - T^n < 0$. When proper end conditions are satisfied, Equations (4.31) and (4.32) define finite rate availabilities (exergies) of engine and heat pump mode, respectively. They simplify to the classical exergy whenever the process is quasistatically in a cascade with infinite number of stages, that is, when $\theta^n \to 0$ and $N \to \infty$.

When optimizing discrete cascades two basic computational schemes can be applied as shown in Figure 4.4. In the first, or standard one, shown in the upper part of Figure 4.4, no distinction between various control variables is made. In the second (lower) scheme a special control θ^n is distinguished. Whenever control θ^n appears linearly in the model its properties are special (similar to those of continuous systems), hence the rationale to treat control θ^n separately in the optimization algorithm. In this case it is also convenient to separate the continuous time t^n from the set of the state variables. It may be shown that whenever the optimization model is linear in control θ^n a *discrete* Hamiltonian is constant and the optimization algorithm is very close to the classical Pontryagin's algorithm.

Bellman's principle of optimality is applied to multistage power production or consumption. In the forward algorithm of the dynamic programming method ellipse-shaped balance areas pertain to sequential subprocesses that evolve by the inclusion of remaining stages.

The solution to the considered variational problem is associated with the optimal choice of Carnot temperatures T'^n and stage size intervals θ^n (also measures of stage transfer areas). This solution is usually obtained for a prescribed final temperature and total transfer area (constant T^N and τ^N) and leads to two discrete exergy functions. The problem represents a discrete counterpart of a corresponding continuous problem (Sieniutycz, 1997a). To solve this problem we exploit the algorithm of a discrete maximum principle, Equations (88)–(95) of an

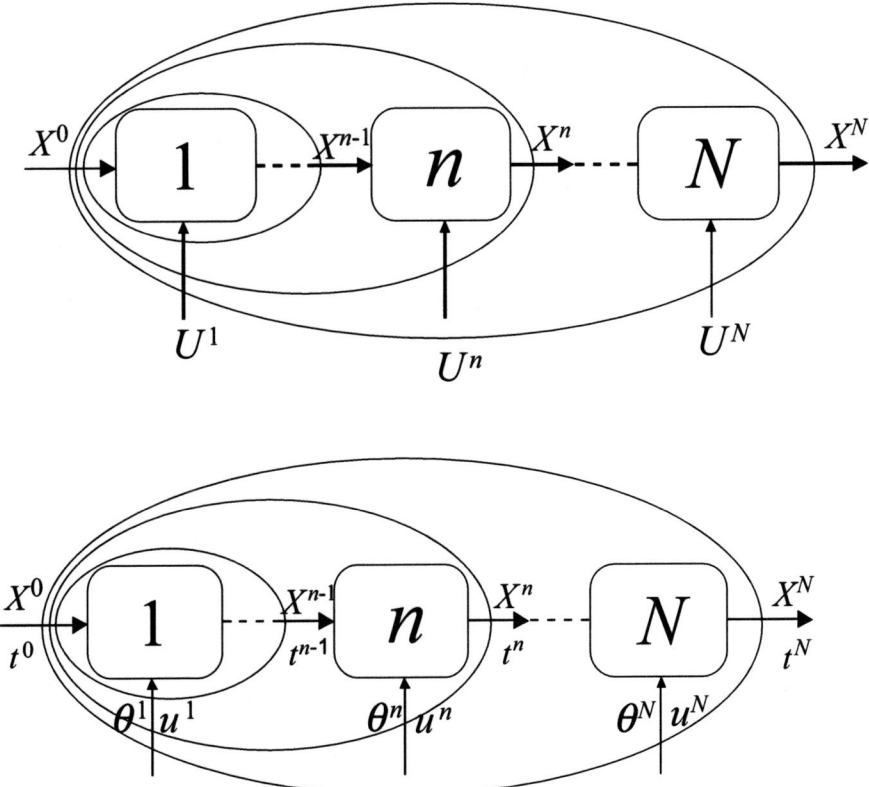

Figure 4.4 Two schemes of a discrete process with stage size control θ^n and other controls $\mathbf{u}^n$.

earlier paper (Sieniutycz, 1999a). Its basic property is constancy of a Hamiltonian function in autonomous systems.

The general definition of a Hamiltonian is

$$H^{n-1}(\mathbf{x}^n, \mathbf{z}^{n-1}, \mathbf{u}^n, t^n) = f_0^n(\mathbf{x}^n, \mathbf{u}^n, t^n) + \sum_{\nu=1}^{s} z_{\nu}^{n-1} f_{\nu}^n(\mathbf{x}^n, \mathbf{u}^n, t^n) \qquad (4.33)$$

where $\mathbf{z}^{n-1}$ is the adjoint vector and $\mathbf{u}^n$ is the control vector, excluding θ^n. The discrete Hamiltonian for the problem of extremum work (4.31) is

$$H^{n-1} = \left(z^{n-1} - c\left(1 - \Phi^n \frac{T^e}{T'^n}\right)\right)(T'^n - T^n) \qquad (4.34)$$

With this function we investigate a subclass of all solutions satisfying the condition

$$H^{n-1} = h \qquad (4.35)$$

This subclass is described by the equations

$$\frac{T^n - T^{n-1}}{\theta^n} = \frac{\partial H^{n-1}}{\partial z^{n-1}} = T'^n - T^n \qquad (4.36)$$

$$\frac{z^n - z^{n-1}}{\theta^n} = -\frac{\partial H^{n-1}}{\partial T^n} = z^{n-1} - c\left(1 - \Phi^n \frac{T^e}{T'^n}\right) \qquad (4.37)$$

$$\left(z^{n-1} - c\left(1 - \Phi^n \frac{T^e}{T'^n}\right)\right)(T'^n - T^n) = h \qquad (4.38)$$

and

$$\frac{\partial H^{n-1}}{\partial T'^n} = z^{n-1} - c\left(1 - \frac{T^e \Phi^n T^n}{(T'^n)^2}\right) = 0 \qquad (4.39)$$

We also should note that under the extremum control condition, that is, when Equation (4.39) is valid, the temperature adjoint, Equation (4.37), becomes

$$\frac{z^n - z^{n-1}}{\theta^n} = -\frac{\partial H^{n-1}}{\partial T^n} = \frac{cT^e \Phi^n}{(T'^n)^2}(T'^n - T^n) \qquad (4.40)$$

The properties of the optimal control T' are illustrated in Figure 4.5 in terms of rate variable $u = T' - T$. The graph also applies to the continuous case considered in Section 4.6.

The parameter Φ^n in Equations (4.34)–(4.39) explicitly shows the influence of internal irreversibilities in a thermal machine on the optimality conditions. The basic effect is caused by the efficiency reduction, although there are also some minor effects such as the influence of Φ^n on the overall operative number of transfer units at each stage. The condition $H^{n-1} = H^n = h$ means restriction to a subclass of the total family of optimal trajectories. A common intensity h and various undetermined durations characterize such a subclass. In our model one deals with the single variable of thermodynamic state T^n and a constant number of stages N. When N is constant the application of the condition $H^n = h$ restricts the solution to a definite trajectory.

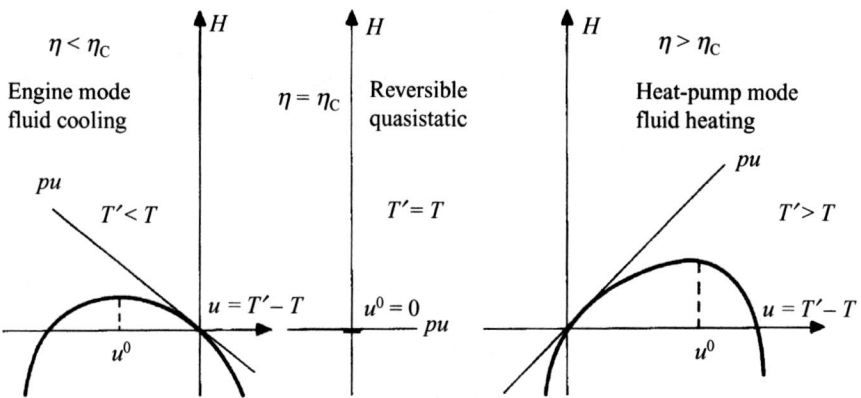

Figure 4.5 Hamiltonian in terms of rate variable $u = T' - T$ for two basic modes of the process.

By eliminating z^{n-1} from Equations (4.38) and (4.39) an integral of discrete motion follows:

$$cT^e \Phi^n \left(1 - \frac{T^n}{T'^n}\right)^2 = h \qquad (4.41)$$

This has two roots, the first referring to cooling processes in engine mode:

$$\sqrt{\Phi^n}\left(1 - \frac{T^n}{T'^n}\right) = -\sqrt{\frac{h}{cT^e}} \qquad (4.42)$$

and the second to heating processes in heat pump mode:

$$\sqrt{\Phi^n}\left(1 - \frac{T^n}{T'^n}\right) = \sqrt{\frac{h}{cT^e}} \qquad (4.43)$$

These formulae are consistent with the corresponding equation for continuous systems without internal dissipation (Sieniutycz, 1997a, 1997b). The Hamilton–Jacobi theory for continuous power-producing processes can be generalized to cascades described by Equation (4.41). The condition $T' > 0$ is sufficient for the stationary maximum of the Hamiltonian function, corresponding with the maximum work delivered in engine modes and minimum work supplied in heat pump modes. For these modes one obtains two possible realizations of Carnot control T'^n which correspond, respectively, to increasing and decreasing temperatures T^n in time τ^n, or with heating and cooling strategies for the resource fluid. For heating processes optimal Carnot temperatures satisfy an equation

$$T'^n = \frac{T^n}{1 - \sqrt{h/cT^e \Phi^n}} \qquad (4.44)$$

whereas for cooling processes the same square root appears with the opposite (plus) sign. When temperatures T^n are higher then those in the environment, heating processes can be identified with heat pump modes and cooling processes with engine modes.

Defining the intensity function $\xi(h, \Phi)$ of heating processes in terms of the numerical Hamiltonian h as

$$\xi^n(h, \Phi^n) = \frac{\sqrt{h(cT^e\Phi^n)^{-1}}}{1 - \sqrt{h(cT^e\Phi^n)^{-1}}} \tag{4.45}$$

accounting for the rule that opposite signs at square roots refer to cooling processes and expressing Equation (4.44) in terms of $u^n = T^{'n} - T^n$, one obtains $u^n = \xi T^n$, or, after using the state equation (4.29),

$$\frac{T^n - T^{n-1}}{\theta^n} = \xi^n(h, \Phi^n)T^n \tag{4.46}$$

This equation describes the discrete rate in an extremum process, which changes proportionally with the temperature, the result analogous to the equation

$$\frac{dT}{d\tau} = \xi(h, \Phi)T \tag{4.47}$$

which is valid for the continuous version. Equation (4.46) is the discrete generalization of the optimality condition (4.47) known if $\Phi = 1$ for the optimal trajectories of heat exchangers, simulated annealing and infinitesimal CNCA sequences (Sieniutycz, 1997a; Salamon et al., 1980; Andresen and Gordon, 1992a, 1992b, 1994; Spirkl and Ries, 1995).

When $\Phi^n = \Phi^{n-1} = \Phi$, a common parameter, that is, for coefficents of internal irreversibilities is the same in all thermal machines of cascade, a difference model yielding equation (4.46) can be solved analytically. For this purpose one uses a second canonical equation with adjoints z^{n-1} and z^n evaluated from the extremum condition of Hamiltonian. Equations (4.40) and (4.42) yield

$$\frac{cT^e\Phi^nT^n}{(T^{'n})^2} - \frac{cT^e\Phi^{n+1}T^{n+1}}{(T^{'n+1})^2} = \frac{cT^e\Phi^n}{(T^{'n})^2}(T^{'n} - T^n)\theta^n \tag{4.48}$$

With state equation (4.29), the above result can be written in terms of the discrete change of the temperature exclusively, and then simplified. This leads to an equation describing the temperature relations for any two neighboring stages, n

and $n + 1$, as

$$\frac{\Phi^n T^{n-1}}{(T'^n)^2} = \frac{\Phi^{n+1} T^{n+1}}{(T'^{n+1})^2} \tag{4.49}$$

Using control expression (4.44) for stages n and $n + 1$ in this formula yields a basic result for heating processes:

$$T^n = \sqrt{T^{n-1} T^{n+1}} \frac{\sqrt{cT^e \Phi^n} - \sqrt{h}}{\sqrt{cT^e \Phi^{n+1}} - \sqrt{h}} \tag{4.50}$$

For cooling processes the square root of h appears with the opposite (plus) sign. This equation shows a relatively weak influence of internal dissipation in thermal machines on a trajectory extremizing work. In particular, the extremal trajectory is unaffected by internal losses whenever the assumption $\Phi^n = \Phi^{n-1} = \Phi$ is allowed. In this special case the interstage temperatures T^n between stages n and $n + 1$ are geometric means of the boundary temperatures for any two-stage subprocess. The special result was first obtained by dynamic programming (Sieniutycz, 1999b) for the endoreversible case $\Phi = 1$. Yet abandoning endoreversibility in the present analysis renders the result more general and valuable. The application of Equation (4.50) for finding the optimal temperature sequence for $\Phi^n = \Phi^{n-1} = \Phi$ is described below.

For a prescribed T^0 and in terms of the temperature T^1 assumed as a "first decision" the recursive use of the formula $T^{n+1} = (T^n)^2 / T^{n-1}$ yields the sequence

$$T^2 = \frac{(T^1)^2}{T^0} \tag{4.51}$$

$$T^3 = \frac{(T^2)^2}{T^1} = \frac{[(T^1)^2 / T^0]^2}{T^1} = \frac{(T^1)^3}{(T^0)^2} \tag{4.52}$$

$$T^4 = \frac{(T^3)^2}{T^2} = \frac{[(T^1)^3 / (T^0)^2]^2}{[(T^1)^2 / T^0]} = \frac{(T^1)^4}{(T^0)^3} \tag{4.53}$$

and so on. Hence the final temperature T^N for the N stage process is

$$T^N = \frac{(T^1)^N}{(T^0)^{N-1}} \tag{4.54}$$

This leads to the temperature T^1 as a function of the initial and final temperatures:

$$T^1 = (T^N)^{1/N}(T^0)^{(N-1)/N} \tag{4.55}$$

With this result one may return to the intermediate temperatures (4.51)–(4.53) to obtain these in terms of the boundary temperatures

$$T^2 = (T^N)^{2/N}(T^0)^{2(N-1)/N-1} \tag{4.56}$$

$$T^3 = (T^N)^{3/N}(T^0)^{3(N-1)/N-2} \tag{4.57}$$

$$T^4 = (T^N)^{4/N}(T^0)^{4(N-1)/N-3} \tag{4.58}$$

$$T^{N-1} = (T^N)^{(N-1)/N}(T^0)^{[(N-1)^2/N-(N-2)]} = (T^N)^{(N-1)/N}(T^0)^{1/N} \tag{4.59}$$

In the case when $\Phi^n = \Phi^{n-1} = \Phi$, we can operate with Equation (4.46) in the form

$$\xi = (\theta^n)^{-1}\left(1 - \frac{T^{n-1}}{T^n}\right) = (\theta^{n+1})^{-1}\left(1 - \frac{T^n}{T^{n+1}}\right) \cdots \tag{4.60}$$

that uses constant $\xi^n(\Phi^n) = \xi^{n-1}(\Phi^{n-1}) \equiv \xi(\Phi)$ in accord with Equation (4.45). Substituting into this result Equation (4.50) in the case $\Phi^n = \Phi^{n-1} = \Phi$ we obtain

$$\xi = (\theta^n)^{-1}\left(1 - \sqrt{\frac{T^{n-1}}{T^{n+1}}}\right) = (\theta^{n+1})^{-1}\left(1 - \sqrt{\frac{T^{n-1}}{T^{n+1}}}\right) \cdots \tag{4.61}$$

This means that the effective (Φ-dependent) stage-size intervals are equal along an extremal

$$\theta^n = \theta^{n+1} \cdots = \theta = \frac{\tau^N - \tau^0}{N} \tag{4.62}$$

With Equations (4.59) and (4.62) and the first equality of Equation (4.60) applied to the Nth stage cascade, the path-independent value of optimal intensity ξ can be expressed in terms of the boundary temperatures (T^0 and T^N), duration $\tau^N - \tau^0$ and total number of stages N:

$$\xi = \frac{N}{\tau^N - \tau^0} \frac{T^N - T^{N-1}}{T^N} = \frac{N}{\tau^N - \tau^0} \left(1 - \left(\frac{T^0}{T^N}\right)^{1/N}\right) \tag{4.63}$$

For an infinite stage process with fixed end thermodynamic states one obtains from Equation (4.63) the corresponding formula for the limiting continuous process

$$\xi = \frac{\ln(T^N/T^0)}{\tau^N - \tau^0} \tag{4.64}$$

which holds for $T^N = T^f$ and $\tau^0 \equiv \tau^i = 0$. As the system is autonomous, we further assume $\tau^0 = 0$ in the relevant formulae. Use of function $\xi(h)$, Equation (4.45), in Equation (4.63) links total duration τ^N with the numerical value of Hamiltonian h for given boundary temperatures. For a fluid's heating processes accomplished in heat pump modes,

$$\tau^N(h, \Phi) = N \frac{\left(1 - \sqrt{h(cT^e\Phi)^{-1}}\right)}{\sqrt{h(cT^e\Phi)^{-1}}} \left(1 - \left(\frac{T^0}{T^N}\right)^{1/N}\right) \tag{4.65}$$

($\tau^0 = 0$). A formula inverse with respect to Equation (4.65) is of importance:

$$h(\Phi, \tau^N) = cT^e\Phi \left(\frac{\tau^N}{N[1 - (T^0/T^N)^{1/N}]} + 1\right)^{-2} \tag{4.66}$$

Equation (4.66) shows that internal irreversibility Φ increases the numerical value of the Hamiltonian in engine modes as the resulting intensity factor when achieving a given final state in a given finite time. Similarly the internal irreversibility Φ decreases the numerical value of the Hamiltonian and attainable process intensity in heat pump modes. The numerical Hamiltonian h defines a cost of the fluid's residence time or the numerical value of the Lagrange multiplier associated with the duration constraint or equipment size constraint. This is also the numerical value of the Hamiltonian constant h that has to be used when Equations (4.34)–(4.39) are solved for a given duration τ^N.

The above results make it possible to obtain the work potential $V^n = \max(W^n)$ and a related exergy function A for multistage processes characterized by the finite number of stages N with an ideal mixing at each stage. Using the dynamic programming method (Aris, 1964; Bellman, 1967; Berry et al., 2000) we find the following work function V^n:

$$V^N = c(T^0 - T^N) + cT^e \Phi N \left[1 - \left(\frac{T^0}{T^N} \right)^{1/N} \right]$$

$$- cT^e \Phi \frac{\{N[1 - (T^0/T^N)^{1/N}]\}^2}{\tau N - N[1 - (T^0/T^N)^{1/N}]}$$

(4.67)

With this formula the exergy of heat pump mode corresponds with $T^0 = T^e$, whereas the negative exergy of engine mode corresponds with $T^N = T^e$. The exergy is the maximal work $Wmax = V(T^i, \tau^i, T^f, \tau^f)$ with $T^i = T^N = T$ and $T^f = T^0 = T^e$ for engine mode, and the negative minimal work $(-W)min = -V(T^i, \tau^i, T^f, \tau^f)$ with $T^i = T^0 = T^e$ and $T^f = T^N = T$ for heat pump mode. For the intensity $\xi = 0$ and in continuous case (N approaching infinity) the change of the classical thermal exergy is recovered: hence the exergy formulae

$$A^N = c(T - T^e) - cT^e \Phi N \left[\left(\frac{T}{T^e} \right)^{1/N} - 1 \right]$$

$$- cT^e \Phi \frac{\left\{ N \left[1 - (T/T^e)^{1/N} \right] \right\}^2}{\tau N + N \left[(T/T^e)^{1/N} - 1 \right]}$$

(4.68a)

for engine mode, and

$$A^N = c(T - T^e) - cT^e \Phi N \left[1 - \left(\frac{T^e}{T} \right)^{1/N} \right]$$

$$+ cT^e \Phi \frac{\left\{ N \left[1 - (T^e/T)^{1/N} \right] \right\}^2}{\tau N + N \left[1 - (T^e/T)^{1/N} \right]}$$

(4.68b)

for heat pump mode. The middle terms in the above equations describe the product of $T^e \Phi$ and two sequences ΦS^N of entropy change, ΔS_{eng}^N and ΔS_{hp}^N, each approaching the change of standard thermodynamic entropy between states T and T^e, $\Delta S = c \ln (T/T^e)$, at the limit of an infinite number of stages, $N \to \infty$. However, an essential difference exists between the changes: whereas standard thermodynamic $\Delta S = c \ln (T/T^e)$ is the *lower* limit for engine sequence ΔS_{eng}^N, it is simultaneously the *upper* limit for the heat pump sequence, ΔS_{hp}^N. This is shown in Figure 4.6. Correspondingly, as shown below, availabilities in Equations (4.68a) and (4.68b) have the following properties: continuous exergy of engine mode A^∞ is the *upper* limit of A^N, whereas continuous exergy of heat pump mode A^∞ is the *lower* limit of A^N.

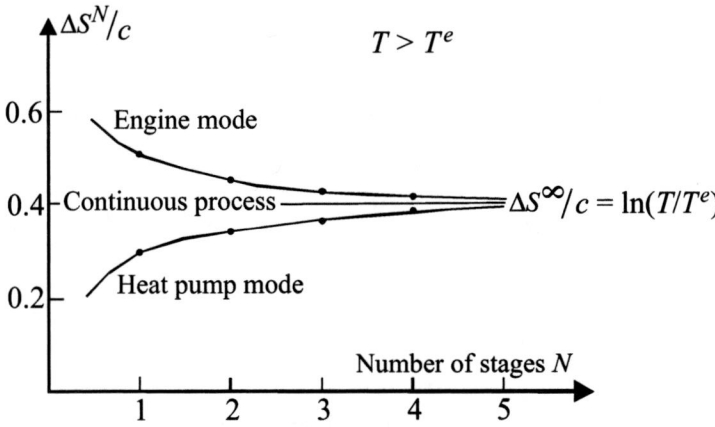

Figure 4.6 Change of discrete entropy function with total number of stages N. When $T > T^e$ the change of thermodynamic entropy in a continuous process is the upper limit of ΔS^N in heat pump modes and the lower limit of ΔS^N in engine modes.

The Carnot temperatures T'^n which assure this exergy and an optimal trajectory of a corresponding irreversible process are

$$
\begin{aligned}
T'^n &= T^n(1 + \xi) \\
&= (T^N)^{n/N}(T^0)^{[n(N-1)/N]-(n-1)}\left(1 + \frac{N}{\tau N}\left[1 - \left(\frac{T^0}{T^N}\right)^{1/N}\right]\right)
\end{aligned}
\qquad (4.69)
$$

Consider now limiting functions for N approaching infinity (continuous process). Let us start with heat pump mode, that is, the case where energy is supplied to the system. We evaluate the limit of middle expression in Equation (4.68b):

$$
\lim_{N \to \infty} cT^e \Phi N\left(1 - \left(\frac{T^e}{T}\right)^{1/N}\right) = \lim_{x \to 0} cT^e \Phi \frac{1 - (T^e/T)^x}{x}
$$

The limit calculated for $a = T^e/T$ yields the change in standard entropy

$$
\lim_{x \to 0}\left(cT^e \Phi \frac{1 - a^x}{x}\right) = \lim_{x \to 0}\left(-cT^e \Phi a^x \ln a\right) = cT^e \Phi \ln\left(\frac{T}{T^e}\right)
$$

A common differential equation holds for extremals of extremum work and minimum entropy production as a manifestation of Gouy–Stodola law (Gouy, 1889; Stodola, 1905; Bejan, 1982). Use of the variational calculus to extremize the endoreversible power yields

$$
V\left(T^{i}, T^{f}, \tau^{f} - \tau^{i}\right) \equiv \max\left(\frac{P}{G_{f}}\right) = -\max \int_{\tau^{i}}^{\tau^{f}} c\left(1 - \frac{T^{e}}{T'}\right)(T' - T)\mathrm{d}\tau
$$

$$
= h(T^{i}) - h(T^{f}) - T^{e}(s(T^{i})) - (s(T^{f})) - T^{e}\min\int_{T^{i}}^{T^{f}} c\frac{(T' - T)^{2}}{T'T}\mathrm{d}\tau
$$

$$(4.70)$$

which leads to an equation

$$
T\frac{\mathrm{d}^{2}T}{\mathrm{d}\tau^{2}} - \left(\frac{\mathrm{d}T}{\mathrm{d}\tau}\right)^{2} = 0
$$

that will be derived later under a weaker assumption of two finite reservoirs (Equation (4.111) below). It is satisfied by function $T(t)$ which solves the simple differential equation $\mathrm{d}T/\mathrm{d}\tau = \xi T$ following from Equation (4.47) for constant intensity coefficient ξ. The coefficient ξ is positive for the fluid's heating (power consumption) and negative for the fluid's cooling (power yield). An unconstrained extremal is an exponential curve. This is connected with Lorentz cycles and their optimal characteristics for the arbitrary law of heat transfer (Kuznecov et al., 1985). The following condition of optimal contact with the source of finite capacity was obtained if $\Phi = 1$ (which is the condition of minimal dissipation of heat transfer):

$$
q^{2}(T_{1}, T_{2}) = \lambda T_{2}^{2}\frac{\partial q}{\partial T_{2}} \tag{4.71}
$$

For the case of Newton's law of heat transfer it follows from this formula that for any moment of time (or any point at the contact surface) the ratio of the temperatures of the working body and the source must be the same. Substitution of this condition into a differential equation for evolution of the source's temperature yields Equations (4.47) and (4.70) and their discrete analogs. One should emphasize the fact that the working body contacts with the sources not in turn but simultaneously affects only the value of the equivalent heat transfer coefficient and does not affect the optimal law of temperature evolution. Equation (4.70) is the "endoreversible" limit for the work yield between two given states and for a given number of transfer units. Even this simple limit is stronger than that predicted by the classical exergy. What can be said about a yet stronger limit which involves an internal dissipation in the thermal machine? We need to recall the hierarchy of limits. For the limit of higher rank, an internal entropy generation $S_{\sigma}^{\mathrm{int}}$ is included in the dissipation model and then Equation (4.70) is replaced by its simple generalization:

$$V(T^i, T^f, \tau^f - \tau^i, p) = h(T^i) - h(T^f) - T^e(s(T^i)$$
$$- s(T^f)) - T^e \min \left(S_\sigma^{int} + \int_{T^i}^{T^f} c \frac{(T' - T)}{T'T} d\tau \right)$$

(4.71′)

This is in complete agreement with the Gouy–Stodola law (Bejan, 1982). For a still better limit, other components of entropy source are included at the expense of a more detailed input of information, but with the advantage that the limit is closer to reality.

At the limit of an infinite N the discrete trajectory converges to the exponential path $T_1(\tau) = T_1^i (T_1^f/T_1^i)^{\tau/\tau^f}$ and yields an exergy function. For the heat pump mode

$$A^\alpha = c(T - T^e) - cT^e \Phi \ln \left(\frac{T}{T^e} \right) + cT^e \Phi \frac{[\ln(T/T^e)]^2}{\tau^f + \ln(T/T^e)}$$

(4.72)

In the case of $\Phi = 1$ this function corresponds with the known results for continuous endoreversible systems (Sieniutycz, 1997a, 1997b, 1997c, 1998a, 1998b, 1998c, 1998d, 1998e, 2001a, 2001b, 2003a, 2003b, 2003c).

We can make explicit enhanced bounds resulting from these generalizations of thermal exergy by transforming the above expressions to suitable forms. We continue our search for a heat pump mode. The inequality

$$\delta A_{hp}^N \equiv cT^e \Phi \left\{ \ln \left(\frac{T}{T^e} \right) - N \left(1 - \left(\frac{T^e}{T} \right)^{1/N} \right) \right\} \geq 0$$

(4.73)

shows that continuous exergy A^∞ of heat pump mode is the lower limit of A^N. Here we have made use of the system's property that the change of discrete entropy ΔS^N (the second term of subtraction) increases with N in a heat pump mode, and thus the change of standard thermodynamic entropy $\Delta S \equiv \Delta S^\infty$ is the upper limit of ΔS_{hp}^N (Figure 4.6). Consequently, functions (4.68b) and (4.72) of work consumption modes can, respectively, be written in a form in which all terms are nonnegative:

$$A^N = A^{class} + c(1 - \Phi)T^e \ln \left(\frac{T}{T^e} \right) + \delta A_{hp}^N$$

$$+ cT^e \Phi \frac{\{N[1 - (T^e/T)^{1/N}]\}^2}{\tau^N + N[1 - (T^e/T)^{1/N}]}$$

(4.74)

$$A^{\infty} = A^{\text{class}} + c(1 - \Phi)T^{e}\ln\left(\frac{T}{T^{e}}\right) + cT^{e}\Phi\frac{[\ln(T/T^{e})]^{2}}{\tau^{f} + \ln(T/T^{e})} \qquad (4.75)$$

where classical thermal availability is defined in the standard way:

$$A^{\text{class}} \equiv c(T - T^{e}) - cT^{e}\ln\left(\frac{T}{T^{e}}\right) \qquad (4.76)$$

Note that inequality $\Phi < 1$ is necessary for positiveness of the Φ term of Equation (4.74) in all heat pump modes. Very generally, in all finite-rate heat pump modes: $A^{N} > A^{\infty} > A^{\text{class}}$. The result obtained means that in heat pump modes the generalized exergies A^{N} and A^{∞} provide higher (and therefore stronger) lower bounds for mechanical energy supply than the lower bound of energy consumption defined by the classical exergy A^{class}. The enhanced nature of bounds for a finite number of finite-size stages and finite-rate systems is a manifestation of the second law of thermodynamics. See Chapter 6 and especially Figure 6.3 for discussion of the influence of internal irreversibilities Φ on limiting finite-rate work generated in engines and consumed in heat pumps, as well as more information about properties of generalized exergy of a limiting continuous process, A^{∞}.

Similarly, for an engine mode, when $\Phi > 1$, an engine counterpart of Equation (4.74) can be derived in which all nonclassical terms are nonpositive. In engine modes mechanical energy is yielded by the system, and the inequality holds:

$$\delta A^{N}_{\text{eng}} = cT^{e}\Phi\left\{\ln\left(\frac{T}{T^{e}}\right) - N\left(\left(\frac{T}{T^{e}}\right)^{1/N} - 1\right)\right\} \leq 0 \qquad (4.77)$$

showing that the continuous exergy of engine mode A^{∞} is the *upper* limit of A^{N}. Here we have made use of the system's property that the change of discrete entropy ΔS^{N} decreases with N in engine modes, and thus the change of standard thermodynamic entropy $\Delta S \equiv \Delta S^{\infty}$ is the lower limit of $\Delta S^{N}_{\text{eng}}$ (Figure 4.6). Engine mode functions (4.68a) and (4.72) can, respectively, be written in the form

$$A^{N} = A^{\text{class}} + c(1 - \Phi)T^{e}\ln\left(\frac{T}{T^{e}}\right) + \delta A^{N}_{\text{eng}}$$

$$- cT^{e}\Phi\frac{\{N[1 - (T/T^{e})^{1/N}]\}^{2}}{\tau^{N} + N[(T/T^{e})^{1/N} - 1]} \qquad (4.78)$$

$$A^{\alpha} = A^{\text{class}} + c(1 - \Phi)T^{e}\ln\left(\frac{T}{T^{e}}\right) - cT^{e}\Phi\frac{[\ln(T/T^{e})]^{2}}{\tau^{f} + \ln(T/T^{e})} \qquad (4.79)$$

where, again, reversible thermal availability is defined in the standard way, Equation (4.76).

As $\Phi > 1$ in all engine modes, each nonclassical term in Equation (4.78) is nonpositive. This means that the generalized exergies A^N and A^∞ of engine mode provide upper bounds for mechanical energy supply that are lower (and therefore stronger) than the upper bound on energy yield defined by the classical exergy. Again, very generally, in all finite-rate engine modes: $A^N < A^\infty < A^{class}$. Enhanced bounds resulting from irreversible exergies A^N and A^∞ are qualitatively similar to those in Chapter 6 (Figure 6.3), where the dashed region is that of possible improvement due to reduction of internal dissipation in thermal machines.

The continuous Carnot-temperature control ensuring extremum of work is

$$T'(\tau) = T(\tau)(1 + \xi(\Phi, \tau^f - \tau^i)) = T^i \left(\frac{T^f}{T^i}\right)^{(\tau - \tau^i)/(\tau^f - \tau^i)} \left(1 + \frac{\ln(T^f/T^i)}{\tau^f - \tau^i}\right)$$

(4.80)

where ξ is rate indicator (4.64), which is positive for the fluid's heating and negative for the fluid's cooling. An unconstrained extremal is an exponential curve satisfying Equation (4.47).

The obtained exergy functions are *discrete* generalizations of the classical exergy for finite-time (dissipative) processes (Ondrechen et al., 1981; Andresen et al., 1983; Sieniutycz, 1997b, 1999a; Berry et al., 2000). As in continuous processes, the generalized exergy of processes departing from the equilibrium (upper sign) is larger than that of those approaching the equilibrium (lower sign). This is because one respectively adds or subtracts the various forms of entropy production in Equations (4.68a, 4.68b). We observe that limits for mechanical energy yield or consumption provided by exergies A^N and A^∞ are always stronger than those defined by the classical exergy. Thus, in both modes the generalized exergies provide enhanced bounds in comparison with those predicted by classical exergy. Equations (4.74) and (4.75) show that both internal and external dissipation increases the minimum work that must be supplied to the system. Likewise, Equations (4.78) and (4.79) show that both internal and external dissipation decreases the maximum work that can be produced by the system. Functions A^N and A^∞ provide work limits which take into account limitations resulting from finite rates of external transports and internal irreversibilities. Function $A^N < A^\infty$ is the work limit of higher rank than A^∞ as it takes into account a finite number of stages, the property ignored in both A^∞ and A^{class}.

4.5. EXAMPLE

Here we solve the following example: determine the critically short residence time t of the fluid in a cascade (discrete system) or in a pipeline (continuous system) for which no work production is possible by the system composed of the cascade or pipeline, engines and the environment. To solve the problem we search for the particular value of nondimensional time τ at which the upper bound for work A vanishes in the considered system. Applying the condition of vanishing A to Equations (4.78) and (4.79) subject to the definitions (4.76) and (4.77) yields, respectively,

$$A^{\text{class}} + c(1 - \Phi)T^e \ln \left(\frac{T}{T^e} \right) + \delta A_{\text{eng}}^N = cT^e \Phi \frac{\{N[1 - (T/T^e)^{1/N}]\}^2}{\tau^N + N[(T/T^e)^{1/N} - 1]} \tag{4.81}$$

$$A^{\text{class}} + c(1 - \Phi)T^e \ln \left(\frac{T}{T^e} \right) = cT^e \Phi \frac{[\ln(T/T^e)]^2}{\tau^f + \ln(T/T^e)} \tag{4.82}$$

These equations describe in an implicit way the vanishing work conditions with respect to the nondimensional time τ (number of heat transfer units in the system). From the first of these equations the critical value of time τ_c in the cascade easily follows as

$$\tau_c^N = cT^e \Phi \frac{\{N[1 - (T/T^e)^{1/N}]\}^2}{A^{\text{class}} + c(1 - \Phi)T^e \ln(T/T^e) + \delta A_{\text{eng}}^N} - N \left[\left(\frac{T}{T^e} \right)^{1/N} - 1 \right] \tag{4.83}$$

It is associated with the critical residence time of the fluid in the Nth stage cascade:

$$t_c^N = \chi cT^e \Phi \frac{[\ln(T/T^e)]^2}{A^{\text{class}} + c(1 - \Phi)T^e \ln(T/T^e)} - \chi \ln \left(\frac{T}{T^e} \right) \tag{4.84}$$

Here $\chi = \rho c/(\alpha/a_v)$ is the time constant which links the nondimensional time τ with the residence time t measured in the usual units.

Similarly, for the pipeline the critical nondimensional number of heat transfer units τ_c is

$$\tau_c^f = cT^e \Phi \frac{[\ln(T/T^e)]^2}{A^{\text{class}} + c(1 - \Phi)T^e \ln(T/T^e)} - \ln(T/T^e) \tag{4.85}$$

and the critical residence time of the fluid follows in the form

$$t_c^f = \chi cT^e \Phi \frac{[\ln(T/T^e)]^2}{A^{\text{class}} + c(1 - \Phi)T^e \ln(T/T^e)} - \chi \ln \left(\frac{T}{T^e} \right) \tag{4.86}$$

As the time constant $\chi = \rho c/(\alpha' a_v)$ decreases with the overall heat transfer coefficient α', the critical time t_c is in each case shorter for media with large thermal conductivities. This also means that for media conducting heat perfectly ("heat superconductors") the critical time equals zero, which means that power generation is always possible in reversible nonequilibrium systems. However, the prediction based on the superconductivity assumption is consistent with using the classical, reversible exergy. In fact, as shown by Equations (4.84) and (4.86), for thermally resistive media (i.e. those with finite heat conductivities) critical times are always greater than zero, which means that no mechanical energy production is possible when the medium's residence time is less than t_c (i.e. when the system's length is too short). This result is of kinetic origin; it is stronger and more realistic than the reversible one consistent with the classical exergy.

4.6. CONTINUOUS IMPERFECT SYSTEM WITH TWO FINITE RESERVOIRS

The analyses developed up to this point describe the dynamical extension of the original CNCA problem (Chambadal, 1957; Novikov, 1958; Curzon and Ahlborn, 1975) under the condition of one finite reservoir. Solution of the continuous problem of maximum power involving two finite reservoirs was recently obtained in Lingen Chen's research group (Li et al., 2007a, 2007b). Application of the optimal control theory in the form of the continuous maximum principle (Pontryagin et al., 1962; Fan, 1966) shows that the temperatures of a linear system change exponentially with respect to the flow rate and process duration, that is, the solution is the same type as in the dynamical CNCA system with one finite reservoir. While the two-reservoir analysis of the Chinese researchers, presented here, is applied to the continuous case, the application of the discrete maximum principle (Fan and Wang, 1964; Poświata, 2005; Sieniutycz, 1991, 2006b) can lead to further extensions involving discrete cascades.

Bejan (1982) derived the reversible limit of an engine extracting maximum work from a continuous hot stream with pure heat transfer. Its endoreversible limit was solved by Sieniutycz (1997a, 1997b, 1997c). Bejan and Errera (1998), Sieniutycz and von Spakovsky (1998), and Sieniutycz (2000a, 2000b, 2000c, 2000d, 2000e) obtained the extremal work from endoreversible heat engine and heat pump systems with one reservoir of a finite thermal capacity. Sieniutycz (2000a, 2000b, 2000c, 2000d, 2000e, 2004a, 2004b, 2004c) also obtained the extremal work from imperfect continuous and discrete thermal systems with one finite reservoir. Li et al. (2007a, 2007b, 2009) studied a multistage endoreversible continuous thermal machine system with two finite thermal capacity reservoirs and obtained the extremal work of the system based on the results of Sieniutycz and von Spakovsky (1998). Yet real thermal machines do not contain Carnot units in practical and industrial systems. Based on recent works (Sieniutycz, 2004a, 2004b, 2004c; Li et al., 2007a, 2007b, 2009), this chapter treats thermal machines with two finite thermal reservoirs and internal irreversibilities to obtain the extremal work and the optimal temperature profile of the driving fluid of the system.

Schemes of imperfect heat engine and heat pump systems are shown in Figures 4.1–4.3. Confining ourselves to the continuous case we focus on Figure 4.2. The first fluid (driving fluid) flows along the x-axis; the infinitesimal imperfect engines (or heat pumps) are located continuously between two separated boundary layers of the fluid. For the engine mode, the driving fluid supplies the pure heat to the infinitesimal imperfect engine at a high temperature T_1 and releases the pure heat to the second fluid at a low temperature T_2. The cumulative power yield is the sum of contributions from all stages. For an inverse process in which work is added (the continuous limit of the process in Figure 4.3), the system works as a heat pump mode. It is assumed that work W delivered in the heat engine mode is positive. In the heat pump mode W is negative, which means that the positive work $(-W)$ must be supplied to the system. For the engine mode, the heat fluxes between imperfect engines and two heat reservoirs are

$$dQ_1 = d\gamma_1(T_1 - T_{1'}), \qquad dQ_2 = d\gamma_2(T_{2'} - T_2) \qquad (4.87)$$

where $d\gamma_1 = \alpha_1 dA_1$ and $d\gamma_2 = \alpha_2 dA_2$ are heat conductivities between engines and two heat reservoirs, respectively. α_1 and α_2 are the heat transfer coefficients and dA_1 and dA_2 are the corresponding high- and low-temperature heat exchanger surface areas. $T_{1'}$ and $T_{2'}$ are (high and low) temperatures of working fluid in infinitesimal imperfect heat engines.

From Equation (4.87) one obtains

$$T_{1'} = T_1 - \frac{dQ_1}{d\gamma_1} \qquad (4.88)$$

whereas the internal entropy balance of an infinitesimal process yields

$$\Phi \frac{dQ_1}{T_{1'}} = \frac{dQ_2}{T_{2'}} \qquad (4.89)$$

As usual, $\Phi = \Delta S_{2'}/\Delta S_{1'}$ is an internal irreversibility factor (where $\Delta S_{1'}$ and $\Delta S_{1'}$ are the entropy changes of the circulating fluid along the two isotherms, $T_{1'}$ and $T_{2'}$). Φ equals the ratio of the entropy fluxes across the thermal machine, $\Phi = J_{S2'}/J_{S1'}$, and $\Phi = 1$ refers to the reversible system. As Φ is a complicated function of the machine's operating variables, an averaged value of Φ over the cycle is applied which is treated as the process constant.

Applying Equations (4.87) and (4.88) into Equation (4.89) yields

$$T_{2'} = \frac{T_2(T_1 - dQ_1/d\gamma_1)}{T_1 - dQ_1/d\gamma_1 - \Phi dQ_1/d\gamma_2} \qquad (4.90)$$

The efficiency of an infinitesimal, imperfect heat engine is

$$\eta = 1 - \frac{dQ_2}{dQ_1} = 1 - \Phi \frac{T_{2'}}{T_{1'}} \tag{4.91}$$

Substituting $T_{1'}$ and $T_{2'}$ into Equation (4.91) yields

$$\eta = 1 - \frac{\Phi T_2}{T_1 - dQ_1/d\gamma_1 - \Phi dQ_1/d\gamma_2} \tag{4.92}$$

For each differential engine we define the overall heat conductivity as

$$\begin{aligned}
d\gamma &= \frac{d\gamma_1 d\gamma_2}{\Phi d\gamma_1 + d\gamma_2} = \frac{\alpha_1 k dA \alpha_2 (1-k) dA}{\Phi \alpha_1 k dA + \alpha_2 (1-k) dA} = \frac{\alpha'_1 \alpha'_2}{\Phi \alpha'_1 + \alpha'} dA \\
&= \alpha' dA
\end{aligned} \tag{4.93}$$

where $\alpha'_1 = k\alpha_1$, $\alpha'_2 = (1-k)\alpha_2$, $A = A_1 + A_2$, $k = A_1/A$, and A_1 and A_2 are cumulative heat exchange surface areas between engines and two heat reservoirs, respectively. With these designations Equation (4.93) can be written as

$$\eta = 1 - \frac{\Phi T_2}{T_1 - dQ_1/d\gamma} \tag{4.94}$$

From the conservation of energy $dP = \eta dQ_1$, one can obtain

$$dP = \left(1 - \frac{\Phi T_2}{T_1 - dQ_1/d\gamma}\right) dQ_1 \tag{4.95}$$

where dP is the power output of an infinitesimal imperfect heat engine. In the engine mode, the temperature of the driving fluid decreases along its path, that is, $dT_1 < 0$. For the driving fluid, one has $dQ_1 = -G_1 c_1 dT_1$, where dT_1 is the differential temperature change of the driving fluid, G_1 is the mass flux of the driving fluid, and c_1 is its specific heat. For the second fluid, one has $dQ_2 = -G_2 c_2 dT_2$, where dT_2 is the differential temperature change of the second fluid, G_2 is the mass flux of the second fluid and c_2 is its specific heat. Quantities c_1 and c_2 are assumed to be constants whenever any integral formulae are derived. From Equation (4.95) and the above analysis, one can obtain

$$dP = -G_1 c_1 \left(1 - \frac{\Phi T_2}{T_1 - dQ_1/d\gamma}\right) dT_1 \tag{4.96}$$

For any process mode, the cumulative power delivered per unit mass flux of driving fluid is obtained by integration of Equation (4.96) between an initial temperature T_{1i} and a final temperature T_{1f} of the fluid. This integration yields the specific work of the flowing fluid in the form of the functional

$$W = \frac{P}{G_1} = -\int_{T_1^i}^{T_1^f} c_1 \left(1 - \frac{\Phi T_2}{T_1 - dQ_1/d\gamma}\right) dT_1 \qquad (4.97)$$

With Equation (4.97), the work maximization problem can be stated for the engine mode:

$$W_{max}^{eng} = \max\left\{ -\int_{T_1^i}^{T_1^f} c_1 \left(1 - \frac{\Phi T_2}{T_1 - dQ_1/d\gamma}\right) dT_1 \right\} \qquad (4.98)$$

whereas for the heat pump mode, the work supplied to the system $(-W)$ must be minimized with respect to various possible finite time paths which lead the system from T_1^f to T_1^i. One states the minimization problem

$$(-W_{min}^{pump}) = \min\left\{ \int_{T_1^f}^{T_1^i} c_1 \left(1 - \frac{\Phi T_2}{T_1 - dQ_1/d\gamma}\right) dT_1 \right\} \qquad (4.99)$$

It is useful to define the length scale of the process (an analog of height of the heat transfer unit in heat exchangers and mass transfer unit in packed columns)

$$\varphi_1 = \frac{G_1 c_1}{\alpha' a_{v1} F_1} \qquad (4.100)$$

where $a_{v1} = A/v_1$ is total specific exchange area per unit volume of the driving fluid and F_1 is the fluid cross-sectional area, perpendicular to x.

Consequently, the nondimensional time of the process is defined as

$$\tau = \frac{x}{\varphi_1} = \frac{\alpha' a_{v1} F_1 x}{G_1 c_1} = \frac{\alpha' a_{v1} F_1 v_1 t_1}{G_1 c_1} \qquad (4.101)$$

where v_1 is the linear velocity of the driving fluid and t_1 is the contact time of this fluid with the heat exchange surface. As in Equation (4.21) the heat flux control variable (with the units of temperature) is defined which satisfies the equation

$$u = \frac{-dQ_1}{d\gamma} = \frac{G_1 c_1 dT_1}{\alpha' dA} = \frac{G_1 c_1 dT_1}{\alpha' a_{v1} F_1 dx} = \frac{G_1 c_1 dT_1}{\alpha' a_{v1} F_1 v_1 dt_1} = \frac{dT_1}{d\tau} = \dot{T}_1$$

$$(4.102)$$

This equation links the unit heat exchange with the temperature of the driving fluid as discussed in a number of references (Sieniutycz, 1997a, 2001a, 2001b, 2006a, 2006b).

For the second fluid, one can obtain

$$\dot{T}_2 = \frac{dT_2}{d\tau} = \frac{dQ_2}{G_2 c_2 d\tau} = \frac{\Phi T_{2'} dQ_1/T_{1'}}{G_2 c_2 d\tau} = -\frac{\Phi G_1 c_1 T_2 dT_1/(T_1 - dQ_1/d\gamma)}{G_2 c_2 d\tau}$$

$$= -\frac{\Phi G_1 c_1 T_2 u}{G_2 c_2 (T_1 + u)} \qquad (4.103)$$

Consequently, Equations (4.98) and (4.99) take the form

$$W_{max}^{eng} = \max\left\{ -\int_{\tau^i}^{\tau^f} c_1 \left(1 - \frac{\Phi T_2}{T_1 + u}\right) u \, d\tau \right\} \qquad (4.104)$$

and

$$(-W_{min}^{pump}) = \min\left\{ \int_{\tau^f}^{\tau^i} c_1 \left(1 - \frac{\Phi T_2}{T_1 + u}\right) u \, d\tau \right\} \qquad (4.105)$$

These optimization problems will be solved below by using Pontryagin's algorithm of the optimal control theory (Pontryagin et al., 1962; Fan, 1966).

Applying the optimal control theory we define the Hamiltonian function

$$H = -c_1 \left(1 - \frac{\Phi T_2}{T_1 + u}\right) u + \lambda_1 u - \lambda_2 \frac{\Phi \psi T_2 u}{T_1 + u} \qquad (4.106)$$

where Lagrange multipliers λ_1 and λ_2 are called adjoint variables and $\Psi = G_1 c_1/(G_2 c_2)$ is a constant positive for a fixed system. The optimal control follows from the equation

$$\frac{\partial H}{\partial u} = 0 \Rightarrow -c_1 + \lambda_1 + \frac{c_1 - \lambda_2 \psi}{(T_1 + u)^2} \Phi T_1 T_2 = 0 \qquad (4.107)$$

The adjoint equations are

$$\dot{\lambda}_1 = -\frac{\partial H}{\partial T_1} = \frac{c_1 - \lambda_2 \psi}{(T_1 + u)^2} \Phi T_2 u \qquad (4.108)$$

and

$$\dot{\lambda}_2 = -\frac{\partial H}{\partial T_2} = \frac{\lambda_2 \psi - c_1}{T_1 + u} \Phi u \qquad (4.109)$$

The derivative of both sides of Equation (4.107) with respect to the nondimensional time τ is

$$
\begin{aligned}
&\dot{\lambda}_1 + \frac{c_1 - \lambda_2 \psi}{(T_1 + u)^2} \Phi T_2 u + \frac{c_1 - \lambda_2 \psi}{(T_1 + u)^2} \Phi T_1 \dot{T}_2 \\
&+ \frac{(T_1 + u)(-\dot{\lambda}_2 \psi) - 2(c_1 - \lambda_2 \psi)(u + \dot{u})}{(T_1 + u)^3} \Phi T_1 T_2 = 0
\end{aligned}
\tag{4.110}
$$

Substituting Equations (4.102), (4.103), (4.108), and (4.109) into Equation (4.110) yields

$$
\dot{T}_1^2 - T_1 \ddot{T}_1 = 0
\tag{4.111}
$$

An optimal path satisfying this equation is the same as for Equation (4.47) describing the case of single finite reservoir. The solution of Equation (4.111) is represented by two expressions:

$$
T_1 = T_1^0 \exp(\xi \tau)
\tag{4.112}
$$

$$
u = \xi T_1^0 \exp(\xi \tau)
\tag{4.113}
$$

where ξ is an arbitrary constant which may be positive (heat pump mode) or negative (engine mode). For the engine mode, T_1^0 is the initial temperature T_1^i. For heat pump mode, T_1^0 is the final temperature T_1^f in the engine mode. Substituting Equations (4.112) and (4.113) into Equation (4.103) yields

$$
T_2 = T_2^i \exp\left[\frac{-\psi \Phi \xi \tau}{1 + \xi}\right]
\tag{4.114}
$$

where T_2^i is the initial temperature of the second fluid in engine mode. For the heat pump mode, T_2^i is the final temperature T_2^f of the engine mode. Thus, for fixed initial and finial states and positive duration $\tau_f - \tau_i$ one obtains with the help of Equation (4.112)

$$
\xi = \pm \frac{1}{\tau^f - \tau^i} \ln \frac{T_1^f}{T_1^i}
\tag{4.115}
$$

The plus sign in Equation (4.115) refers to the engine mode and the minus sign refers to the heat pump mode. The optimal temperature profile of the driving fluid is

$$
T_1 = T_1^0 \exp\left(\pm \frac{\tau}{\tau^f - \tau^i} \ln \frac{T_1^f}{T_1^i}\right)
\tag{4.116}
$$

By using the Legendre condition it may be proven that the profile of T_1 is the extremal profile. The boundary conditions of the engine mode are assumed: $T_1^i = T^i$, $T_2^i = T^e$, $T_1^f = T^f$, and T_2^f (unknown). The process duration is τ^f ($\tau^i = 0$). Substituting the boundary conditions into Equations (4.104) and (4.105) yields

$$(-W_{\max}^{\text{eng}}) = c_1(T^i - T^f) - \frac{c_1 T_e}{\psi} \left\{ \exp\left[-\frac{\Phi\psi\ln(T^f/T^i)}{1 + \ln(T^f/T^i)/\tau^f}\right] - 1 \right\}$$

$$= c_1(T^i - T^f) - \frac{G_2 c_2 T_e}{G_1} \left\{ \exp\left[-\frac{\Phi G_1 c_1 \ln(T^f/T^i)/G_2 c_2}{1 + \ln(T^f/T^i)/\tau^f}\right] - 1 \right\}$$

$$(4.117)$$

and

$$(-W_{\min}^{\text{pump}}) = c_1(T^i - T^f) - \frac{c_1 T_e}{\psi} \left\{ \exp\left[-\frac{\Phi\psi\ln(T^f/T^i)}{1 - \ln(T^f/T^i)/\tau^f}\right] - 1 \right\}$$

$$= c_1(T^i - T^f) - \frac{G_2 c_2 T_e}{G_1} \left\{ \exp\left[-\frac{\Phi G_1 c_1 \ln(T^f/T^i)/G_2 c_2}{1 - \ln(T^f/T^i)/\tau^f}\right] - 1 \right\}$$

$$(4.118)$$

When the second fluid is an infinite reservoir ($T_2 \equiv T^e$, $\Psi \to 0$) and $T^f = T^e$ exergy type functions follow which simplify to the classical thermal exergy when $\tau \to \infty$:

$$W_{\max}^{\text{eng}} = c_1(T^i - T^f) - \lim_{\psi \to 0} \frac{c_1 T^e}{\psi} \left\{ \exp\left[-\frac{\Phi\psi\ln(T^f/T^i)}{1 + \ln(T^f/T^i)/\tau^f}\right] - 1 \right\}$$

$$= c_1(T^i - T^e) - \frac{c_1 T^e \Phi}{1 - \ln(T^i/T^e)/\tau^f} \ln\frac{T^i}{T^e} \qquad (4.119)$$

and

$$(-W_{\min}^{\text{pump}}) = c_1(T^i - T^f) - \lim_{\psi \to 0} \frac{c_1 T^e}{\psi} \left\{ \exp\left[-\frac{\Phi\psi\ln(T^f/T^i)}{1 - \ln(T^f/T^i)/\tau^f}\right] - 1 \right\}$$

$$= c_1(T^i - T^e) - \frac{c_1 T^e \Phi}{1 + \ln(T^i/T^e)/\tau^f} \ln\frac{T^i}{T^e} \qquad (4.120)$$

The second lines of Equations (4.119) and (4.120) represent the results of Sieniutycz (2004a, 2004b, 2004c). When $\Phi = 1$, Equations (4.117) and (4.118) along with Equations (4.119) and (4.120) describe the results obtained by Sieniutycz and von Spakovsky (1998), and Li et al. (2007a, 2007b, 2009), respectively.

In conclusion it can be said that the extremal works from two reservoirs of finite thermal capacity are different for engine mode, where work is released, and heat pump mode, where work is supplied. The results obtained can be compared with those of Sieniutycz (1997a, 1997b, 1998a, 1998b, 1998c, 1998d, 1998e), Sieniutycz and von Spakovsky (1998), Sieniutycz (2004a, 2004b, 2004c), Li et al. (2007a, 2007b, 2009), and also with earlier works by Ondrechen et al. (1981, 1983) on maximum work from a finite reservoir by sequential Carnot cycles. The similarities and differences of the results can be discussed for fixed initial and final states and process durations: if one controls the driving fluid, whether the second fluid is finite or not and thermal machines are reversible or not, the optimal temperature profiles and optimal controls are described by extremals of the same type, Equations = (4.112) and (4.113). The extremal works in two process modes are different. Kinetic bounds on work delivered or supplied are stronger than reversible bounds predicted by the classical thermodynamics.

4.7. FINAL REMARKS

For irreversible processes bounds on mechanical energy can be obtained either by dynamic programming or by the maximum principle. Based on equations of nonequilibrium work, enhanced bounds are found for real processes. The hysteretic properties of the finite-rate work functions are important. An essential decrease of the maximum work received from an engine system and an increase of the minimal work added to a heat pump system occurs in the high-rate regimes and for short durations. These works are quantified in terms of finite-rate exergies, which exclude some evolutions allowed by the classical exergy (Szargut and Petela, 1965; Kotas, 1985). In thermostatics, the two bounds on work (the upper bound on the work produced and the lower bound on the work consumed) do coincide. But the static limits are often too far from reality to be really useful. Generalized exergies provide bounds stronger than classical. They do not coincide for work production and work consumption; they are "thermokinetic" rather than "thermostatic" bounds. Only for infinitely long durations or for perfect processes with excellent transfer (infinite number of transfer units, τ) and without internal losses ($\Phi = 1$) do thermokinetic bounds reduce to thermostatic ones. The divergence of bounds means that the second law excludes some processes allowed by thermostatics. The exclusion region grows when conductances are smaller, which causes higher mean process rates and higher entropy production. For more information see Chapter 6, where the issue of excluded processes is illustrated in Figure 6.3.

The practical value of multistage solutions refers to energy limits enhanced due to the inclusion of imperfect thermal machines with a small number of stages and internal dissipation, where endoreversible bounds on continuous systems, obtained earlier, are too weak to be sufficiently useful. The classical exergy bounds are weaker even, as this function yields an exact estimate of bounds on extremal work only for excellent transfer conditions, or for infinitely long times of energy exchange.

For short durations the heat pump mode exergy $A = (-W)$min, which defines the lower bound on the work consumption, can be significantly higher than the minimal work of classical thermodynamics. It explains the restrictive applicability of classical thermodynamic bounds when they are applied to real processes, and it shows that these bounds should be replaced by stronger bounds obtained from nonequilibrium thermodynamics (Hoffmann, 1990; Sieniutycz, 2000c). It is essential that in either of two methodological versions of the thermodynamic approach, of which the first gives up internal irreversibilities (endoreversible approach) whereas the second one estimates the internal irreversibilities from a model, the thermokinetic limits on energy consumption or production are stronger then those predicted by the classical exergy. In short, this results from the "process rate penalty" that is taken into account in irreversible approaches. In the hierarchy of limits resulting from more and more detailed models, limits of the second and higher rank are stronger than limits of the first rank (endoreversible) version. The weakest are limits of zeroth rank: those of classical thermodynamics, obtained from the classical exergy. In the described scheme any consideration of the relation between the irreversibility and economic costs is unnecessary.

Extensions for the radiation conversion, separation processes and chemical reactions will be the topics of the next chapters.

5 Maximum power from solar energy

5.1. INTRODUCING CARNOT CONTROLS FOR MODELING SOLAR-ASSISTED OPERATIONS

In recent years the use of solar energy has become more and more attractive. Engineering analyses are available for heat transfer in liquid flat plates and concentrating solar collectors, convective heat transfer effects within honeycomb structures in flat plate collectors, solar air heaters, solar ponds, furnaces and photovoltaic converters in both terrestrial and celestial systems. Developments in solar energy and in solar collector design have been summarized by Duffie and Beckman (1974), Kreider and Kreith (1975), Sayigh (1977), and Imre (1990). Bejan (1982, 1996a, 1996b, 1996c, 1996d, 1996e, 1988a, 1988b) initiated local analyses of solar units based on entropy generation minimization (EGM) which is the approach equivalent to power extremization under the conditions of the Guy–Stodola law. Thermodynamic aspects of conversion of solar energy into electric energy are reviewed by Alaphilippe et al. (2007), where the coupling of a parabolic trough solar concentrator with a hot air Ericsson engine in open cycle is modeled. These analyses have been used in collector design, solar energy storage, solar powered refrigeration and air conditioning, heating and cooling of homes (cooling absorption systems), and other areas concerning the economics of solar energy.

de Vos (1992) developed analyses of energy limits in solar collectors based on second-law analyses and "finite time thermodynamics" (FTT). de Vos and co-workers have investigated power limits of many steady endoreversible operations in thermal, solar, photothermal, photovoltaic, chemical and hybrid systems. Simultaneously configuration aspects of solar collectors have been investigated. A rigorous derivation of the relationships between effectiveness and the number of transfer units (Lund, 1986, 1989) indicates that the thermal performance of serpentine-flow flat-plate solar collectors is superior to that of parallel-flow collectors for the same number of transfer units; this result is in contrast to a number of previous conclusions. A maximum of the net overall efficiency arises in the FTT of solar collectors driving heat engines, since the (first-law) solar collector efficiency decreases with increasing solar collector temperature while the heat engine efficiency increases. Gordon (1988) has developed an optimization of solar-driven heat engines based on FTT. Lund (1990) has used FTT for analysis of the differences between the optimal strategies for solar thermal engines working in terrestrial and celestial environments. The approach to solar energy conversion based on thermodynamics provides a unified quantitative treatment of solar engines, solar concentrators, photosynthesis engines, and photothermal and photovoltaic conversion in a manner which is simultaneously concise and

Energy Optimization in Process Systems
©2013 Elsevier Ltd. All rights reserved.

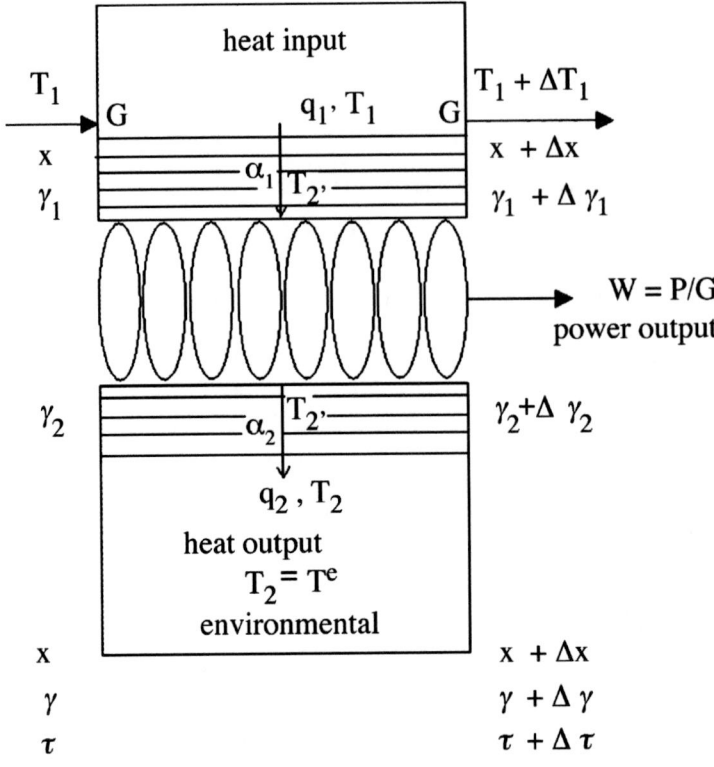

Figure 5.1 Sequential power generation in a flow system with radiation, as a resource fluid. The scheme is a tool for evaluation of generalized exergy of radiation fluid.

deep in physical meaning (de Vos, 1992). A Stefan–Boltzmann engine has been introduced (de Vos, 1992) in analogy to the CCNA engine.

In the context of dynamic limits, the piston paths which maximize the output of work in light-driven engines have been found by optimal control theory (Watowich et al., 1986). Upper and lower limits for the coefficient of performance of solar absorption cooling cycles have been derived from the first and second laws (Mansoori and Patel, 1979). These limits depend not only on the environmental temperatures of the cycle components but also on the thermodynamic properties of refrigerants, absorbents and mixtures thereof. Comparative studies of different refrigerant–absorbent combinations are now possible.

Here, however, we shall limit ourselves to considering the utilization of solar thermal energy for power generation. Estimating energy limits from radiation proceeds with analyses of power production in radiation engines. The scheme of power production is principally the same as for more traditional processes (Figure 5.1). Using this scheme we summarize here our previous derivations leading to Carnot variables (Sieniutycz, 2003a, 2003c) and then show peculiarities of their use for fluids governed by nonlinear kinetics (e.g. radiation fluid).

Figure 5.1 depicts a scheme of sequential power generation in which the upper fluid may equally well be classical Newtonian fluid or radiation fluid. For brevity we restrict ourselves to a single steady endoreversible cycle characterized by reservoir temperatures T_1 and T_2 and temperatures of circulating fluid $T_{1'}$ and $T_{2'}$. In engine mode $T_1 > T_{1'} > T_{2'} > T_2$. Evaluating entropy production σ_s as the difference of outlet and inlet entropy fluxes yields

$$\sigma_s = \frac{q_2}{T_2} - \frac{q_1}{T_1} = \frac{(1-\eta)q_1}{T_2} - \frac{q_1}{T_1} = \frac{q_1}{T_2}\left(1 - \eta - \frac{T_2}{T_1}\right) \tag{5.1}$$

where η is the first-law efficiency. Since $\eta = 1 - T_{2'}/T_{1'}$ we obtain in terms of temperatures of circulating fluid,

$$\sigma_s = \frac{q_1}{T_2}\left(\frac{T_{2'}}{T_{1'}} - \frac{T_2}{T_1}\right) \tag{5.2}$$

Therefore after introducing an effective temperature called Carnot temperature

$$T' \equiv T_2\frac{T_{1'}}{T_{2'}} \tag{5.3}$$

endoreversible entropy production (5.2) takes the following simple form:

$$\sigma_s = q_1\left(\frac{1}{T'} - \frac{1}{T_1}\right) \tag{5.4}$$

This form is identical to the familiar expression obtained for the process of purely dissipative heat exchange between two bodies with temperatures T_1 and T'.

The endoreversible efficiency $\eta = 1 - T_{2'}/T_{1'}$ takes in terms of T' the Carnot form

$$\eta = 1 - \frac{T_2}{T'} \tag{5.5}$$

which substantiates the name "Carnot temperature" for T'. Moreover, power produced in the endoreversible system also takes the classical form

$$p = \eta q_1 = \left(1 - \frac{T_2}{T'}\right)q_1 \tag{5.6}$$

It is essential to realize that the derivation of Equations (5.1)–(5.6) does not require any specific assumptions regarding the nature of heat transfer kinetics. In terms of T' description of endoreversible cycles is broken down to formally "classical" equations which contain T' in place of T_1. In fact, Carnot temperature T' efficiently represents temperature of the upper reservoir, T_1, in irreversible situations. Yet, at the reversible Carnot point, Equation (5.3) yields $T' = T_1$, thus returning to the classical reversible theory. Equations (5.4)–(5.6) then yield $\sigma_s = 0$, $\eta = \eta_C$, and $p = \eta_C q_1$, where heat flux $q_1 = 0$ for all systems in which q_1 is a function of temperature differences or the difference of some temperature functions (as, e.g., in the Stefan–Boltzmann equation).

The above properties of Carnot temperature render descriptions of endoreversible and reversible cycles very similar. They also make the variable T' a very suitable control variable in both static and dynamic situations. We shall also see soon that the descriptions using Carnot temperature can be generalized to cases with internal irreversibilities.

Kinetic aspects of Carnot temperature, transfer of this notion to the chemical potential, and abandoning endoreversibility are discussed in the literature (Sieniutycz, 2003a, 2003c). Here we shall make just a few remarks based on the results obtained in previous chapters.

For the purpose of this chapter it is worth knowing that in terms of T' the linear (Newtonian) heat transfer is described by a simple kinetic equation:

$$q_1 = g(T_1 - T') \tag{5.7}$$

For an unsteady-state relaxation of a resource to the thermodynamic equilibrium a kinetic equation associated with Equation (5.7) has the linear form

$$\frac{dT_1}{d\tau} = T' - T_1 \tag{5.8}$$

where the nondimensional time τ represents the number of heat transfer units and is related to the overall conductance g of Equation (5.7). Subscript 1 may be neglected in dynamical equations of this sort. The resource (or a finite "upper reservoir") is upgraded when Carnot temperature T' is higher than the resource's temperature T_1; whereas the resource is downgraded (relaxes to the thermodynamic equilibrium with an infinite "lower reservoir" of temperature T_2) when Carnot temperature T' is lower than the resource's temperature T_1. In linear systems the optimal, power-maximizing temperature T' is proportional to the resource's temperature T_1 at each time instant (Sieniutycz and Szwast, 2003). For nonlinear systems, for example those with radiation, an equation generalizing (5.8) will be derived in this chapter (see Equation (5.139)).

Now consider a steady radiation engine, also called the Stefan–Boltzmann engine, in which both the upper and lower heat exchange takes place in accordance with the laws of radiation:

$$q_1 = g_1(T_1^4 - T_{1'}^4) \tag{5.9}$$

and

$$q_2 = g_2(T_{2'}^4 - T_2^4) \tag{5.10}$$

In spite of the model's simplicity, its two "resistive parts" rigorously take into account the entropy generation caused by simultaneous emission and absorption of black-body radiation, the model's property which some FTT adversaries seem not to be aware of. This entropy generation follows as the "classical" sum:

$$\sigma_s = q_1(T_{1'}^{-1} - T_1^{-1}) + q_2(T_2^{-1} - T_{2'}^{-1}) \tag{5.11}$$

where each q_i is given by the Stefan–Boltzmann law, Equation (5.1).

Along with the two efficiency expressions we have at our disposal the equality $\eta = 1 - T_{2'}/T_{1'} = 1 - T_2/T'$ whence $T_{2'} = T_{1'}T_2/T'$ which agrees, of course, with Equation (5.3). Substituting this expression into the continuity equation for the entropy flux,

$$g_1\frac{T_1^4 - T_{1'}^4}{T_{1'}} = g_2\frac{T_{2'}^4 - T_2^4}{T_{2'}} \tag{5.12}$$

we obtain a single equation for $T_{1'}$. Then, after using this $T_{1'}$ and applying $T_{2'} = T_{1'}T_2/T'$ again, we obtain an analogous equation for $T_{2'}$. Next, with Stefan–Boltzmann Equations (5.9) and (5.10) we calculate heat fluxes, q_1 and q_2. The obtained thermal characteristic that describes heat flux q_1 in terms of T' has the form

$$q_1 = g_1g_2\frac{T_1^4 - T'^4}{g_1(T'/T_2)^3 + g_2} \tag{5.13}$$

In a similar way an equation is obtained that describes the second heat flux, q_2:

$$q_2 = g_1g_2\frac{T''^4 - T_2^4}{g_1 + g_2(T''/T_1)^3} \tag{5.14}$$

Note that the second Carnot temperature, T'', is applied here, for which $\eta = 1 - T''/T_1$. Observe that, corresponding with the equality $1 - \eta = T_2/T' = T''/T_1$, the heat flux q_1 was expressed in terms of its corresponding temperature T' whereas the flux q_2 was expressed in terms of the temperature T''. The equality $T'T'' = T_1T_2$, resulting from the efficiency invariance, can be exploited in order to get the second Carnot temperature. Yet, because T' and T'' are dependent, only one Carnot temperature is sufficient in the description of power systems, so in the following only quantity T' will be used.

At the Carnot point the following equalities are satisfied: $T' = T_1 = T_{1'}$, $T'' = T_2 = T_{2'}$, $\eta = 1 - T_2/T_1$, $q_1 = q_2 = p = 0$. Therefore, the heat fluxes expressed by their own Carnot temperatures (T' for q_1 and T'' for q_2) satisfy equations which are similar but not identical to equations of a corresponding process without work. Only at the short circuit point, where $T' = T_2$ and $T'' = T_1$ and $q_1 = q_2$,

Equations (5.13) and (5.14) yield

$$q_1 = g[T_1^4 - T'^4] = g[T''^4 - T_2^4] \qquad (5.15)$$

Clearly Equations (5.13) and (5.14) rather than Equation (5.15) should be used in evaluating the work limits for solar-assisted operations. The results, Equations (5.13)–(5.15), confirm the reservation which states that the overall kinetics, expressed in terms of temperatures T' or T'', may be identical in processes with work and without work only in linear cases. For nonlinear models, benefits resulting from the description of both processes in terms of the Carnot variables manifest themselves due to common expressions describing the entropy source and the work potential. However, as long as we are limited to processes without chemical reactions and those transports in which the linear approximations are acceptable, the models and optimization in systems with work and without work are the subject of a simple mutual mapping.

It may be observed that the main idea of the method that uses the Carnot intensity variables is based on the identity of *thermodynamic* equations describing the entropy production in processes with and without work in terms of the control variables T' and (possibly) μ' (Sieniutycz, 2003a, 2003c, 2007a, 2007b, 2007c, 2007d). This is caused by the abstract nature of the thermodynamic equations that are free from the time variable and materials characteristics. Yet, the complete identity of *kinetic* expressions that describe the entropy production and one of the heat fluxes in processes, with and without work, holds only for the linear kinetic models. In general nonlinear cases the identity is not satisfied; thus the use of the method based on the control variables T' and μ' requires a modification. This is shown for solar-assisted operations that involve nonlinear models due to the radiative energy transfer. In nonlinear cases the equations which describe the overall kinetics of processes with and without work are different even if they are both expressed in terms of the controls T' and μ'. While full coincidence is still attained in the "short circuit point" of the system, beyond that point the coefficients of the overall kinetics become dependent on the controls (T' and μ') and a form of divergence is observed in the structure of both equations. Our process description must then take into account this divergence. This divergence is restricted to assure the Carnot point the satisfaction of the basic equalities: $T' = T_1 = T_{1'}$, $T'' = T_2 = T_{2'}$, $\eta = 1 - T_2/T_1$, $q_1 = q_2 = p = 0$.

Below we apply thermodynamic analysis in modeling, simulation and optimization of radiation engines as nonlinear energy converters. We also perform critical analysis of available data for photon flux and photon density that leads to the exact numerical value of the photon flux constant. Basic thermodynamic principles lead to expressions for the converter's efficiency and generated work in terms of driving energy flux in the system. Steady and dynamical processes are investigated. In the latter, associated with an exhaustion of radiation resource and its temperature decrease along the process path, real work is a cumulative effect obtained in a system composed of a radiation fluid, sequence of engines, and an infinite bath.

Variational calculus can be applied in the trajectory optimization of relaxing radiation described by a pseudo-Newtonian model. The principal performance function that expresses optimal work depends on thermal coordinates and a dissipation index, h; in fact a Hamiltonian of the optimization problem for extremum power or minimum entropy production. As an example of work limit in the radiation system under pseudo-Newtonian approximation the generalized exergy of radiation fluid will be estimated in terms of finite rates quantified by the Hamiltonian h. Essential results are also dynamical equations of state for radiation temperature and work output in terms of process control variables. In Chapters 6 and 7 these equations and their discrete counterparts will serve to derive efficient algorithms for power optimization in the form of Hamilton–Jacobi–Bellman equations and dynamic programming equations.

The contemporary theory of work flux in energy systems includes both energy generators (engines) and refrigerators or heat-pumps. Locally, work flux is an indirect effect of energy transfer between two reservoirs and a thermal machine producing or consuming power. Energy systems can be described by considering the behavior of efficiency, energy flux, entropy production and mechanical power, in steady and unsteady operations. Quite often, a quantitative description of nonlinear energy transfer in various parts of the system assumes that the energy flux is proportional to the difference of temperature in a certain power, T^m. Of special importance is the case when $m = 4$, which refers to radiation engines. In our paper (Sieniutycz and Kuran, 2005) we have analyzed the effect of nonlinear laws on efficiency of power production and entropy generation in systems composed of a resource, the environment and an imperfect (nonCarnot) thermal machine. Observing that a finite flow of a resource fluid in steady systems is consistent with a finite reservoir in unsteady ones, we suggested that common dynamical equations describing changes of the driving fluid in time (spatial or chronological) can be obtained. Such equations will be derived here for systems with radiation that produce or consume work. They serve as differential constraints in problems of dynamical optimization, in particular as constraints in work optimization problems. In the optimal control theory such differential constraints are customarily called the state equations.

In this chapter we derive state equations describing the temperature in power yield systems with radiation. These equations can be applied to formulate optimization algorithms, thus leading to optimal controls, optimal trajectories and optimal performance functions. Each optimal performance function represents a potential. In the second part of this chapter we will use a part of our earlier results (Sieniutycz and Kuran, 2005) and new findings obtained here to formulate efficient algorithms for work optimization in nonlinear energy systems. With the help of these algorithms work potentials can be found as principal functions of related optimization problems. When suitable boundary conditions are assumed in a nonlinear system of interest (Sieniutycz, 2001a, 2001b) the work potential becomes a finite-rate generalization of the classical, reversible exergy (Kotas,

1985). This results in enhanced bounds on work delivered from or supplied to the system in comparison with those implied by the classical exergy, which is the reversible work potential (Ondrechen et al., 1981; Bejan and Errera, 1998; Sieniutycz, 2001a, 2001b, 2004a). In particular, we focus on the thermal behavior and work flux from a sequence of radiation engines (Figure 5.1) for which we develop several (exact or approximate) state equations.

Also, with basic formulae of black radiation thermodynamics and the Stefan–Boltzmann law (to describe effects of emission and adsorption of radiation), a general formula for a converter's efficiency follows, applicable when estimating an irreversible work limit as an extension of the classical work potential. The real work is a cumulative effect obtained from a system composed of: radiation fluid at flow, a set of sequentially arranged engines, and an infinite bath. To set a maximum work problem for this system (Figure 5.1) the concept of a multistage process is used in which each elementary stage is a Chambadal–Novikov–Curzon–Ahlborn (CNCA) operation (Ondrechen et al., 1981; de Vos, 1992; Sieniutycz, 2001a, 2001b). Each stage can be illustrated on the T–S diagram, and irreversibilities in thermal machines can be considered by using internal irreversibility factor Φ. Figure 3.1 of Chapter 3 interprets the role of Φ as a synthetic measure of the thermal machine's imperfection. By definition, $\Phi = \Delta S_{2'}/\Delta S_{1'}$ (where $\Delta S_{1'}$ and $\Delta S_{2'}$ are respectively entropy changes of the circulating fluid along the two isotherms $T_{1'}$ and $T_{2'}$), which equals the ratio of the entropy fluxes across the thermal machine, $\Phi = J_{s2'}/J_{s1'}$. The quantity Φ satisfies inequality $\Phi > 1$ for engine mode and $\Phi < 1$ for heat-pump mode of the system.

Black radiation is a specific fluid governed by its characteristic statistical mechanics, thermodynamics and kinetics (Section 5.2). The mechanism of energy transfer between that fluid and medium circulating in the engine has a significant influence on the efficiency of power production. In the radiation case the energy transfer is strongly nonlinear. Instead of Newton's linear law, the radiative energy transfer obeys the law, $q \propto \Delta(T^m)$, that is, energy flux is proportional to the difference in T^m for $m = 4$. The performance index or work W delivered in the radiation engine mode is positive by assumption. In the heat-pump mode W is negative, which means that the positive work $(-W)$ must be supplied to the system.

Performance bounds for thermal machines governed by the transfer law $q \propto \Delta(T^m)$ are known from the works of de Vos (1992), Gordon and Ng (2000), Chen and Yan (1989), Chen et al. (1990, 1999), Wu (1989, 1992) and co-workers of these researchers. Some treatments (Ondrechen et al., 1981; Bejan and Errera, 1998; Gordon and Ng, 2000; Sieniutycz, 2001a, 2001b, 2004a, 2004b, 2004c; Sieniutycz and Kuran, 2005) relax the limitation to steady systems (associated with infinite reservoirs), and take into account the effect of internal irreversibilities within energy generators. Carnot engines are then replaced by more realistic thermal machines. Consequently, a contemporary theory is nonlinear and treats imperfect processes subject to the assumption of finite resources (finite hot reservoirs).

The problem of a generalized exergy is associated with work production by a finite resource interacting with the environment in a finite time. To find the extremum work at flow and associated exergy, optimization problems are considered, for a maximum of work delivery (max W) and for a minimum of the work supply (min (−W)). The generalized exergy is the maximum work that refers to a minimally irreversible, finite-rate process. It is quantified in terms of the states of the resource and environment, a process rate index (Hamiltonian h) and imperfection factor of the thermal machine, Φ. Yet most post-classical terms of generalized exergies were evaluated under the assumption of the exponential relaxation, consistent with linear dynamics (Sieniutycz, 1997a, 1997b, 1997c, 2004a, 2004b, 2004c).

However, in radiation fluids, which are nonlinear thermodynamic systems, a fluid's properties vary along the path, and the optimal relaxation curve is nonexponential. Nevertheless, the shape of the optimal curve can be determined from the optimum power conditions. In this chapter, various differential models of controlled relaxation dynamics are studied, differing with the degree of accuracy of the process description. While simpler models are easier to solve, more complicated ones may describe the related physics in a more exact way, and this is why they may be preferred. Modifications of relaxation models are also considered, depending on the mode of energy exchange with the environment.

5.2. THERMODYNAMICS OF RADIATION

A part of this section contains material familiar to a physicist, but we shall briefly adduce useful formulae to make the chapter self-contained. However, we shall also discuss relatively novel information related to the proper adjustment of the so-called photon flux constant essential for consistent description of photon flows.

Free energy of radiation can be derived from boson statistics:

$$F = E - TS = -\frac{1}{3}aVT^4 = -\frac{4}{3c}\sigma VT^4 \tag{5.16}$$

(Gumiński, 1982). The universal coefficient a in Equation (5.16) is related to the Stefan–Boltzmann constant $\sigma = \Pi^2 k_B^4 (60h^3c^2)^{-1}$ by the direct and inverse formula

$$a = \frac{4\sigma}{c}, \quad \sigma = \frac{ac}{4}$$

The entropy of homogeneous radiation occupying the volume V is then

$$S = -\frac{\partial F}{\partial T} = \frac{4}{3}aVT^3 = \frac{16}{3c}\sigma VT^3 \tag{5.17}$$

whereas the radiation pressure P is

$$P = -\frac{\partial F}{\partial V} = \frac{1}{3}aT^4 = \frac{4}{3c}\sigma T^4 \tag{5.18}$$

and the energy of radiation is

$$E = F + TS = aVT^4 = \frac{4}{c}\sigma VT^4 \tag{5.19}$$

This expression corresponds with the heat capacity at the constant volume

$$C_v(T, V) = 4aT^3V = 16c^{-1}\sigma T^3V \tag{5.20}$$

It follows (see below) that the number of photons in a black box is also proportional to the product T^3V. Therefore, the useful conclusion stemming from Equations (5.17) and (5.20) is that both entropy and heat capacity per one black photon are constant.

Clearly, the following relations are valid:

$$PV = \frac{1}{3}aVT^4 = \frac{4}{3c}\sigma VT^4 = \frac{E}{3}$$

Now we will make use of the fact that the free energy (5.16) does not contain explicitly the number of particles N. From Equation (5.16) and free energy differential the chemical potential of black radiation is

$$\mu = \left(\frac{\partial F}{\partial N}\right)_{T,V} = 0$$

This is consistent with the vanishing Gibbs function for the black radiation:

$$G = H - TS = F + PV = -\frac{1}{3}aVT^4 + \frac{1}{3}aVT^4 = 0 \tag{5.21}$$

Associated with the vanishing chemical potential μ is the equality

$$H = TS = -T\frac{\partial F}{\partial T} = \frac{4}{3}aVT^4 = \frac{16}{3c}\sigma VT^4 \tag{5.22}$$

When a general formula for the first differential of the pressure (grand) potential

$$d\Omega = -P\,dV - S\,dT - N\,d\mu$$

is applied to the function

$$\Omega \equiv -PV = -\frac{1}{3}aVT^4 = -\frac{4}{3c}\sigma VT^4 = -\frac{E}{3}$$

correct pressure and entropy are obtained by partial differentiation of Ω with respect to the volume and temperature. Yet, differentiation of Ω with respect to chemical potential (to get the particle number N) cannot be effective due to the constancy of (vanishing) μ.

Nonetheless, statistical mechanics calculations of quantum theory show that N is proportional to VT^3, or $N/(VT^3) = p^0$ is a constant (Tolman, 1949; Landau and Lifshitz, 1974; Massa, 1986, 1989; Gumiński, 1982; Kirwan, 2004). This also means that the photon flux, $\dot{N}$, the product of the density of photons N/V and their mean flow velocity $c/4$, is proportional to T^3. A numerical value of related proportionality constant, also called the constant of the photon flux, is approximately

$$p = 1.52 \times 10^{11} \text{ photons cm}^{-2} \text{ K}^{-3} \text{ s}^{-1} \qquad (5.23)$$

In terms of this quantity, the flux density of photons is $J_N = pT^3$ photons cm^{-2} s^{-1}. Of course, $\dot{N} = pT^3 F$; that is, the flux itself is the product of the quantity J_N and the cross-sectional area F. The literature values of p fluctuate depending on approximations made in statistical calculations. From the statistical evaluation by Landau and Lifshitz (1974) of the number of photons is:

$$N = 0.244(k_B T \hbar^{-1} c^{-1})^3 V \qquad (5.24)$$

the value of $p = 1.17 \times 10^{11}$ photons cm^{-2} K^{-3} s^{-1} is obtained, which differs from that given by Equation (5.23). As we shall show soon, both considered values may be regarded as approximate, but there is a way to find the precise value of p or p^0 based on the requirement that photons should satisfy the state equation for an ideal gas.

Although the numerical value of p in Equation (5.23) is not exact, we shall use it to show how this approximation influences the form of the state equation and values of some thermodynamic quantities attributable to one photon. The proportionality of N to VT^3 means that the entropy per one black photon $S/N = $ constant and that the energy of a single black photon is proportional to the absolute temperature, T. With p of Equation (5.23) the entropy per one black photon is

$$s_{ph} = \frac{S}{N} = \frac{J_s}{J_N} = \frac{(4/3)\sigma T^3}{pT^3} = \frac{4\sigma}{3p}$$

$$= \left(\frac{4}{3}\right) \frac{5.6696 \times 10^{-5} \, erg \, cm^{-2} \, s^{-1} \, K^{-4}}{1.52 \times 10^{11} \, cm^{-2} \, K^{-3}}$$

$$= 4973 \times 10^{-16} \, erg \, K^{-1}$$

Comparing this value with the numerical value of the universal Boltzmann constant

$$k_B = 1.3807 \times 10^{-16} \, erg \, K^{-1}$$

yields an approximate result

$$s_{ph} = 3.60\,204 k_B \tag{5.25}$$

The energy per one black photon is approximately

$$e_{ph} = \frac{E}{N} = \frac{\sigma T^4}{pT^3} = \frac{\sigma}{p}T$$

$$= \frac{5.6696 \times 10^{-5} \, erg \, cm^{-2} \, s^{-1} \, K^{-4}}{1.52 \times 10^{11} \, cm^{-2} \, K^{-3}}T \tag{5.26}$$

$$= 3.73 \times 10^{-16} T \, [erg]$$

or as the multiplicity of $k_B T$

$$\frac{e_{ph}}{k_B T} = \frac{3.73 \times 10^{-16}}{1.3807 \times 10^{-16}} = 2.70 \tag{5.27}$$

As $PV = E/3$, the energy density of photons may be written as

$$\frac{E}{V} = 3P = \frac{2.70 N k_B T}{V} \tag{5.28}$$

whence

$$PV = 0.9 N k_B T \tag{5.29}$$

As this is a perfect gas equation but with coefficient 0.9 instead of 1, it may be claimed that the coefficient value is caused by approximations in statistical mechanics calculations of $p^0 = N/(VT^3)$. In his approach, Massa (1986) expresses the average energy of a photon as $3k_B T$ rather than $2.7 k_B T$ of Equation (5.18). This gives a Wien coefficient that differs from the Planck coefficient by only 6%.

This also agrees with Massa's analysis when applied to one photon. His results lead also to reasonable data of gravitational constraints on black-body radiation and the maximum (Planck) temperature (Massa, 1989).

We observe that, with Massa's adjustment, photons are particles satisfying the perfect gas formula in its classical form

$$PV = Nk_{\rm B}T \tag{5.30}$$

Inversion of this formula and use of the photon pressure in terms of temperature, $P = (1/3)aT^4 = (4/3c)\sigma T^4$, allows for precise results for the photon number and density flux to be obtained in terms of their temperature and volume:

$$N = \frac{P(T)V}{k_{\rm B}T} = \frac{a}{3k_{\rm B}}T^3V = \frac{4\sigma}{3ck_{\rm B}}T^3V \tag{5.31}$$

$$J_N = \frac{N}{V}\frac{c}{4} = \frac{ac}{12k_{\rm B}}T^3 = \frac{\sigma}{3k_{\rm B}}T^3 \tag{5.32}$$

This result is the particle counterpart of the Stefan–Boltzmann formula for density of flowing energy

$$\frac{\dot{E}}{V} = \sigma T^4 \tag{5.33}$$

Applying in Formula (5.26) the readjusted value $3k_{\rm B}T$ for the energy per one black photon

$$e_{\rm ph} = \frac{E}{N} = \frac{\sigma}{p}T = 3k_{\rm B}T \tag{5.34}$$

yields the following value of the photon flux constant

$$p = \frac{\sigma}{3k_{\rm B}} = \frac{ac}{12k_{\rm B}} \tag{5.35}$$

This yields a "possibly exact" numerical value of $p = 1.369 \times 10^{11}$ photons cm^{-2} K^{-3} s^{-1}, roughly in the middle of the two values considered above. This is the value that assures satisfaction of the state equation of perfect gas in terms of photons and sets the value of single photon entropy at the level $s_{\rm ph} = 4k_{\rm B}$ (Equation (5.37) below). Equation (5.32) thus satisfies the expression $J_N = pT^3$ for the above value of p. The related result for the photon number N, Equation (5.31), is in terms of p

$$N = \frac{P(T)V}{k_{\rm B}T} = \frac{4\sigma}{3ck_{\rm B}}T^3V = \frac{4p}{c}T^3V \tag{5.36}$$

The photon density formula involves the constant $p^0 = 4p/c$. The corresponding value of the entropy per one black photon is

$$s_{ph} = \frac{J_s}{J_N} = \frac{(4/3)\sigma T^3}{pT^3} = \frac{4\sigma}{3p} = 4k_B \qquad (5.37)$$

We can also speak about mass of black photons which is the ratio of their energy E and c^2. In particular, the division of energy (5.36) by c^2 yields the mass of a single black photon as the quantity increasing linearly with the absolute temperature T. Since the number of photons increases with T^3 and the energy per one photon increases with T, their total mass in the enclosed system, $M = aV\,T^4/c^2$, increases proportionally to T^4.

5.3. CLASSICAL EXERGY OF RADIATION

The work potential of radiation is particularly important in applications. A number of ways to derive exergy of enclosed radiation understood in the classical sense can be advocated. First, the exergy of radiation can be derived from a general expression consistent with the exergy definition. As the classical exergy is the maximum reversible work obtained from the system (radiation) and the environment (Spanner, 1964; Kotas, 1985), the first differential of the radiation exergy satisfies a general relationship

$$dB = dE - T_0\,dS + P_0\,dV - \mu_0\,dN = (T - T_0)\,dS - (P - P_0)\,dV \\ + (\mu - \mu_0)\,dN \qquad (5.38)$$

with $\mu = 0$ and $\mu_0 = 0$. The integration of this expression yields the familiar formula

$$B = E - E_0 - T_0(S - S_0) + P_0(V - V_0) + \mu_0(N - N_0) \qquad (5.39)$$

where the last term vanishes for the black radiation. This is the first formula from which the radiation exergy follows in the so-called Petela's form (Equation (5.46) below). On the other hand, the formula can easily be transformed to a simpler form after applying the well-known thermodynamic relation

$$E = TS - PV + \mu N \qquad (5.40)$$

for the two states considered, where the first state is the current state and the second one is the zeroth (equilibrium) state

$$E_0 = T_0 S_0 - P_0 V_0 + \mu_0 N_0 \qquad (5.41)$$

The result is

$$B = E - T_0 S + P_0 V + \mu_0 N \tag{5.42}$$

or equivalently, after repeated use of Formula (5.40),

$$B = (T - T_0)S - (P - P_0)V + (\mu - \mu_0)N \tag{5.43}$$

For the black radiation its chemical potentials μ and μ_0 equal zero, hence

$$B = (T - T_0)S - (P - P_0)V \tag{5.44}$$

and its first differential satisfies Equation (5.38) with $\mu = 0$.

Equation (5.44) provides probably the simplest way to obtain the radiation exergy enclosed in the volume V. Applying in Equation (5.44) the formulae describing entropy S and pressure P in terms of temperature T and volume V, Equations (5.17) and (5.18) yield a familiar Petela's formula for the exergy density

$$b_v \equiv \frac{B}{V} = \frac{a}{3}(3T^4 - 4T^3 T_0 + T_0^4) \tag{5.45}$$

(Petela, 1964). Its alternative form is

$$B = aT^4 V \left(1 - \frac{4}{3}\frac{T_0}{T} + \frac{1}{3}\left(\frac{T_0}{T}\right)^4\right) \tag{5.46}$$

The large brackets of this equation contain Petela's efficiency of the energy conversion, η_P.

Let us transform exergy equation (5.44) using the condition of vanishing Gibbs function, $G = 0$. The condition assures the equalities $TS = H$ and $T_0 S_0 = H_0$. Clearly from Equation (5.44)

$$B = H - T_0 S - (P - P_0)V \tag{5.47}$$

But in view of the equality $T_0 S_0 = H_0$ one may subtract from this result H_0 and simultaneously add $T_0 S_0$. This yields

$$B = H - H_0 - T_0(S - S_0) - V(P - P_0) \tag{5.48}$$

This result proves that the exergy of enclosed radiation, satisfying Petela's formula (5.46), can also be found from the enthalpy counterpart of standard equation (5.39) subject to the condition $\mu = \mu_0 = 0$, valid for black photons. One can observe a simple connection between Equation (5.48) and that describing exergy flux per unit photon flux (see further text). While Formula (5.46) complies with other results (Petela, 1964; Press, 1976; Landsberg and Mallinson, 1976), it is not nondebatable since some authors advocated different equations for the exergy of enclosed radiation (Spanner, 1964; Jeter, 1981). A discussion of these issues can be found in papers quoted above and in reviews (Bejan, 1982, 1987a, 1987b; Petela, 2003; Badescu, 2004, 2008). The results here confirm that Equation (5.46) is valid for enclosed radiation but, otherwise, they seem to imply that its flow counterpart cannot satisfy Petela's formula despite some claims in the literature. This would mean that Petela's result is limited to the enclosed radiation and fails for a steady photon flux. A possible explanation of this controversy has been achieved (Badescu, 2008 and Section 5.4). Also, the incorrectness of the supposition that the entropy flux of thermal radiation is the same as that for heat conduction, the heat flux divided by the local temperature, q/T, has been discussed. An analysis shows that the net entropy flux should rather be put in the form $n(q/T)$, where n is a coefficient unique to the radiative fluxes involved (Wright, 2007).

A suitable way to evaluate and interpret the classical exergy of radiation is provided by an equation given recently by Özturk and his co-workers, who developed a general scheme of thermodynamic transformations involving availability rather than entropy (Özturk, 1997; Ozekmekci et al., 2000).

For a *heat-pump mode* (process departure from the equilibrium) their equation reads

$$dB = \left(1 - \frac{T_0}{T}\right) C_v(T, V)\, dT + \left[(T - T_0)\left(\frac{\partial P}{\partial T}\right)_V - (P - P_0)\right] dV$$

(5.49)

In this equation the expressions for pressure (5.18) and heat capacity (5.20) yields the perfect differential of exergy B in the form

$$dB = 4aVT^3 \left(1 - \frac{T_0}{T}\right)\, dT + \frac{a}{3}(3T^4 - 4T^3 T_0 + T_0^4)\, dV \qquad (5.50)$$

where

$$b_v \equiv \frac{B}{V} = \frac{a}{3}(3T^4 - 4T^3 T_0 + T_0^4)$$

is the exergy of unit volume in Petela's form (5.45). Equation (5.50) replaces in a sense a similar one with particle number N as a variable known in the classical thermodynamics of open systems. In the radiation case the use of the particle number variable simultaneously with T would be inappropriate since N is a function of T rather than an independent coordinate of state. On the other hand, Equation (5.50) contains two independent variables T and V that properly characterize the physical state of the system.

An integral form of Equation (5.50) for a process starting at (T_0, V_0) and terminating at (T, V) is

$$B = \int_{T_0, V_0}^{T, V} \left(1 - \frac{T_0}{T}\right) 4aVT^3 \, dT + \frac{a}{3}(3T^4 - 4T^3T_0 + T_0^4) \, dV \quad (5.51)$$

To calculate the integral we start at the point T_0, V_0 and first integrate with respect to temperature along the horizontal line of the constant volume $V = V_0$. After the temperature achieves its upper limit T, the integration is with respect to volume, along a vertical of a constant T, until the upper limit of volume V is attained. The result of integration is

$$B = aV_0(T^4 - T_0^4) - T_0 \frac{4}{3}aV_0(T^3 - T_0^3) + \frac{a}{3}(3T^4 - 4T^3T_0 \\ + T_0^4)(V - V_0) \quad (5.52)$$

and the rearrangements yield

$$B = \frac{a}{3}(3T^4 - 4T^3T_0 + T_0^4)V_0 + \frac{a}{3}(3T^4 - 4T^3T_0 + T_0^4)(V - V_0) \quad (5.53)$$

Thus finally

$$B = \frac{a}{3}(3T^4 - 4T^3T_0 + T_0^4)V \quad (5.54)$$

in accordance with Petela's formula (5.45).

For the *engine mode* of the process (approach to the equilibrium) the sign of the right-hand side of Equation (5.49) is inverted, and for the radiation engine

$$dB = -\left(1 - \frac{T_0}{T}\right) 4aVT^3 \, dT - \frac{a}{3}(3T^4 - 4T^3T_0 + T_0^4) \, dV \quad (5.55)$$

Integration of this equation for the inverse reversible process starting at (T, V) and terminating at (T_0, V_0) requires evaluation of the integral

$$B = \int_{T, V}^{T_0, V_0} -\left(1 - \frac{T_0}{T}\right) 4aVT^3 \, dT - \frac{a}{3}(3T^4 - 4T^3T_0 + T_0^4) \, dV \quad (5.56)$$

Because of the absence of derivatives with respect to T and V and the perfect differential property of the integrand, the integration result is the same; that is Equations (5.45) and (5.54) are obtained again. This shows that reversible work produced in a certain process equals that consumed in its inverse. Such a property is associated with the potential nature of classical exergy, and does not hold when a residual dissipation is admitted in cases with finite rates.

5.4. FLUX OF CLASSICAL EXERGY

Consider now the flow of black photons. The total time derivative of the exergy of enclosed volume

$$\dot{B} = \left(1 - \frac{T_0}{T}\right) 4aVT^3\dot{T} + \frac{a}{3}(3T^4 - 4T^3T_0 + T_0^4)\dot{V} \qquad (5.57)$$

does not represent the exergy flux. The situation is similar to that for energy at flow when the time derivative of the enclosed energy is not the energy flux because the latter incorporates the work against the pressure forces. As the energy flux is the product of the enthalpy density and the volume flux, the flux of radiation exergy $\dot{B}_f$ satisfies the general thermodynamic formula

$$\dot{B}_f \equiv \dot{B} + (P - P_0)\dot{V}$$

Using this formula and Equation (5.48) one obtains for radiation fluid

$$\dot{B}_f \equiv \dot{B} + (P - P_0)\dot{V} = \{h_v - h_{v0} - T_0(s_v - s_{v0})\}\dot{V} \qquad (5.58)$$

where subscript v refers to the respective quantity per unit volume (density). The same result can be obtained from Equation (5.39). Equation (5.58) thus complies with the results of general thermodynamics. It is also consistent with an expression for reversible power production associated with exergy flux $\dot{B}_f$, represented by the integral

$$\dot{B}_f = -\int_{T,P}^{T_0,P_0} \dot{V}\left(c_v(T)\left(1 - \frac{T_0}{T}\right) + \frac{dP}{dT}\right) dT \qquad (5.59)$$

By applying heat capacity density at the constant volume c_v we omit here familiar difficulties associated with an infinite heat capacity of photons at constant pressure. As for black radiation $dP = (4/3)aT^3\,dT$, we calculate the enthalpy-related power $\dot{B}_f$ using a *substitutional* heat capacity

$$c_h(T) \equiv c_v(T) + \frac{dP}{dT} = \left(4 + \frac{4}{3}\right)aT^3 = \left(\frac{16}{3}\right)aT^3 \qquad (5.60)$$

It may be noted that $c_h/c_v = 4/3$ and that the entropy contribution to power is still governed by c_v. We recall that in each black radiation system the value 4/3 is the power coefficient in an equation for an isentropic, adiabatic process, $PV^{4/3} = f(S)$. In fact, equations describing radiation in this process are formally identical to equations of perfect gases in which the ratio c_p/c_v equals 4/3. In the case of black radiation the analogy is only formal because the heat capacity c_p (partial derivative of enthalpy with respect to T at a constant P) is infinite (P is a function of T for the black radiation).

Substitution of Equation (5.60) into (5.59) yields an integral of reversible power produced by black radiation at flow:

$$\dot{B}_f = -4a\dot{V} \int_T^{T_0} \left\{ \left(\frac{4}{3}\right) T^3 - T^2 T_0 \right\} dT \tag{5.61}$$

Its integration between a variable initial temperature T and a constant final temperature T_0 yields

$$\dot{B}_f = \frac{4}{3} a\dot{V}((T^4 - T_0^4) - T_0(T^3 - T_0^3)) \tag{5.62}$$

The reader can quickly generalize this result to the power formula applicable when both integration limits are arbitrary, that is when the temperature of the environment is not necessarily an integration limit.

Below we verify that Equations (5.59) and (5.62) indeed describe the enthalpy-based exergy flux of black radiation; that is, that they satisfy the general thermodynamic formula $\dot{B}_f \equiv \dot{B} + (P - P_0)\dot{V}$. From Equation (5.62),

$$\dot{B}_f = \left(\frac{4}{3}a(T^4 - T_0^4) - T_0\frac{4}{3}a(T^3 - T_0^3)\right)\dot{V} = (h_v - h_{v0} - T_0(s_v - s_{v0}))\dot{V}$$

$$= \dot{B} + (P - P_0)\dot{V} \tag{5.63}$$

Yet some specific formulae hold for radiation caused by the condition $\mu = 0$ for black photons. The upper line of Equation (5.63) yields after simplification

$$\dot{B}_f = \frac{4}{3}a(T^4 - T^3 T_0)\dot{V} = \frac{4}{3}aT^4\left(1 - \frac{T_0}{T}\right)\dot{V} \tag{5.64}$$

But since the enthalpy density should be consistent with Formula (5.22) we conclude that the photons' exergy flux is equal to the product of the enthalpy flux and the Carnot efficiency

$$\dot{B}_f \equiv \{h_v - h_{v0} - T_0(s_v - s_{v0})\}\dot{V} = \frac{4}{3}aT^4\dot{V}\left(1 - \frac{T_0}{T}\right) = h_v\left(1 - \frac{T_0}{T}\right)\dot{V} \tag{5.65}$$

In brief, the flux of the classical exergy of radiation satisfies the formula

$$\dot{B}_f = \dot{H}_f\left(1 - \frac{T_0}{T}\right) = \dot{S}_f(T - T_0) \tag{5.66}$$

The energy flux is expressed above as in the case of a usual substance, that is in terms of enthalpy flux, $h_v \dot{V}$, and the entropy flux equals $s_v \dot{V}$.

In conclusion, the calculation of the exergy flux of radiation is as exact as that of a substance, and the same general formulae can be applied provided that the constraint $P = P(T)$ is incorporated for radiation.

5.5. EFFICIENCIES OF ENERGY CONVERSION

It should be realized that while the first line of Equation (5.65) seems unquestionable, its second line (and hence the Carnot efficiency it contains) depends on formulae used for thermodynamic functions. In the paper by Badescu (2008), which gives new perspectives to earlier results, four statistical microscopic models are used to derive the radiation exergy and its flux. They consist of combinations of quantum and classical descriptions of the state occupation number and entropy, respectively. In all four cases the black body exergy (or exergy flux density) is given by the internal energy (or energy flux density) times an efficiency-like factor containing the ambient temperature and the black-body radiation temperature.

The enthalpy base of the energy flux is in agreement with the energy flux formula obtained as the formal component of the energy momentum tensor for isotropic radiation (Tolman, 1949). The enthalpy-based exergy efficiency is that of Carnot

$$\eta_J \equiv \frac{\dot{B}_f}{\dot{H}_f} = \frac{(4/3)a(T^4 - T^3 T_0)\dot{V}}{(4/3)aT^4 \dot{V}} = 1 - \frac{T_0}{T} \qquad (5.67)$$

in agreement with Jeter's (1981) result for the radiation conversion. This result can be compared with Petela's efficiency

$$\eta_P \equiv \frac{B}{E} = 1 - \frac{4}{3}\frac{T_0}{T} + \frac{1}{3}\left(\frac{T_0}{T}\right)^4 \qquad (5.68)$$

This expression is often called the Petela–Landsberg–Press efficiency as it was derived independently by each of these authors. It predicts efficiencies lower than Carnot. Badescu (2008), who claims that the same efficiency factor is valid for the enclosed and flowing radiation, shows that Petela–Landsberg–Press efficiency is the "exact" result while the Jeter's (Carnot) efficiency corresponds to the classical approximation under which the black-body radiation behaves like heat and exhibits the vanishing pressure. The reader is referred to Badescu (2008) for more details.

Numerous efficiency formulae are available in references (Petela, 2003; Bejan, 1982, 1987a; Wright et al., 2002; Badescu, 1998a, 1998b, 1999, 2000, 2008). In particular, Badescu claims that efficiency (5.58) takes into account "more irreversibilities" than efficiency. Spanner's (1964) efficiencies can additionally be quoted whose values lie between the values of the efficiencies of Carnot and Petela:

$$\eta_{Sp} \equiv \frac{B}{E} = 1 - \frac{4T_0}{3T} \qquad (5.69)$$

The discrepancies between various efficiencies were explained (Bejan, 1987a, 1987b). Using his procedure under the framework of endoreversible thermodynamics Badescu (1998a, 1998b, 1999) found relationships generalizing Spanner, Jeter and Petela efficiencies. For example, the equation generalizing the last efficiency is

$$\eta_B \equiv 1 - \frac{4}{4-n}\left(\frac{T_0}{T}\right)^n + \frac{n}{4-n}\left(\frac{T_0}{T}\right)^4 \qquad (5.70)$$

where n is a parameter characterizing an endoreversible thermal engine. For $n = 1$ (the case of Carnot engine) Petela's efficiency is recovered from Equation (5.70). As already stated, Petela's efficiency is lower then that of Carnot. Badescu (2004) argues that this is so because it takes into account two irreversibilities, filling the system with and emptying it of radiation. "More irreversibilities – less efficiency" is a simple rule stemming from his discussion. Yet, apart from Sieniutycz and Kuran (2005), these works ignore (the factor of) internal irreversibilities Φ, which also contribute to the efficiency decrease in practical systems.

5.6. TOWARDS A DISSIPATIVE EXERGY OF RADIATION AT FLOW

Further analysis is directed towards generalization of the radiation exergy for finite rates. As dissipative components are present in real reservoirs and within energy generators, any finite rates involve an inevitable minimum of dissipation. To define a rate-dependent exergy that extends the classical exergy for processes with dissipation a sequence of Chambadal–Novikov–Curzon–Ahlborn (CNCA) thermal machines is the basic theoretical tool. During the approach to equilibrium the so-called *engine mode* of the system takes place in which work is released; during the departure the so-called *heat-pump mode* occurs in which work is supplied. The work W delivered in the engine mode is positive by assumption ("engine convention").

It is advisable to recognize the quantities that do not vary along paths of flow processes. In sequential processes with constant cross-sectional area F that are pertinent for exergy evaluation one of the constants is the flux density of the radiation volume, J_V:

$$J_V = \frac{\dot{V}}{F} \equiv u \qquad (5.71)$$

The constancy of $J_V = u$ means that an average macroscopic velocity of the photon mixture in the direction of its flow, u, is the same for each point of the system. (For a special case of photons leaving the black box the constant value of $u = J_V$ equals $c/4$.) For sequential processes, in which cross-sectional area F perpendicular to the photon flow is a constant quantity, the constancy of the volumetric density of photons, $u = J_V$, implies the constancy of their volume flux through the stages

$$\dot{V} = FJ_V = Fu \qquad (5.72)$$

The constancy of $\dot{V}$ in a steady flow system is the counterpart of the condition of constant volume V in the enclosed system. Yet for traditional particles the particle flux is conserved along a flow; for photons this is not a case. As the photon flux satisfies the equality

$$\dot{N} = \frac{N}{V}\dot{V} \qquad (5.73)$$

and the density of black photons N/V is a function of T which varies along the flow, the flux of black photons cannot be constant in systems with a constant $\dot{V}$. Using perfect gas formula (5.30) associated with Massa's (1986) formula $e_{ph} = 3k_B T$ we evaluate variation of photon flux along the process path in terms of current temperature T and volume flow $\dot{V}$

$$\dot{N} = \frac{a}{3k_B}T^3\dot{V} = \frac{4\sigma}{3ck_B}T^3\dot{V} \qquad (5.74)$$

The corresponding flux density of photons is

$$J_N = \frac{\dot{N}}{F} = \frac{4\sigma}{3ck_B F}T^3\dot{V} \qquad (5.75)$$

For constant $\dot{V}$ these formulae ensure a decrease of $\dot{N}$ and J_N in the engine mode of the process when T decreases along the path and energy is delivered from the radiation engine. On the other hand, the formulae ensure an increase of N and J_N when energy is consumed in a "radiation-utilizing" heat-pump mode when radiation temperature T increases along the path. For constant volume flux, $\dot{V}$, both N and $\dot{N}$ depend on T in the same way, and Equation (5.73) implies constancy of photons volume along the path of the considered process.

In the irreversible finite-rate situations quasistaticity is lost and any extension of exergy to irreversible situations is nontrivial. Because of finite rates instantaneous efficiencies η are different from those of Carnot at each time instant. Therefore, before any formulation of a work integral, prior evaluation of a proper efficiency η should be made. In optimization approaches based on the variational calculus, η has to be evaluated as a function of state T and rate $dT/d\tau$, to assure the functional property (path dependence) of a work integral. As an exergy is a limiting work, its evaluation must be associated with the optimization that maximizes work W and assures an optimal path. The optimal work follows in the form of a potential function that depends on the end states and duration. This function is a finite-rate exergy when the final state of engine mode is that of equilibrium with the environment. Another function, also exergy type, is obtained when the initial state of heat-pump mode is that of equilibrium with the environment. While the reversibility property is lost for both these exergies, their kinetic bounds are stronger and hence more useful than classical thermostatic bounds. This substantiates the role of extended exergies in evaluation of energy limits in practical systems.

For the exergy evaluation the finiteness of the resource (radiation fluid) is essential; this makes the power production process unsteady in time. The analysis of a single CNCA unit is insufficient in this case, rather the treatment of a complex sequential system with (finite or infinite number of) CNCA units is necessary. The work-production process involves the active energy exchange between two fluids through finite "conductances" (products of the effective transfer coefficient and the area). In the case of the radiation exergy the first fluid in Figure 5.1 is the radiation fluid. As follows from the Stefan–Boltzmann law, the transfer coefficient of radiation fluid α_1 is necessarily temperature dependent, $\alpha_1 = 4\sigma\varepsilon T_1^3$. The second fluid is a low-temperature fluid being either a low-temperature radiation or the usual environment composed of the common substances of the Earth, as defined in the exergy theory. The second fluid possesses a boundary layer as its own dissipative component, so that the corresponding exchange coefficient is α_2. (The second coefficient of the energy exchange, α_2, can also be temperature-dependent.) In the physical space, the direction of the radiation flow is along a horizontal coordinate x.

However, the use of transfer coefficients α_i is unnecessary in radiation problems. In fact, functions $\alpha_1 = \alpha_{10} T_1^3$ and $\alpha_2 = \alpha_{20} T_2^3$ are applied only in the so-called pseudo-Newtonian approach when, by assumption, a temperature dependent conductance is attributed to the driving force defined as the simple temperature difference. As we shall see soon, the virtue of the pseudo-Newtonian approach is its potential of getting an analytical solution under an approximation of an overall coefficient for energy transfer. While this takes temperature dependence into account, more exact approaches to the energy flux are preferred that involve differences of temperature T in power $m = 4$, that is use the Stefan–Boltzmann law. Yet, these approaches do not lead to analytical solutions, so that numerical optimization techniques must be developed. They are briefly characterized below (Kuran, 2006).

We begin with considerations of the symmetric nonlinear case in which the energy transfer rate is proportional to the difference of absolute temperatures in power m. The case of $m = 4$ refers to the radiation, $m = -1$ to the Onsagerian kinetics and $m = 1$ to the Fourier law of heat exchange. (In the Onsagerian case quantities g_i are negative in the common formalism.)

Next we adduce the "hybrid nonlinear case" in which kinetics in the lower reservoir is Newtonian. The upper-temperature fluid is still governed by the kinetics proportional to the difference in T^m. Still other cases are possible, as for example, the case with the environmental kinetics of the natural convection (where $q \propto (\Delta T)^m$), and some "mixed" cases. Consequently, a variety of physical models and related optimization algorithms can be applied, each model leading to its own generalized exergy function.

5.7. BASIC ANALYTICAL FORMULAE OF STEADY PSEUDO-NEWTONIAN MODEL

First we focus on a single infinitesimal CNCA engine as a one-stage component of the sequential system in Figure 5.1. Next, an analysis is developed to model cumulative power output (input) from (to) an infinite number of infinitesimal steps that model thermal behavior and exergy of the sequential system at the continuous limit. As the theory of the elementary CNCA process is well known, we briefly present here its counterpart called the Stefan–Boltzmann engine under pseudo-Newtonian approximation and then pass to main formulae associated with cumulative power and entropy production of the sequential system at its continuous limit.

A single engine in Figure 5.1 depicts an infinitesimal stage of the system. The location of this stage in the system is between x and $x + dx$, where x is the geometric coordinate in the direction of the radiation flow. In a steady situation the state changes of the fluid in the differential engine (two isotherms and two irreversible adiabates) are stationary loops in the space. The radiation fluid (subscript 1) flows as a whole in the direction of the axis x with a finite volume flux, $\dot{V}$. The unknowns $T_{1'}$ and $T_{2'}$ (explicit in Figure 5.1) are upper and lower temperatures of the fluid circulating in each engine. Between x and $x + dx$ is located the circulation loop of each small engine and the differential conductances $g_1 = d\gamma_1$ and $g_2 = d\gamma_2$, where $d\gamma_1 = \alpha_1(T_1) dA_1$ and $d\gamma_2 = \alpha_2(T_2) dA_2$. The differentials dA_1 and dA_2 are two exchange areas at the infinitesimal stage. They are components of the composite area A whose differential satisfies $dA = dA_1 + dA_2$. The overall conductance γ is defined in terms of $g_1 = d\gamma_1$ and $g_2 = d\gamma_2$ in the traditional way; $(d\gamma)^{-1} = (d\gamma_1)^{-1} + (d\gamma_2)^{-1}$. Consequently, $d\gamma$ is the product of an overall coefficient of heat transfer, α', and total differential area dA.

The differential flux, $d\dot{Q}_1$, is the energy flux subtracted from the radiation fluid when its state changes from T_1 to $T_1 + dT_1$. The radiative energy exchange $d\dot{Q}_1$ occurs by the emission and adsorption of radiation in the temperature range T_1 and $T_{1'}$. In the pseudo-Newtonian modeling the (nonNewtonian) flux of the exchanged radiation energy is the product of the variable coefficient $g_1 = d\gamma_1 = \alpha_1(T_1^3)\,dA$ and the temperature difference $T_1 - T_{1'}$. When the exact Stefan–Boltzmann law is used its energy exchange model rigorously takes into account entropy generation caused by simultaneous emission and absorption of black-body radiation. The entropy generation is the "classical" sum: $d\dot{Q}_1(T_{1'}^{-1} - T_1^{-1}) + d\dot{Q}_2(T_2^{-1} - T_{2'}^{-1})$, where $d\dot{Q}_1$ is given by the Stefan–Boltzmann law.

The low-T part of the engine releases the heat proportional to $T_{2'} - T_0$ to the environment (or fluid 2) through conductance $d\gamma_2$ that also may be temperature dependent.

For an imperfect engine the first-law efficiency of each infinitesimal unit is given by the pseudo-Carnot formula, $\eta = 1 - \Phi T_{2'}/T_{1'}$, where $\Phi = \Delta S_{2'}/\Delta S_{1'} = J_{S_{2'}}/J_{S_{1'}}$ is the parameter of internal irreversibility. $T_{1'}$ and $T_{2'}$ are not independent but connected by entropy balance

$$\frac{g_2(T_2)(T_{2'} - T_2)}{T_{2'}} - \Phi\frac{g_1(T_1)(T_1 - T_{1'})}{T_{1'}} = 0 \tag{5.76}$$

We invert pseudo-Carnot formula, $\eta = 1 - \Phi T_{2'}/T_{1'}$, to substitute $T_{2'} = (1 - \eta)\Phi^{-1}T_{1'}$ into the above entropy balance. We then obtain an equation for $T_{1'}$

$$T_{1'} = (g_1 + \Phi^{-1}g_2)^{-1}(g_1 T_1 + (1 - \eta)^{-1}g_2 T_2) \tag{5.77}$$

With this result the flux of the radiation energy,

$$q_1 = d\dot{Q}_1 \cong g_1(T_1^3)(T_1 - T_{1'}) \tag{5.78}$$

exchanged by simultaneous emission and adsorption follows in the pseudo-Newtonian formalism as

$$q_1 = g'\left(T_1 - \frac{\Phi T_2}{1 - \eta}\right) \tag{5.79}$$

In Equation (5.79) operational overall conductance was defined with all g_i as functions of respective bulk temperatures of reservoirs

$$g' \equiv g_2 g_1(\Phi g_1 + g_2)^{-1} = (g_1^{-1} + \Phi g_2^{-1})^{-1} \tag{5.80}$$

In fact, this is an overall conductance of the usual heat transfer that is suitably modified by the coefficient of internal dissipation. The work flux (power) follows in the form

$$p = \eta q_1 = \eta g'(\Phi, T_1, T_2) \left(T_1 - \frac{\Phi T_2}{1 - \eta} \right) \tag{5.81}$$

The expression

$$T' = \Phi T_2 (1 - \eta)^{-1} \tag{5.82}$$

appearing in Equations (5.79) and (5.81) describes the so-called Carnot temperature in terms of the efficiency. (We recall the thermodynamic definition of Carnot temperature, $T' \equiv T_2 T_{1'}/T_{2'}$.) We note that despite temperature-dependent conductances g_1 and g_2, Equation (5.81) yields the same maximum power efficiency $\eta_{mp} = 1 - \sqrt{\Phi T_2/T_1}$ as for engines driven by usual fluids where heat flux is governed by (linear) Newtonian law of cooling.

The differential of total entropy produced has a kinetics-independent form

$$dS_\sigma = \frac{dQ_1}{T_2} \left(\Phi \frac{T_{2'}}{T_{1'}} - \frac{T_2}{T_1} \right) = dQ_1 \left(\frac{\Phi - 1}{T'} + \left(\frac{1}{T'} - \frac{1}{T_1} \right) \right) \tag{5.83}$$

In a pseudo-Newtonian process conductances g_i are only functions of temperatures of respective reservoirs. This is an approximation as, in fact, they are also influenced by temperatures of the circulating fluid. A way to improve this situation is described below.

5.8.   STEADY NONLINEAR MODELS APPLYING STEFAN–BOLTZMANN EQUATION

In a more exact modeling of radiation engines we abandon the transfer coefficients and exploit the Stefan–Boltzmann equation for energy transfer in its exact form.

In the *symmetric nonlinear case* (Sieniutycz and Kuran, 2005) we assume that the energy fluxes in both reservoirs depend on the difference of temperatures in the same power m

$$q_1 = g_1(T_1^m - T_{1'}^m) \tag{5.84}$$

($m = 4$ for radiative energy exchange and 1 for the Newtonian one). Conductances g_i are now constants, different from those in the previous section, yet, as usual, they are proportional to the areas $g_i = \sigma_i A_i$, where each σ_i is the product of Stefan–Boltzmann constant σ and emission coefficient ε_i. We can still use the notion of Carnot temperature $T' \equiv T_1 T_2/T_{2'}$ as a suitable control variable. From this definition temperature $T_{2'}$ satisfies the inverse formula

$T_{2'} \equiv T_{1'}T_2/T'$. Substituting this formula into the internal balance equation of entropy

$$\Phi g_1 \frac{T_1^m - T_{1'}^m}{T_{1'}} = g_2 \frac{T_2^m - T_{2'}^m}{T_{2'}} \tag{5.85}$$

and solving the result obtained with respect to $T_{1'}$ yields

$$T_{1'} = \left(T_1^m - g_2 \frac{T_1^m - T'^m}{\Phi g_1 (T'/T_2)^{m-1} + g_2} \right)^{1/m} \tag{5.86}$$

From Equation (5.86) and Equation (5.44) written in the form

$$T_{1'} = \left(T_1^m - \frac{q_1}{g_1} \right)^{1/m} \tag{5.84'}$$

energy flux $q_1(T')$ follows:

$$q_1 = g_1 g_2 \frac{T_1^m - T'^m}{\Phi g_1 (T'/T_2)^{m-1} + g_2} \tag{5.87}$$

This formula represents "thermal characteristics" of the system. An expression for $T_{2'}$ corresponding with Equation (5.86) follows from the thermodynamic definition of the Carnot variable, whence $T_{2'} \equiv T_{1'}T_2/T'$. Also, one may calculate heat flux $q_2 = q_1(1 - \eta) = \Phi T_2 q_1/T'$. The power yield related to Equation (5.87) is

$$p = g_1 g_2 \frac{T_1^m - T'^m}{\Phi g_1 (T'/T_2)^{m-1} + g_2} \left(1 - \Phi \frac{T_2}{T'} \right) \tag{5.88}$$

The maximization of p can be performed analytically or graphically, using Carnot temperature T' as a free control.

The entropy generation caused by simultaneous emission and absorption of black-body radiation is the external part of the total entropy production that follows as "classical" sum:

$$\sigma_s^{ext} = q_1(T_{1'}^{-1} - T_1^{-1}) + q_2(T_2^{-1} - T_{2'}^{-1}) \tag{5.89}$$

where each q_i incorporates the Stefan–Boltzmann law. Yet, this is only a part of the entropy production in the system. For the "symmetric" kinetics governed by the differences in T^m the T'-representation of the total entropy production follows from Equations (5.83) and (5.87):

$$\sigma_s = g_1 g_2 \frac{T_1^m - T'^m}{\Phi g_1 (T'/T_2)^{m-1} + g_2} \left(\frac{\Phi - 1}{T'} + \left(\frac{1}{T'} - \frac{1}{T_1} \right) \right) \qquad (5.90)$$

In this model no explicit formula exists for mechanical power or entropy production in terms of energy flux, q_1. A hybrid model, considered below, offers such an opportunity.

In the *nonsymmetric or hybrid nonlinear case* (Sieniutycz and Kuran, 2005) the radiative exchange ($m = 4$) takes place in the upper reservoir only, whereas the energy exchange in the lower one is Newtonian:

$$q_2 = g_2 (T_{2'} - T_2) \qquad (5.91)$$

As before, to get $T_{1'}$ in terms of T' we substitute the expression $T_{2'} \equiv T_{1'} T_2 / T'$ into the suitable (internal) balance equation for the entropy

$$\Phi g_1 \frac{T_1^m - T_{1'}^m}{T_{1'}} = g_2 \frac{T_{2'} - T_2}{T_{2'}} \qquad (5.92)$$

Yet the procedure leads now to T' explicit in terms of $T_{1'}$ rather than $T_{1'}$ in terms of T'

$$T' = T_{1'} - \Phi g_1 \frac{T_1^m - T_{1'}^m}{g_2} \qquad (5.93)$$

The mechanical power p in terms of $T_{1'}$ is

$$p = q_1 \eta = g_1 (T_1^m - T_{1'}^m) \left(1 - \frac{\Phi T_2}{T_{1'} - \Phi g_1 (T_1^m - T_{1'}^m)/g_2} \right) \qquad (5.94)$$

The large bracket of this formula contains the pseudo-Carnot efficiency expressed in terms of $T_{1'}$ rather than in T'. With Equation (5.84), the energy flux representation of power (Equation (5.94)) follows:

$$p = q_1 \eta = q_1 \left(1 - \frac{\Phi T_2}{(T_1^m - q_1/g_1)^{1/m} - \Phi q_1/g_2} \right) \qquad (5.95)$$

The corresponding entropy production satisfies Equation (5.83) with T' defined by Equation (5.93). Hence the power of entropy generation in terms of temperature $T_{1'}$ is

$$\sigma_s = g_1 (T_1^m - T_{1'}^m) \left(\frac{\Phi}{T_{1'} - \Phi g_1 (T_1^m - T_{1'}^m)/g_2} - \frac{1}{T_1} \right) \qquad (5.96)$$

For the upper reservoir $T_{1'} = (T_m^1 - q_1/g_1)^{1/m}$, and Equation (5.93) yields the following expression for Carnot temperature T' in terms of q_1:

$$T' = T_{1'} - \Phi g_1 \frac{T_1^m - T_{1'}^m}{g_2} = \left(T_1^m - \frac{q_1}{g_1}\right)^{1/m} - \Phi \frac{q_1}{g_1} \tag{5.97}$$

Equations (5.96) and (5.97) now lead to the representation of the entropy source in terms of q_1:

$$\sigma_s = q_1 \left(\frac{\Phi}{(T_1^m - q_1/g_1)^{1/m} - \Phi q_1/g_2} - \frac{1}{T_1} \right) \tag{5.98}$$

Equation (5.88), (5.94), or (5.95) allows analytical or graphical maximization of power with respect to a single control variable, T', $T_{1'}$ or q_1. Because of the reversible component of power that persists for vanishing rates, these equations have nontrivial optimal solutions even in the absence of constraints on the rates. Optimization leads to the steady limits on power production in imperfect units. Yet Equations (5.90), (5.96), and (5.98) have nontrivial optimization solutions for entropy source only when their control variables are constrained. Otherwise they imply vanishing σ_s at the reversible Carnot point as unconstrained minimum of σ_s. In dynamical problems below, constraints result from the energy balance.

5.9. DYNAMICAL THEORY FOR PSEUDO-NEWTONIAN MODELS

In dynamical problems the temperature of at least one reservoir changes (decreases in engine mode) due to the reservoir's finite capacity. This is the case with a finite resource, appropriate to define an exergy. By integration of power expressions in the previous section, functionals of power generation (consumption) and related exergies are obtained.

Our exergy-directed analysis extends those previous ones by considering the sequential operation with internal irreversibilities (within thermal machines of each stage). The factor of internal irreversibilities, Φ, satisfies inequality $\Phi > 1$ for engine mode and $\Phi < 1$ for heat-pump mode of the system. In terms of Φ suitable formulae follow for generalized work and exergy. Use of the substitutional quantity $c_h(T)$, Equation (5.60), leads to enthalpy density of photons and allows us to overcome difficulties resulting from an infinite value of c_p of the photon gas.

As already remarked, several physical models can be applied, each leading to its own generalized exergy. Below we shall consider these models in order. First we focus on the pseudo-Newtonian model and corresponding exergies for two modes considered.

We begin with the energy exchange formula written in terms of quantities cumulative along the process path

$$d\dot{Q}_1 \equiv d\gamma'\left(T_1 - \Phi\frac{T_2}{1-\eta}\right) \tag{5.99}$$

Its inversion yields the first-law efficiency of the imperfect process in the form

$$\eta = 1 - \Phi\frac{T_0}{T - d\dot{Q}_1/d\gamma'} = 1 - \Phi\frac{T_0}{T - d\dot{Q}_1/(\alpha'\,dA)} \tag{5.100}$$

where T stands for any T_1 on the path. The derivative term $v = -d_1/d\gamma'$ is a control with $\dot{Q}$ units of temperature itself. It may be written in the form of several alternative expressions

$$\frac{d\dot{Q}}{d\gamma'} \equiv -v = -\dot{V}c_h(T)\frac{dT}{\alpha'(T)a_v F\,dx} = -c_h(T)\frac{dT}{\alpha'(T)a_v\,dt}$$

$$= -\chi\frac{dT}{dt} = -\frac{dT}{d\tau} \tag{5.101}$$

of which the two last ones are the most suitable. In Equation (5.101) $\chi = c_h/(\alpha'a_v)$ is a time constant for the energy exchange process. Two other useful quantities can also be selected in Equation (5.101). The first one is a spatial scale for the overall transfer, H_{TU},

$$\frac{\dot{V}c_v}{\alpha'a_v F} = H_{TU}$$

whereas the second is a nondimensional time, τ,

$$\tau \equiv \frac{x}{H_{TU}} = \frac{\alpha'a_v F}{\dot{V}c_v}x$$

H_{TU} has the units of length and is known as the "height of the heat transfer unit". By definition H_{TU} introduced above refers to the radiation fluid at state 1. The independent variable τ is a nondimensional length, $\tau = x/H_{TU}$ called the "number of transfer units." Clearly τ measures the system extent, and is a measure of the fluid's holdup time t. Due to the similar type of dependence of α' and c_h on T, the time constant $\chi \equiv c_s/(\alpha'a_v)$ linking t and τ is practically temperature independent. This substantiates the usefulness of τ.

For ignored thermal resistance of ambient fluid (entropy production caused only by emission and adsorption of radiation), overall coefficient of radiation energy transfer, α', varies proportionally to T^3, and so does c_h. In this limiting case χ is exactly constant, in other cases its constancy is only an approximation. Assuming a nondissipative environment one can admit constancy of χ in the last two expressions of Equation (5.101) as a suitable property. With Equation (5.101) efficiency formula (5.100) becomes a simple modification of the Carnot formula:

$$\eta = 1 - \Phi \frac{T_0}{T + \chi \, dT/dt} = 1 - \Phi \frac{T_0}{T + dT/d\tau} \tag{5.102}$$

Yet, this result is not as universal as its quasistatic (zero rate) limit; in fact its denominator contains Carnot temperature operator $T'(T, \dot{T})$ in the form restricted to pseudo-Newtonian models. Primarily, mostly classical fluids satisfy Equation (5.102); its applicability to radiation is due to the approximate constancy of χ, discussed above.

The dynamical process is the passage of the vector $\mathbf{T} = (T, \tau)$ from its initial state $\mathbf{T}^i$ to its final state $\mathbf{T}^f$. In the absence of frictional effects the power functional corresponding to efficiency (5.102) is the following generalization of reversible functional (5.59):

$$\dot{W}_f = -\dot{V} \int_T^{T_0} \left(c_v(T) \left(1 - \Phi \frac{T_0}{T + \chi \, dT/dt} \right) + P_T \right) dT$$
$$= -\dot{V} \int_T^{T_0} \left(c_h(T) - c_v(T) \Phi \frac{T_0}{T + \chi \, dT/dt} \right) dT \tag{5.103}$$

where $P_T \equiv dP/dT = (4/3)aT^3$ and $c_h(T) \equiv c_v(T) + P_T + (16/3)aT^3$. A more transparent form of the above power integral is obtained after transforming it so as to extract from it the effect of the reversible power (5.59) and the associated efficiency term. For the pseudo-Newtonian model we obtain

$$\dot{W}_f = -\dot{V} \int_T^{T_0} \left(c_h(T) - c_v(T) \frac{T_0}{T} \right) dT$$
$$- T_0 \dot{V} \int_T^{T_0} \left(c_v(T) \left(\frac{\chi (dT/dt)^2}{T(T + \chi \, dT/dt)} + (1 - \Phi) \frac{dT/dt}{T + \chi \, dT/dt} \right) \right) dT \tag{5.104}$$

Associated entropy production per unit flowing volume can be evaluated as the difference between the outlet and inlet entropy fluxes. In terms of Carnot temperature $T' = T_1 + \chi \, dT_1/dt$ and after using $q_2 = q_1 \Phi T_0/T'$ we find for the pseudo-Newtonian model

$$\sigma_s = \frac{q_2}{T_2} - \frac{q_1}{T_1} = \frac{g'(T_1 - T')^2}{T_1 T'} + q_1 \frac{\Phi - 1}{T'} \tag{5.105}$$

This is consistent with general Equation (5.83) and the model-related Equation (5.99). Comparison of Equations (5.104) and (5.105) shows that the term multiplied by T_0 in the power expression (5.104) is the entropy production of the pseudo-Newtonian model. It is here split into the sum of two nonnegative terms. The first term, related to the approximate description of the effect of emission and adsorption of radiation, is obviously positive. The second term, or the product $q_1(\Phi - 1)/T'$, is always nonnegative as the signs of q_1 and $\Phi - 1$ are the same (positive for engine and negative for heat-pump). A concise form of power functional (5.104)

$$\dot{W}_f = -\dot{V} \int_T^{T_0} \left(c_h(T) - c_v(T)\frac{T_0}{T} \right) dT - T_0 \dot{V} \int_T^{T_0} \sigma_v \, dt \qquad (5.106)$$

(for photons $c_v(T) = 4aT^3$) is in agreement with the Gouy–Stodola law. This form is quite general and not restricted to the pseudo-Newtonian model. However, the (first) reversible term of this equation apparently shows a disagreement between the resulting, reversible efficiency and the Carnot efficiency. Therefore we stress that the reversible *thermal* efficiency of the radiation conversion is always Carnot. Indeed, comparison of Equations (5.58) and (5.59) shows that the apparent disagreement is caused by the additive, work-related term $P_T \equiv dP/dT$ in the exergy formulae, Equations (5.59) and (5.103) and the like. For radiation fluids, the pressure contribution to the exergy in the form of term $P_T \equiv dP/dT$ is masked by the dependence of P on T.

Integration of the first (reversible) part of integral (5.106) and calculation of $\dot{W}_f^{rev}/\dot{V}$ yields the classical exergy of flowing radiation fluid per unit volume, Equations (5.58)–(5.63). When the environment temperature is not necessarily an integration limit, the specific work of flowing radiation between two arbitrary states is obtained as the exergy difference. From Equation (5.61),

$$\frac{\dot{W}_f^{rev}}{\dot{V}} = -\int_{T^i}^{T^f} \left\{ \left(\frac{16}{3}\right) aT^3 - 4aT^2T_0 \right\} dT = (h_v^i - h_v^f) - T_0(s_v^i - s_v^f) = \Delta b_v$$

$$(5.107)$$

These results are in agreement with general thermodynamics. They confirm that the first term of power functional (5.104) is path-independent. Thus, whenever a power extremum is sought, only the second, irreversible term contributes to the optimization solution.

In terms of nondimensional time $\tau = t/\chi$ and the per unit system volume entropy production functional of the pseudo-Newtonian model, Equation (5.104), is

$$s_{v_{gen}} \equiv \frac{\dot{S}_{gen}}{\dot{V}} = \int_T^{T_0} \left(c_v(T) \left(\frac{\dot{T}^2}{T(T+\dot{T})} + (1 - \Phi)\frac{\dot{T}}{T+\dot{T}} \right) \right) d\tau \qquad (5.108)$$

The additive structure of two parts in Equation (5.104) is an important property that causes the two problems of extremum work (5.104) and the associated problem of minimum entropy generation (5.108) to have the same solutions whenever end states are fixed. The optimization problem can thus be stated as a variational problem for either functional of work or of entropy production. When work W is an optimization criterion, the problem is that of maximum W for engine mode and of minimum $(-W)$ for heat-pump mode. When optimization of the entropy production is considered, a minimum is sought for each process mode. A generalized exergy is an extremum of W with appropriate integration limits ($T^i = T$ and $T^f = T_0$ for engine mode and $T^i = T_0$ and $T^f = T$ for heat-pump mode). In the quasistatic limit (zero rates, $\dot{T} = 0$), Equation (5.104), always leads to *classical exergy*. Moreover, it leads to the same classical exergy for each mode when proper integration limits, stated above, are used. The absolute value of work (5.104) describes a change of generalized exergy of radiation in operations with imperfect thermal machines and when dissipative phenomena due to the radiation emission and adsorption are essential.

We focus here on the minimum entropy production formulation for functional (5.108). An equation for the optimal temperature follows from the condition $\varepsilon = h$, where $\varepsilon = (\partial L / \partial \dot{T}) \dot{T} - L$ is the energy-like integral for Lagrangian L contained in equations of power or entropy production, and h is a constant value of ε determined from the boundary conditions for T and τ. The present h has units of entropy density or specific heat per unit volume, and should be distinguished from Hamiltonians used occasionally with respect to the energy. Our $h = H/VT_0$, where H is the Hamiltonian expressed in the energy units and V is the volume. For any rate independent Φ the first integral for L of Equation (5.108) is

$$\varepsilon(T, \dot{T}) = \frac{\partial L}{\partial \dot{T}} \dot{T} - L = \Phi c_v(T) \frac{\dot{T}^2}{(T + \dot{T})^2} = h \qquad (5.109)$$

We obtain an optimal trajectory from Equation (5.99). After introducing the function

$$\xi\left(\frac{h}{\Phi c_v(T)}\right) \equiv \pm \sqrt{\frac{h}{\Phi c_v(T)}} \left(1 - \pm\sqrt{\frac{h}{\Phi c_v(T)}}\right)^{-1}$$
$$= \left(\pm\sqrt{\frac{\Phi c_v(T)}{h}} - 1\right)^{-1} \qquad (5.110)$$

(upper sign refers to the heat-pump mode, lower one to the engine mode) a pseudo-exponential extremal follows in the form

$$\dot{T} = \xi(h, \Phi, T)T \qquad (5.111)$$

Here the slope of the logarithmic rate $\xi = \mathrm{d}\ln T/\mathrm{d}\tau$ is a state dependent quantity. The slope ξ is the rate indicator, positive for the fluid's heating and negative for the fluid's cooling.

Application of extremal (5.111) in Equation (5.108) leads to the minimum entropy production in the form

$$
s_{v_{\text{gen}}} = \int_{T^i}^{T^f} \frac{c_v(T)}{T} \left(\pm\sqrt{\frac{h}{\Phi c_v(T)}} + (1 - \Phi)\left(1 - \pm\sqrt{\frac{h}{\Phi c_v(T)}}\right) \right) \, dT
$$

$$(5.112)$$

With this result and the Gouy–Stodola law we obtain the density of generalized exergy for the fluid at flow

$$
a_v(T, T_0, h) = b_v(T, T_0, 0) \pm T_0 s_{v_{\text{gen}}} = \frac{4}{3}aT^4\left(1 - \frac{T_0}{T}\right)
$$

$$
\pm T_0 \int_T^{T_0} \frac{c_v(T)}{T} \left(\pm\sqrt{\frac{h}{\Phi c_v(T)}} + (1 - \Phi)\left(1 - \pm\sqrt{\frac{h}{\Phi c_v(T)}}\right) \right) \, dT
$$

$$(5.113)$$

where the second-line term is nonclassical. For radiation $c_v = 4aT^3$, the expression $h = (4/3)aT^4$ is the enthalpy of the radiation. The classical term in the above exergy equation is the reversible flow exergy of black radiation per unit volume which is recovered at the reversible limit when Hamiltonian $h = 0$. This classical exergy satisfies the formula

$$
b_v \equiv \frac{\dot{B}_f}{\dot{V}} = h_v - h_{v0} - T_0(s_v - s_{v0}) \tag{5.114}
$$

consistent with Equation (5.65) for the radiation fluid. For that fluid the optimal trajectory which solves Equations (5.110) and (5.111) is

$$
\pm\frac{4}{3}a^{1/2}\Phi^{1/2}h^{-1/2}(T^{3/2} - T^{i3/2}) - \ln\frac{T}{T^i} = \tau - \tau^i \tag{5.115}
$$

The integration limits refer to initial (i) and current state (no index) of the radiation fluid, that is to temperatures T^i and T, corresponding with τ^i and τ. Figure 5.2 shows an example of optimal paths for radiation in both process modes. Relaxation to the equilibrium occurs in engine mode whereas utilization – or escape from the equilibrium – occurs in heat-pump mode. Clearly, radiation does not relax exponentially. A qualitative difference in relaxation curves from those describing exponential relaxation in linear processes is observed.

Equation (5.115) also allows curves to be drawn describing nondimensional durations $t^f - \tau^i$ in terms of Hamiltonian h for various internal irreversibilities

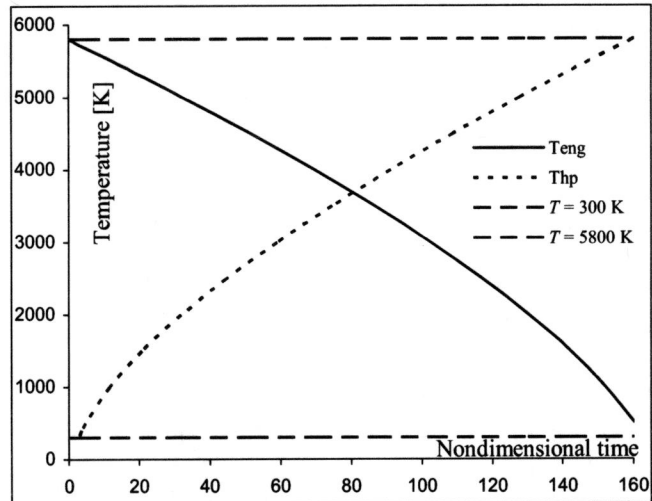

Figure 5.2 Decreasing temperature of radiation relaxing in engine mode and increasing temperature of radiation utilized in heat-pump mode in terms of nondimensional time, for a constant value of Hamiltonian $h = 1 \times 10^{-8}$ ($J K^{-1} m^{-3}$), and the same coefficient $\phi = 1,0$ in both modes ("endoreversible" relaxation and utilization of radiation).

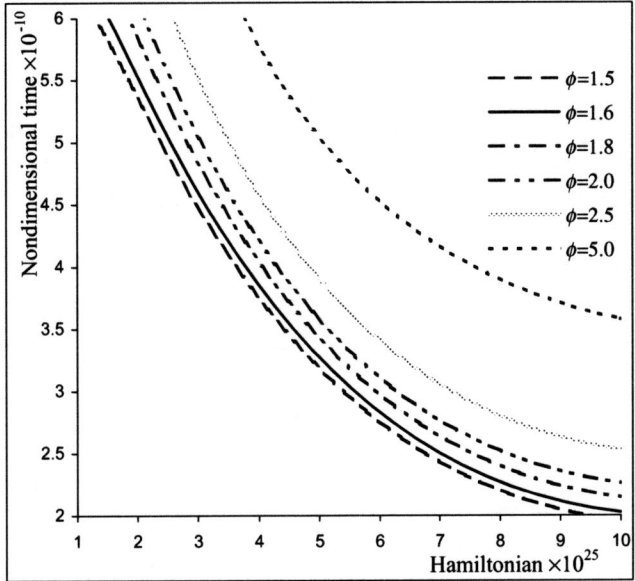

Figure 5.3 Nondimensional duration of engine mode in terms of Hamiltonian h ($J K^{-1} m^{-3}$) as an intensity index, for various values of coefficient of internal reversibilities ϕ, at prescribed boundary temperatures, $T^{i} = 300 \, K$ and $T^{f} = 5800 \, K$.

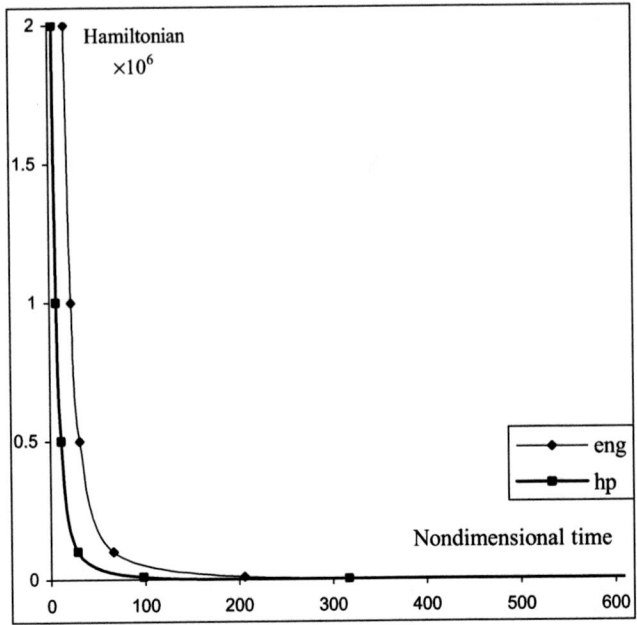

Figure 5.4 Hamiltonian h ($\mathrm{J\,K^{-1}\,m^{-3}}$) as a function of nondimensional duration of engine mode (h_{eng}) or heat-pump mode (h_{hp}) at prescribed boundary temperatures, $T^{\mathrm{i}} = 300\,\mathrm{K}$ and $T^{\mathrm{f}} = 5800\,\mathrm{K}$.

ϕ and at fixed end temperatures T^{i} and $T = T^{\mathrm{f}}$. This is illustrated in Figure 5.3, for $T^{\mathrm{i}} = 300\,\mathrm{K}$ and $T^{\mathrm{f}} = 5800\,\mathrm{K}$.

After inverting these data to obtain curves for which duration $\tau^{\mathrm{f}} - \tau^{\mathrm{i}}$ is an independent variable one can solve the problem of the numerical value of Hamiltonian needed for a prescribed duration, as shown in Figure 5.4.

Equations (5.114) and (5.115) are associated with the entropy production (5.108) and the generalized availability of radiation

$$
\begin{aligned}
a_v(T, T_0, h) = b_v(T, T_0, 0) &\pm \frac{4}{3} a^{1/2} h^{1/2} \Phi^{1/2} T_0 (T^{3/2} - T_0^{3/2}) \\
&+ \frac{4}{3} a T_0 (1 - \Phi)(T^3 - T_0^3)
\end{aligned}
\tag{5.116}
$$

The classical availability of *radiation at flow* resides in the above equation in Jeter's form

$$
\begin{aligned}
b_v(T, T_0, 0) = h_v - h_{v0} - T_0(s_v - s_{v0}) &= h_v \left(1 - \frac{T}{T_0} \right) \\
&= \frac{4}{3} a T^4 \left(1 - \frac{T}{T_0} \right)
\end{aligned}
\tag{5.117}
$$

(see Equation (5.65) and Jeter's (1981) result).

As two modes are included, the common symbol T in generalized availability (5.116) refers to the initial temperature of engine mode or the final temperature of heat-pump mode. The qualitative properties of this availability function with respect to nondimensional duration or the overall number of heat transfer units $\Delta\tau$ (cf. Figure 2 in Sieniutycz and Kuran, 2005) are similar to those described on Figure 6.3 of the next chapter.

The slope ξ is constant in traditional Newtonian fluids with a constant c_v. ξ is positive for a fluid's heating (heat-pump mode) and negative for a fluid's cooling (engine mode). The numerical value of Newtonian ξ characterizes a constant logarithmic intensity satisfying the dynamical equation for the temperature logarithm

$$\mathrm{d}\ln T = \xi\,\mathrm{d}\tau$$

Its integration for the fixed-end boundary conditions leads to exponential dynamics

$$T(\tau) = T^{\mathrm{i}}\exp\left(\xi(\tau - \tau^{\mathrm{i}})\right) \tag{5.118}$$

The associated Carnot temperature control ensuring extremum work is

$$T'(\tau) = T(\tau)(1 + \xi(\Phi)) = T^{\mathrm{i}}\left(\frac{T^{\mathrm{f}}}{T^{\mathrm{i}}}\right)^{(\tau-\tau^{\mathrm{i}})/(\tau^{\mathrm{f}}-\tau^{\mathrm{i}})}\left(1 + \ln\frac{T^{\mathrm{f}}/T^{\mathrm{i}}}{\tau^{\mathrm{f}} - \tau^{\mathrm{i}}}\right) \tag{5.119}$$

It corresponds with the power formula (5.95) in the case $m = 1$. The exponential decay of T, implied by Equation (5.118) for the relaxation in the Newtonian process (negative ξ), can be compared with the temperature decrease in the radiation relaxation process described by Equation (5.115). ξ can be determined from the boundary conditions of the fixed-end problem

$$\xi = (\tau^{\mathrm{f}} - \tau^{\mathrm{i}})^{-1}\ln\frac{T^{\mathrm{f}}}{T^{\mathrm{i}}} \tag{5.120}$$

In the Newtonian process the density of entropy generated is described by the formula

$$s_{v\mathrm{gen}} = c_v\left(\left(\pm\sqrt{\frac{h}{\Phi c_v}}\right) + (1 - \Phi)\left(1 - \pm\sqrt{\frac{h}{\Phi c_v}}\right)\right)\ln\frac{T^{\mathrm{f}}}{T^{\mathrm{i}}} \tag{5.121}$$

For endoreversible processes ($\Phi = 1$), we recover from the above equation the special formula

$$s_{v_{\text{gen}}} = \pm c_v \int_{T^i}^{T^f} \sqrt{\frac{h}{c_v}} \, \mathrm{d} \ln T = \pm c_v \sqrt{\frac{h}{c_v}} \ln \frac{T^f}{T^i} \qquad (5.121')$$

This result is exploited to obtain the generalized exergy of a compressible Newtonian fluid in which viscous friction is ignored

$$b_v(T, T_0, h) = b_v(T, T_0, 0) + c_{p_v} T_0 \left(\frac{\xi}{1 + \xi} \right) \ln \frac{T}{T_0}$$

$$= c_{p_v} T_0 \left[\left(\frac{T}{T_0} - 1 \right) - \ln \frac{T}{T_0} + \ln \left(\frac{P}{P_0} \right)^{(k-1)/k} \right]$$

$$+ c_{p_v} T_0 \left(\left(\pm \sqrt{\frac{h}{\Phi c_{p_v}}} \right) + (1 - \Phi) \left(1 - \pm \sqrt{\frac{h}{\Phi c_{p_v}}} \right) \right) \ln \frac{T}{T_0}$$

$$(5.122)$$

The last line describes the rate-related term. In the classical case the pressure contributes to the exergy with a separate term. This is not a surprise because, in this case, T and P are two independent variables, whereas in the case of radiation the specification of T already defines the pressure P.

5.10. DYNAMICAL MODELS USING THE STEFAN–BOLTZMANN EQUATION

Again, the optimization task is to find an optimal profile of the driving temperature T' along the path of the radiation resource (fluid 1) that assures a minimum of the integral entropy production and—simultaneously—an extremum of work consumed or delivered. However, nondimensional time τ of the previous section related to *overall* number of transfer units cannot now be used as it is no longer effective. Rather a time variable τ_1 associated solely with the resource properties (Equation (5.130)) will be applied.

Exact modeling of mechanical power yield from radiation, that uses the Stefan–Boltzmann equation, involves several difficulties (irreducible nonlinearities). They may be attributed to the temperature dependence of quantity $G_c \equiv GC_m$ or the product of the molar fluid's flow and its molar heat capacity. From Equation (5.16) capacities per unit volume are $c_v(T) = 4aT^3$ and $c_h(T) \equiv c_v(T) + \mathrm{d}P/\mathrm{d}T = (16/3)aT^3$. From Equation (5.36), the ratio of the photon volume to their number or a single photon volume (reciprocal of number density) is

$$v_{\text{ph}} \equiv \frac{V}{N} = \frac{k_B T}{P(T)} = \frac{3k_B}{a} T^{-3} = \frac{c}{4p} T^{-3} \tag{5.123}$$

The molar volume, V_m, is the product of this quantity and Avogadro number, A_v. This leads to evaluation of products $c_{v_{\text{ph}}} \equiv v c_v(T) = 12k_B$ and $c_{v_{\text{ph}}} \equiv v c_v(T) = 16k_B$. This also means that molar heat capacities of black photons are, respectively, $C_{mv} = 12R$ and $C_{mb} = 16R$, where R is the universal gas constant.

As shown by the state equation (5.36), products $\dot{V} c_v(T)$ and $\dot{V} c_b(T)$ in power functionals (5.59), (5.103), and (5.104) vary in time. For a constant $\dot{V}$, Equation (5.36) implies that both particle and molar flows, $\dot{N}$ and $\dot{G}_m = \dot{N}/A_v$, are proportional to T^3.

$$\dot{G}_m = \frac{4p}{cA_v} T^3 \dot{V} = \frac{\dot{V}}{V_m} \tag{5.124}$$

(Note that in some works the symbol G is used for the molar flux.) Equation (5.124) is, in fact, a simple transformation of Equation (5.26). For a constant $\dot{V}$ Equation (5.124) proves the decrease of the photon flux in the engine mode and the increase of this flux in the heat-pump mode. These effects are associated with a corresponding decrease and increase of T along the process path. The products $\dot{G}_m c_{mv}$ and $\dot{G}_m c_{mb}$ are, respectively

$$GC_{mv} = \frac{48k_B p}{c} T^3 \dot{V}, \qquad GC_{mb} = \frac{64k_B p}{c} T^3 \dot{V} \tag{5.125}$$

These formulae serve to accomplish effective calculations of the work integrals.

However, there are also nonlinearities associated with the analytical structure of differential constraints, different from those in the pseudo-Newtonian model of Section 5.9. In fact, the differential constraint of that model was contained in the formula linking the local heat control v with the temperature change, $v = -\chi \, dT/dt = -dT/d\tau$. The simplicity of that constraint caused its easy imbedding in the power functional. Yet, in nonlinear models in the present section constraints are nonlinear counterparts of Equation (5.101), thus they are more involved; an example is Equation (5.139) below. Their imbedding into power functionals (to express these in terms of T and dT/dt) is not always possible, so they reside in the mathematical model as separate entities. Thus, a typical control scheme is Pontryagin's maximum principle where controls are not the simple derivatives of state coordinates with respect to time.

Based on local power expression $p_1 = \eta q_1$ the cumulative power is the integral over $\eta\,d\dot{Q}_1$ or

$$\dot{W} = \int_0^{\dot{Q}_1} \left(1 - \Phi\frac{T_2}{T'}\right) d\dot{Q}_1 \tag{5.126}$$

From the energy balance the differential energy flux q_1 corresponding with infinitesimal changes of dT_1, dx and dt equals $d\dot{Q}_1 = -\dot{G}_m(T_1)C_m\,dT_1$. The related functional of cumulative power is

$$\dot{W} = -\int_{t^i}^{t^f} \dot{G}_m(T_1)C_m \left(1 - \frac{\Phi T_2}{T'}\right) \dot{T}_1\,dt \tag{5.127}$$

In hybrid models it is possible to express T' in the form $T' = T'(T_1, \dot{T}_1)$, as we shall soon see. In power formulae like the above, the dots over symbols refer either to flow quantities (e.g. $\dot{G}_m(T_1)$) or to time derivatives of thermal potentials (e.g. T_1) with respect to physical time, t, which is the contact time of the driving fluid with the power generator. However, it should be noted that, due to the invariance $\dot{T}_1^{(t)}\,dt = \dot{T}_1^{(\xi)}\,d\xi$, the product $\dot{T}_1\,dt$ in the above functional describes the same differential dT_1 for any definition of time variable. Such definition can involve nondimensional times τ and τ_1, cumulative area A, or a fluid's contact time t. In effect, dots over the temperature symbol in formulae like Equation (5.127) may be referred to as arbitrary independent variables, whereas dots over the symbols of flow quantities (e.g. $\dot{G}_m(T_1)$) must be reserved to flows in physical time, t.

With Equation (5.90) the cumulative power of the entropy production describing lost work in equations of extended availabilities is an integral

$$\sigma_s = -\int_{t^i}^{t^f} \dot{G}_m(T)C_m \left(\frac{\Phi}{T'} - \frac{1}{T_1}\right) \dot{T}_1\,dt \tag{5.128}$$

For fixed end states the limiting production or consumption of mechanical energy is associated with extremum work (5.127) or minimum of entropy production (5.128). Whenever an operator describing Carnot temperature T' in terms of radiation temperature T_1 and its time derivative can be found, variational calculus can be applied to solve the optimization problem of extremum work. In the opposite case Equation (5.128) and the related work functionals must be written in the form in which T' and T_1 are two distinct variables in an algorithm of the optimal control. This is Pontryagin's algorithm in which a differential equation constraints' changes of T_1, dT_1/dt, and T' are shown (see Equation (5.139) below).

We shall now deal specifically with the *symmetric nonlinear case*. It involves the radiative heat transfer ($m = 4$) in both reservoirs and corresponds with form

(5.90) of the local entropy production. We exploit both the variable heat-capacity flux $\dot{G}_{ch}(T)$ and the effective heat coefficient $\alpha_1(T_1)$ to define the nondimensional time τ_1 by the equality

$$
\frac{q_1}{g_{1N}} = g_1 \frac{T_1^m - T_{1'}^m}{g_{1N}} = -\frac{\dot{G}_{ch}(T_1)\,dT_1}{\alpha_1(T_1)a_v F_1\,dx}
$$

$$
\equiv -\frac{dT_1}{d\tau_1} = \sigma \frac{T_1^m - T_{1'}^m}{\alpha_1(T_1)} \cong T_1 - T_{1'}
\tag{5.129}
$$

where the ratio g_1/g_{1N} above equals $\sigma/\alpha_1 = T^{-3}$ and $\dot{G}_{ch} \equiv \dot{G}_m c_{mh}$ is the product of the fluid's molar flow and its molar heat capacity, the second expression in Equation (5.125). The nondimensional time τ_1 is defined by the equality

$$
d\tau_1 = \frac{\alpha_1(T_1)a_v F_1}{\dot{G}_{ch}(T_1)}\,dx
\tag{5.130}
$$

Since both quantities $\dot{G}_{ch}$ and $\alpha_1(T_1)$ vary as T^3, the effect of T cancels out and nondimensional time τ_1 is a suitable quantity proportional to physical residence time t. Equation (5.129) describes the energy balance for the radiation fluid. The nondimensional time τ_1 is simultaneously the number of energy transfer units related to the fluid in state 1. Equation (5.129) shows that the driving energy flux can be measured in terms of the temperature drop of radiation fluid per unit of nondimensional time τ_1. Equation (5.129) is, in fact, a mixed structure: while it uses the Stefan–Boltzmann law, it also bears some properties of pseudo-Newtonian systems due to the presence the transfer coefficient $\alpha_1(T_1)$ in the definition of nondimensional time τ_1. After using Equation (5.86) for $T_{1'}$ in Equation (5.129) we obtain the state equation

$$
-\frac{dT_1}{d\tau_1} = T_1 - \left(T_1^m - g_2 \frac{T_1^m - T'^m}{\Phi g_1(T'/T_2)^{m-1} + g_2} \right)^{1/m}
\tag{5.131}
$$

We may also proceed in another way. We exploit energy exchange formula (5.87) with conductances $g_i = \sigma A_i$ based on universal Stefan–Boltzmann constant, σ. Comparing two boundary expressions of Formula (5.129) with Equation (5.87) we obtain an equation

$$
\frac{dT_1}{d\tau_1} = -\frac{g_1 g_2}{g_{1N}} \frac{T_1^m - T'^m}{\Phi g_1(T'/T_2)^{m-1} + g_2}
\tag{5.132}
$$

which corrects Equation (5.45) of Sieniutycz and Kuran (2005) by inclusion of ratio g_2/g_{1N}. For radiation the ratios g_2/g_{1N} and g_1/g_{1N} appearing in Formula (5.132) equal respectively $\sigma/\alpha_1 = T_1^{-(m-1)} = T_1^{-3}$ and $\sigma/\alpha_2 = T_2^{-3}$.

While each of Equations (5.131) and (5.132) can be used as a constraint in the optimization of work (5.127) their transfer-coefficient base leads us to a further search towards an exact equation containing only universal (Stefan–Boltzmann) constants. This exact equation can be obtained from an alternative form of energy balance (5.129) which is

$$-G_{cb}(T_1)\, dT_1 = \sigma_1(T_1^m - T_{1'}^m)a_v F_1 u\, dt = \sigma_1(T_1^m - T_{1'}^m)a_v \dot{V} dt \qquad (5.133)$$

Note that the time variable used is the contact time of the radiation fluid with the energy generator, that is $t = t_1$, but, for simplicity, we shall further neglect index 1 when designating any property of the fluid. This will also serve to point out that such properties can be variable quantities rather than constants. Expressing in Equation (5.133) molar flux $\dot{G}_m$ as a function of volume flux $\dot{V}$ and molar volume V_m

$$\dot{G}_m = \frac{\dot{V}}{V_m} = \frac{4p}{cA_v}T^3\dot{V} = p_m^0 T^3 \dot{V} \qquad (5.134)$$

where $p_m^0 \equiv 4p/(cA_v)$ is the molar constant of a photon's density we obtain

$$-c_{mb}\frac{dT_1}{dt} = \sigma_1(T_1^m - T_{1'}^m)a_v V_m \qquad (5.135)$$

This basic result describes, in fact, heating of one mole of photons with volume V_m and effective capacity c_{mb} in an energy exchange governed by the Stefan–Boltzmann law. Writing molar volume V_m in terms of the universal constant, $V_m = T^{-(m-1)}/p_m^0$, we obtain

$$\frac{dT_1}{dt} = -\beta\frac{T_1^m - T_{1'}^m}{T_1^{m-1}} \qquad (5.136)$$

where

$$\beta \equiv \frac{\sigma_1 a_v}{c_{bm}p_m^0} \qquad (5.137)$$

As $T_{1'}$ is not an independent control variable, it is suitable to express this equation in terms of variables such as T' or η that are independent controls. In terms of Carnot temperature the differential constraint is obtained by using Equation (5.86) in Equation (5.136); the result is

$$\frac{dT_1}{dt} = -\left(\frac{\sigma_1 a_v}{c_{bm}p_m^0}\right)\frac{T_1^m - T'^m}{(\Phi(g_1/g_2)(T'/T_2)^{m-1} + 1)T_1^{m-1}} \qquad (5.138)$$

Now we can ignore subscript 1 in the variable temperature of the radiation fluid and lump coefficients into a single constant β. In terms of the fluid's variable temperature $T = T_1(t)$ and Carnot temperature T' the state equation (5.138) is

$$\frac{dT}{dt} = -\beta \frac{T^m - T'^m}{(\Phi'(T'/T_2)^{m-1} + 1)T^{m-1}} \tag{5.139}$$

where coefficient β is defined by Equation (5.137) and

$$\Phi' \equiv \Phi \frac{g_1}{g_2} \tag{5.140}$$

and time variable t is the contact time of the radiation fluid with the engine.

We shall now consider the *hybrid nonlinear case* in which the radiative energy transfer ($m = 4$) occurs only in the first (upper) reservoir; whereas, in the second reservoir, the energy exchange is governed by the mechanism of convective heat exchange described by Newton's law (5.91). In order to obtain a power functional in this case we use Equation (5.94) as the particular representation of power p in terms of T_1'. From the energy balance the differential energy flux q_1 corresponding with changes of dT_1, dx and dt equals $d\dot{Q}_1 = -\dot{G}_m(T_1)C_{bm}\,dT_1$. Calculating cumulative power as the integral over $\eta\,d\dot{Q}_1$ with the help of Equations (5.93), (5.94) and (5.127) yields

$$\dot{W} = -\int_{t^i}^{t^f} \dot{G}_m(T_1)C_{bm}\left(1 - \frac{\Phi T_2}{T_{1'} - \Phi g_1(T_1^m - T_{1'}^m)/g_2}\right)\dot{T}_1\,dt \tag{5.141}$$

With Equation (5.136), temperature $T_{1'}$ of Equation (5.141) can be expressed in terms of variables T and dT_1/dt. Consequently, the production (consumption) of mechanical energy is described by the following power integral:

$$\dot{W} = -\int_{\tau^i}^{t^f} \dot{G}_c(T_1)\left(1 - \frac{\Phi T_2}{\left(T_1^m + \dot{T}_1^m\right)^{1/m} + \Phi\dot{T}_1^m g_1/g_2}\right)\dot{T}_1\,d\tau_1 \tag{5.142}$$

The temperature function $\dot{G}_{cb}(T)$ in Equation (5.132) and in power equations below is defined as

$$\dot{G}_{cb}(T) \equiv \dot{G}_m C_{bm} = \frac{64 k_B p}{c}T^3\dot{V}C_{bm} = p_m^0 T^3\dot{V}C_{bm} \tag{5.143}$$

This formula follows from the second expression in Equations (5.125) and (5.134). Equation (5.142) uses operator representation of the temperature of the upper circulating fluid

$$T_{1'}^m = T_1^m + \frac{\mathrm{d}T_1^m}{a\beta\,\mathrm{d}t} \equiv T_1^m + \frac{\mathrm{d}T_1^m}{\mathrm{d}\tau_1} \equiv T_1^m + \dot{T}_1^m \tag{5.144}$$

which results from Equation (5.136). This operator representation also defines the dimensionless time of the problem, $\tau_1 \equiv \beta a t$.

The related expression for the total entropy production is

$$\sigma_s = -\int_{\tau^i}^{\tau^f} \dot{G}_{ch}(T_1) \left(\frac{\Phi}{\left(T_1^m + \dot{T}_1^m\right)^{1/m} + \dot{T}_1^m \Phi g_1/g_2} - \frac{1}{T_1} \right) \dot{T}_1\,\mathrm{d}\tau_1 \tag{5.145}$$

While an equation of this form was suggested earlier (Sieniutycz and Kuran, 2005), only the present analysis defines the relation between dimensionless time and physical or contact time: $\tau_1 \equiv \beta m t$. As it is our policy here to prefer formulae using physical time t, we write below power formula (5.142) in terms of t rather than τ.

$$\dot{W} = -\int_{\tau^i}^{t^f} \dot{G}_{ch}(T_1) \left(1 - \frac{\Phi T_2}{(T_a^1 + \chi \dot{T}_1^m)^{1/m} + \Phi \chi \dot{T}_1^m g_1/g_2} \right) \dot{T}_1\,\mathrm{d}t \tag{5.146}$$

where $\chi = (m\beta)^{-1}$ and $\dot{T}_1^m \equiv \mathrm{d}T_1^m/\mathrm{d}t = mT_1^{m-1}\,\mathrm{d}T/\mathrm{d}t$. Hence, after the omission of unnecessary subscript 1 at T_1

$$\dot{W} = -\int_{\tau^i}^{t^f} \dot{G}_{ch}(T) \left(1 - \frac{\Phi T_2}{(T^a + \beta^{-1}T^{m-1}\dot{T})^{1/m} + \Phi\beta^{-1}T^{m-1}\dot{T}g_1/g_2} \right) \dot{T}\,\mathrm{d}t \tag{5.147}$$

Equations (5.142) and (5.145)–(5.147) contain Carnot temperature operator T' expressed in terms of the temperature of the upper reservoir and its appropriate time derivative. As these equations are Lagrange functionals, classical method of calculus of variations can be applied for their optimization. Yet, this property only refers to the hybrid model, because the Lagrange structures do not appear in the symmetric model.

Optimal control approaches are also possible for hybrid models. After identifying the temperature derivative as the possible control $\dot{T} = u$ we obtain

$$W = -\int_{\tau^i}^{t^f} \dot{G}_{ch}(T) \left(1 - \frac{\Phi T_2}{(T^m + \beta^{-1}T^{m-1}u)^{1/m} + \Phi\beta^{-1}T^{m-1}ug_1/g_2}\right) u \, dt$$

(5.148)

The differential constraint for the above integral has a trivial form resembling that in (5.101)

$$\frac{dT}{dt} = u$$

(5.149)

Analogously one can treat the integral of the entropy production, Equation (5.145). In terms of u the procedure yields the functional

$$\sigma_s = -\int_{t^i}^{t^f} \dot{G}_{ch}(T) \left(\frac{\Phi}{(T^m + \beta^{-1}T^{m-1}u)^{1/m} + \Phi\beta^{-1}T^{m-1}ug_1/g_2} - \frac{1}{T}\right) u \, dt$$

(5.150)

Again, it should be minimized subject to Equation (5.149).

As in the case of the symmetric problem, integrals of power and entropy production of the hybrid problem can always be treated by the algorithm of Pontryagin's maximum principle. In that case Equations (5.148) and (5.150) are optimized subject to constraint (5.149). However, the most suitable way to optimize hybrid models is to write down and then solve the Euler–Lagrange equation of the variational problem. For this purpose functionals of T and $\dot{T}$ are relevant, such as Equation (5.145) or (5.146). Analytical solutions are rare, thus one has to rely on numerical techniques.

5.11. TOWARDS THE HAMILTON–JACOBI–BELLMAN APPROACHES

We shall now describe some benefits resulting from the derived differential models. Equations (5.131), (5.132), (5.136), (5.138), (5.139) and (5.149) are differential constraints in problems extremizing power or total entropy production treated by Pontryagin's maximum principle. The extremization leads to optimal profiles $T'(\tau_1)$ and $T_1(\tau_1)$ that assure extremum of work produced in a sequential engine system (Figure 5.1) or consumed in a sequential heat-pump system. Both systems are dynamic, with an infinite number of infinitesimal stages. The extremum work thus-obtained is a finite-time exergy of the resource working in the continuous system. An example is the extended exergy referred to in

Equation (5.116). For practical reasons, its discrete counterparts for finite stages are also of interest.

With power functionals at our disposal we can formulate the Hamilton–Jacobi–Bellman theory (HJB theory) for extremum work and related extended exergy. Hamiltonian and Lagrangian formalisms are not really suitable as they do not yield directly the extremum work expression. Yet in some cases the principal function can be found also within these formalisms, by finding optimal paths and evaluating the optimal work along these paths.

The HJB theory of the principal function is the basic ingredient in variational calculus (Elsgolc, 1960; Rund, 1966) and optimal control (Bellman, 1961; Pontryagin et al., 1962; Leitman, 1966; Fan, 1966; Findeisen et al., 1980; Crandall and Lions, 1983; Sieniutycz, 1991; Berry et al., 2000; Crandall et al., 1992; Barles, 1994). Standard equations of the HJB type are partial differential equations yet discrete structures of the HJB type also exist (Sieniutycz, 2000c). The former are associated with ordinary differential equations (such as those in this chapter), the latter with difference equations. Bellman's recurrence equation can be regarded as a discrete HJB equation, yet there are also other discrete relationships that are structurally closer to HJB equations than Bellman's equations (Sieniutycz, 2006b). An HJB equation generalizes the classical Hamilton–Jacobi equation (Rund, 1966) by inclusion of extremum conditions for control variables.

In the two next chapters we shall formulate and solve HJB equations for some continuous models of work production and consumption. We shall also develop discrete counterparts of these equations describing cascade processes with stages of finite size. This will serve to develop numerical methods in complex cases with state dependent coefficients, when an HJB equation cannot be solved analytically or does not allow the classical solution. Because of the link between the HJB theory and the method of dynamic programming, associated numerical approaches will make use of Bellman's recurrence equation (Aris, 1964; Berry et al., 2000).

5.12. FINAL REMARKS

In this chapter dynamical results have been presented that show the complex (nonexponential) nature of the radiation relaxation. State equations have been obtained for continuous processes with radiation. They constitute ordinary differential equations which describe the response of a power-producing radiation system to external controls. Under the pseudo-Newtonian approximation difficulties stemming from use of the exact Stefan–Boltzmann theory are to a large extent abandoned; it is possible to use analytical methods and to find explicit formulae for generalized exergies, Equations (5.116) and (5.122). Yet, use of the exact Stefan–Boltzmann theory causes essential difficulties in obtaining analytical solutions. Therefore, in the next chapters optimization algorithms (HJB equations and dynamic programming equations) will be obtained and solved for continuous processes and their discrete counterparts describing cascades with finite stages.

Generalized energy limits stem from irreversible extensions of the classical exergy by including minimally irreversible processes. Limiting work estimates made with the help of classical exergies are too weak and often insufficient; generalized exergies assure stronger work limits that take into account extra limitations coming from the process kinetics (Ondrechen et al., 1981; Bejan and Errera, 1998; Sieniutycz, 1998a, 1998b, 1998c, 1998d, 1998e, 1999c, 2002a, 2002b, 2002c, 2003a, 2003b, 2003c, 2007c; Sahin, 2002; Berry et al., 2000; Sieniutycz and Szwast, 2003; Sieniutycz and Kuran, 2006; Kuran and Sieniutycz, 2007). Considering the classical component of the radiation exergy it has been shown that the well-known Petela's (1964) equation for the density of radiation exergy is restricted to the exergy of enclosed radiation, whereas its direct application to the radiation flux ceases to be valid. The exact equation for the radiation exergy flux agrees with basic thermodynamics and proves that the efficiency of reversible transformation of solar flux equals the Carnot efficiency provided that the radiation energy flux is determined by the Stefan–Boltzmann equation.

A generalized exergy of processes departing from the equilibrium (a fluid's utilization) is larger than that in processes approaching the equilibrium (a fluid's relaxation). This is because one respectively adds or subtracts the product of T_0 and entropy production in a formula of the generalized exergy. By inserting the entropy production, limits for energy yield or consumption set by generalized exergies are stronger than those defined by the classical exergy. It fact, finite rates increase the minimum work that must be supplied to the system and decrease the maximum work that can be produced by the system.

6 Hamilton–Jacobi–Bellman theory of energy systems

6.1. INTRODUCTION

In this chapter we treat power limits in dynamic energy systems driven by nonlinear fluids that are restricted in their amount or magnitude of flow, and as such, play the role of resources. A resource is a valuable substance or energy used in a process; its value can be quantified by specifying its exergy: the maximum work that can be obtained when the resource relaxes to the equilibrium. A power limit is an upper (lower) bound on power produced (consumed) in the system. To find the power limit cumulative power is maximized in the system with a nonlinear fluid, an engine or a sequence of engines, and an infinite bath. Reversible relaxation of the resource leads to classical exergy: when dissipation is allowed, generalized exergies are obtained.

The thermodynamic theory developed in Chapters 2–5 is treated here in the context of the *classical* Hamilton–Jacobi–Bellman (HJB) theory of optimal systems; nonclassical aspects are discussed in Chapter 7. The reader is referred to previous chapters of the book for details regarding mathematical modeling and simulation of nonlinear energy converters, in particular radiation engines, in steady and dynamic situations. Dynamic state equations and expressions for efficiency of imperfect converters obtained in previous chapters are used to derive and solve Hamilton–Jacobi equations describing resource upgrading and downgrading. Various mathematical tools are applied in the trajectory optimization, with special attention given to the relaxing radiation. In agreement with the findings of Chapter 5, the radiation relaxation curve is nonexponential, characteristic of a nonlinear system. Provided that nonclassical solutions are admitted (Chapter 7), Hamilton–Jacobi–Bellman equations have the potential of power optimization algorithms which lead directly to work limits and generalized availabilities. The performance functions of energy converters depend on end thermodynamic coordinates and a process intensity index, h; that is, the Hamiltonian of the power optimization problem. As an example of the limiting work from radiation, a finite-rate exergy of radiation fluid is estimated in terms of finite rates quantified by the Hamiltonian h.

In fact, generalized exergies described by Hamilton–Jacobi–Bellman equations quantify deviations of the system's efficiency from the Carnot efficiency. In the context of the HJB theory, an exergy is the principal component (value function) of the solution to the variational problem for extremum work under suitable boundary conditions. Other components are the optimal trajectory and optimal control. In purely thermal systems, due to the absence of chemical changes, the trajectory is characterized by the temperature of the resource fluid, $T(t)$. Whereas the control is Carnot temperature $T'(t)$ defined in Chapter 3, in agreement with references in Sieniutycz (2003a, 2003c), whenever $T'(t)$ differs from $T(t)$, the resource relaxes with a finite rate, and with an efficiency different from the Carnot efficiency. Only when $T'(t) = T(t)$ is the efficiency Carnot, but this corresponds to an infinitely slow relaxation rate of the resource to thermodynamic equilibrium with the environmental fluid.

The structure of this chapter is as follows. Section 6.2 discusses various aspects of steady and dynamic optimization of power yield. Quantitative analysis of processes downgrading a finite resource (contained in the first reservoir) and issues regarding the generalization of the classical exergy for finite rates are presented in Section 6.3. Sections 6.4–6.6 display various Hamilton–Jacobi–Bellman equations and Hamilton–Jacobi equations for power production problems. Extensions highlighting systems with nonlinear kinetics (e.g. radiation) and internal dissipation are treated in Section 6.7. Generalized exergies based on classical smooth solutions are discussed in Section 6.8, while Section 6.9 summarizes the results obtained.

The size limitation of the present chapter does not allow for inclusion of all derivations to make the chapter self-contained, thus the reader may need to refer to some previous chapters as well as to some previous works (Sieniutycz, 2003a, 2003c; Sieniutycz and Kuran, 2005, 2006).

In view of the difficulties either in getting classical analytical solutions in complex systems or because the HJB equations may not admit classical solutions at all in favor of viscosity solutions (Bardi and Capuzzo-Dolcetta 1997), discretized equations and numerical approaches to HJB solutions are considered in Chapter 7. This chapter also discusses selected convergence issues of discrete numerical algorithms to viscosity solutions of HJB equations and the role of Lagrange multipliers in dimensionality reduction (linked with considerations related to the Hamiltonian constancy preservation).

6.2. DYNAMIC OPTIMIZATION OF POWER IN A FINITE-RESOURCE PROCESS

Restricted amount or flow of a working resource causes a decrease of the resource potential in time (chronological or spatial), and consequently processes with resource downgrading are dynamic ones, and any studies of their extrema require dynamic optimization methods. From the optimization viewpoint the dynamic process is any process in which one can distinguish sequential changes of state, either in the chronological time or in (spatial) holdup time. The first group of dynamic processes, evolving in chronological time, occurs in nonstationary systems; the second group, evolving in spatial time, in steady-state systems.

In a typical energy production system two resting reservoirs do interact through an energy generator (engine). In this system the power flow is steady only when two reservoirs are infinite. The energy balance shows that when one, say upper, reservoir is finite, its thermal potential must decrease in time. Any finite reservoir is thus a resource reservoir. It is the resource property of the fluid that leads to the dynamic behavior of the fluid and its relaxation to the equilibrium with an infinite lower reservoir (usually the environment).

Alternatively, fluid at a steady flow can replace the resting upper reservoir. The process of resource downgrading is then arranged as a steady-state process in which the resource fluid flows through a pipeline or stages of a cascade. The fluid's state then changes along a steady trajectory. As in the previous case the trajectory is a curve describing the relaxation process towards the equilibrium between the fluid and the lower reservoir (the environment). This relaxation is sometimes called "active relaxation" as is associated with the simultaneous work production. It should be contrasted with "dissipative relaxation," a well-known natural process between a body or a fluid and the environment without any power production.

Relaxation (either active or dissipative) leads to a decrease of the resource potential (i.e. temperature) in time. An inverse of the relaxation process is the one in which a body or a fluid abandons the equilibrium or increases its distance from the equilibrium. This process cannot be spontaneous, that is, it cannot occur outside an external sustainable flux; rather, it needs a supply of external power. The process, referred to as thermal upgrading of the resource, can be accomplished with a heat pump. The inverse process is also considered here.

Much of the research on power limits published to date deals with stationary systems, in which both reservoirs are infinite. To this case refer to steady-state analyses of the Chambadal–Novikov–Curzon–Ahlborn engine (CNCA engine). In their original paper Curzon and Ahlborn (1975) considered the case in which the energy exchange is described by the Newtonian law of cooling. A more general model (Chapter 5) may be the Stefan–Boltzmann engine for a system with radiation fluids and the energy exchange governed by the Stefan–Boltzmann law (de Vos, 1992). ("Radiation fluid" is an isotropic mixture of a large number of photons obeying laws of thermodynamics.) As a result of their stationarity (caused by the infiniteness of both reservoirs), controls maximizing power are represented by a fixed point in the state space. In fact, for the CNCA engine, the maximum power point may be related to the optimum value of a free (unconstrained) control variable which can be efficiency η or Carnot temperature T'. In terms of the reservoirs, temperatures T_1 and T_2 and the internal irreversibility factor Φ one finds $T'_{opt} = (T_1 \Phi T_2)^{1/2}$ (Sieniutycz and Kuran, 2005). An exact expression for the optimal power point of the Stefan–Boltzmann engine cannot be determined analytically, yet this temperature can be found graphically from the power diagram $W = f(T')$. Moreover, the method of Lagrange multipliers can be successfully applied (Kuran, 2006). As the elimination of these multipliers from a set of resulting equations is quite easy, the problem is broken down to the numerical solving of a single nonlinear equation for the optimal control T'. Finally, the so-called pseudo-Newtonian model (Sieniutycz and Kuran, 2005, 2006), which uses the state or temperature-dependent heat exchange coefficient

$\alpha(T^3)$, omits, to a considerable extent, analytical difficulties associated with the use of the Stefan–Boltzmann equation. Applying this model in the so-called symmetric case, where both reservoirs are filled up with radiation, one shows that the optimal (power maximizing) Carnot temperature of the steady radiation engine is that for the CNCA engine, that is, $T'_{opt} = (T_1 \Phi T_2)^{1/2}$ (Sieniutycz and Kuran, 2005, Chapter 5). The result is, in fact, a good approximation, admissible when the transfer coefficients depend solely on bulk temperatures of reservoirs. With this assumption, the equation also holds for a nonsymmetric or hybrid model, where the radiation fills only the upper reservoir, whereas the lower one contains a Newtonian fluid.

The application of nonlinear models in the prediction of dynamic energy yield requires the evaluation of an extremal curve rather than an extremum point. This is associated with the application of variational methods (to handle functional extrema) in place of static optimization methods (to handle extrema of functions). For example, the use of the pseudo-Newtonian model to quantify the dynamic energy yield from radiation gives rise to an extremum curve describing the radiation relaxation to the equilibrium. This curve is nonexponential, the consequence of the nonlinear properties of the relaxation dynamics. Other curves describing the radiation relaxation are also nonexponential, for example, those following from exact models using the Stefan–Boltzmann equation, symmetric and hybrid (Sieniutycz and Kuran, 2005, 2006).

Analytical difficulties associated with dynamic optimization of nonlinear systems may be severe; this is why diverse models of power yield and numerical approaches are applied. Various control variables may be used in modeling as the process analysis using a particular control can be substantially easier than the analysis in terms of another one.

Optimal (i.e. power-maximizing) relaxation curve $T(t)$ is associated with the optimal control curve $T'(t)$; they are both components of the dynamic optimization solution to a continuous problem. In the corresponding discrete problem, formulated for numerical purposes via a suitable discretization, one searches for optimal temperature sequences $\{T^n\}$ and $\{T'^n\}$. Various discrete optimization methods leading to optimal sequences $\{T^n\}$ and $\{T'^n\}$ may be used. They comprise: direct search, dynamic programming, discrete maximum principle, and combinations of these methods.

Minimum power supplied to the system is described in a suitable way by function sequences $R^n(T^n, t^n)$, whereas maximum power produced is described by functions $V^n(T^n, t^n)$. We stress that the profit-type performance function V and cost-type performance function R simply differ by sign, that is, $V^n(T^n, t^n) = -R^n(T^n, t^n)$. The beginner may find the change from symbol V to symbol R and back as unnecessary and confusing. Yet each function is positive in its own, natural regime of working (V is positive in the engine range and R is positive in the heat pump range). Moreover, symbol R is preferred for the quantity most popular in analytical mechanics, where minima of cost-type functions described by Lagrangians convex in rates are considered. (Using R instead V should help the physicist to read this chapter.) As the systems considered are usually autonomous, elimination of the duration variable t^n is possible, which results in improved accuracy of calculations. The elimination of the duration variable is associated with the introduction of the Lagrange multiplier and use of primed or modified power cost functions $R'(T^n, \lambda)$ (Chapter 7). In these algorithms time intervals θ^n are chosen optimally to participate in the power optimization.

Importantly, energy limits of dynamic processes are inherently connected with the exergy functions, the classical exergy and its rate-dependent extensions. To obtain the classical exergy from power functions it suffices to assume that the thermal efficiency of the system is identical to the Carnot efficiency. On the other hand, nonCarnot efficiencies lead to generalized exergies. The latter depend not only on classical thermodynamic variables but also on their rates. These generalized exergies refer to state changes in a finite time, and can be contrasted with the classical exergies that refer to reversible quasistatic processes evolving in time infinitely slowly. The benefit obtained from generalized exergies is that they define stronger energy limits than classical exergies. A systematic approach to exergies (classical or generalized) based on work functionals leads to several original results in thermodynamics of energy systems. The approach especially allows one to explain some unknown properties of exergy of black-body radiation or solar radiation, and to expect that the efficiency of the solar energy flux transformation is equal or close to the Carnot efficiency. Staying on classical ground, one can also show that the familiar Petela's (1964) equation for the radiation exergy is restricted to the enclosed radiation (Sieniutycz and Kuran, 2006), yet a direct application of this equation to the flowing radiation (when evaluating the exergy flux) may be incorrect.

6.3. TWO DIFFERENT WORKS AND FINITE-RATE EXERGIES

Whenever resources are finite (e.g. influx of solar energy or its amount is finite), dynamic analysis replaces the steady analysis, and the formalism is in the realm of functionals. For appropriate boundary conditions, the principal function of the variational problem of extremum work coincides with the notion of an exergy, the function that characterizes the quality of finite resources.

Two different works – the first associated with the resource downgrading during its relaxation to the equilibrium, and the second with the reverse process of resource upgrading – are essential (Figure 6.1). During the approach to the equilibrium, the *engine mode* of the system takes place in which work is released, and during the departure, *heat pump mode* occurs in which work is supplied. Work W delivered in the engine mode is positive by assumption ("engine convention"). A sequence of irreversible engines (CNCA or Stefan–Boltzmann) serves to determine a rate-dependent exergy extending the classical exergy for irreversible, finite-rate processes. Before maximization of a work integral, process efficiency η has to be expressed in terms of state T and a control variable (energy flux q or rate $dT/d\tau$) to assure the functional property (path dependence) of the work integral. The integration must be preceded by the maximization of power or

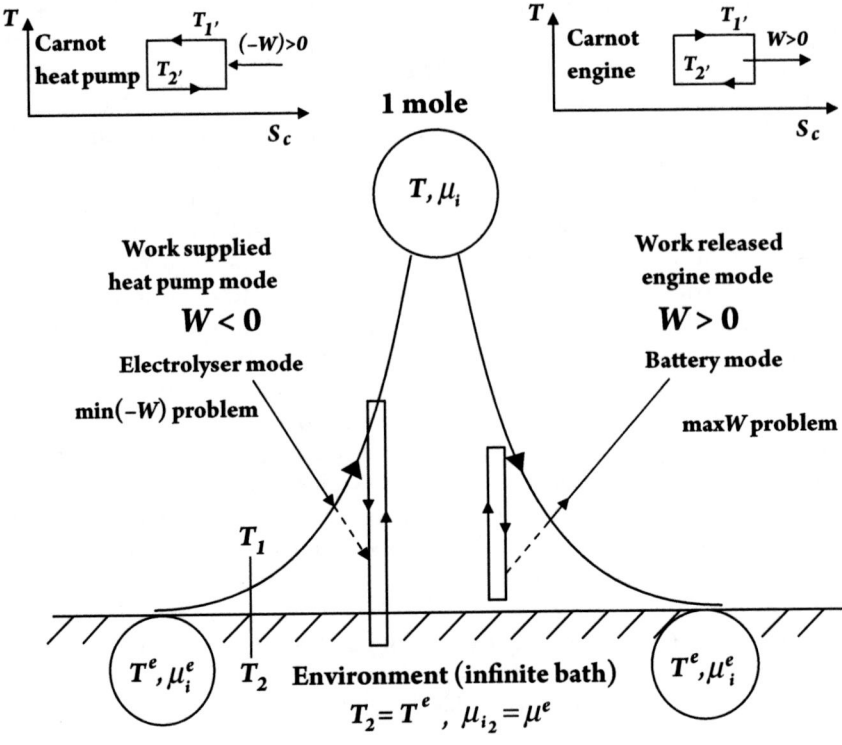

Figure 6.1 Two works: limiting work produced and limiting work consumed are different in an irreversible process.

work at flow w (the ratio of power and flux of driving substance) to assure an optimal path. The optimal work is sought in the form of a potential function that depends on the end states and duration.

Figure 6.1 illustrates the sequential idea of an infinite number of infinitesimal CNCA steps necessary in exergy calculation. Each step is a work-producing (consuming) stage with the energy exchange between two fluids and the thermal machine through finite "conductances" (products of transfer coefficients and related areas). For a radiation engine it follows from the Stefan–Boltzmann law that the effective transfer coefficient α_1 in the "driving" (radiation) fluid is necessarily temperature-dependent, $\alpha_1 = \propto T_1^3$. The second or low-T fluid represents the usual environment, as defined in the exergy theory. This fluid possesses its own boundary layer as a dissipative component, and the corresponding exchange coefficient is α_2. In the physical space, the flow direction of the resource fluid is along the horizontal coordinate x. The optimizer's task is to find an optimal temperature of the resource fluid along the path that extremizes the work consumed or delivered. For traditional fluids (constant c_v) an optimal path is known to be exponential (Sieniutycz 2003a, 2003c, and Chapter 4). Yet no exponential decay of temperature occurs for nonlinear fluids, for example, radiation fluid, during its relaxation to the equilibrium.

Total power obtained from an infinite number of infinitesimal engines is determined as the Lagrange function of the following structure:

$$\dot{W}[\mathbf{T}^i, \mathbf{T}^f] = \int_{t^i}^{t^f} f_0(T, T')\, dt = -\int_{t^i}^{t^f} \dot{G}_c(T)\eta(T, T')\, \dot{T}\, dt \qquad (6.1)$$

where f_0 is the power generation intensity, $\dot{G}$ is the resource flux, $\dot{G}_c(T)$ is the specific heat, $\eta(T, T')$ is the efficiency in terms of state T and control T'; further, $\mathbf{T}$ is the enlarged state vector comprising state and time, and t is the time variable (residence time or holdup time) for the resource contacting the heat transfer surface. Sometimes one uses a nondimensional time τ, identical with the so-called number of the heat transfer units. Note that, for constant mass flow $\dot{G}$ of a resource, one can extremize power per unit mass flux (the quantity of work dimension). In this case Equation (6.1) describes a problem of extremum work. Integrand f_0 is common for both modes, yet the numerical results it generates differ by sign (positive for engine mode; "engine convention"). Power generation function f_0 can always be replaced by power consumption function $l_0 = -f_0$. Formally, l_0 plays the role of a Lagrangian of the process, whereas f_0 is the negative of the Lagrangian. When the resource flux is constant the following work functional can be obtained from Equation (6.1) for the *thermal* exergy flux per unit flux of resource:

$$w_{\max_{dT/dt}} = -\int_{T=T^i}^{T^e=T^f} c(T)\left(1 - \frac{T^e}{T'(T, dt/dT)}\right)\, dT \qquad (6.2)$$

Note that the independent variable in this equation is T, that is, it is different from that in Equation (6.1).

The function f_0 in Equation (6.1) contains a thermal efficiency function, η, described by a practical counterpart of the Carnot formula. When $T > T^e$, efficiency η decreases in the engine mode above η_c and increases in the heat pump mode below η_c. At the limit of vanishing rates, $dT/dt = 0$ and $T' \to T$. Then work of each mode simplifies to the common integral of the classical exergy. For the specific *thermal* exergy,

$$w_{\max_{dT/dt \to 0}} = -\int_{T=T^i}^{T^e=T^f} c(T)\left(1 - \frac{T^e}{T}\right)\, dT = h - h^e - T^e(s - s^e) \qquad (6.3)$$

Thus, with appropriate boundary conditions, Equation (6.1) leads to a generalization of thermal exergy for finite time processes. A form of this functional has two additive components: the classical (potential) part and a nonpotential part which depends on the history of the process.

Nonlinearities can have a thermodynamic or kinetic origin; the former refers, for example, to state-dependent heat capacity, $c(T)$, the latter to nonlinear energy exchange. Problems with linear kinetics (Newtonian heat transfer) are an important subclass. In such problems the fluid's specific work at flow, w, is described by an equation

$$w[T^i, T^f] = \frac{\dot{W}}{\dot{G}} = - \int_{T^i}^{T^f} c(T)\left(1 - \frac{T^e}{T}\right) dT - T^e \int_{t^i}^{t^f} c(T) \frac{(T' - T)^2}{T'T} d\tau$$

$$(6.4)$$

where

$$\tau \equiv \frac{x}{H_{TU}} = \frac{\alpha' a_v F}{\dot{G}c} x = \frac{\alpha' a_v F v}{\dot{G}c} t = \frac{t}{\chi} \tag{6.5}$$

is *nondimensional* time of the process. Equation (6.5) assumes that a resource fluid flows with velocity v through cross-section F and contacts with unit heat transfer exchange surface a_v (Chapter 3). Quantity τ is identical to the so-called *number of the heat transfer units*.

Solutions to work extremum problems can be obtained by

(a) Variational methods, that is, via the Euler–Lagrange equation of variational calculus

$$\frac{\partial L}{\partial T} - \frac{d}{dt}\left(\frac{\partial L}{\partial \dot{T}}\right) = 0 \tag{6.6}$$

1. In the example considered above, that is, for a thermal system with linear kinetics we find

$$T\frac{d^2 T}{d\tau^2} - \left(\frac{dT}{d\tau}\right)^2 = 0 \tag{6.7}$$

2. which corresponds with the optimal trajectory

$$T(\tau, \tau^f, T^i, T^f) = T^i \left(\frac{T^f}{T^i}\right)^{\tau/\tau^f} \tag{6.8}$$

($\tau^i = 0$ is assumed in Equation (6.8)). However, the solution of the Euler–Lagrange equation does not contain any information about the optimal work function. This is assured by solving the Hamilton–Jacobi–Bellman equation.

(b) The HJB equation for "principal function" (V or R), also called value function or extremum work function. For the example considered,

$$\frac{\partial V}{\partial \tau} + \min_{T'}\left\{\left(\frac{\partial V}{\partial T} + c\left(1 - \frac{T^e}{T'}\right)\right)(T' - T)\right\} = 0 \tag{6.9}$$

The extremal work function V is a function of the final state T and total duration. After evaluation of optimal control and its substitution to Equation (6.9), one obtains a nonlinear equation

$$\frac{\partial V}{\partial \tau} - c\left\{ \sqrt{T^e} - \sqrt{T\left(1 + c^{-1}\frac{\partial V}{\partial T}\right)} \right\}^2 = 0 \qquad (6.10)$$

which is the Hamilton–Jacobi equation of the problem. Its solution can be found by the integration of work intensity along an optimal path, between limits T^i and T^f, as shown in Section 6.6. A reversible (path independent) component of V is the classical exergy $A(T, T^e, 0)$.

(c) Discretization and use of the dynamic programming method (DP) which converges to the nonclassical (viscosity) solution for V (Chapter 7).

Details of models of multistage power production in sequences of infinitesimal engines are known from previous publications (Sieniutycz, 1997a, 1997b, 2000c, 2003a, 2003c; Sieniutycz and Kuran, 2005, 2006). These models provide power generation functions f_0 or thermal Lagrangians $l_0 = -f_0$ and dynamic constraints.

6.4. SOME ASPECTS OF CLASSICAL ANALYTICAL HJB THEORY FOR CONTINUOUS SYSTEMS

While the next chapter offers discrete algorithms whose solutions converge to (nonclassical) solutions of continuous HJB equations, classical differentiable solutions have their own analytical theory. This theory deals with differentiable value functions $V(\mathbf{x}, t)$ and $R(\mathbf{x}, t)$, and is outlined here. It may be derived by continuous dynamic programming (Bellman, 1961; Fan, 1966; Findeisen et al., 1980; Sieniutycz, 1991). Here, for brevity, a simple approach exploiting Caratheodory's idea of potentiality of optimal performance function (Rund, 1966) is applied. We also show the relation of HJB equations to the Hamiltonian function

$$H = f_0(\mathbf{x}, \mathbf{u}, t) + \sum_{i=1}^{s} \left(\frac{\partial R}{\partial x_i}\right) f_i(\mathbf{x}, \mathbf{u}, t) \qquad (6.11)$$

that contains state vector $\mathbf{x}$, control vector $\mathbf{u}$, rates f_0 and f_i and state adjoints $z_i = \partial R/\partial x_i$. The scalar f_0 is the growth rate of a generalized profit (intensity of power yield in our case). Hamiltonian H is an energy-type quantity, constant along an optimal path of an autonomous system. An associated quantity is the momentum-type variable; for the kth coordinate

$$z_k = \frac{\partial R}{\partial x_k} \qquad (6.12)$$

In terms of H and $z = \partial R/\partial x$ a general form of the HJB equation is

$$\frac{\partial R}{\partial t} + \max_{\mathbf{u}(t)} H\left(\frac{\partial R}{\partial \mathbf{x}}, \mathbf{x}, \mathbf{u}, t\right) = 0 \qquad (6.13)$$

Let us derive the above equation for the engine mode of an energy process. A control problem of maximum delivery of cumulative power is governed by the characteristic function

$$V(T^i, t^i, T^f, t^f) \equiv \max_{T'(t)} \dot{W}[T^i, T^f] = \max_{T'(t)} \left(\int_{t^i}^{t^f} f_0(T, T')\, dt\right) \qquad (6.14)$$

As it follows from the definition of the maximum performance function V,

$$\max_{T'(t)} \left(\int_{t^i}^{t^f} f_0(T, T')\, dt - V(T^i, t^i T^f, t^f)\right) = 0 \qquad (6.15)$$

Here $f_0 = -l_0$ is the profit generation rate, or, in our case, intensity of power production. Differentiation of Equation (6.15) with respect to the upper limit of the integral, t^f, yields

$$\begin{aligned}
\max_{T'(t)} &\left(f_0^f(T, T') - \frac{dV(T^i, t^i, T^f, t^f)}{dt^f}\right) \\
&= \max_{T'(t)} \left(f_0^f(T, T') - \frac{\partial V}{\partial t^f} - \frac{\partial V}{\partial T^f} f^f(T, T')\right) = 0
\end{aligned} \qquad (6.16)$$

Observe that all rates (f_0 and f) and partial derivatives of V are evaluated at the final state (the so-called "forward equation"). In the second expression, the total time derivative is expanded in terms of all rates. Now, the partial derivative of the characteristic function V with respect to time can be taken off this equation and the superfluous index f can be omitted (variable final states). Then, after replacing V by $-R$, an HJB equation is obtained, consistent with Equation (6.13):

$$\frac{\partial R}{\partial t} + \max_{T'(t)} \left(\frac{\partial R}{\partial T} f(T, T') - l_0(T, T')\right) = 0 \qquad (6.17)$$

For resources relaxing with *linear kinetics* (Newtonian heat exchange) a standardized Lagrangian $l_0 = c(1 - T^e T'^{-1})(T' - T)$ and the related kinetics refer to nondimensional time τ rather than to the usual time t. Equation (6.17) then yields

$$\frac{\partial R}{\partial \tau} + \max_{T'(t)} \left\{\left(\frac{\partial R}{\partial T} - c\left(1 - \frac{T^e}{T'}\right)\right)(T' - T)\right\} = 0 \qquad (6.18)$$

Systems governed by nonlinear kinetics are discussed in the following section.

Numerical methods apply suitable discrete models, for given rates f_0 (or l_0) and f. The important issue of convergence conditions of discrete solutions to nonclassical (viscosity) solutions is discussed in Chapter 7. With the discrete models the theory can be restated and have own life in the realm of difference equations, sums, recurrence relations, two-stage criteria, and so on, often achieving a form strongly dissimilar to the original HJB theory.

6.5. HJB EQUATIONS FOR NONLINEAR POWER GENERATION SYSTEMS

Here we shall display various Hamilton–Jacobi–Bellman equations for power systems described by nonlinear kinetics.

6.5.1. Arbitrary Nonlinear Kinetics

For an arbitrary work generation function f_0 and nonlinear kinetics $dT/dt = f(T, T)$, an HJB equation in terms of work generation function V, final state and final time is

$$-\frac{\partial V}{\partial t} + \max_{T'(t)} \left(f_0(T, T') - \frac{\partial V}{\partial T} f(T, T') \right) = 0 \qquad (6.19)$$

6.5.2. Radiation Engine Approximated by Pseudo-Newtonian Model

A suitable example is a radiation engine whose power integral is expressed as

$$\dot{W} = -\int_{t^i}^{t^f} \dot{G}_m c_m \left(1 - \Phi' \frac{T^e}{T'} \right) \upsilon(T', T) \, dt \qquad (6.20)$$

with $\upsilon = \alpha(T^3)(T' - T)$. An alternative form uses Carnot temperature T' explicit in υ (Sieniutycz and Kuran, 2006). Then

$$\dot{W} = -\int_{T}^{T_0} \dot{G}_m \left(c_{hm}(T) - c_{vm}(T) \frac{T^e}{T} \right) \upsilon \, dt$$
$$-\int_{T}^{T_0} T^e \dot{G}_m \left(c_{vm}(T) \left(\frac{\chi \upsilon^2}{T(T + \chi \upsilon)} \right) + (1 - \Phi) \frac{\upsilon}{T + \chi \upsilon} \right) dt \qquad (6.21)$$

This defines f_0 in terms of υ, with the entropy production singled out in the second term. The optimal power function of this problem can be referred to each of these integrals, for example

$$V(T^i, t^i, T^f, t^f) \equiv \max_{T'(t)} \dot{W} = \max_{T'(t)} \left(\int_{t^i}^{t^f} -\dot{G}_m c_m \left(1 - \Phi' \frac{T^e}{T'}\right) u \, dt \right) \quad (6.22)$$

In this case the HJB Equation (6.19) applies with f_0 defined as integrand of Equation (6.20).

6.5.3. Stefan–Boltzmann Engine

For *the symmetric model* of radiation conversion (both reservoirs composed of radiation)

$$\dot{W} = \int_{t^i}^{t^f} \dot{G}_c(T) \left(1 - \frac{\Phi T^e}{T'}\right) \beta \frac{T^a - T'^a}{(\Phi'(T'/T^e)^{a-1} + 1)T^{a-1}} \, dt \quad (6.23)$$

The coefficient $\beta = \sigma a_v c_h^{-1}(p_m^0)^{-1}$ is related to the molar constant of the photons density p_m^0 and the Stefan–Boltzmann constant of radiation σ. In the physical space, power exponent $a = 4$ for radiation and $a = 1$ for a linear resource. The integrand of this integral is the generalized profit f_0. With the state equation

$$\frac{dT}{dt} = -\beta \frac{T^a - T'^a}{(\Phi'(T'/T^e)^{a-1} + 1)T^{a-1}} \quad (6.24)$$

(Sieniutycz and Kuran (2006) applied in general Equation (6.19)), we obtain an HJB equation

$$-\frac{\partial V}{\partial t} + \max_{T'(t)} \left\{ \left(\dot{G}_c(T)\left(1 - \Phi'\frac{T^e}{T'}\right) + \frac{\partial V}{\partial T}\right) \right.$$
$$\left. \beta \frac{T^a - T'^a}{(\Phi'(T'/T_2)^{a-1} + 1)T^{a-1}} \right\} = 0 \quad (6.25)$$

Dynamical Equation (6.24) is the characteristic equation for Equation (6.25).

For *a hybrid model* of radiation conversion where the upper reservoir is composed of the radiation and the lower one of a Newtonian fluid, the power is

$$\dot{W} = -\int_{\tau^i}^{t^f} G_c(T) \left(1 - \frac{\Phi T^e}{(T^a + \beta^{-i} T^{a-i} u)^{1/a} + \Phi \beta^{-i} T^{a-i} u g_1/g_2}\right) u \, dt$$

$$(6.26)$$

(Sieniutycz and Kuran, 2006) and the corresponding Hamilton–Jacobi–Bellman equation is

$$
-\frac{\partial V}{\partial t^f} + \max_{T'(t)} \left\{ \left(-\dot{G}_c(T) \left(1 - \frac{\Phi T^e}{(T^a + \beta^{-1}T^{a-1}u)^{1/a} + \Phi\beta^{-1}T^{a-1}ug_1/g_2)} \right) \right. \right.
$$
$$
\left. \left. + \frac{\partial V}{\partial T^f} \right) u \right\} = 0 \tag{6.27}
$$

6.6. ANALYTICAL SOLUTIONS IN SYSTEMS WITH LINEAR KINETICS

In all HJB equations, extremized expressions are Hamiltonian expressions. In fact, they are Pontryagin-type H in a state space, related to nonextremal processes, rather than extremum H that are energy-like functions resembling the energy of classical mechanics. Moreover, in the HJB formalism Hamiltonians reside in the enlarged state space (T, τ) or (T, t) rather than in the phase space (T, z, τ) or (T, z, t). For example, Pontryagin's Hamiltonian of the linear system (Newtonian energy flow in time τ rather than t) is

$$
H = \left[z - c \left(1 - \frac{T^e}{T'} \right) \right] [T' - T] = \left[z - c \left(1 - \frac{T^e}{T + u} \right) \right] u \tag{6.28}
$$

Clearly, its first expression (in terms of T') is precisely that in Equation (6.18). Comparison of both equations shows that $z = \partial R/\partial T$, that is, the temperature adjoint is the gradient of R (or negative gradient of V). By applying the feedback control, optimal driving temperature T' is implemented as a quantity maximizing Hamiltonian (6.25) or its HJB counterpart (6.18) with respect to T' at each point of the path.

The maximization of Equation (6.18) implies the one in Equation (6.28) with respect to the same control T'. This leads to two equations. The first expresses optimal control T' in terms of T and z or $\partial R/\partial T$:

$$
\frac{\partial R}{\partial T} - \frac{\partial l_0(T, T')}{\partial T'} = \frac{\partial R}{\partial T} = c \left(1 - \frac{T^e T}{T'^2} \right) = 0 \tag{6.29}
$$

whereas the second is the original Equation (6.18) without maximizing operation:

$$
\frac{\partial R}{\partial \tau} + \frac{\partial R}{\partial T}(T' - T) - c \left(1 - \frac{T_2}{T'} \right) (T' - T) = 0 \tag{6.30}
$$

To obtain optimal control function $T'(z, T)$ one should solve the second equality in Equation (6.29) in terms of T'. The result is Carnot control T' in terms of T and $z = \partial R/\partial T$:

$$T' = \left(\frac{T^e T}{1 - c^{-1}\partial R/\partial T} \right)^{1/2} \tag{6.31}$$

This is next substituted into Equation (6.30); the result is the nonlinear Hamilton–Jacobi equation:

$$\frac{\partial R}{\partial \tau} + cT \left(\sqrt{1 - c^{-1}\frac{\partial R}{\partial T}} - \sqrt{\frac{T^e}{T}} \right)^2 = 0 \tag{6.32}$$

which contains the energy-like (extremum) Hamiltonian of the extremal process:

$$H \left(T, \frac{\partial R}{\partial T} \right) = cT \left(\sqrt{1 - c^{-1}\frac{\partial R}{\partial T}} - \sqrt{\frac{T^e}{T}} \right)^2 \tag{6.33}$$

This is also an optimal H for Equation (6.28) and $z = \partial R/\partial T$. For a positively defined H, each Hamilton–Jacobi equation for optimal work preserves a general form of autonomous equations known from analytical mechanics and theory of optimal control:

$$\frac{\partial R}{\partial \tau} + H \left(T, \frac{\partial R}{\partial T} \right) = 0 \tag{6.34}$$

(In this example H does not contain time explicitly.) Equation (6.34) differs from the HJB equations as it refers to an extremal path only and H is the extremum Hamiltonian.

Expressing the extremum Hamiltonian (Equation (6.33)) in terms of state variable T and Carnot control T' yields an energy-like function satisfying the following relations:

$$E(T, u) = \frac{\partial l_0}{\partial u}u - l_0 = f_0 - u\frac{\partial f_0}{\partial u} = cT^e\frac{(T' - T)^2}{T'^2} \tag{6.35}$$

E is the Legendre transform of the work Lagrangian $l_0 = -f_0$ with respect to the rate $u = dT/d\tau$.

Assuming a numerical value of the Hamiltonian, say h,

$$cT \left(\sqrt{1 - c^{-1}z} - \sqrt{\frac{T^e}{T}} \right)^2 = h \tag{6.36}$$

one can exploit the constancy of autonomous H to eliminate adjoint z. Next, combining Equation (6.36) with optimal control Equation (6.31), or with an equivalent equation using energy flow control $u = T' - T$,

$$u = \left(\frac{T^e T}{1 - c^{-1}z} \right)^{1/2} - T \tag{6.37}$$

yields optimal rate $u = \dot{T}$ in terms of temperature T and the Hamiltonian constant h:

$$\dot{T} = \left\{ \pm \sqrt{\frac{h}{cT^e}} \left(1 - \pm \sqrt{\frac{h}{cT^e}} \right)^{-1} \right\} T \tag{6.38}$$

A general form of this result is

$$\dot{T} = \xi(h, T^e) T \tag{6.39}$$

where

$$\xi(h, T^e) \equiv \pm \sqrt{\frac{h}{cT^e}} \left(1 - \pm \sqrt{\frac{h}{cT^e}} \right)^{-1} \tag{6.40}$$

is an intensity index. Positive ξ refers to heating of the resource fluid in the heat pump mode, and the negative value refers to cooling of this fluid in the engine mode.

Equation (6.39) describes the optimal trajectory in terms of state variable T and constant h. The corresponding optimal (Carnot) control is

$$T' - (\xi(h, T^e) + 1) T \tag{6.41}$$

Now one can find the (solution to the problem of) *Hamiltonian representation of extremum work*. Substituting temperature control (Equation (6.41)) into work functional (Equation (6.4)) and integrating work intensity along an optimal path yields the extremal work function

$$\begin{aligned} V(T^i, T^f, h) &= c(T^i - T^f) - cT^e \ln \frac{T^i}{T^f} + cT^e \frac{\xi(h)}{1 + \xi(h)} \ln \frac{T^i}{T^f} \\ &= c(T^i - T^f) - cT^e \ln \frac{T^i}{T^f} - cT^e \sqrt{\frac{h}{cT^e}} \ln \frac{T^i}{T^f} \end{aligned} \tag{6.42}$$

This expression is valid for every process mode. Integration of Equation (6.39) subject to boundary conditions $T(\tau^i) = T^i$ and $T(\tau^f) = T^f$ allows one to express

Equation (6.42) in terms of the process duration

$$V(T^i, T^f, \tau^i, \tau^f) = c(T^i - T^f) - cT^e \ln\frac{T^i}{T^f} - \frac{cT^e[\ln(T^i/T^f)]^2}{\tau^f - \tau^i - \ln(T^i/T^f)} \qquad (6.43)$$

6.7. EXTENSIONS FOR SYSTEMS WITH NONLINEAR KINETICS AND INTERNAL DISSIPATION

In systems governed by nonlinear kinetics, the intensity index ξ is no longer a state-independent constant. For a class of "pseudo-Newtonian" systems with an internal irreversibility Φ,

$$\dot{T} = \xi(h, T, T^e, \Phi)T \qquad (6.44)$$

(Sieniutycz and Kuran, 2006). The presence of resource temperature T in function ξ proves that the relaxation curve is no longer exponential. An example is the pseudo-Newtonian model of a radiative engine with

$$\dot{W} = -\dot{V} \int_T^{T^e} \left(c_h(T) - c_v(T)\Phi \frac{T^e}{T + \chi \, dT/dt} \right) dT \qquad (6.45)$$

where $c_h(T) \equiv c_v(T) + P_T = (16/3)a_0 T^3$, $P_T \equiv dP/dT = (4/3)a_0 T^3$, and $\chi = \rho c_v(a'a_v)^{-1}$. Note that the ratio dt/χ is the differential of the nondimensional number of transfer units, $d\tau$. An alternative form of the above integral is Equation (6.21), which has singled out the term with the entropy production. In terms of volumetric flow $\dot{V}$ and Carnot temperature $T' = T + \chi v$,

$$\dot{W} = -\dot{V} \int_T^{T^e} \left(c_h(T) - c_v(T)\frac{T^e}{T} \right)(T' - T') \, dt$$
$$-T^e \dot{V} \int_t^{t^f} \left(c_v(T) \left(\frac{(T' - T)^2}{\chi T T'} + (1 - \Phi)\left(\frac{T' - T}{\chi T'} \right) \right) \right) dt \qquad (6.46)$$

As the Carnot temperature expression for this case has Newtonian origin, that is, $T' = T + \chi v$, this particular expression is appropriate for Newtonian and pseudo-Newtonian models, but is inapplicable for more rigorous models based on Stefan–Boltzmann equations.

The application of variational calculus to nonlinear radiation fluids with the temperature-dependent heat capacity $c_v(T) = 4a_0 T^3$ yields a pseudo-exponential extremal (6.44). (Constant a_0 is related to the Stefan–Boltzmann constant; $a_0 = 4\sigma/c$.) To single out the effect of environmental temperature from

$\xi(h,T,T^e,\varPhi)$, entropy production Hamiltonian $h_\sigma = h/T^e$ is introduced in place of h. The optimal relaxation curve then has the form

$$\dot{T} = \left(\pm\sqrt{\frac{h_\sigma}{\varPhi c_v(T)}} \left(1 - \pm\frac{h_\sigma}{\varPhi c_v(T)}\right)^{-1} \right) T \equiv \xi(h_\sigma, \varPhi, T)T \qquad (6.47)$$

Here the slope of the logarithmic rate $\xi = d \ln T/d\tau$ is a state-dependent quantity. Operative Carnot control assuring the extremum power yield along an optimal path is

$$T' = (1 + \xi(h_\sigma, \varPhi, T))T \qquad (6.48)$$

The slope $\xi(h_\sigma, \varPhi, \tau)$ is a rate indicator, positive for the resource's heating and negative for cooling. ξ is constant in Newtonian energy exchange for resources with a constant c_v. When thermal resistance of the environment is ignored, for example, when the entropy production is only due to emission and adsorption of radiation, the overall coefficient of energy transfer, α', varies proportionally to T^{a-1} (T^3 for radiation), and so does c_v. In this case $\chi = dt/d\tau$ is constant, or times t and τ are proportional. In terms of Hamiltonian of the minimum entropy production, $h_\sigma = h/T^e$, instantaneous values of ξ are described by a generalization of Equation (6.40) for $\varPhi \neq 1$,

$$\xi(h_\sigma, \varPhi, T) \equiv \left(\pm\sqrt{\varPhi c_v(T)h_\sigma^{-1} - 1} \right)^{-1} \qquad (6.49)$$

Like ξ, Hamiltonian h_σ is a rate indicator that vanishes for quasistatic processes.
For the radiation $c_v(T) = 4a_0 T^3$, where a_0 is a constant, an optimal trajectory solving Equations (6.48) and (6.49) has the form

$$\pm\frac{4}{3}a_0^{1/2}\varPhi^{1/2}h_\sigma^{-1/2}\left(T^{3/2} - T^{i3/2}\right) - \ln\frac{T}{T^i} = \tau - \tau^i \qquad (6.50)$$

The integration limits refer to the initial state (i) and a current state of the radiation fluid, that is, temperatures T^i and T corresponding with τ^i and τ.
Optimal curve (6.50) is illustrated in Figure 6.2. It refers to the case when the radiation relaxation is subject to an operative constraint imposed on T (resulting in nonvanishing partial derivatives of V versus T). This causes a significant deviation of the relaxation curve from the exponential curve.
Equations (6.49) and (6.50) are associated with the entropy production term in Equations (6.21) and (6.46). The corresponding extremal work function per

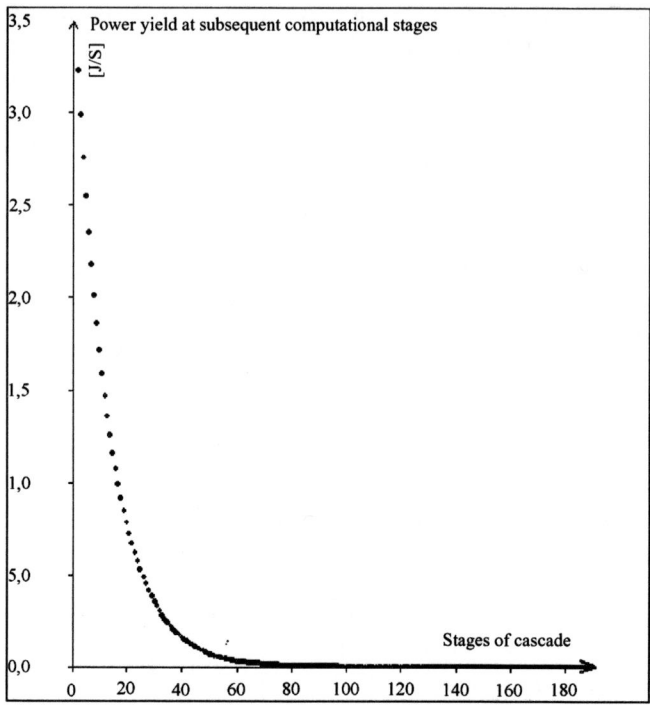

Figure 6.2 Power relaxation curve for black radiation without constraint on the final temperature.

unit volume of flowing radiation is

$$V \equiv h_\nu^i - h_\nu^f - T^e(s_\nu^i - s_\nu^f) - \frac{4}{3}a_0^{1/2}h_\sigma^{1/2}\Phi^{1/2}T^e(T^{i3/2} - T^{f3/2})$$

$$+\frac{4}{3}a_0T^e(1 - \Phi)(T^{i3} - T^{f3}) \tag{6.51}$$

Also, the corresponding exergy function, Equation (6.52) below, has an explicit analytical form.

6.8. GENERALIZED EXERGIES FOR NONLINEAR SYSTEMS WITH MINIMUM DISSIPATION

6.8.1. Radiation as a Pseudo-Newtonian Resource

By using pertinent boundary conditions, a formula for the finite-rate exergy of radiation is obtained from Equation (6.51). The particular extremal work leading to an exergy (generalized or classical) should contain the environment temperature as one of the boundary states. The finite-rate exergy is the maximal work $W_{max} = V(T^i, t^i, T^f, t^f)$ with $T^i = T$ and $T^f = T^e$ for the engine mode, and the negative minimal work $(-W)_{min} = -V = R(T^i, t^i, T^f, t^f)$ with $T^i = T^e$ and $T^f = T$ for the heat pump mode. Equations (6.49)–(6.51) yield a finite-rate exergy (also called generalized availability):

$$A_v(T, T^e, b_\sigma) = A_v^{class}(T, T^e, 0) \pm \frac{4}{3} a_0^{1/2} b_\sigma^{1/2} \Phi T^e (T^{3/2} - T^{e3/2})$$
$$+ \frac{4}{3} a_0 T^e (1 - \Phi)(T^3 - T^{e3}) \tag{6.52}$$

The upper sign refers to the heat pump mode, the lower one to the engine mode. The classical availability of *radiation at flow* resides in the above equation in Jeter's (1981) form:

$$A_v^{class}(T, T^e, 0) = b_v - b_v^e - T^e(s_v - s_v^e) = b_v \left(1 - \frac{T^e}{T}\right)$$
$$= \frac{4}{3} a_0 T^4 \left(1 - \frac{T^e}{T}\right) \tag{6.53}$$

This result proves that the well-known Petela's equation (Petela, 1964, 2003) should be restricted to the "enclosed radiation." As two modes are described, the common symbol T in Equations (6.52) and (6.53) refers to the initial temperature of engine mode or the final temperature of heat pump mode of the process. Likewise, variable duration of engine or heat pump modes, measured in terms of nondimensional time, is denoted in Figure 6.3 by the common symbol $\tau = \tau^f - \tau^i$.

6.8.2.   Two Finite Reservoirs, First One Filled up with Radiation

The solution to the maximum power problem involving two finite reservoirs, the first one filled up with radiation, was obtained by Chen's research group (Li et al., 2009). The system consists of two reservoirs of finite thermal capacity, in which the energy transfer between the heat source and the working fluid obeys the radiation law described by the pseudo-Newtonian model, and the energy transfer between the working fluid and heat sink obeys the linear law. The function describing maximum cumulative work (which is now *not* an exergy)

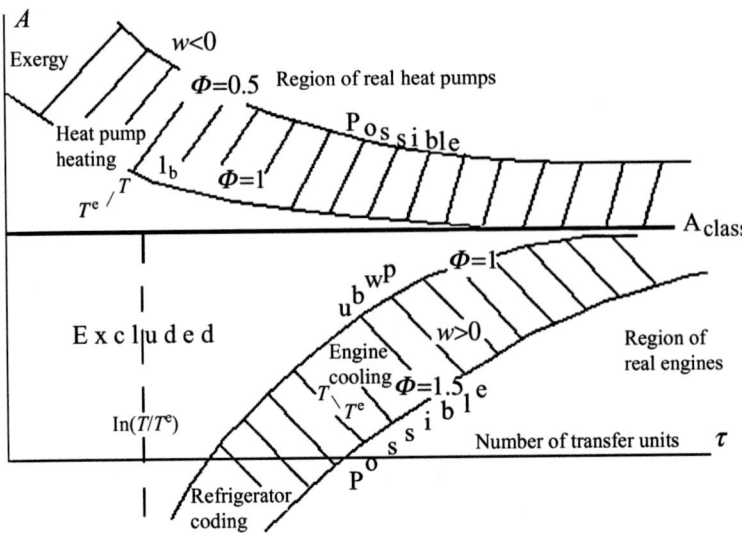

Figure 6.3 Influence of internal irreversibilities Φ on limiting finite-rate work generated in engines and consumed in heat pumps ($T_0 = T^e$). Generalized exergy prohibits processes from operating below the heat pump mode line for $\Phi = 0.5$ (lower bound for work supplied) and above the engine mode line for $\Phi = 1.5$ (upper bound for work produced). Weaker limits of classical exergy are represented by the straight line $A = A_{class}$. Dashed lines mark regions of possible improvements when imperfect thermal machines are replaced by those with better performance coefficients. Note a lower limit for the number of transfer units, below which no power production is possible in the system.

follows in the form of an equation

$$W_{max} = -4a \int_{T1i}^{T_{if}} 4T_1^3/3 \, dT_1 + 4aT_{2i} \int_{T_{1i}}^{T} (\xi^{-1} T_1^{1/2} - T_1^2)$$

$$\exp\left[8a\psi(T_1^{3/2} - T_{1i}^{3/2})/(3\xi) - 4a\psi(T_1^3 - T_{1i}^3)/3\right] dT_1 \quad (6.54)$$

$$= -4a(T_{1f}^4 - T_{1i}^4)/3 - T_{2i}\{\exp[8a\psi(T_{1f}^{3/2} - T_{1i}^{3/2})/(3\xi)$$

$$- 4a\psi(T_{1f}^3 - T_{1i}^3)/3] - 1\}/\psi$$

This corresponds with the optimal temperature curve (6.50) for the first fluid and the following temperature curve for the second fluid:

$$\ln\frac{T_2}{T_{2i}} = \left(\frac{8a\psi}{3\xi}\right)\left(\frac{T_1^{3/2}}{T_{1i}^{3/2}}\right) - \left(\frac{4a\psi}{3}\right)(T_1^3 - T_{1i}^3) \quad (6.55)$$

6.8.3. Compressible Newtonian Resource Without Viscous Friction

In this case, which returns to systems with an infinite second reservoir, integration can be performed analytically, and leads to the generalized exergy of unit volume in the form

$$A_v(T, T^e h_\sigma) = A_v(T, T^e, 0) + c_{p_v} T^e \left(\frac{\xi}{1+\xi}\right) \ln \left(\frac{T}{T^e}\right)$$

$$= c_{p_v} T^e \left[\left(\frac{T}{T^e} - 1\right) - \ln \left(\frac{T}{T^e}\right) + \ln \left(\frac{P}{P^e}\right)^{(k-1)/k}\right]$$

$$+ c_{p_v} T^e \left(\left(\pm \frac{h}{\Phi c_{p_v}}\right) + (1 - \Phi)\left(1 - \pm \frac{h}{\Phi c_{p_v}}\right)\right) \ln \frac{T}{T^e}$$

$$(6.56)$$

Compressibility is represented by the pressure (P) term. The last line term is nonclassical. For vanishing intensities h or ξ the classical thermal exergy is recovered. In fact, the result in the last line of Equation (6.56) agrees with the second law in the Gouy–Stodola form (Kotas, 1985).

Generalized exergies are irreversible extensions of the reversible work potential including minimally irreversible processes. Limits for energy yield or consumption provided by generalized exergies are stronger than those defined by the classical exergy. This property is caused by finite rates that increase a minimum work that must be supplied to the system and decrease a maximum work that can be produced by the system. These results help engineers with a better evaluation of energy limits in practical processes with radiation, especially in those with solar engines, solar driven heat pumps and solar cells.

6.9.   FINAL REMARKS

We considered energy limits in dynamic energy systems driven by nonlinear fluids that are restricted in their amount or flow, and, as such, play the role of resources. We discussed the main aspects of classical analytical HJB theory (differentiable V) and various HJB equations for a nonlinear power generation system. Applications of HJB theory subject to appropriate boundary conditions led to finite-rate generalizations of the standard availability (exergy), influenced by imperfect phenomena. In modes departing from the equilibrium this generalized exergy is different from their inversions approaching the equilibrium. Analytical value functions were obtained for systems with linear kinetics and their nonlinear extensions. Specific results show a complex, nonexponential form of the radiation relaxation.

The hysteretic property of power limits, which follows from the nonequivalence of the two modes discussed, is important. Both modes refer to different values of the extremum work obtained when a process which leaves the equilibrium is compared with its inverse which approaches the equilibrium. While in the reversible case both processes can be accomplished with exactly the same magnitude of work, in the dissipative case works consumed and produced are no longer equal. An essential decrease of the maximal work received from an engine system and an increase of the minimal work added to a heat pump system occurs in the high-rate regimes and for short durations (Figure 6.3).

Consequently, bounds for the mechanical energy yield and consumption provided by generalized exergies are mode-dependent and stronger than those defined by the classical exergy. Dashed lines in Figure 6.3 mark regions of improvement when imperfect thermal machines are replaced by those with better performance coefficients (note also endoreversible limits for Carnot machines). The results show that for an energy generator with thermally resistive media (i.e. media with finite heat conductivities) there exists a positive critical time t_c below which no power production is possible when the holdup time of a driving resource is less than t_c (or when the system's length is too short). This result limits to a significant degree the possibility of reducing investment costs in a converter's design; it is of kinetic origin, and is not predicted by approaches based on the classical exergy.

7 Numerical optimization in allocation, storage and recovery of thermal energy and resources

7.1. INTRODUCTION

This chapter describes selected numerical tools that are applied in the trajectory optimization of dynamic power production problems, with special attention paid to nonlinear systems. Numerical methods often provide the only way to solve the problem because in many cases the governing Hamilton–Jacobi–Bellman (HJB) equations neither can be solved analytically nor do they admit classical solutions. In the latter case the viscosity solutions remain the only option (Crandall and Lions, 1983). Systems with nonlinear kinetics, for example, radiation, are particularly difficult; therefore, discrete counterparts of continuous HJB equations and numerical approaches are useful. Discrete algorithms of dynamic programming (DP), which lead to work limits and generalized availabilities, are especially effective.

We consider convergence of discrete algorithms to viscosity solutions of HJB equations, solutions by discrete approximations, and the role of Lagrange multiplier λ associated with duration constraint. In the analytical discrete schemes, the Legendre transformation is a significant tool leading to the original work function. We also describe numerical algorithms of dynamic programming and consider dimensionality reduction in these algorithms. Finally, we outline some applications of the method for other systems, in particular chemical energy systems.

Analytical and computational aspects of energy limits refer to fluids that are restricted in their amount or flow, and, as such, play the role of resources. In the practical processes of engineering and technology the resource is a useful, valuable substance of a limited amount or flow. The value of the resource can be quantified thermodynamically by specifying its exergy or a maximum work that can be delivered when the resource is downgraded to equilibrium with the environment. Reversible relaxation of the resource is associated with the classical exergy. When some dissipative phenomena are allowed, generalized exergies are obtained. These exergies incorporate both the limited availability and a minimum work supply for the resource's production. In the classical exergy only the first property is taken into account (Sieniutycz, 2003a).

Knowledge of a power integral is required when calculating the exergy or exergy-like functions. A generalized exergy follows as the extremum of this power integral subject to the exergy initial conditions. In thermal systems this integral involves the product of thermal efficiency and the differential of exchanged energy. Various process models lead to diverse formulae for thermal efficiencies, which show how the efficiency of a practical system deviates from the Carnot efficiency. In fact, the generalized exergies somewhat quantify these deviations. Formally, an exergy follows from the principal function of the variational problem of extremum work (maximum of work delivered or minimum of work supplied) subject to the boundary condition of equilibrium with the environment. Variational solution to the work extremum problem contains an exergy-type function as its principal component. Other components are the optimal trajectory and optimal control. In thermal systems the trajectory is characterized by the temperature of the resource fluid $T(t)$, whereas the control variable may be efficiency η or Carnot temperature $T'(t)$. The latter quantity, defined as $T' \equiv T_2 T_{1'} / T_{2'}$ (see previous chapters and references, e.g. Sieniutycz, 2003c, 2007d), is particularly suitable for describing driving forces in energy systems. Whenever $T'(t)$ differs from $T(t)$, the resource relaxes to the environment with a finite rate determined by the efficiency deviation from the Carnot efficiency. Only when $T'(t) = T(t)$ the efficiency is Carnot, but this corresponds with an infinitely slow relaxation rate of the resource to the thermodynamic equilibrium.

In Chapter 6 we presented analytical aspects of the Hamilton–Jacobi–Bellman (HJB) and Hamilton–Jacobi equations for nonlinear thermal systems with power generation. Solutions to these equations were obtained for systems with linear kinetics. Rate-dependent exergies, generalizing classical exergies for systems with finite durations, were evaluated. Processes bounded by these generalized exergies involve various imperfect phenomena such as, for example, heat conduction or nonideal compression and expansion. In process modes departing from the equilibrium, generalized exergies are larger than in their inversions approaching the equilibrium. Corresponding bounds for the mechanical energy yield or consumption are stronger than those defined by the classical exergy (enhanced bounds).

In order to obtain generalized exergies (or corresponding functions describing energy limits) one has to solve appropriate HJB equations. The problem is not only that most of the optimal solutions cannot be obtained in the form of explicit analytical formulae. In fact, the HJB equations may not permit classical solutions at all, and one has to work with viscosity solutions (Section 7.3). This difficulty especially refers to systems with nonlinear kinetics (e.g. radiation systems). To overcome the difficulty, discrete counterparts of continuous equations are solved in numerical approaches to HJB equations. A few forms of discrete dynamic programming algorithms are especially efficient at solving continuous HJB equations.

Discrete Hamiltonians H^n are generally not constants of discrete autonomous paths. The motivation behind this chapter is to show that even if the constancy of a H^n is violated (e.g. by complex discretization) the Lagrange multiplier λ of the duration constraint still exhibits the constancy property. Therefore, it is the constant λ that is capable of decreasing the discrete problem dimensionality and thus simplifying the numerical optimization procedure. In fact, in the examples presented, the Lagrange multiplier λ takes over the role of the energy integral, and this is what makes the problem interesting.

Section 7.2 shows an example of a continuous model of power production from the black radiation and difference equations following from the model's discretization. Section 7.3 discusses the constancy preservation conditions for discrete Hamiltonians and convergence conditions of discrete numerical schemes to viscosity solutions of HJB equations, whereas Section 7.4 presents a dynamic programming equation for power yield from radiation. Section 7.5 elucidates a solving method by discrete approximations and introduces the time adjoint as a Lagrange multiplier. Mean and local intensities of discrete processes are discussed in Section 7.6. Section 7.7 shows the significance of the Legendre transform in recovering original work functions. Section 7.8 describes numerical procedures using dynamic programming, while Section 7.9 discusses dimensionality reduction in numerical DP algorithms. Section 7.10 presents the most essential conclusions.

7.2. A DISCRETE MODEL FOR A NONLINEAR PROBLEM OF MAXIMUM POWER FROM RADIATION

In the previous chapter we considered HJB equations for continuous, active (work producing) systems working with finite rates. Here we shall focus on discrete schemes solving these continuous equations, either analytically, via discrete approximations, or numerically, with the help of a computer.

Let us first recall a representative problem of minimum work consumed in a system subject to constraints imposed on dynamics and duration. Consider a dynamical radiation system, characterized by highly nonlinear kinetics (Sieniutycz and Kuran, 2005, 2006). For a *symmetric model* of power yield from radiation (both reservoirs consist of radiation) a suitable power integral (Sieniutycz and Kuran, 2006) is

$$\dot{W} = \int_{t^i}^{t^f} \dot{G}_c(T) \left(1 - \frac{\Phi T^e}{T'}\right) \beta \frac{T^a - T'^a}{(\Phi'(T'/T^e)^{a-1} + 1)T^{a-1}} \, dt \qquad (7.1)$$

In the physical space, power exponent $a = 4$ for radiation and $a = 1$ for a linear resource. The integrand of Equation (7.1) represents the intensity of generalized profit, f_0. In the engine mode integral (7.1) has to be maximized subject to the dynamical constraint ("state equation")

$$\frac{dT}{dt} = -\beta \frac{T^a - T'^a}{(\Phi'(T'/T^e)^{a-1} + 1)T^{a-1}} \qquad (7.2)$$

As it follows from the general theory (Chapter 2) the extremum conditions for the optimization problem involving Equations (7.1) and (7.2) are contained in the HJB equation

$$-\frac{\partial V}{\partial t} + \max_{T'(t)} \left\{ \left(\dot{G}_c(T) \left(1 - \Phi \frac{T^e}{T'} \right) + \frac{\partial V}{\partial T} \right) \beta \frac{T^a - T'^a}{(\Phi'(T'/T^e)^{a-1} + 1)T^{a-1}} \right\} = 0$$

(7.3)

where $V = \max \dot{W}$. Since it is impossible to solve this equation analytically, except for the case when $a = 1$, we describe here a way of numerical solving based on Bellman's method of dynamic programming (DP).

Considering computer needs we introduce a related discrete scheme

$$\dot{W}^N = \sum_{k=1}^{N} \dot{G}_c(T^k) \left(1 - \frac{\Phi T^e}{T'^k} \right) \beta \frac{T^{ka} - T'^{ka}}{(\Phi'(T'^k/T^e)^{a-1} + 1)T^{ka-1}} \theta^k \quad (7.4)$$

$$T^k - T^{k-1} = \theta^k \beta \frac{T'^{ka} - T^{ka}}{(\Phi'(T'^k/T^e)^{a-1} + 1)T^{ka-1}} \quad (7.5)$$

$$\tau^k - \tau^{k-1} = \theta^k \quad (7.6)$$

We search for the maximum of the sum (7.4) subject to discrete constraints (7.5) and (7.6). Our task is to define conditions when the numerical schemes of dynamic programming for the set (7.4)–(7.6) converge to solutions of the Hamilton–Jacobi–Bellman equation (7.3). The following section analyzes this problem in general terms (i.e. for arbitrary profit functions and constraints).

7.3. NONCONSTANT HAMILTONIANS AND CONVERGENCE OF DISCRETE DP ALGORITHMS TO VISCOSITY SOLUTIONS OF HJB EQUATIONS

Conditions determining when *discrete* optimization schemes converge to solutions of Hamilton–Jacobi–Bellman equations (HJB equations) are quite involved. Moreover, in the present approach they have to be linked with systematic studies of the problem of the Hamiltonian constancy preservation (Jordan and Polak, 1964a, 1964b; Sieniutycz, 1973a, 1973b, 1973c, 1973d; Szwast, 1979, 1988; Marsden, 1992; McLachlan, 1993; Quispel, 1995; Berry et al., 2000; Sieniutycz, 2006b). For discrete autonomous models the Hamiltonian constancy means that $H^n = H^{n-1}$, for nonautonomous ones the constancy refers to the enlarged Hamiltonian involving the time adjoint. The problem is quite old and started with traditional mathematics but before viscosity solutions of HJB equations were discovered and worked out. The theory of viscosity solutions (Crandall and Lions, 1983; Crandall et al., 1992) developed in the last two decades is nowadays the main theory concerning HJ and HJB equations. Its basic premise is that, in general, value functions are nonsmooth and hence solutions are sought in the viscosity sense. The computation typically amounts to suitable discretization and the use of discrete dynamic programming. The literature on this subject is wide, see for example, books by Barles (1994) and Bardi and Capuzzo-Dolcetta (1997). Nowadays the theory for a single Hamilton–Jacobi equation is quite complete and ensures existence and uniqueness results for nondifferentiable and even discontinuous solutions. In parallel with the strong theoretical activity, many approximation techniques have been proposed in the last 20 years so that the numerical analysis of Hamilton–Jacobi equations has also experienced fast growth (Barles and Souganidis, 1991; Falcone and Makridakis, 2001; Falcone and Ferretti, 2006). This progress has also had an impact on the rectification and improvement of the optimization algorithms of the discrete maximum type for finite-difference control systems (Mordukhovich, 1988, 1995; Mordukhovich and Shvartsman, 2004). Moreover, starting with the work of Ishii (1985) on discontinuous Hamiltonians it has been shown that HJ and HBJ equations with discontinuous ingredient may arise in some continuous systems, as, for example, propagations of fronts in inhomogeneous media, geometric optics in the presence of layers and shape-from-shading problems (Coclite and Risebro, 2003, 2005, 2007). Hamilton–Jacobi equations are now considered whose coefficients have some kind of singularity and the Hamiltonian depends discontinuously on both the spatial and temporal location. The main results are the existence and well-posedness of a viscosity solution to the Cauchy problem (Tourin, 1992; Ishii and Ramaswamy, 1995). These works show that if the Hamiltonian H satisfies some structure conditions, essentially amounting to discontinuous jumps "in one direction", there exists a unique solution to the initial value problem (Capuzzo-Dolcetta and Perthame, 1996; Stromberg, 2003). In particular, for so-called shape-from-shading applications, Ostrov (1999, 2000) shows that a sequence of approximations defined by smoothing the coefficients converges to the unique viscosity solution defined as the solution of an auxiliary control problem. The technique of using the associated control problem (for convex Hamiltonians) is also exploited (Dal Maso and Frankowska, 2002; Karlsen et al., 2003) when studying some Hamilton–Jacobi equations with a discontinuous Hamiltonian. Connection with the conservation laws is also shown, and a quite general theory for conservation laws with discontinuous fluxes is established (Jin and Wen, 2005) with an entropy condition obtained for nonlinear parabolic equations with discontinuous coefficients.

Yet whenever H is the total energy of an autonomous system it has often been argued that its constancy must be preserved for physical reasons. In a series of works, Jin and co-workers have considered Hamiltonian-preserving schemes for the Liouville equation with discontinuous potentials (Jin and Wen, 2005, 2006a, 2006b, 2006c; Jin and Liao, 2006). Their designing principle is to build the behavior of waves at the interface – either cross over with a changed velocity according to a constant Hamiltonian, or be reflected with a negative velocity (or momentum) – into the numerical flux. These schemes are the Hamiltonian-preserving schemes. In general, they build into the numerical flux the wave-scattering information at the interface, and use the Hamiltonian-preserving principle to couple the wave numbers at both sides of the interface. The constancy of H and the canonical structure of some discrete optimization algorithms (Jordan and Polak, 1964a, 1964b; Szwast, 1988; Sieniutycz, 2006a, 2006b) also conform to present tendencies in physics in constructing numerical integration schemes for ordinary differential equations (ODEs) in such a way that a qualitative property of the solution of the ODE is exactly preserved. For example, they are Poisson-structure-preserving integration schemes (symplectic integrators), symmetries and related invariants, and so on. The optimal choice of free time intervals θ, necessary for the constancy of a discrete H, may be compared with the group of special-purpose integration methods collectively called the structure-preserving integrators (also called mechanical or geometric integrators). In these methods a local discretizing structure may be established without explicit recourse to an optimization criterion although it has to preserve (exactly) a number of important properties known for OD equations (Marsden, 1992; McLachlan, 1993; Quispel, 1995; Quispel and Dyt, 1998; McLachlan and Quispel, 1998; McLachlan et al., 1999).

The preference for differential schemes with continuous H follows also from the fact that the classical theory for viscosity solutions of Hamilton–Jacobi equations (Crandall and Lions, 1983; Crandall et al., 1992) does not cover the case where the Hamiltonian H is discontinuous. This is because the straightforward method of comparing sub- and supersolutions does not work if H is discontinuous in x or t. Therefore, the H-preserving models contain the very core that may be pursued nowadays in the modern context. This core refers to the energy conservation in discontinuous systems, the property that is important when H has the meaning of physical energy (or steady energy flux) of an autonomous system, where the energy must be constantly on an optimal path. This problem also appears in the context of variable resolution discretizations for high-accuracy solutions of optimal control problems (Munos and Moore, 2001; Falcone and Ferretti, 1994; Fialho and Georgiu, 1999). The latter work develops a variational approach involving the theory of viscosity solutions of Hamilton–Jacobi equations (Crandall and Lions, 1983) to establish the convergence of the Euler approximation schemes via discrete dynamic programming. Fialho and Georgiu (1999) also provide an algorithm to compute upper bounds for value functions and give the proof of convergence of approximation schemes for the control problem. Their case studies assess the robustness of a feedback system and the quality of trajectory tracking in the presence of uncertainty.

The approach of Hamiltonian preservation has generated both followers and adversaries. The former claim that in differential systems H must be constant in an autonomous case because of physical symmetry requirements and resulting conservation laws, the latter admit violations of H preservation as the mathematical possibility supported by analyses of solutions. Sometimes even the followers of Hamiltonian preservation admit systems with discontinuous Hamiltonians. For example, in a recent paper by Jin et al. (2007) the Hamiltonian is smooth in parts but with many discontinuities. The problem arises when classical particles move through media that contain barriers or interfaces at which the Hamiltonian is discontinuous. It also arises in the Lagrangian description of geometrical optics, or in the propagation of linear high frequency waves through interfaces. In fact a vast amount of research material has been collected to date showing that the constancy of Hamiltonians for both continuous and discrete models may not be preserved.

One may ask the question: how does the physics allow for and manage the violation of the Hamiltonian constancy? The answer, stemming from the present reasoning and earlier papers (Sieniutycz, 2006a, 2006b) is simple: the Hamiltonian describes the system energy only when the HJB and HJ equations exist; whenever they do not (as in systems with discontinuous solutions to the value function) the role of the physical energy is taken over by another quantity. This quantity is the Lagrange multiplier of the process duration constraint, the time adjoint which is constant in both discrete and continuous autonomous systems. In other words, Hamiltonians represent the energy only in limited situations. Hamiltonians of discrete processes (even Pontryagin's of energy-like structure) are particularly susceptible to violation of the Hamiltonian constancy. An arbitrary discretization mode of an original differential model violates the constancy preservation of the Hamiltonian (Sieniutycz, 2006a, 2006b), and only discrete models linear with respect to the time interval $\Delta t^n \equiv \theta^n$ and without local constraints imposed on θ^n can admit constant discrete Hamiltonians. Importantly, when these conditions are not satisfied, the Lagrange multiplier λ of the duration constraint still exhibits the constancy property. Therefore, the motivation behind this chapter is to show how to implement this Lagrange multiplier λ to decrease the problem's dimensionality and to simplify the numerical procedure of optimization.

To outline discrete structures and extremum conditions that may lead to either constant or nonconstant Hamiltonians we consider a family of optimization models obtained by discretization of original continuous ones. In this case one has to determine the necessary optimality conditions of a general discrete process governed by a work criterion W^N:

$$W^N = \sum_{n=1}^{N} f_0(\mathbf{x}^n, t^n, \mathbf{u}^n, \theta^n)\theta^n \tag{7.7}$$

subject to constraints resulting from difference equations

$$x_i^n - x_i^{n-1} = f_i(\mathbf{x}^n, t^n, \mathbf{u}^n, \theta^n)\theta^n \equiv f_i(\tilde{\mathbf{x}}^n, \mathbf{u}^n, \theta^n)\theta^n \tag{7.8}$$

The scalar f_0 is the rate of the profit generation. Superscripts refer to stages and subscripts to coordinates. The integer n ($n = 1, \ldots, N$) is usually called discrete time, the entity that should be distinguished from continuous time t. The latter is usually the physical time (t is the chronological time in unsteady-state operations and holdup or residence time in steady cascade operations). Both n and t are monotonously increasing. The s-dimensional vector $\mathbf{x} = (x_1, \ldots, x_s)$ is the state vector, and the r-dimensional vector $\mathbf{u} = (u_1, \ldots, u_r)$ is the control vector, where $\mathbf{x}^n \in E^s$, $\mathbf{u}^n \in E^r$ and rate functions f_0^n and f_i^n are continuously differentiable always in $\mathbf{x}$ and θ, but not always in $\mathbf{u}$. The rate change of state coordinate x_i in time t is the ith component of s-dimensional vector of rates, $\mathbf{f}$. The change of time t through the stage n defined as $\theta^n = t^n - t^{n-1}$ is called the time interval.

Various discretization schemes for constraining differential equations (Equation (7.2)) lead to discrete models (e.g. Equation (7.8)) either linear or nonlinear in time interval θ^n. While θ^n is a control-type quantity, it is excluded from the coordinates of vector $\mathbf{u}$, that is, it is treated separately in the optimization model. The $s + 1$-dimensional vector $\tilde{\mathbf{x}} = (\mathbf{x}, t)$ is also used which is the enlarged state vector describing the space–time. Usually, one assumes that a control sequence $\{\mathbf{u}^n\}$ and the corresponding trajectory $\{\mathbf{x}^n, t^n\}$ are admissible, that is, that they satisfy the control constraint $u^n \in U^n$ and the state–space constraint $\tilde{x}^n \in \tilde{X}^n$.

For work production problems criterion (7.7) is maximized; for work consumption a minimum of (7.1) is sought. In the optimization problems with constrained duration $t^N - t^0$ (the so-called fixed-horizon problem), a discrete model must explicitly include an equation defining time interval θ^n, either as the increment of a monotonously increasing state coordinate satisfying an equation $x_{s+1}^n - x_{s+1}^{n-1} = \theta^n$ or as the increment of usual time

$$t^n - t^{n-1} = \theta^n. \tag{7.6}$$

The monotonic increase of the time-like coordinate, implying nonnegative θ at each stage n, is crucial for many properties of model (7.7)–(7.9). Throughout the whole chapter models with free (unconstrained) intervals θ^n are considered as only those that are able to achieve their own continuous limit.

Two classes of discrete models, linear and nonlinear in free θ^n, should be distinguished when considering convergence of their optimality conditions to viscosity solutions of continuous Hamilton–Jacobi–Bellman (HJB) equations. In the first class, HJB equations follow straightforwardly from the optimality conditions. In the second class, a condition of weak nonlinearity of the discrete rates with respect to θ^n (discussed below) must be satisfied.

In the second class, the Hamiltonian of an autonomous system is not constant (as in the first class) but satisfies the condition

$$H^{n-1} + \theta^n \left(\frac{\partial H^{n-1}}{\partial \theta^n} \right) = H_1^{n-1} = \lambda \tag{7.9}$$

where H_1^{n-1} is the Hamiltonian of the first class. Below we outline the optimization scheme that substantiates the above condition and its further use in the dimensionality reduction of the dynamic programming algorithms (Bellman, 1961; Aris, 1964).

In each case some locally sufficient optimality criteria are valid in discrete systems (CBS criteria: Rund, 1966; Boltyanski, 1969; Sieniutycz, 1991). For a value function V^n (maximum work function in radiation example)

$$V^n \equiv \max \sum_{k}^{n} f_0^k(\mathbf{x}^k, t^k, \mathbf{u}^k, \theta^k)\theta^k \tag{7.10}$$

a general optimality criterion at stage n that uses the θ-dependent rates has the form

$$0 = \max_{\mathbf{u}^n, \theta^n, \mathbf{x}^n, t^n} \{f_0^n(\mathbf{x}^n, t^n, \mathbf{u}^n, \theta^n)\theta^n$$

$$-(V^n(\mathbf{x}^n, t^n) - V^{n-1}(\mathbf{x}^n - \mathbf{f}^n(\mathbf{x}^n, t^n, \mathbf{u}^n, \theta^n)\theta^n, t^n - \theta^n))\} \tag{7.11}$$

In what follows we shall also use functions of cost type, local costs $l_0 = -f_0$ and integral optimal costs $R = -V$.

For a special mode of discretization with θ-free rates f criterion, Equation (7.11), refers to the "standard" or "canonical" model associated with all participating equations linear in θ^n. It may then be shown that these models are characterized by a constant Hamiltonian that satisfies a discrete HJB equation whose discrete value function converges under the reasonable assumption of absence of local constraints on θ^n to the unique viscosity solution which is the value function of the original problem. Criterion (7.11) can be applied to derive a set of usable (Hamiltonian-based) optimality conditions including those with respect to $\mathbf{x}^n$ and time t^n. Using Equation (7.11) one can pass to an algorithm of discrete maximum principle and related canonical equations. This issue is only briefly discussed here; for our purposes a dynamic programming equation (see below) is most essential as it provides direct information about the extremum performance function. Importantly for applications in power systems, Bellman's equation of dynamic programming follows in "forward" form from criterion (7.11) for fixed final states and times.

Below we present a set of special optimality conditions for the process described by Equations (7.7)–(7.11) and characterized by differentiable V. Convexity of rate functions and constraining sets is assumed. From these conditions those for subsystems linear in θ^n easily follow for θ-independent rates and a Hamiltonian.

Optimizing time intervals θ^n and controls $\mathbf{u}^n$ of Equation (7.11) within the admissible control set leads respectively to two extremum conditions:

$$
\frac{-\partial V^{n-1}}{\partial t^{n-1}} + f_0^n(\mathbf{x}^n, t^n, \theta^n, \mathbf{u}^n) - \left(\frac{\partial V^{n-1}}{\partial \mathbf{x}^{n-1}}\right) \cdot \mathbf{f}^n(\mathbf{x}^n, t^n, \mathbf{u}^n, \theta^n)
$$

$$
+ \frac{\theta^n \partial \{f_0^n(\mathbf{x}^n, t^n, \theta^n, \mathbf{u}^n) - (\partial V^{n-1}/\partial \mathbf{x}^{n-1}) \cdot \mathbf{f}^n(\mathbf{x}^n, t^n, \mathbf{u}^n, \theta^n)\}}{\partial \theta^n} = 0 \tag{7.12}
$$

and

$$
\frac{\partial \{f_0^n(\mathbf{x}^n, t^n, \theta^n, \mathbf{u}^n) - (\partial V^{n-1}/\partial \mathbf{x}^{n-1}) \cdot \mathbf{f}^n(\mathbf{x}^n, t^n, \mathbf{u}^n, \theta^n)\}}{\partial \mathbf{u}^n} = 0 \tag{7.13}
$$

For convex functions and constraining set $u^n \in U^n$, stationarity condition (7.13) allows us to find optimal control $\mathbf{u}^n$ from the maximum condition for a Hamiltonian expression

$$
\mathbf{u}^n = \underset{\mathbf{u}^n}{\mathrm{argmax}} \left\{ \frac{-\partial V^{n-1}}{\partial t^{n-1}} + f_0^n(\mathbf{x}^n, t^n, \theta^n, \mathbf{u}^n) - \left(\frac{\partial V^{n-1}}{\partial \mathbf{x}^{n-1}}\right) \cdot \mathbf{f}^n(\mathbf{x}^n, t^n, \theta^n, \mathbf{u}^n) \right\}
$$

$$\tag{7.14}$$

Discrete HJB Equations

It follows from Equations (7.12) and (7.13) that a discrete HJB equation for the optimum profit function

$$
-\frac{\partial V^{n-1}}{\partial t^{n-1}} + \max_{\mathbf{u}^n} \left\{ f_0^n(\mathbf{x}^n, t^n, \mathbf{u}^n) - \left(\frac{\partial V^{n-1}}{\partial \mathbf{x}^{n-1}}\right) \cdot \mathbf{f}^n(\mathbf{x}^n, t^n, \mathbf{u}^n) \right\} = 0 \tag{7.15}
$$

or an equivalent equation, in terms of the cost functions $R = -V$ and $l_0 = -f_0$,

$$
\frac{\partial R^{n-1}}{\partial t^{n-1}} + \max_{\mathbf{u}^n} \left\{ -l_0^n(\mathbf{x}^n, t^n, \mathbf{u}^n) + \left(\frac{\partial R^{n-1}}{\partial \mathbf{x}^{n-1}}\right) \cdot \mathbf{f}^n(\mathbf{x}^n, t^n, \mathbf{u}^n) \right\} = 0 \tag{7.16}
$$

can be obtained for convex models linear in θ^n (this is why variable θ^n is absent in rates of Equations (7.15) and (7.16)). However, Equation (7.12) proves that the classical Hamilton–Jacobi–Bellman structure is not attained when the discrete model is nonlinear in time intervals θ^n. Equation (7.14) describes, in fact, a maximum principle with respect to $\mathbf{u}^n$ for a "Hamiltonian", that is, the expression in braces of this equation. The principle is here written for the so-called enlarged Hamiltonian that includes the partial time derivative of V. The above findings are summarized by the following lemma.

Lemma (maximum principle)

Let control constraint $u^n \in U^n$ and state–space constraint $\tilde{x}^n \in \tilde{X}^n$ be convex sets. Then the necessary condition for the maximum of criterion W^N with respect to the control sequence $\{u^n\}$ is the maximum for the Hamiltonian expression

$$H^{n-1} \equiv -l_0^n(\mathbf{x}^n, t^n, \mathbf{u}^n) + \left(\frac{\partial R^{n-1}}{\partial \mathbf{x}^{n-1}}\right) \cdot \mathbf{f}^n(\mathbf{x}^n, t^n, \mathbf{u}^n) \qquad (7.17)$$

that appears in the discrete Hamilton–Jacobi–Bellman equations of the optimization problem, Equations (7.15) and (7.16). The proof is the immediate consequence of Equations (7.12), (7.13) and (7.14) (the latter follows from Equations (7.11) and (7.13) under the convexity assumptions). Note that while the maximum of H^{n-1} holds in the state space $\mathbf{x}^n$, the lemma leads directly to the maximum principle in the phase space, where phase–space Hamiltonian is defined as follows:

$$H^{n-1}(\mathbf{x}^n, z^{n-1}, t^n, \mathbf{u}^n, \theta^n) \equiv -l_0^n(\mathbf{x}^n, t^n, \mathbf{u}^n, \theta^n) + \sum_{i=1}^{s} z_i^{n-1} f_i^n(\mathbf{x}^n, t^n, \mathbf{u}^n, \theta^n)$$

$$(7.18)$$

The phase–space Hamiltonian (of energy type) is the sum of profit intensity $f_0^n = -l_0^n$ and the scalar product of state adjoints

$$z_k^{n-1} \equiv \frac{\partial R^{n-1}}{\partial x_k^{n-1}}, \quad k = 1, 2, \ldots s \qquad (7.19)$$

and corresponding rates f_k^n. Within the continuous limit and for problems of analytical mechanics state adjoints are identical with the generalized momenta, yet the extremum Hamiltonian does not represent the system's energy when rates f_i^n contain explicitly time intervals θ^n. This is the price of arbitrary discretization.

Discrete Hamilton–Jacobi Equations

When optimal controls θ^n and u^n are evaluated from Equations (7.12) and (7.13) and next substituted into Equation (7.12), the resulting equation constitutes a discrete counterpart of the Hamilton–Jacobi equation of continuous processes.
   For models linear with respect to θ^n one obtains

$$\frac{\partial R^{n-1}}{\partial t^{n-1}} + H^{n-1}\left(x_1^n, \ldots, x_s^n, t^n, \frac{\partial R^{n-1}}{\partial x_{x_1}^{n-1}}, \ldots, \frac{\partial R^{n-1}}{\partial x_s^{n-1}}\right) = 0 \qquad (7.20)$$

whereas for those nonlinear in θ^n Hamiltonian H^{n-1} is replaced by partial derivative

$$h_\theta \equiv \frac{\partial(\theta^n H^{n-1})}{\partial \theta^n}$$

$$\frac{\partial R^{n-1}}{\partial t^{n-1}} + h_\theta{}^{n-1}\left(x_1^n, \ldots, x_s^n, t^n, \frac{\partial R^{n-1}}{\partial x_1^{n-1}}, \ldots, \frac{\partial R^{n-1}}{\partial x_s^{n-1}}\right) = 0 \qquad (7.21)$$

In both cases optimal controls are expressed in terms of state coordinates and state adjoints. However strange, the replacement of H^{n-1} by the derivative $h_\theta \equiv \partial(\theta^n H^{n-1})/\partial \theta^n$ in the second case leads, in fact, to a valid structure that is explained in the enlarged space–time in terms of the (vanishing) partial derivative of an enlarged Hamiltonian function with respect to free control, θ^n (Sieniutycz, 2006b). Note that both equations are nonlinear in terms of $\partial R^{n-1}/\partial x^{n-1}$.

In the general case of nonsmooth solutions the results discussed above lead to convergence conditions of discrete computational schemes to viscosity solutions of continuous Hamilton–Jacobi equations of physical processes. To formulate these convergence conditions for discrete models whose rates contain explicitly intervals θ a notion of the weak dependence of the discrete Hamiltonian on θ is important. Note that the Hamiltonian is a weighted measure quantifying dependence of the discrete rate vector on time interval θ.

Definition

The θ-differentiable Hamiltonian function H^{n-1} is said to be weakly dependent on θ^n in the vicinity of 0+ if for any positive number ε a positive number η exists such that for sufficiently small positive $\theta^n \le \eta$ the absolute value of the product $I\theta^n\partial(H^{n-1})/\partial\theta^n$, $I \le \varepsilon$, for $n = 1, 2, \ldots, N$. This means that in the vicinity of 0+ the reciprocal of derivative $\partial(H^{n-1})/\partial\theta^n$ tends to zero slower than θ^n itself, or that $\lim(\theta^n\partial(H^{n-1})/\partial\theta^n) = 0$ for $\theta^n \to 0+$ regardless of the form of rate functions f_k. In fact, many popular discretization schemes lead to discrete rates and Hamiltonians that are weakly dependent on θ^n for positive θ in the vicinity of the point $\theta = 0$.

For Hamiltonians weakly dependent on θ^n the following corollary holds.

Corollary

Assume fixed end states (x^0, x^N), end times (t^0, t^N) and an arbitrarily large number M. Observe that for a sufficiently large total number of stages $N > M$ each free interval θ^n is sufficiently close to zero, that is, $\lim\theta^n = 0+$ for $N \to \infty$ as the consequence of the monotonic property of time coordinate t. Then for unconstrained intervals θ^n and each Hamiltonian function $H^{n-1}(x_1^n, \ldots, x_s^n, t^n, -(\partial V^{n-1}/\partial x_1^{n-1}) \ldots (\partial V^{n-1}/\partial x_s^{n-1}), \theta^n)$ weakly dependent on θ^n in the vicinity of 0+, the classical Hamilton–Jacobi equation holds in the limit of $N \to \infty$.

Proof. For Hamiltonians H^{n-1} weakly dependent on θ^n in the vicinity of 0+ and a sufficiently large N the derivative $\partial(\theta^n H^{n-1})/\partial\theta^n$ is sufficiently close to H^{n-1} for each $n = 1, \ldots, N$. In the limiting case of $N \to \infty$ the sequence of free θ^n satisfies the conditions: $\lim \theta^n = 0+$ and $\lim \partial(\theta^n H^{n-1})/\partial\theta^n = \lim H^{n-1} = H(\mathbf{x}, t, \partial R/\partial\mathbf{x})$. Equation (7.21) then goes over into the Hamilton–Jacobi equation of a continuous process

$$\frac{\partial R}{\partial t} + H\left(x_1, \ldots, x_s, t, \frac{\partial R}{\partial x_1}, \ldots, \frac{\partial R}{\partial x_s}\right) = 0 \tag{7.22}$$

The above corollary assures that the limiting Hamilton–Jacobi equation of a continuous system is obtained in a unique form (7.22), regardless of the discretization mode and no matter if the underlying discrete equation has the form of Equations (7.20) or (7.21). Discrete Hamiltonians weakly dependent on θ^n in the vicinity of 0+ are common in thermodynamic systems described in spaces of arbitrary state variables rather than extensive thermodynamic coordinates.

The discrete dynamics described in terms of the Hamiltonian are omitted in this chapter, where the theoretical schemes of dynamic programming leading to optimal performance functions V and R are essential. The reader interested in the first issue is referred to the literature (Sieniutycz, 2006b).

In the general case of nonsmooth solutions for the value function V, the discretization modes that satisfy the condition of the weak dependence of process rates on free θ^n in the vicinity of 0+ have solutions convergent to the viscosity solutions of continuous Hamilton–Jacobi and HJB equations.

7.4.   DYNAMIC PROGRAMMING EQUATION FOR MAXIMUM POWER FROM RADIATION

The previous section has shown that, regardless of the discretization mode, equations modeling the continuous processes (including, of course, equations of power systems considered) can be solved numerically by the discrete dynamic programming (DP) algorithm associated with stage criterion (7.11). Consequently, one can use the DP method associated with Bellman's recurrence equation

$$\begin{aligned} R^n(\mathbf{x}^n, t^n) = \min_{\mathbf{u}^n, \theta^n} \{ l_0^n(\mathbf{x}^n, t^n, \mathbf{u}^n, \theta^n)\theta^n \\ + R^{n-1}(\mathbf{x}^n - \mathbf{f}^n(\mathbf{x}^n, t^n, \mathbf{u}^n, \theta^n)\theta^n, t^n - \theta^n) \} \end{aligned} \tag{7.23}$$

The enlarged Hamiltonian is not constant as in the linear, free-θ theory but satisfies the condition (7.9), where λ is the Lagrange multiplier of the duration constraint and H_1^{n-1} is the Hamiltonian of the discrete model whose rates do not involve time interval. Therefore, difference models linear in θ^n (those with θ-independent rates f_k) are the primary candidates to accomplish the efficient solution of continuous equations characterized by their own Hamilton–Jacobi–Bellman equations and Hamilton–Jacobi equations.

We can now return to the difficult radiation problem described by Equations (7.4)–(7.6). Applying Equation (7.23) to this problem, the following recurrence equation is obtained:

$$
R^n(T^n, t^n) = \min_{\mathbf{u}^n, \theta^n} \left\{ \dot{G}_c(T^n) \cdot \left(1 - \frac{\Phi T^e}{T'^n}\right) \beta \frac{T'^{n^a} - T''^{n^a}}{(\Phi(T'^n/T^e)^{a-1} + 1)T'^{n^{a-1}}} \theta^n \right.
$$

$$
\left. + R^{n-1}\left(T^n - \theta^n \beta \frac{T'^{n^a} - T^{n^a}}{(\Phi'(T'^n/T^e)^{a-1} + 1)T'^{n^{a-1}}}, t^n - \theta^n\right)\right\}
$$

(7.24)

While the analytical treatment of Equations (7.1) and (7.2) is a tremendous task, it is quite easy to solve recurrence equation (7.24) numerically. Low dimensionality of the state vector for Equation (7.24) assures reasonable accuracy of the DP solution. Moreover, the original accuracy can be significantly improved by performing the so-called dimensionality reduction associated with the elimination of time t^n as the state variable. In the transformed problem, without coordinate t^n, accuracy of DP solutions is high. Section 7.9 discusses related computational issues.

7.5. DISCRETE APPROXIMATIONS AND TIME ADJOINT AS A LAGRANGE MULTIPLIER

In this section we consider another discrete example whose DP solution converges to the viscosity solution of a continuous problem of power generation. Yet, for brevity of formulas, we restrict ourselves to systems in which nonlinearities are absent in process kinetics although they are still present in the power expression. We consider solutions of HJB equations by discrete approximations (produced by the method of dynamic programming) in association with state dimensionality reduction (elimination of time coordinate) by using a Lagrange multiplier λ.

First we shall outline the generation of certain suboptimal costs in terms of the Lagrangian multiplier λ associated with the duration constraint as the time adjoint λ is constant in autonomous systems.

Consider a minimum of consumed work with constraints imposed on discrete dynamics and process duration

$$R^n(T^n, \tau^n) = \sum_{k=1}^{n} c\left(1 - \Phi' \frac{T^e}{T'^k}\right)(T^k - T^{k-1})$$

$$= \sum_{k=1}^{n} c\left(1 - \Phi' \frac{T^e}{T'^k}\right)(T'^k - T^k)\theta^k$$

$$\tag{7.25}$$

$$T^k - T^{k-1} = \theta^k(T'^k - T^k) \tag{7.26}$$

$$\tau^k - \tau^{k-1} = \theta^k \tag{7.27}$$

Observe that the above difference equations model a continuous problem of minimum work subject to the kinetic constraint $\dot{T} = T' - T$ (Sieniutycz, 2007c, 2007d).

Exploiting the constancy of λ we eliminate state variable τ by introducing a (primed) criterion of modified work:

$$R'^n(T^n, \lambda) = \min \sum_{k=1}^{n} \left\{ c\left(1 - \Phi' \frac{T^e}{T'^k}\right)(T^k - T^{k-1}) + \lambda\theta^n \right\} \tag{7.28}$$

or, in view of state equation (7.26),

$$R'^n(T^n, \lambda) = \min \sum_{k=1}^{n} \left\{ c\left(1 - \Phi' \frac{T^e}{T'^k}\right) + \frac{\lambda}{T'^k - T^k} \right\}(T^k - T^{k-1})$$

$$\tag{7.29}$$

In this problem, the idea of parametric representations for the principal performance function, Lagrange multiplier and process duration has proven its usefulness. While these representations are unnecessary for linear optimization problems, they are quite effective for describing solutions of nonlinear problems, where the optimal work, Lagrange multiplier and optimal duration are obtained in terms of an optimal control variable as a parameter.

To begin with we determine optimality conditions from Equation (7.29). We consider two initial stages of the process, 1 and 2. A procedure leading to parametric representations is defined below.

The equation of work modified by the presence of the Lagrange multiplier λ, yet without a minimization sign,

$$R'^1(T^1, T'^1, \lambda) = \left\{ c \left(1 - \varPhi' \frac{T_2}{T'^1} \right) + \frac{\lambda}{T'^1 - T^1} \right\} (T^1 - T^0) \qquad (7.30)$$

constitutes a component of the parametric representation of $R'^1(T^1, \lambda)$ provided that the following procedure is implemented:

1. λ is determined from the extremum condition of work function R'^1 with respect to a control variable, here with respect to Carnot control:

$$\lambda = c\varPhi'T^e \frac{(T'^1 - T^1)^2}{(T'^1)^2} \qquad (7.31)$$

2. Extremum λ is substituted into the work function R'^1, and the result of this substitution

$$R'^1(T^1, T'^1) = \left\{ c \left(1 - \varPhi' \frac{T^e}{T'^1} \right) + c\varPhi'T^e \frac{T'^1 - T^1}{(T'^1)^2} \right\} (T^1 - T^0) \qquad (7.32)$$

is taken together with the stationary λ.

In view of the above, parametric representation of work function $R'^1(T^1, \lambda)$ in terms of Carnot control as a parameter is given by the set of Equations (7.31) and (7.32), or, after simplification of the work equation

$$R'^1(T^1, T'^1) = c \left(1 - \varPhi'T^e \frac{T^1}{(T'^1)^2} \right) (T^1 - T^0) \qquad (7.33a)$$

$$\lambda = c\varPhi'T^e \left(1 - \frac{T^1}{T'^1} \right)^2 \qquad (7.33b)$$

In this example it is possible to eliminate the parameter T'^1 which leads to an explicit function of work consumption for $n = 1$:

$$R'^1(T^1, T^0, \lambda) = c(T^1 - T^0) - c\varPhi'T^e \left(1 - \pm\sqrt{\frac{\lambda}{c\varPhi'T^e}} \right) \left(\frac{T^1 - T^0}{T^1} \right)$$

$$(7.34)$$

The corresponding optimal control satisfies an equation

$$T'^1 = T^1 \left(1 - \pm \sqrt{\frac{\lambda}{c\Phi'T^e}} \right)^{-1} \tag{7.35}$$

obtained by solving Equation (7.31) with respect to T'^1.

It should be kept in mind that elimination of the parameter is not always possible, and then parameter-dependent functions are the only representation of the solution. It is in just this case that the parametric representations are inevitable and helpful.

Let us proceed further. Optimal work supply to two-stage system $R'^2(T^2, \lambda)$ is described by the equation

$$R'^2(T^2, \lambda) = \min_{T^1} \left\{ c(T^2 - T^1) - c\Phi'T^e \left(1 - \pm \sqrt{\frac{\lambda}{c\Phi'T^e}} \right)^2 \left(\frac{T^2 - T^1}{T^2} \right) \right.$$
$$\left. + c(T^1 - T^0) - c\Phi'T^e \left(1 - \pm \sqrt{\frac{\lambda}{c\Phi'T^e}} \right)^2 \left(\frac{T^1 - T^0}{T^1} \right) \right\}$$

whence, after making simplifications

$$R'^2(T^2, \lambda) = \min_{T^1} \left\{ c(T^2 - T^0) \right.$$
$$\left. - c\Phi'T^e \left(1 - \pm \sqrt{\frac{\lambda}{c\Phi'T^e}} \right)^2 \left(\frac{T^2 - T^1}{T^2} + \frac{T^1 - T^0}{T^1} \right) \right\} \tag{7.36}$$

Optimal interstage temperature between stages 1 and 2, T^1, satisfies the stationarity condition for expression in the large bracket of the above equation

$$\frac{\partial}{\partial T^1} \left(\frac{T^2 - T^1}{T^2} + \frac{T^1 - T^0}{T^1} \right) = \frac{T^0}{(T^1)^2} - \frac{1}{T^2} = 0 \tag{7.37}$$

Therefore optimal interstage temperature T^1 is the geometric mean of boundary temperatures of both considered stages:

$$T^1 = (T^0 T^2)^{1/2} \tag{7.38}$$

The minimum value of optimized expression (7.37) is

$$\min_{T^1} \left(\frac{T^2 - T^1}{T^2} + \frac{T^1 - T^0}{T^1} \right) = \frac{T^2 - \sqrt{T^0 T^2}}{T^2} + \frac{\sqrt{T^0 T^2} - T^0}{\sqrt{T^0 T^2}} \tag{7.39}$$
$$= 2(1 - \sqrt{T^0/T^2})$$

This leads to the optimum work function for $n = 2$:

$$R'^2(T^2, \lambda) = c(T^2 - T^0) - 2c\Phi'T^{\mathrm{e}} \left(1 - \pm\sqrt{\frac{\lambda}{c\Phi'T^{\mathrm{e}}}} \right)^2 \left(1 - \left(\frac{T^0}{T^2} \right)^{1/2} \right) \tag{7.40}$$

For $n = 3$ we apply an expression for local work

$$K'^3(T^3, T^2, \lambda) = c(T^3 - T^2) - c\Phi'T^{\mathrm{e}} \left(1 - \pm\sqrt{\frac{\lambda}{c\Phi'T^{\mathrm{e}}}} \right)^2 \left(\frac{T^3 - T^2}{T^3} \right) \tag{7.41}$$

which has the same structure as the one-stage function of Equation (7.34), but the indices are shifted ahead by one. In the recurrence equation for $n = 3$, the explicit work function $R'^3(T^3, \lambda)$ is the result of optimization described by the following expression:

$$R'^3(T^3, \lambda) = \min_{T^2} \left\{ c(T^3 - T^2) - c\Phi'T^{\mathrm{e}} \left(1 - \pm\sqrt{\frac{\lambda}{c\Phi'T^{\mathrm{e}}}} \right)^2 \left(\frac{T^3 - T^2}{T^3} \right) \right.$$
$$\left. + c(T^2 - T^0) - 2c\Phi'T^{\mathrm{e}} \left(1 - \pm\sqrt{\frac{\lambda}{c\Phi'T^{\mathrm{e}}}} \right)^2 \left(1 - \left(\frac{T^0}{T^2} \right)^{1/2} \right) \right\}$$

After simplifying we obtain

$$R'^3(T^3, \lambda) = \min_{T^2} \left\{ c(T^3 - T^0) - c\Phi'T^{\mathrm{e}} \left(1 - \pm\sqrt{\frac{\lambda}{c\Phi'T^{\mathrm{e}}}} \right)^2 \left[\left(\frac{T^3 - T^2}{T^3} \right) \right. \right.$$
$$\left. \left. + 2 \left(1 - \left(\frac{T^0}{T^2} \right)^{1/2} \right) \right] \right\}$$
$$\tag{7.42}$$

Consequently, optimal interstage temperature T^2 satisfies the stationarity condition

$$\frac{\partial}{\partial T^2}\left(\frac{T^3 - T^2}{T^3} + 2\left(1 - \left(\frac{T^0}{T^2}\right)^{1/2}\right)\right) = 0 \qquad (7.43)$$

Performing the differentiation with respect to T^2 one obtains in terms of T^0 and T^3

$$T^2 = (T^0)^{1/3}(T^3)^{2/3} \qquad (7.44)$$

and

$$T^1 = (T^0 T^2)^{1/2} = (T^0)^{1/2}[(T^0)^{1/3}(T^3)^{2/3}]^{1/2} = (T^0)^{2/3}(T^3)^{1/3} \qquad (7.45)$$

Let us eliminate T^0 and determine T^2 in terms of T^1 and T^3. We obtain

$$T^2 = (T^0)^{1/3}(T^3)^{2/3} = (T^1)^{1/2}(T^3)^{-1/6}(T^3)^{2/3} = (T^1)^{1/2}(T^3)^{1/2} \qquad (7.46)$$

Therefore, as one would expect, optimal interstage temperature T^2 is also the geometric mean of boundary temperatures of two considered stages

$$T^2 = (T^1 T^3)^{1/2} \qquad (7.47)$$

Substitution of optimal temperature $T^2 = (T^0)^{1/3}(T^3)^{2/3}$ into work function (7.42) yields

$$R'^3(T^3, \lambda) = c(T^3 - T^0) - c\Phi'T^e\left(1 - \pm\sqrt{\frac{\lambda}{c\Phi'T^e}}\right)^2\left[\left(1 - \left(\frac{T^0}{T^3}\right)^{1/3}\right)\right.$$

$$\left. +2\left(1 - \left(\frac{T^0}{T^3}\right)^{1/3}\right)\right]$$

which can be simplified to the form

$$R'^3(T^3, \lambda) = c(T^3 - T^0) - 3c\Phi'T^e\left(1 - \pm\sqrt{\frac{\lambda}{c\Phi'T^e}}\right)^2\left(1 - \left(\frac{T^0}{T^3}\right)^{1/3}\right)$$

$$(7.48)$$

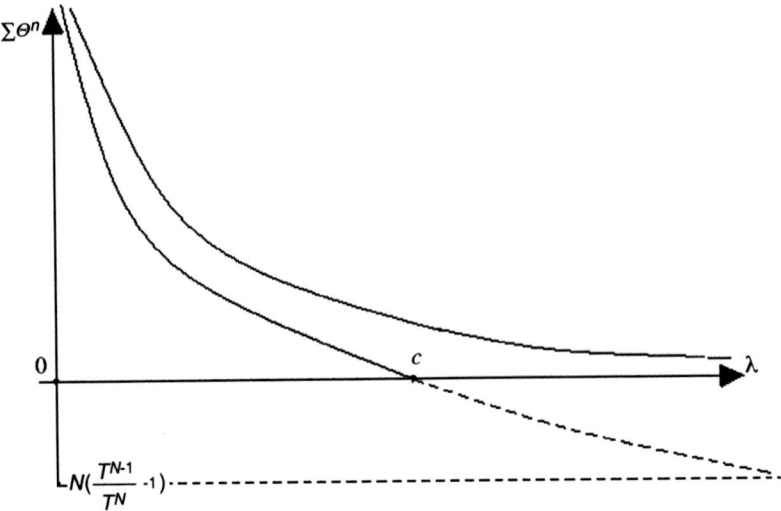

Figure 7.1 Two cases of dependence of Lagrange multiplier λ on total optimal duration $\tau = \sum \Theta''$ in a cascade of power generation systems.

Comparing this expression with corresponding ones for $n = 1$ and $n = 2$, Equations (7.24) and (7.40), lead to the optimal work function for an arbitrary n:

$$R'''(T^n, \lambda) = c(T^n - T^0) - nc\Phi'T^e\left(1 - \pm\sqrt{\frac{\lambda}{c\Phi'T^e}}\right)^2\left(1 - \left(\frac{T^0}{T^n}\right)^{1/n}\right)$$

$$(7.49)$$

The corresponding optimal duration is the partial derivative of optimal work function with respect to Lagrangian multiplier λ:

$$\tau^n = \frac{\partial R'''(T^n, \lambda)}{\partial \lambda} = \frac{1 - \pm\sqrt{\lambda/(c\Phi'T^e)}}{\pm\sqrt{\lambda/(c\Phi'T^e)}}n\left(1 - \left(\frac{T^0}{T^n}\right)^{1/n}\right) \qquad (7.50)$$

Qualitative properties of the duration function are illustrated in Figure 7.1. Equation (7.50) refers to linear kinetics, in which case the curve intersects the axis of λ for $\lambda = c$. In nonlinear systems with a variable c the intersection point may move to large values of λ. This case is also shown in Figure 7.1. In any process, linear or not, λ is a monotonically decreasing function of duration.

Knowledge of the partial derivative of optimal work function with respect to Lagrangian multiplier λ is essential when one wishes to return to the original work function $R^n(T^n, \tau^n)$ (without the Lagrange multiplier term). In this operation the Legendere transformation plays an essential role (Section 7.7). Consider first some properties of intensity parameter ξ.

7.6.   MEAN AND LOCAL INTENSITIES IN DISCRETE PROCESSES

Further transformations are easier if the following intensity criterion is introduced:

$$\xi \equiv \frac{n}{\tau^n}\left(1 - \left(\frac{T^0}{T^n}\right)^{1/n}\right) \tag{7.51}$$

To identify the physical meaning of ξ let us calculate its limiting value for n approaching infinity (continuous process). Since

$$\lim_{n\to\infty} n\left(1 - \left(\frac{T^0}{T^n}\right)^{1/n}\right) = \lim_{x\to 0}\frac{1 - (T^0/T^n)^x}{x} = \ln\left(\frac{T^n}{T^0}\right)$$

we obtain from Equation (7.51)

$$\lim_{n\to\infty}\xi \equiv \lim_{n\to\infty}\frac{n}{\tau^n}\left(1 - \left(\frac{T^0}{T^n}\right)^{1/n}\right) = \frac{\ln T^n - \ln T^0}{\tau^n} \tag{7.52}$$

Therefore ξ defined by Equation (7.51) is a discrete counterpart of the *mean* relaxation rate of the temperature logarithm for the n-stage process.

For an arbitrary stage n we can also introduce a local intensity of a discrete process

$$\xi^n \equiv \frac{T^n - T^{n-1}}{T^n\theta^n} \tag{7.53}$$

This quantity can be obtained after careful use of Equation (7.51) for $n = 1$ and appropriate change of symbols. Its limit for $n \to \infty$ is instantaneous logarithmic rate of state change in a continuous process:

$$\lim_{n\to\infty}\frac{T^n - T^{n-1}}{T^n\theta^n} = \frac{d\ln T}{d\tau} = \xi \tag{7.54}$$

Applying the geometric sequence property for optimal path in the considered example

$$\frac{T^0}{T^n} = \frac{T^0}{T^1}\frac{T^1}{T^2}\cdots\frac{T^{n-1}}{T^n} = \left(\frac{T^{n-1}}{T^n}\right)^n \tag{7.55}$$

we obtain

$$\frac{T^{n-1}}{T^n} = \left(\frac{T^0}{T^n}\right)^{1/n} \tag{7.56}$$

Thus we can easily prove the equality of mean and local rates

$$\xi \equiv \frac{n}{\tau^n}\left(1 - \left(\frac{T^0}{T^n}\right)^{1/n}\right) = \frac{n}{\tau^n}\left(1 - \frac{T^{n-1}}{T^n}\right) = \frac{T^n - T^{n-1}}{T^n\theta^n} \tag{7.57}$$

Local and mean intensities ξ are in general different quantities. Yet, in processes with linear kinetics, considered in the present example, there is no need to distinguish between them and the same symbol for both may be used in equations.

Writing the duration formula, Equation (7.50), in the form

$$\xi^{-1} \equiv \frac{\tau^n}{n}\left(1 - \left(\frac{T^0}{T^n}\right)^{1/n}\right)^{-1} = \pm\left(\sqrt{\frac{\lambda}{c\Phi'T^e}}\right)^{-1} - 1 \tag{7.58}$$

we find a useful equality determining the Lagrange multiplier in terms of the process intensity (mean or instantaneous):

$$\lambda = c\Phi'T^e\left(\frac{\xi}{\xi+1}\right)^2 \tag{7.59}$$

Two values of ξ for a given λ correspond with heating and cooling of the resource fluid in heat-pump and engine modes (upgrading and downgrading of the resource). Both λ and ξ vanish in reversible quasistatic processes.

As the following equality is valid

$$1 - \pm\sqrt{\frac{\lambda}{c\Phi'T^e}} = \frac{1}{\xi+1} \tag{7.60}$$

optimal work function in terms of ξ assumes the form

$$R'''(T^n, \xi) = c(T^n - T^0) - \frac{c\Phi'T^e}{(1+\xi)^2}n\left(1 - \left(\frac{T^0}{T^n}\right)^{1/n}\right) \tag{7.61}$$

Taking into account the limiting value of the expression

$$\lim_{n \to \alpha} n \left(1 - \left(\frac{T^0}{T^n} \right)^{1/n} \right) = \ln \left(\frac{T^n}{T^0} \right)$$

we find that the limiting value of function $R'''(T^n, \xi)$ in a quasistatic $(\xi = 0)$ and reversible process $(\Phi = 1)$ represents the change of classical thermal exergy:

$$R'''(T^n, 0) = c(T^n - T^0) - cT^e \ln \left(\frac{T^n}{T^0} \right) \qquad (7.62)$$

Therefore, optimal work function (7.61) is a finite-rate exergy of the considered discrete process. In the following section other functions of this kind are obtained.

7.7. LEGENDRE TRANSFORM AND ORIGINAL WORK FUNCTION

The minimum of consumed work (7.25) is described by the original principal function $R^n(T^n, \tau^n)$. This function is the Legendre transform of $R'''(T^n, \lambda)$ with respect to λ:

$$R^n(T^n, \tau^n) = R'''(T^n, \lambda) - \lambda \tau^n = R'''(T^n, \lambda) - \lambda \frac{\partial R'''(T^n, \lambda)}{\partial \lambda} \qquad (7.63)$$

For our example we obtain

$$R^n(T^n, \lambda) = c(T^n - T^0)$$
$$-nc\Phi'T^e \left(1 - \pm \sqrt{\frac{\lambda}{c\Phi'T^e}} \right)^2 \left(1 - \left(\frac{T^0}{T^n} \right)^{1/n} \right)$$
$$-\lambda \frac{1 - \pm\sqrt{\lambda/(c\Phi'T^e)}}{\pm\sqrt{\lambda/(c\Phi'T^e)}} n \left(1 - \left(\frac{T^0}{T^n} \right)^{1/n} \right) \qquad (7.64)$$

which should be transformed to the space of variables T^n and τ^n. In transformations we use intensity ξ as an intermediate variable to increase the lucidity of formulae. Applying the equality

$$\frac{1 - \pm\sqrt{\lambda/(c\Phi'T_2)}}{\pm\sqrt{\lambda/(c\Phi'T_2)}} = \frac{1/(1+\xi)}{\xi/(1+\xi)} = \frac{1}{\xi} \qquad (7.65)$$

we obtain

$$\tau^n = \frac{\partial R'^n(T^n, \lambda)}{\partial \lambda} = \frac{n}{\xi}\left(1 - \left(\frac{T^0}{T^n}\right)^{1/n}\right) \tag{7.66}$$

and

$$R^n(T^n, \xi) = c(T^n - T^0) - c\Phi'T^e\left(\frac{1}{1+\xi}\right)^2 n\left(1 - \left(\frac{T^0}{T^n}\right)^{1/n}\right)$$
$$- \frac{\lambda(\xi)}{\xi} n\left(1 - \left(\frac{T^0}{T^n}\right)^{1/n}\right) \tag{7.67}$$

where the function describing λ in terms of ξ is given by Equation (7.50). A complementary formula expressing λ in terms of duration τ follows from Equation (7.50):

$$\lambda = c\Phi'T^e\left(\frac{\tau^n}{n[1 - (T^0/T^n)^{1/n}]} + 1\right)^{-2} \tag{7.68}$$

Monotonic decrease of λ with τ is a general property of both linear and nonlinear processes. Since λ is numerically equal to the Hamiltonian of the optimization problem, Figure 5.4 in Chapter 5 shows the behavior of the Lagrange multiplier λ in terms of nondimensional duration at prescribed boundary temperatures.

Using Equations (7.67) and (7.59) we find

$$R^n(T^n, \xi) = c(T^n - T^0) - \frac{c\Phi'T^e}{1+\xi} n\left(1 - \left(\frac{T^0}{T^n}\right)^{1/n}\right) \tag{7.69}$$

This is a finite-rate exergy, yet it differs from function $R'^n(T^n, \xi)$ of Equation (7.67) in the previous section by the structure of the ξ term. To single out from this equation a ξ-independent term we write $(1+\xi)^{-1}$ as $1 - \xi/(1+\xi)$ and then

$$R^n(T^n, \xi) = c(T^n - T^0) - c\Phi'T^e n\left(1 - \left(\frac{T^0}{T^n}\right)^{1/n}\right)$$
$$+ \frac{\xi}{1+\xi} c\Phi'T^e n\left(1 - \left(\frac{T^0}{T^n}\right)^{1/n}\right) \tag{7.70}$$

Equation (7.51) can next be applied to express the last (rate-dependent) term of Equation (7.70) in terms of process duration. We obtain

$$\frac{\xi}{1+\xi} \equiv \frac{n(1-(T^0/T^n)^{1/n})}{\tau^n + n(1-(T^0/T^n)^{1/n})} \tag{7.71}$$

and

$$R^n(T^n, \tau^n) = c(T^n - T^0) - c\Phi'T^e n\left(1 - \left(\frac{T^0}{T^n}\right)^{1/n}\right)$$

$$+ c\Phi'T^e \left\{ n\left(1 - \left(\frac{T^0}{T^n}\right)^{1/n}\right)\right\}^2 \left(\tau^n + n\left(1 - \left(\frac{T^0}{T^n}\right)^{1/n}\right)\right)^{-1} \tag{7.72}$$

The first two components of this work function are of static origin. The function describes the minimum work supplied to a resource to upgrade it from T^0 to T^n in a finite (nondimensional) time τ^n. As in the case of primed function R' a limiting value of $R^n(T^n, \tau^n)$ in a reversible and quasistatic process ($\Phi' = 1$, $\tau^n \to \infty$) describes a change of classical thermal exergy (7.62).

This approach can also be organized in the entropy representation, where the principal function R''_σ is the minimum entropy production modified by the Lagrange multiplier term. Then

$$R'^1_\sigma(T^1, \lambda) = c\left(1 - \frac{T^0}{T^1}\right)\left(1 - \left(1 - \pm\sqrt{\frac{\lambda}{c\Phi'T^e}}\right)^2\right) \tag{7.73}$$

and

$$R''^n_\sigma(T^n, \lambda) = nc\left(1 - \left(\frac{T^0}{T^n}\right)^{1/n}\right)\left(1 - \left(1 - \pm\sqrt{\frac{\lambda}{c\Phi'T^e}}\right)^2\right) \tag{7.74}$$

Further analysis can be developed in a way analogous to that performed for work functions.

7.8. NUMERICAL APPROACHES APPLYING DYNAMIC PROGRAMMING

Numerical aspects of DP algorithms are only briefly mentioned here as there is a vast literature available that discusses these issues. The "curse of dimensionality" and problems of grid expansion are the main difficulties (Aris, 1964; Bellman, 1961; Sieniutycz, 1991).

Optimal performance functions are a direct outcome of numerical methods which apply the dynamic programming. Optimal control problems of both continuous and discrete processes with a single independent variable (time or length) can be treated in the framework of a common discrete formalism. As computer needs require a discrete set of equations, in the continuous case prior discretizing of ordinary differential equations is required to obtain a set of difference equations. Thus, it is appropriate to focus on numerical multistage optimization, so we describe generation of optimal function $V^n = \max W$.

Assume that at the stage n (duration $\Delta t^n = \theta^n$) a profit function $D^n = D^n(\mathbf{x}^n, t^n, \mathbf{u}^n, \theta^n, n)$ is given. In a maximum work problem $D^n = f_0^n \theta^n$, where D^n is the work produced at stage n. The total profit at the n-stage subprocess equals $\sum D^n = \sum f_0^m \theta^m$. Along with discrete state equations and local constraints, various data of D^n – analytic, graphic or tabular – are sufficient to develop computational principles for cascade processes with an arbitrary number of stages. It is, however, important that D^n is properly expressed at the stage n as a function of state $\mathbf{x}^n$, time t^n, and controls $(\mathbf{u}^n, \theta^n)$. Data of optimal work functions V^1, ..., V^n, ..., V^N should be generated over subprocesses composed respectively of stage 1, stages 1 and 2, ..., stages 1, 2, ..., n and, finally, stages 1, 2, ..., N.

For a given set of difference constraints (7.8) each profit function, for example, optimal work potential V^n, is found from Bellman's equation of dynamic programming. A typical form of this equation in terms of enlarged state vector $\tilde{\mathbf{x}}^n \equiv (\mathbf{x}^n, t^n)$ and one-stage profit D^n is

$$V^n(\tilde{\mathbf{x}}^n) = \max_{\mathbf{u}^n, \theta^n} \{D^n(\tilde{\mathbf{x}}^n, \mathbf{u}^n, \theta^n) + V^{n-1}(\tilde{\mathbf{x}}^n - \tilde{\mathbf{f}}^n(\tilde{\mathbf{x}}^n, \mathbf{u}^n, \theta^n)\theta^n)\} \quad (7.75)$$

Gauged profits $\tilde{D}^n$ obtained by addition (subtraction) of a difference of a scalar function $G(\mathbf{x}^n, t^n) - G(\mathbf{x}^{n-1}, t^{n-1})$ to original one-stage profit D^n can also be considered. They are associated with Bolza functionals and necessarily nonlinear in θ^n:

$$\tilde{D}^n(\tilde{\mathbf{x}}^n, \mathbf{u}^n, \theta^n) \equiv \tilde{f}_0^n \theta^n \equiv \left(f_0^n(\tilde{\mathbf{x}}^n, \mathbf{u}^n, \theta^n) + \frac{G(\tilde{\mathbf{x}}^n) - G(\tilde{\mathbf{x}}^n - \tilde{\mathbf{f}}^n(\tilde{\mathbf{x}}^n, \mathbf{u}^n, \theta^n)\theta^n)}{\theta^n} \right) \theta^n$$

$$(7.76)$$

Tildas over symbols refer to extended quantities; thus, a tilda over the profit symbol refers to extended profit including the effect of the gauging function G. As both profit functions D^n and $\tilde{D}^n$ differ by the path independent increment of state function G, simultaneous generation of optimal data for D^n and $\tilde{D}^n$ is possible. Since one has to generate computational data within a definite domain of the variables $\tilde{\mathbf{x}}$, one can conveniently omit in Equation (7.75) stage superscript

n at the (enlarged) state vector

$$V^n(\tilde{x}) = \max_{u^n, \theta^n} \{D^n(\tilde{x}, u^n, \theta^n) + V^{n-1}(\tilde{x} - \tilde{f}^n(\tilde{x}, u^n, \theta^n)\theta^n)\} \qquad (7.75')$$

The solution to Equations (7.75) and (7.75') is obtained iteratively in the form of tables for $n = 1, 2, \ldots, N$, which describe the sequence of functions $V^n(\tilde{x})$, $u^n(\tilde{x})$ and $\theta^n(\tilde{x})$. The iterative procedure starts with $V^0 = 0$. Potential $V^n(\tilde{x})$ is in general time-dependent even if the process is autonomous.

Organization of calculations requires a suitable computational grid. In the nodes of this grid data of optimal functions and optimal controls are computed and stored. A total number of stages, N, is assumed. The numerical DP algorithm generates a potential function $V^n(\tilde{x})$ from function $V^{n-1}(\tilde{x})$ of $n - 1$ stage sub-process and state transformations, Equation (7.8). In agreement with Bellman's optimality principle, consistent with Equation (7.75), a computer maximizes the sum of optimal cost of all previous $n - 1$ stages (optimal function V^{n-1}) and nonoptimal profit $\tilde{D}^n$ at the stage n. To determine $V^n(\tilde{x})$ exactly for a definite n, the computer would have to numerically determine values of this function for every value of $\tilde{x}$, an impossible task. Therefore, these values are determined on a discrete subset of $\tilde{x}$, and the data are used in the way that makes possible evaluation of $V^n(\tilde{x})$ everywhere. If $\langle \tilde{x}_I, \tilde{x}_{II} \rangle$ is the interval of interest, one can take $\tilde{x}_I = A\delta$ and $\tilde{x}_{II} = B\delta$, where δ is a small accepted value. The vectors A and B are, respectively, computed as $\tilde{x}_I/\delta$ and $\tilde{x}_{II}/\delta$. The quantity δ cannot be too large, because the accuracy of results is then poor. It cannot be too small either, as the computation time becomes very long. The discrete subset of $\tilde{x}$, for which the values $V^n(\tilde{x})$ are computed for various n, has the form

$$\tilde{x} = A\delta, (A + 1)\delta, \ldots, (B - 1)\delta, B\delta \qquad (7.77)$$

This refers to the linear grid of values $\tilde{x} = a\delta$, where $a = A, A + 1, \ldots, B$. Other values of $V^n(\tilde{x})$, for example, those in an interval $\langle a\delta, (a + 1)\delta \rangle$, are defined by accepting $V^n(a\delta)$ or $V^n[(a + 1)\delta]$ depending on the location of $x\delta$ with respect to $(a + 1/2)\delta$, or by using interpolation, for example

$$V^n(\tilde{x}) = V^n(A\delta) + \frac{V^n((A + 1)\delta) + V^n(A\delta)}{\delta}(\tilde{x} - a) \qquad (7.78)$$

The discrete subset of admissible controls u^n is defined in a similar way. For example, when constrains imposed on u^n are described by inequality $u^* \leq u^n \leq u^*$, variable u^n may assume only the discrete values

$$u^* = E\gamma + (E + 1)\gamma, \ldots, (F - 1)\gamma, \quad F\gamma = u^* \qquad (7.79)$$

for an appropriately small value γ. This refers to a linear grid of controls $u = b\delta$, where $b = E, E + 1, \ldots, F - 1, F$.

Optimal controls are determined from the formula

$$\{\mathbf{u}^n(\tilde{\mathbf{x}}), \theta^n(\tilde{\mathbf{x}})\} = \underset{\mathbf{u}^n \theta^n}{\operatorname{argmax}} \{\tilde{D}^n(\tilde{\mathbf{x}}, \mathbf{u}^n, \theta^n) + V^{n-1}(\tilde{\mathbf{x}} - \tilde{\mathbf{f}}^n(\tilde{\mathbf{x}}, \mathbf{u}^n, \theta^n)\theta^n)\} \quad (7.80)$$

along with the sequence of the optimal functions V^n. The first optimal function, $V^1(\tilde{\mathbf{x}})$, and the corresponding optimal controls for $n = 1$ follow from the application of the initial condition $V^0(\tilde{\mathbf{x}}) = 0$ in Equations (7.75) and (7.80); this yields

$$V^1(\tilde{\mathbf{x}}) = \max_{\mathbf{u}^1, \theta^1} \{\tilde{D}^1(\tilde{\mathbf{x}}, \mathbf{u}^1, \theta^1)\} \quad (7.81)$$

and

$$\{\mathbf{u}^1(\tilde{\mathbf{x}})\theta^1(\tilde{\mathbf{x}})\} = \underset{\mathbf{u}^n \theta^n}{\operatorname{argmax}} \{\tilde{D}^1(\tilde{\mathbf{x}}, \mathbf{u}^1, \theta^1)\} \quad (7.82)$$

To find these functions, the computer chooses the first point $\tilde{\mathbf{x}} = \mathbf{A}\delta = (A_1\delta, A_2\delta)$ and compares $\tilde{D}^1(\tilde{\mathbf{A}}\delta, E_1\gamma, E_2\gamma)$ with $\tilde{D}^1(\mathbf{A}\delta), (E_1 + 1)\gamma, E_2\gamma)$. The larger of these values is stored and compared with $\tilde{D}^1(\mathbf{A}\delta, (E_1 + 2)\gamma, E_2\gamma)$, and so on. This process is continued until the whole discrete set of controls $(\mathbf{u}^1, \theta^1)$ is exhausted. The largest of the thus-obtained values is always a maximum of $\tilde{D}^1$ with respect to $(\mathbf{u}^1, \theta^1)$ for a fixed discrete point $\tilde{\mathbf{x}}$. The coordinates of $\mathbf{u}^1$ and θ^1 which maximize $\tilde{D}^1$ are stored. Analogous operations are next performed for $\tilde{\mathbf{x}} = ((A_1 + 1)\delta, A_2\delta), ((A_1 + 2)\delta, A_2\delta)$, and so on. Again, this leads to a maximum of $\tilde{D}^1$ and optimal values of $\mathbf{u}^1$ and θ^1. The data for the same quantity differ at various points $\tilde{\mathbf{x}}$ (different nodes of the grid). The computer outputs are DP tables which contain only optimal data: $V^1(\tilde{\mathbf{x}})$, $\theta^1(\tilde{\mathbf{x}})$ and $\mathbf{u}^1(\tilde{\mathbf{x}})$.

For $n = 2$ (two-stage process), and for larger n, the procedure is analogous but uses Equations (7.75) and (7.80) in their complete form. Data of V^{n-1} are found in tables describing previous computations, for cascade with $n - 1$ stages. When using these data a difficulty can appear which is called "the danger of grid expansion". It follows from the fact that the values $V^{n-1}(\tilde{\mathbf{x}})$ were computed within the range $\tilde{\mathbf{x}}_I \le \tilde{\mathbf{x}} \le \tilde{\mathbf{x}}_{II}$, but the computation of $V^n(\tilde{\mathbf{x}})$ from Equation (7.75) requires the knowledge of $V^{n-1}(\tilde{\mathbf{x}}^T)$ for the transformed state $\tilde{\mathbf{x}}^T \equiv \tilde{\mathbf{x}} - \tilde{\mathbf{f}}^n(\tilde{\mathbf{x}}, \mathbf{u}^n, \theta^n)\theta^n$. This means that for some forms of rate functions $\tilde{\mathbf{f}}$ the computation of $V^n(\tilde{\mathbf{x}})$ requires the knowledge of values $V^{n-1}(\tilde{\mathbf{x}})$ for $\tilde{\mathbf{x}}$ located outside of the range $\tilde{\mathbf{x}}_I \le \tilde{\mathbf{x}} \le \tilde{\mathbf{x}}_{II}$. Therefore, to evaluate $V^n(\tilde{\mathbf{x}})$ within the range satisfying $\tilde{\mathbf{x}}_I \le \tilde{\mathbf{x}} \le \tilde{\mathbf{x}}_{II}$ it may be necessary to determine $V^{n-1}(\tilde{\mathbf{x}})$ within a boundary which is larger than that described by the inequality $\tilde{\mathbf{x}}_I \le \tilde{\mathbf{x}} \le \tilde{\mathbf{x}}_{II}$.

The procedure leads to the optimal values of V^n, θ^n and $\mathbf{u}^n$ stored at each node of the grid of $\tilde{\mathbf{x}}$, for each n. These values constitute discrete representations of optimal functions $V^n(\tilde{\mathbf{x}})$, $\theta^n(\tilde{\mathbf{x}})$ and $\mathbf{u}^n(\tilde{\mathbf{x}})$. Additionally, values of coordinates of the transformed state, $\tilde{\mathbf{x}}^T \equiv \tilde{\mathbf{x}} - \tilde{\mathbf{f}}^n(\tilde{\mathbf{x}}, \mathbf{u}^n, \theta^n)\theta^n$, can be stored. Data of $\tilde{\mathbf{x}}^T$

describe optimal inlet states to the stage n in terms of outlet states from this stage, $\tilde{x}$.

Backward reading of the solution

Dynamic programming tables, which describe all computed data, can be used to find the solution of a particular (N-stage) problem in which final values of $\tilde{x}^N = \tilde{x}^f$ and N are prescribed. This is a backward procedure in which we first identify in DP tables the final point $\tilde{x}^f$ for $n = N$, and, next, in these tables, we read off data of optimal controls $\theta^N(\tilde{x}^f)$ and $u^N(\tilde{x}^f)$. In the tables we also find transformed outlet states $\tilde{x}^{TN} \equiv \tilde{x} - \tilde{f}^N(\tilde{x}, u^N, \theta^N)\theta^N$, which are inlet states to the stage N.

Now we pass to the $N-1$ stage subprocess. It has its own outlet state $\tilde{x}^{N-1}$ which was just found as $\tilde{x}^{TN}$. By interpolating in the tables for $n = N - 1$ we find all suitable data for the state $\tilde{x} = \tilde{x}^{N-1}$. We thus find optimal profit $V^{N-1}(\tilde{x}^{N-1})$, optimal controls $\theta^{N-1}(\tilde{x}^{N-1})$, $u^{N-1}(\tilde{x}^{N-1})$, and inlet states to the stage $N-1$, $\tilde{x}^{N-2} = \tilde{x}^{TN-1}$. Continuing the procedure on the computer, we obtain an optimal solution as a sequence of optimal controls $u^N, u^{N-1}, \ldots, u^1$ and $\theta^N, \theta^{N-1}, \ldots, \theta^1$, and an optimal discrete trajectory, $\tilde{x}^N, \tilde{x}^{N-1}, \ldots, \tilde{x}^1, \tilde{x}^0$. The sequence of optimal costs for all related subprocesses $V^N, V^{N-1}, \ldots, V^1$ also follows.

The optimality principle excludes investigation of (the huge number of) all possible $n - 1$ stage subprocesses, which causes extremely large savings in computational time. Another virtue of the DP method is that it always leads to absolute maximum, and, as opposed to other methods, the increase of the number of constraints simplifies the computer search (fewer points to be tested). Functions $\tilde{D}^n$ and f^n describing profit and state transformation need not be continuous or analytical; they may be in graphic or tabular form. Two-point boundary values do not cause problems either, as the recurrence equation is not influenced by end conditions. Only evaluation of first function, $V^1(\tilde{x})$, changes depending on initial conditions. Large dimensionality of the control vector does not cause essential difficulties.

It is very fortunate that many dynamical energy production problems (especially those without mass transfer) are of low state dimensionality. It is just the property that causes the easy solving procedure for Equation (7.24).

7.9. DIMENSIONALITY REDUCTION IN DYNAMIC PROGRAMMING ALGORITHMS

In energy problems with multicomponent mass transfer and/or chemical reactions a very serious difficulty arises connected with the use of dynamic programming. This is the so-called "curse of dimensionality", referring to the large dimensionality of state vector, $\tilde{x}$. Indeed, for $s = \dim \tilde{x} = 1$, a single column of discrete set of $\tilde{x}$ is sufficient, for $s = 2$ the computational grid must constitute, say, a rectangle. For $s = 3$, however, the grid must be cubic, for $s = 4$ the data must be obtained (and stored) as a set of cubes, and so on. Clearly, the number of computational points, and hence the computer memory requirements, increase tremendously with state dimensionality s. Problems with $s = 1$ and $s = 2$ are quite easy to solve numerically, problems with higher $s = 3$ are troublesome or serious, and problems with $s \geq 5$ are practically intractable if good accuracy is required. Therefore, numerical dynamic programming can effectively be applied only for problems characterized by the small dimensionality of the state vector $\tilde{x}$; problems of large dimensionality, such as those encountered in the static optimization, are excluded. Fortunately, many dynamical problems of energy production are of low dimensionality. In the case of high s, other methods, especially those associated with maximum principles, must be applied. These are based on canonical equations, state adjoints and a Hamiltonian function involving these variables, and are described in a number of sources (e.g. Aris, 1964; Boltyanski, 1969, 1973; Pontryagin et al., 1962; Sieniutycz, 1991). Sometimes, however, the dimensionality reduction is possible in DP problems. For the energy problems considered here dimensionality reduction is possible, for example, in autonomous systems due to the constancy of the time adjoint λ along an optimal path. This is described below.

For V^n regarded as energy production profit, the net economic-type profit, or the difference between V^n and the "time penalty cost" λ $(t^n - t^0)$, can be defined. λ is the Lagrange multiplier associated with time t. We shall designate by an asterisk subscript thus-modified profits (or costs), and will focus on their properties in autonomous processes. When the discrete process is autonomous and time interval θ^n is not explicitly present in rates f_k, λ is identical to a constant Hamiltonian $H^{n-1} = H$. Under a weaker assumption of autonomous process and intervals θ^n explicit in rates f_k, the constancy property refers only to λ. To analyze both cases simultaneously we deal with modified optimal functions, net profits $V_*^n \equiv V^n - \lambda(t^n - t^0)$ or net costs $R_*^n \equiv R^n + \lambda(t^n - t^0)$, both criteria being equivalent because the second is obtained by multiplication of the first by the minus unity. The quantity λ describes the decrease of the original profit when the process duration is increased by one unit.

Local profits and costs are defined in a similar way. For single-stage profit $\tilde{D}^n$, which appears in Equation (7.75), a net profit is $\tilde{D}_*^n \equiv \tilde{D}^n - \lambda\theta^n$, where $\theta^n \equiv \Delta t^n$. Similarly total cost at the stage n is $\tilde{K}_*^n \equiv \tilde{K}^n + \lambda\theta^n$, where $\tilde{K}^n \equiv -\tilde{D}^n$.

Given net profit $\tilde{D}_*^N$, the optimal process is governed by a sequence of asterisk functions: $V_*^1, \ldots, V_*^n, \ldots, V_*^{N-1}$ and V_*^N. The sequence of these optimal functions obeys the equation:

$$V_*^n(\mathbf{x}^n, \lambda) = \max_{\mathbf{u}^n, \theta^n} \{\tilde{D}_*^n(\mathbf{x}^n, \mathbf{u}^n, \theta^n, \lambda) + V_*^{n-1}(\mathbf{x}^n - \mathbf{f}^n(\mathbf{x}^n, \mathbf{u}^n, \theta^n)\theta^n, \lambda)\}$$

$$(7.83)$$

This differs from Equation (7.75) by the presence of vector $\mathbf{x}$ rather than $\tilde{x} \equiv (\mathbf{x}, t)$. Because of the constancy of λ along a discrete optimal path, state dimensionality of the problem described by Equation (7.83) is decreased by one in comparison with that for Equation (7.75). In fact, the continuous limit of Equation (7.83) is a HJB equation:

$$\max_{\mathbf{u}} \left\{ \tilde{f}_{0*}(\mathbf{x}, \mathbf{u}, \lambda) - \frac{\partial V_*(\mathbf{x}, \lambda)}{\partial \mathbf{x}} f(\mathbf{x}, \mathbf{u}, \lambda) \right\} = 0 \qquad (7.84)$$

Again, this is similar to the Hamilton–Jacobi equation of classical mechanics but contains the maximizing sign before a Hamiltonian expression. The equation refers to an optimal duration $T = (t^f - t^i)$ equal to $\partial V_*(\mathbf{x}, \lambda)/\partial \lambda$.

Optimal functions of work production, V^n and V_*^n, preserve a number of basic qualitative properties of economic production profits and total economic profits. The same remark refers to functions of work consumption, $R^n = - V^n$ and $R_*^n = V_*^n$. For multistage control processes, optimal data generated by DP always have the form of sequence functions $V^n(\mathbf{x}, t)$ or their duals $V_*^n(\mathbf{x}, \lambda)$, where $\mathbf{x}$ is the process state, t is a time variable and n is the number of stages. Profit functions $V^n(\mathbf{x}, t)$ and $V_*^n(\mathbf{x}, \lambda)$, or cost functions $R^n(\mathbf{x}, t)$ and $R_*^n(\mathbf{x}, \lambda)$, are related by the Legendre transformation with respect to their independent variable (Leitman, 1966; Landau and Lifshitz, 1971; Sieniutycz, 1973a). The limiting case of a continuous process is characterized by functions $V(\mathbf{x}, t)$ and $V_*^n(\mathbf{x}, \lambda)$, which are mathematical equivalents of Hamilton's principal action and abbreviated action in classical mechanics or related phase functions in optics (Landau and Lifshitz, 1971). The relation between the optimal cost functions generated by dynamic programming and Pontryagin's maximum principle is now well understood (Leitman, 1966). The optimal paths of a control problem are equivalent to mechanical paths in mechanics or light rays in optics. The use of dynamic programming in constructing finite-time potentials for discrete and continuous control separation processes has been summarized (Berry et al., 2000). A computational example showing wave-path duality is available, dealing with a separation process in which a volatile component is evaporated from a porous, fluidizing solid by a hot gas (Sieniutycz, 1973b; Berry et al., 2000).

The mathematics of these approaches is independent of the specific applications. This is why they can be conducted first in frames of thermodynamic models and the experience gained serves to formulate and solve more involved problems of economics.

Again, in the above discussion we have omitted procedures of discrete maximum principle (Boltyanski, 1973) as these do not generate data of optimal performance function.

7.10.   CONCLUDING REMARKS

In this chapter we have presented a basic formulation for maximum power in dynamical power systems and considered convergence of discrete value functions to viscosity solutions of HJB equations. Lagrangian multipliers associated with

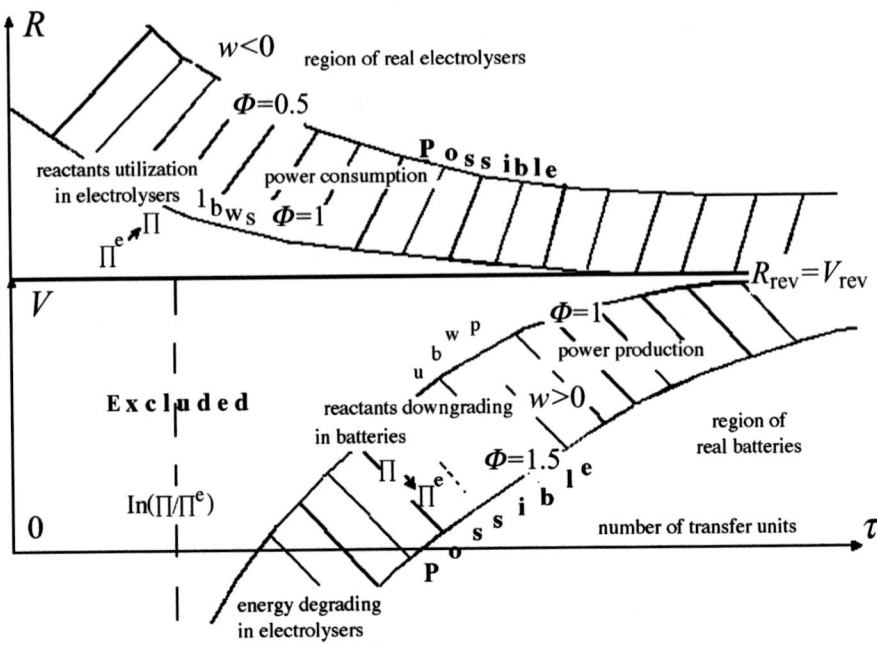

Figure 7.2 Work limits for reversible and real electrolysers and batteries. System with work production are described by function $V(T, \tau)$, systems with work consumption by function $R(T, \tau)$. Endoreversible limits correspond with curves for $\Phi = 1$; weaker reversible limits are represented by the straight line $R_{rev} = V_{rev}$. Dashed lines mark regions of possible improvements when imperfect thermal machines are replaced by those with better performance coefficients, terminating at endoreversible limits for Carnot machines.

duration constraint have been used to reduce dimensionality of some power production problems. Analytical and numerical approaches to power generation problems, applying the dynamic programming method, have been described. Legendre transform was applied to recover optimal work as function $R^n(T^n, \tau)$.

It was shown that while arbitrary discretization of the original model violates in general the constancy of discrete autonomous Hamiltonians H^n, one may preserve this constancy either by working in the coordinate frames in which discrete rates do not contain time intervals θ^n or by applying the Lagrange multiplier of the duration constraint which still exhibits the constancy property.

Generalized (time-dependent) work potentials can be found for nonlinear systems. They lead to the thermodynamic bounds on the power produced (consumed) with a finite rate. These are stronger than the classical bounds as they involve constraints coming from the process kinetics.

Another important application of the approach considered involves chemical energy systems, and, especially, fuel cells. Figure 7.2 shows the comparison of optimal work production and work consumption functions, V and R, in terms of slowness variable τ, for a fixed change of system state. Reversible upper bound V_{rev} achieved in production modes is equal to reversible lower bound

R_{rev} achieved in consumption modes. For the irreversible bounds this equality does not hold, and a lower bound of R is larger than upper bound of V. Note a qualitative similarity of this plot of work limits to charts characterizing generalized exergies (Figure 6.3). This similarity is a suitable starting point to investigate energy yield in chemical and electrochemical systems (Chapters 9 and 10, respectively).

8 Optimal control of separation processes

8.1. GENERAL THERMOKINETIC ISSUES

The exergy and heat consumed in separation units can now be treated in general terms without reference to any specific process, whether it be distillation, desorption, or drying. This leads to limits on the performance of separation processes (Orlov and Berry, 1991b). For a given separation effect, the lowest bound for heat consumption is determined by thermostatics and is given by the ratio of the minimum work of separation to the related Carnot efficiency. However, this limit is unrealistically low, and, more importantly, it does not correspond to any real feed flow. An irreversible bound on the heat consumed in separation processes has been determined as a function of feed flow (Orlov and Berry, 1991b) and gives a more realistic limit which includes the effect of entropy production σ and simplifies to the classical result in the limit of vanishing σ. These results show some resemblance to those known for the efficiency of thermal engines evaluated at the maximum power point.

For any finite rate separation process with a given nonvanishing mass flow (average mass flow in the case of cyclic processes) the exergy consumption is larger than the corresponding reversible consumption. Since the constraint on the feed flow (and any other constraints on, e.g., boundary concentrations) is operative, only a part of the entropy produced can be reduced through an optimal choice of an operational parameter. For a given feed flow such a reduction causes a related decrease in the valuable heat; hence, the minimum of the heat consumed corresponds to the minimum of σ. Thus, there exists a more realistic lower bound, greater than the classical value, on the valuable heat consumed. This bound is a function of the flow F. Any real separation process with a given feed flow will consume an amount of heat that cannot be lower than this limit. This value is still just a lower bound and is not the economically optimal heat consumption for any particular separation unit. Whatever the economical heat consumption is, for a given operational situation, this consumption cannot be less than this lower bound. Knowledge of this bound is of value for design.

A chemical engineer realizes, however, that it is not necessary to speak about entropy production at all in order to determine the lower bound on heat consumption. One could simply minimize the heat as a function of a parameter at given operational constraints. However, entropy production (or exergy dissipation) is a convenient common measure of the imperfection of very diverse processes. Yet the rate-constrained thermodynamic bounds of FTT may implicitly contain a constant vector of certain nonthermodynamic quantities (e.g., design parameters). Care is necessary when their application to a design, where this vector may change, is in question. The difference between FTT bounds and the actual heat consumption can be illustrated by an example of the economic design of a typical rectification column (Ciborowski, 1965). The tradeoff between the operational and investment costs results in the economical reflux, usually several times larger than the lowest possible one. In the rectification column the consumption of heat supplied to the boiler grows linearly with the reflux R; $Q = D(R + 1)r$, where r is the average heat of evaporation and D is the flow of distillate. The actual heat is then several times larger than the lowest possible one corresponding to the minimum of entropy production. In conclusion, real columns should never be designed to operate at the bound of minimum heat (or minimum σ) even if this is not a thermostatic limit. However, for existing plants where the investment is fixed or its variation can be ignored, FTT, EGM and their analyses of limiting possibilities are useful.

Distillation plants have been analyzed with the goal of reducing overall plant irreversibility and exploitation costs through adjustment of the principal influencing factors. Earlier applications have been reviewed by Kotas (1985). Linnhoff et al. (1988) have discussed the merits and disadvantages of the use of exergy efficiencies in distillation. Mullins and Berry (1984) have evaluated methods for reducing entropy production by finding the optimal location of intermediate heat exchangers. Fritzmoriss and Mach (1980) have calculated exergy losses and efficiencies for the column and for the system as a whole; by varying the reflux ratio and the number of plates they have demonstrated the effects of these two parameters on efficiencies. Fonyo and Rev (1981) have carried out a detailed analysis of the internal and external exergy losses in distillation columns and reached conclusions about the appropriateness of interboilers and intercondensers. Distillation design can be rigorously treated in the framework of process synthesis (Umeda et al., 1979; Naka et al., 1982). Naka et al. (1982) have optimized a complex multicomponent distillation system with heat integration with the maximum of exergy recovery from condensers and reboilers. The role of energy limits and second law analyses is substantial in Le Goff's research; in particular Ranger et al. (1990) have stressed the role of a combination of rectification and reverse rectification processes in the design of a new type of absorption heat-pump (see next section). The reader is referred to the synthesis of results in distillation in the context of thermodynamic geometry (Andresen and Salamon, 2000) and to the review on separation processes by Tsirlin and Kazakov (2002a).

Below we shall consider in detail the important issue of quantitative theory of thermodynamic limits on irreversible separation processes and associated analysis of entropy production following a systematic approach developed by Tsirlin's group (Tsirlin and Kazakov, 2002a). Limits on the performance of irreversible separation operations can be derived including work of separation in finite time (generalization of Van't Hoff reversible work), maximum productivity of heat-driven separations (an analog of maximum power of an irreversible engine), and minimum dissipation in an irreversible distillation column along with the corresponding operating line and column productivity. These issues are important because separation processes are those that consume the most energy per unit amount of the product or they are least energy-efficient. In industry diverse sorts of separation operations are applied, such as: distillation, evaporation, desorption, drying, membrane separation, centrifuging, and so on.

Classical thermodynamics provides reversible estimates for thermodynamic limits that are, however, too weak to be very useful. These classical limits do not take into account kinetic components of process equations, that contribute to limits of higher rank, for example to "endoreversible" limits or overall limits corresponding with minimum entropy generation, S_σ^{min}. When process productivity is specified, the degree of thermodynamic perfection is evaluated by comparing actual dissipative losses with those resulting from S_σ^{min}. In the approach of Tsirlin and Kazakov (2002a) a minimum rate of entropy generation is found in terms of heat and mass transfer coefficients for some irreversible gas and liquid separation processes. Thermodynamic balancing is next used to get the minimum possible work or energy of separation.

The benefits following from irreversible thermodynamic analysis are numerous. First, gas separation can be treated. The lower bound on the power consumption in gas separation of a prescribed productivity and fixed compositions of the input and output fluxes can be derived.

Second, a general heat-propelled irreversible separation process, or the so-called thermal separation, can be analyzed. It may be shown that the productivity (rate) of the thermal separation has an upper limit. This means that for heat fluxes exceeding a certain critical value the productivity of thermal separation decreases with the heat flux.

Third, an irreversible binary distillation can be treated to derive a new estimate minimum possible entropy generation along with the corresponding ideal operating line. For the maximum productivity of thermal separation, the stronger bound can be obtained than the one for a general thermal separation.

8.2. THERMODYNAMIC BALANCES TOWARD MINIMUM HEAT OR WORK

We shall now consider the thermodynamic balance of a general separation process. We shall describe formally the separation scheme shown in Chapter 3, where Figure 3.7 depicts input and output fluxes for computation of separation operations. The stream of a mixture is characterized by the flux g_0, composition x_0, temperature T_0 and pressure P_0; this stream is separated into two streams with parameters g_i, x_i, T_i and P_i, where $i = 1, 2$. The power of mechanical work supplied to the system is p. The heat flux q_+ with the temperature T_+ is supplied to the system, whereas the heat flux q_- with the temperature T_- leaves the system.

Assuming the steady-state condition of the operation, balance equations for mass and energy are formulated. Following Tsirlin and Kazakov (2002a) we consider a mixture composed of m constituents j, where $j = 1, k, \ldots, m$. The mass balance is

$$g_0 x_{0j} - g_1 x_{1j} - g_2 x_{2j} = 0 \tag{8.1}$$

or

$$g_0 x_{0j} = \sum_{i=1}^{i=2} g_i x_{ij} \tag{8.2}$$

$$\sum_{j=1}^{m} x_{ij} = 1, \quad i = 0, 1, 2 \tag{8.3}$$

where index $j = 1, k$ refers to species and index $i = 0, 1, 2$ to streams. The energy balance is

$$p + q_+ - q_- + g_0 h_0 - g_1 h_1 - g_2 h_2 = 0 \tag{8.4}$$

where h_i is the molar enthalpy of the ith stream. The corresponding entropy balance is

$$\frac{q_+}{T_+} - \frac{q_-}{T_-} + g_0 s_0 - g_1 s_1 - g_2 s_2 + \sigma = 0 \tag{8.5}$$

where s_i is the molar enthalpy of the ith stream and $\sigma \equiv \tilde{S}_\sigma$ is the rate of the entropy generation within the system. As $g_0 = g_1 + g_2$, assured by Equations (8.1) and (8.3), g_0 can be eliminated from Equations (8.4) and (8.5). When the heat flux q_- is eliminated from the so-transformed Equation (8.4) and the result is substituted into Equation (8.5) the following general expression is obtained for the entropy production:

$$\sum_{i=1}^{2} g_i \left((s_0 - s_i) - \frac{h_0 - h_i}{T_-} \right) + q_+ \left(\frac{1}{T_+} - \frac{1}{T_-} \right) - \frac{p}{T_-} + \sigma = 0 \tag{8.6}$$

The case of purely thermal separation corresponds with $p = 0$; then

$$\sum_{i=1}^{2} g_i \left(\frac{T_-(s_0 - s_i) - (h_0 - h_i)}{T_-} \right) + q_+ \left(\frac{T_- - T_+}{T_- T_+} \right) + \sigma = 0 \tag{8.7}$$

Consequently the heat consumption in irreversible operations of thermal separation is

$$q_+ = \frac{T_+}{T_+ - T_-} \left(\sum_{i=1}^{2} g_i(h_i - h_0 - T_-(s_i - s_0)) + T_-\sigma \right) \tag{8.8}$$

On the other hand, for mechanical processes of separation, that is when $q_+ = q_- = 0$,

$$p = \sum_{i=1}^{2} g_i(h_i - h_0 - T_-(s_i - s_0)) + T_-\sigma \tag{8.9}$$

We shall now consider the question of assumptions under which conditions the reversible expressions in Equations (8.8) and (8.9) can be described in terms of exergy-type functions. To answer this question we shall test the applicability of availability function A^{class} that resembles the classical exergy, but contains the stream flux fractions y_i in place of the usual concentrations. For the environmental parameters h_i^e, s_i^e and μ_i^e,

$$A^{class} = \sum_{i=1}^{s} (h_i - h_i^e - T^e(s_i - s_i^e))y_i \tag{8.10}$$

This equation applies to the separation operation (with $s = 2$) whenever:

(i) $T^e = T_- \equiv T_0$ (i.e. if the low temperature at which the heat is rejected, T_-, is equal to both the inlet temperature of the separated mixture, T_0, and the temperature of the environment, T^e);
(ii) each $h_i^e \equiv h_0$, each $s_i^e \equiv s_0$; and
(iii) $y_i = g_i/g_0$ play formally the role of concentrations.

Equation (8.10) is in agreement with the well-known thermodynamic rule which states that every specific property of the solution is the sum of products of partial specific properties and mass fractions. With Equation (8.10), the reversible separation work for a definite system follows from the input–output analysis applied to A^{class}. A generalized expression complementing the above formula in finite-time thermodynamics takes into account irreversible losses of work stemming from the entropy generation $S_\sigma = \sigma/g_0$. The role of stream fractions $y_i = g_i/g_0$ in the separation model is explicit in an expression describing the power per unit mass of inlet flow p/g_0:

$$\frac{p}{g_0} = \sum_{i=1}^{2} y_i(h_i - h_0 - T_-(s_i - s_0)) + \frac{T_-\sigma}{g_0} \tag{8.11}$$

With the stream flux fractions $y_i = g_i/g_0$ and "chemical potentials of streams" (in fact, Gibbs functions of all output streams)

$$\mu_i = h_i - T_i s_i \tag{8.12}$$

one can transform Equations (8.9) or (8.10) into suitable well-known structures in which the reversible work follows as an increment of the exergy-like function (8.10). For example, with the help of Equation (8.12) one can transform Equation (8.10) into a suitable exergy-like form to refer the specific work of separation at flow, $W = -p/g_0$, to the increment of the function A^{class} as follows:

$$
\begin{aligned}
(-W) &= \Delta \sum_{i=1}^{s} (h_i - T^e s_i - \mu^e) y_i + T^e S_\sigma \\
&= \Delta h - T^e \Delta s - \sum_{i=1}^{s} \mu^e \Delta y_i + T^e S_\sigma = (-W)_{\min}^{\text{rev}} + T^e S_\sigma
\end{aligned}
\tag{8.13}
$$

(note the engine sign convention). The nonflow counterpart of function (8.13) was used to describe separation processes in polymeric solutions (Sieniutycz, 2001a, 2001b).

Equations describing power or work at flow always have the same structure, and Equation (8.13) is typically representative. The first term (describing the static effect) is reversible, and can be related to change of a potential, whereas the second term is irreversible and refers to process kinetics and entropy production. Knowledge of the residual or minimal entropy production obtained, with the help of optimization techniques for the processes of interest, leads to irreversible estimates of minimum energy consumption:

$$q_+ \geq q_+^{\text{rev}} + \frac{T_- \sigma_{\min}^{\text{T}}}{\eta_C} \tag{8.14}$$

$$p \geq p^{\text{rev}} + T_- \sigma_{\min}^{\text{M}} \tag{8.15}$$

where $\sigma_{\min}^{\text{T}}$ and $\sigma_{\min}^{\text{M}}$ are, respectively, the estimates of minimum entropy production for both thermal and mechanical separation and $\eta_C = 1 - T_+/T_-$ is the appropriate Carnot efficiency.

Explicit formulae can be given for the reversible estimate of separation heat and power in mixtures that are close to ideal solutions. In this case the molar enthalpies and entropies h_i and s_i can be described by the expressions

$$h_i(T_i, P_i, x_i) = \sum_{j=1}^{k} x_{ij} h_j(T_i, P_i) \tag{8.16}$$

$$s_i(T_i, P_i, x_i) = \sum_{j=1}^{k} x_{ij}[s_j^0(T_i, P_i) - R \ln x_{ij}] \tag{8.17}$$

where the index j refers to the species and i to the streams ($i = 0, \ldots, 2$). The reversible estimate of separation heat, obtained from Equation (8.8) in the special case of $\sigma = 0$, is

$$q_+^{\text{rev}} = \frac{1}{\eta_C} \sum_{i=1}^{2} g_i(T_-(s_0 - s_i) - (h_0 - h_i)) \tag{8.18}$$

For mixtures described by Equations (8.16) and (8.17) the estimate takes the form

$$q_+^{\text{rev}} = \frac{1}{\eta_C} \sum_{i=1}^{2} g_i T_-$$

$$\left(\sum_{j=1}^{k} x_{0j} s_j^0(T_0, P_0) - x_{ij} s_j^0(T_i, P_i) - R(x_{0j} \ln x_{0j} - x_{ij} \ln x_{ij}) \right) \tag{8.19}$$

$$- \frac{1}{\eta_C} \sum_{i=1}^{2} g_i \sum_{j=1}^{k} (x_{0j} h_j(T_0, P_0) - x_{ij} h_j(T_i, P_i))$$

On the other hand, for reversible mechanical processes of isothermal and isobaric separation, that is, those with $\sigma = q_+ = 0$ and $T_+ = T_- = T$, one gets from Equations (8.9), (8.16) and (8.17),

$$p^{\text{rev}} = g_0 \sum_{i=1}^{2} y_i(T_-(s_0 - s_i) - (h_0 - h_i)) = g_0 RT \sum_{i=1}^{2} y_i \sum_{j=1}^{k} (x_{ij} \ln x_{ij} - x_{0j} \ln x_{0j})$$

$$\tag{8.20}$$

In an irreversible operation this power is increased by the term $T\sigma$ in agreement with Equation (8.97) and the Gouy–Stodola law.

Now we turn the reader's attention to an important quantity

$$a_i(x_i) = -RT \sum_{j=1}^{k} x_{ij} \ln x_{ij}, \quad i = 0, 1, 2 \tag{8.21}$$

which is the reversible work of isothermal and isobaric separation of one mole of ith stream into pure components. The usefulness of this quantity is immediately seen; since $\Sigma y_i = 1$, Equation (8.20) can be given the form

$$p^{\text{rev}} = -g_0 RT \left(\sum_{j=1}^{k} x_{0j} \ln x_{0j} - \sum_{i=1}^{2} y_i \sum_{j=1}^{k} x_{ij} \ln x_{ij} \right) \qquad (8.20')$$

Thus, the reversible power of separation p^{rev} is equal to the difference between the reversible power used for complete separation of the input flux and combined reversible power for separation of output fluxes. As in the case of availability function A^{class}, stream fractions y_i are essential to calculate the combined reversible power for output fluxes.

One may regard the work of reversible separation as the trivial integral of Equation (8.20') in which the integration of the flux $g_0 = dN_0/dt$ takes place over the time t. Then for the complete separation of N_0 moles of zeroth stream into pure components

$$A_0 = -N_0 RT \sum_{j=1}^{k} x_{0j} \ln x_{0j} \qquad (8.22)$$

consistent with Equation (8.21). Whence in a special case of a binary mixture the reversible work of the complete separation has the form

$$A_0 = -N_0 RT(\mathbf{x} \ln \mathbf{x} + (1 - \mathbf{x})\ln(1 - \mathbf{x})) \qquad (8.23)$$

where $\mathbf{x} \equiv x_{01}$ is the molar fraction of the first component in the initial mixture. The separation of a mixture of k substances can be made as a sequence of $k - 1$ binary separations. No transfer coefficients are present in the reversible formulae such as Equation (8.23). In a more general problem with an incomplete separation an inlet binary mixture (0) with concentrations x_{0j} is separated in a batch process into two outlet mixtures (1 and 2) with final concentrations x_{1j} and x_{2j}. Equation (8.20') then leads to the usable result

$$A_0 = -N_0 RT$$

$$\left(\sum_{j=1}^{k} x_{0j} \ln x_{0j} - \frac{N_1}{N_0} \sum_{j=1}^{k} x_{1j} \ln x_{1j} - \left(1 - \frac{N_1}{N_0} \right) \sum_{j=1}^{k} x_{2j} \ln x_{2j} \right) \qquad (8.24)$$

A classical model of mechanical separation involves a binary gas mixture in a chamber equipped with two semi-permeable pistons that move toward each other. The left piston is permeable for first gas only, whereas the right one is permeable for the second gas only. When two pistons touch and pressures in both parts become equal the separation process is completed. The reversible works (8.23) and (8.24) constitute the worst estimates of the lower bound for work consumption; any irreversible generalizations of these provide better (higher) estimates. Due to the various irreversible processes associated with finite separation rates, the actual work of separation includes the product TS_σ and can be significantly higher than the reversible work. Models are therefore designed that lead to residual (minimal) entropy production to obtain irreversible estimates for mechanical energy consumption in accordance with Equation (8.15). These models take into account a finite duration of separation, finite size of separator or finite rates necessary to assure finite productivity of the operation. The estimates for power, work or heat, that are enhanced in this way, depend explicitly on the transfer coefficients and size characteristics of the separator.

8.3. RESULTS FOR IRREVERSIBLE SEPARATIONS DRIVEN BY WORK OR HEAT

In the approach by Tsirlin and Kazakov (2002a), these estimates are defined by choosing a particular distribution of transferred mass fluxes along the system's length that minimize the entropy production and the separation work. Equivalently, the additional separation work caused by dissipation

$$\Delta A = \int_0^\tau (g_1(P_{01}, P_1)\Delta\mu_1 + g_2(P_{02}, P_2)\Delta\mu_2)\mathrm{d}t \tag{8.25}$$

is minimized subject to the conditions fixing the average rates through membranes

$$\int_0^\tau g_1(P_{01}, P_1)\mathrm{d}t = N_0 x_1(0) \tag{8.26}$$

$$\int_0^\tau g_2(P_{02}, P_2)\mathrm{d}t = N_0 x_2(0) \tag{8.27}$$

and other constraints that are characteristics of the process

$$V_0(t) + V_1(t) + V_2(t) = 1 \tag{8.28}$$

$$N(t)x_i(t) + N_i(t) = N_0 x_i(0) \tag{8.29}$$

$$P_{0i}(t) = \frac{RT}{V_0(t)} N(t)x_i(t) \tag{8.30}$$

$$P_i(t) = \frac{RT}{V_i(t)}N_i(t) \tag{8.31}$$

$$\frac{dN_i}{dt} = g_i(P_{0i}, P_i) \tag{8.32}$$

$$N_i(0) = 0 \tag{8.33}$$

$$N_i(t) = N_0 x_i(0) \tag{8.34}$$

This formulation represents quite complex problems of optimal control in which the controls are $V_1(t)$ and $V_2(t)$; their sum must not exceed the total volume of the system. It may be noted that the optimality criterion ΔA does not depend on the state variables N_i. Another feature is that both g_i and $\Delta \mu_i$ depend on the same variables P_{0i} and P_i. Some simplifications are therefore possible (Tsirlin and Kazakov, 2002a).

If one-to-one dependence between the increments of chemical potentials and fluxes is assumed the irreversible losses (8.25) take the form

$$\Delta A = \int_0^\tau (g_1 f_1(g_1) + g_1 f_2(g_1)) dt \tag{8.35}$$

This criterion can be minimized only with respect to the integral conditions (8.27) and subject to the sign constraints $g_1 \geq 0$ and $g_2 \geq 0$. A feasibility check of the remaining constraints is recommended (Tsirlin and Kazakov, 2002a). The optimization problem can be decomposed into two subproblems; to each the so-called averaged nonlinear programming is applied (Berry et al., 2000). Their solutions for optimal functions $g_i(t)$ are the piecewise functions of time that take no more than two values (hence no more than three values for the vector solution of the problem). The minimal work of separation is described by an equation of the structure

$$A_{min} = A_0 + \Delta A_{min} = A_0(x_1, x_2) + \frac{N_0^2}{t^f - t^i}\left(\frac{x_1^2}{\alpha_1} + \frac{x_2^2}{\alpha_2}\right) \tag{8.36}$$

where $x_2 = 1 - x_1$ and α_i ($i = 1, 2$) are permeabilities linking g_i with μ_i in the formula $g_i = \alpha_i \Delta \mu_i$.

In nonlinear generalizations products of $g_i = \beta$ $(P_0 - P_i)$ and $\Delta \mu_i = RT \ln(P_0/P_i)$ are involved. The optimal rates of flows through membranes are constant and equal $N_0 x_i/(t^f - t^i)$. The minimal consumption of mechanical energy in the irreversible process is

$$A_{\min} = A_0 + \Delta A_{\min} = A_0(x_1, x_2)$$

$$+ \frac{RTN_0}{t^f - t^i} \left(\sum_{i=1}^{2} x_i \ln \frac{P_{0\max}}{P_{0\max} - N_0 x_i / \beta_i (t^f - t^i)} \right) \qquad (8.37)$$

where $x_2 = 1 - x_1$ and $P_{0\max}$ represents a maximum allowable pressure of the operation. See Tsirlin and Kazakov (2002a) and the discussion below for formulae describing incomplete separation.

For short durations the quantity $A_{\min}$ increases rapidly, which is associated with the fact that at $x_1(0) = 0$ and $x_2(0) = 1$ the irreversible estimate of the separation work has the discontinuity. This explains why the real separation work of mixtures with low concentration of one component is much higher than the reversible estimate. For example, according to Chambadal (1963), the actual work required to separate uranium isotopes may exceed the reversible estimate by the order of 10^5.

Consider now briefly the incomplete separation of a binary mixture. Designate by x the initial concentration of the key component and by x_1 and x_2 its final concentrations in the left and right chamber achieved in the finite time $t^f - t^i$. From mass balances it follows that the ratio of N_0 moles of mixture in the central chamber to the total amount of mixture that is separated equals

$$b(x, x_1, x_2) = \frac{N_0}{N} = \frac{(x - x_2)(x_1 - x)}{x(1 - x)(x_1 - x_2)} \qquad (8.38)$$

This quantity is called the completeness of the separation. Indeed, when $b = 0$ there is no separation at all, whereas $b = 1$ refers to the complete separation (on the pure components). Using the expression for reversible work of incomplete separation, Equation (8.24), referred to as one mole, one may determine the unit reversible work $a_n^0(x) \equiv A_0/N_0$. With this quantity the work estimate that has to be spent for incomplete separation of the binary mixture is

$$A_{\min} = A_0 + \Delta A_{\min}$$

$$= N_0 a_n^0(x) + \frac{N_0^2}{t^f - t^i} \left(\frac{x^2}{\alpha_1} + \frac{(1 - x)^2}{\alpha_2} \right) b^2(x, x_1, x_2) \qquad (8.39)$$

and the corresponding power estimate (lower limit)

$$p_{\min} = \frac{A_0 + \Delta A_{\min}}{t^f - t^i} = g a_n^0(x) + g^2 \left(\frac{x^2}{\alpha_1} + \frac{(1 - x)^2}{\alpha_2} \right) b^2(x, x_1, x_2) \qquad (8.40)$$

where b is defined by Equation (8.38) and the ratio $g = N_0/(t^f - t^i)$ is the flux of the input mixture or a measure of the productivity. The quantity $p_{\min}$ increases with

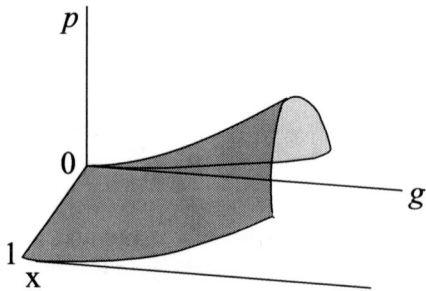

Figure 8.1 Power supply p for mechanical separation has a maximum with respect to initial concentration x of the mixture and grows monotonically with the productivity g of operation (Tsirlin and Kazakov, 2002a).

g and has a maximum with respect to x. Figure 8.1 illustrates the power supply for mechanical separation as a function of the productivity g and concentration x of the input flux.

This information supplements information related to thermal separation processes presented in Chapter 3 (Section 3.8; Equations (3.64)–(3.68) and the corresponding text).

As a rule, diffusion engines and separation systems are based on membranes. Tsirlin and Leskov (2007) quantify the limiting performance of membrane systems in the separation processes with fixed rates, focusing on maximizing the power of diffusion engines and minimizing the power necessary for the separation of a prescribed feed mixture. The latter problem is the inverse with respect to the first, and is associated with maintaining a nonequilibrium state by consuming power. The authors consider engines with periodic and constant contact between the working fluid and sources. The solution of both optimization problems depends on whether the feed mixture used by the engine is gaseous or liquid because this determines the form of the chemical potentials of components and, hence, the driving forces of the process. Sieniutycz (1999a) develops a thermodynamic scheme for optimal control of multistage engines with heat and mass transfer, characterized by the vector of efficiencies.

Below we pass to examples of optimization and energy limits in some separation operations, thermal or mechanical.

8.4. THERMOECONOMIC OPTIMIZATION OF THERMAL DRYING WITH FLUIDIZING SOLIDS

8.4.1. General Remarks

Drying and evaporation are separation operations leading to the removal of an active component (moisture) from solid to gas. Drying is one of the unit operations requiring the largest quantity of exergy. Its exergy consumption is closely linked with the technology used. While the general principles of energy conservation have been described in monographs (Reay, 1977; Grant, 1979), their application to drying is still insufficient. Energy policy in drying has been reviewed by Strumillo (1983), Strumillo and Lopez-Cacicedo (1984), and Strumillo and Kudra (1987). Exergy analyses of drying began with the work of Bosniakovic (1965) and Opman (1967); for recent results in the framework of exergy costs and related optimizations see Zawlocki (1987), Szwast (1990) and Sieniutycz (1985, 1991), Sieniutycz and Szwast (1982a). Many industrial data are, however, still reported in terms of energy rather than exergy efficiency and suffer from the lack of a strong link with well-established kinetic models. Indeed, reasonable modeling of dried bodies naturally involves sets of coupled partial differential equations for temperature and moisture content. They are derived from the irreversible thermodynamic formalism using Onsager's theory (Luikov and Mikhailov, 1963). However, the solvable models of drying are most often lumped ones and deal with the space-averaged properties (Sieniutycz, 1973b, 1984). Graphical and numerical solutions of the balance and kinetic equations are effective (Ciborowski, 1965; Poświata, 2005).

Bejan's constructal theory (Bejan, 2000) has been used in porous systems to show analytically and numerically how an originally uniform flow structure transforms itself into a nonuniform one when the objective is to minimize global flow losses (Ordonez et al., 2003). The flow connects one point (source, sink) to a number of points (sinks, sources) distributed uniformly over a two-dimensional domain. It is shown that flow "maldistribution" and the abandonment of symmetry are necessary for the development of flow structures with minimal resistance. The flow medium is continuous and permeated by Darcy's flow. The flow structures generated based on the pressure gradient criterion have consistently smaller flow resistances. As flow systems become smaller and more compact, the flow systems themselves become "designed porous media." Constructal theory has also been applied to find the geometry and operation of adsorption processes. Moreover, particle agglomeration and design of air-cleaning devices have been modeled by the constructal theory (Reis et al., 2006), and the dendritic geometry has been predicted along with the existence of asymmetric bifurcation (Wechsatol et al., 2006).

The potential for exergy saving in drying processes through changes in design and operation is significant. Possible process and design modifications involve waste heat recovery from solids, energy recuperation from gas, application of heat-pumps for waste energy upgrading, recycling of the exhausted drying agent, combinations of mechanical and thermal drying, and use of solar energy. Żyłła and Strumiłło (1991) have advocated the use of heat-pumps to decrease energy consumption in drying. The multiple effects of evaporation from solution have been treated (Itahara and Stiel, 1966), in particular with the exploitation costs evaluated on the basis of exergy input (Sieniutycz, 1991). The combination of solar and dehumidification drying offers advantages not found when either of these methods are used alone; reviews of the economic aspects of solar drying are available (Imre et al., 1986; Imre, 1990). Since the efficiency of flat-plate

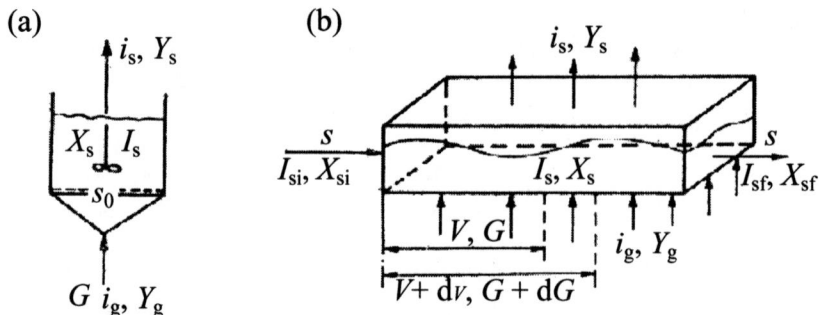

Figure 8.2 Batch fluidized drying and steady fluidized drying in a horizontal exchanger.

collectors is correlated with the exit temperature of the warmed working agent, an appropriate efficiency of drying can be achieved only at a moderate temperature level (Bejan, 1982; Imre et al., 1986). A prerequisite for both exergy and economic efficiency is therefore the application of solar energy to situations allowing moderate temperatures of the driving agent. Second law analyses of solar drying imply that the system should be integrated into the power system of an agricultural plant, and the exergy received used for other operations during periods when the basic process is not running. A number of researchers have advocated a combined solar energy-heat-pump dryer and the combination of wind and solar energies (see references in the review by Strumillo and Lopez-Cacicedo (1984)). Exergy saving methods of drying using dielectric heating which use radio and microwave frequencies are under development (Cross et al., 1982; Bednarek et al., 1981); they offer the further advantages of improved product quality and decreased equipment size.

Interestingly, only a few of the types of industrial dryers in use consume most of the total exergy used in drying; five types—flash, tower, rotary, cylinder and fluidized bed dryers—account for approximately 99% of the total exergy usage (Strumillo and Lopez-Cacicedo, 1984). A meaningful reduction of drying exergy could be achieved by the optimal control of these particular dryers; on average, the potential for exergy reduction is more than 20%. However, only the exergy optimization of the fluidized bed dryers has received considerable attention (Sieniutycz and Szwast, 1982a; Sieniutycz, 1982, 1985; Szwast, 1990).

Thermal drying of fine particles in fluidized beds is an effective operation because of good contact of phases and a large interfacial surface. In this section we consider the optimal transition of solid phase from initial thermodynamic state (I_{si}, X_{si}) to final thermodynamic state (I_{sf}, X_{sf}) in the fluidized bed. The dynamic programming method is used to find optimal decisions and optimal trajectories of a drying process corresponding with the minimum of costs required to run the process in a finite time τ.

Figure 8.2a depicts the first objective which is the dynamical process of batch fluidization drying. Its model also describes the process of steady-state fluidization drying in a horizontal exchanger, Figure 8.2b, which is the second objective. In the first process, the solid state changes as a function of time; in the second process, it changes as a function of distance when ideal mixing in the first apparatus and ideal mixing in vertical cross-section of the second apparatus are assumed. In an isobaric and adiabatic process, the thermodynamic state of the solid is described by enthalpy I_s and moisture content X_s, the inlet gas by enthalpy i_g and absolute humidity Y_g. The gas phase is the controlling phase and the solid phase is the controlled phase. The solid enthalpy I_s and solid moisture content X_s are state variables in the original formulation of the optimization problem. The inlet gas enthalpy i_g, the inlet gas humidity Y_g and the inlet gas flow G are decision variables in the most general case. In this book, the optimization problem is considered only for the most realistic case where the enthalpy (or temperature) of the inlet gas is the only decision variable with the required gas humidity Y_g and gas flow constant and equal respectively to Y_{g0} and G_0.

The usual way of conducting the processes considered is with a constant inlet gas state. In such a case the inlet gas enthalpy i_g and absolute humidity Y_g do not change with solid residence time in either of the processes. It is known that driving forces of such processes are large at the beginning and small at the end of the process and that this fact makes processes uneconomical, that is, cause large production costs. Therefore, it should be expected that a variable inlet gas state should make the processes considered more economical. Finding a technique which minimizes production costs in a finite time or the total costs in a free (optimization determined) time is a task of the optimization in question. For processes with a single controlling phase, such as those considered here, where the gaseous phase is controlling the process, the exergy cost optimization and the economic optimization (based on data of prices and the sum of the production and investment costs) give the same family of optimal trajectories and optimal decisions if the unit economic price of the controlling (gaseous) phase is assumed to be proportional to the unit available energy of gas. Note, however, that the optimal final states of a process with some free (nonprescribed) coordinates depend on the relation between the prices of the controlling and controlled phase, and hence the optimal final states of the thermodynamic and economic optimization differ. (This does not disprove the conclusion, already stated, that the *families* of trajectories in both optimizations are identical.) Since the dynamic programming method leads to the family of the optimal trajectories, the difference between thermodynamic and economic optimization is broken down to the difference between the particular trajectories that correspond with the thermodynamically and economically optimal final states. This substantiates the use of the present analysis for economic purposes (Figure 8.3).

8.4.2. Drying in a Quasi-Homogeneous Fluidized Bed

When the fluidization process is quasi-homogeneous, sets of equations describing the energy and mass exchange in the batch process as well as the process in a horizontal fluidized exchanger can be derived under the assumption that the

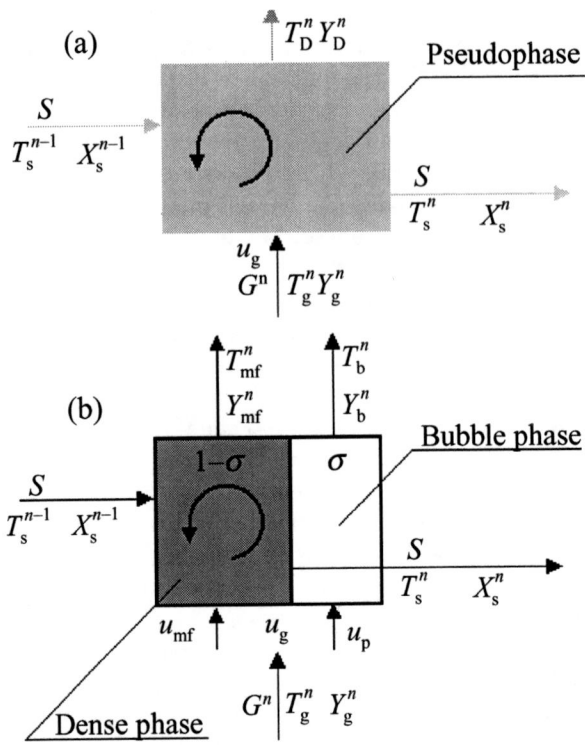

Figure 8.3 Comparison of two schemes of fluidized bed modeling to describe stage n of the cascade. In case A gas and solid constitute a perfectly mixed nonequilibrium pseudophase (Sieniutycz, 1972, 1973b, 1973d), in case B bubble and dense phase are distinguished (Poświata, 2005; Poświata and Szwast, 2003, 2007). The scheme allows the simultaneous treatment of the batch fluidization and fluidization in a horizontal exchanger as the continuous limit of a cascade.

gas–solid system in the fluidized bed is a well-mixed disequilibrium pseudophase (Sieniutycz, 1972, 1973b, 1973d). This refers to case A in Figure 8.3 where the outlet gas state (T_0^n, Y_0^n) is not in equilibrium with the solid (T_s^n, X_s^n). The disequilibrium is associated with a finite mass transfer unit number, N_g.

The model with an ideal mixing in both phases can only be a very rough approximation of the inhomogeneous gas fluidization, applicable in the close vicinity of the thermodynamic equilibrium between outlet gas and outlet solid. In order to take channeling and bubbling effects into account a two phase model of fluidization has to be exploited, Section 8.4.3. In this section only the pseudo-homogeneous model and its optimization solution for the equilibrium between outlet gas and outlet solid ($N_g \to \infty$) is discussed. Analogous solutions can easily be found for a finite N_g, using Equations (8.41)–(8.43) given below for this purpose.

We display below the resulting equations of the nonequilibrium pseudophase model using in principle Poświata's (2005) notation rather than the notation in the original papers (Sieniutycz, 1973b, 1973d). This should facilitate comparison of both models. Under the assumption of an ideal mixing in the nonequilibrium pseudophase (finite N_g) the differential state equations are

$$\frac{dI_s}{d\tau} = \frac{N_g}{Le} \left\{ i_g - i_s[I_s, X_s] + \frac{(i_p(Le - 1)(Y_g - Y_s[I_s, X_s]))/(N_g + 1)}{N_g/Le + 1} \right\} \quad (8.41)$$

$$\frac{dX_s}{d\tau} = \frac{N_g(Y_g - Y_s[I_s, X_s])}{N_g + 1} \quad (8.42)$$

The variable τ is a dimensionless time of the process defined as:

$$\tau = \begin{cases} Gt/S_0 \text{ for batch fluidization process} \\ G/S \text{ for the process in the horizontal exchanger} \end{cases} \quad (8.43)$$

Square brackets are reserved for functional relations which are often implicit. We omit here derivations of these equations, referring the reader to the original publications (Sieniutycz, 1973a, 1973b). The single pseudophase model constitutes an approximate, insufficient description of fluidization, often too approximate. However, it may be used with an admissible accuracy (ca. $\pm 12\%$) in situations where the bed arrangements (baffles, mixers, etc.) cause a significant mixing of the fludized bed, making the dispersion contribution to the plug flow prevailing. In the opposite limiting case, of the negligible dispersion in the gas phase, that is, when an ideal plug flow of gas can be assumed, the two-phase model of fluidization (Kunii and Levenspiel, 1969, 1991) is operative. The state equations of this two-phase model are in the most general case implicit. However, for the majority of real fluidizing apparatus, where the bed height is of the order 0.5 m, practical thermodynamic equilibrium is observed between outlet gas and solid due to the very large specific surface of the solid and very small height of the transfer units (HTU). In this limiting case, $N_{g \to \infty}$, the state equations involve exclusively equilibrium balances, and, as such, they are independent of the specific fluidization model. The equilibrium state equations are

$$\frac{dI_s}{d\tau} = i_g - i_s[I_s, X_s] \quad (8.44)$$

$$\frac{dX_s}{d\tau} = Y_g - Y_s[I_s, X_s] \quad (8.45)$$

Clearly, these continuous equations follow immediately from the energy and mass balances of the fluidized bed. In this volume only the optimization solution in the case of equilibrium between outlet gas and solid, described by Equations

(8.44) and (8.45), is discussed. If necessary the reader can easily find the analogous solution for the more general case, using Equations (8.41) and (8.42) given above for this purpose. (They complement equations of the two-phase model of fluidization (Kunii and Levenspiel, 1969, 1991; Wagialla et al., 1991).) The state equations have on their right-hand sides controls (i_g, Y_g) and thermodynamic functions of the state variables $i_s[I_s, X_s]$ and $Y_s[I_s, X_s]$. The latter are always possible to find for a definite gas–solid system from the data of gas–solid equilibria (sorption equilibria). For the silica gel–air–water system equilibrium data in terms of functions $i_s[I_s, X_s]$ and $Y_s[I_s, X_s]$ are available (Sieniutycz, 1973c; Sieniutycz and Szwast, 1982b).

Industrial processes should usually ensure maximum profit defined as the difference between the market value of products and operational cost. As the market value of products depends on current market condition, operational cost may decide net process profit, and then also the process attractiveness. Engineers may then restrict themselves to taking this operational cost into consideration in their design. Optimization of drying can often be reduced to minimization of the process operational cost. Such optimization should make drying cheaper and more effective.

The performance index of the batch fluidization process is accepted as the difference between the total input value of substances delivered to the system and the total output value of substances leaving it, increased by the difference between the value of substances in the system before and after the process. Such a criterion is referred to here as the unit mass of the solid S_0. Without modification, it may be used for processes undergone in a given equipment (fixed investment costs). For the fluidization in the horizontal exchanger a corresponding criterion involves the difference between the flows of value of substances delivered to the system and those leaving it.

The performance index of both processes expressed as a cost, referred to as the mass unit of dry solid, can be written in the common form as

$$K^p = \int_0^{t^f} (c_g - \mu c_s)dt + C_{si} - C_{sf} \tag{8.46}$$

where c_g and c_s are, respectively, gas prices per unit mass of inlet and outlet gas; C_{si} and C_{sf} are unit prices of solid at the initial and final state of the process. Exergy tariff of prices can be used (Szargut and Petela, 1965); then a counterpart of the above formula will be defined in terms of specific exergies (Sieniutycz, 1973a, 1973b). The performance index formula is valid given the same assumption under which the process equations were derived. The quantity μ appearing in Equation (8.46) is the so-called coefficient of usability of the outlet gas. When $\mu = 0$ then the outlet gas is being released into the atmosphere. In such a case the effective exergy of outlet gas is equal to zero. When $\mu = 1$ then the outlet gas is being fully exploited in other operations. The value of μ can be a decision variable when complex system links are considered.

Now we consider the case of an equilibrium fluidized drying ($N_g \to \infty$) with a constant inlet gas humidity X_{g0}. The corresponding equations are

$$\frac{dI_s}{d\tau} = i_g - i_s[I_s, X_s] \tag{8.44}$$

$$\frac{dX_s}{d\tau} = Y_g - Y_s[I_s, X_s] \tag{8.45}$$

$$\frac{dK^P}{d\tau} = c_g[i_g, Y_{g0}] - \mu c_s[I_s, X_s] - \frac{dC_s}{d\tau} \tag{8.46$'$}$$

Equation (8.46$'$) describes the changes of cost K^P along the process path. It was obtained by differentiating the integral describing these costs, Equation (8.46), with respect to the variable upper limit of integration and assuming the equilibrium model (in the sense already described).

In order to exploit the autonomous nature of the model with respect to time τ it is advisable to transform the set of Equations (8.44)–(8.46$'$) into such a form in which τ is a dependent variable. We then obtain an equivalent set of process equations which do not contain dependent variable τ explicitly on their right-hand sides. This dependent variable will allow the use of a Lagrangian multiplier to reduce process dimensionality. In order to obtain the set of equations with time as a dependent variable, we use the solid moisture content X_s as an independent variable. The set of Equations (8.44)–(8.46$'$) then takes the form

$$\frac{dI_s}{d\tau} = \frac{i_g - i_s[I_s, X_s]}{Y_{g0} - Y_s[I_s, X_s]} \tag{8.47}$$

$$\frac{d\tau}{dX_s} = \frac{1}{Y_{g0} - Y_s[I_s, X_s]} \tag{8.48}$$

$$\frac{dK^P}{dX_s} = \frac{c_g[i_g, Y_{g0}] - \mu c_s[I_s, X_s]}{Y_{g0} - Y_s[I_s, X_s]} - \frac{dC_s}{dX_s} \tag{8.49}$$

In this set, state variables are solid enthalpy I_s and time τ; the control is inlet gas enthalpy i_g (X_{g0} = constant). We shall distinguish the original optimization problem (without Lagrangian multiplier) from the transformed problem (after the Lagrangian multiplier is introduced).

The Original Optimization Problem

It is useful now to turn our attention to the original optimization problem. We are investigating the minimum of a functional obtained by the integration of Equation (8.49):

$$K^p\{i_g[X_s]\} = \frac{c_g[i_g, Y_{g0}] - \mu c_s[I_s, X_s]}{Y_{g0} - Y_s[I_s, X_s]} dX_s + C_s[I_{si}, X_{si}] - C_s[I_{sf}, X_{sf}] \quad (8.50)$$

with differential Equations (8.47) and (8.48) plus additional conditions imposed on temperatures and concentrations of the gas and material. Solid moisture content X_s is specified at the initial and final stages, whereas final enthalpy I_{sf} (final temperature T_{sf}) and the overall time τ_f may be prescribed or arbitrary. We recognize that this variational problem has the Bolza form.

The Lagrangian Multiplier and the Transformed Optimization Problem

Considering two constraining equations, Equations (8.47) and (8.48), the original optimization problem is imbedded in three-dimensional space of variables I_s, X_s and τ. Since multidimensional problems are difficult to solve, the possibility of dimensionality reduction should be explored. Because the dependent variable is not explicitly present on the right-hand sides of Equations (8.47) through (8.49), there exists the possibility of using the Lagrangian multiplier λ (associated with the process time τ) and then transforming the original problem to one with a single restrictive equation. This procedure gets rid of the constraint, Equation (8.48), and relies on determining the functional extremum

$$K = K^p + \lambda(t_f - t_i) = K^p + \lambda \int_{X_{si}}^{X_{sf}} \frac{dX_s}{Y_{g0} - Y_s[I_s, X_s]} \quad (8.51)$$

or, on the basis of Equation (8.50),

$$K\{i_g[X_s], \lambda[X_s]\} = \int_{X_{si}}^{X_{sf}} \frac{c_g[i_g, Y_{g0}] - \mu c_s[I_s, X_s] + \lambda}{Y_{g0} - Y_s[I_s, X_s]} dX_s$$
$$+ C_s[I_{si}, X_{si}] - C_s[I_{sf}, X_{sf}] \quad (8.52)$$

with the differential restriction from Equation (8.47), appropriate initial and final conditions, as well as any extra local constraints imposed on the state and decision variables. It should be pointed out that the final time in the transformed problem cannot be prescribed.

The role of Lagrangian multipliers in dynamic programming is clarified by Bellman (1957) and Bellman and Dreyfus (1967). Properties of multiplier λ used here are discussed further (Figure 8.4).

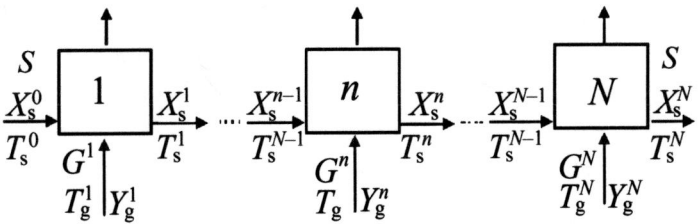

Figure 8.4 Scheme of the fluidized drying calculation as a multistage process.

Discretization of the Process Equations for the Purpose of Computation of Optimal Control

Since thermodynamic relations in Equations (8.47)–(8.49) and (8.52), that is, functions $i_s[I_s, X_s]$ and so on are too complex in their analytical form, it is necessary to solve the optimization problem numerically. Because of the process restrictions and the necessity of numerical methods, it is most appropriate to apply the discrete version of dynamic programming as the method of optimization. Although the processes considered are continuous, in order to make use of computer facilities, Equation (8.47) and performance index, Equation (8.52), should be written in an approximate difference form. We designate $\Delta = -\Delta X_s$. The process criterion is

$$
K[i_g^1, i_g^2, \ldots, i_g^N, \lambda] = \sum_{n=1}^{N} \left\{ \frac{c_g[i_g^n, Y_{g0}] - \mu c_s[I_s^n, X_{si} - n\Delta] + \lambda}{Y_s[I_s^n, X_{si} - n\Delta] - Y_{g0}} \Delta \right.
$$
$$
\left. + C_s[I_s^{n-1}, X_{si} - (n-1)\Delta] - C_s[I_s^n, X_{si} - n\Delta] \right\}
$$

(8.53)

Thus, we should find the extremum of discrete functional (8.53) with the difference constraint

$$
\frac{I_s^n - I_s^{n-1}}{\Delta} = \frac{i_g^n - i_s[I_s^n, X_{si} - n\Delta]}{Y_s[I_s^n, X_{si} - n\Delta] - Y_{g0}}
$$

(8.54)

and appropriate initial and final conditions as well as some additional local constraints on the state and decision variables (see below).

The Recurrence Equation of Dynamic Programming

With the help of Equations (8.53) and (8.54) and Bellman's principle of optimality, it is possible to derive a basic recurrence equation for the transformed problem. Designating

$$
\min_{\{i_g^n\}} K[i_g^1, i_g^2, \ldots, i_g^N, \lambda] \equiv F^n[I_s^N, \lambda]
$$

(8.55)

and taking advantage of the restrictive Equation (8.54) to express the inlet gas enthalpy i_g^n as a function of the material enthalpies before and after the stage (I_s^{n-1} and I_s^n respectively)

$$i_g^n = i_s[I_s^n, X_{si} - n\Delta] + \frac{I_s^n - I_s^{n-1}}{\Delta}(Y_s[I_s^n, X_{si} - n\Delta] - Y_{g0}) \qquad (8.54')$$

the following recurrence equation is obtained:

$$F^n[I_s^n, \lambda] = \min_{I_s^{n-1}} \left\{ \left(\frac{c_g[i_s[I_s^n, X_{si} - n\Delta] + \Delta^{-1}(I_s^n - I_s^{n-1})(Y_s[I_s^n, X_{si} - n\Delta] - Y_{g0}), Y_{g0}]}{Y_s[I_s^n, X_{si} - n\Delta] - Y_{g0}} \right. \right.$$

$$\left. - \frac{\mu c_s[I_s^n, X_{si} - n\Delta] + \lambda}{Y_s[I_s^n, X_{si} - n\Delta] - Y_{g0}} \right) \Delta + C_s[I_s^{n-1}, X_{si} - (n-1)\Delta] - C_s[I_s^n, X_{si} - n\Delta]$$

$$\left. + F^{n-1}[I_s^{n-1}, \lambda] \right\}. \qquad (8.56)$$

This equation starts with $F^0[I_s^0, \lambda] = 0$. The enthalpy I_s^n of the solid leaving stage n is the state variable and the solid enthaply before stage I_s^{n-1} is the new decision variable. Using decision I_s^{n-1} instead of the original decision i_g^n makes computations simpler.

The recurrence equation, Equation (8.56), can be written in a general form

$$F^n[I_s^n, \lambda] = \min_{I_s^{n-1}} \{P^n(I_s^n, I_s^{n-1}, \lambda) + F^{n-1}[I_s^{n-1}, \lambda]\} \qquad (8.57)$$

The quantity P^n in Equation (8.57) is the cost consumed at the nth process stage. The term $F^{n-1}[I_s^{n-1}, \lambda]$ represents the results of all previous computations of the optimal costs for the $n - 1$ stage process. Obtaining the optimization solution relies on recursive minimization of the right-hand side of Equation (8.56). The standard procedure for solving Equation (8.57) is shown in many books on optimization, for example, Bellman and Dreyfus (1967).

At first, function $F^1[I_s^1, \lambda]$ is obtained for an assumed constant λ by substituting the initial values $I_s^{n-1} = I_{si}$ and $n = 1$ into the right-hand side of Equation (8.57). As there is the possibility to choose only a single value for $I_{si} = $ constant, the minimization P^1 with respect to I_{si} does not take place in the present problem. This leads to the function being equal to $P^1[I_s^1, I_{si}, \lambda]$.

When $n = 2$ the optimization procedure relies on finding the minimum of the sum

$$S^2 = P^2(I_s^2, I_s^1, \lambda) + F^1[I_s^1, \lambda]$$

with respect to the enthalpy I_s^1 but at a constant enthalpy I_s^2. Iterating minimization for various discrete values I_s^2 leads to optimal functions $I_s^1[I_s^2, \lambda]$ and $F^2[I_s^2, \lambda]$.

An identical procedure holds in the case of $n = 3, 4, \ldots, N$. The procedure is applied to solve Equation (8.56). Following the minimization of the right-hand side of the recurrence Equation (8.56) and storage of optimal thermodynamic parameters of solid before and after every stage, the results of the optimal gas inlet enthalpy and optimal process time are computed and stored. This is accomplished, respectively, by means of Equation (8.54) and the following formula:

$$\tau^n[I_s^n, \lambda] = \frac{\Delta}{Y_s[I_s^n, X_{si} - n\Delta] - Y_{g0}} + \tau^{n-1}\{I_s^{n-1}[I_s^n, \lambda], \lambda\} \qquad (8.58)$$

which represents the difference form of Equation (8.48). All of the optimization results depend upon the assumed value of the parameter λ as well as upon the state of the process (I_s^n, X_s^n).

A complete flow diagram of the program used in the computations of the optimal decisions and optimal trajectories as well as a sample of the computational data are available (Sieniutycz, 1972, 1973a, 1973b; Sieniutycz and Szwast, 1982a).

Equation (8.56) has been solved for the constant inlet solid state $I_{si} = -4.2\,\text{kJ kg}^{-1}$, $X_{si} = 0.1\,\text{kg kg}^{-1}$ ($T_{si} = 22.6\,°\text{C}$). Constant inlet gas humidity was accepted as that found in the atmospheric air; $X_{g0} = 0.008\,\text{kg kg}^{-1}$. It was assumed that $\mu = 0$, that is, the outlet gas is not exploited. The maximum admissible inlet gas temperature t_{gmax} was assumed as equal to $375\,°\text{C}$. A constraint on the solid parameters was assumed in the form

$$Y_s[I_s^n, X_{si} - n\Delta] \geq Y_{g0} \qquad (8.59)$$

that limits equilibrium gas humidities. This inequality establishes the working regime of solid states, since in the case of the drying process, the recurrence relationship, Equation (8.56), must be solved within the boundary of the variables (I_s, W_s) where the evaporation direction is from solid to gas. This boundary is known from the drying equilibrium data. These data as well as thermodynamic functions of gas and solid are known (Sieniutycz, 1973c).

The Properties of the Lagrangian Multiplier

The form of expression for the performance index K, Equation (8.51), shows that the quantity K plays the role of total process cost, which is the sum of the production cost K^p and the investment cost K^i, the latter expressed here as $\lambda\tau_f$. All process costs can be expressed either in economic (price) units or in exergy units. The production costs follow from the exergy balance, analogous to the economic balance. By definition, the exergy investment costs are, in our case, the economic investment costs divided by the economic value of the unit exergy of the gaseous phase.

A comparison of economic and thermodynamic results (Sieniutycz and Szwast, 1982a) shows that the (economic) Lagrangian multiplier λ is related to the

Table 8.1 The relationship between the Lagrangian multiplier (in exergy units) and the optimal process time (for $\mu = 0$ and $\tau_1 = 0$)

τ_f	13.5	5.7	3.45	2.62
λ (kJ kg^{-1})	0.42	4.2	21.0	105

fluidized apparatus price per unit area of sieve bottom p as follows:

$$\lambda = \left(\frac{z}{T} + \beta\right) \frac{p}{\tau_u v_g e_g} \tag{8.60}$$

The symbols z, T, β, and τ_u pertain to the economic quantities; z is the nondimensional factor describing freezing of the capital costs, τ is the maximum acceptable payout time (in years), β is the coefficient taking into account renovations (years^{-1}), τ_u is the utilization time of the dryer (h year^{-1}). Rules for assuming these coefficients are available (Szargut, 1970). The quantity e_g in Equation (8.60) is the economic value of unit exergy for gaseous phase, discussed in various sources (Szargut and Petela, 1965; Szargut, 1970).

In the fluidized bed, superficial gas velocity v_g is, in fact, a constant multiple of the critical velocity. Therefore, the velocity v_g in Equation (8.60) is a nearly constant quantity, if a definite diameter of the solid grain is assumed. Also, other quantities in Equation (8.60), except the apparatus price p, can be assumed as constant. Under the above assumptions, the Lagrangian multiplier λ should be interpreted as the unit apparatus price expressed in appropriate units. The formula, Equation (8.60), is important in applications, since it makes possible estimation of how the numerical value of λ is related to a definite fluidized apparatus. Typical values of λ are of the order of 0.5–20 kJ kg^{-1}. With this information a subclass of the whole family of optimal trajectories is obtained for a constant λ as a result of the solution of the transformed optimization problem. Such a solution gives information about the optimal process time directly, on the basis of Equation (8.58).

Iterating optimization computations within a region of the state variables (I_{sf}, X_{sf}) and changing only the values of parameter λ, tables describing the relationship between the multiplier λ and the optimal process time τ_f are obtained ($\tau_s = 0$). An example of this link for $\mu = 0$ is given in Table 8.1. The results pertain to initial solid state $I_{si} = -4.2$ kJ kg^{-1}, $X_{si} = 0.1$ kg kg^{-1} ($t_{si} = 22.6\,°C$) and the final solid state $I_{sf} = -21$ kJ kg^{-1}, $X_{sf} = 0.06$ kg kg^{-1} ($t_{sf} = 40\,°C$).

The results show that when λ increases, τ_f decreases. Since the final solid state (I_{sf}, X_{sf}) is fixed, average process intensity described by the vector (I_{sf}/τ_f, X_{sf}/τ_f) increases with λ. Therefore the Lagrangian multiplier has dual interpretation: it represents both the apparatus price (expressed in appropriate units) and the optimal process intensity index. In the optimal process the changes of the apparatus price are balanced by the changes of the process intensity. Here we encounter a manifestation of Tondeur and Kvaalen's (1987) principle of equipartition of the entropy production and Bejan's (1987b) conclusion that the exchange area should be concentrated in places where the process is most intense.

It is useful to compare applications of two types of solution to our optimization problems. The most general solution of the transformed optimization problem describes the dependence of overall costs, optimal decision and optimal time upon the final state $[I_{sf}, X_{sf}]$ and the Lagrangian multiplier λ (functions $F[I_{sf}, X_{sf}, \lambda]$, $i_g[I_{sf}, X_{sf}, \lambda]$, $\tau_f [I_{sf}, s_f, \lambda]$). (The moisture content dependence in these relations is the consequence of the explicit presence of the stage number n in the transformed optimization problem.) The corresponding solution of the original optimization problem leads to relations which express the production costs K^p, multiplier and the last decision as a function of the final state and the final time τ_f (functions $K^p[I_{sf}, X_{sf}, \tau_f]$, $i_g[I_{sf}, X_{sf}, \tau_f]$, $\lambda[I_{sf}, X_{sf}, \tau_f]$).

The solution of the transformed optimization problem is useful when a numerical value of the unit apparatus price p is prescribed. In such a case one can solve the transformed optimization problem for definite values of λ, evaluated from Equation (8.60). Furthermore, one can obtain direct information about the optimal overall time of the process from Equation (8.58).

On the other hand, the solution of the original optimization problem is useful when the final process time is assumed. Knowledge of such a solution enables one to find the optimal multiplier value λ, and hence the optimal apparatus price p, from the condition

$$\lambda = -\frac{\partial K^p[I_{sf}, X_{sf}, \tau_f]}{\partial \tau_f} \tag{8.61}$$

which expresses a necessary condition for the stationary minimum of the total costs $K = K^p + \lambda \tau_f$ with respect to the final time τ_f. By computing λ from Equation (8.61) and using Equation (8.60), one can find the optimal price p of a designed apparatus.

Since the case when the unit apparatus price p is a prescribed quantity is very common, the multiplier of investment costs, λ, most often has a synonymously specified value. Therefore the solution of the transformed optimization problem is usually the most convenient to use. In addition, the solution of such a problem is easier to obtain since the related equation of dynamic programming is one-dimensional. One may realize that the (transformed) optimization problem with given p or λ is simultaneously the problem of a constant process intensity or a constant rate penalty (in terms of Hamiltonian or λ). For the above reasons most of the optimization calculations are performed for the transformed problem. Since the case when outlet gas is being released to atmosphere is most popular in practice, the results discussed in the following pertain to the case when $\mu = 0$. Some related qualitative results in the case $\mu = 1$ are presented elsewhere (Sieniutycz, 1972). Since silica gel is a popular industrial adsorbent, the silica gel–water–air system was studied in optimization computations.

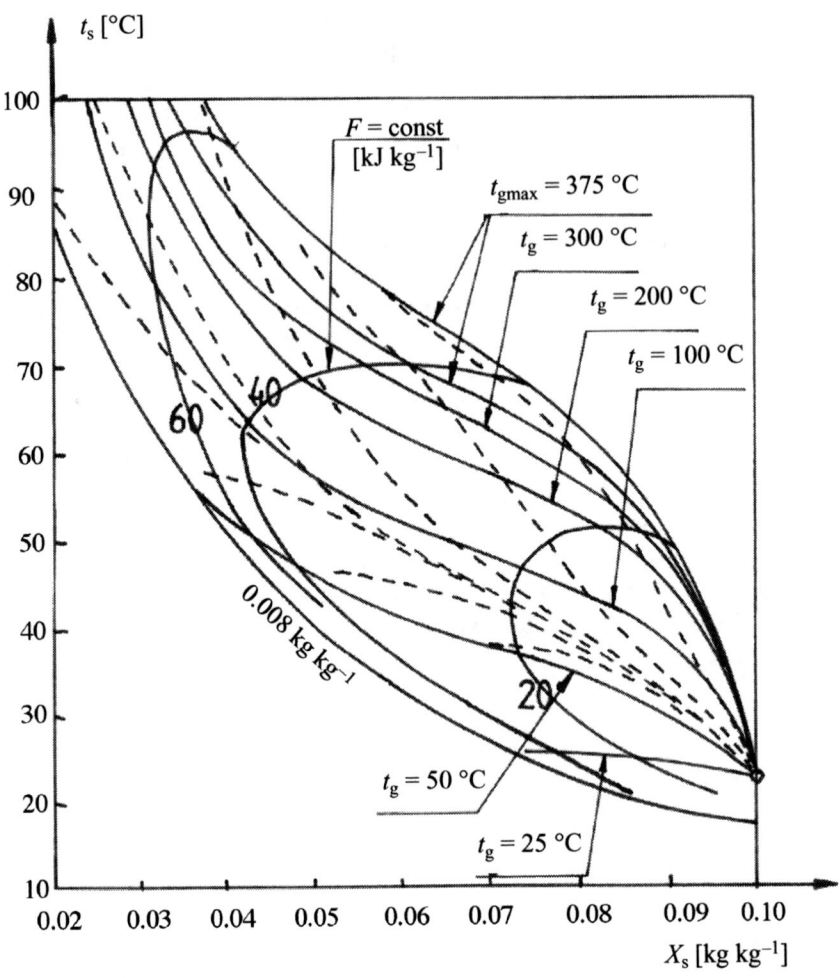

Figure 8.5 Rays (trajectories) and lines of constant costs. The equilibrium problem (N_g). Results for silica gel–air–water system, $\mu = 0$, $t_{si} = 22.6\,°C$, $I_{si} = -4.2\,kJ\,kg^{-1}$, $X_{si} = 0.1\,kg\,kg^{-1}$, $\Delta I_s = 0.84\,kJ\,kg^{-1}$, $\Delta = -\Delta X_s = 0.01\,kg\,kg^{-1}$, $\lambda_B = 4.2\,kJ\,kg^{-1}$, $X_{g0} = 0.008\,kg\,kg^{-1}$, $t_{gmax} = 375\,°C$.

Results of Computations

The following values—the optimal costs F^n, the optimal solid enthalpy I_s^{n-1}, the optimal gas enthalpy i_g^n (or the optimal gas temperature t_g^n) and the optimal time τ^n—are found by computations as functions of variables I_s^n, X_s^n, and λ.

The basic curves characterizing the optimal process are shown in Figure 8.5 on the temperature–moisture content diagram for the solid phase (silica gel). The graph contains "rays" (broken lines of optimal paths) and "wave-fronts" (continuous lines of constant costs). Other continuous lines on the graph are constant decision curves, t_g. One can see that no specific curve exists corresponding

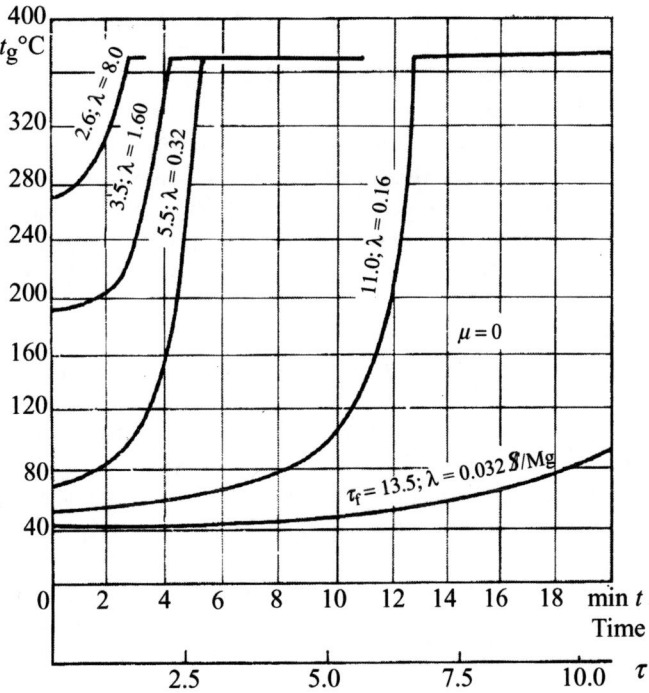

Figure 8.6 Optimal drying gas temperature t_g versus dimensionless time τ and Lagrangian multiplier λ.

with the maximum admissible temperature t_{gmax}, but boundaries do exist for the solid-state variables (T_s, X_s). Inside these boundaries the state change is connected with the constant gas temperature policy.

A graph of the optimal inlet gas temperature as a function of a dimensionless time τ and multiplier λ is shown in Figure 8.6. The results in this graph indicate that an expensive apparatus (large λ) is characterized by high inlet gas temperatures and short final process times (i.e. large process intensities). These results confirm earlier conclusions that the multiplier λ is the process intensity index or rate penalty index. Data on Figure 8.6 shows that in order to attain any final state of material in definite apparatus (λ = constant) the optimal gas inlet temperature should be increased monotonically in time until the maximum admissible gas temperature t_{gmax} is achieved. Then the optimal policy remains constant in time and the optimal gas temperature t_g is equal to the maximum temperature t_{gmax}. When there are no restrictions on the inlet gas temperature the difference between the temperatures of both phases, that is, a process driving force, remains nearly equal in time.

The common method of conducting the fluidization processes is the use of the constant inlet gas temperature technique. The results of optimization show that the optimum of costs corresponds with an increasing gas temperature policy and that the constant inlet gas temperature technique can only be a final part of the optimal policy when an upper limit of the gas temperature is assumed. The optimal temperature policy can simply be adjusted in the batch fluidization case when the inlet gas stream is heated with the intensity which is variable in time (variable flow of heating medium or variable intensity of electric heating with the use of an autotransformer). In the case of the horizontal exchanger the optimal policy can be accomplished only approximately when the inlet gas is divided into several streams flowing to different sections of the apparatus and when each gas stream is heated to a definite temperature.

Some Summarizing Remarks

In pure thermodynamic analysis the constraint on the total process time or an average rate of the process is most often neglected. Such an analysis often leads to the trivial conclusion that the optimal process is a quasistatic one. Since a quasistatic process has a rate approaching zero, the practical value of this conclusion is very low.

In view of the above, it is advisable to point out several important qualitative conclusions which hold in the case when the apparatus price is considered as a variable parameter, the inlet gas humidity X_g varies, and when $\mu = 1$, that is, when all the process products (e.g., outlet gas and outlet solid streams) are fully exploited. Such a consideration deals with various optimal processes having the same initial and final states which take place in different apparatus (different values of λ). The optimal time of such processes increases when λ decreases. In the case where λ approaches zero (very inexpensive apparatus) the optimal final time τ_f approaches infinity. Corresponding to this case there is a limiting numerical value of optimal exergy costs K^p and F equal to zero when $\mu = 1$. This value of costs is the result of conducting the process in which the solid final state (I_{sf}, X_{sf}) is attained from the initial state (I_{sf}, X_{sf}) without the existence of driving forces between both phases $(i_g = i_s, Y_g = Y_s)$. Such a process has a rate approaching zero and is known in thermodynamics as quasistatic. Thus, a quasistatic process with variable X_g can be optimal only in the case where the apparatus price approaches zero and all of the process products are fully exploited $(\mu = 1)$.

In the case where the apparatus price is greater than zero (positive λ), the optimal final time has a definite value for arbitrary value of coefficient μ. The driving forces $i_g - i_s$ and $Y_g - Y_s$ are then different from zero. In this case, only an irreversible process is optimal. The relation between apparatus price, process intensity (rate penalty), and the process time is quite general, and has been confirmed by other researchers (Bejan, 1987b; Tondeur and Kvaalen, 1987; Szwast, 1990).

For expensive apparatus, only intensive processes can be optimal; they are characterized by a large irreversibility index, that is, large rate penalty. This means that the design of expensive apparatus must be associated with the design of intense optimal processes to assure short process time. The above results help avoid basic mistakes in the design of new equipment.

8.4.3. Drying in an Inhomogeneous Fluidized Bed Described by a Bubble Model

Rationale for Using a Bubble Bed Model in Fluidized Drying

For inhomogeneous fluidization it is quite realistic to assume that a part of the gas is in plug flow ("bubble phase") whereas its remaining part (in "dense phase") is ideally mixed with the solid (Kunii and Levenspiel, 1969, 1991). Applications of the model to fluidized drying are known (Poświata, 2005; Poświata and Szwast, 2003, 2007). The cascade approach to fluidized bed drying provides a common "multistage context" capable of treating both batch fluidization and fluidization in a steady horizontal exchanger; Figure 8.2, cases a and b. In the two-phase model proposed by Kunii and Levenspiel (1969) a gas excess which flows in the form of bubbles is considered with reference to gas flow at minimum fluidization. The heat and mass transfer between solid and gas and between a dense phase (with gas and solid) and bubble phase determines the values of optimal gas parameters.

While the use of relatively high velocities of inlet gas improves conditions of heat and mass transfer, a portion of gas flows as bubbles or canals and thus the energy carried by this gas is not exploited efficiently. This is the reason for searching for the best operating parameters which minimize total cost or maximize related efficiency. Optimization leads to optimal parameters of drying gas (flow rate, temperature and humidity). Total exergy supplied to the drying process is the process performance index. This function consists of three parts, the first describing thermal exergy of gas, the second describing its chemical exergy, and the third one referring to investment costs expressed in exergy units. The chemical exergy of gas takes into account the difference between gas humidity and the ambient humidity.

An effective approach involves an algorithm of maximum principle for multistage optimization. Earliest algorithms of this sort are those by Katz (1962) and Fan and Wang (1964); their more rigorous forms came a decade later (Boltyanski, 1969, 1973). They are all less suitable because of their undefined symplectic structure and properties of the Hamiltonian along an optimal path. Therefore we apply here Poświata's (2005) extension of Sieniutycz and Szwast's algorithm, which is formally similar to the well-known continuous algorithm of Pontryagin (Sieniutycz and Szwast, 1982b, 1983). Poświata's extension admits nonlinear intervals of time in the optimization model, a property which is useful in the optimization of imperfect cascades. Optimization calculations take into consideration bed hydrodynamics and kinetics of transport processes. These considerations are focused here on the optimization of fluidized drying. Their main purpose is the presentation of relationships between optimal parameters of drying gas caused by complex hydrodynamic and kinetic phenomena in the fluidized bed.

Problem Formulation

The optimization problem deals with the drying process in which the solid is transformed from a fixed initial state to a final one. The initial values of solid coordinates (the solid's moisture content and its temperature or enthalpy) are fixed, whereas some final values of solid coordinates can be undetermined. While the final solid moisture content is usually fixed in accordance with technology requirement, the solid final temperature is often free. (The opposite situation is observed in the heating or cooling process of wet solid in which final solid moisture content is free whereas its final temperature is fixed.) In Poświata's (2005) study the controls are inlet gas temperature, humidity and gas flow rate. The task is to determine optimal parameters of gas and solid in the drying process associated with minimum exergy consumption.

Figure 8.4 depicts a multistage scheme of the fluidized drying calculation. For N-stage cascade of fluidized dryers ($n = 1, \ldots, N$) the exergy consumption per unit stream of dried material is

$$I = NC_0 + \sum_{n=1}^{N} \left[\frac{1}{2}A\left(T_g^n - T_a\right)^2 + \frac{1}{2}B\zeta\left(Y_g^n - Y_a\right)^2 + \kappa \right] \theta_g^n \quad (8.62)$$

(Sieniutycz and Szwast, 1982b; Berry et al., 2000). In this formula C_0 is a fixed cost (assumed the same for every stage of the cascade) that is independent of drying gas properties. θ_g is a dimensionless gas flow rate or the ratio of gas flow rate, G, to solid flow rate, S. The first and second terms in square brackets of Equation (8.62) approximate, respectively, thermal exergy of inlet gas and chemical exergy of this gas. The parameter ζ is the ratio of chemical exergy price to the thermal exergy one. The presence of ζ underlines that unit prices of various kinds of exergy may differ. Change of thermal exergy is connected with heating of drying gas whereas that of chemical exergy is connected with preliminary drying of this gas. As the heating process is usually much cheaper than the drying one, the values of parameter ζ are greater than 1. Parameters A and B in Equation (8.62) are found by Taylor expansion of the gas exergy formula and calculated with the help of expressions

$$A = \frac{c_g + Y_a c_v}{T_a} \quad (8.63)$$

and

$$B = \frac{RT_a}{Y_a(M_v + Y_a M_g)} \quad (8.64)$$

Quantities subscripted by index a refer to the ambient air. Coefficient κ is the so-called exergy coefficient of investment and gas pumping cost. This coefficient describes the cost related to unit dimensionless flow rate of gas θ^n, and is expressed in exergy terms. Its value can be computed on the basis of an expression given in the literature (Sieniutycz and Szwast, 1982b; Berry et al., 2000). For a fixed number of stages in the cascade, N, the term NC_0 is constant and can be omitted in optimization calculations. Thus, the following working form of performance index is accepted (Poświata and Szwast, 2003, 2007):

$$I = \sum_{n=1}^{N} \left[\frac{1}{2} A (T_g^n - T_a)^2 + \frac{1}{2} B \zeta (Y_g^n - Y_a)^2 + \kappa \right] \theta_g^n \qquad (8.65)$$

Hydrodynamics of Fluidized Bed

In fluidized beds various disturbances of gas flow can appear; bubbles, channels or plugs are often observed. This causes density inhomogeneities within the volume of fluidized bed. The disturbance forms (bubbles, channels, and plugs) contain inside them a small number of solid particles. Gas velocity within these forms is higher than in the dense phase of the bed. In addition, spaces with a higher density of solid particles are present. All this causes serious difficulties in the accurate descriptions of bed hydrodynamics. A number of mathematical models of bed hydrodynamics have been formulated (Kunii and Levenspiel, 1969, 1991; Kato and Wen, 1969; Mori and Wen, 1975).

Poświata's (2005) studies apply the Kunii and Levenspiel two-phase model to the optimization of the fluidized bed drying (Figure 8.3, case B). Here we recapitulate the main results of this optimization. The following assumptions are made:

1. fluidized bed consists of two phases: dense phase and bubble phase;
2. dense phase is ideally mixed;
3. gas flows through the dense phase with the minimum fluidization velocity; gas flow in the bubble phase is a plug phase;
4. each bubble has the same diameter considered as a constant parameter of the bed;
5. solid particles are absent in the bubble phase.

Volume fraction of bubble phase, σ, in fluidized bed is described by the ratio

$$\sigma = \frac{u_g - u_{mf}}{u_p - u_{mf}} \qquad (8.66)$$

where u_g, u_{mf}, and u_p represent appropriate gas velocities (superficial, through dense phase at minimum fluidization, and through the bubble phase, respectively). In the literature it is assumed (Kunii and Levenspiel, 1969, 1991) that small bubbles satisfy the following relationship:

$$\frac{u_{br}}{u_{mf}} \leq \frac{1}{\varepsilon_{mf}} \qquad (8.67)$$

The velocity of gas flowing through the bubble phase is then calculated from the equation

$$u_p = u_b + 3u_{mf} \tag{8.68}$$

for small bubbles, and the equation

$$u_p = u_b \tag{8.69}$$

for large bubbles, that is, for which

$$\frac{u_{br}}{u_{mf}} \geq \frac{5}{\varepsilon_{mf}} \tag{8.70}$$

where

$$u_b = u_g - u_{mf} + u_{br} \tag{8.71}$$

and

$$u_{br} = 0.711\sqrt{gd_b} \tag{8.72}$$

The symbol u_b describes the velocity of rising bubbles in a fluidized bubble bed, whereas u_{br} describes the rising velocity of an isolated single bubble. For ratios u_{br}/u_{mf} located between $1/\varepsilon_{mf}$ and $5/\varepsilon_{mf}$ the linear interpolation is applied.

The volume or mass fraction of gas flowing as the bubble phase is calculated from the following relationship (Poświata and Szwast, 2007):

$$\sigma_g = 1 - u_{mf}\frac{(1 - \sigma)}{u_g} \tag{8.73}$$

where σ is the volume fraction of bubble phase in the bed, and σ_g is the mass fraction of gas flowing through a bubble.

Drying Description

For the drying in a two-phase fluidized bed the moisture balance for stage n takes the form

$$X_s^n - X_s^{n-1} = \theta_g^n[(1 - \sigma_g)(Y_g^n - Y_{mf}^n) + \sigma_g(Y_g^n - Y_b^n)] \tag{8.74}$$

where the contents of moisture in solid and in gas are expressed in kilograms of moisture per kilogram of dry inert. For the moisture transfer from solid to gas in the dense phase the kinetic equation can be expressed in the following form:

$$X_s^n - X_s^{n-1} = \frac{Y_{mf}^n - Y_e^n}{W_{mf}}\theta_g^n \tag{8.75}$$

where Y_e is equilibrium humidity at the solid surface.

We shall first focus on the first drying period in which this quantity is only a function of the temperature at the solid surface. A useful coefficient in the development of parametric studies is a dimensionless measure of resistance of mass transfer W_{mf} defined as follows:

$$W_{mf} = \frac{u_g \rho_g}{k_{mf} h_{mf}} \tag{8.76}$$

Moisture transfer between the dense phase and the bubble phase is described by the equation

$$Y_g^n - Y_b^n = \frac{\Delta Y_b^n}{W_b} \tag{8.77}$$

where the coefficient W_b is the following resistance measure:

$$W_b = \frac{u_g \rho_g \sigma_g (1 - \sigma)}{k_b h_{mf} \sigma} \tag{8.78}$$

The driving force in Equation (8.77), ΔY_b^n, is defined as the logarithmic average of humidity differences between the dense phase and the bubble phase.

The state transformation equation describing change of solid moisture content at the nth process stage in the first drying period (Poświata and Szwast, 2003) is

$$X_s^n - X_s^{n-1} = \theta_g^n(Y_g^n - Y_e^n)\frac{1 - \sigma_g e^{-1/W_b}}{1 + W_{mf}(1 - \sigma_g e^{-1/W_b})} \tag{8.79}$$

When the solid is in the second drying period some equations should replace those relevant for the first period. For example, under the piston flow assumption, a state transformation describing the change of solid moisture content at the nth stage follows from integration of the familiar rate formula

$$\frac{dX_s}{dt} = -K(X_s - X_e)$$

between the inlet moisture content X_s^{n-1} and outlet moisture content X_s^n. This yields

$$\frac{X_s^{n-1} - X_e^n}{X_s^n - X_e^n} = \exp(K^n \theta_s^n) \qquad (8.80)$$

where $\theta_s^n = V_s^n / S_v$ is the solid holdup time expressed as the ratio of solid volume V_s to volumetric flow rate of solid, S_v. Time θ_s^n may also be expressed by some bed parameters and dimensionless gas flow rate θ_g^n, and can be written in the following form:

$$\theta_s^n = \frac{h_{mf} \rho_s (1 - e_{mf})}{\rho_g u_g} \theta_g^n \equiv a\theta_g^n \qquad (8.81)$$

Poświata and Szwast (2007). These authors also cite the literature showing the calculation of drying coefficients K. The state transformation Equation (8.80) in the second drying period can thus be written as follows:

$$X_s^n - X_s^{n-1} = (X_s^n - X_e^n)\exp(1 - K^n a\theta_g^n) \qquad (8.82)$$

The enthalpy balance for the nth stage takes a form analogous to the mass balance:

$$I_s^n - I_s^{n-1} = \theta_g^n [\sigma_g (i_g^n - i_b^n) + (1 - \sigma_g)(i_g^n - i_{mf}^n)] \qquad (8.83)$$

The first term in square brackets contains parameters of gas flowing through the bubble phase, whereas the second term contains parameters of gas flowing through dense phase.

Ambient temperature and liquid moisture are taken as the reference state for enthalpy calculation; then enthalpy of solid is calculated from the following equation:

$$I_s = (c_s + Xc_w)(T - T_a) \qquad (8.84)$$

whereas enthalpy of gas is calculated from equation

$$i = (c_g + Yc_v)(T - T_a) + YL_t \qquad (8.85)$$

where L_t is heat of vaporization in the reference temperature.

A kinetic equation for energy transfer between the solid and gas in the dense phase is

$$I_s^n - I_i^n = -\frac{c_g(T_{mf}^n - T_s^n)}{Z_{mf}} \theta_g^n + (X_s^n - X_s^{n-1})i_{vmf} \qquad (8.86)$$

The left-hand side of the above equation describes changes of solid enthalpy, the first term on the right-hand side describes heat transferred from gas to solid by convection, whereas the second one describes enthalpy transferred from solid to gas by the moisture flux. Nondimensional, nonnegative coefficient Z_{mf} represents a heat transfer resistance defined as

$$Z_{mf} = \frac{u_g \rho_g c_g}{\alpha_{mf} h_{mf}} \qquad (8.87)$$

where i_{vmf} is enthalpy of moisture transported between solid and gas. For $Z_{mf} = 0$ the heat resistance is equal to zero, which means that heat transfer is rapid, and solid temperature is equal to the temperature of gas in the dense phase.

An equation can be written for enthalpy of gas in the bubble phase:

$$i_g^n - i_b^n = \frac{c_g \Delta T_b^n}{Z_b} - (Y_b^n - Y_g^n) i_{vb}^n \qquad (8.88)$$

where ΔT_b^n is the logarithmic average of differences of temperature between bubble phase and dense phase, i_{vb} is enthalpy of moisture transported between these phases, and nondimensional resistance coefficient is defined as

$$Z_b = \frac{c_g u_g \rho_g \sigma_g (1 - \sigma)}{\alpha_b h_{mf} \sigma} \qquad (8.89)$$

This coefficient is a measure of the resistance of heat transfer from bubble phase to dense phase. Coefficients Z_b and W_b take positive values or zero, similar to coefficient Z_{mf}. If each of these coefficients equals zero then the thermodynamic equilibrium is achieved between gas in the bubble phase and gas in the dense phase. One may note that the ratio of respective coefficients Z_b and W_b defines the Lewis number. Equality of the coefficients Z_b and W_b can be accepted for the water–air system occurring commonly in drying processes.

The set of Equations (8.74)–(8.89) cannot be solved analytically; thus numerical methods are required.

Optimization Algorithm

Exergy consumption in the N-stage drying process is related to a discrete functional that appears on the right-hand side of Equation (8.65). This functional is minimized by an optimal choice of three decision variables which are gas temperature, gas humidity and gas flow rate. As in Section 8.4.2, it should be noted that in the formulation with the Lagrange multiplier the final time cannot be constrained. This is associated with only two (not three) state transformations, one for solid moisture content and one for solid temperature. They are described below by the following implicit equations:

$$X_s^n - X_s^{n-1} = f_X(X_s^n, T_s^n, T_g^n, Y_g^n, \theta_g^n)\theta_g^n \tag{8.90}$$

$$T_s^n - T_s^{n-1} = f_T\left(X_s^n, T_s^n, T_g^n, Y_g^n, \theta_g^n\right)\theta_g^n \tag{8.91}$$

(No explicit analytical form exists for these equations.) State transformations (8.90) and (8.91) are valid for fluidized drying processes considered without assumption of equilibrium between output streams. These formulae are structurally different from the respective state transformations for processes with thermodynamic equilibrium between the outlet phases (Berry et al., 2000; Sieniutycz and Szwast, 1982b). In Equations (8.90) and (8.91) dimensionless gas flow rate θ_g^n appears as the argument of functions f_T and f_X, which is not the case for the processes linear with respect to the interval θ_g^n. For operations considered earlier (e.g., those in Chapter 4) the linearity of the optimization model with respect to decision θ_g^n allowed us to apply an optimization algorithm with a constant Hamiltonian (Sieniutycz, 1973a, 1974; Sieniutycz and Szwast, 1982b, 1983; Berry et al., 2000). For the drying operations considered here the constant H algorithm cannot be applied. Therefore, Poświata (2005) has constructed an extended version of the original algorithm which does not require linearity of state equations with respect to decision variable θ_g (Poświata, 2005; Sieniutycz, 2006b). In this extended version of the algorithm the Hamiltonian function is defined identically as in the original version of discrete algorithm with a constant Hamiltonian. Therefore the Hamiltonian of the drying process considered has the form

$$\begin{aligned} H^{n-1}&\left(T_g^n, Y_g^n, \theta_g^n, T_s^n, X_s^n, z_T^{n-1}, z_X^{n-1}\right) \\ &= [\frac{1}{2}A(T_g^n - T_a)^2 + \frac{1}{2}B\zeta(Y_g^n - Y_a)^2 + \kappa] \\ &\quad + z_T^{n-1}f_T^n(X_s^n, T_s^n, T_g^n, Y_g^n, \theta_g^n) \\ &\quad + z_X^{n-1}f_X^n(X_s^n, T_s^n, T_g^n, Y_g^n, \theta_g^n), \end{aligned} \tag{8.92}$$

where z_T and z_X are costate variables, adjoint with solid temperature and moisture content.

However the above Hamiltonian is not constant along the optimal trajectory. The equation describing the constancy of the Hamiltonian in the algorithm linear in θ_g^n is here replaced by Equation (8.93), which refers to models nonlinear in θ_g^n. This equation constitutes the necessary condition for the stationary optimality of decision θ_g^n,

$$H^{n-1} + \theta_g^n \frac{\partial H^{n-1}}{\partial \theta_g^n} = 0 \tag{8.93}$$

The necessary conditions for the stationary optimality of inlet gas temperature and inlet gas humidity are given by equations

$$\frac{\partial H^{n-1}}{\partial T_g^n} = 0 \tag{8.94}$$

and

$$\frac{\partial H^{n-1}}{\partial Y_g^n} = 0 \tag{8.95}$$

Moreover, equations describing changes of adjoint variables at stage n are of the form

$$z_T^n - z_T^{n-1} = \frac{\partial H^{n-1}}{\partial T_s^n} \theta_g^n \tag{8.96}$$

and

$$z_X^n - z_X^{n-1} = \frac{\partial H^{n-1}}{\partial X_s^n} \theta_g^n \tag{8.97}$$

Of course, the algorithm is completed by state transformations (8.90) and (8.91).

Sieniutycz (2006b) has explained that in discrete models nonlinear with respect to the time interval the control variable θ_g^n ceases to have its distinguished properties and becomes "usual decision" (such as T_g^n and Y_g^n in the present example). This led him to the conclusion that the present theory bridges the standard algorithm of the discrete maximum principle (Katz, 1962; Fan and Wang, 1964; Halkin, 1966; Boltyanski, 1969, 1973) and the algorithm with a constant Hamiltonian within a different canonical formalism that uses Halkin's (1996) definition of the discrete Hamiltonian (Sieniutycz, 2006b).

The set of Equations (8.90), (8.91) and (8.93)–(8.97) must be solved numerically for every $n = 1, \ldots, N$ starting from the last stage, $n = N$. At the beginning of computations values of final solid temperature T_s^N and final adjoint variable z_X^N have to be assumed (note that final solid humidity X_s^N is fixed whereas final adjoint variable z_T^N equals zero, as final solid temperature is free). After the computations are performed along the cascade, values of inlet solid temperature, T_s^0, and concentration X_s^0, are computed. For correctly assumed values of T_s^N and

z_X^N computed values of T_s^0 and X_s^0 are the same as those defined by the initial conditions.

Results of Calculations

Using the bubble (two-phase) model Poświata and Szwast (2003, 2007) have performed optimization calculations of multistage fluidized drying for both drying periods. For the first drying period they investigated the case where the Lewis number (ratio of coefficients Z and W) equals 1. This case refers to the following data: initial solid moisture content $X_s^0 = 0.8$, initial solid temperature $T_s^0 = 290$ K, final solid moisture content $X_s^3 = 0.4$, and free final solid temperature.

Optimization calculations for the first drying period show that the main part of the drying process occurs in the first stage, where solid moisture content decreases above 95%, whereas the rest of the moisture (less than 5%) is removed during the following stages. A similar but weakened tendency of this kind is observed for the second drying period (Figure 8.7).

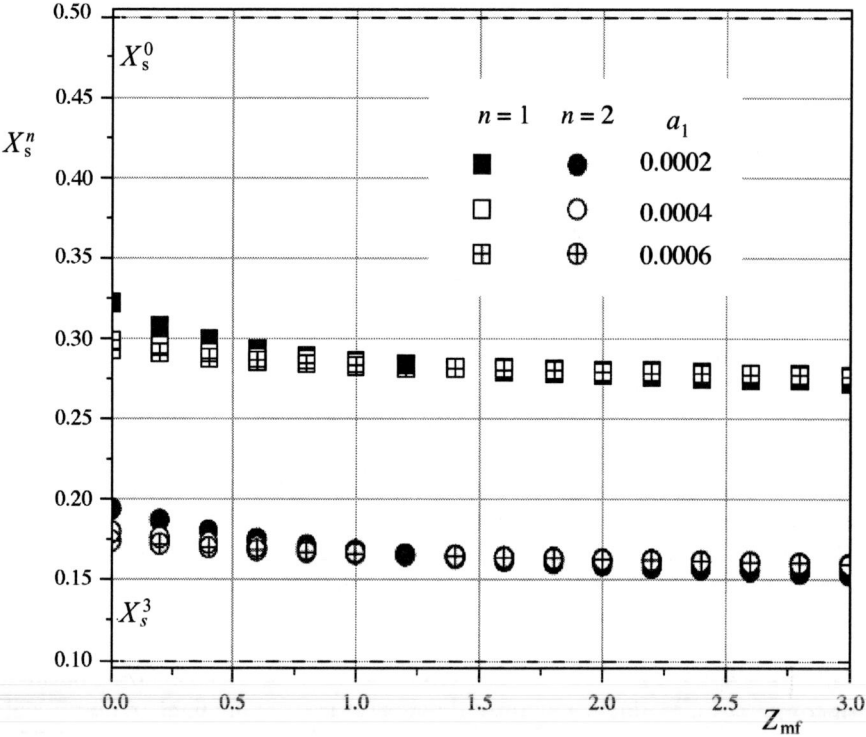

Figure 8.7 Interstage moisture content of solid (X_s^1 and X_s^2) at various stages of cascade versus coefficient Z_{mf} in the second drying period. Dashed lines mark initial and final concentrations, X_s^0 and X_s^3 respectively (Poświata and Szwast, 2007, with permission).

The results obtained, especially those in Figure 8.7, show that an interstage moisture content in solid ($n = 1$ or $n = 2$) is weakly dependent on (or—in the first drying period—independent of) bed hydrodynamics and kinetics described by coefficients Z_{mf}, Z_b, W_{mf} and W_b. Interstage moisture content increases in an optimal process with coefficient κ, whereas the effect of parameter ζ is the opposite, that is, the interstage moisture content decreases with ζ.

Poświata and Szwast (2003) also discuss driving forces of heat transfer in the fluidized drying process. They define a thermal driving force as the difference between the (optimal) temperature of inlet gas and the outlet temperature of solid. Consistently, the mass transfer driving forces are defined as the difference between the equilibrium gas humidity calculated at outlet solid temperature and inlet gas humidity. Their results show that in the first drying period (with $Le = Z/W = 1$) optimal driving forces are independent of bed kinetics and hydrodynamics. In the second drying period a weak dependence of driving forces on coefficient Z_{mf} is observed. Optimal driving forces for heat transfer attain the greatest values at the first stage, and then decrease with the stage number. A more complicated situation is observed for optimal driving forces of mass transfer. For small values of parameter ζ, these driving forces increase with stage number n, whereas for large values of ζ they decrease with n. To explain why in mass transfer such a complicated situation may occur, observe that driving forces depend on inlet gas humidity and temperature of solid inside the dryer. For low ζ optimal inlet gas humidity takes relatively low values because a preliminary drying of inlet gas is relatively cheap, whereas for high ζ inlet gas humidity is nearly the same as the ambient humidity because the preliminary drying of inlet gas is expensive. Moreover, it is solid temperature which decides the value of equilibrium gas humidity, Y_e. Thus, for low ζ driving forces increase with n because optimal inlet gas humidity decreases with n. Otherwise, driving forces decrease for large ζ because lowered values of equilibrium gas humidity (caused by decreasing solid temperature) are not compensated well enough by slowly decreasing humidity of inlet gas.

Poświata and Szwast's (2003) calculations of the first drying period show that the optimal solid temperature T_s^n increases at the first stage (heating of cold solid) whereas it decreases at the next ones (prevailing evaporation effect). As the final solid temperature is free, there is never a need to heat the solid to a high temperature. The solid heating at the first stage secures sufficiently high driving forces, whereas, at the next stages, solid enthalpy partially covers the energy requirement for vaporization with a decrease of solid temperature. In the second drying period a pronounced increase of T_s^n with n may be observed.

Figure 8.8 shows optimal temperatures of solid in the second drying period. Corresponding changes of controlling parameters of drying gas—temperatures T_g^n and humidities Y_g^n—are illustrated in Figures 8.9 and 8.10. Whereas quantities T_g^n and Y_g^n represent "intensive control variables" weakly affected by hydrodynamics, gas flow rate θ_g^n represents "extensive control" whose changes caused by bed imperfections are significant (Figure 8.11).

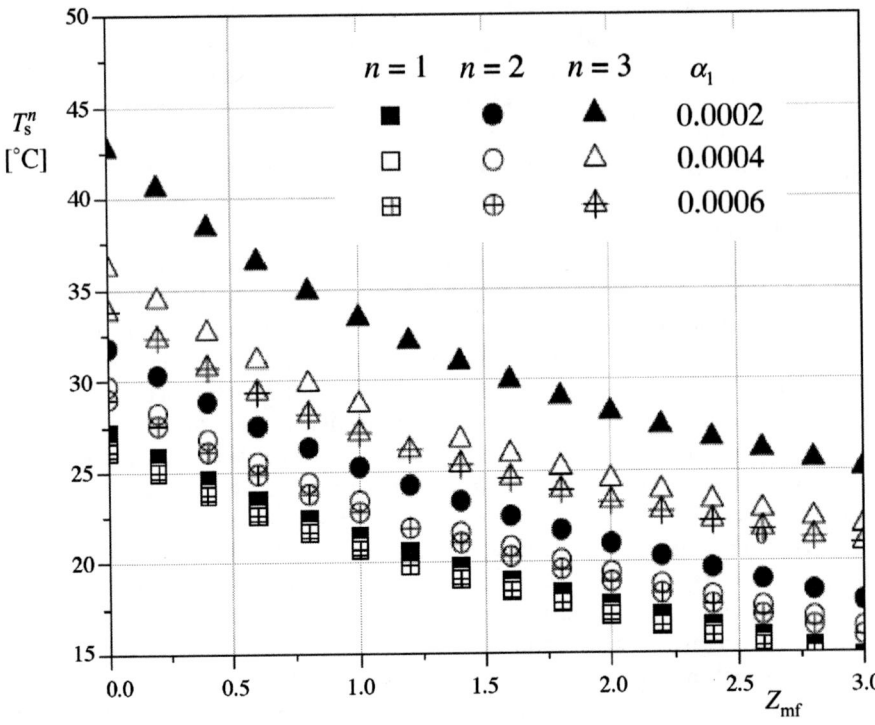

Figure 8.8 Optimal solid temperature at various stages versus coefficient Z_{mf} in the second drying period (Poświata and Szwast, 2007, with permission).

In fluidized drying optimal driving forces are weakly dependent on hydrodynamics and kinetics. Yet this is not observed with respect to optimal gas flow rates θ_g^n (Figure 8.11). They strongly increase when hydrodynamics get worse (larger values of coefficients Z and W, in both periods of drying). Optimal gas flow rates θ_g^n also strongly depend on coefficient of investment costs κ (θ_g^n decreases with κ). One can see that the parameter ζ does not influence gas flow rate, it only changes driving forces of mass and heat transport processes.

Ratios of costs incurred at the second and at the third stage to total process cost, I^2/I and I^3/I, are discussed by Poświata and Szwast (2003). The results for the first drying period show that the main part of the process cost is incurred at the first stage. The sum of costs incurred at the second stage and at the third does not exceed 5% of total drying cost. Thus, the contribution of stages other than the first one is small, both in regard to the drying advancement and its cost. In the economic calculation of number of stages performance index, Equation (8.62) should be applied instead of Equation (8.65). This is because criterion (8.62) contains the term describing fixed costs, NC_0. It is the value of C_0 multiplied by the number of stages N which decides the rationale for using a cascade in a practical drying operation.

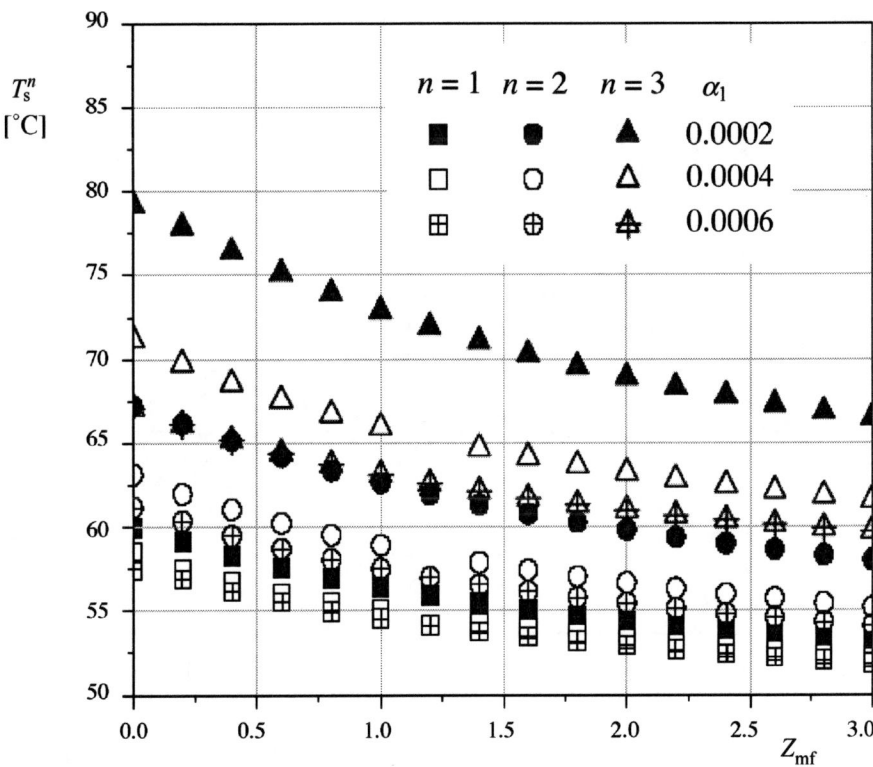

Figure 8.9 Optimum gas temperature versus coefficient Zmf for three-stage cascade of fluidized drying in the second drying period (Poświata and Szwast, 2007).

Summarizing Remarks

Multistage fluidized drying in the first drying period proceeds mainly at the first stage. Fixed costs defined by the number of stages in the cascade decide the advisability of applying the cascade in practical drying. When the Lewis number or the ratio of resistance coefficients of mass and heat transfer equals 1, optimal driving forces are virtually independent of hydrodynamics and kinetics. Only optimal gas flow rate strongly depends on hydrodynamics and kinetics in the bed. Optimal driving forces of transfer processes depend on parameter ζ which describes the ratio of prices of thermal and chemical exergies, whereas the optimal gas flow rate is independent of ζ. For low values of ζ driving forces increase with n because the optimal inlet gas humidity decreases with n. For large ζ driving forces decrease with n because the lowered values of equilibrium gas humidity caused by decreasing T_s are not compensated by a very slow decrease of inlet gas humidity. Optimal gas flow rate θ_g rapidly decreases with coefficient of investment costs κ. One may observe that in practice parameter ζ does not

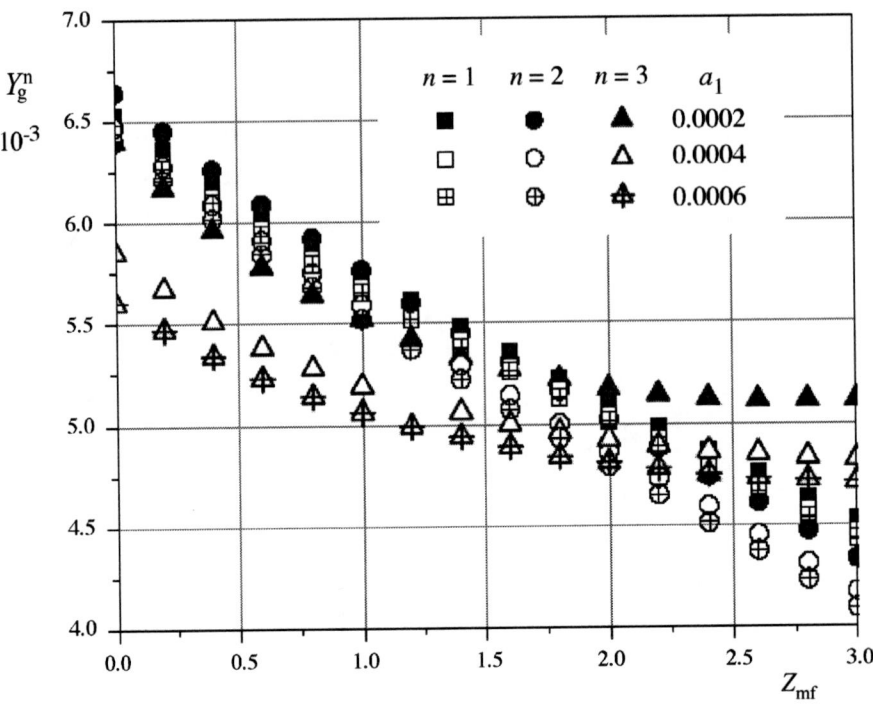

Figure 8.10 Optimum gas humidity versus coefficient Z_{mf} for three-stage fluidized drying in the second drying period (Poświata and Szwast, 2007, with permission).

influence gas flow rate, and only influences driving forces of mass and heat transport processes.

8.5. SOLAR ENERGY APPLICATION TO WORK-ASSISTED DRYING

In this section we consider dryers driven by the thermal energy of the gas upgraded in a heatpump. We focus on applications in which the optimal control theory plays a role.

8.5.1. A Formula for Performance Coefficient

In Chapters 3–5 we have considered devices of engine type (power generators) and of heat-pump type (power consumers), each driven either by fluxes of heat or matter, including solar radiation. The unit where such operations can be accomplished is the generalized Chambadal–Novikov–Curzon–Ahlborn system (Novikov, 1958; Curzon and Ahlborn, 1975), which takes into account the internal dissipation (Chapter 3, Figure 3.1). In the present section we shall focus on

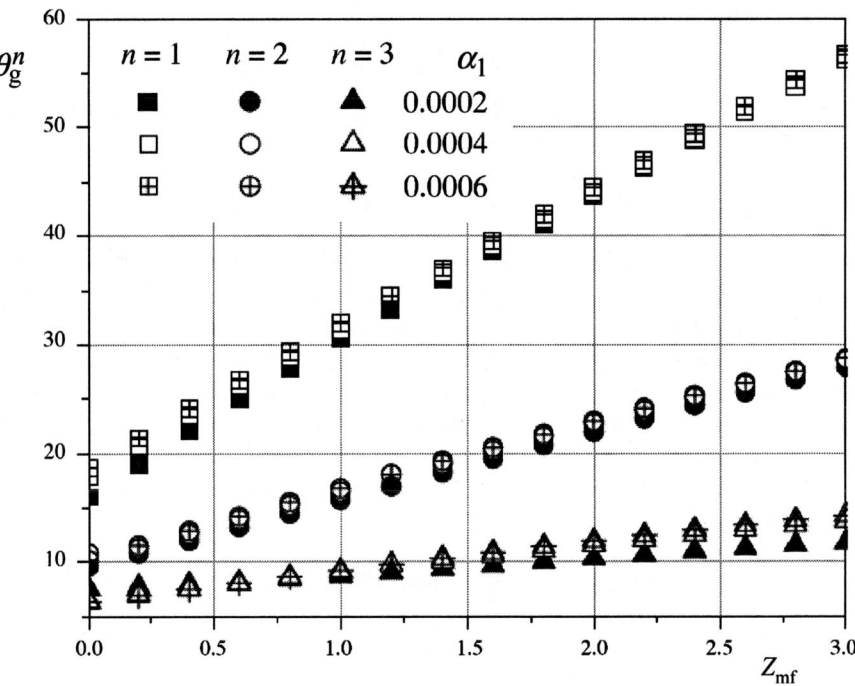

Figure 8.11 Optimum gas flow rate as a function of coefficients Z_{mf} for three-stage fluidized drying in the second drying period (Poświata and Szwast, 2007, with permission).

such dissipative units applied to the work-assisted drying. The unit can work either in the heat-pump mode (larger, external loop in Figure 3.1) or in the engine mode (smaller, internal loop in Figure 3.1). For the present purposes the heat-pump mode is essential.

As in Chapter 3, internal imperfections are quantified by internal irreversibility coefficient, Φ, which equals $\Delta S_{2'}/\Delta S_{1'}$ along isotherms $T_{1'}$ and $T_{2'}$ in Figure 3.1. Φ is also the ratio of the entropy fluxes in the thermal machine, $\Phi = J_{s2'}/J_{s1'}$. The second law implies $J_{s2'}/J_{s1'} > 1$ for engines and $J_{s2'}/J_{s1'} < 1$ for heat-pumps at the steady states. For heat-pump mode the quantity Φ satisfies inequality $\Phi < 1$. Data of internal entropy production can serve to estimate Φ.

For a heat-pump its coefficient of performance c_{op} (reciprocal of efficiency η) can be expressed in terms of the temperatures of the circulating fluid:

$$c_{op} = (\eta)^{-1} = \frac{dQ_1}{dQ_1 - dQ_2} = (1 - \Phi T_{2'}/T_{1'})^{-1} \qquad (8.98)$$

Both temperatures are here mutually dependent (linked with the internal entropy balance (8.101)). Using Equation (8.98), the Carnot temperature T'

satisfying the usual definition

$$T' \equiv T_2 T_{1'}/T_{2'} \tag{8.99}$$

the performance coefficient for the heat-pump is expressed in terms of T' as free control

$$c_{op} = q_1/p = (1 - \Phi T_2/T')^{-1} \tag{8.100}$$

This formula shows that the performance coefficient of an endoreversible heat pump is lower than that of Carnot heat pump. Equations (8.98) or (8.100) are suitable to evaluate the power consumption in systems with heat pumps. Yet, to get an explicit heat flux, one must apply a special model of the heat exchange. As explained in Chapter 5 for thermal machines driven by nonlinear fluids, one may use a pseudo-Newtonian model with conductances and transfer coefficients that are functions of T_1 and T_2.

The entropy balance constraint for the internal part of the single thermal machine

$$g_2(T_2)(T_{2'} - T_2)T_{2'}^{-1} - \Phi g_1(T_1)(T_1 - T_{1'})T_{1'}^{-1} = 0 \tag{8.101}$$

proves that the system has only one degree of freedom (one independent control, T'). We use Equation (8.99) to insert $T_{2'} \equiv T_{1'}T_2/T'$ into Equation (8.101). We then obtain $T_{1'}$ in terms of T':

$$T_{1'} = (\Phi g_1 T_1 + g_2 T')(\Phi g_1 + g_2)^{-1} \tag{8.102}$$

and a corresponding equation for $T_{2'} \equiv T_{1'}T_2/T'$. The flux of upgraded energy (negative in the engine sign convention) follows as

$$q_1 = g_1(T_1 - T_{1'}) = g'(T_1 - T') \tag{8.103}$$

The second flux is $q_2 = q_1(1 - \eta)$, where η follows from the pseudo-Carnot expression (8.100). In Equation (8.103) an operational overall conductance, g', appears, defined as

$$g'(\Phi, T_1, T_2) \equiv g_2 g_1(\Phi g_1 + g_2)^{-1} = (g_1^{-1} + \Phi g_2^{-1})^{-1} \tag{8.104}$$

As in other chapters, this is a suitably modified *overall* conductance of an inactive heat transfer in which use of the operative (Φ and T dependent) heat conductance, g', is required. From Equations (8.100) and (8.103) mechanical power $p = \eta q_1$ follows.

Equation (8.103) leads to the familiar working expression for Carnot temperature

$$T' = T_1 - q_1/g' \tag{8.105}$$

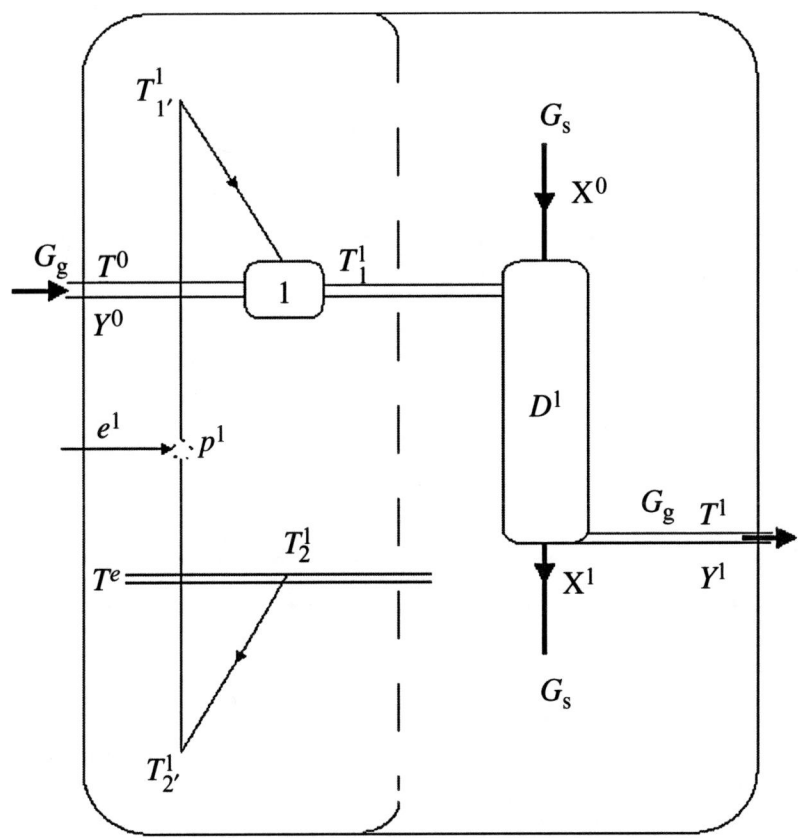

Figure 8.12 A scheme of one-stage drying with a heatpump.

With this expression and Equation (8.100), a working formula for the performance coefficient follows:

$$c_{\mathrm{op}} = \left(1 - \Phi\frac{T_2}{T_1 - q_1/g'}\right)^{-1} \tag{8.106}$$

8.5.2. Two-Stage Optimization of Drying with Heat-Pumps

As an example of application of Formula (8.105) let us consider a drying operation in which a drying agent is heated before drying in order to achieve a sufficiently high temperature. The heating process is accomplished in the condenser of a heat-pump. The principle of the one-stage operation, which may refer to the first stage of a multistage system, is shown in Figure 8.12. The typical state changes of gas in a related multistage system are illustrated in the enthalpy–concentration diagram of Figure 8.13.

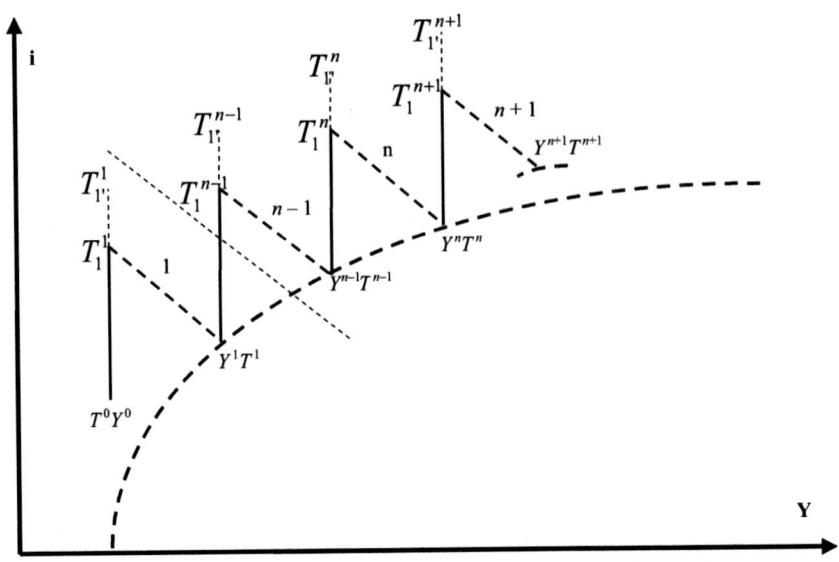

Figure 8.13 Changes of gas states in a multistage work-assisted drying operation. Primed states refer to temperatures of circulating fluids which heat gases supplied to dryers 1, 2, ..., n.

Formulation of a Two-Stage Problem

The gas leaving the first stage flows to the heat-pump and dryer of the second stage. Similarly the outlet solid from the first stage flows to the dryer of the second stage. Thus, the drying may be continued; all following stages repeat the one-stage operation for their own inputs. We assume that the drying takes place in an apparatus in which a pulverized solid makes contact concurrently with the gas. For simplicity we assume that the process occurs during the first drying period, so that solid temperature changes are then negligible. For simplicity we assume that outlet solid and outlet gas are in thermodynamic equilibrium due to a large specific area of solid.

Our purpose is to minimize the power consumption in the two-stage operation by a suitable choice of the intermediate moisture content between the first and second stage.

The balance equations of mass and sensible heat can be written in the joint form

$$\frac{rG_s}{cG_g}(X^0 - X^1) = T_1^1 - T^1 = -\frac{r}{c}(Y_1^1 - Y_s(T^1)) \qquad (8.107)$$

where r is specific evaporation heat, c is mean heat capacity of the flowing gas, and $Y_s(T)$ is a function describing humidity of gas at equilibrium with solid.

It is essential to understand that the addition of a dryer at a stage changes the state variable. The state variable must be the temperature of gas from the dryer, T^1, not that of gas leaving the heat-pump. An equation which describes the specific work supplied to the heat-pump should match this new definition of state in agreement with designations shown in Figure 8.12. This matching requires us to identify the temperature T^1 of Equation (8.108) with the temperature T_1^1 in Figure 8.12. Therefore power consumed at a single stage per unit flow of gas, e^1, follows as

$$
\begin{aligned}
e^1 &\equiv -\frac{p}{G_g} = c\left(1 - \Phi\frac{T^e}{T^1 - q^1/g'}\right)\left(-\frac{q^1}{g'}\right)\left(\frac{g'}{cG_g}\right) \\
&= c\left(1 - \Phi\frac{T^e}{T_1^1 + u^1}\right)u^1\theta^1
\end{aligned}
\tag{8.108}
$$

where q^1 is the energy supply to the drying gas in the condenser of the heat-pump, $\theta = g'(cG_g)^{-1}$ and $u^1 = -q^1/g > 0$ is a measure of the energy supply in the temperature units.

Substituting into the above equation the temperature T_1^1 following from Equation (8.20),

$$
T_1^1 = T^1 + \frac{r}{c}(Y_s(T^1) - Y_1^1)
\tag{8.109}
$$

and taking into account that $Y_1^1 = Y^0$ (also $Y_1^n = Y^{n-1} = Y_s(T^{n-1})$), for $n = 2, \ldots,$ N) we find the unit consumption of mechanical energy at the stage

$$
e^1 = c\left(1 - \frac{T^e}{T^1 + r/c(Y_s(T^1) - Y^0) + u^1}\right)u^1\theta^1
\tag{8.110}
$$

This is transformed further in view of the link between u^1 and θ^1 (consider the difference constraint describing $\Delta T^n = u^n\theta^n$ for $n = 1$):

$$
\begin{aligned}
e^1 = c\bigg(1 - \frac{T^e}{T^1 + rc^{-1}(Y_s(T^1) - Y^0) + ((T^1 + rc^{-1}(Y_s(T^1) - Y^0) - T^0)/\theta^1)}\bigg) \\
(T^1 + rc^{-1}(Y_s(T^1) - Y^0) - T^0)
\end{aligned}
\tag{8.111}
$$

In terms of the "adiabatic temperature function"

$$
T^a(T^1) \equiv T^1 + rc^{-1}Y_s(T^1)
\tag{8.112}
$$

the work expression takes the final form

$$
\begin{aligned}
e^1 = c\bigg(1 - \frac{T^e}{T^a(T^1) - rc^{-1}Y^0 + ((T^a(T^1) - rc^{-1}Y^0 - T^0)/\theta^1)}\bigg) \\
(T^a(T^1) - rc^{-1}Y^0) - T^0)
\end{aligned}
\tag{8.113}
$$

An analogous function, but with the shifted superscripts, is valid for the second stage

$$
e^2 = c \left(1 - \frac{T^e}{T^a(T^2) - rc^{-1}Y_s(T^1) + ((T^a(T^2) - rc^{-1}Y_s(T^1) - T^1)/\theta^2)} \right)
$$
$$
(T^a(T^2) - rc^{-1}Y_s(T^1) - T^1)
$$

(8.114)

The constraint $Y^1 = Y_s(T^1)$ resulting from the equilibrium between both outlet phases from the first dryer is incorporated in the second work expression. Note that exploiting the definition of the adiabatic temperature at the second stage yields

$$
e^2 = c \left(1 - \frac{T^e}{T^a(T^2) - rc^{-1}Y_s(T^1) + ((T^a(T^2) - T^a(T^1))/\theta^2)} \right)
$$
$$
(T^a(T^2) - T^a(T^1))
$$

(8.115)

The sum of both works yields total work consumed. This is the thermodynamic cost that should be minimized. For a fixed τ^2 there are two independent controls: θ^1 and T^1.

8.5.3. Outline of Work Minimization Procedure

The minimizing procedure searches for an optimal interstage temperature T^1 and an optimal heat transfer area in the first heat-pump, a^1, contained in the control variable

$$
\theta^1 = \frac{g^1}{cG_g} = \frac{\alpha' a^1}{cG_g}
$$

(8.116)

First we assume the constancy of the sum $\tau^2 = \theta^1 + \theta^2$ or

$$
\tau^2 = \frac{\alpha'(a^1 + a^2)}{cG_g}
$$

(8.117)

which is the constraint associated with fixing the cumulative area $A^2 = a^1 + a^2$. At the end of the optimization procedure we relax this constraint (see below), thus determining an optimal value of A^2.

The requirement of sufficiently low final moisture content in solid, X^2, defines the amount of the evaporated moisture per unit time, $\mathcal{M} \equiv G_s(X^0 - X^2) = G_g(Y^2 - Y^0)$. This, in turn, defines the final humidity of gas leaving the second dryer

$$Y^2 = Y^0 + \frac{G_s}{G_g}(X^0 - X^2) \tag{8.118}$$

From the saturation line on a humid air chart one can find the solid temperature T^2 such that $Y_s(T^2) = Y^2$. Alternatively, knowing the function $Y_s(T)$ in an
analytical form, $T^2(Y^2)$ can be evaluated.

The minimization of the sum $e^1 + e^2$ leads to two optimal controls, a^1 and
T^1. The optimal areas are usually close to each other. The optimal temperatures constitute an increasing sequence. The optimal work (power per unit
flux of gas) supplied to the two-stage system decreases significantly with the
total transfer area A^2 allocated between the two heat-pumps. Otherwise, the
cost of the transfer area increases linearly with A^2. Minimization of the sum
$R_* = R(A^2) + (\kappa_a/\kappa_e)A^2$, in which κ_e and κ_a are unit prices of the energy and
transfer area, leads to the optimal choice of the total area A^2.

The optimization can be extended to the N-stage cascade system. In this case
a suitable optimization algorithm is a recurrence equation of dynamic programming

$$R^n(T^n, \tau^n) = \min_{T^{n-1}, \theta^n} \{e^n(T^n, T^{n-1}, \theta^n) + R^{n-1}(T^{n-1}, \tau^n - \theta^n)\} \tag{8.119}$$

which holds for $n = 2, \ldots, N$. There are two controls T^{n-1} and θ^n at the stage n.
The function e^n (T^n, T^{n-1}, θ^n) in the above equation represents the unit work
input to the stage n:

$$e^n = c\left(1 - \frac{T^e}{T^a(T^n) - rc^{-1}Y_s(T^{n-1}) + ((T^a(T^n) - T^a(T^{n-1}))/\theta^n)}\right)$$
$$(T^a(T^n) - T^a(T^{n-1})) \tag{8.120}$$

Again, the sum $R_*^n = R(A^n) + (\kappa_a/\kappa_e)A^n$ can be minimized to determine the optimal value of total transfer area A^n in the finite-size system.

An interesting aspect of the optimization algorithm represented by the
autonomous model of Equation (8.119) (stage cost independent of time τ) is the
possibility of state dimensionality reduction in the recurrence Equation (8.119).
This is associated with the elimination of state variable τ from this equation by
introducing the Lagrange multiplier associated with the constraint for the total
time $\tau^n = \Sigma\theta^n$. The resulting recurrence equation then becomes one-dimensional,
that is, it contains temperature T^n as the only state variable

$$R'^n(T^n) = \min_{T^{n-1}, \theta^n} \{e^n(T^n, T^{n-1}, \theta^n) + \lambda\theta^n + R'^{n-1}(T^{n-1})\} \tag{8.121}$$

Since the discrete time variable θ^n does not appear in the optimal cost function $R'^{n-1}(T^{n-1})$, the optimality condition with respect to θ^n has a local nature, that is, can only be determined by the local cost function e^n. The minimization with respect to θ^n yields for every stage n,

$$-\partial e^n(T^n, T^{n-1}, \theta^n)/\partial \theta^n = \lambda \qquad (8.122)$$

Thus the partial derivative of local cost e^n is a constant of an optimal path or it represents a discrete integral of motion. In optimization formalism λ is a time adjoint or, in most cases, a constant Hamiltonian of the optimization problem. It should be realized that local cost at a stage must be properly expressed: in terms of boundary temperatures and time increment θ^n.

The multistage heat-pump example, presented above, is not the only one that constitutes an application of the theory developed. Other examples of practical applications could be considered, such as dryers working on solar-energy-concentrated, solar-energy-driven desalination systems, solar fields heat exchangers, and so on.

8.6. CONCLUDING REMARKS

In this chapter we developed the thermodynamic theory for irreversible separation operations driven by work or heat, and presented examples of energy limits and optimization solutions for some separation operations.

In particular, we have considered various fluidized drying operations as examples of separation processes. We have shown that descriptions of fluidized drying may use quasi-homogeneous models of the fluidized bed or inhomogeneous models which assume that a part of the gas is in plug flow ("bubble phase") whereas its remaining part (in "dense phase") is ideally mixed with the solid. Optimization computations show that the intense parameters of optimal drying operation are little influenced by these phenomena, yet their effect on the extensive process parameters (e.g., optimal gas flow rate) is essential. Computations also show that optimal gas flow rate rapidly decreases with coefficient of investment costs κ.

We have also considered dryers driven by the thermal energy of the gas which is upgraded in a heatpump before entering the dryer. Minimization of power supply for two-stage drying with heatpumps has been outlined. The developed example shows how to optimize a sequence of heatpumps by the method of dynamic programming. The optimization solution shows that the optimal areas are close to each other. The optimal temperatures constitute an increasing sequence. The optimal work supplied to the two-stage system decreases significantly with the total transfer area. With some modification, the approach can be extended to complex nonsequential systems.

9 Optimal decisions for chemical reactors

9.1. INTRODUCTION

In this and next chapter we outline various tools coming from fields of thermodynamics, kinetics, and optimization that are useful in dynamic modeling and optimal control of chemical, electrochemical, and biological reactors, in particular fuel cells.

In the first part we analyze nonlinear electrochemical and thermal transport processes, where transport steps are treated as peculiar chemical reactions described by appropriate affinities. Rate and transport equations contain terms exponential with respect to Planck's potentials and temperature reciprocal. In each elementary step, of transport or rate nature, kinetic mass action law leads to the identification of two competing unidirectional fluxes. While they are equal in the thermodynamic equilibrium, their difference out of the equilibrium constitutes the observed flux that represents the resulting rate of the process. A generalized affinity emerges as a suitable driving force. Correspondence with the classical Butler–Volmer kinetics is secured for electrochemical systems. Near the thermodynamic equilibrium the theory converges to the standard linear kinetics described by Onsager's equations.

In the second part we transfer to the chemical realm the method of thermodynamic optimization that was developed in earlier chapters for thermal machines. In the present context, the method is aimed at the maximum production of power in chemical systems, and, after suitable extension, it may efficiently treat complex electrochemical flow systems, in particular fuel cells. With the thermodynamic knowledge and methods of dynamic optimization (especially the method of dynamic programming) kinetic limits are estimated for the optimal work function W_{max} that describes an integrated power output and generalizes the familiar reversible work W_{rev} for finite rates. Optimization results lead to the energy limits, which, in the context of fuel cells (next chapter), depend on the overvoltage effects caused by rate processes (electrochemical reactions) and associated transport phenomena.

9.2. DRIVING FORCES IN TRANSPORT PROCESSES AND CHEMICAL REACTIONS

The aim of this section is a general macrokinetics of transport and rate processes (chemical or electrochemical reactions) and its application to the theory of optimal power yield in chemical or electrochemical systems. Analyzing two competing directions in elementary chemical or transport steps, we investigate equations of nonlinear kinetics of Marcelin–Kohnstamm–de Donder type that contain terms exponential with respect to the Planck potentials and temperature reciprocal. The accepted approach distinguishes in each elementary process, diffusive or chemical, two competing unidirectional fluxes, which are equal at the thermodynamic equilibrium. The observed (resulting) flux, which describes an elementary rate, is the difference of these unidirectional fluxes off equilibrium. We regard the kinetics of this sort as the potential representation of a generalized law of mass action that comprises the effect of rate processes, transfer phenomena, and external fields. Important are physical consequences of these results close and far from equilibrium, which show how diverse processes can be described, and how the basic equation of electrochemical kinetics can be obtained within the general kinetic description. In these considerations we point out the significance of a generalized affinity and correspondence with the standard kinetic mass action law and Onsager's theory near the thermodynamic equilibrium.

We begin with a discussion of the driving forces in generalized kinetic schemes. The general equations of macrokinetics are formulated for a set of chemical reactions, $j = 1, 2, \ldots, N$

$$\sum_{i=1}^{n} v_{ij}^{\mathrm{f}} B_i^{\mathrm{f}} \leftrightarrow \sum_{i=1}^{n} v_{ij}^{\mathrm{b}} B_i^{\mathrm{b}} \tag{9.1}$$

between species $i = 1, 2, \ldots, n$ and also for the transport (diffusion) processes involving the same species. The absolute stoichiometric coefficients v_{ij} always assume positive values. Both groups of processes are described by the same basic equation; the differences appear in the way the stoichiometric coefficients are treated, so to secure their equality on both sides of the energy barrier for transport processes.

The absolute rates are expressed as functions of potentials F_k, which are components of the vector $\mathbf{F} = (T^{-1}, -\mu_1 T^{-1}, \ldots, -\mu_n T^{-1})$. The classical definition of the chemical affinity is here transferred to the entropy representation (superscript s) and extended so that it can take into account the effect of transport phenomena

$$A_j^s = \sum_{i=0}^{n} (v_{ij}^{\mathrm{b}} F_i^{\mathrm{b}} - v_{ij}^{\mathrm{f}} F_i^{\mathrm{f}}) \tag{9.2}$$

In the kinetic regime a genuine chemical step takes place. In this case the transport thermodynamic force $-X_i = F_i^{\mathrm{f}} - F_i^{\mathrm{b}} = 0$ or $F_i^{\mathrm{f}} = F_i^{\mathrm{b}}$. As the resulting stoichiometric coefficient $v_{ij} = v_{ij}^{\mathrm{b}} - v_{ij}^{\mathrm{f}}$ is nonvanishing for a genuine reaction (it is positive for products and negative for substrates), an isothermal reaction in the kinetic regime is driven by the classical affinity

$$A_j^s = \sum_{i=1}^{n} (v_{ij}^{\mathrm{b}} - v_{ij}^{\mathrm{f}}) F_i = \sum_{i=1}^{n} v_{ij} F_i = (\mathbf{v}^T \mathbf{F})_j \tag{9.3}$$

For a diffusion step $v_{ij}^{b} = v_{ij}^{f} = v_{ij}^{f,b}$, that is, the *resulting* stoichiometric coefficient vanishes, or $v_{ij} = 0$. In this case the process is driven by the classical driving forces or the differences $X_i = F_i^b - F_i^f$. Then

$$A_j^s = \sum_{i=1}^{n} (v_{ij}^b F_i^b - v_{ij}^f F_i^f) \rightarrow \sum_{i=1}^{n} v_{ij}(F_i^b - F_i^f) = \sum_{i=1}^{n} v_{ij} X_i \qquad (9.4)$$

Thus, in the diffusion regime quantities X_i emerge naturally, meaning that the diffusional fluxes are related to Onsagerian driving forces.

For any elementary step the generalized affinity (9.2) can be written down in the form

$$A_k^s = \sum_i v_{ik}^b F_i^b - \sum_i v_{ik}^f F_i^f = R \ln \left(\frac{k_k^f \prod_{i=0}^{s} \left(c_i^f \right)^{v_{ik}^f}}{k_k^b \prod_{i=0}^{s} \left(c_i^b \right)^{v_{ik}^b}} \right) = R \ln \left(\frac{r_k^f}{r_k^b} \right) (9.5)$$

The quantity $A_k^s = A_k/T$ is the affinity of kth exchange process in the entropy representation. In imperfect systems affinities a_i should replace concentrations c_i. It may be shown that the rates can be given the form of Ohm's law with appropriately defined resistance (see Section 9.4).

After calculating net fluxes and forces we observe that in the conventional Onsager's description of irreversible processes one deals with net fluxes and net thermodynamic forces. Traditional rate equations postulated in the Onsager's theory have the following structure:

$$\mathbf{J} = f(\mathbf{X}), \qquad \mathbf{J} = -\Delta \mathbf{I}, \qquad \mathbf{X} = \Delta \mathbf{F} \qquad (9.6)$$

Here I_k is the vector of "absolute" or unidirectional rates, exemplified by Equation (9.7) or such like. In the Onsager's theory, the typical notation for Equation (9.6) is

$$J_i = \sum_k L_{ik} X_k$$

where L_{ik} is the Onsager's phenomenological coefficient. The link of the present formalism with the Onsager's theory can be interpreted graphically (Oláh, 1997). Each unidirectional flux I_j is a direction-independent function of potentials F_k (Figure 9.1).

9.3.   GENERAL NONLINEAR EQUATIONS OF MACROKINETICS

Consequently, the observed kinetic mechanism follows as the net result of two opposite steps. At the thermodynamic equilibrium both absolute rates I_k and potentials F_k equalize, although they are different at nonequilibrium states. In any ith step the quantities I_i^0 ("exchange currents") are the same for both

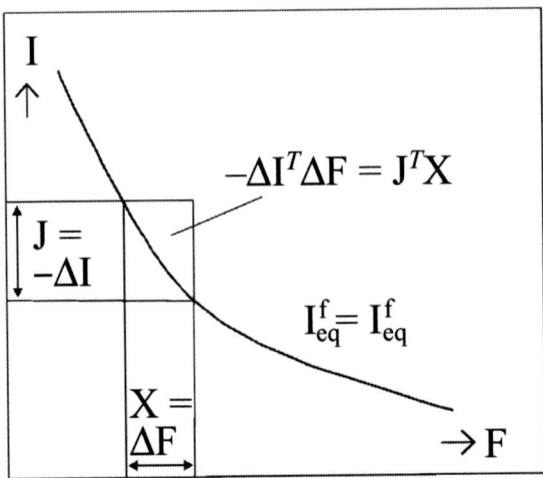

Figure 9.1 Interpretation of absolute fluxes **I**, potentials **F**, net fluxes **J**, and classical thermal forces **X**.

directions. The unidirectional chemical flux in the forward direction (a component of unidirectional vector $\mathbf{I}^f$) satisfies an equation

$$r_j^f = r_j^0 \exp\left(-\sum_i v_{ij}^f F_i^f\right), \qquad j = 0, 1, \ldots, N \tag{9.7}$$

The same structure holds for the backward process with only the difference in indices (b instead of f). Yet, there are two competing processes, direct and reverse (or forward and backward). In the discrete model describing exchange processes between two different subsystems the resulting rate of an elementary transport or chemical step subsumes the competition of forward and backward rates

$$r_j = r_j^f - r_j^b = r_j^0 \left(\exp\left(-\sum_i v_{ji}^f F_i^f\right) - \exp\left(-\sum_i v_{ji}^b F_i^b\right)\right) \tag{9.8}$$

In these equations, potentials $F_i = (1/T, -\mu_i/T)$ are the partial derivatives of the entropy with respect to the extensive variables appearing in the fundamental Gibbs equation for the entropy differential. In the kinetic regime (fast transport; $F_i^f = F_i^b = F_i^{eq}$) classical chemical kinetics is recovered in the Marcelin–Kohnstamm–de Donder form; see Section 9.4. (As implied by Equation (9.8), this is the kinetics in terms of the potential quantities instead of concentrations.) In the diffusive regime (fast reactions) the kinetic system represents a nonlinear diffusion process.

9.4. CLASSICAL CHEMICAL AND ELECTROCHEMICAL KINETICS

We shall now show the correspondence of the present theory with classical kinetics of rate processes governed by the mass action law. We consider the competition of forward and backward chemical steps. Using the usual structure of chemical potentials

$$\mu_i(\mathbf{c}, T) = \mu_{i0}(T) + RT \ln c_i$$

we substitute concentrations c_i into the classical mass action kinetics

$$r_j = r_j^{\mathrm{f}} - r_j^{\mathrm{b}} = k_j^{\mathrm{f}}(T) \prod_{i=1}^{n} c_i^{v_{ij}^{\mathrm{f}}} - k_j^{\mathrm{b}}(T) \prod_{i=1}^{n} c_i^{v_{ij}^{\mathrm{b}}} \qquad (9.9)$$

The result is the chemical rate in the Marcelin–de Donder form

$$r_j(\mathbf{c}, T) = r_j^{\mathrm{f}}(\mathbf{c}, T) - r_j^{\mathrm{b}}(\mathbf{c}, T) = r_j^{0}(T) \left(\exp \sum_{i=1}^{n} v_{ij}^{\mathrm{f}} \frac{\mu_i}{RT} - \exp \sum_{i=1}^{n} v_{ij}^{\mathrm{b}} \frac{\mu_i}{RT} \right)$$

$$(9.10)$$

This equation has the typical (yet not the most general) nonlinear structure of macrokinetic equations. Its virtue is a single rate constant, r_j^0, common for both directions, representing the so-called *exchange current*, and the explicit satisfaction of the principle of microscopic reversibility at the state of equilibrium. The exchange current expressed by "usual" rate constants has the form

$$r_j^{0}(T) \equiv k_j^{\mathrm{f}}(\mathbf{c}, T) \left(\exp \sum_{i=1}^{n} -v_{ij}^{\mathrm{f}} \frac{\mu_{i0}}{RT} \right) = k_j^{\mathrm{b}}(\mathbf{c}, T) \left(\exp \sum_{i=1}^{n} -v_{ij}^{\mathrm{b}} \frac{\mu_{i0}}{RT} \right)$$

$$(9.11)$$

In the case of ionic systems the chemical potentials are replaced by electrochemical potentials.

Electrochemistry is, in fact, the realm where the notion of the exchange current is best known. In fact, it is the condition of the vanishing electrochemical affinity at equilibrium, the condition that makes it possible to define the universal rate constant r_j^0 or the exchange current of electrochemistry (Figure 9.2).

In the isothermal case, the result of the procedure involving the potential description of kinetics is the Butler–Volmer equation that describes the electric current J as the difference between the anodic and cathodic currents. Oláh (1997) has given clear interpretation of anodic and cathodic currents as well as of the exchange current in terms of voltage. In his diagrams the abscissa of crossing point of both currents describes the exchange current. He has also interpreted logarithms of anodic and cathodic currents as well as of the exchange current

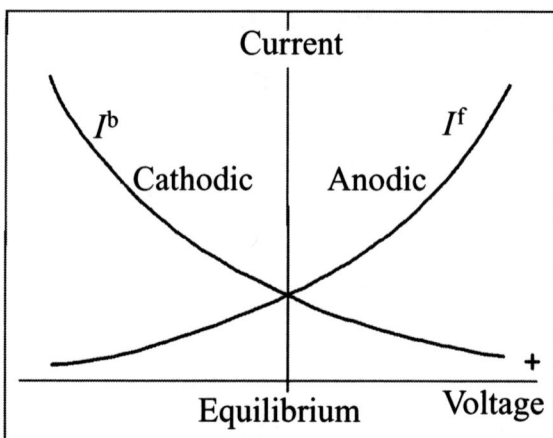

Figure 9.2 Oláh's interpretation of anodic and cathodic currents as well as of the exchange current in terms of voltage. Abscissa of crossing point of both currents describes the exchange current.

in terms of electrochemical potential. The potential form of the Butler–Volmer equation is

$$J = j^{anod}(\mathbf{c}, T) - j^{cathod}(\mathbf{c}, T) = j^0(T)\left(\exp\sum_{i=1}^{n} v_i^f \frac{\tilde{\mu}_i}{RT} - \exp\sum_{i=1}^{n} v_i^b \frac{\tilde{\mu}_i}{RT}\right)$$

(9.12)

whereas a customary, equivalent form of the Butler–Volmer equation describes the electric current of a cell in terms of the overvoltage η and universal Faraday's constant F

$$J = j^0\left(\exp\left(\frac{(1-\alpha)F\eta}{RT}\right) - \exp\left(\frac{-\alpha F\eta}{RT}\right)\right)$$

(9.13)

The coefficient α characterizes the symmetry of the energy barrier and is near to 1/2. Equation (9.13) is commonly used to describe various electrochemical systems including fuel cells. Its consequence is the Tafel chart presented in Figure 9.3.

9.5. INCLUSION OF NONLINEAR TRANSPORT PHENOMENA

We shall return to the general nonlinear equations of macrokinetics described in Section 9.3. Using these general equations one may consider coupled transfer of heat (h) and mass (m). Introduced are potentials $F_i = (1/T, -\mu_i/T)$, $i = 0, 1, \ldots, n$,

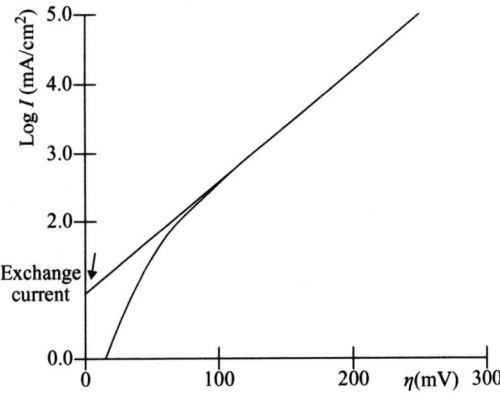

Figure 9.3 Tafel chart.

which are the thermodynamic conjugates of the extensive variables in the Gibbs equation for the system's entropy

$$dS = T^{-1}\,dE - T^{-1}\mu_\alpha\,dc_\alpha \equiv \sum_{i=0}^{s} F_i\,dC_i \equiv \mathbf{F}\cdot d\mathbf{C} \tag{9.14}$$

The process kinetics is described by the *general exchange equation* for the net flux J_i

$$J_i = I_j^0 \left\{ \exp\left(\frac{-\sum_i v_{ik}^* F_k^f}{R}\right) - \exp\left(\frac{-\sum_i v_{ik}^* F_k^b}{R}\right) \right\} \tag{9.15}$$

whose equivalent form in terms of deviations from equilibrium is

$$J_i = I_i^{eq} \Delta\exp\left(\frac{-\sum_i v_{ik}^*(F_k^f - F_k^{eq})}{R}\right) \tag{9.16}$$

with

$$I_i^{eq} \equiv I_i^0 \exp\left(\frac{-\sum v_{ik}^* F_k^{eq}}{R}\right) \tag{9.17}$$

as the common value of the absolute current at equilibrium.

Consider now as a simplest possible example an energy generation process driven by the pure heat transfer. In this case the corresponding kinetic set reduces to just one equation that describes the nonlinear heat flux. The set of equations describing the exchange of heat between each reservoir and the thermal machine involves two equations. The first one describes the heat flux driving the engine

$$q_1 = I_1^{eq} \Delta\exp\left(\frac{-E(T^{-1} - T^{-1eq})}{R}\right) \tag{9.18}$$

where

$$I_1^{eq} \equiv I_1^0 \exp\left(\frac{-ET^{-1eq}}{R}\right) \tag{9.19}$$

and the symbol Δ in Equation (9.18) refers to the suitable difference of temperatures T_1 and $T_{1'}$. The constant E is the activation energy for the thermal transfer. Analogous equation holds for the second (low temperature) fluid. Close to equilibrium, the linear approximation of the nonlinear exchange equation may be considered

$$q_1 = I_1^{eq} E R^{-1}(T_{1'}^{-1} - T_1^{-1}) \cong I_1^{eq} \frac{ER^{-1}(T_1 - T_{1'})}{T_1 T_{1'}} \tag{9.20}$$

Comparing this expression with the classical Newton's structure $q_1 = g_1(T_1 - T_{1'})$, we find that the conductance in the present model can be evaluated in accordance with the equation

$$g_1 = I_1^{eq} \frac{ER^{-1}}{T_1 T_{1'}} = I_1^0 \exp\left(\frac{-ET^{-1eq}}{R}\right) \frac{ER^{-1}}{T_1 T_{1'}} \tag{9.21}$$

This also shows that the correspondence with classical equation of heat exchange can be assured. The above equation may serve to estimate the value of the exchange current provided that the heat exchange coefficient α, constant E, and the common equilibrium temperature of two bodies are known. It should be realized that E characterizes the temperature dependence of the heat conductivity.

We can now write down the kinetic equations for both reservoirs. For the known process coefficients the heat exchange equation for the first reservoir has the form

$$q_1 = I_1^0 \left\{ \exp\left(\frac{-E_1}{RT_1}\right) - \exp\left(\frac{-E_1}{RT_{1'}}\right) \right\} \tag{9.22}$$

whereas that for the second one is

$$q_2 = I_2^0 \left\{ \exp\left(\frac{-E_2}{RT_{2'}}\right) - \exp\left(\frac{-E_2}{RT_2}\right) \right\} \tag{9.23}$$

On this ground, one can develop the nonlinear theory in which thermal conductivities and related conductances are variable, that is, are state functions.

The basic form (9.8) is, in fact, a generalized mass action law in terms of potentials F_i. It is not restricted to the kinetic regime. One may write down Equation (9.8) in a condensed form

$$r_j = r_j^0(\exp(\Pi_j^f) - \exp(\Pi_j^b)) \tag{9.24}$$

where

$$\Pi_j^f \equiv -\sum_{i=1}^n (v_{ij}^f F_i^f), \qquad \Pi_j^b \equiv -\sum_{i=1}^n (v_{ij}^b F_i^b) \tag{9.25}$$

are, respectively, one-directional components of affinity in forward and backward directions. Consider now the jth reaction step characterized by a *constant* specific resistance ρ_j and the (electro)chemical vector $\mathbf{J}_j$ through the surface area A. Equations (9.5), (9.14), (9.15), and (9.25) show that the chemical or electrochemical kinetics can be given the form of Ohm's law

$$J_j = R_j^{-1}(\Pi_j^f - \Pi_j^b) \tag{9.26}$$

in which both quantities Π_j are the boundary values of potentials of the reaction j. However, this form of the law is exact only for a constant specific resistance ρ_j. For a complicated motion of the chemical complex through the energy barrier, knowledge of the local resistance formula and its integration is required. This shows the restrictive ingredient of the model of discrete energy barrier that—in spite of many successes—is not capable of avoiding mean quantities characteristic of the whole barrier connected with finite affinities or driving forces, such as chemical resistance R_j.

9.6. CONTINUOUS DESCRIPTION OF CHEMICAL (ELECTROCHEMICAL) KINETICS AND TRANSPORT PHENOMENA

The nonlinear theory described above has its own counterpart in the field description outlined in this section.

To describe a chemical transformation as a motion through the energy barrier treated as a continuum an effort can be made to replace the logarithmic resistance (the mean quantity R_j associated with a finite affinity) by its local counterpart. The result is a continuous description governed by a principle of Fermat type with an infinite number of infinitesimal refractions of the ray (Tan and Holland, 1990; Sieniutycz, 2000b, 2002a, 2002b, 2007a). The results show that the path of chemical complex bends into a direction that ensures its shape associated with longest residence time in regions of lower resistance. These results make possible prediction of shapes of chemical or electrochemical paths and related kinetic formulae.

Considering a thermal conduction process coupled with mass diffusion a quasilinear formula is obtained

$$\mathbf{J}_q = L_{11}(\mathbf{F})\text{grad}(-T^{-1}) + L_{12}(\mathbf{F})\text{grad}(\mu T^{-1}) \tag{9.27}$$

The quasilinearity is with respect to the potential vector $\mathbf{F}$. In Equation (9.27) L_{11} is Onsager's conductivity connected with the usual one by the relation $L_{11} = \lambda T^2$. Similarly, as the result of the integration along the gradient direction of the vector relationship

$$r_j = -\lambda_j(\mathbf{F})\text{grad}\,\Pi_j \equiv \lambda_j(\mathbf{F})\text{grad}\left\{\sum_{i=1}^{n}(Q_{ij}F_i)\right\} \tag{9.28}$$

the chemical Ohm's law follows. The above is the continuous form of Ohm's law, replacing the discrete structure

$$r_j = \Lambda_j A_j \tag{9.29}$$

or its equivalent counterpart, Equation (9.26). Equation (9.28) is integrated between the limits $Q_{ij}^f = v_{ij}^f$ and $Q_{ij}^b = v_{ij}^b$. Here the vector of resulting chemical rate is considered in the physical space. This approach also includes electrochemistry whenever F_i involve electrochemical potentials.

The field representation of the nonlinear theory of macrokinetics implies that the diffusion coefficients D are not constants but are exponential functions of intensive state coordinates (temperature reciprocal and Planck potentials). The exponential dependence of D on $1/T$ was confirmed in a large number of experiments, especially in those describing diffusion in metals and melts and drying of solids.

We shall now pass to the applicative sections of the chapter in which we outline how thermodynamics can contribute to the theory of efficiency and power generation in chemical systems. In particular, the method can be extended to irreversibly working fuel cells. As in the heat systems, considered earlier, the method is based on the optimal control theory and thermodynamics. First (Section 9.7–9.16), we shall analyze simple chemical systems. Complex chemistries are treated in Section 9.17.

9.7.   TOWARD POWER PRODUCTION IN CHEMICAL SYSTEMS

Power optimization in chemical processes undergoing in irreversible, power-producing systems bears the difficulty caused by multiple (vectorial) efficiencies. Whenever power limits are considered, optimal efficiency formulae refer to chemical machines working at the maximum power. Steady-state modeling describes a chemical system in which two reservoirs are infinite, whereas an unsteady model treats a dynamic system with finite upper reservoir and gradually decreasing chemical potential of a key fuel component. In the chemical systems considered total power output is maximized at constraints that take into account dynamics of mass transport and efficiency of power generation. Dynamic optimization methods, in particular variational calculus, lead to optimal functions that describe integral power limits and extend reversible chemical work W_{rev} to finite-rate situations. Optimization data quantify effects of chemical rates and transport phenomena. Legendre transform of a local power function is an effective tool to obtain an optimal path in a dynamic process of power yield.

Here we analyze the performance of a nonisothermal chemical engine in terms of heat and mass fluxes flowing from a fuel reservoir to the power generator. By assumption, a fuel mixture that drives the power generator is composed of an inert and an active component. Efficiency, power yield, and fuel flux are essential variables determining the performance of the chemical system (de Vos, 1992;

Sieniutycz, 1999a, 1999d, 2007b). The problem of finite-rate limits, which was treated in our earlier papers for thermal processes (Sieniutycz, 2003a, 2003b, 2003c, 2004a, 2004b, 2004c, 2006a, 2006b), is applied here to chemical systems, steady and unsteady. Enhanced limits caused by finite rates are evaluated for power released from an engine system or added to a power-consuming system.

We discuss two basic models of chemical units producing power. Steady-state model, originated by de Vos (1992), refers to the situation when both reservoirs are infinite, whereas a new, unsteady model treats a dynamic case with finite upper reservoir and gradually decreasing chemical potential of the active component of fuel. As opposed to these earlier works we extend here the original problem to situations with nonisothermal generation of power. In the dynamic case Lagrangian and Hamiltonian approaches to power functionals and optimization algorithms using canonical equations are effective. Finite-rate models incorporate a minimum entropy production caused by irreversible diffusion phenomena.

Modeling a power-assisted chemical operation for the purpose of dynamic limits is a difficult task. Evaluation of dynamic limits requires sequential operations (Sieniutycz, 1999a, 1999b, 1999c, 1999d, 1999e, 2003b, 2004a, 2006a), where total power yield is maximized at constraints that describe dynamics of energy and mass exchange. The dynamic model can be continuous or discrete; the latter are frequent for computational purposes. The results are limiting work functions in terms of end states, duration, and (in discrete processes) number of stages (Sieniutycz, 2004a). Modeling of power generation processes is consistent with general philosophy of optimization (Berry et al., 2000; Sieniutycz and de Vos, 2000; Sieniutycz and Farkas, 2005). Constraints take into account dynamics of mass transport and rates of fuel consumption. Finite-rate, endoreversible modeling includes irreducible losses of classical exergy caused by resistances. Optimal performance functions, which describe extremum power and incorporate a residual minimum entropy production, are determined in terms of end states, duration, and (in discrete processes) number of stages (Sieniutycz, 2004a, 2006a). Similarly as in thermal systems enhanced power limits follow from the constrained optimization of total power.

In chemical engines mass transport processes participate in the transformation of differences of chemical potentials into mechanical power. However, a more precise statement refers not to the individual chemical potentials μ_k, but to their linear combination in the form of unidirectional component of chemical affinity, Π_j (Sieniutycz, 2004b). This affinity component is, in fact, the potential Π_j of the reagents that decreases along the reaction path. The quantity Π_j (which reduces to the chemical potential of a single component only in the case of a simplest isomerization reaction) plays in chemical engines role analogous to that played by the temperature in heat engines. Yet, in chemical systems, generalized reservoirs are present that are capable of providing heat and substance. Heat conductors of the thermal engine theory became conductors of both matter and substance in the chemical engine theory.

Infinite reservoirs are capable of keeping constant the process potentials (T, μ_k, Π_j, etc.). For such reservoirs problems of extremum power (maximum of power

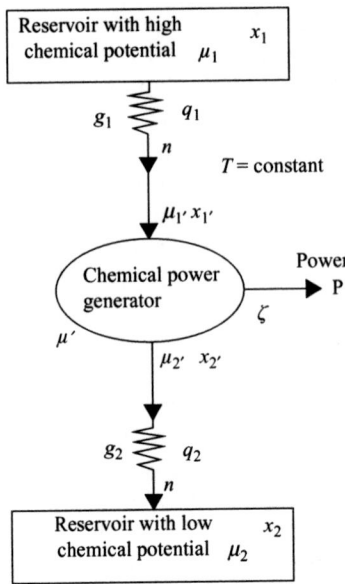

Figure 9.4 Principle of a chemical engine driven by mass transfer of an active component through an inert.

produced and minimum of power consumed) are static optimization problems. On the other hand, for finite reservoirs, in which the amount of an active reactant is gradually reduced and its chemical potential decreases in time, the considered problems are those of dynamic optimization.

To deal with steady situations we consider a single "endoreversible" chemical machine, an engine or power consumer, in Figure 9.4. The engine is propelled by a high-μ active reactant of a fluid mixture (fuel mixture) that is supplied to the power generator from an infinitely large reservoir. The fluid mixture is composed of the active reactant and an inert. The flux of this active reactant, designated by n, is a basic quantity influencing the intensity of power yield.

Yet, tackling a dynamic situation requires considering a multistage production (consumption) of power in a sequence of chemical engines. This case is associated with a finite upper reservoir and a gradual exhaust of the active reactant in the fuel mixture. In the multistage engine the chemically active reactant drives at each stage the chemical power generator from which power is released. In the multistage power consumer the fuel mixture is upgraded in the system to which power is supplied. In each case the second fluid is an infinite reservoir. The fluids are of finite thermal and mass conductivity; hence, there are finite resistances in the system. In a multistage engine operation the driving chemical potential decreases at each stage; the whole operation is described by the sequence $\mu^0, \mu^1, \ldots, \mu^N$. The popular "engine convention" is used: work generated in an engine, W, is positive, and work generated in a power consumer is negative; this implies a positive work $(-W)$ consumed in the power consumer.

The sign of an optimal work function $V^N = \max W^N$ defines working mode for an optimal sequential process as a whole. In engine modes, $W > 0$ and $V > 0$. In power consumption modes, $W < 0$ and $V < 0$; therefore, working with a function $R^N = -V^N = \min(-W^N)$ is more convenient. Of special attention are two processes: the one that starts with the state $X^0 = X^e$ and terminates at an arbitrary $X^N = X$ and the one that starts at an arbitrary $X^0 = X$ and terminates at X^e. For these processes functions V^N and R^N are counterparts of the classical exergy in state changes with finite durations.

The topological nature of the state space and its influence on the system dynamics are determined by both the state of the finite-resource fuel flowing through stages of the cascade and the properties of the heat bath (infinite thermal reservoir). Without bath the system topology would correspond to a simple sequence of stages (a cascade) with the state dimensionality defined by the number of independent coordinates of the fuel mixture. The presence of an infinite bath, the intensive parameters of which, that is, its temperature T^e and chemical potentials μ_i^e, do not change along the process path, introduces into the mathematical model constant parameters T^e and μ_i^e. In fact, it is the condition of an infinite bath that enables us to treat all power functions p^n as bath-history independent. The potential function of extremum work, obtained via optimization of a work integral, is of exergy type, that is, it contains intensive parameters of the bath and state coordinates of the fuel mixture. This property makes the (parameter dependent) function of extremum work different from the (parameter independent) thermodynamic potentials.

The range of optimization in this chapter is restricted to thermodynamic limits, or, more specifically, to a generalized quantity of exergy type attributed to the single stream of fuel. This quantity constitutes a generalized potential of extremum work that depends on end states of the fuel mixture and its holdup time in the system (time of fuel consumption). Alternatively, a generalized potential could be expressed in terms of end states of the fuel stream and the optimal Hamiltonian that is a measure of the process intensity along an optimal path.

Generalized (dissipation-affected) power limits for a finite time degrading of the fuel mixture are the main task of this chapter. To find these limits the system unit must contain a chemically active part or a "reaction zone." Only in "endoreversible" systems this zone is a purely reversible part of the system, that is, its efficiency of energy production is given by a reversible formula. This reversible formula constitutes a chemical counterpart of the familiar Carnot formula $\eta_C = 1 - T_2/T_1$, valid for thermal engines. In chemical engines, however, the reversible formula refers to the chemical affinity, which has nothing in common with the Carnot efficiency, η_C.

Yet, the restriction to external irreversibilities is unnecessary; in fact, thermodynamic models can go beyond "endoreversible limits," that is, they can treat internal irreversibilities as well; see references (Berry et al., 2000; Chen et al., 2001; Sieniutycz, 2004a, 2006a, 2007c, 2007d). It is most essential, however, that in either of two methodological versions of the thermodynamic approach, of which the first gives up internal irreversibilities, whereas the second one estimates these from a model, the obtained power limits are stronger than those predicted by the classical exergy. In short, this results from the "process rate penalty" that is taken into account in every version of the approach.

In classical Carnot-like analyses the resource and environment reservoirs are insensitive to the effect of dissipators (boundary layers, resistances, etc.) because the reversible situation requires the spatial homogeneity of each reservoir. In the irreversible analysis, performed here, which admits dissipative transports, inhomogeneities in transport potentials play essential role.

When calculating power limits, we search for *purely physical* extrema with no regard to economic optima.

9.8. THERMODYNAMICS OF POWER GENERATION IN NONISOTHERMAL CHEMICAL ENGINES

This section analyzes a single-stage chemical process depicted in Figure 9.4. To obtain a power yield formula associated with a single isomerization reaction

$$B_1 \Leftrightarrow B_2 \qquad (9.30)$$

we apply balances of energy, entropy, and mass. We assume that an active reagent 1 and the sensible heat q_1 are transferred to the reaction zone through an inert gas. As the result of the chemical transformation the reaction zone yields product 2 that is transferred through the same inert gas to the environment. The energy balance

$$\varepsilon_1 = \varepsilon_2 + p \qquad (9.31)$$

and the mass balance in terms of molar fluxes

$$n_1 = n_{1'} = n_{2'} = n_2 \qquad (9.32)$$

are combined with an equation describing the continuity of the entropy flux

$$\frac{\varepsilon_1 - \mu_{1'} n_{1'}}{T_{1'}} = \frac{\varepsilon_2 - \mu_{2'} n_{2'}}{T_{2'}} \qquad (9.33)$$

Equation (9.33) holds in the chemically active, reversible part of the system where the chemical reaction runs.

Equations (9.31)–(9.33) yield the power expression

$$p = \varepsilon_1 - \varepsilon_2 = \left(1 - \frac{T_{2'}}{T_{1'}}\right)\varepsilon_{1'} + T_{2'}\left(\frac{\mu_{1'}}{T_{1'}} - \frac{\mu_{2'}}{T_{2'}}\right)n_{1'} \qquad (9.34)$$

The energy flux in the dissipative parts of the system is continuous, for example

$$\varepsilon_1 = q_1 + h_1 n_1 = q_{1'} + h_{1'} n_{1'} = \varepsilon_{1'} \qquad (9.35)$$

Using the primed part of this equation in the power formula (9.34) yields

$$p = \left(1 - \frac{T_{2'}}{T_{1'}}\right)q_{1'} + [(h_{1'} - h_{2'}) - T_{2'}(s_{1'} - s_{2'})]n_{1'} \qquad (9.36)$$

Thus, the combination of the second law and the reversible balance of the entropy leads to power expression (9.36) in which the Carnot efficiency of an endoreversible process

$$\eta = 1 - \frac{T_{2'}}{T_{1'}} \qquad (9.37)$$

is the thermal component of a two-dimensional efficiency vector. The second component of this vector is the exergy-like function of the active component evaluated for the primed state 2 as the reference state

$$\beta' = h_{1'} - h_{2'} - T_{2'}(s_{1'} - s_{2'}) \qquad (9.38)$$

Observe that in the considered case (where molar flux n is the efficiency basis) the chemical component of efficiency is not nondimensional. Other cases of this sort will be observed later when we shall also note the identity of efficiencies with active driving forces. The primed quantities and equations are often applied in a transformed form expressing all physical quantities in terms of the bulk state variables of both fluids and certain controls. The latter are related to fluxes of heat and matter. Applying the energy flux continuity and mass flux continuity in Equation (9.36) to eliminate fluxes $q_{1'}$ and $n_{1'}$ on account of q_1 and n_1 yields

$$p = \left(1 - \frac{T_{2'}}{T_{1'}}\right) q_1 + \beta n_1 \qquad (9.39)$$

where coefficient β is defined as

$$\beta = \left[c_p(T_1 - T_{1'}) \left(1 - \frac{T_{2'}}{T_{1'}}\right) + (h_{1'} - h_{2'}) - T_{2'}(s_{1'} - s_{2'}) \right] n_1 \qquad (9.40)$$

Note bilinear power structures in Equations (9.36) and (9.39) (Sieniutycz, 1999a, 1999d). Both these equations are useful measures of chemical efficiency, yet they differ in the heat flux (q_1 or $q_{1'}$) accepted as the control variable. The second component of power expressions (9.36) and (9.39) is associated with the work production (consumption) due to the mass transfer. Its interpretation is the product of mass flow n and exergy-like functions β' or β whose structure follows from combination of the energy and (conservative) entropy balances.

Yet, we may eliminate exergy-like functions β and β' by passing to the process description in terms of different fluxes. Leaving mass flux n unchanged we introduce a new energy flux, $Q_{1'}$, called total heat flux, which is defined by an equation

$$Q_{1'} \equiv q_{1'} + T_{1'}s_{1'}n_{1'} \qquad (9.41)$$

Clearly, $Q_{1'}$ is the product of temperature $T_{1'}$ and the total entropy flux, the latter being the sum of entropy transferred with heat $q_{1'}/T_{1'}$ and with substance

$s_1 n_1$. The virtue of flux $Q_{1'}$ is that the power production $p = \dot{w}$ assumes in terms of $Q_{1'}$ and n_1 an intuitively natural form that contains the Carnot efficiency and chemical potential difference of the active component

$$\dot{w} = p = \left(1 - \frac{T_{2'}}{T_{1'}}\right) Q_{1'} + (\mu_{1'} - \mu_{2'})n \tag{9.42}$$

In this case the chemical efficiency component is just the chemical affinity for the single isomerization reaction considered.

Alternatively we may base the efficiency on a certain mass analogue of heat flux $Q_{1'}$ defined as follows:

$$G_{1'} \equiv \mu_{1'} n_{1'} \tag{9.43}$$

The quantity $G_{1'}$ represents, in fact, the flux of Gibbs free energy of the active component of fuel. The suitability of this quantity, which we call Gibbs flux, follows from its capability of measuring quality of mass flux $n_{1'}$. In fact, flux $G_{1'}$ measures the quality of $n_{1'}$ in the same way as heat $Q_{1'}$ measures the quality of entropy flux, the associated "quality potentials" being, respectively, $\mu_{1'}$ and $T_{1'}$. With Equation (9.43), power expression (9.42) is

$$p = \left(1 - \frac{T_{2'}}{T_{1'}}\right) Q_{1'} + \left(1 - \frac{\mu_{2'}}{\mu_{1'}}\right) G_{1'} \tag{9.44}$$

The entropy and mass balances of an endoreversible machine take in terms of variables $Q_{1'}$ and $G_{1'}$ the following simple forms:

$$\frac{Q_{1'}}{T_{1'}} = \frac{Q_{2'}}{T_{2'}} \tag{9.45}$$

$$\frac{G_{1'}}{\mu_{1'}} = \frac{G_{2'}}{\mu_{2'}} \tag{9.46}$$

When the Gibbs flux is the efficiency basis the structure of the mass transfer balance (9.46) is formally the same as that of the heat transfer balance (9.45). Note that the endoreversible thermal efficiency in terms of properties of the circulating fluid is the same as in the process of pure heat transfer, that is, it has the Carnot form.

The power expression in "the flux representation" conforms with the general structure

$$p = \dot{w} = \eta(Q_{1'}, G_{1'})Q_{1'} + \omega(Q_{1'}, G_{1'})G_{1'} \tag{9.47}$$

Total heat flux $Q_{1'}$ and Gibbs energy flux $G_{1'}$ are two independent fluxes driving the process of power production.

In a traditional analysis power (9.44) may be considered in terms of two unknown ("primed") temperatures and two unknown ("primed") chemical potentials of circulating fluid, these variables being linked by entropy and mass balances across the reversible part of the system. Such balance constraints make equations quite difficult to use, and a method to overcome the difficulty should be designed. The Carnot quantities play in this matter an essential role (Sieniutycz, 1999a, 1999b, 1999c, 1999d, 1999e, 2003a, 2003b, 2003c, 2007a, 2007b, 2007c, 2007d).

9.9.   NONISOTHERMAL ENGINES IN TERMS OF CARNOT VARIABLES

We shall now pass to corresponding relationships in terms of the Carnot variables (Sieniutycz, 1999b, 2003a, 2003b, 2003c, 2007d). The rationale for their use follows from the fact that temperatures and chemical potentials of circulating fluid are not independent variables but they are constrained by balances of entropy and substance through the engine. Application of Carnot variables leads to the process description (consistent with the entropy and mass balances) in terms of free controls that assure suitable formulae for efficiency, heat and mass fluxes, and power. Carnot temperature was comprehensively discussed earlier (Sieniutycz, 1999b, 2003a, 2003b, 2003c). The present text gives the good opportunity to discuss various definitions of the Carnot chemical potential, and the choice of the most suitable one.

An easiest formal way to set a suitable definition of a Carnot variable is to assume the invariance of efficiency, η or ω. Considering, for example, $1 - \omega$ we obtain the equality

$$\frac{\mu_{2'}}{\mu_{1'}} = \frac{\mu_2}{\mu'} \tag{9.48}$$

(possible coefficients of internal irreversibilities cancel in these expressions). Definition of the Carnot chemical potential then follows as

$$\mu' \equiv \frac{\mu_{1'}}{\mu_{2'}}\mu_2 \tag{9.49}$$

Thus, when the chemical efficiency is based on the Gibbs energy flux (and only in this case) the chemical efficiency expression is fully analogous to that describing the Carnot temperature

$$T' \equiv \frac{T_{1'}}{T_{2'}}T_2 \tag{9.50}$$

For the latter quantity, see Sieniutycz's work (1999b, 2007d). The definition of Carnot temperature T' remains unchanged when the mass transfer accompanies the heat transfer.

The thermal efficiency is the first component of two-dimensional efficiency vector. For the thermal efficiency in terms of Carnot temperature we find

$$\eta = 1 - \frac{T_2}{T'} \tag{9.51}$$

The second efficiency component is related to the difference of chemical potentials

$$\omega = 1 - \frac{\mu_2}{\mu'} \tag{9.52}$$

As the Gibbs flux is the efficiency basis and Equations (9.43), (9.44), (9.46), and (9.47) hold, the chemical efficiency (9.52) has the Carnot form.

However, in terms of variables $Q_{1'}$ and $n_{1'}$ the chemical efficiency is neither nondimensional nor of the Carnot structure; rather it follows in the affinity units

$$\zeta \equiv \frac{p}{n} \equiv \mu_{1'} - \mu_{2'} \tag{9.53}$$

whereas the associated Carnot chemical potential of the active component is

$$\mu' \equiv \mu_2 + \mu_{1'} - \mu_{2'} \tag{9.54}$$

(Sieniutycz, 2007b). Of course, the chemical efficiency (9.53) has dimensional units, those of energy per mole. Like Equation (9.50), Equation (9.54) may be seen as the trivial consequence of the efficiency invariance, $\mu_{1'} - \mu_{2'} = \mu' - \mu_2$. In the reversible case, that is, when $\mu_{1'} = \mu_1$ and $\mu_{2'} = \mu_2$, expression (9.54) reduces μ' to μ_1; thus, in this case, the chemical component of the efficiency vector equals $\mu_1 - \mu_2$. Note the direct link between efficiencies and driving forces. For example, efficiency (9.53) is the driving force describing the chemical affinity of the isomerization reaction considered.

Carnot chemical potential, μ', should be redefined when a more usual energy flux $\varepsilon = q + \sum h_i n_i$ is applied instead of Q. In this case Equation (9.34) leads to the following definition of Carnot chemical potential μ' in terms of parameters of circulating fluid ($T_{1'}$, $T_{2'}$, $\mu_{1'}$, and $\mu_{2'}$):

$$\mu' = T' \left(\frac{\mu_2}{T_2} + \left(\frac{T_{2'}}{T_2} \right) \left(\frac{\mu_{1'}}{T_{1'}} - \frac{\mu_{2'}}{T_{2'}} \right) \right) \tag{9.55}$$

This quantity, which was occasionally used in some of our previous papers (Sieniutycz, 2003a, 2003c, 2007b), is equivalent to μ' of Equation (9.54) only in an isothermal case. Chemical potentials in Equations (9.49), (9.52), and (9.53) are all correct; still they differ because each refers to a different pair of fluxes. In any process with mass transfer the work yield is characterized by the vector of thermal efficiencies, (η, ω), such that its first component is the Carnot efficiency.

Thermodynamic functions of state referred to Carnot temperatures and chemical potentials are called Carnot functions. One may always pass to corresponding relationships in terms of "thermodynamic Carnot quantities." In this way one may deal with Carnot energy, enthalpy, entropy, etc. Importantly, in terms of thermodynamic Carnot variables, reversible structure of basic equations is preserved in irreversible cases, and prediction is possible of irreversible equations on the basis of well-known or easily derived equations of reversible processes.

At the "short circuit point" (de Vos, 1992) the equalities $T' = T_2$ and $\mu' = \mu_2$ hold and all components of efficiency vector (η, ω) do vanish. Yet, at the "Carnot point" efficiencies refer to the quasistatic process (de Vos, 1992). In this special case the efficiency vector in Equation (9.44) has the reversible components

$$\eta_C = 1 - \frac{T_2}{T_1} \tag{9.56}$$

and

$$\omega_C = 1 - \frac{\mu_2}{\mu_1} \tag{9.57}$$

Leaving aside other special cases and returning to the problem in terms of $Q_{1'}$ and $G_{1'}$, we can state that a general optimization task is to seek for optimal T' and μ' that maximize power p in "Carnot variables representation"

$$p \equiv \dot{w} = \eta(T', T_2)Q_{1'}(T', T_1, \mu', \mu_1) + \omega(\mu', \mu_2)G_{1'}(T', T_1, \mu', \mu_1)$$

$$= \left(1 - \frac{T_2}{T'}\right)Q_{1'} + \left(1 - \frac{\mu_2}{\mu'}\right)G_{1'} = \left(1 - \frac{T_2}{T_1}\right)Q_{1'} + \left(1 - \frac{\mu_2}{\mu_1}\right)G_{1'}$$

$$- T_2 Q_{1'}\left(\frac{1}{T'} - \frac{1}{T_1}\right) + \mu_2 G_{1'}\left(\frac{1}{\mu_1} - \frac{1}{\mu'}\right) \tag{9.58}$$

The first equality of the second line is the irreversible power expression that was split into the reversible part (the second equality in the second line, without Carnot controls) and an irreversible part (the expression in third line which is the negative product of T_2 and entropy production). Note that Carnot temperature and Carnot chemical potential are independent variables and the power is extremized with respect to these variables as free process controls. The reversible balances of entropy and mass across the (perfect) thermal machine are included; thus, the extremizing procedure works without constraints.

9.10. ENTROPY PRODUCTION IN STEADY SYSTEMS

Let us determine an expression for the entropy production from the entropy balance of a steady system. From the entropy balance of an overall system composed of reservoirs and reactor the intensity of the entropy production σ_s follows in terms of the reservoir parameters and system fluxes

$$\sigma_s = \frac{q_2}{T_2} - \frac{q_1}{T_1} + (s_2 - s_1)n \tag{9.59}$$

Using in this equation the energy conservation law

$$q_1 + h_1 n = q_2 + h_2 n + p \tag{9.60}$$

to eliminate flux q_2 we obtain

$$\sigma_s = \frac{q_1 + h_1 n - h_2 n - p}{T_2} - \frac{q_1}{T_1} + (s_2 - s_1)n \tag{9.61}$$

whence, in terms of fluxes q_1 and n

$$\sigma_s = q_1 \left(\frac{1}{T_2} - \frac{1}{T_1} \right) + \frac{(\mu_1 - \mu_2)n + (T_1 - T_2)s_1 n}{T_2} - \frac{p}{T_2}$$

$$= \frac{(\eta_C - \eta)q_1 + (\beta_C - \beta)n}{T_2}$$

$$(9.62)$$

Equations (9.41) and (9.44) yield in terms of other fluxes and corresponding efficiencies:

$$\sigma_s = Q_1 \left(\frac{1}{T_2} - \frac{1}{T_1} \right) + \frac{(\mu_1 - \mu_2)n}{T_2} - \frac{p}{T_2} = \frac{(\eta_C - \eta)Q_1 + (\zeta_C - \zeta)n}{T_2}$$

$$= \frac{(\eta_C - \eta)Q_1 + (\omega_C - \omega)G_1}{T_2}$$

$$(9.63)$$

Equations (9.62) and (9.63) generalize to the nonisothermal case an earlier result limited to isothermal situations (Sieniutycz, 2007b). They show that, modulo the multiplier $(T_2)^{-1}$, the isothermal component of the entropy production is the product of the reactant flux and the deviation of chemical efficiency ζ from the corresponding efficiency of the reversible process, ζ_C. Yet, as shown by Equation (9.63), the entropy production formula of a nonisothermal process contains the second (thermal) component equal to the product of total (i.e., mass transfer including) heat flux $Q_1 \equiv q_1 + T_1 s_1 n$ and the deviation of thermal efficiency η from the Carnot efficiency.

The subscript C points out that the reversible efficiencies η_C, ζ_C, and ω_C refer to the so-called Carnot point of the system, also called "open circuit point" (de Vos, 1992). This point is associated with vanishing currents and upper, reversible limits for thermal and chemical efficiencies.

9.11. DISSIPATIVE AVAILABILITIES IN DYNAMIC SYSTEMS

As resources are finite by definition, in real systems only the second fluid constitutes an infinite constant reservoir (e.g., an environment), whereas the first fluid (a resource) changes its properties when it proceeds through stages in time. We have to attribute changes in state variables to each considered flux.

For the mass flux, which equals $G_1{'}/\mu_1{'}$, the suitable state coordinate is the invariant amount of reacting substance $N = N_1{'} = N_2{'}$ or the concentration $X = N/I$. For the entropy flux, $Q_1{'}/T_1{'}$, the state variable is the entropy S. In the formulae below we use the symbols S and T for the variable entropy $S_1(t)$ and temperature $T_1(t)$ in the bulk of the resource fluid. For an endoreversible process the yield of the specific work at flow in terms of the Carnot controls is

$$W \equiv \frac{P}{I} = - \int_{T^i, \mu^i}^{T^f, \mu^f} \{(T' - T_2)dS + (\mu' - \mu_2)dN\}$$

$$= - \int_{T^i, \mu^i}^{T^f, \mu^f} \{(T - T_2)dS + (\mu - \mu_2)dN\} \qquad (9.64)$$

$$- \int_{T^i, \mu^i}^{T^f, \mu^f} \{(T' - T)dS + (\mu' - \mu)dN\}$$

The first integral in the second line of this equation describes the reversible work, whereas the second integral is the lost work. When the subscript 2 refers to the state of environment, Equation (9.64) leads to a generalized or finite time exergy $A = \max W$ satisfying the formula

$$A = \max W \equiv \max \left(\frac{P}{I} \right) = \max \int_{T^i, \mu^i}^{T^f, \mu^f} \{-(T - T_2)dS - (\mu - \mu_2)dN\}$$

$$= H^i - H^f - T_2(S^i - S^f) - \mu_2(N^i - N^f) \qquad (9.65)$$

$$-\min \int_{T^i, \mu^i}^{T^f, \mu^f} \{(T' - T)dS + (\mu' - \mu)dN\}$$

Equation (9.65) expresses—in terms of Carnot controls—the Gouy–Stodola law for the endoreversible system (Sieniutycz, 1997a). In terms of the Carnot controls the thermodynamic form of the lost work expression (without kinetics incorporated) is classical, and it is described by the sum of products of thermodynamic fluxes and forces

$$T_2 S_\sigma = \int_{T^i}^{T^f} \{(T' - T)\dot{S} + (\mu' - \mu)\dot{N}\}d\tau \qquad (9.66)$$

As shown by Equation (9.65) the maximum work produced by the engine equals to the change of the classical exergy reduced by the lost work or the product of the reservoir's temperature and the minimum entropy production. This expresses, of course, the Gouy–Stodola law that links the lost work with the entropy generation. An analogous equation is obtained for the power consumer, but then the effect of entropy production is added to the classical exergy change, that is, an increase of work input is necessary to assure the required state change in a finite time.

Equation (9.65) defines the "endoreversible" limit for the work production accomplished between two given states and for a given number of transfer units. Even this simple limit is stronger than the one predicted by the classical exergy. What can be said about a yet stronger limit that involves an internal dissipation in the participating thermal machine? We need to consider the hierarchy of limits. For limits of higher rank, an internal entropy generation is included in the dissipation model and then Equation (9.65) is replaced by its simple generalization that contains the sum of the "endoreversible" and "internal" productions of the entropy, $S_\sigma^{endo} + S_\sigma^{int}$, in agreement with the Gouy–Stodola law. In general, extra components of total entropy source are included at the expense of a more detailed information input, but with the advantage that the limit is closer to reality.

For a sufficiently high rank of the limit, it approaches the real work quite closely, but the cost of the related information is very large. A proper compromise associated with the accepted limit of a finite rank is important. For limits of various ranks, inequalities are related to A and real work W_{real} that are valid in the form $W_{real} > \Delta A_k > \Delta A_1 > \Delta A_0$, where ΔA_1 refers to the change of "endoreversible exergy" and ΔA_0 pertains to the change of the classical exergy. The classical exergy change constitutes then the weakest or the worst standardized limit on the real work. In the described scheme any considerations of relations between the irreversibility and costs are unnecessary.

9.12. CHARACTERISTICS OF STEADY ISOTHERMAL ENGINES

Expressing chemical potentials

$$\mu_k = \mu_{0k} + RT \ln x_k = \mu_{0k} + RT \ln \left(\frac{X_k}{1 + X_k} \right) \tag{9.67}$$

in terms of molar fraction of the active component in "upper" and "lower" parts of the system ($k = 1, 2$), we obtain the affinity-related efficiency (9.53) in the form

$$\zeta = \mu_{1'} - \mu_{2'} = \zeta_0 + RT \ln \left(\frac{x_{1'}}{x_{2'}} \right) \equiv \zeta_0 + RT \ln \left(\frac{X_{1'}(1 + X_{2'})}{X_{2'}(1 + X_{1'})} \right) \tag{9.68}$$

where

$$\zeta_0 = \mu_{01'} - \mu_{02'} = \mu_{01} - \mu_{02} \tag{9.69}$$

The last equation applies the property of chemical passivity for each component in the chemically inactive parts of the system, where only diffusive transport takes place. The equation describes, in fact, the standard Gibbs energy for the isomerization reaction considered. The constant ζ_0 involves chemical potentials of substrate and product in their reference states. The reference value ζ_0 vanishes only if both components are identical. In general, however, the constant ζ_0 is nonvanishing.

When both reservoirs are infinite, the process is at the steady state. The mass transfer between each reservoir and the production section of the system is described by certain kinetic equations. For simplicity we assume that these equations are linear. The mass balances for the substance transferred, produced, and consumed are contained in the equations

$$n = g_1(x_1 - x_{1'}) \tag{9.70}$$

and

$$n = g_2(x_{2'} - x_2) \tag{9.71}$$

In order to determine work characteristics of the chemical engine at the steady state (unlimited stock of fuel) one searches for concentrations $x_{1'}$ and $x_{2'}$ expressed in terms of a control variable. For the chemical engine a suitable quantity can be efficiency ζ (Equations (9.68) and (9.69)). This means that the following system of equations should be solved:

$$\zeta = \zeta_0 + RT \ln \left(\frac{x_{1'}}{x_{2'}} \right) \tag{9.72}$$

$$g_1(x_1 - x_{1'}) = g_2(x_{2'} - x_2) = n \tag{9.73}$$

From the first equation of this set we find

$$\frac{x_{2'}}{x_{1'}} = \exp \left(-\frac{\zeta - \zeta_0}{RT} \right) \tag{9.74}$$

Substituting $x_{2'}$ from this equation to the first equality of Equation (9.73) written in the form

$$g_1 x_{1'} + g_2 x_{2'} = g_1 x_1 + g_2 x_2 \tag{9.75}$$

one obtains

$$x_{1'} = \frac{g_1 x_1 + g_2 x_2}{g_1 + g_2 \exp(-(\zeta - \zeta_0)/RT)} \tag{9.76}$$

and

$$\begin{aligned} x_{2'} &= \frac{g_1 x_1 + g_2 x_2}{g_1 + g_2 \exp(-(\zeta - \zeta_0)/RT)} \exp \left(-\frac{\zeta - \zeta_0}{RT} \right) \\ &= \frac{g_1 x_1 + g_2 x_2}{g_1((\zeta - \zeta_0)/RT) + g_2} \end{aligned} \tag{9.77}$$

Each of the last two equations can be used in the balance-kinetic formula (9.73). This leads to an equation describing the feed flux of the active component of fuel in terms of chemical efficiency

$$n = g_1 x_1 - g_1 x_{1'} = g_1 x_1 - g_1 \frac{g_1 x_1 + g_2 x_2}{g_1 + g_2 \exp(-(\zeta - \zeta_0)/RT)} \tag{9.78}$$

This equation can be yet simplified into the form

$$\begin{aligned} n = g_1(x_1 - x_{1'}) &= g_1 g_2 \frac{x_1 \exp(-(\zeta - \zeta_0)/RT) - x_2}{g_1 + g_2 \exp(-(\zeta - \zeta_0)/RT)} \\ &= g_1 g_2 \frac{x_1 - x_2 \exp((\zeta - \zeta_0)/RT)}{g_1 \exp((\zeta - \zeta_0)/RT) + g_2} \end{aligned} \tag{9.79}$$

Thus, the feed rate of the system by the active reactant can be described by two equivalent expressions

$$n = \frac{x_1 - x_2 \exp((\zeta - \zeta_0)/RT)}{(g_1)^{-1} + (g_2)^{-1}\exp((\zeta - \zeta_0)/RT)} \tag{9.80}$$

or

$$\frac{n}{g_1} = \frac{x_1 - x_2 \exp((\zeta - \zeta_0)/RT)}{1 + (g_1/g_2)\exp((\zeta - \zeta_0)/RT)} \tag{9.81}$$

The function inverse to the above defines the chemical efficiency in terms of the reactant's feed rate n

$$\zeta = \zeta_0 + RT \ln\left(\frac{x_1 - ng_1^{-1}}{ng_1^{-1}(g_1/g_2) + x_2}\right) \tag{9.82}$$

This equation shows that an effective concentration of the reactant in the upper reservoir $x_{1\mathrm{eff}} = x_1 - g_1^{-1}n$ is decreased, whereas an effective concentration of the product in the lower reservoir $x_{2\mathrm{eff}} = x_2 - g_2^{-1}n$ is increased due to the finite mass flux. Consequently, the efficiency ζ decreases nonlinearly with n. When the effect of the resistance g_k^{-1} is ignorable or the flux n is very small, reversible efficiency, ζ_C, is attained. Quite generally the power function described by the product $\zeta(n)n$ exhibits the maximum power for a finite value of the flux n.

Equations (9.34) and (9.80) yield power in terms of efficiency ζ of an isothermal process

$$p = (\mu_{1'} - \mu_{2'})n = \zeta\frac{x_1 - x_2 \exp((\zeta - \zeta_0)/RT)}{(g_1)^{-1} + (g_2)^{-1}\exp((\zeta - \zeta_0)/RT)} \tag{9.83}$$

This power function exhibits a maximum for a certain efficiency, ζ, this efficiency being a chemical analogue of the well-known Chambadal–Novikov–Curzon–Ahlborn efficiency (CNCA efficiency) (de Vos, 1992; Chen et al., 2001). The maximum power can be considered with respect to control variables n and ζ, for example. Yet, other control variables can be considered (Figure 9.5).

Similarly like in heat processes, for which the Carnot temperature was applied as a suitable control (Sieniutycz, 1999b, 2003a, 2003b, 2003c), we can control the considered mass transfer process by using the Carnot chemical potential, μ'. For an isothermal process Carnot chemical potential μ' may be obtained from the invariance of the chemical efficiency

$$\mu_{1'} - \mu_{2'} = \mu' - \mu_2 \tag{9.84}$$

Therefore

$$\mu' = \mu_2 + \mu_{1'} - \mu_{2'} = \mu_2 + \zeta \tag{9.85}$$

The structure of μ' may be more involved in nonisothermal processes when the classical definition of heat flux is applied. However, for the *generalized* heat flux, $Q_{1'} \equiv q_{1'} + T_{1'}s_{1'}n$, Carnot chemical potential μ' represented by Equation

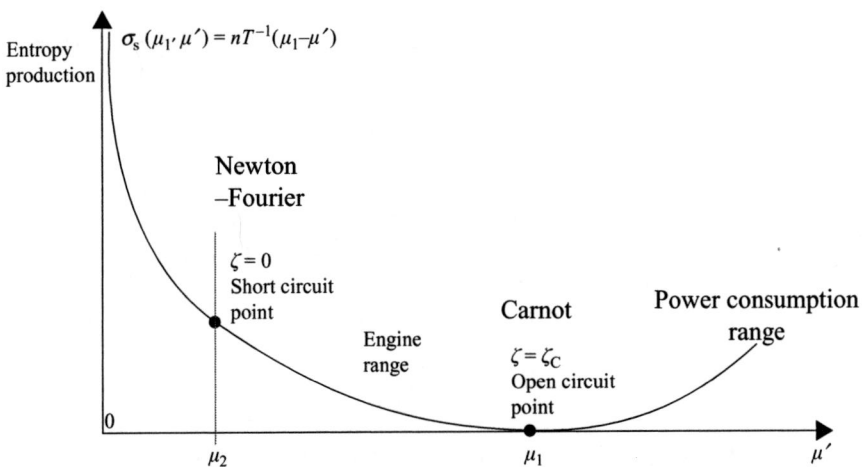

Figure 9.5 Entropy generation in the engine in terms of Carnot chemical potential μ' as a control variable.

(9.85) remains valid even in nonisothermal cases. The chemical efficiency of any (isothermal or not) process of energy generation expressed in terms of μ' is

$$\zeta = \frac{p}{n} = \mu' - \mu_2 \tag{9.86}$$

The quantity μ' can thus be a process control suitable in considerations of the connection between the rate change of the resource potential and the process driving force. Whenever $\mu' = \mu_1$, the process rate vanishes, and equations of classical thermodynamics are sufficient to characterize the system. For $\mu' < \mu_1$ the driving reactant (fuel) is consumed in the upper reservoir, its chemical potential μ_1 decreases gradually, and the system produces power. In the opposite case, when $\mu' > \mu_1$, the system consumes power for the purpose of the resource utilization, or to increase its chemical potential. These properties are valid regardless of particular structure of kinetic equations (linear or not).

At the Carnot point (also called "open circuit point of the system"; de Vos, 1992), rates and power vanish, and the system efficiency ζ attains its upper limit. Only then is this efficiency identical with the reversible chemical affinity of the reaction. For our model

$$\zeta_C = \frac{p}{n} = \mu_1 - \mu_2 \tag{9.87}$$

Putting Equation (9.80) or (9.81) to zero yields

$$x_1 = x_2 \exp\left(\frac{\zeta_C - \zeta_0}{RT}\right) \tag{9.88}$$

This formula leads to the limiting reversible efficiency in the form determined by the chemical affinity of reaction at the reversible Carnot point

$$\zeta_C = \zeta_0 + RT \ln \left(\frac{x_1}{x_2} \right) \equiv A_C \tag{9.89}$$

However, the power produced at the "open circuit" state equals zero, corresponding with vanishing feed flux of the driving reactant.

Intensity of the entropy generation in the system (Equation (9.63)) can be presented in several forms. For the isothermal system considered three forms of σ_s can be discussed. In the first, the controlling quantity is the efficiency ζ

$$\sigma_s = \frac{(\zeta_C - \zeta)n}{T} = \frac{\zeta_C - \zeta}{T} \left(\frac{x_1 - x_2 \exp((\zeta - \zeta_0)/RT)}{(g_1)^{-1} + (g_2)^{-1} \exp((\zeta - \zeta_0)/RT)} \right) \tag{9.90}$$

In the second, the control variable is Carnot chemical potential

$$\begin{aligned} \sigma_s &= \frac{((\mu_1 - \mu_2) - (\mu' - \mu_2))n}{T} \\ &= \frac{\mu_1 - \mu'}{T} \frac{x_1 - x_2 \exp((\mu' - \mu_2 - \zeta_0)/RT)}{(g_1)^{-1} + (g_2)^{-1} \exp((\mu' - \mu_2 - \zeta_0)/RT)} \end{aligned} \tag{9.91}$$

In the third form the control is the fuel flux n; then, from Equation (9.82):

$$\sigma_s = \frac{(\zeta_C - \zeta)n}{T} = R \ln \left(\left(\frac{x_1}{x_2} \right) \left(\frac{ng_2^{-1} + x_2}{x_1 - ng_1^{-1}} \right) \right) \tag{9.92}$$

In the so-called "short circuit point," there is no power production for any value of ζ_0, despite possible chemical reaction. Only entropy is then produced with an intensity

$$\begin{aligned} \sigma_{s_{zw}} &= \frac{\zeta_C}{T} \left(\frac{x_1 - x_2 \exp(-\zeta_0/RT)}{(g_1)^{-1} + (g_2)^{-1} \exp(-\zeta_0/RT)} \right) \\ &= \frac{\mu_1 - \mu_2}{T} \frac{x_1 - x_2 \exp(-\zeta_0/RT)}{(g_1)^{-1} + (g_2)^{-1} \exp(-\zeta_0/RT)} \end{aligned} \tag{9.93}$$

that is the maximum intensity of the engine range. When $\zeta_0 = 0$ the entropy production at the "short circuit point" corresponds with the situation without chemical reaction and energy generation

$$\begin{aligned} (\sigma_s)_{\zeta=0} &= \frac{\zeta_C n_{sc}}{T} = \frac{\zeta_C}{T} \left(\frac{x_1 - x_2}{(g_1)^{-1} + (g_2)^{-1}} \right) \\ &= T^{-1} \frac{(\mu_1 - \mu_2)(x_1 - x_2)}{(g_1)^{-1} + (g_2)^{-1}} \end{aligned} \tag{9.94}$$

Clearly, only at the short circuit point and for the associated absence of power the entropy production at this point is proportional to the product of the reaction rate and its chemical affinity. Of course, the proportionality coefficient is the temperature reciprocal, T^{-1}. This is the classical result, which, however, does not hold when the system produces power (i.e., belongs to the class of "active systems").

Assuming a finite ζ_0, we determine the rate of the system feed by the active reactant in two alternative forms. The first one describes the feed flux n in terms of ζ, whereas the second—in terms of μ'

$$
\begin{aligned}
n &= \frac{x_1 - x_2 \exp((\zeta - \zeta_0)/RT)}{(g_1)^{-1} + (g_2)^{-1}\exp((\zeta - \zeta_0)/RT)} \\
&= \frac{x_1 - x_2 \exp((\mu' - \mu_2 - \zeta_0)/RT)}{(g_1)^{-1} + (g_2)^{-1}\exp((\mu' - \mu_2 - \zeta_0)/RT)}
\end{aligned}
\tag{9.95}
$$

Power produced in the reactor may be determined in three forms, as a function of ζ, in terms of μ', or as a function of n. The first two forms are described by an equation

$$
\begin{aligned}
p &= (\mu_{1'} - \mu_{2'})n = \zeta\frac{x_1 - x_2 \exp((\zeta - \zeta_0)/RT)}{(g_1)^{-1} + (g_2)^{-1}\exp((\zeta - \zeta_0)/RT)} \\
&= \frac{(\mu' - \mu_2)(x_1 - x_2 \exp((\mu' - \mu_2 - \zeta_0)/RT))}{(g_1)^{-1} + (g_2)^{-1}\exp((\mu' - \mu_2 - \zeta_0)/RT)}
\end{aligned}
\tag{9.96}
$$

Using inverse function $\zeta(n)$ (Equation (9.82)), one may describe power in terms of the reagent flux n

$$
p = \zeta(n)n = \zeta_0 n + RTn \ln\left(\frac{x_1 - ng_1^{-1}}{ng_2^{-1} + x_2}\right)
\tag{9.97}
$$

Equation (9.82) proves that efficiency ζ decreases nonlinearly with feed flux n. The consequence of this property is the maximum of the product $\zeta(n)n$ describing the power yield. The location of the maximum power point is determined by maximization of power with respect to a free control variable, ζ, μ', or n. Analytical methods are seldom effective. However, one can use diagrams describing functions $p(\zeta)$, $p(\mu')$, or $p(n)$ to determine the maximum point graphically. Information regarding the location of maximum point can also be obtained by a numerical search.

System characteristics in terms of control μ' are presented in Figure 9.6.

Equation (9.97) shows that a maximum of power p is attained for a fuel flux n satisfying an equation

$$
\frac{dp}{dn} = \zeta_0 + RT \ln\left(\frac{x_1 - ng_1^{-1}}{ng_2^{-1} + x_2}\right) - RTn\frac{g_1^{-1}x_2 + g_2^{-1}x_1}{(x_1 - ng_1^{-1})(ng_2^{-1} + x_2)} = 0
\tag{9.98}
$$

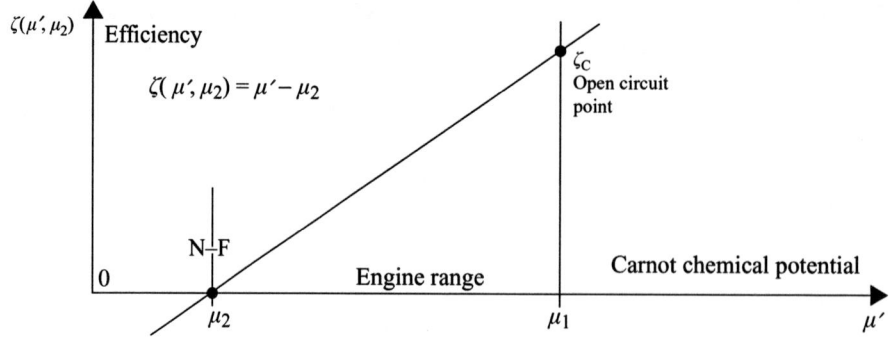

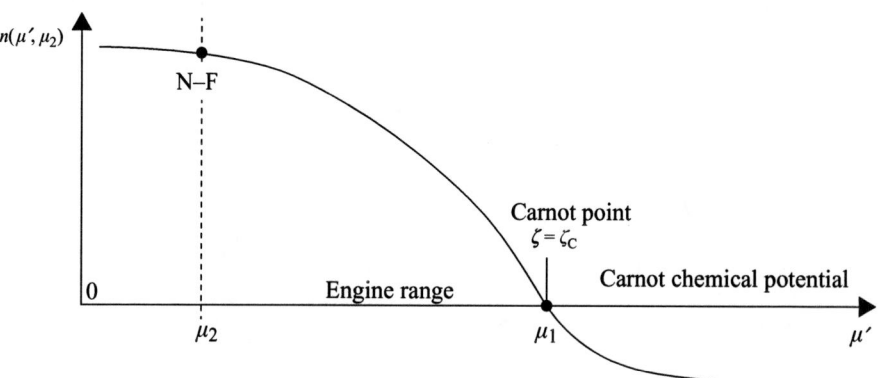

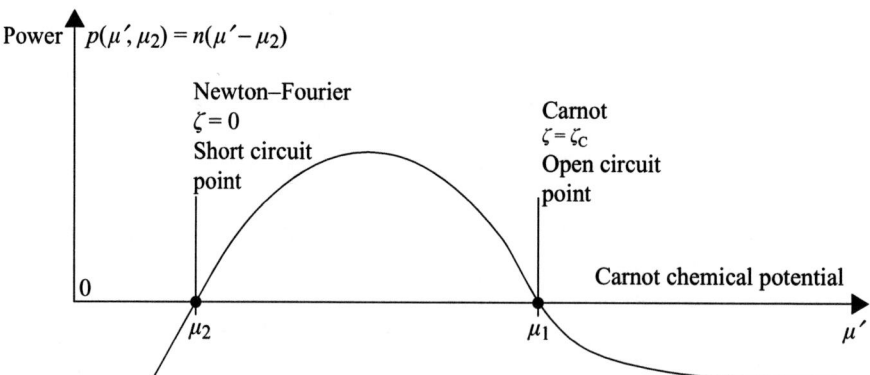

Figure 9.6 Engine characteristics: efficiency, driving flux, and power yield vs. Carnot chemical potential μ'.

Its numerical solution generates a function describing the optimal feed of the system by the active reactant

$$n_{mp} = f(\zeta_0, T, x_1, x_2, g_1, g_2) \tag{9.99}$$

Substituting this result into Equation (9.82) leads to the chemical counterpart of the CNCA efficiency (de Vos, 1992; Chen et al., 2001). It describes the efficiency ζ_{mp} of the chemical engine at the maximum power point in terms of the system properties

$$\zeta_{mp} = f(\zeta_0, T, x_1, x_2, g_1, g_2) \tag{9.100}$$

The curve of produced power has two zero points. The first is the "short circuit point" (also the point of vanishing efficiency), and the second is the "open circuit point" or Carnot point at which the feed of the system with the active component is infinitely slow. Equation (9.80) proves that the feed rate of the active reagent at the short circuit point (where $\zeta = 0$) is

$$n_{sc} = \frac{x_1 - x_2 \exp(-\zeta_0/RT)}{(g_1)^{-1} + (g_2)^{-1}\exp(-\zeta_0/RT)} \tag{9.101}$$

For a nonvanishing $\zeta_0 = \mu_{01'} - \mu_{02'} = \mu_{01} - \mu_{02}$, the short circuit point corresponds with a particular, "purely dissipative" state of the system at which lossy elements predominate so significantly that the chemical reaction does not produce any power, despite nonvanishing reaction rate and existing finite potential of ζ_0 (but not ζ) for the production of this power. For $\zeta_0 = 0$ the discussed equation describes the situation in which the reactor does not exist, the fuel stream flows by two sequentially connected conductors, and molar flux of the reactant, n, is determined by the specification of the overall conductance $g = ((g_1)^{-1} + (g_2)^{-1})^{-1}$. Consequently, for $\zeta_0 = 0$ and at the short circuit point of the system

$$n_{sc} = \frac{x_1 - x_2}{(g_1)^{-1} + (g_2)^{-1}} \tag{9.102}$$

9.13. SEQUENTIAL MODELS FOR DYNAMIC POWER GENERATORS

When resources are limited, their quality decreases in time, and dynamic processes arise. Power optimization requires then variational methods to maximize power integrals subject to process differential constraints. Taking into account analytical difficulties, we often apply methods of discrete optimization, for example, the dynamic programming method. An extremum problem for a power integral is then broken down into an optimization problem for a cascade with a finite number of stages.

9.13.1. Discrete Models

Regarding computer needs, we shall now introduce a suitable discrete model of the chemical engine. We consider the optimal fuel consumption in a cascade of K engines, with efficiency ζ^k or local feed n^k as control variables at kth stage. We also test Carnot chemical potentials μ'^k as possible controls. To describe fuel consumption we exploit the mass balance. Some suitable notions are introduced first.

Let us introduce a cumulative flux of active reactant over the first k stages of the cascade, $N^k = \sum n^l$, where $l = 1, 2, \ldots, k$. The sequence of local fluxes n^l (characterized by "upper" potentials μ_1^k in the kth stage engine) describes allocations of N^k between stages $1, 2, \ldots, k$. Each local flux n^k of the active reagent at stage k equals the change of cumulative mass flux $N^k - N^{k-1}$.

The mass balance at the stage k

$$n^k \equiv N^k - N^{k-1} = -I(X_1^k - X_1^{k-1}) = -Ic_m(\mu_1^k - \mu_1^{k-1}) \qquad (9.103)$$

shows that the local molar flux at the stage k (an interval of the cumulative mass flux N^k) can be evaluated as the (negative) product of molar flux of inert, I, and the change of reactant's concentration, ΔX^k. Mass capacity c_m can be introduced as the partial derivative of the concentration X with respect to the chemical potential at the constant temperature. Then, in the case of an isothermal mass exchange, any change in the cumulative coordinate N can be evaluated in terms of the reactant's chemical potential.

The reactant's flux n^k can be eliminated on account of difference $X_1^k - X_1^{k-1}$ with the help of Equation (9.103). Simultaneously it is convenient to have at our disposal a separate expression for efficiency ζ^k in terms of concentrations X_1^k and X_1^{k-1}. Both these needs are satisfied in one stroke below. Mass balance (9.103) and the current–efficiency characteristic (9.81) yield an expression

$$-I(X_1^k - X_1^{k-1}) = N_1^k - N_1^{k-1}$$
$$= \frac{x_1 - x_2 \exp((\zeta - \zeta_0)/RT)}{1 + (g_1/g_2)\exp((\zeta - \zeta_0)/RT)} \beta_i'^k (A^k - A^{k-1}) \qquad (9.104)$$

where $\beta'_1 \equiv \beta_1 y = \beta_1 a_1/(a_1 + a_2)$ and $\beta_1'^k(A^k - A^{k-1}) = g_1^k$, whence

$$-\frac{I(X_1^k - X_1^{k-1})}{\beta_1'^k(A^k - A^{k-1})} \equiv \frac{n^k}{g_1^k} = \frac{x_1 - x_2 \exp((\zeta - \zeta_0)/RT)}{1 + (g_1/g_2)\exp((\zeta - \zeta_0)/RT)} \qquad (9.105)$$

Since the inversion of Equation (9.81) or the right-hand side of Equation (9.105) is Equation (9.82):

$$\zeta = \zeta_0 + RT \ln \left(\frac{x_1 - ng_1^{-1}}{ng_1^{-1}(g_1/g_2) + x_2} \right) \qquad (9.82)$$

the chemical efficiency ζ^k in terms of X_1^k and X_1^{k-1} is

$$\zeta^k = \zeta_0^k + RT \ln \left(\frac{(X_1^k/(1 + X_1^k)) + ((I(X_1^k - X_1^{k-1}))/(\beta_1'^k a_v^k F^k(l^k - l^{k-1})))}{-((I(X_1^k - X_1^{k-1}))/(\beta_1'^k a_v^k F^k(l^k - l^{k-1})))(g_1/g_2) + x_2} \right)$$

(9.106)

In the engine mode the concentration of the active reactant can only decrease along a path; thus, the term with the discrete slope $\Delta X^k/\Delta l$ in Equation (9.106) is negative. Consequently, the efficiency of a stage working in the engine mode is lower than the Carnot efficiency.

The quantity

$$\frac{I}{\beta'_1 a_v F} \equiv H_{TU_1}$$

(9.107)

has units of length, and is known from the mass transfer theory as the "height of the mass transfer unit" (H_{TU}). In Equation (9.107) it is referred to partial mass transfer coefficient of active reactant, β'_1, although an analogous quantity could be defined for the product. The nondimensional length

$$\tau_1 \equiv \frac{l}{H_{TU_1}}$$

(9.108)

is identical with the "number of transfer units" N_{TU} for mass transfer. Since it is proportional to the system's extent l and hence to the contact time of active reactant with mass exchange area, it also plays the role of a nondimensional time, and this is why it is designated by τ_1

$$\frac{p^k}{I} = -\left\{ \zeta_0 + RT \ln \left(\frac{(X_1^k/(1 + X_1^k)) + ((X_1^k - X_1^{k-1})/(\tau_1^k - \tau_1^{k-1}))}{x_2 - j((X_1^k - X_1^{k-1})/(\tau_1^k - \tau_1^{k-1}))} \right) \right\}$$
$$(X_1^k - X_1^{k-1})$$

(9.109)

The conductance ratio is defined as coefficient $j \equiv g_1/g_2$.

The total power delivered from the N-stage process per unit flux of the inert is the sum of contributions of stages. This sum is a discrete functional that is maximized by the suitable choice of the interstage concentrations and allocation of time intervals between the stages:

$$W^N \equiv \sum_{k=1}^{N} w^k$$

$$= -\sum_{k=1}^{N} \left\{ \zeta_0 + RT \ln \left(\frac{(X_1^k/(1 + X_1^k)) + ((X_1^k - X_1^{k-1})/(\tau_1^k - \tau_1^{k-1}))}{x_2 - j((X_1^k - X_1^{k-1})/(\tau_1^k - \tau_1^{k-1}))} \right) \right\}$$
$$(X_1^k - X_1^{k-1})$$

(9.110)

9.13.2. Continuous Models

Models of continuous systems producing power are obtained as limits of models of suitable discrete systems for number of stages approaching infinity. An example is the continuous limit describing an integral of total power per unit molar flow of the inert

$$W = -\int_{\tau_1^i}^{\tau_1^f} \left\{ \zeta_0 + RT \ln \left(\frac{(X/(1+X)) + (dX/d\tau_1)}{x_2 - j(dX/d\tau_1)} \right) \right\} \frac{dX}{d\tau_1} d\tau_1 \quad (9.111)$$

The continuous variable X plays here the same role as the discrete variable X_1^k in Equation (9.110). Equations (9.110) and (9.111) represent, respectively, discrete and continuous Lagrange descriptions of the optimized process.

We can also apply controls that are certain state variables, and have no interpretation of rates. For example, using Carnot chemical potential $\mu' = \mu_2 + \zeta$ in integral (9.111), we obtain

$$W = -\int_{\tau_1^i}^{\tau_1^f} \zeta \frac{dX}{d\tau_1} d\tau_1 = -\int_{\tau_1^i}^{\tau_1^f} (\mu' - \mu_2) \frac{dX}{d\tau_1} d\tau_1 \quad (9.112)$$

where μ' satisfies an equation

$$\mu' = \mu_2 + \zeta_0 + RT \ln \left(\frac{(X/(1+X)) + (dX/d\tau_1)}{x_2 - j(dX/d\tau_1)} \right) \quad (9.113)$$

Its solution with respect to the time derivative yields

$$\frac{dX}{d\tau_1} = -\frac{(X/(1+X)) - x_2 \exp((\mu' - \mu_2 - \zeta_0)/RT)}{1 + j \exp((\mu' - \mu_2 - \zeta_0)/RT)} \quad (9.114)$$

Using in this formula an expression linking molar fraction $x = X/(1+X)$ with chemical potential

$$\frac{X}{1+X} = \exp \frac{\mu - \mu_{0_1}}{RT} \quad (9.115)$$

and introducing mass capacity $c_m(\mu)$

$$c_m(\mu) \equiv \frac{dX}{d\mu} = \frac{1}{RT} \frac{\exp((\mu - \mu_{0_1})/RT)}{(1 - \exp((\mu - \mu_{0_1})/RT))^2} \quad (9.116)$$

we obtain a nonlinear equation of process dynamics in terms of the chemical potential of the active component of fuel

$$c_m(\mu) \frac{d\mu}{d\tau_1} = -\frac{\exp((\mu - \mu_{0_1})/RT) - x_2 \exp((\mu' - \mu_2 - \zeta_0)/RT)}{1 + j \exp((\mu' - \mu_2 - \zeta_0)/RT)}$$

$$(9.117)$$

A corresponding power formula has the form

$$W = \int_{\tau_1^i}^{\tau_1^f} (\mu' - \mu_2) \frac{\exp((\mu - \mu_{0_1})/RT) - x_2 \exp((\mu' - \mu_2 - \zeta_0)/RT)}{1 + j \exp((\mu' - \mu_2 - \zeta_0)/RT)} d\tau_1$$

(9.118)

Dynamic optimization tackles integral (9.118) subject to differential constraint (9.117). In a chemical problem, the mathematical structure of the power integral and constraint(s) is more complicated than in the heat problem. In the latter the use of the Carnot temperature assures a simplest mathematical structure of the problem (Sieniutycz, 1997c, 2000c, 2000d; Kubiak, 2005; Kuran, 2006). This also substantiates the use of numerical methods.

9.14. A COMPUTATIONAL ALGORITHM FOR DYNAMIC PROCESS WITH POWER MAXIMIZATION

We shall now outline a computational algorithm in which the discrete problem of maximum power is treated as an optimal control problem, and a suitable control variable is chosen to assure a relatively simple model. The simplicity condition is satisfied by a simple link of control with rate change of reactant's concentration in the fuel, $dX/d\tau_1$. In the discrete version this rate is replaced by the difference ratio $\Delta X/\Delta \tau_1$.

We introduce a control variable describing the fuel consumption

$$v^k \equiv -\frac{n^k}{g_1}$$

(9.119)

It is negative in engine modes and positive in power consumption modes. For the process investigated, we find

$$v \equiv -\frac{n}{g_1} = -\frac{N^k - N^{k-1}}{\gamma_1^k - \gamma_1^{k-1}} = \frac{I(X_1^k - X_1^{k-1})}{\beta_1'^k(A_1^k - A_1^{k-1})} = \frac{X_1^k - X_1^{k-1}}{\tau_1^k - \tau_1^{k-1}}$$

(9.120)

$$u \equiv -\frac{q_1}{g_1} = -\frac{Q^k - Q^{k-1}}{\gamma_1^k - \gamma_1^{k-1}} = \frac{I_c(T_1^k - T_1^{k-1})}{\alpha_1'^k(A_1^k - A_1^{k-1})} = \frac{T_1^k - T_1^{k-1}}{\tau_1^k - \tau_1^{k-1}}$$

(9.121)

The optimization problem searches for a maximum of the performance index

$$W^N = -\sum_{k=1}^{N} \left\{ \zeta_0 + RT^k \ln \left(\frac{X_1^k(1 + X_1^k)^{-1} - v^k}{x_2 - jv^k} \right) \right\} v^k \theta^k$$

(9.122)

($j \equiv g_1/g_2$) subject to difference constraints

$$X_1^k - X_1^{k-1} = v^k \theta^k$$

(9.123)

$$\tau^k - \tau^{k-1} = \theta^k$$

(9.124)

In a nonisothermal problem an extra equation has to be included to treat temperature changes

$$T^k - T^{k-1} = u^k \theta^k \tag{9.125}$$

and the power function has to be enlarged to include the thermal component of power yield. Some information related to the thermal component is given in Section 9.17.

Change in the sign of performance function (9.122) transforms the problem into a discrete problem of power minimization, whose numerical solution is outlined below.

We apply the method of dynamic programming that searches for a solution of Bellman's recurrence equation. A general form of this equation is

$$R^n(\mathbf{x}^n, t^n) = \min_{\mathbf{u}^n, \theta^n} \{l_0^n(\mathbf{x}^n, t^n, \mathbf{u}^n, \theta^n)\theta^n$$

$$+ R^{n-1}(\mathbf{x}^n - \mathbf{f}^n(\mathbf{x}^n, t^n, \mathbf{u}^n, \theta^n)\theta^n, t^n - \theta^n)\} \tag{9.126}$$

where $R^n(\mathbf{x}^n, t^n) = \min(-W^n)$ is the function describing the minimum of power consumed. This is a function of optimal cost type. In an isothermal case $\mathbf{x} = X_1$, $\mathbf{u} = v$, and $t = \tau$.

Applying Equation (9.126) to the isothermal problem described by Equations (9.122)–(9.124), the following recurrence equation is obtained:

$$R^k(X^k, \tau^k) = \min_{\mathbf{u}^k, \theta^k} \left\{ \left\{ \zeta_0 + RT\ln\left(\frac{X_1^k(1 + X_1^k)^{-1} - v^k}{x_2 - jv^k}\right) \right\} v^k \theta^k \right.$$

$$\left. + R^{n-1}((X_1^k - \theta^n v^k, \tau^k - \theta^k) \right\} \tag{9.127}$$

The forward dynamic programming algorithm corresponding with this equation is associated with the optimality principle interpreted in Figure 9.7.

While the analytical solving of the discrete problem (9.122)–(9.124) is a difficult task, it is quite easy to solve recurrence equation (9.127) numerically. Low dimensionality of state vector in Equation (9.127) assures a decent accuracy of DP solution. Moreover, an original accuracy can significantly be improved after performing the so-called dimensionality reduction associated with the elimination of time t^k as the state variable by using a Lagrange multiplier λ. In the transformed problem, without coordinate t^k, accuracy of DP solutions is high. Section 2.3 of a review (Sieniutycz, 2000c) discusses related computational issues with more detail.

The block scheme of the computational process is also available (Sieniutycz, 2007b), where the system state is represented by the concentration of an active component of fuel, X_1^k, as in Equation (9.127).

For a continuous isothermal process with constant coefficients g_1, g_2, j, etc., an analytical condition associated with the power optimum or Equation (9.127) can be determined in the form of constancy of an energy-like function along

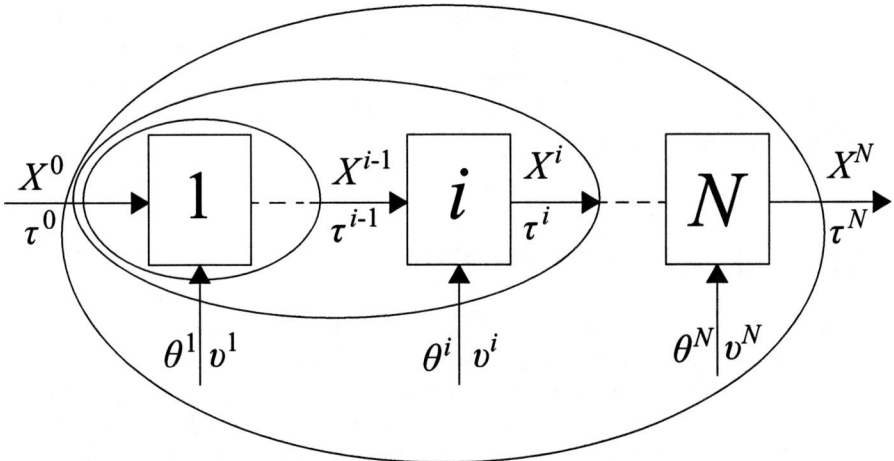

Figure 9.7 Principle of multistage power generation in a dynamic chemical engine. The system stage comprises: fuel mixture, engine, and environment; at each stage mechanical power is generated. Maximized is total flux of resulting power. Control θ^n characterizes stage extent or holdup time of the fluid at the stage n, whereas control $\mathbf{u}^n$ is the fuel supply control. X^n is concentration of the active component in the fuel at stage n. In complex reaction systems coordinates of the vector of Carnot chemical potentials μ'^n may be applied in place of concentrations $\mathbf{X}^n$.

an optimum path. This energy-like function is, in fact, the Hamiltonian of the power optimization problem that is expressed here in terms of rates rather than adjoint variables (the latter being the natural variables of the Hamiltonian). The optimality condition with respect to the rates v^k and u^k proves that the state adjoints are equal to the partial derivatives of the following Lagrangian:

$$
\begin{aligned}
L &= \zeta_0 \frac{dX}{d\tau_1} + RT \frac{dX}{d\tau_1} \ln\left(\frac{(X/(1+X)) + (dX/d\tau_1)}{x_2 - j(dX/d\tau_1)}\right) \\
&\equiv \zeta_0 \dot{X} + RT\dot{X} \ln\left(\frac{x_1(X) + \dot{X}}{x_2 - j\dot{X}}\right)
\end{aligned}
\tag{9.128}
$$

with respect to its rates $\dot{X} = dX/d\tau_1$ and $\dot{T} = dT/d\tau_1$. We have defined $x_1(X) \equiv X(1+X)^{-1}$. Equation (9.128) leads to the energy-like function as the Legendre transform of Lagrangian (9.128):

$$
H(X, \dot{X}) \equiv \frac{\partial L}{\partial \dot{X}} \dot{X} - L = RT\dot{X}^2 \frac{x_2 + jx_1(X)}{(x_1(X) + \dot{X})(x_2 - j\dot{X})}
\tag{9.129}
$$

In a close vicinity of the thermodynamic equilibrium where relaxation rates are very low (close to quasistatic ones) the above optimality condition simplifies to the form

$$
H(X, \dot{X}) = RT\dot{X}^2 \frac{x_2 + jx_1(X)}{x_1(X)x_2} = RT\dot{X}^2 \left(\frac{1+X}{X} + \frac{j}{x_2}\right)
\tag{9.130}
$$

We observe that for low rates and large concentrations X (mole fractions x_1 close to the unity) optimal relaxation rate is approximately constant in time. In an arbitrary situation, however, optimal rates $\dot{X}$ are state dependent so as to preserve the constancy of Hamiltonian (9.130). No constancy of chemical driving forces along the optimal path is observed.

9.15. RESULTS OF COMPUTATIONS

Enhanced limits are obtained for dynamic extrema of power production or consumption in sequential chemical systems, for a finite number of steps or a finite time of resource exploitation. New results refer to the multistage production (consumption) of power in the chemical process that is driven by the diffusive transport step (Stefan diffusion through the inert) and takes into account the effect of drying out of the active component of fuel. The results of calculations for cost function $R^n(X^n, t^n) = \min(-W^N)$ show that the power limits differ for power generated and consumed, and that they depend on global working parameters of the system, for example, total number of mass transfer units, factor of internal irreversibilities within the power generator, average process rate, number of process steps, etc. These solutions provide the design bounds for energy generators that are stronger than the familiar thermostatic bounds (i.e., classical limits for energy transformation).

As shown by the results of calculations, optimal process efficiency increases with the intensity of fuel feed and decreases with the numerical value of Lagrange multiplier λ, an adjoint variable of the process duration. The numerical value of λ is associated with the global constraint imposed on the total holdup time or the size of the equipment. The multiplier λ is also the intensity factor whose numerical value is equal to the Hamiltonian of the power optimization problem. As such, λ is constant along an optimal trajectory of the process in which a prescribed final state is attained in a prescribed time. This constant value defines a cost of increase of the process duration by one unit. This is finally a number that should be used in calculations when the process equations are solved for an unspecified time, τ^N.

In the classical thermostatics, both limiting lines of thermodynamic bounds (upper bound for power produced and lower bound for power consumed) do coincide. However, the lines of classical bounds are often too far from the working regimes of real processes to make these bounds fully useful. The results obtained show the divergence of limiting lines for two basic process modes, mode of power production and mode of power consumption. They prove that the second law of thermodynamics does exclude some processes that are otherwise allowed by thermostatics. The region of the excluded processes grows for shorter process durations τ, associated with faster mean rates and larger entropy production. Fuel holdup time and number of transfer units play an essential role. Nonlinearities and state dependence of Hamiltonian determine complex changes of chemical and thermal efficiencies and driving forces along the optimal paths of power production process.

9.16. SOME ADDITIONAL COMMENTS

Analyzing the mechanical energy yield in nonlinear chemical systems, we have developed an approach that uses kinetic integrals of power and provides the energy limits stronger than those known in classical thermodynamics. Whenever process rates and durations are finite in systems with power production (engine modes), the power optimization approach leads to decreased upper bounds, that is, it provides lower and hence more realistic bounds for power yield in comparison with classical thermodynamics. In systems with power consumption the power optimization approach leads to increased lower bounds, meaning that these bounds are higher and hence more realistic than the lower bounds determined in classical thermodynamics (thermostatics). The enhanced nature of power limits in chemical and other systems constitutes the practical consequence of nontrivial implications of the second law of thermodynamics.

The hierarchical structure of power limits should be observed, in which the "endoreversible" limits (assuming a perfect chemical generator) are one step better than those derived from the classical thermodynamics. In the analyzed model, the endoreversible step is just the step forward sufficient to incorporate the entropy production caused by the transport phenomena. Further step includes imperfections within the chemical generator causing the effect of internal entropy production.

Also, we point out that the *economic* problem of the system optimization and the *physical* problem of work limits for a resource, considered here, are two different problems. The real work delivered from a chemical power generator per unit time at economically optimal conditions may sometimes be much lower than the maximum limit associated with an appropriate consumption of fuel. In the context of work limits, the trade-off between exploitation and investment costs and the problem of investment reduction by admission of exergy losses are, in fact, irrelevant issues. Also, the entropy production minimization for the chemical system considered has no relevance to the economic optimum of fuel consumption.

9.17. COMPLEX CHEMICAL POWER SYSTEMS WITH INTERNAL DISSIPATION

Consider first the comparison of chemical and thermal operations of power production. To compare the chemical process described by Equations (9.122)–(9.124) with a typical thermal process we adduce first relationships describing a dynamic thermal engine.

In a discrete thermal problem the performance index and constraints are

$$W^N \equiv \sum_{k=1}^{N} w^k = -c \sum_{k=1}^{N} \left\{ 1 - \frac{T_2}{T_1^k + u^k} \right\} u^k \theta^k \qquad (9.131)$$

$$T_1^k - T_1^{k-1} = u^k \theta^k \qquad (9.132)$$

$$\tau^k - \tau^{k-1} = \theta^k \qquad (9.133)$$

where W^N is the total power per unit mass flux and the bracketed expression in Equation (9.131) is the first-law efficiency. The continuous limit of Equation (9.131) is the work integral

$$\begin{aligned} W &= -\int_{\tau^i}^{\tau^f} c \left\{ 1 - \frac{T_2}{T + dT/d\tau} \right\} \left(\frac{dT}{d\tau} \right) d\tau \\ &= -\int_{\tau^i}^{\tau^f} c \left\{ 1 - \frac{T_2}{T + \chi\, dT/dt} \right\} \left(\frac{dT}{dt} \right) dt \end{aligned} \qquad (9.134)$$

The associated integral of entropy production is

$$S_\sigma = \int_{\tau^i}^{\tau^f} c \frac{(dT/d\tau)^2}{T(T + dT/d\tau)} d\tau = \int_{t^i}^{t^f} c \frac{\chi(dT/dt)^2}{T(T + \chi(dT/dt))} dt \qquad (9.135)$$

The temperature derivative and slope coefficients are taken here either with respect to the nondimensional time τ (identical with the *overall* number of heat transfer units) or with the resource's holdup time t. In the periodic process of energy upgrading the latter variable represents the utilization time of a thermal resource. Both time variables are linked with length coordinate, x, by the equation

$$\tau \equiv \frac{\alpha' a_v F}{Gc} x = \frac{\alpha' a_v}{\rho c} t = \frac{t}{\chi} \qquad (9.136)$$

where α' is an overall heat transfer coefficient, a_v is the specific area, F is the cross-sectional area for fluid's flow, and $\chi = \rho c/(\alpha' a_v)$ plays the role of a time constant for the system (Sieniutycz, 1997a, 1997b, 1997c). An optimal, power maximizing, relaxation process is described by a family of exponential curves (Sieniutycz, 1997a, 1997b; Kubiak, 2005). However, in systems with nonlinear transport, for example, in radiation engines, relaxation curves are no longer exponential (Tsirlin et al., 1998; Kuran, 2006; Sieniutycz and Kuran, 2006; see Chapter 6. Other models of dynamic chemical systems are available in the literature (Tsirlin et al., 1998).

For an imperfect thermal process the efficiency equation can be adopted in several forms (consider, e.g., Equations $((4.94))$ and $((5.79))$ inverted to the efficiency form)

$$
\begin{aligned}
\eta &= 1 - \Phi\frac{T_2}{T + dT/d\tau} = 1 - \Phi\frac{T_2}{T + \chi\, dT/dt} \\
&= 1 - \Phi\frac{T_2}{T - Q_1/g'} = 1 - \Phi\frac{T_2}{T'}
\end{aligned}
$$

(9.137)

where the effective thermal conductance g' is defined by Equation (5.80). When mass transfer is involved in power yield, Equation (9.137) can still be adopted either in terms of pure heat flux q_1 or in terms of total heat flux Q_1. Its thermal efficiency formula holds in pseudo-Carnot form if Carnot temperature T' is applied. Yet, whenever affinity-based chemical efficiency is adopted as the second efficiency component (see below), total heat flux Q_1 is the only choice as the driving heat flux of the process.

Power evaluation in practical chemical generators calls for the relaxation of the assumption about reversibility of chemical reaction within the active zone with power yield. This requires introducing internal entropy production within the active zone. Moreover, single chemical efficiency, ζ, defined as power yield per 1 mole of an invariant molar flux n, is insufficient in nonisothermal generators. In these units, multiple reactions and extra component of power, associated with the temperature difference between the reservoirs, play a role.

In complex reaction systems power yield formula is a vector generalization of Equation (9.47) or the like for the case of multiple reactions. When reaction mass fluxes n_j are used rather than Gibbs fluxes (such as Gibb's flux in Equation (9.47)), a power equation has the following structure:

$$
p = \dot{w} = \eta(Q_{1'}, n_j)Q_{1'} + \sum_{j=1}^{R}\xi_j(Q_{1'}, n_j)n_j
$$

(9.138)

or, in terms of bulk variables

$$
p = \eta(Q_1, n_j)Q_1 + \sum_{j=1}^{R}\xi_j(Q_1, n_j)n_j
$$

(9.139)

In the above equations efficiencies of thermal and chemical conversions appear, η and ξ_j, respectively. Our purpose below will be the derivation of suitable forms for efficiency functions contained in power equations.

An example of a multireaction system is illustrated in Figure 9.8. This scheme assumes a set of chemical reactions, $j = 1, 2, \ldots, R$, undergoing between species B_i $(i = 1, 2, \ldots, m)$ in the system. Arbitrary reaction j of Figure 9.8 can be written in the form

$$
\begin{aligned}
\nu_{1j}^{f}B_1 + \nu_{2j}^{f}B_2 + \nu_{3j}^{f}B_3 + \cdots + \nu_{mj}^{f}B_m &= \nu_{1j}^{b}B_1 \\
&+ \nu_{2j}^{b}B_2 + \nu_{3j}^{b}B_3 + \cdots + \nu_{mj}^{b}B_m
\end{aligned}
$$

(9.140)

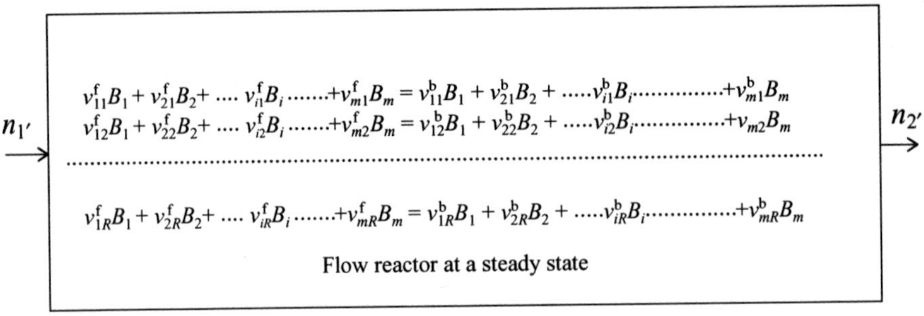

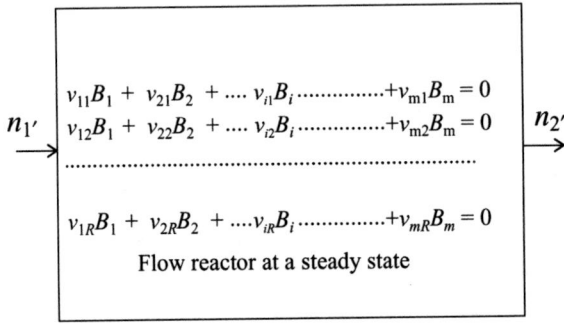

$$n_{1'} \quad \begin{aligned} v_{11}^f B_1 + v_{21}^f B_2 + \ldots v_{i1}^f B_i \ldots + v_{m1}^f B_m &= v_{11}^b B_1 + v_{21}^b B_2 + \ldots v_{i1}^b B_i \ldots + v_{m1}^b B_m \\ v_{12}^f B_1 + v_{22}^f B_2 + \ldots v_{i2}^f B_i \ldots + v_{m2}^f B_m &= v_{12}^b B_1 + v_{22}^b B_2 + \ldots v_{i2}^b B_i \ldots + v_{m2}^b B_m \\ \ldots\ldots\ldots\ldots\ldots\ldots\ldots\ldots & \\ v_{1R}^f B_1 + v_{2R}^f B_2 + \ldots v_{iR}^f B_i \ldots + v_{mR}^f B_m &= v_{1R}^b B_1 + v_{2R}^b B_2 + \ldots v_{iR}^b B_i \ldots + v_{mR}^b B_m \end{aligned} \quad n_{2'}$$

Flow reactor at a steady state

Figure 9.8 Scheme of steady, multireaction system.

$$n_{1'} \quad \begin{aligned} v_{11} B_1 + v_{21} B_2 + \ldots v_{i1} B_i \ldots + v_{m1} B_m &= 0 \\ v_{12} B_1 + v_{22} B_2 + \ldots v_{i2} B_i \ldots + v_{m2} B_m &= 0 \\ \ldots\ldots\ldots\ldots\ldots\ldots\ldots\ldots & \\ v_{1R} B_1 + v_{2R} B_2 + \ldots v_{iR} B_i \ldots + v_{mR} B_m &= 0 \end{aligned} \quad n_{2'}$$

Flow reactor at a steady state

Figure 9.9 A scheme of a multireaction system described by the traditional stoichiometry.

where B_i are species participating in R chemical reactions $j = 1, 2, \ldots, R$, and $v_{i,j}$ is the *absolute* stoichiometric coefficient of component i in reaction j. The reaction description in terms of Equation (9.140) is equivalent to the reaction scheme presented in Figure 9.8. The absolute stoichiometric coefficients $v_{i,j}^f$ and $v_{i,j}^b$ appearing therein are always positive. They should be distinguished from relative stoichiometric coefficients, traditional quantities defined in the physical chemistry as $v_{ij} = v_{ij}^b - v_{ij}^f$. For the latter, convention usually states that $v_{i,j} > 0$ if component i is a product, $v_{i,j} < 0$ if it is a reactant, and $v_{i,j} = 0$ if component i is an inert (Figure 9.9).

Depending on the problem at hand, the considered states may refer to the following "pairs of cross-sections" (also referred to as "boundary states of an endoreversible zone"):

$$1 = f \text{ and } 1' = b, \qquad 1' = f \text{ and } 2' = b, \qquad 2' = f \text{ and } 2 = b \quad (9.141)$$

The thermodynamic equation for power yield, which results from the energy balance of the internal reacting chamber and is expressed in terms of heat fluxes

$(q$ or $Q)$, has the form

$$
\begin{aligned}
p &= q_{1'} - q_{2'} + h_{1'}n_{1'} - h_{2'}n_{2'} \cdots \\
&= Q_{1'} - Q_{2'} + \mu_{1_1'}n_{1_1'} + \cdots \mu_{i_1'}n_{i_1'} \cdots + \mu_{m_1'}n_{m_1'} \\
&\quad -\mu_{1_2'}n_{1_2'} - \cdots \mu_{i_2'}n_{i_2'} \cdots - \mu_{m_2'}n_{m_2'}
\end{aligned} \tag{9.142}
$$

where we have introduced "total heat flux" Q_1 as the sum of the sensible heat q_1 and the mass transfer–related heat flux

$$
Q_1 \equiv q_1 + T_1 s_{1_1} n_{1_1} + \cdots + T_1 s_{i_1} n_{i_1} + \cdots + T_1 s_{m_1} n_{m_1} \tag{9.143}
$$

The virtue of using total heat flux Q_1 is its additive contribution to power yield in power formulas containing chemical efficiencies in the form of driving forces or affinities.

For a nonisothermal process with a complete conversion, Equation (9.142) yields an expression

$$
p = Q_{1'} - Q_{2'} + \mu_{1_1'}n_{1_1'} + \cdots \mu_{i_1'}n_{i_1'} - \cdots \mu_{i+1_2'}n_{i+1_2'} \cdots - \mu_{m_2'}n_{m_2'} \tag{9.144}
$$

where molar fluxes may be linked by the chemical stoichiometry.

For a simple isomerization reaction, $A_1 - A_2 = 0$, and under the assumption of the complete conversion, the application of flux continuity for entropy and matter yields

$$
p = Q_{1'} - Q_{2'} + \mu_{1'}n_{1'} - \mu_{2'}n_{2'} = \left(1 - \frac{T_{2'}}{T_{1'}}\right)Q_{1'} + (\mu_{1'} - \mu_{2'})n_{1'} \tag{9.145}
$$

Consider now a general while still idealized case of an endoreversible engine with many components and chemical reactions, assuming that the boundary states of the endoreversible zone refer to the cross-sections $1'$ and $2'$. In such an engine fluxes of entropy and matter are continuous through the active (endoreversible) zone, because there is no entropy production within reversible part of the generator, and the mass conservation principle is valid. In terms of the total heat flux Q, entropy flux n_s, and Gibb's fluxes G_j, the continuity equations are

$$
\frac{Q_{1'}}{T_{1'}} = \frac{Q_{2'}}{T_{2'}} = n_s \tag{9.146}
$$

$$
\frac{G_{j1'}}{\Pi_{j1'}} = \frac{G_{j2'}}{\Pi_{j2'}} = n_j, \qquad j = 1, 2, \ldots, R \tag{9.147}
$$

In the second equation, reaction potentials Π_j appear at the boundaries of the endoreversible system (cross-sections $1'$ and $2'$). In terms of the chemical potentials and *absolute* stoichiometric coefficients $v_{ij}^{abs} \equiv v_{ij}$ reaction potentials are defined by the following equation:

$$\Pi_j \equiv \sum_{i=1}^{n} (v_{ij}\mu_i) \tag{9.148}$$

Depending on case considered, the potential of each reaction changes continuously between its boundary values, either between Π_j^f and Π_j^b as in Figure 9.8, where

$$\Pi_j^f \equiv \sum_{i=1}^{n} (v_{ij}^f \mu_i^f) \tag{9.149}$$

and

$$\Pi_j^b \equiv \sum_{i=1}^{n} (v_{ij}^b \mu_i^b) \tag{9.150}$$

or between $\Pi_{j1'}$ and $\Pi_{j2'}$, as in Equation (9.142). (Compare the above equations with Equation (9.25) of the entropy representation.)

Applying continuity Equations (9.146) and (9.147) in the general power formula of an endoreversible chemical engine driven by heat fluxes and R chemical reactions

$$p = Q_{1'} - Q_{2'} + \sum_{j=1}^{R} (G_{j1'} - G_{j2'}) \tag{9.151}$$

power expression of an idealized (endoreversible) engine follows in several alternative forms

$$
\begin{aligned}
p &= Q_{1'} - Q_{2'} + \sum_{j=1}^{R} (G_{j1'} - G_{j2'}) \\
&= \left(1 - \frac{Q_{2'}}{Q_{1'}}\right) Q_{1'} + \sum_{j=1}^{R} \left(1 - \frac{G_{j2'}}{G_{j1'}}\right) G_{j1'} \\
&= \left(1 - \frac{T_{2'}}{T_{1'}}\right) Q_{1'} + \sum_{j=1}^{R} \left(1 - \frac{\Pi_{j2'}}{\Pi_{j1'}}\right) G_{j1'} \\
&= \left(1 - \frac{T_{2'}}{T_{1'}}\right) T_{1'} n_{s1'} + \sum_{j=1}^{R} \left(1 - \frac{\Pi_{j2'}}{\Pi_{j1'}}\right) \Pi_{j1'} n_{j1'} \\
&= \left(1 - \frac{T_{2'}}{T_{1'}}\right) Q_{1'} + \sum_{j=1}^{R} (\Pi_{j1'} - \Pi_{j2'}) n_{j1'}
\end{aligned}
\tag{9.152}
$$

The last term of this equation contains endoreversible chemical efficiencies expressed as changes of reaction potentials, for all reactions involved, $j = 1, 2, \ldots, R$.

We recall that, when the reaction flux n_j is the process variable and the efficiency basis, the endoreversible chemical efficiency of jth reaction, ξ_j, is identical with the related chemical affinity, A_j, which involves boundary states of the chemically active zone or the states in the cross-sections $1'$ and $2'$. Indeed, in terms of the reaction potentials and absolute stoichiometric coefficients, an endoreversible chemical efficiency (or associated affinity) is the difference of forward and backward reaction potentials, Π_j^f and Π_j^b in the active boundary states $1'$ and $2'$

$$\xi_j = A_j = \sum_{i=1}^{n} (v_{ij}^f \mu_i^f)_{1'} - \sum_{i=1}^{n} (v_{ij}^b \mu_i^b)_{2'} = \Pi_{j1'}^f - \Pi_{j2'}^b \tag{9.153}$$

In terms of the relative stoichiometry and for the absence of diffusion (a classical case with the same chemical potentials for each species in states f and b) the reversible chemical efficiency or the classical chemical affinity satisfies the familiar formulae

$$\xi_j = -\sum_{i=1}^{n} v_{ij} \mu_i = -(\mathbf{v}^T \mu)_j = \sum_{i=1}^{n} (v_{ij} \mu_i)^{\text{subs}} - \sum_{i=1}^{n} (v_{ij} \mu_i)^{\text{prod}} = \Pi_{j1'}^f - \Pi_{j2'}^b \tag{9.154}$$

In the classical case the affinity A_j is also equal to the difference $\Pi_j^f - \Pi_j^b$, although, in general, Π_j^f is not the substrate component of the classical affinity and Π_j^b is not the product component of the classical affinity.

The scheme in Figure 9.8 and the above equations describe a general case of the system with m components and R reactions. However, for simple didactic illustration of physical effects, we shall now consider a single-reaction system with an arbitrary chemical reaction.

As an example consider the application of this formalism to the simple isomerization reaction A $\Leftrightarrow$ B in the case when feed stream can contain the product and outlet stream can contain the substrate. That reaction process can be written in the form

$$v_1^f A + v_2^f B = v_1^b A + v_2^b B \tag{9.155}$$

The boundary potentials of the reaction are

$$\Pi^f = v_A^f \mu_A^f + v_B^f \mu_B^f \tag{9.156}$$

and

$$\Pi^b = v_A^b \mu_A^b + v_B^b \mu_B^b \tag{9.157}$$

For this reaction the affinity-related chemical efficiency is

$$\xi = \sum_{i=1}^{2} (\nu_i^f \mu_i^f - \nu_i^b \mu_i^b) = \Pi^f - \Pi^b = \nu_A^f \mu_A^f + \nu_B^f \mu_B^f - \nu_A^b \mu_A^b - \nu_B^b \mu_B^b$$

$$(9.158)$$

This is a "general formula" for the efficiency of the considered reaction. When mixing in the reactor is perfect, chemical potentials of the same species in states f and b are identical, that is, $\mu_A^f = \mu_A^b \equiv \mu_A$ and $\mu_B^f = \mu_B^b \equiv \mu_B$. Then, the following efficiency is obtained:

$$\begin{aligned}\xi &= \Pi^f - \Pi^b = \nu_A^f \mu_A^f + \nu_B^f \mu_B^f - \nu_A^b \mu_A^b - \nu_B^b \mu_B^b \\ &= (\nu_A^f - \nu_A^b)\mu_A + (\nu_B^f - \nu_B^b)\mu_B = -\nu_A \mu_A - \nu_B \mu_B\end{aligned}$$

$$(9.159)$$

In terms of the traditional physiochemical stoichiometry, the reaction considered reads

$$\nu_A A + \nu_B B = 0 \qquad (9.160)$$

where $\nu_A = -1$ and $\nu_B = 1$. In this case efficiency formula (9.158) yields the endoreversible efficiency (or the related affinity) of the reaction A $\leftrightarrow$ B in the familiar form

$$\xi = \Pi^f - \Pi^b = \mu_{A1'} - \mu_{B2'} \qquad (9.161)$$

(see Equation (9.53)). The resulting equation contains the difference of end chemical potentials of substrate A and product B, and it appears to be a special case of Equation (9.158).

Consider now *a general nonideal case*, characterized by the internal production of entropy and incomplete conversion of substrates into products. Since identification of active affinity force with chemical efficiency is valid only for thermodynamically reversible reactions, we shall explain how internal entropy production and incomplete conversion affect forms of efficiency expressions.

The second law applied to the internal entropy production in the engine mode dictates that $n_{s2'} > n_{s1'}$, or $\Phi > 1$. Moreover, $n_{j2'} < n_{j1'}$, or $\Xi_j \leq 1$, for each reaction j, which refers to the reduction of the outgoing chemical flow of product j by incomplete conversion. Thus, the following relations hold in an imperfect system:

$$\frac{Q_{2'}}{T_{2'}} = \Phi \frac{Q_{1'}}{T_{1'}} = \Phi n_{s1'} \qquad (9.162)$$

$$\frac{G_{j2'}}{\Pi_{j2'}} = \Xi_j \frac{G_{j1'}}{\Pi_{j1'}} = \Xi_j n_{j1'} \qquad (9.163)$$

where imperfection coefficients satisfy the inequalities $\Phi > 1$ and $\Xi_j < 1$. The first inequality follows from the second law, whereas the second is the consequence of the incomplete conversion that dictates that the outlet chemical flux $n_{j2'}$ can only be less than the inlet chemical flux $n_{j1'}$.

With Equations (9.162) and (9.163), a nonideal power formula is obtained in the form

$$
\begin{aligned}
p &= Q_{1'} - Q_{2'} + \sum_{j=1}^{R} (G_{j1'} - G_{j2'}) = \left(1 - \frac{Q_{2'}}{Q_{1'}}\right) Q_{1'} + \sum_{j=1}^{R} \left(1 - \frac{G_{j2'}}{G_{j1'}}\right) G_{j1'} \\
&= \left(1 - \Phi \frac{T_{2'}}{T_{1'}}\right) Q_{1'} + \sum_{j=1}^{R} \left(1 - \Xi_j \frac{\Pi_{j2'}}{\Pi_{j1'}}\right) \Pi_{j1'} n_{j1'} = \left(1 - \Phi \frac{T_{2'}}{T_{1'}}\right) Q_{1'} \\
&\quad + \sum_{j=1}^{R} (\Pi_{j1'} - \Xi_j \Pi_{j2'}) n_{j1'}
\end{aligned}
\tag{9.164}
$$

When diffusion processes can be ignored, a simple approximation of the above formula for the simple isomerization reaction a $\Leftrightarrow$ b (or A $-$ B $\Leftrightarrow$ 0) has the form

$$
p \cong \left(1 - \Phi \frac{T_{2'}}{T_{1'}}\right) Q_{1'} + (\mu_{1'} - \Xi \mu_{2'}) n_{1'}
\tag{9.165}
$$

This result contains reduced thermal and chemical efficiencies, and it represents an "imperfect counterpart" of the idealized power formula (9.145) that is limited to reversible conversions.

Using the above equation we can reinterpret imperfection factors and provide extra substantiation for the structure of imperfect power formula. The balances of entropy and mass for the imperfect part of the engine are

$$
\frac{Q_{2'}}{T_{2'}} - \frac{Q_{1'}}{T_{1'}} = \sigma_s^{\text{int}}
\tag{9.166}
$$

$$
n_{b2'} - n_{a1'} = \sigma_p^{\text{int}}
\tag{9.167}
$$

The above set contains a negative source σ_p^{int} that points to incomplete conversion of substrate a when product p is created in a chemical process between sections 1' and 2' of the chemical chamber.

Remember that σ_b^{int} describes only the insufficiency of the product yield, not the actual yield of this product in the chemical reaction.

Transforming Equation (9.166) we can write the entropy balance in the form

$$
\frac{Q_{2'}}{Q_{1'}} = \frac{T_{2'}}{T_{1'}} + \frac{T_{2'} \sigma_s^{\text{int}}}{Q_{1'}} = \left(1 + \frac{T_{1'} \sigma_s^{\text{int}}}{Q_{1'}}\right) \frac{T_{2'}}{T_{1'}} \equiv \Phi \frac{T_{2'}}{T_{1'}}
\tag{9.168}
$$

In terms of the entropy flux n_s the above result reads

$$
\frac{n_{s2'}}{n_{s1'}} = 1 + \frac{\sigma_s^{\text{int}}}{n_{s1'}} = \Phi
\tag{9.169}
$$

whereas the transformed mass balance equation (9.167), which describes the incomplete yield of the product b (characterized by the negative insufficiency σ_b^{int}), can be presented in the form

$$\frac{n_{b2'}}{n_{a1'}} = 1 + \frac{\sigma_b^{int}}{n_{a1'}} \equiv \Xi_b \tag{9.170}$$

Since the inequality $n_{b2'} < n_{a1'}$ holds for an incomplete transformation of the substrate a in the product b, the imperfection function Ξ_b satisfying the inequality $\Xi_b \leq 1$ and assuring the conversion insufficiency $\sigma_b^{int} < 0$ can be introduced and then applied.

Summing up, two internal imperfection functions $\Phi > 1$ and $\Xi_b < 1$ can be applied in the balance equations and power formula (9.165). These functions are defined in terms of internal sources and the system's operational parameters, in accordance with the expressions

$$\Phi \equiv 1 + \frac{T_{1'}\sigma_s^{int}}{Q_{1'}} \tag{9.171}$$

$$\Xi_b \equiv 1 + \frac{\sigma_b^{int}}{n_{a1'}} \tag{9.172}$$

It is essential that the inequalities $\Phi > 1$ and $\Xi_b < 1$ are restricted to engine modes.

With chemical imperfection factor Ξ_b we obtain from Equation (9.170) the equality $n_{b2'} = \Xi_b n_{a1'}$, or, briefly, $n_{2'} = \Xi n_{1'}$. Consequently, the considered example leads to the power formula

$$p = Q_{1'} - Q_{2'} + \mu_{1'} n_{1'} - \mu_{2'} n_{2'} = Q_{1'}\left(1 - \frac{Q_{2'}}{Q_{1'}}\right) + (\mu_{1'} - \Xi\mu_{2'})n_{1'}$$

$$= \left(1 - \Phi\frac{T_{2'}}{T_{1'}}\right)Q_{1'} + (\mu_{1'} - \Xi\mu_{2'})n_{1'} \tag{9.173}$$

(Compare with Equation (9.145).) Functions of internal sources, Φ and Ξ, satisfy Equations (9.171) and (9.172).

Power formulae (9.165) or (9.173) generalize idealized power of Equation (9.145). Imperfect efficiency vector has two components. The first or thermal component describes power generated by total heat flux $Q_{1'}$ with lowered (non-Carnot) efficiency, $\eta = 1 - \Phi T_{2'}/T_{1'}$, whereas the second or chemical component describes power generated with nonideal chemical efficiency $\xi = \mu_{1'} - \Xi\mu_{2'}$. The fractional nature of coefficient Ξ is consistent with the negative sign of chemical potentials (Baierlein, 2001). In the literature, where this property is not always respected, imperfection coefficients different than Ξ are occasionally applied.

In the engine mode, where $\Phi > 1$ and $\Xi < 1$, imperfect chemical system (the one with internal imperfections) behaves as it would work with an increased temperature of lower reservoir ($\Phi T_{2'}$ instead of $T_{2'}$) and with a decreased chemical affinity of an effective value $\mu_{1'} - \Xi\mu_{2'}$. Of course, these imperfections reduce power production.

Let us now attempt to derive equations of *incomplete conversion from the first principles* that involve general mass balances in multireaction and multicomponent systems (Figure 9.8). Considering this general system with arbitrary inlet and outlet molar flows, we shall take into account the balance of mole numbers for *j*th chemical reaction

$$\frac{(n_{1_{1'}} - n_{1_{2'}})_j}{v_{1j}} = \frac{(n_{2_{1'}} - n_{2_{2'}})_j}{v_{2j}} \cdots = \frac{(n_{i_{1'}} - n_{i_{2'}})_j}{v_{ij}} \cdots =$$
$$\frac{(n_{k_{1'}} - n_{k_{2'}})_j}{v_{kj}} = \frac{(n_{l_{1'}} - n_{l_{2'}})_j}{v_{lj}} = \frac{(n_{m_{1'}} - n_{m_{2'}})_j}{v_{mj}} \equiv -n_j \qquad (9.174)$$

In treating flow systems we introduce into considerations chemical flux of *j*th reaction, n_j, defined as the product of the homogeneous reaction rate and the volume. (When the reaction is heterogeneous, analogous product of its rate and the surface is introduced in Chapter 10.) For the reaction *j*, its chemical flux can be defined as

$$n_j = \frac{(n_{i2'} - n_{i1'})_j}{v_{i,j}} \qquad (9.175)$$

where n_j is the product of reaction rate r_j and the reaction volume, $(n_{i2'} - n_{i1'})_j$ is the molar flux of component *i* reacting in reaction *j*, and $v_{i,j}$ is the relative stoichiometric coefficient of *i*th component in reaction *j*, defined in physical chemistry as $v_{ij} = v_{ij}^b - v_{ij}^f$ (traditional stoichiometric coefficient).

Instead of the absolute stoichiometry of Equation (9.140), we shall now use traditional stoichiometric coefficients $v_{ij} = v_{ij}^b - v_{ij}^f$. Then

$$(v_{1j}^b - v_{1j}^f)B_1 + (v_{2j}^b - v_{2j}^f)B_2 + (v_{3j}^b - v_{3j}^f)B_3 + \cdots + (v_{mj}^b - v_{mj}^f)B_m = 0$$

$$(9.176)$$

or

$$v_{1j}B_1 + v_{2j}B_2 + \cdots v_{ij}^b B_i + v_{i+1j}B_{i+1} + \cdots + v_{mj}B_m = 0 \qquad (9.177)$$

Interpretation of the above equation is illustrated in Figure 9.9. Negative v_{ij} refer to substrates and positive v_{ij} refer to products. Vanishing v_{ij} refer to inert species.

Let us consider first an idealized case, the one with a complete conversion. For convenience, components are numbered such that species 1, 2, ..., *i* are substrates and species $i+1, i+2, ..., m$ are products. We shall take advantage of mass conservation in each chemical reaction (Equation (9.174)). That formula uses (characteristic for each reaction) molar fluxes of active reagents, n_j, identical with products of reaction volume V and derivatives of reaction extents, ε_j, with respect to time t (i.e., reaction rates, r_j).

For a complete conversion, there are no substrate fluxes at the outlet; thus, in each reaction j

$$(n_{1_{2'}} = n_{2_{2'}} \cdots = n_{i_{2'}})_j = 0 \tag{9.178}$$

(species $1, 2, \ldots, i$ are substrates). When products are absent in the feed stream

$$(n_{i+1_{1'}} = n_{i+2_{1'}} \cdots = n_{m_{1'}})_j = 0 \tag{9.179}$$

(species $i+1, i+2, \ldots, m$ are products). Therefore, in this case Equation (9.174) takes the form

$$-\frac{(n_{1_{1'}})_j}{v_{1j}} = -\frac{(n_{2_{1'}})_j}{v_{2j}} \cdots = -\frac{(n_{i_{1'}})_j}{v_{ij}} = \frac{(n_{i+1_{2'}})_j}{v_{i+1j}} = \frac{(n_{l_{2'}})_j}{v_{lj}} = \frac{(n_{m_{2'}})_j}{v_{mj}} \equiv n_j \tag{9.180}$$

$(j = 1, 2, \ldots, R)$. Since for each i and j

$$n_i = \sum_{j=1}^{R} (n_i)_j \tag{9.181}$$

and Equations (9.178)–(9.180) are valid, total power yield of an isothermal multireaction process

$$p = \mu_{1} n_{1_{1'}} + \mu_{2_{1'}} n_{2_{1'}} + \cdots \mu_{i_{1'}} n_{i_{1'}} \cdots + \mu_{m_{1'}} n_{m_{1'}} \\ -\mu_{1} n_{1_{2'}} - \mu_{2_{2'}} n_{2_{2'}} \cdots - \mu_{i_{2'}} n_{i_{2'}} \cdots - \mu_{m_{2'}} n_{m_{2'}} \tag{9.182}$$

(a special, "isothermal" case of Equation (9.142)) takes the form

$$p = \sum_{j=1}^{R} \{\mu_1 (n_{1_{1'}})_j + \mu_{2_{1'}} (n_{2_{1'}})_j + \cdots \mu_{i_{1'}} (n_{i_{1'}})_j \\ -\mu_{i+1_{2'}} (n_{i+1_{2'}})_j \cdots - \mu_{l_{2'}} (n_{l_{2'}})_j \cdots - \mu_{m_{2'}} (n_{m_{2'}})_j\} \tag{9.183}$$

and, finally (keep in mind that each $n_j > 0$ and traditional stoichiometric coefficients v_{ij} of substrates are negative)

$$p = \sum_{j=1}^{R} \{p_j\} \\ = \sum_{j=1}^{R} -\{\mu_{1_{1'}} v_{1j} + \mu_{2_{1'}} v_{2j} + \cdots \mu_{i_{1'}} v_{ij} \\ +\mu_{i+1_{2'}} v_{i+1j} \cdots + \cdots \mu_{m-1_{2'}} v_{m-1j} \cdots + \mu_{m_{2'}} v_{mj})\} n_j \tag{9.184}$$

n_j is the chemical flux of jth reaction equal to the product of reaction rate and volume, $n_j = r_j V$. Equation (9.184), equality $n_j = r_j V$, and the affinity definition

$$A_j = \Pi_j^f - \Pi_j^b = -\sum_{i=1}^{m} (v_{ij} \mu_i) = -\mathbf{v}^T \mu \tag{9.185}$$

prove that (for a complete conversion) power yield from the unit volume of the reactor is equal to the sum of products of affinity driving forces and rates of chemical reactions

$$p = \sum_{j=1}^{R} \{\tilde{A}_j n_j\} = V \sum_{j=1}^{R} \{\tilde{A}_j r_j\} \tag{9.186}$$

Note that the affinity-like quantity $\tilde{A}_j$, which appears in Equation (9.186), is not precisely identical with classical chemical affinity A_j. The former is a "distributed quantity" involving states in cross-sections $1'$ and $2'$, whereas the latter is the state function. Still the distributed affinity $\tilde{A}_j$ may be used in calculations of chemical efficiencies and power yield.

For a single reaction the chemical efficiency ζ may be defined as power yield per 1 mole of the invariant molar flux of reagents; for many reactions, vector of efficiencies may be defined. These efficiencies are identical with affinities $\tilde{A}_j$ in the chemically active part of the system. While they are not dimensionless, they can describe correctly the system.

With these general formulas one can propose a multicomponent *extension of the present approach to the case of imperfect conversion*. Actually, the approach introduces fictitious negative flows of products at the inlet $n_{i+1_{1'}} \cdots n_{i+2_{1'}} \cdots = n_{m_{1'}} < 0$ in order to subsume the effect of incomplete conversion in the reactor described by standard balance equations. This results in an imperfection coefficient Ξ_b of the same type as other coefficients Ξ. For a product b the imperfection coefficient Ξ_b is now defined as

$$\Xi_b = \frac{n_{b2'}}{n_{b_{2'}} - n_{b_{1'}}} \equiv \frac{n_{b2'}}{\Delta n_b} \tag{9.187}$$

($b = i+1, i+2, \ldots, m$). For an engine, $\Phi > 1$ and $\Xi_b < 1$, where the chemical inequality refers to a fictitious negative flux of product b at the reactor inlet, $n_{b1'}$.

This approach assumes that a basic equality of the complete conversion model

$$n_{1_{2'}} = n_{2_{2'}} \cdots = n_{i_{2'}} = 0 \tag{9.188}$$

is valid (i.e., substrates $1, 2, \ldots, i$ do not appear in cross-section $2'$), but that

$$n_{i+1_{1'}} = n_{i+2_{1'}} \cdots = n_{m_{1'}} \leq 0 \tag{9.189}$$

that is, there are negative input fluxes (outputs) of products $i+1, i+2, \ldots, m$ at the inlet cross-section $1'$. Then instead of "perfect equality" (9.180), applied here for a single reaction, $R = 1$, we obtain

$$-\frac{n_{1_{1'}}}{\nu_1} \cdots = -\frac{n_{i-1_{1'}}}{\nu_{i-1}} = -\frac{n_{i_{1'}}}{\nu_i} = \frac{n_{i+1_{2'}} - n_{i+1_{1'}}}{\nu_{i+1}} = \cdots = \frac{n_{m_{2'}} - n_{m_{1'}}}{\nu_m} \equiv n$$

$$\tag{9.190}$$

This result can be written in the form

$$
-\frac{n_{1_{1'}}}{\nu_1} \cdots = -\frac{n_{i_{1'}}}{\nu_i} \cdots = \frac{n_{i+1_{2'}}}{\nu_{i+1} n_{i+1_{2'}} / (n_{i+1_{2'}} - n_{i+1_{1'}})}
$$
$$
+ \cdots \frac{n_{m_{2'}}}{\nu_m n_{m_{2'}} / (n_{m_{2'}} - n_{m_{1'}})} \equiv n \tag{9.191}
$$

An equivalent form of this equation contains factors Ξ_p defined as the values of ratios $\Xi_p = n_{p2'}/\Delta n_p$ for products $(p = i+1, \ldots, m)$, each satisfying property $\Xi_p < 1$. For a common $\Xi_p = \Xi$

$$
-\frac{n_{1_{1'}}}{\nu_1} \cdots = -\frac{n_{i_{1'}}}{\nu_i} = \frac{n_{i+1_{2'}}}{\nu_{i+1}\, \Xi} \cdots = \cdots \frac{n_{m_{2'}}}{\nu_m\, \Xi} \equiv n \tag{9.192}
$$

Nonisothermal power yield (9.144) is now an apparent formula. After using Equation (9.192) in Equation (9.144) power generated by a single complex reaction assumes the form

$$
p = Q_{1'}\left(1 - \frac{\Phi T_{2'}}{T_{1'}}\right)
$$
$$
-\{\mu_{1_{1'}}\nu_1 + \cdots \mu_{i_{1'}}\nu_i + \Xi\mu_{i+1_{2'}}\nu_{i+1} + \Xi\mu_{m-1_{2'}}\nu_{m-1_{2'}} + \Xi\mu_{m_{2'}}\nu_{m_{2'}}\}n \tag{9.193}
$$

A formal interpretation of this power formula is based on effective increase of chemical potentials of products and Gibbs flux of these products, which causes reduction of the chemical efficiency. The approach presented here is similar to that known in the theory of imperfect heat machines that operates with the factor of internal irreversibilities Φ (satisfying the inequality $\Phi > 1$ and acting multiplicatively on the temperature of fluid contacting with the second reservoir, $T_{2'}$). The effective temperature of this fluid $T_{2'}^{\mathrm{eff}}$ equals then $\Phi T_{2'}$, as shown in the thermal efficiency term of Equation (9.193). Of course, this effect causes a decrease of the thermal efficiency. In the chemical case the role similar to Φ is played by coefficient Ξ (or family of such coefficients, Ξ_p, where each member pertains to a different product). The effective chemical potential of product p becomes $\Xi\mu_p$ at $2'$. Since each $\Xi_p < 1$ (consistent with negative signs of chemical potentials), the effective chemical potentials of products that reside in the chemical efficiency are higher ("less negative") than actual ones. This effect decreases chemical efficiencies (driving forces and affinities).

The power formula (Equation (9.193)) generalizes the idealized power of Equation (9.144). Vector nature of efficiency manifests itself in nonisothermal chemical systems. The first or thermal component of the power yield describes power generated by the total heat flux with the decreased (nonCarnot) thermal efficiency $\eta = 1 - \Phi T_{2'}/T_{1'}$, whereas the second or chemical component describes the power yield caused by the chemical reaction. Chemical power is generated with the lowered (nonideal) chemical efficiency

$$\zeta_\Xi \equiv -\{\mu_{1_{1'}}\nu_1 + \cdots + \mu_{i_{1'}}\nu_i + \Xi\mu_{i+1_{2'}}\nu_{i+1} + \Xi\mu_{m-1_{2'}}\nu_{m-1_{2'}} + \Xi\mu_{m_{2'}}\nu_{m_{2'}}\}$$

$$(9.194)$$

This is equivalent to efficiency described by "reduced affinity" $A_\Xi = \Pi_{1'} - \Xi\Pi_{2'}$ and consistent with the general formula describing imperfect efficiency of jth reaction in terms of the reaction potentials ("one-way affinities," Π_j)

$$\zeta_{j\Xi} \equiv \Pi_{j1'} - \Xi\Pi_{j2'} \qquad (9.195)$$

The chemical term in power formulae predominates in systems that work closely to isothermal conditions. For a simple isomerization reaction ($i = 1$), $\nu_1 = -1$ in the cross-section $1'$ in Figure 9.9 and $\nu_{i+1} = \nu_2 = 1$ in the cross-section $2'$. Equation (9.194) then yields efficiency $\zeta_\Psi \equiv \mu_{1_{1'}} - \Xi\mu_{2_{2'}}$, in agreement with the previous results and Equation (9.195). Effectively, in the engine mode where $\Phi > 1$ and $\Xi < 1$, an imperfect chemical system (with internal irreversibilities) behaves as it would work with the higher temperature of circulating fluid in its lower branch ($\Phi T_{2'}$ instead of $T_{2'}$) and with a decreased affinity of the effective value (9.195). Of course, power production is decreased by these imperfections. An example of the engine power reduction caused by the imperfect chemical efficiency is presented by Lin et al. (2004).

In the next chapter, this approach will be applied to estimate power output in imperfect fuel cells described by the formalism of neutral components.

10 Fuel cells and limiting performance of electrochemobiological systems

10.1. INTRODUCTION

In this chapter power maximization approach is developed for steady and dynamic fuel cells (FCs) and some biosystems. FCs are flow engines driven by fluxes of chemical reagents and electrochemistry of electric current generation. Faraday's law determines intensity of the electric current generated in an FC. The first-principle theory combines Faraday's law with the efficiency formalism established from balance equations for energy, entropy, and mass. Operating voltage is the departure from the ideal voltage in terms of polarizations. Power output formula follows from transport laws for electrons and ions, irreversible polarizations, and efficiency of energy generation. Some models and experiments for FCs are considered, including those obtained for systems with incomplete conversions. In particular, a steady-state model of a high-temperature solid oxide fuel cell (SOFC) is considered that refers to constant chemical potentials of incoming hydrogen fuel and oxidant. An approach is developed that attributes lowering of the cell voltage below its reversible value to polarizations and imperfect conversions of reactions. Performance curves of the FC and the effect of typical design and operating parameters on the FC performance are analyzed. Biological fuel cells (BFCs) are introduced. Finally we consider energy and size limits of organisms in biological systems.

10.2. ELECTROCHEMICAL ENGINES

The future of power production will certainly include FC systems. An FC is an electrochemical engine or a power converter that directly transforms a part of chemical energy into electrical energy by consuming fuel and oxidant. It operates continuously, generating electricity as long as the fuel and oxidant are supplied. FC system is a thermodynamic device that converts the energy of chemical reaction directly into electricity and heat, thus producing efficiently power with a finite rate. An FC energy source is not stored internally but is continuously provided in the form of fuel such as hydrogen and an oxidant such as oxygen. The process modeling involves individual cells grouped into stacks (modules), that is, a modular system in which stages are connected electrically to ensure a practical voltage and power output.

Energy Optimization in Process Systems
©2013 Elsevier Ltd. All rights reserved.

FCs have recently attracted great attention by virtue of their inherently clean, efficient, and reliable performance. For all types of vehicles FCs represent a developing technology that could potentially replace the internal combustion engine (Lior and Dunbar, 1991; Lior et al., 1991). FC systems operate with significantly higher fuel efficiency and greatly reduced emissions, and have the capability of running on a variety of conventional and alternative fuels (such as hydrogen, methanol, ethanol, and natural gas). The vast use of FC vehicles could have a major impact on reducing petroleum consumption and on improving air quality in cities and large urban areas. FC vehicles in widespread use would create a new growth industry along with jobs for a stronger economy. A major advantage of FCs compared to heat engines is that their efficiency is not a major function of device size. High-temperature SOFCs, which are of main attention here, began with Nernst's 1899 discovery of the still-used yttria-stabilized zirconia solid-state ionic conductor; however, little additional practical development occurred until the 1960s. While the electronic and ionic transfers are necessary to sustain power generation in FCs, it is the overall chemical reaction that is the source of power, and it is the chemical engine property that constitutes the first essential component of the theory of power generation in FCs. The second essential component involves the kinetics of electronic, ionic, and thermal transfer phenomena.

Although internal combustion engines can use hydrogen, it is the high-efficiency, zero-emission hydrogen FC that has largely captured the attention of auto producers. Several automakers have recognized FCs as a superior zero-emission technology, and have large development and commercialization programs. General Motors, Daimler, Honda, Hyundai, and Toyota have presented plans to commercialize FC vehicles sometime between 2015 and 2020. Many auto producers recognize complementary roles for hydrogen FCs and battery electric vehicles and are pursuing both technologies. FCs are highly efficient electrochemical engines that combine hydrogen and oxygen in air to produce electricity to power the vehicle. They work without combustion or emissions of pollutants or greenhouse gases; the only tailpipe emission is water. Nowadays FC vehicles have fuel economies twice that of comparable gasoline cars, and 35–65% higher than gasoline hybrids. FC vehicles use electric drive trains but have a longer range, a faster refueling time, and potential for lower cost than battery electric cars. In addition to the FC stack, other key components of a hydrogen FC vehicle include hydrogen storage, electric motors and power controllers, and batteries for hybrid operation and cold start support (most FC vehicles today are hybrids). A key technology for automotive applications is the proton exchange membrane (PEM). The current focus in academia is the research, laboratory, and educational programs in support of the design and implementation of FC systems technology in advanced vehicles. The standard research focuses on the performance and system integration of FC stacks and associated components in electric and hybrid vehicles. The components are subsystems that include fuel storage, fuel processing and reforming, air delivery units, heat exchangers for thermal integration with other vehicle systems, humidification and water management, DC power processing, and sensors and controls. FC propulsion systems are optimized not only from the standpoint of power, efficiency, and performance but also with respect to criteria involving manufacturing, recycling, safety, cost, reliability, durability, and aging of FC stacks and components.

SOFCs are regarded most efficient energy conversion devices for steady power yield. Out of many types of FCs the SOFCs have a number of distinctive characteristics such as high efficiency of energy conversion, modularity, and environmental compatibility. They do not need noble materials. The high operating temperatures of SOFCs (873–1273K) allow cogeneration and heat reuse. There is a flexibility of fuel choice, for instance, both hydrogen and carbon monoxide can be used as fuel, either directly or indirectly through the water–gas shift reaction. The main virtues of such systems are high electrical and thermal efficiencies, low pollutant emissions, and reliability (Calise et al., 2006a, 2006b, 2007; Dincer and Rosen, 2007; Hussain et al., 2007). Performance of a high-temperature SOFC combined with a conventional recuperative gas turbine (GT–SOFC) plant is analyzed (Haseli et al., 2008). Nowadays, various types of SOFC are investigated, including tubular, planar, and microtubular. Tubular SOFC (TSOFC), developed by Siemens-Westinghouse since the 1970s, can be regarded as most reliable (Li, 2006). Before their commercialization, however, a number of problems need to be solved, such as cost targets, operating life, optimal properties, and integration with traditional devices (Calise et al., 2007). In the last few years many researchers have been involved in the investigation of SOFCs. In the literature numerous papers are available that deal with time-consuming steady and transient CFD simulations of the SOFC stack and of power plants. Yet, zero-dimensional (0D) simulations are much simpler and faster even if they cannot allow to determine profiles of state variables (temperature, pressure, current density, reagents concentrations, etc.). In order to overcome problems associated with high-temperature operation of SOFC and enhance its performance, much of the present research is focused on developing new materials and configurations (Milewski, 2004; Molenda, 2006). These developments usually provide improved performance at reduced temperatures. Another consequence of the temperature reduction is a decrease of ionic conductivity of the electrolyte, which causes increased ohmic polarization.

Methods based on thermodynamics extend to FC systems the optimization approaches belonging to family of those that were recently worked out for power generators. Constraints take into account dynamics of heat and mass transport and rate of real work production. The basic structure of FCs includes electrolyte layer that is in contact with a porous anode and cathode on either side. In a typical FC (Figure 10.1a), gaseous fuels are fed continuously to the anode (negative electrode) compartment and an oxidant (i.e., oxygen from air) is fed continuously to the cathode (positive electrode) compartment; the electrochemical reactions take place at the electrodes to produce an electric current. The fuel or oxidant gases flow past the surface of the anode or cathode opposite the electrolyte and generate electrical energy by the electrochemical oxidation of fuel, usually hydrogen, and the electrochemical reduction of the oxidant, usually oxygen. This makes FC similar to a chemical engine.

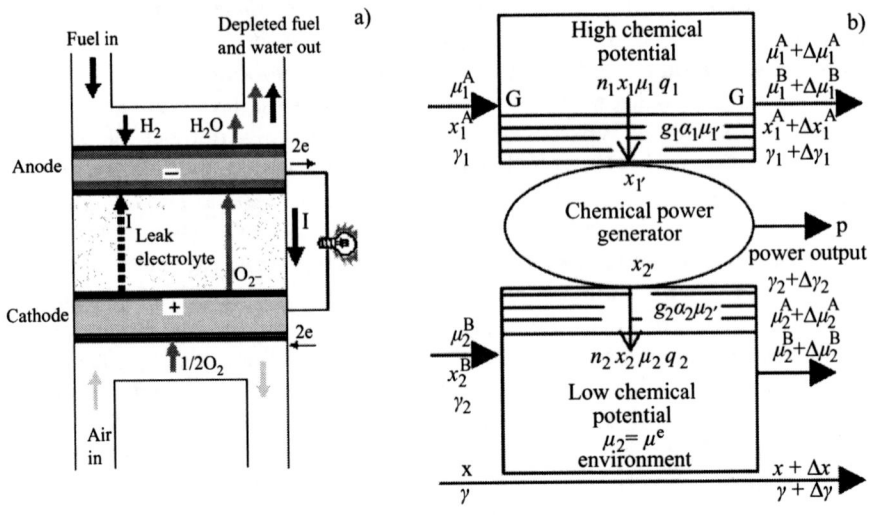

Figure 10.1 Principle of a solid oxide fuel cell (a) as compared with a flow chemical engine (b).

In chemical engines (Figure 10.1b), mass and energy transports propel the transformation of chemical affinities into mechanical power (Sieniutycz, 2009c; Chapter 9); in FCs the transformation leads directly to electrical power. Highly valued initial chemical potentials of reagents decrease in engine mode systems. This situation also occurs in engines based on phase changes (Sieniutycz, 2009b). As opposed to thermal machines (de Vos, 1992), in chemical and electrochemical devices versatile streams (reservoirs) are present, capable of providing both heat and substance. Infinite flows (reservoirs) assure constancy of driving potentials (temperatures and chemical potentials) along stream paths, and problems of extremum power (maximum of power produced and minimum of power consumed) are those of static optimization. Yet, for finite flows (reservoirs), causing limited fluxes of propelling energy and substances, process potentials change along the stream paths, so that the considered problems are those of dynamic optimization and variational calculus. This refers to any potential (thermal or chemical) and is valid for both thermal and chemical engines (de Vos, 1992; Sieniutycz, 1999a, 1999d, 2006a, 2009a; Sieniutycz and de Vos, 2000).

As the FC is a power generator, it is reasonable to ask about an upper bound on power production or, if the process is reversed, a lower bound on power consumption. In reversible evaluations of classical thermodynamics these two bounds are, in fact, absent. Classical reversible thermodynamics is capable of determining reversible cell voltage E and associated power $P = IE$ in terms of the molar Gibbs free energy change for the FC reaction, $E = -\Delta g/(n_e F)$. The voltage E is the equilibrium cell potential obtained at thermodynamically reversible condition. However, power P estimated from the expression $P = IE$ is usually much larger than that observed in experiments (especially for large currents), and,

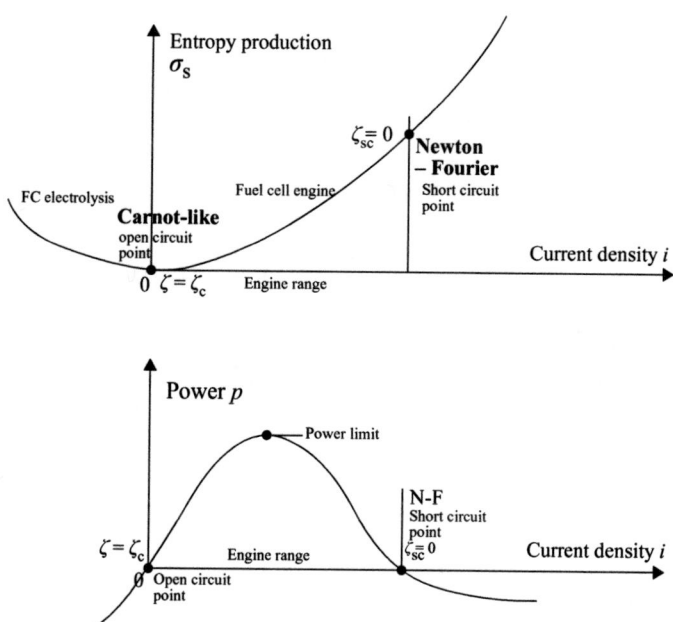

Figure 10.2 Qualitative picture illustrating entropy production and power yield in a fuel cell in terms of density of electric current.

moreover, the experiments show the existence of power limits (upper bounds for FCs) that are not predicted by the formula $P = IE$. Thus, because of their inherent reversibility, reversible evaluations lead to the paradox of unlimited power and results that are in disagreement with reality. By introducing rate-dependent approaches and quantification of various irreversible "polarizations," irreversible thermodynamics is capable of predicting voltage drops, power limits, and power disappearance for sufficiently large electric currents, so the results are much closer to the observed reality.

Power limits in FCs refer to either depletion of basic resources in the FC engine mode or their upgrading in the process undergoing in inverse direction (FC electrolysis mode). For an isothermal process these limits can be determined as the change of the Gibbs function for the flow process in the FC. The electrolysis mode corresponds to inversion of the electric current and power consumption instead of power yield. The range of both processes is illustrated in Figure 10.2.

Power output of an FC is the product of electric current and operational voltage. Power estimation is only apparently simpler than in thermal machines where one must first find a formula for the imperfect converter's efficiency and then multiply it by the driving heat in order to determine a power expression. In FCs the actual voltage is a good indicator of their efficiency, yet this indicator is not easy to calculate because of many voltage-lowering dissipative phenomena that emerge in the FC system. The distinctive role of the operational voltage is not disproved by the observation that, as itself, the voltage measure of the efficiency is not a nondimensional quantity.

Indeed, knowing the operational voltage it is easy to define a nondimensional efficiency as the ratio $\chi = V/E^0$, where E^0 is the reversible cell voltage or the equilibrium cell potential. For the power density expression in terms of χ we find $p = iE^0\chi$ or $p = \chi p_{rev}$, wihich means that the so-defined efficiency equals the ratio of the actual power to the maximum reversible power. This definition links the FC efficiency with the second law, and stresses substantial role of the operational voltage. The latter may be calculated as the difference between equilibrium cell potential E and cumulative contribution of all lossy voltage drops. In a similar way one may evaluate "overvoltages" that have to be overcome in the electrolysis mode of the system. However, one can consider also other definitions of FC efficiencies, in particular those which involve both laws of thermodynamics (first and second laws, Section 10.9).

In dynamic systems power limits may often be linked with finite amounts of resources, similarly to limits in the flow systems that work with finite streams of reagents. The limits for finite resources may often be associated with exergy and either sequential depletion of resource(s) to the environment (engine mode) or their upgrading in the process undergoing in inverse direction (electrolysis mode). Related power integral (total work yield) is a cumulative effect obtained from a dynamic system composed of flowing resources, an unsteady stack, and the external environment. To accomplish the task of limit evaluation, optimization problems are solved, the one for the maximum integral work generated and the one for the minimum of integral work produced. The result is an optimal process trajectory and optimal controls assuring limiting values of total work in a finite time. This leads to dynamic limits for work generation. In the present section, however, only steady-state limits associated with static optimization of power yield in FCs are considered.

A successful result consists in the development of analytical expressions quantifying power produced in steady FCs, P, in terms of current j, and predicting a maximum of P in the regime of moderate j before the dissipation begins to prevail. Maximum power effects have been confirmed in experiments (El-Genk and Tournier, 2000; Ando et al., 2001; Li, 2006). Fundamentals of FCs are associated with thermodynamics of flow systems (Oláh, 1997; Hoffmann et al., 1997; Bejan and Errera, 1998; Hirschenhofer et al., 1998; Fellner and Newman, 2000; Ro and Sohn, 2007; Sieniutycz, 2003b). Starting with basic thermodynamic ideas of the second-law analyses for chemical processes (Denbigh, 1956), exergy-type analyses have been developed for batteries and FCs (Bejan and Dan, 1996, 1997; Bedringas et al., 1997; Cownden et al., 2001; Lior, 2002; Radcenco et al., 2001). Analyses of mass transport and other kinetic phenomena in FCs are available (Singh et al., 1999). Power maximization and design for power problems are pursued (Haynes and Wepfer, 2000; Shi et al., 2002; Spakovsky and Olsommer, 2002; Li, 2006).

A variety of FCs are in different stages of development. They can be attributed to diverse categories, depending on the combination of type of fuel and oxidant, whether the fuel is processed outside of the FC (external reforming) or inside of it (internal reforming), the electrolyte type, the temperature of operation, whether the reactants are fed to the cell by internal or external manifolds, etc.

The most common classification of FCs is by the type of electrolyte used in the cells and operating temperature, for example: (1) polymer electrolyte fuel cell (PEFC) ~ 80 °C, (2) alkaline fuel cell (AFC) ~ 100 °C, (3) phosphoric acid fuel cell (PAFC) ~ 200 °C, (4) molten carbonate fuel cell (MCFC) ~ 605 °C, (5) intermediate-temperature solid oxide fuel cell (ITSOFC) ~ 800 °C, (6) TSOFC ~ 1000 °C, etc.

In this chapter, we focus on power production characteristics for an overall FC reaction (reaction between uncharged components), theoretical estimates of maximum power from the FC engine, and some experimental data confirming power limits in FCs. We also suggest a consistent thermodynamic methodology for the calculation of basic performance characteristics of SOFCs.

The goals of the present chapter include: (a) formulation of an advanced thermoelectrochemical model of imperfect FCs, especially those with incomplete chemical conversions, (b) implementation of the model to high-T SOFCs, (c) prediction of various losses and their influence on cell performance, and (d) application of FC characteristics in determining power limits. The final part of the chapter deals with BFCs and energy/size limits of organisms in biological systems.

10.3. THERMODYNAMICS OF ENTROPY PRODUCTION AND POWER LIMITS IN FUEL CELLS

Irreversible thermodynamics provides reliable methodologies for optimizing performance and improving design of various power systems (de Vos, 1992; Sieniutycz, 2009c; Zhao et al., 2008). This field also constitutes an efficient tool in evaluating the impact of resource exploitation and upgrading on the environment.

Knowledge of operational voltage helps to define a cell efficiency as the ratio $\chi = V/E^0$, where E^0 is the reversible cell voltage. For power density in terms of χ one finds $p = iE^0 \chi$ or $p = \chi p^{rev}$, which means that the efficiency equals the ratio of actual power to maximum reversible power. This definition links FC efficiency with the second law, and stresses the substantial role of the operational voltage.

Both the experiments and the theory show that an FC operates between two extreme zero-power limits (Figure 10.2). One extreme is the reversible, a Carnot-like limit with the highest voltage and voltage-related efficiency. This voltage limit is at the open circuit state, and is associated with the zero current. This is also the state of zero power, where the cell works extremely slowly (quasistatically). For finite, yet small currents the power of an FC engine becomes greater than zero. Yet, for too large currents the FC operates inefficiently because the dissipative effects prevail and the operational voltage decreases significantly. Hence, another extreme arises (a counterpart of the Newton–Fourier point in thermal machines

Characteristics of SOFC at various temperatures

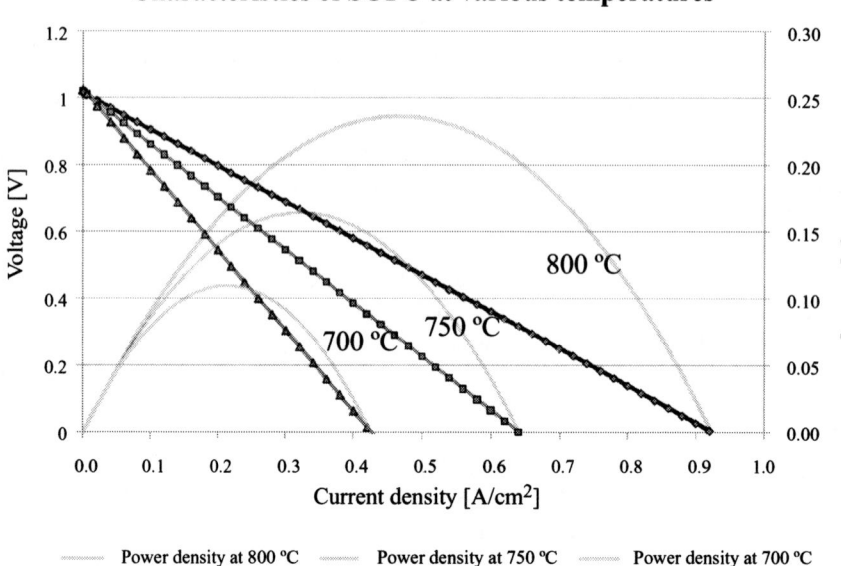

Power density at 800 °C ──── Power density at 750 °C ──── Power density at 700 °C

Figure 10.3 Voltage and power data vs. electric current density for a SOFC at various temperatures. Continuous lines represent the Aspen Plus™ calculations testing the model consistency with the experiments. The lines were obtained in Wierzbicki's MsD thesis supervised by S. Sieniutycz and J. Jewulski (Wierzbicki, 2009). Points refer to experiments in Warsaw Institute of Energetics (Wierzbicki, 2009, and reference [18] therein).

theory) with vanishing power and voltage. For some intermediate currents a power maximum is observed (Figure 10.3), which is the upper limit of power. Working FCs frequently operate somewhere between the point of maximum efficiency and that of maximum power.

In order to analyze the performance of a class of hydrogen–oxygen FCs, the following basic assumptions are often accepted (Zhao et al., 2008):

(1) FCs operate in stationary regime.
(2) In the domain of interest temperature and pressure are uniform and constant ($P = 1$ atm).
(3) Fuel composition in the anode inlet consists of hydrogen and water vapor mixture, and air composition in the cathode inlet is at the typical value of 21% O_2 and 79% N_2.
(4) Reactants are compressible ideal gases having constant thermodynamic and electrochemical properties.
(5) Complete chemical conversion takes place and no reactants are left after the reaction.
(6) No side or electrode reactions (e.g., corrosion) occur and no gas leakage is taken into account.
(7) No current leakage is possible.

Interestingly, assumption (5) can be relaxed, and the present considerations present some ways showing how this could be done for FCs.

Assume that all incoming streams (those with "higher" input of Gibbs flux $G_{in} = G_{1'}$) represent a common phase of "substrates" (all system components in the state before the chemical transformation, index $1'$). All outgoing streams (those with "lower" Gibbs flux $G_{out} = G_{2'}$) represent a common phase of "products" (all system components in the state after the transformation, index $2'$). Power p follows from entropy conservation and energy balance in the reversible part of the system (primed indices). For an isothermal reactor described in terms of neutral components (Sundheim, 1964; Ekman et al., 1978)

$$p = \mu_{1_1'}n_{1_1'} + \mu_{2_1'}n_{2_1'} + \cdots \mu_{i_1'}n_{i_1'} \cdots + \mu_{m_1'}n_{m_1'}$$
$$-\mu_{1_2'}n_{1_2'} - \mu_{2_2'}n_{2_2'} \cdots - \mu_{i_2'}n_{i_2'} \cdots - \mu_{m_2'}n_{m_2'}$$

$$(10.1)$$

This formula states that, in a steady and isothermal process, power yield of an engine system is the difference between the input and output flux of the Gibbs function (Zhao et al., 2008).

For the important special case of a complete conversion, we can transform Equation (10.1) to the pronounced form of Equation (10.2). In this case components are numbered in such a way that species 1, 2, ..., i are substrates and species $i + 1$, $i + 2$, ..., m are products. Thus, total power yield of an isothermal multireaction process takes the form

$$p = \sum_{j=1}^{R} \{p_j\}$$
$$= \sum_{j=1}^{R} -\{\mu_{1_1'}\nu_{1j} + \mu_{2_1'}\nu_{2j} + \cdots \mu_{i_1'}\nu_{ij}$$
$$+ \mu_{i+1_2'}\nu_{i+1j} \cdots + \cdots \mu_{m-1_2'}\nu_{m-1j} \cdots + \mu_{m_2'}\nu_{mj}\}n_j$$

$$(10.2)$$

Quantities n_j are molar chemical fluxes of reagents, that is, products of the electrode surface area F and heterogeneous rates, r_j. For a complete conversion, power yield from the electrode unit area equals the sum of products of affinities and reaction rates

$$p = \sum_{j=1}^{R} \{\tilde{A}_j n_j\} = F \sum_{j=1}^{R} \{\tilde{A}_j r_j\}$$

$$(10.3)$$

Yet, the assumption of a complete transformation of substrates into products can be too restrictive. Therefore, it is advisable to show how the assumption can be relaxed simultaneously with the affinity formalism preserved. By considering chemistry of systems with power production and transport phenomena, effects of incomplete conversions in homogeneous reactions were evaluated in Zhao et al., 2008. For a single heterogeneous reaction, such as FC reaction considered here, an isothermal power formula that replaces Equation (10.3) to include the effect of incomplete conversions can be written in the form

$$p = (\Pi_{1'} - \Xi\Pi_{2'})n_{1'} = -\frac{iA}{n_e F}\Delta g^{\text{eff}}(T, p) = -\Delta \dot{G}^{\text{eff}} \qquad (10.4)$$

where primed quantities refer to inputs and outputs of the chemically active zone. Quantity $\Pi_{1'}$ (Chapter 9) is "one-way chemical affinity" attributed to reactants with known chemical potentials and $n_{1'}$ is the (positive) chemical flux defined as the product of heterogeneous reaction rate and electrode area. Incomplete conversion is described by the fraction Ξ that is close to the reciprocity of "less correct" coefficient Ψ introduced earlier (Sieniutycz, 2009c). In the literature several quantities such as Ξ, Ψ, etc. appear, which characterize detrimental increase of chemical potentials of reaction products caused by their dilution by remaining reactants.

Equation (10.4) generalizes the idealized power of an "endoreversible" system ($\Xi = 1$), in which case the potential difference $\Pi_{1'} - \Pi_{2'}$ is the chemical affinity or $-\Delta g$. In fact, the equation describes the chemical component of power caused by chemical flux $n_{1'}$. Power is generated with a nonideal chemical efficiency $\xi = \Pi_{1'} - \Xi\Pi_{2'}$. For the simplest reaction, $1 \leftrightarrow 2$, $\xi = \mu_{1'} - \Xi\mu_{2'}$, which is lower than $\mu_{1'} - \mu_{2'}$ (note that chemical potentials are negative). Effectively, in the engine mode where $\Xi \leq 1$, a system with internal imperfections behaves as if it operated with a decreased affinity of an effective value $\Pi_{1'} - \Xi\Pi_{2'}$.

By generalizing the first equality in Equation (10.4) to nonisothermal chemical systems with power production one should take into account thermal effects in the power production formula and internal entropy production within these systems. For a single chemical reaction the corresponding power formula can be written in the form

$$P = \left(1 - \Phi\frac{T_{2'}}{T_{1'}}\right)\dot{Q}_{1'} + (\Pi_{1'} - \Xi\Pi_{2'})\dot{n}_{1'} \qquad (10.5)$$

where $\dot{Q}_{1'}$ is the total heat flux entering the active chemical zone (the flux involving the sensible heat flux and the sum of products of partial entropies and fluxes of species multiplied by the temperature T). Internal imperfection functions, Φ and Ξ, are, respectively, related to internal entropy production and incomplete conversion. A working counterpart of Equation (10.5) for an isothermal FC process is given by Equation (10.32).

Vector nature of efficiency is the rule in nonisothermal and multireaction systems. The first (thermal) component of power yield (10.5) describes power generated by total heat flux with the imperfect (nonCarnot) thermal efficiency $\eta = 1 - \Phi T_{2'}/T_{1'}$, whereas the second or chemical component represents power yield caused by chemical flux $n_{1'}$. In the engine mode where $\Phi > 1$ and $\Xi \leq 1$, the system with internal imperfections behaves as it would operate with the higher temperature of working fluid in its lower branch ($\Phi T_{2'}$ instead of $T_{2'}$) and with a decreased affinity of an effective value $\Pi_{1'} - \Xi\Pi_{2'}$. Of course, power production is decreased by these imperfections.

In the considerations below, we incorporate the present approximate approach to the FC modeling in order to get rid of the assumption that no reactants remained after the FC reaction. An alternative account of the issue of incomplete conversions can be found in the literature (Zhao et al., 2008; Wierzbicki, 2009).

Limited power yield in FCs, that is, impossibility to achieve arbitrarily large power, is caused by nonvanishing entropy production. Thermal and electrochemical irreversibilities, essential in the FC operation, decrease its performance. Many researchers have focused on experimental and theoretical studies of irreversible FCs; see, for example, literature reviews and books (Larminie and Dinks, 2000; Żbikowski, 2004; Zhao et al., 2008; Chen and Zhao, 2009; Wierzbicki, 2009; Mench et al., 2009; Sieniutycz, 2010c).

Yet, it is still appropriate to validate the accuracy of generally accepted descriptions that ignore some possible effects. In particular, issues such as incomplete conversion and current leakage still require further considerations. Computational models and simulation tools can provide valuable guidance for design, performance optimization, and cost reduction of FCs. They may be coupled with a computational fluid dynamics model that treats diffusive transport in the electrodes and convective transport in the reactant flow channels. Computer simulations can also test quantitative role of effects ignored in earlier researches. For instance, cell performance is usually calculated under an assumption of no leakage conductance that, however, cannot be ignored in all FC operations (Zhao et al., 2008).

10.4. CALCULATION OF OPERATIONAL VOLTAGE

10.4.1. Some Introductory Issues

A single cell can be made to achieve whatever current and power is required, simply by scaling up the size of the active electrode area. However, for realistic operating conditions, the output voltage of a single H_2 polymer electrolyte membrane fuel cell (PEMFC), DMFC, or SOFC is less than 1 V. Therefore, for most applications and for compact design, we use an FC stack of several individual cells connected in series.

The FC operates continuously generating electricity as long as hydrogen and oxygen are supplied. Its basic principle is illustrated in Figure 10.1a by the SOFC using hydrogen as fuel and air as oxidant. The cell contains an electrolyte in contact with the anode and cathode on either side. The electrolyte acts as a barrier between the anode and the cathode allowing only certain types of ions to pass through. The electrons travel around the external circuit from the anode to the cathode producing useful current.

At the cathode, oxygen is reduced by the incoming electrons to produce the oxygen anions. These are conducted through the electrolyte to the anode where they electrochemically combine with the adsorbed hydrogen to form water and heat as a by-product and release electrons to the external circuit. The electrons return from the external circuit to participate in the electrochemical reduction at the cathode.

The electrochemical reactions in the cell can be outlined as follows:

Anode reaction:

$$H_2 + O^{2-} \rightarrow H_2O + 2e \tag{10.6}$$

Cathode reaction:

$$\frac{1}{2}O_2 + 2e \rightarrow O^{2-} \tag{10.7}$$

Overall reaction:

$$H_2 + \frac{1}{2}O_2 \rightarrow H_2O + \text{heat} + \text{electricity} \tag{10.8}$$

The intensity of the electric current generated in the FC is the effect of the chemical reaction of fuel oxidation. In the reaction four electrons flow through the external circuit per single molecule of the oxygen that participates in the reaction, consistent with the fact that two electrons are carried through the electrolyte by the single ion O^{2-}. Therefore, in agreement with Faraday's law

$$I = iA = n_e F \dot{n}_{\text{fuel}} = 4F\dot{n}_{O_2} = 2F\dot{n}_{O^{2-}} \tag{10.9}$$

where n is the number of moles, $\dot{n}_{\text{fuel}} = dn/dt$ is the molar consumption rate of fuel, n_e is the number of electrons transferred in reaction, $F = 96\ 485$ C/mol is Faraday's constant, I is current intensity, i is current density, and A is electrode surface area.

For a steady-state cell driven by a single reaction the mole balance of species k is

$$\frac{n_k^{\text{in}} - n_k^{\text{out}}}{v_k} = -n \tag{10.10}$$

where the balance is in terms of fluxes and we have introduced a k-independent quantity, n, called the electrochemical flux defined as the product of the heterogeneous reaction rate and the electrode surface area. In Equation (10.10) n_k is the mole number of species k and v_k is the stoichiometric coefficient of species k in the reaction.

The notion of electrochemical flux can easily be generalized for multiple reaction systems. For the heterogeneous reaction j, its electrochemical flux is defined as

$$n_j = -\frac{(n_k^{\text{in}} - n_k^{\text{out}})_j}{v_{k,j}} \tag{10.11}$$

where n_j is the product of jth reaction rate and the electrode surface area, $n_k^{\text{in}} - n_k^{\text{out}}$ is the change of molar flux of component k reacting in reaction j, and $v_{k,j}$ is the stoichiometric coefficient of component k in reaction j. Equations like (10.10) and (10.11) serve to derive the definition of the imperfection coefficient Ξ in Equation (10.5). Ξ refers to the detrimental effect of incomplete conversion on power (see also Equation (10.32)).

For the charge transfer via the hydrogen reaction, $H_2 + O^{2-} \rightarrow H_2O + 2e$, $\nu_{H_2} = -1$, $\nu_{H_2O} = 1$, and $n_e = 2$. For the transfer via oxygen $1/2 O_2 + 2e \rightarrow O^{2-}$ the same n_e follows.

10.4.2. Equilibrium Cell Potential

The perfect FC achieves its maximum reversible voltage when no current is flowing by the external load. The reversible voltage is the equilibrium cell potential obtained from the reversibility condition

$$E^0 = -\frac{\Delta g(T, p)}{n_e F} \tag{10.12}$$

The above expression contains the molar Gibbs free energy of the cell reaction, Δg. This quantity depends on the temperature T and partial pressures of reactants p_k:

$$\Delta g(T, p) = \Delta g^0(T) - RT \ln \left(\frac{p_{H_2} p_{O_2}^{1/2}}{p_{H_2O}} \right) \tag{10.13}$$

where $R = 8.314 \, J/(mol \, K)$ is the universal gas constant and $\Delta g^0(T)$ is the molar Gibbs free energy of reaction at $p_0 = 1$ atm. The temperature-dependent quantity Δg^0 is the standard molar Gibbs free energy of the reaction.

10.4.3. Leakage and Internal Resistances

The operating voltage of an FC can be expressed as a departure from the ideal voltage in terms of "polarizations." This is because the FC output voltage is always less than the reversible cell voltage since some voltage drops across the cell are caused by irreversible losses. Multiple phenomena contribute to irreversible losses in an actual FC.

The losses include three main sources: activation polarization (V_{act}), ohmic polarization (V_{ohm}), and concentration polarization (V_{conc}). To treat these losses in a unified and consistent way it is advisable to introduce associated activation resistance (R_{act}), ohmic resistance (R_{ohm}), and concentration resistance (R_{conc}). These resistances, in general, are current and temperature dependent. The total internal resistance of the FC (R_{int}) is composed of these partial resistances, and can be determined as

$$R_{int} = \frac{V_{int}}{I} = \frac{V_{act} + V_{conc} + V_{ohm}}{I} = R_{act} + R_{conc} + R_{ohm} \tag{10.14}$$

where V_{int} is the voltage on the internal resistance. Activation and concentration polarization occurs at both anode and cathode locations, while the resistive polarization represents ohmic losses throughout the FC.

As pointed out in a number of sources, for example (Zhao et al., 2008; Mench, 2008; Mench et al., 2009; Wierzbicki, 2009), the measured open circuit potential in an FC is usually lower than the ideal reversible potential. A way to manage this effect, proposed by Zhao et al. (2008), is to assume some electronic current leakage through the electrolyte. In this case, a leakage conductance $(R_{leak})^{-1}$ is introduced, parallel with the external load (Figure 10.1a). When electrons flow through the internal and external FC circuit, local behavior of irreversible losses is described as an equivalent circuit, which is a series of an ideal (reversible) voltage defined by Equation (10.12), internal resistance represented by the sum of three polarization contributions, and leakage resistance in parallel with the load. In this approach, the operating voltage of the FC is represented as a departure from the ideal voltage:

$$V = E^0 - V_{int} = E^0 - V_{act} - V_{conc} - V_{ohm}$$
$$= E^0 - I(R_{act} + R_{conc} + R_{ohm}) \tag{10.15}$$

Polarization losses are generally dependent on gas thermodynamic parameters (partial pressures, temperature) and current density in the FC.

Equation (10.15) can include the electrolysis mode in which the sign of the electric current is negative. In terms of total resistance R both cases can be described by the common formula

$$V = E^0 - RI \tag{10.16}$$

This equation shows that in electrolysis operating voltage V must be greater than the reversible voltage E^0. Also, power $P = VI$ consumed in the electrolysis mode is greater than reversible power $E^0 I$ evaluated for the same absolute value of current intensity I. For the total resistance R independent of current direction, the difference between power for running the FC electrolysis and that released from the FC engine equals the double of power dissipated in one system's mode:

$$\Delta P = I(V_{electrol} - V_{eng}) = 2RI^2 \tag{10.17}$$

10.4.4. Activation Polarization

Activation polarization, which dominates losses at low current density, is the manifestation of voltage overpotential required to overcome the activation energy of the electrochemical reaction on the electrode surface, and is thus similar to the activation energy of chemical reactions. This polarization can be represented by the Tafel equation at each electrode; it may be regarded as an extra potential necessary to overcome the energy barrier of the rate-determining step of the reaction to a value such that the electrode reaction proceeds at a desired rate. An assumption made in FC modeling assumes that the occurrence of each reaction is a one-step, single-electron transfer process. The kinetic expression for the oxygen reduction rate can be described by using the Butler–Volmer equation

$$i = i_0 \left\{ \exp\left(\beta \frac{FV_{act}}{RT}\right) - \exp\left(-(1-\beta)\frac{FV_{act}}{RT}\right) \right\} \tag{10.18}$$

where β is the transfer coefficient and i_0 is the exchange current density. The transfer coefficient is the fraction of the change in polarization that leads to a change in the reaction rate constant, and usually is 0.5 for FCs (Larminie and Dinks, 2000; Mench et al., 2009). With $\beta = 0.5$, polarizations for the anode and cathode can be obtained from Equation (10.18) in the form

$$V_{act;a} = 2RTF^{-1} \sinh^{-1} \left(\frac{i}{2i_{0,a}} \right) \tag{10.19}$$

and

$$V_{act;c} = 2RTF^{-1} \sinh^{-1} \left(\frac{i}{2i_{0,c}} \right) \tag{10.20}$$

It follows that the activation polarization depends principally on the exchange current density. The exchange current density is defined as the current density flowing equally in each direction at the reversible potential. Several research reviews, for example, Zhao et al. (2008) and Larminie and Dinks (2000), show that the anode and cathode exchange current densities can be found from semi-empirical correlations

$$i_{0,a} = \gamma_a \left(\frac{p_{H_2}}{p_0} \right) \left(\frac{p_{H_2O}}{p_0} \right) \exp \left(-\frac{E_{act,a}}{RT} \right) \tag{10.21}$$

and

$$i_{0,c} = \gamma_c \left(\frac{p_{O_2}}{p_0} \right)^{1/4} \exp \left(-\frac{E_{act,c}}{RT} \right) \tag{10.22}$$

where γ_a and γ_c are preexponential coefficients, for the anode and cathode, respectively, p_0 is the reference pressure (1 atm), and $E_{act,a}$ and $E_{act,c}$ are activation energies for anode and cathode, respectively. These parameters and other suitable data that have become now common in many FC calculations are collected by Zhao et al. (2008), as shown in Table 10.1.

10.4.5. Ohmic Polarization

For a properly designed cell, ohmic polarization is typically dominated by electrolyte conductivity, which is primarily a function of water content and temperature in PEM FCs, and operating temperature in SOFCs. Ohmic polarization losses occur in an FC because of resistance to the flow of ions in the electrolyte and resistance to the flow of electrons through the electrode. In the cells produced recently, however, the resistance exhibited by the anode and cathode materials is negligible in comparison with the electrolyte resistance. Thus, with a good approximation, only the electrolyte contribution to resistance can be considered, which can be evaluated as

$$R_{ohm} = \frac{L_{el}}{\sigma_{el}A} \tag{10.23}$$

Table 10.1 Standardized data of operating conditions and performance-related parameters for SOFCs, as listed by Zhao et al. (2008)

Parameter	Value
Operating temperature, T (K)	1073
Operating pressure, p_0 (atm)	1
Fuel composition, p_{H_2}, p_{H_2O}	0.97; 0.03
Air composition, p_{O_2}, p_{N_2}	0.21; 0.79
Charge transfer coefficient, β	0.5
Number of electrons, n_e	2
Prefactor for anode exchange current density, γ_a (A/m^2)	5.5×10^8
Activation energy of anode, $E_{act,a}$ (J/mol)	1.0×10^5
Prefactor for cathode exchange current density, γ_c (A/m^2)	7.0×10^8
Activation energy of cathode, $E_{act,c}$ (J/mol)	1.2×10^5
Electrode thickness, L_{el} (μm)	20
Activation energy of O^{2-}, E_{el} (J/mol)	8.0×10^4
Prefactor of O^{2-}, σ_0 (S/m)	3.6×10^7
Ratio of the internal resistance to the leakage resistance, k	1/100
Anode limiting current density, $i_{L,a}$ (A/m^2)	2.99×10^4
Cathode limiting current density, $i_{L,c}$ (A/m^2)	2.16×10^4
Faraday's constant, F (C/mol)	96 485
Universal gas constant, R (J/(mol K))	8.314

where L_{el} is the electrolyte thickness and σ_{el} is the ionic electrolyte conductivity. The latter can be defined in terms of temperature as follows:

$$\sigma_{el} = \frac{\sigma_0}{T} \exp\left(-\frac{E_{el}}{RT}\right) \tag{10.24}$$

(Zhao et al., 2008; Mench et al., 2009; Larminie and Dinks, 2000). In this formula, E_{el} is the activation energy for ion transport and σ_0 is the reference conductivity. With the above equations ohmic polarization of the FC can be expressed as the product IR_{ohm} or

$$V_{ohm} = \frac{iL_{el}T}{\sigma_0} \exp\left(\frac{E_{el}}{RT}\right) \tag{10.25}$$

Even a small reduction in ohmic losses through advanced materials, thinner electrolytes, or optimal temperature distribution can significantly improve FC performance and its power. Yet, there are still other polarizations, such as concentration polarization, described below.

10.4.6. Concentration Polarization

At very high current densities, mass transport limitation of fuel or oxidizer to the electrode causes a sharp decline in the output voltage. This effect is referred to as the concentration polarization. The related region of the polarization curve is a mass transport controlled regime; thus, any creative means of facilitating species transport to the electrode surface results in improved fuel utilization at high current densities. Concentration polarization prevents the surrounding material to maintain the initial concentration of the bulk fluid as the reactant is consumed at the electrode. This polarization loss occurs over the entire range of current density, but becomes significant at high currents when it is difficult to provide enough reactant flow to the reaction sites. The Damköler number (Da), which is the ratio of the characteristic electrochemical reaction rate to the rate of mass transport to the reaction surface, is a related dimensionless parameter. In the limiting case of infinite kinetics (high Da), one derives the Tafel-based formulae

$$V_{conc,a} = \frac{RT}{n_e F} \ln \left(1 - \frac{i}{i_{L,a}} \right) \tag{10.26}$$

and

$$V_{conc,c} = \frac{RT}{n_e F} \ln \left(1 - \frac{i}{i_{L,c}} \right) \tag{10.27}$$

In these formulae $i_{L,a}$ and $i_{L,c}$ are the limiting current densities of related electrodes, which represent maximum currents produced when the reactant surface concentration is reduced to zero at the reaction site. The limiting current density is affected by transports in the electrodes, and depends on various parameters such as cell temperature, chemical composition, operating pressure, etc. Yet, the previous research shows that the limiting current is only slightly dependent on T, especially when the current is far from its limit (Zhao et al., 2008; Mench et al., 2009; Larminie and Dinks, 2000). Therefore, the calculation of the overall concentration polarization can be simplified by attributing constant values for the limiting current densities at the electrodes, as listed in Table 10.1.

Mathematical models for an FC including feed channels can be developed to simulate cell performance. A set of conservation equations and species concentration equations can be solved numerically in a coupled gas channel and porous media domain. A basic idea in modeling is to construct an algorithm with the programming potential for the whole FC system. Improvement in efficiency and reliability of economic optimization proclaim the importance of such mathematical models (Żbikowski, 2004).

10.5. THERMODYNAMIC ACCOUNT OF CURRENT-DEPENDENT AND CURRENT-INDEPENDENT IMPERFECTIONS

10.5.1. Effective Enthalpy and Effective Gibbs Free Energy of Reaction

To evaluate power production and effects of current-independent losses on the FC performance, we apply below some effective thermodynamic quantities and mimic the approach used in describing the reversible operation. Let us focus on the isothermal case.

With Equations (10.5) and (10.9)–(10.11), we introduce effective enthalpy change and Gibbs free energy change between the products p and reactants s that take into account incomplete conversion in the overall electrochemical reaction. In Equations (10.28)–(10.31), fraction Ξ is the coefficient of internal imperfections caused by incomplete conversion (in multiple reactions, 1, 2, ..., N, different coefficients Ξ_j may be introduced for each reaction), h_k is the molar enthalpy of the species k at temperature T, p is the partial pressure of the species, μ_k is the partial molar Gibbs free energy of species k usually called the chemical potential of species k, and subscripts p and s represent the pth product and sth reactant of the reaction, respectively.

For the steady-state fuel cell at temperature T, the effective enthalpy change per unit time can, respectively, be expressed as

$$\Delta \dot{H}^{\text{eff}} = \sum_p \Xi \dot{n}_p h_p(T) - \sum_s \dot{n}_s h_s(T) = \frac{iA}{n_e F} \Delta h^{\text{eff}}(T) \qquad (10.28)$$

and the effective or reduced Gibbs flux change of the FC reaction is

$$\Delta \dot{G}^{\text{eff}} = \sum_l \Xi \dot{n}_p \mu_p(T) - \sum_k \dot{n}_s \mu_s(T) = \frac{iA}{n_e F} \Delta g^{\text{eff}}(T, p)$$

$$= -\frac{iA}{n_e F}(\Pi_s - \Xi \Pi_p) \qquad (10.29)$$

where the following definitions of effective quantities are introduced:

$$\Delta h^{\text{eff}}(T) = \sum_p v_p \Xi \mu_p(T) - \sum_s v_s h_s(T) \qquad (10.30)$$

and

$$\Delta g^{\text{eff}}(T, p) = \sum_p v_p \Xi \mu_p(T, p) - \sum_s v_s \mu_s(T) \qquad (10.31)$$

The difference $\Pi_s - \Xi \Pi_p$ in Equation (10.29) is the reduced affinity of the FC reaction.

Clearly, for $\Xi = 1$ the formulae describing perfect systems are recovered. These perfect-system formulae constitute familiar equations of reversible thermodynamics of FCs.

10.5.2.   Current-Independent Imperfections

We can now consider current-independent imperfections and attribute these non-idealities to the lowering of the open circuit potential below the ideal reversible potential.

Thanks to introducing Ξ, effective chemical potential of a product p becomes $\Xi \mu_p$ at the outlet; because each $\Xi_j < 1$, and μ_k are negative, effective chemical potentials of products in the chemical affinity are higher than actual ones. This effect decreases chemical efficiencies (driving forces and affinities).

As the change of Gibbs flux in an electrochemical reaction is a maximum electrical energy obtained as work (Zhao et al., 2008; Sieniutycz, 2009c), the largest power output generated by an imperfect, isothermal FC reaction can be determined as

$$P_{irr} = -\Delta\dot{G}^{eff} = \sum_s \dot{n}_s \mu_s(T,p) - \sum_p \dot{n}_p \Xi\mu_p(T,p) = -\frac{iA}{n_e F}\Delta g^{eff}(T,p)$$

(10.32)

Clearly, the calculations of $\Delta h^{eff}(T)$ and $\Delta g^{eff}(T)$ mimic those of $\Delta h(T)$ and $\Delta g(T)$ given, for example, by Zhao et al. (2008), and are based on the tabulated values at operating temperature T and pressure $p_0 = 1$ atm, where Δh is independent of pressure. For an example of the latter data, see Table 2 in Zhao et al. (2008).

In nonisothermal systems, the first term of power formula (10.5) serves to generalize the isothermal power of Equation (10.32). When temperature difference $T_1' - T_2'$ is actively exploited in power generation, thermal component of power (10.5) describes extra power generated by total heat Q_1' with the lowered (nonCarnot) thermal efficiency $\eta = 1 - \Phi T_2'/T_1'$. In fact, thermal management is one of the major issues in many FCs. The energy released by the electrochemical reaction is split between electrical and thermal components. The waste heat rate can be as much as 60% or more of the output, and is roughly equal to the current density, i, multiplied by the departure from actual open circuit voltage:

$$P_{waste} = niA(E_{oc} - V_{cell})$$

(10.33)

where n is the number of cells in a series in the stack, V_{cell} is the single cell operating voltage, and E_{oc} is the theoretical open circuit voltage (Mench et al., 2009). For discussion of associated cooling issues, see Larminie and Dinks (2000) and Mench et al. (2009).

When no current is required by the external load, an internally imperfect FC achieves its largest voltage. This voltage is also the imperfect cell potential at the open circuit point:

$$E^{eff} \equiv E_0 = -\frac{\Delta g^{eff}(T,p)}{n_e F}$$

(10.34)

For ignorable internal imperfections of power conversion ($\Xi = 1$), the above equation reduces to the formula describing the equilibrium cell potential, Equation (10.12).

As explained by Sieniutycz (2010c), the use of effective quantities in the presented approach is an approximation whose simplicity and practical value is of some importance. Due to losses resulting from undesired species crossover from one electrode through the electrolyte and internal currents, the actual open circuit voltage is usually below the theoretical value (Larminie and Dicks, 2000; Mench et al., 2009).

Fuel utilization efficiency is never 100%, especially because of concentration polarization limitation on performance. Unused hydrogen in the case of SOFC (or liquid and vapor-phase methanol in the case of DMFC) must be actively recycled, utilized, or converted prior to exhaust to the environment. Potential schemes include the use of condensers, recyclers, secondary burners, or catalytic converters.

10.5.3. Current-Dependent Imperfections

We shall now pass to current-dependent imperfections and attribute them to a finite rate of entropy generation. Equation (10.32) is an indicator leading to the largest power obtained from the reaction in an imperfect FC and associated losses resulting from incomplete conversion and undesired species crossover from one electrode through the electrolyte and internal currents. In practice, even this part of potential power is never completely utilized because of the current-dependent irreversibilities. They are attributed to the system's entropy production by heat transfer, friction, mixing, and chemical reactions, and cause polarizations that we have discussed earlier. When the FC works normally and produces useful power through the external load, the corresponding entropy production resulting from the internal resistance and leakage resistance can be determined by an expression

$$\dot{S}_{tot} = \dot{S}_{int} + \dot{S}_{leak} = \frac{I_{int}^2 R_{int}}{T} + \frac{I_{leak}^2 R_{leak}}{T} \qquad (10.35)$$

where T is the ambient temperature, I_{int} is the electric current through the equivalent internal resistance R_{int}, and I_{leak} is the electric current through the leakage resistance R_{leak} (Zhao et al., 2008). Equation (10.35) includes the thermal contribution of the power degradation to the environment. The (positive) dissipation rate, $\dot{S}_{tot}$, is that of the FC and environment. Some formulations use this quantity for optimization of performance parameters in thermal units and FCs.

10.6. EVALUATION OF MASS FLOWS, POWER OUTPUT, AND EFFICIENCY

10.6.1. Mass Flow Rate of Reactants

Several factors determine the appropriate mass flow rate of reactants. These factors are related to requirements such as the minimum requirement for electrochemical reaction, maintaining proper water balance, and thermal management (Mench et al., 2009). Water management concerns may dictate the need for increased flow rate, for example. However, the minimum flow requirements for all FCs are determined by the requirements of the electrochemical reaction. An expression for the flow rate of a reactant required for electrochemical reaction follows from Equation (10.9) in the form

$$\dot{n}_{react} = \frac{I}{n_e F} = \frac{iA}{n_e F} \tag{10.36}$$

where i and A represent the current density and total electrode area, respectively (Mench et al., 2009). The stoichiometric ratio for an electrode reaction is defined as the ratio of reactant supplied to that needed for the electrochemical reaction. Mench et al. (2009), for example, consider a hydrogen-fed polymer membrane FC (H_2 PEMFC) stack with 20 plates, an active area of $100\,cm^2$ per plate, operating at a current density of $1\,A/cm^2$. This gives a total active area of $2000\,cm^2$. For the anode reaction, n_e is 2, and one can determine the minimum required flow rate of hydrogen from Equation (10.36). If this minimum rate is doubled, then the anode stoichiometric ratio, ξ_a, is 2. On the cathode side, n_e is 4 (Figure 10.1a) and the required molar oxygen flow rate is simply half of the hydrogen molar flow rate. Since air is typically used instead of pure oxygen, the resulting value must be converted into an air flow rate. Considering Equation (10.36), useful expressions can be developed for the required mass flow rates as a function of FC electrical power and operating conditions. For example, for the direct methanol FC cathode and anode, Mench et al. (2009) show that

$$\dot{m}_{air} = 3.56 \times 10^{-4} \frac{P_{stack}}{V_{cell}} \xi_c \tag{10.37}$$

and

$$\dot{m}_{fuelsolution} = 1.71 \times 10^{-3} \frac{P_{stack}}{V_{cell} Mol} \xi_a \tag{10.38}$$

where P_{stack} is the FC stack electrical power output, V_{cell} is the operating cell voltage, and Mol is the fuel solution molarity. These expressions can be useful to approximate the flow requirements for any sized system. For example, a 2 kW DMFC stack operating at an average of 0.4 V per plate and stoichiometry of 3 will require ca 5.34 g/s of air. Similar data are obtained for other FC systems (Mench et al., 2009).

10.6.2. Power Output and Efficiency

An analytical method for predicting the performance of the FC can play an important role in assessing the effect of different parameters. From Equations (10.15), (10.19), (10.20), (10.25)–(10.27), (10.32), and (10.34), an expression of the cell terminal voltage can be presented in the form resembling that of Zhao et al. (2008), namely:

$$
\begin{aligned}
V &= E^{\text{eff}} - V_{\text{act,a}} - V_{\text{act,c}} - V_{\text{conc,a}} - V_{\text{conc,c}} - V_{\text{ohm}} \\
&= \frac{1}{n_e F} \left[-\Delta g^{0\text{eff}}(T, P) + RT \ln \left(\frac{x_{H_2} x_{O_2}^{1/2}}{(x_{H_2O})^{\Xi}} \right) - RTd_1 \right]
\end{aligned}
\tag{10.39}
$$

where, in accordance with the original definition of these authors

$$
\begin{aligned}
d_1 = {} & 2n_e \sinh^{-1} \left(\frac{i}{2i_{0,a}} \right) + 2n_e \sinh^{-1} \left(\frac{i}{2i_{0,c}} \right) \\
& -\ln \left(1 - \frac{i}{i_{L,a}} \right) - \ln \left(1 - \frac{i}{i_{L,c}} \right) + \frac{i n_e F L_{el}}{\sigma_0 R} \exp \left(\frac{E_{el}}{RT} \right)
\end{aligned}
\tag{10.40}
$$

As $\Xi < 1$, chemical affinities are reduced and, considering all the irreversibilities involved, the power output deduced from Equations (10.32)–(10.35) and (10.39) is

$$
\begin{aligned}
P_{\text{irr}} &= -\Delta \dot{G}^{\text{eff}} - T\dot{S}_{\text{tot}} = -\Delta \dot{G}^{\text{eff}} - iA V_{\text{int}} - \frac{(E^{\text{eff}} - V_{\text{int}})^2}{R_{\text{leak}}} \\
&= \frac{iA}{n_e F} \left(m^{\text{eff}} - \frac{k}{RTd_1} m^{\text{eff}2} \right)
\end{aligned}
\tag{10.41}
$$

where $k = R_{\text{int}}/R_{\text{leak}}$ and

$$
m^{\text{eff}} = -\Delta g^{\text{eff}}(T, P) + RT \ln \left(\frac{x_{H_2} x_{O_2}^{1/2}}{(x_{H_2O})^{\Xi}} \right) - RTd_1
\tag{10.42}
$$

Thus, an approach is organized that attributes a drop of open circuit voltage below the reversible cell voltage to imperfect conversions of cell reactions without excluding voltage drops caused by the current leakage mechanism proposed in Zhao et al. (2008).

For an energy conversion device that absorbs energy to produce work, the conventional definition of thermal efficiency is its power output divided by the energy input (fuel-to-electricity efficiency, $\eta = -\dot{W}(\Delta \dot{H})^{-1}$). The customary efficiency definition for an irreversible FC is

$$
\eta = \frac{P_{\text{irr}}}{-\Delta \dot{H}^{\text{eff}}} = \frac{P_{\text{irr}}}{-(iA/n_e F)\Delta h^{\text{eff}}(T)} = \frac{1}{-\Delta h^{\text{eff}}(T)} \left(m^{\text{eff}} - \frac{k}{RTd_1} m^{\text{eff}2} \right)
\tag{10.43}
$$

Here the energy input of the FC is the effective reaction enthalpy change, evaluated at temperature T. This property allows comparison of different electrochemical devices and the comparison of efficiency (10.43) with the voltage-based efficiency.

In many works numerical predictions are studied to outline how to improve the cell performance. It is desirable to reduce electrolyte thickness to reduce internal ohmic losses. However, the electrolyte must also be relatively impermeable to fuel and oxidizer to minimize reactant crossover, stable in oxidizing and reducing environments over time, and maintain structural integrity at operating conditions. Permeability of electrolyte to reactants results in mixed potentials at the electrodes and reduced performance. The requirements for the electrode include low activation losses, long-term stability, and acceptable ionic/electronic conductivity (Larminie and Dinks, 2000; Li, 2006; Mench, 2008; Mench et al., 2009).

10.7. QUALITY CHARACTERISTICS AND FEASIBILITY CRITERIA

Calculations that characterize FC performance are based on Equations (10.39)–(10.43) and input data of Table 10.1, which are derived from the data available in literature (see, e.g., references [2, 16, 36–38, 42] in the paper by Zhao et al., 2008). The fuel composition is taken as 97% H_2 + 3% H_2O, and the typical oxygen composition in the ambient air, that is, 21% O_2 + 79% N_2, is used as oxidant. Enthalpy and Gibbs free energy in Equations (10.39)–(10.43) are obtained from Table 2 of Zhao et al. (2008). Equation (10.39) is applied to draw polarization curves, which relate the cell voltage to its current density. Equations (10.41)–(10.43) and relevant input parameters also lead to the curves of the power density (Figure 10.3) and traditional efficiency.

For a real FC, there exist a maximum power output P_{max} and a maximum of traditional efficiency η_{max} with respect to the current density, with the corresponding current densities i_P and i_η, respectively. These results can further lead to curves of power density versus FC efficiency as shown by Zhao et al. (2008), where η_m and P_m refer to the efficiency at the maximum power output and the power output at the maximum efficiency, respectively. Such power draft defines the feasibly operating region of an FC situated in the part of the P–η curve that has a negative slope, that is, the power output decreases with the efficiency. In other words, Zhao et al. (2008) have determined feasible ranges of power output and efficiency that are defined by the inequalities $P_m \leq P \leq P_{max}$ and $\eta_m \leq \eta \leq \eta_{max}$. P_{max}, η_{max}, P_m, and η_m are four important parameters of the FC. P_{max} and η_{max} define upper bounds of the power output and efficiency, whereas P_m and η_m define feasible values of the lower bounds of the power output and efficiency.

As pointed out by Zhao et al. (2008), the analytical model represented by Equations (10.41)–(10.43) is applied to define an "optimum region" of the current density i in an FC by the inequality $i_\eta \leq i \leq i_p$. In fact, this is, rather, a feasible region that can be explained by charts describing power density and efficiency versus current density (Figure 10.3 in the present chapter and Figures

3b and 3c in Zhao et al., 2008). In the region of $i < i_\eta$, both power output and efficiency of power yield increase with current density, while in the region of $i > i_P$, both power output and efficiency decrease with current density. The regions of $i < i_\eta$ and $i > i_P$ are not substantiated by thermodynamics, yet the cell can work in these regions. Thus, i_P and i_η are two feasibility parameters of an FC, which determine, respectively, upper and lower bounds of the current density regarded as a control parameter. In practical operation, the researcher should choose a plausible current density within limits defined by the considered inequalities, which locate the FC operation in the feasible region. Power optimization or cost optimization can then be accomplished with the current density as the control variable constrained within the limits i_P and i_η.

10.8. SOME EXPERIMENTAL RESULTS

Modeling of irreversible cell can be used jointly with some experimental work to investigate the effect of various operating and design parameters on the performance of the FC.

There is an abundance of research papers on FC systems, in particular SOFC power systems (Campanari, 2001; Roy-Aikins, 2002; Lemański, 2003; Li and Chyu, 2004; Ordonez et al., 2007). Zero-dimensional modeling of a TSOFC with internal reforming of hydrocarbons is a typical structure that has been both numerically implemented and parametrically analyzed (Lemański and Badur, 2004, 2005). Hybrid power systems containing turbines or microturbines and FCs are treated particularly often (Roy-Aikins, 2002; Rao and Samuelsen, 2002; Chan et al., 2003; Lemański, 2007). Reviews and PhD theses are available (Bessette, 1994; Stambouli and Traversa, 2002; Milewski, 2004; Lemański, 2007). An example of detailed modeling is that for a system of a SOFC tube equipped with a tube-and-shell preformer unit (Calise et al., 2007). Both SOFC tube and preformer are discretized along their axes. Descriptions of the kinetics of shift and reforming reactions are introduced. Balances of energy and substances are set for each slice of the components, allowing the calculation of temperature profiles. Friction factors and heat coefficients are calculated by using experimental correlations. Models are introduced to evaluate SOFC polarizations. With a model, temperatures, pressures, concentrations, and electrical parameters are found for each slice. Finally, the effect of design parameters on system performance is determined. The FC temperature is an important factor that influences power output because it affects the reversible potential as well as the activation, ohmic, and concentration resistances. Higher temperature results in a reduction of activation polarization.

Figure 10.3 shows the effect of the operating temperature on the polarization curve and the performance characteristic curves of the FC with various current densities. Observe that the reversible cell potential decreases with temperature, whereas the operational cell voltage increases with temperature. The power output increases quickly as temperature increases, while the efficiency is not a monotonic function of temperature (Zhao et al., 2008). The efficiency decreases with T when $i < i_\eta$, while it increases with T when $i > i_\eta$. Consequently, the maximum power output P_{max} and the related efficiency η_m increase with T, and the maximum efficiency η_{max} will decrease while the corresponding power output P_{min} increases with T. Finally, Figure 10.3 and efficiency data also show that both i_P and i_η increase with T. The outlined discussion can be extended to more complex electrochemical systems.

At present, many models of irreversible FCs are available, one of the simplest being the model proposed by Wierzbicki (2009), whose treatment of the voltage without load is similar to that presented here. Standardly, the cell voltage in an ideal situation (no losses) is defined by the Nernst equation. Yet, while the first term of the voltage expression defines the voltage without load, it nonetheless includes losses of the so-called idle run. The losses are the effect of flaws in electrode constructions and other imperfections that reduce the open circuit voltage below its theoretical value. Activation polarization V_{act} is neglected in this model; the losses include ohmic polarization and concentration polarization. The first loss term in the model proposed by Wierzbicki (2009) quantifies ohmic losses associated with electric resistance of electrodes and flow resistance of ions through the electrolyte. The second term refers to mass transport losses. Quantity i_L emerges as a particular current arising when the fuel is consumed in the reaction with the maximum possible feed rate.

Many experimental and theoretical data have been published showing power maxima in FCs and proving an analogy of the thermal machines theory to the one describing electrochemical systems. This line of research provides data for power production bounds (limits) enhanced in comparison with those predicted by classical thermodynamics. In fact, these bounds depend not only on changes of the thermodynamic state of participating resources but also on process irreversibilities, ratios of stream flows, stream directions, and mechanism of heat and mass transfer. Yet, the main progress in FC technology is determined primarily by that in the material science of FC systems. An overview (Molenda, 2006) assesses basic SOFC components in terms of their basic functional parameters such as chemical stability, transport, as well as catalytic and thermomechanical properties under operational conditions in an FC.

Validation of the thermodynamic modeling in the context of FCs is based on the organization of FC power experiments and identification of power maxima in terms of control variables. These issues are described in the report (Sieniutycz et al., 2011), paper (Sieniutycz et al., 2012), and Wierzbicki's (2009) MsD work. In these experiments a number of voltage and power measurements were made in terms of the following control variables of SOFC:

- working cell temperature
- various flows of fuels
- different components of fuel

A complete content of the theory and experiments also includes application of the Aspen Plus^{TM} software for simulation purposes, and is presented in the

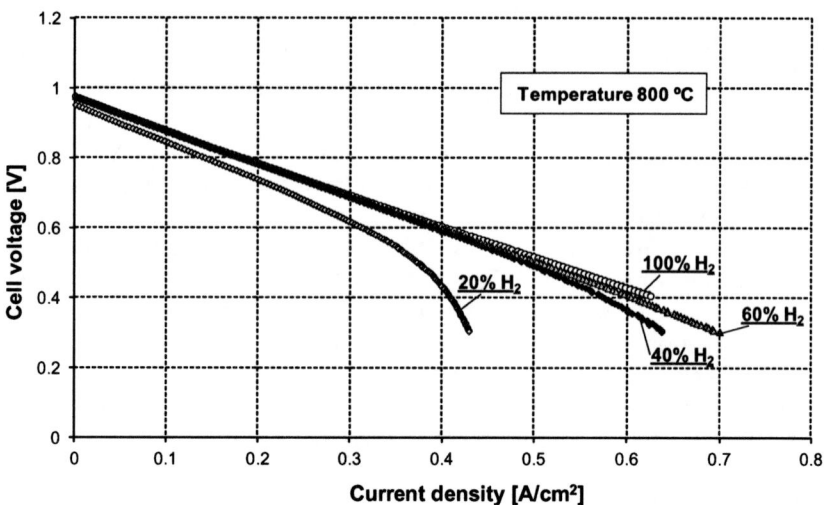

Figure 10.4 Curves of voltage–current density for a SOFC for various fuels at the temperature 800 °C. Data refer to experiments of M. Błesznowski, A. Zieleniak, and J. Jewulski in the Department of Heat Processes at the Warsaw Institute of Energetics. These data are applied for the purpose of the validation of the SOFC thermodynamic model in the forthcoming Błesznowski's PhD thesis (Sieniutycz et al., 2011).

report (Sieniutycz et al., 2011). Diverse aspects of simulations of a TSOFC stack using Aspen Plus^{TM} unit operation models are described by Zhang et al. (2005).

Voltage lowering in FCs below the reversible value is a good measure of their imperfection. Reversible cell voltage E^0 is usually a reference basis calculated from the Nernst equation. Yet, in more general cases, actual voltage without load must take into account losses of the idle run, which are the effect of flaws in electrode constructions and other imperfections. In Wierzbicki's (2009) thesis the operating voltage of a cell is evaluated as the departure from the idle run voltage E_0

$$V = E_0 - V_{int} = E_0 - V_{act} - V_{conc} - V_{ohm} \qquad (10.44)$$

Losses, which are called polarization, include three main sources: activation polarization (V_{act}), ohmic polarization (V_{ohm}), and concentration polarization (V_{conc}). As it follows from Equation (10.44), ratio of the actual voltage and the reversible voltage E^0 can be a suitable measure of cell efficiency, whereas the corresponding voltage decrease is the quantitative measure of the dissipation effects in the FC systems (Figure 10.4).

Power density is the product of voltage V and current density i. A large number of approaches for calculating polarization losses have been presented in the literature, as reviewed by Zhao et al. (2008). The effect of transport phenomena can be evaluated with the help of data and models involving various transfer mechanisms in porous SOFC (Li and Chyu, 2005; Hussain et al., 2007; Kakaç et al., 2007). Activation and concentration polarizations occur at both anode and cathode locations, while the resistive polarization represents ohmic losses throughout the cell. Experiments show power maxima in FCs (Wierzbicki, 2009; Li, 2006, and many others).

There are subtle voltage effects associated with the geometries of current collectors on the power density in a SOFC (Li et al., 2011). To test these effects, as a rule, an analytical model is proposed and an analysis is conducted for SOFCs with the goal of higher power densities. The analytical models are designed to help optimize the size, spacing, and geometric shapes of current collectors.

As the voltage losses increase with current, and power is the product of voltage and electric current, the initially increasing power begins finally to decrease for sufficiently large currents, so that maxima of power are observed. The data include the losses of the idle run attributed to the flaws in electrode constructions and other imperfections (Wierzbicki, 2009).

Entropy and power generations in FCs are compared in Figure 10.2. The power curve has two points at which power production vanishes in the system. The first point, corresponding to reversible behavior of the system, is also called the open circuit point; the electric current and entropy production vanish at this point. The second point at which power vanishes is called Newton–Fourier point or short circuit point. It corresponds with the situation when currents are so large that only irreversible phenomena are present in the system.

In the "short circuit" case all currents flow only by resistances, and there is no power production for any value of efficiency ζ_0, in spite of any possible chemical reaction. Only entropy is then produced. In the special case when $\zeta_0 = 0$, the entropy production at the "short circuit point" corresponds with the situation without chemical reaction and power generation. Only at the short circuit point and in absence of power yield, the entropy production at this point (modulo to multiplier T^{-1}) is equal to the product of the reaction rate and its chemical affinity. This is a classical result, which, however, does not hold when the system produces power (i.e., belongs to the class of "active systems") (Figures 10.5 and 10.6; Table 10.2).

Table 10.2 Summary of limiting parameters of a reference fuel cell at the temperature 800 °C (Sieniutycz et al., 2011)

Gas flow	Maximum power density of fuel cell (mW/cm^2)	Current density at MDP[*] (A/cm^2)	Voltage of fuel cell (V)
(a) 200 ml/min H$_2$	260	0.55	0.47
(b) 120 ml/min H$_2$ + 80 ml/min N$_2$	254	0.53	0.48
(c) 80 ml/min H$_2$ + 120 ml/min N$_2$	246	0.49	0.50
(d) 40 ml/min H$_2$ + 160 ml/min N$_2$	191	0.34	0.56

[*]MDP: maximum density of power.

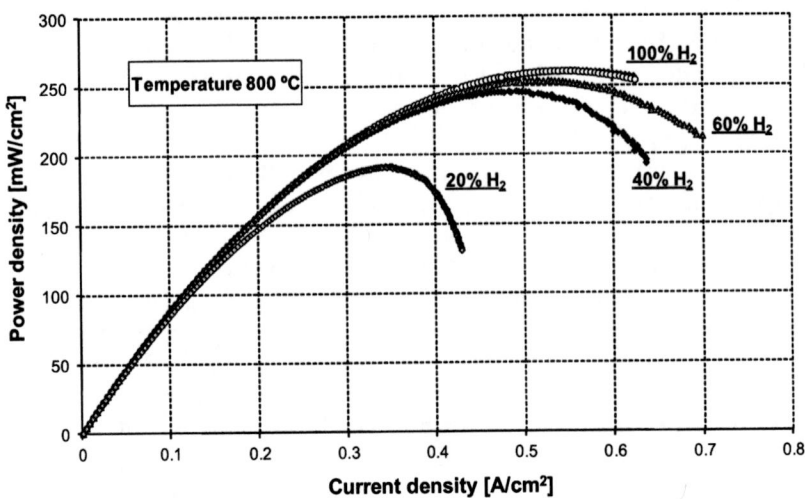

Figure 10.5 Curves of power density of a SOFC for various hydrogen contents in fuel at the temperature 800 °C. Points refer to experiments of M. Blesznowski, A. Zieleniak, and J. Jewulski in the Department of Heat Processes at the Institute of Power Engineering. The data are applied for the purpose of the validation of the SOFC thermodynamic model in the forthcoming Blesznowski's PhD thesis (Sieniutycz et al., 2011).

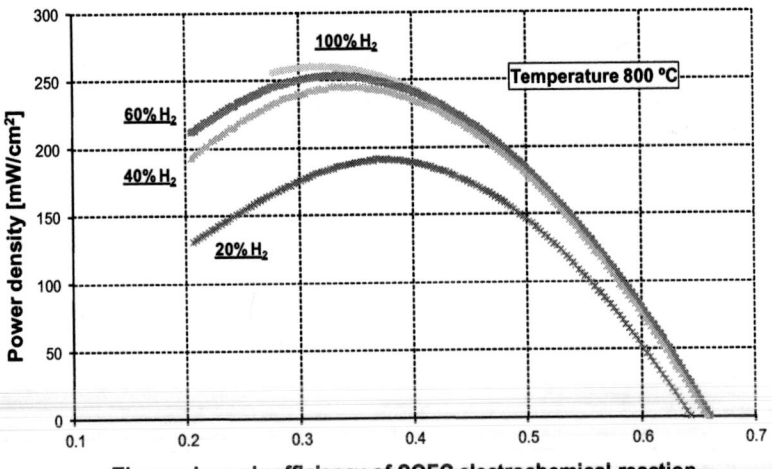

Figure 10.6 Example of data describing power density of an SOFC in terms of the first-law efficiency $\eta = \Delta G(\Delta H)^{-1}$ at temperature 800 °C (Sieniutycz et al., 2011).

10.9. ASSESSING POWER LIMITS IN STEADY THERMOELECTROCHEMICAL ENGINES

Validity of FC models in the thermodynamic framework allows for a rough assessment of power limits in thermoelectrochemical systems of a simple, standardized topology corresponding to Figure 10.1b, as outlined below. The reason why this assessment can only be rough is explicit in Equation (10.45) or the like, which has ignored information about topological structure of many various flows in the system.

Let us focus on FCs described by the formalism of inert components (Sundheim, 1964; Ekman et al., 1978) rather than the ionic description (Newman, 1973). Assume, for simplicity, that the active (power producing) driving forces involve only one temperature difference $T_{1'} - T_{2'}$, single chemical affinity $\mu_{1'} - \mu_{2'}$, and the operating voltage $\phi_1 - \phi_{2'}$. Total power production is the sum of thermal, substantial, and electric components, that is

$$
\begin{aligned}
P &= (T_{1'} - T_{2'})I_s + (\mu_{1'} - \mu_{2'})I_n + (\phi_{1'} - \phi_{2'})I_e \\
&= (T_1 - T_2)I_s + (\mu_1 - \mu_2)I_n + (\phi_1 - \phi_2)I_e \\
&\quad - R_{ss}I_s^2 - R_{nn}I_n^2 - R_{ee}I_e^2 - R_{sn}I_sI_n - R_{se}I_sI_e - R_{ne}I_nI_e
\end{aligned}
\tag{10.45}
$$

Equation (10.45) constitutes the simplest account of linear thermoelectrochemical systems. In general, it is an approximate formula; indeed it does not contain any "topology parameter." Complex configurations of flows contacting such as countercurrent contacting that may exist in FCs are not described well by Equation (10.45). Linear systems described by this equation are those with constant (current independent or flux independent) resistances or conductances. They satisfy Ohm-type or Onsager-type laws linking thermodynamic fluxes and thermodynamic forces (dissipative driving forces which are represented by products $R_{ik}I_k$ in Equation (10.45)). While many thermal power systems and FCs are nonlinear, that is, possess current-dependent resistances, the dependence is often weak, so a linear model can be a good approximation. Below, by applying Equation (10.45) we shall attempt to develop a simple theory of power limits for these systems.

After introducing the enlarged vector of all driving potentials $\tilde{\mu} = (T, \mu, \phi)$, the flux vector $\tilde{I}$ of all currents, and the overall resistance tensor $\tilde{R}$, Equation (10.45) can be written in a simple matrix–vector form

$$
P = (\tilde{\mu}_1 - \tilde{\mu}_2) \cdot \tilde{I} - \tilde{R} : \tilde{I}\tilde{I}
\tag{10.46}
$$

Maximum power corresponds with the vanishing of the partial derivative vector

$$
\frac{\partial P}{\partial \tilde{I}} = \tilde{\mu}_1 - \tilde{\mu}_2 - 2\tilde{R} \cdot \tilde{I} = 0
\tag{10.47}
$$

Therefore, the optimal (power-maximizing) vector of currents at the maximum point of the system can be written in the form

$$\tilde{\mathbf{I}}_{mp} = \frac{1}{2}\tilde{\mathbf{R}}^{-1} \cdot (\tilde{\mu}_1 - \tilde{\mu}_2) \equiv \frac{1}{2}\tilde{\mathbf{I}}_F \tag{10.48}$$

This result means that the power-maximizing current vector $\tilde{\mathbf{I}}_{mp}$ in strictly linear systems equals one half of the purely dissipative current at the Fourier–Onsager point, $\tilde{\mathbf{I}}_F$, at which no power production occurs. Moreover, we note that Equations (10.46) and (10.48) yield the following result for the maximum power limit of the system:

$$P_{mp} = \frac{1}{4}(\tilde{\mu}_1 - \tilde{\mu}_2) \cdot \tilde{\mathbf{R}}^{-1} \cdot (\tilde{\mu}_1 - \tilde{\mu}_2) \tag{10.49}$$

In terms of the purely dissipative flux vector at the Fourier–Onsager point, $\tilde{\mathbf{I}}_F$, the above limit of maximum power is represented by an equation

$$P_{mp} = \frac{1}{4}\tilde{\mathbf{R}} : \tilde{\mathbf{I}}_F\tilde{\mathbf{I}}_F \tag{10.50}$$

Of course, the power dissipated at the Fourier–Onsager point equals

$$P_F = \tilde{\mathbf{R}} : \tilde{\mathbf{I}}_F\tilde{\mathbf{I}}_F \tag{10.51}$$

Equations (10.50) and (10.51) prove that, in linear systems with the simplest flow configurations (no countercurrent flows), only at most 25% of power (10.51), which is dissipated in the natural transfer process, can be transformed into the noble form of the mechanical power. This is probably the lowest (most disadvantageous) evaluation of upper limit of power. Such evaluation cannot be easily generalized to real FCs and other systems where significant deviations from Equations (10.46) and (10.50) are expected depending on the nature of diverse nonlinearities and topology variations, and also topology improvements to include countercurrent contacting. Despite the limitation of the result (10.50) to systems with the simplest linear transfers, the result is of some significance because it shows explicitly the order of magnitude of largest thermodynamic limitations in power production systems.

The analysis presented here proves that a link exists between the mathematics of the thermal engines and FCs, and also that the theory of FCs can be unified with the theory of thermal and chemical engines. Yet the topological differences of both systems may occasionally render both of them quite dissimilar.

Explanation of related physical effects is in order. While power ratios involving Equations (10.50) and (10.51) can be regarded as efficiency measures, they should not be confused with commonly used, popular efficiencies, especially first-law efficiencies. There are a number of definitions of FC efficiencies, based on first or second law, proposed for measuring and comparing the performance of electrochemical processes. Only second-law efficiencies, which show how close the process approaches a reversible process, are entirely correct efficiency measures. Efficiencies based on the first law, such as the theoretical reversible efficiency $\eta = \Delta G (\Delta H)^{-1}$ or the fuel-to-electricity efficiency $\eta = -W(\Delta H)^{-1}$ (often found in the literature), can generate efficiency values greater than 100% for certain systems depending on whether the change in entropy for the overall chemical reaction involved in the process is positive or negative. See, for example, paper by Rao et al. (2004) on various definitions of FC efficiencies.

The FC efficiencies, $\eta = \Delta G / \Delta H$ or $\eta = -W(\Delta G)^{-1}$, which are often applied to many FC systems, can easily achieve numerical values much higher than 1/4 (power ratio of Equations (10.50) and (10.51)). They are first-law efficiencies defined in a different way than the power ratios P_{mp}/P_F satisfying Equations (10.50) and (10.51). In fact, the considered power ratios represent some specific, second-law efficiencies of the simplest standardized thermoelectrochemical process. They are not equivalent, neither theoretically nor numerically, to the most common fuel-to-electricity efficiencies of FCs, $\eta = -W(\Delta H)^{-1}$.

In practice, for FCs, the most commonly used definition is the fuel-to-electricity efficiency, $\eta = -W/\Delta H$, defined as the ratio of produced electricity to input chemical energy of fuels and oxidants. This practical fuel-to-electricity efficiency of an FC is usually between 40% and 60% depending on the type of FCs (refer to http://en.wikipedia.org/wiki/Fuel_cell, and *Fuel Cell Handbook*: EG&G Technical Services Inc., 2004). The high efficiency is one of the major advantages of FCs compared with internal combustion engines and turbines. Still that definition is limited to situations with negative ΔH; otherwise it loses its meaning by implying values higher than 100% if power is produced. Other than the power ratios P_{mp}/P_F second-law efficiencies can also be defined. One of the most correct, simple, and practical definition of efficiency for an FC operating near ambient temperature is the ratio of the actual voltage to the reversible voltage. The virtue of this definition is recognized (Li, 2006).

Numerous theoretical and applicative aspects of FC efficiencies may be reviewed (Arteaga-Pereza et al., 2009; Barbir and Gomez, 1997; Calise et al., 2006a, 2006b; Cenusa et al., 2003, 2004; Chen and Zhao, 2009; Costea et al., 1999; Delsman et al., 2006; Dincer and Rosen, 2007; Durmayaz et al., 2004; Feidt, 2001, 2006; Haseli et al., 2008; Haynes, 2001; Hernandez and Kafarov, 2009; Hou et al., 2007; Hotz et al., 2006; Kuchonthara et al., 2003; Jaluria, 2008; Larminie and Dinks, 2000; Lemański, 2007; Lemański et al., 2004; Lems et al., 2003; Li, 2006; Li et al., 2008a, 2008c; Li and Suzuki, 2004; Li and Chyu, 2005; Lior and Dunbar, 1991; Lior et al., 1991; Logan, 2008; Lovley, 2008; MacLean and Lave, 2003; Mench, 2008; Milewski et al., 2009, 2010; Milewski and Miller, 2012; Morosuk et al., 2004; Ołdak, 2011; Palmore, 2004; Rao et al., 2004; Scholtz and Schroder, 2003; Sidwell and Grover Coors, 2005; Sieniutycz, 2008, 2009a, 2009c, 2010a, 2010c, 2010d, 2011a, 2011b, 2011d, 2012a, 2012b; Sieniutycz et al., 2011, 2012; Sieniutycz and Kuran, 2011; Sieniutycz and Poświata, 2012; Xue and Dong, 1998; Zhang et al., 2005; Zhao et al., 2008; Zhao and Chen, 2009; Zhu and Kee, 2006; Zoulias and Lymberopoulos, 2007, and many others).

10.10.   HYBRID SYSTEMS

Hybrid vehicles have the potential to increase fuel economy by using a primary engine operating at a constant power to supply average power requirements and a surge power unit for peak power demands, and to recover braking energy (Newman, 1973). A hybrid vehicle may combine a primary power plant with an energy storage device. It merges the benefits of conventional and electric vehicle designs since the conventional design offers limited possibilities for improvement, and electric vehicles are expensive, lacking a recharging infrastructure, and have limited range and long recharging times.

To date, no detailed system optimization has been performed for hybrid vehicles. Some 1995 studies propose high-power batteries for use in hybrid vehicles, to combine a version of the lithium-ion battery model with a vehicle model that determines battery power requirements for a given driving cycle. Batteries are designed for either the highest vehicle mileage or minimal acceptable battery dimensions. Hybrid vehicles have potential to more than double mileage as compared to conventional vehicles and have a limited electric vehicle range. It is suggested that a combination of both vehicle and battery is required to determine the complex interaction between hybrid vehicle weight and battery power.

Many automakers see complementary roles for hydrogen FCs and battery electric vehicles, and are pursuing both technologies. FC engines and hybrid vehicles efficiently combine hydrogen and oxygen in air to produce electricity powering the vehicle.

In hybrid systems FC engines operate jointly with thermal engines, in particular GTs. Kuchonthara et al. (2003) treat energy recuperation in SOFC combined with a GT. Benelmir and Feidt (1998) outline an energy management strategy for general energy cogeneration systems. Lior et al. (1991) combine FCs with fuel-fired power plants for improved energy efficiency. Lemański's (2007) thesis proposes an analysis of energy cycles with FC and gas–vapor turbine. Lemański et al. (2004) analyze some feasible strategies for GT–SOFC hybrid cycles. Their results of analysis for hybrid cycles and Inverted Brayton Cycle show that the SOFC electrical efficiency increases with the operation pressure. That analysis also shows that the highest electrical efficiencies are reached by hybrid cycles with the pressurized SOFC. Hybrids of SOFC with a GT, H(SOFC+GT), applied to produce electrical and thermal energy, are competitive for traditional hybrid cycles consisting of a GT and a heat recovery steam generator, because their electrical efficiency attains even 70% in pressurized systems. The high temperature of SOFC is in favor of the solution in which a GT is effectively combined with a SOFC in a hybrid manner to form a single unit for a self-sustainable distributed energy system (Li and Suzuki, 2004; Li, 2006).

Zhao and Chen (2009) and Ołdak (2011) consider modeling and optimization and design criteria of a typical FC–heat engine hybrid system. Their modeling approach involves various sources of irreversible losses, such as the electrochemical reaction, electric resistances, finite-rate heat transfer between the FC and the heat engine, and heat leak from the FC to the environment.

The analysis is based on an existing SOFC model, similar to that considered in Sections 10.4–10.6. This analysis provides design characteristics that help to understand the behavior of a typical FC–heat engine hybrid system at the steady-state operating conditions. Energy and entropy balances yield relationships describing multiirreversible losses and work potentials of the hybrid system. Expressions for the hybrid power output and efficiency are derived along with system's performance characteristics. Effects of the design parameters and operating conditions on the system performance are numerically simulated (Zhao and Chen, 2009; Ołdak, 2011). It is shown that there are certain optimum criteria for important parameters. The results obtained provide the theoretical basis for both the optimal design and operation of real FC–heat engine hybrid systems. These approaches can be extended to hybrid systems with different topologies (Milewski and Miller, 2012). Based on mathematical modeling and numerical simulations, these authors present a control strategy for high-temperature fuel cell hybrid systems (HTFC-HSs). Performance maps with three independent parameters are shown. The independent parameters are stack current, fuel mass flow, and turbine–compressor shaft speed. Those parameters are controlled by external load, fuel valve, and turbine compressor shaft speed, respectively. A control system is directed to satisfy constraints imposed on stack temperature, steam-to-carbon ratio, compressor surge, etc. The research aims to achieve maximum efficiency of power generation subject to constraints. An adequate simulator for the HTFC module is described and a control strategy is obtained on this basis. The performance of the HTFC-HS is shown in various working environments. The presented results indicate that the hybrid possesses a high operation and control flexibility while simultaneously maintains stable thermal efficiency. Operation of these systems is possible over a wide range of parameter changes. Milewski et al. (2009) propose reduction of CO_2 emissions from a coal-fired power plant by using a MCFC. The FC is placed in the flue gas stream of the coal-fired boiler. The main virtues of this idea are higher total power yield of a hybrid system, reduced CO_2 emissions, and higher system efficiency. The results show that use of an MCFC could reduce CO_2 emissions by 61%.

Exergy optimization is the frequent approach applied to thermochemical units and FCs. It provides universal data independent of time and location, and, if applied carefully, it can also be helpful in design (Jaluria, 2008). Applications are vast and refer to a virtually whole spectrum of systems in chemical and mechanical engineering (Grosu et al., 1999; Feidt and Lottin, 2004; Feidt et al., 2002, 2007; Dincer and Rosen, 2007). For example, exergy optimization of solar collector and thermal energy storage system is considered by Aghbalou et al. (2006). Cenusa et al. (2004) consider exergy optimization of the heat recovery steam generators by imposing the total heat transfer area. Cogeneration systems are effectively treated by exergy approaches (Radulescu et al., 2006a, 2006b). Bedringas et al. (1997) present an exergy analysis of SOFC systems; see also further text concerning these systems.

Thin film deposition as applied to fabrication of micro-solid oxide fuel cell (μSOFC) is an emerging and highly active field of research that is currently attracting great attention. In miniaturized SOFCs the active FC components (anode, electrolyte, and cathode) may exist as thin films forming a freestanding membrane on a substrate.

Exergy analyses of FC micropower plants fed by methanol are performed by Hotz et al. (2006). Beckel et al. (2007) review the field of thin films for μSOFCs. The review is focused on thin film (thickness $\leq 1\ \mu$m) deposition techniques and components relevant to SOFCs, including current research on nanocrystalline thin film electrolyte and thin-film-based model electrodes. Calculations showing geometric limits of μSOFCs and first results toward fabrication of μSOFCs are also described. The future goal is to integrate these μSOFCs in a system with reformer, postcombustor, insulation, etc.

Exergy-related optimizations of hybrid systems based on finite-time thermodynamics and thermoeconomics are next stages of classical exergy optimization. In addition to classical thermodynamics, they include the fundamental concepts of heat and mass transfer, fluid mechanics, and some components of industrial economics. Of special importance are irreversibilities originating from finite-time and finite-size constraints (Durmayaz et al., 2004). For the thermodynamic assessment of hybrid systems exergy balances and exergy optimizations are particularly suitable because different system components are treated by a common, unifying measure.

In a series of papers Calise et al. (2006a, 2006b, 2007) study the simulation and exergy analysis of a hybrid SOFC–GT system. For the SOFC reactor model Calise et al. (2006a) assume that only hydrogen participates in the electrochemical reaction and that the high temperature of the stack leads the internal steam reforming to completion; by another assumption the unreacted gases are fully oxidized in the combustor downstream of the SOFC stack. Compressors and GTs are modeled on the basis of their isentropic efficiency. For the heat exchangers and the heat recovery steam generator (a tube-in-tube counterflow arrangement), the simulations use the thermal efficiency-NTU approach. To evaluate local irreversibilities and thermodynamic inefficiencies, energy and exergy balances are performed for each component. Simulations search regimes with various values of operating pressure, fuel utilization factor, fuel-to-air and steam-to-fuel ratios, and current density. The results show that electrical efficiency close to 60% can be achieved for typical conditions; when heat loss recovery is also taken into account, a global efficiency of about 70% can be achieved. Calise et al. (2006b) use the simulating bulk-flow model for synthesis and design optimization of the hybrid SOFC–GT power plant. They determine performance of the centrifugal compressors and radial turbine by using maps, properly scaled to match the values required for mass flow rate and pressure ratio. Their simulations of compact heat exchangers use Colburn and friction factor correlations. For the SOFC, they find curves of cell voltage versus current density from the Nernst potential and polarizations. Both the steam–methane prereforming and internal reforming processes are modeled assuming the water–gas shift reaction to be equilibrium-controlled and the demethanization reactions to be kinetically controlled. A thermoeconomic model is developed by introducing capital cost functions for each plant component. The whole plant is first simulated for a fixed configuration. Then, a synthesis/design optimization procedure is carried out by using the traditional single-level approach. Calise et al. (2007) develop and study a one-dimensional finite-volume simulation model of a TSOFC and a prereformer, with detailed kinetics of the shift and reforming reactions. A related paper by Cenusa et al. (2003) treats GTs and combined cycles. Delsman et al. (2006) develop exergy analysis of an integrated fuel processor and fuel cell (FP–FC) system, applied for a methanol processor integrated with a PEM FC, for use as a portable power generator. They conclude that the calculated overall exergy efficiency of the FP–FC system is higher than the one describing typical combustion engines and rechargeable batteries.

FC hybrid architectures have expanded greatly in recent years. Reliable design of a hybrid system may require complex optimization of the whole plant containing FC stack, prereformer, turbine, etc., and prior analyses may be necessary for local balances of plant components with respect to efficiency, economics, and material issues. Molenda (2006) considers essential material aspects in FC technologies. Chan et al. (2003) develop a multilevel modeling of SOFC–GT. Lin and Hong (2006) model a hybrid FC/micro-gas turbine (MGT) system in order to investigate its transient behavior. Reviews and theses focused on hybrid systems are available (Roy-Aikins, 2002; Stambouli and Traversa, 2002; Bessette, 1994; MacLean and Lave, 2003; Lemański, 2007; Milewski et al., 2010). Clearly, a rapid grow of the research in this field is observed.

10.11. UNSTEADY STATES, DYNAMIC UNITS, AND CONTROL PROBLEMS

Understanding the dynamic behavior is essential to predict the performance and limitations of FC power systems. The development of an FC dynamic model and a feedback control scheme can maintain output voltage despite load changes. Dynamic responses can be determined as solutions of coupled partial differential equations derived from conservation laws of charges, mass, momentum, and energy.

Dynamic SOFC systems and control strategies are reviewed by Chaisantikulwat et al. (2008) and Wu et al. (2008a, 2008b). Dynamic modeling and validation studies for a TSOFC are developed by Bhattacharyya et al. (2009) in terms of characteristics of transient responses and nonlinearities. SOFC is a key component in the new concept of distributed power generation. For connecting SOFC reliably to a load varying grid, its transient behavior needs to be studied in detail with a thoroughly validated dynamic model. Dynamic models are also important for synthesizing efficient controllers (Bhattacharyya et al., 2009). From such models, in particular, the voltage responses to step changes in the fuel concentration and load current can be determined.

Kharton et al. (1999) describe a method of testing TSOFCs at unsteady-state conditions. Testing single cells with externally switched pulse load demonstrates a possibility to optimize the SOFC operating mode at a given resistance of the closing circuit. The test is accomplished by varying the pulse period-to-pulse duration ratio of the pulses that open the circuit. No effect of the pulse load frequency on SOFC performance is observed in the working frequency range (2–50 kHz). The results of testing SOFCs in unsteady-state conditions suggest applicability of the externally switched pulse load to match resistances of single cells in the SOFC stacks. Lin and Hong (2006) simulate the start-up transients of a turbo FC system, which is an important issue in a turbo FC system design, and develop a general dynamic model of the FC–gas micro-turbine hybrid system to investigate the transient behavior during cold start. Obara (2007a) presents a dynamic optimization approach for a combined FC cogeneration, solar module, and geothermal heat pump system using genetic algorithm. Three objective functions – (1) minimization of operation cost, (2) minimization of the error of demand-and-supply balance, and (3) minimization of the amount of greenhouse gas discharge – are proposed. Performance characteristics of the system subject to each objective function are described. Obara (2007b) also proposes an arrangement plan for distributed FCs installed in urban areas.

Methods based on the optimal control theory and thermodynamics extend to electrical systems the optimization approaches belonging to family of those that were already worked out for heat power generators. Constraints take into account dynamics of heat and mass transport and rate of work production. Durmayaz et al. (2004) analyze various approaches in optimization of thermal systems based on finite-time thermodynamics and thermoeconomics.

Below we shall briefly outline a variational approach to the thermodynamic theory of efficiency and work generation in dynamically working FCs.

A performance criterion acceptable in the assessment of a dynamic FC is its integral power output. Functionals describing cumulative power released from FC stacks can be derived and then maximized. In this way limits on the cumulative power output can be determined by optimization. These functionals describe energetics of systems from which power delivery takes place with a finite rate. Power functionals or their discrete analogues (sums) quantify, respectively, continuous and discrete models of power production. This production occurs due to the chemical transformation coupled with the transfer of components and heat. By maximizing the integral power output, limits on work generation and cell voltage are determined. The methods of dynamic programming (DP) and maximum principle are the main tools applied in computations in which total power yield, P, is the performance criterion.

An alternative performance criterion suitable for the thermodynamic analysis of an FC is its entropy production S_σ. Also this criterion must be used in a functional form. It describes the entropy generation in a real cell, from which power delivery takes place with a finite rate. Integrals of P or S_σ, and their discrete analogues (sums), quantify, respectively, continuous and discrete models of power production due to chemical reactions and mass transfer coupled with transfer of heat. By minimizing S_σ limits on the work generation are determined.

Network thermodynamics, finite-time thermodynamics, and exergy analysis prove their potential when evaluating limits on power production in dynamic systems. A vast set of time-dependent energy units and FCs, which work with finite resources, can be analyzed. In dynamic situations these systems are modeled as multistage devices that convert the energy of heat and chemical reaction into mechanical and/or electrical energy, thus producing efficiently power with a finite rate. These investigations are performed within two distinct groups of energy systems. The first group involves energy consumption devices (separators, chillers, and electrolyzers), whereas the second group the energy production units (FCs and engines).

For the latter group (systems with energy generation), affinities and driving forces should appear in the entropy representation. An efficient scheme of the description involves the reaction potentials or "one-way affinities" Π_j^f and Π_j^b for jth reaction, as described in Chapter 9 and in the earlier part of the present chapter.

Whenever an explicit expression for instantaneous power production by the stack is known, a power optimization algorithm can be formulated and then solved. When the affinity component of the reaction substrates, Π^f, is greater than that of the products, Π^b, in the "engine mode" one-way affinity Π decreases along the process path, and the system delivers power. Methods of DP and maximum principle (Chapter 2) are useful to accomplish the multistage optimization (Sieniutycz, 1997a, 1998a, 1998b, 2000c). While the original continuous problem is governed by the Hamilton–Jacobi–Bellman (HJB) theory, its discrete counterpart is described by DP (Chen et al., 2011; Xia et al., 2010a, 2010b, 2011c; Xiao and Basar, 1997, 1999). DP offers a description in terms of wave fronts that, in the present case, are surfaces of constant specific work or power per unit flow of an active reactant. On the other hand, the method of maximum principle (or a similar method of variational calculus) constitutes a description in terms of process trajectories, which characterize state changes of reagents. For an FC stack a nonlinear model describing chemical evolution leads to an optimal work function. This function is determined in terms of a number of transfer units τ or a Hamiltonian h, the latter quantity being a common measure of optimal process intensity (the same for each point of the path). While the number τ is a measure of the residence time of flowing reagents, the quantity h quantifies a minimal irreversibility in the system.

The block scheme of Figure 9.7 applies here, where $\mathbf{X}$ represents the state vector of an energy resource at flow (i.e., hydrogen) and $\mathbf{u}$ refers to a set of control variables, which often include currents. The computational block scheme in Figure 9.7 constitutes an abstract (multistage) representation of a dynamic power production process. For the entropy performance function, S_σ, a cost-like criterion defined as the sum $\sum (l_0^n + h)\theta^n$ is minimized, where l_0^n is a Lagrangian describing the intensity of *original* costs and h is the Lagrange multiplier or time penalty. On the other hand, for the power performance criterion, profit-like criterion defined as the sum $V_* = \sum (p_0^n - h)\theta^n$ is maximized, where p_0^n is the physical, one-stage power yield and θ^n is the time interval at the nth stage.

When the total power is the optimization criterion, a computer generates tables of optimal controls and optimal yields by solving a recurrence equation for the optimal performance function V_*:

$$V_*^n(T^n, Y^n) = \max_{u^n, v^n, \theta^n} \{(p_0^n(T^n, Y^n, u^n, v^n) - h)\theta^n + V_*^n(T^n - u^n\theta^n, Y^n - v^n\theta^n)\}$$
(10.52)

The state vector $X^n = (T^n, Y^n)$ in this formula comprises the temperature and concentrations, whereas the controls u^n and v^n are rates corresponding with change of state variables in the electrochemical system. The presence of the Lagrange multiplier h is associated with the absence of time variable τ^n in the set of state coordinates (dimensionality reduction). Some of end coordinates (T^0, Y^0) and (T^N, Y^N) may be fixed, but the total duration, τ^N, must be free, consistent with the dimensionality reduction. For an assumed h the optimal duration follows as a function of fixed end values of state X^n and total number of stages, N. Accuracy of DP results is much better after the state variable τ^n is eliminated, that is, when the problem is described by only state variables T^n and Y^n. The recurrence equation (10.52) also serves to generate numerical generalizations of function R when both transfer coefficients and heat capacities vary along the process path, and an analytical solution cannot be obtained.

Enhanced limits on power production in FCs (power consumption in electrolyzers) mimic many properties of systems described by Figure 7.2 in terms of internal irreversibility factor Φ. Figure 7.2 shows a comparison of optimal work production and work consumption functions, V and R, in terms of slowness variable τ, for a fixed change of system state. The reversible upper bound of production modes (ubwp), V_{rev}, is the same as the reversible lower bound achieved in consumption modes (lbws), R_{rev}. For irreversible bounds the coincidence does not hold, and a lower bound of R is larger than an upper bound of V. Note a qualitative similarity of this plot of work limits to the one characterizing generalized exergies (Figure 6.3).

Other optimal control problems with FCs are described by Chaisantikulwat et al. (2008) who review dynamic modeling and control of planar anode-supported SOFCs. Aimed at achieving maximum efficiency of power generated within certain constraints (stack temperature, steam-to-carbon ratio, compressor surge limitation, etc.) Milewski et al. (2010) evaluate a control strategy for high-temperature FC hybrid systems. Modeling, analysis, and feedback design are treated in the framework of industrial control of FC power systems by Pukrushpan et al. (2004). Carnot controls, satisfying identically entropy constraint, can be applied in power systems with a number of controls reduced to unconstrained ones (Sieniutycz, 2010a).

Bejan's constructal theory (Bejan, 2000) has been used in optimizing of dimensions, spacings, and the distribution (allocation) of transfer surfaces for a number of systems including FCs (Bejan, 2001b). Trees have been deduced from the maximization of flow access between a point and a volume, a point and an area, and a point and a curve (Lorente and Bejan, 2006). In particular, Lorente (2007) proposes to apply the constructal theory to optimization of ionic transfer by electrokinetics through porous media subjected to an external electrical source. The optimal location of the electrodes is determined from the constructal law. Also, the constructal theory is applied to explore transport problems involving flows governed by Ohm's law (Lewins, 2003). Importantly, the constructal theory is used to optimize the internal structure (relative sizes, spacings) of an FC with fixed total volume, so that the total power is maximized (Vargas and Bejan, 2004). System architecture follows from constrained optimization. The approach (illustrated for an AFC) consists in the optimization of flow geometry starting at the elemental level, where the FC is modeled as a unidirectional flow system. Power maximization also leads to the internal structure (relative sizes, spacings), single cell thickness, and external shape of a PEMFC stack (Vargas et al., 2005). There are two levels of optimization: (i) the internal structure and (ii) the external shape, which accounts for the external aspect ratios of the PEMFC stack. The power, efficiency, and polarization curve are obtained in terms of temperature, pressure, geometry, and operating parameters. Free decisions consist of the relative thicknesses of two diffusion layers, two reaction layers, and the space occupied by the electrolyte solution. The available volume is distributed optimally maximizing the total power. Temperature and pressure gradients are important. The results show that the (twice maximized) stack net power increases monotonically with total volume raised to the power 3/4, similarly to metabolic rate and body size in animal design. Ordonez et al. (2007) develop a structured procedure to optimize the internal structure of a single SOFC so that its power is maximized. The model is developed by using a control volume approach, in which all relevant thermal and electrochemical interactions are residing. Resulting optima are sharp and, therefore, important to be identified in actual SOFC design. The polarization curve and power density are obtained in terms of temperature, geometry, and operating parameters. From optimal balances between the thicknesses of anode and cathode optimal internal structure follows, which accounts for the thicknesses of the two electrodes and the electrolyte, and the flow channel geometry.

For sufficiently high values of currents and rate indices, work consumed in FC electrolyzers may far exceed the classical work, whereas work produced in FC engines can be much lower or even vanish. As opposed to thermal machines, functions of optimal power obtained for imperfect (dissipation-involving) downgrading and utilization of resources in FC systems are not generalizations of the classical exergy. This is because of inherent flow nature of FC systems. Nonetheless, the limits for electrical energy yield or consumption provided by the thermokinetic theory of FCs are stronger than those predicted by the classical thermodynamics. In fact, with thermokinetic models, we can confront and surmount limitations of classical thermodynamic bounds to real processes. This is the direction with many open opportunities. Electrochemical systems, and particularly FCs, are especially important in this context by virtue of their efficient and reliable performance.

We have applied methods of thermodynamics and optimization to FCs, and evaluated static and dynamic power limits. They are useful in the performance analysis and design of FC systems. A dynamic FC system has been treated as a multistage thermodynamic device that converts the energy of chemical reaction directly into electricity and heat, thus producing efficiently power with a finite rate. DP and the discrete maximum principle have been the main tools applied in computations.

In the thermodynamic analysis of modular FC system the performance criterion may be its entropy production S_σ in a form that describes the real stack from which power delivery takes place with a finite rate. The functional S_σ in an integrated form constitutes a performance criterion of the operation occurring due to the chemical reactions and mass transfer coupled with transfer of heat. The minimization of S_σ eliminates all controls from S_σ, thus generating a cost-like function $R_\sigma(X_A, X_B, \tau_B - \tau_A) = \min S_\sigma$, which depends only on initial and final states and the extensive transport parameter called the number of the transfer units. Yet, instead of S_σ, the criterion of total work produced in a given time, W, can be applied to FCs. The results of optimizations for S_σ and W are related by the familiar Gouy–Stodola law that links the lost work with the entropy production.

Because of finite rates, actual cell voltage V and finite-rate work W are smaller than those in an ideal cell. In general, no simple rule exists for optimal control of an unsteady-state system subject to external adjustable decisions. Yet, the optimal solution for total work W often implies a nearly constant intensity of the entropy production along an optimal dynamic path. Such a simple strategy is, however, valid only when no constraints are imposed on control variables. Postquadratic terms and nonlinearities in kinetic equations usually cause violation of this strategy.

The main methodological novelty of this chapter lies in the synthesizing nature of its approach that includes FCs in the class of thermodynamic power yield systems. In the thermodynamic optimization, that is, optimization applying thermodynamic constraints and performance criteria, thermodynamic synthesis stands for combining various optimization models into a "synthesizing" (not always "generalizing") model from which performances of all component units can be predicted. With irreversible thermodynamics, we can predict performance behavior and power limits for quite diverse practical systems.

Canonical (Hamiltonian) treatment of optimal dynamic processes constitutes an approach that may be contrasted with DP approaches. Since the analytical treatment of differential optimization models is most often a very difficult task, solving techniques often treat numerically Bellman's recurrence equation (DP algorithm). However, DP algorithms are effective only for low dimensionality of the state vector. As the number of state variables in chemical systems may be large (catalyst activities may accompany temperature and concentrations), DP algorithms become inefficient and inaccurate in real systems.

Therefore, in many dynamic power systems we need to abandon DP approaches and to apply the Pontryagin-type (Hamiltonian-based) approaches

(Chapter 2). As opposed to the DP algorithms, Hamiltonian algorithms, which involve discrete or difference equations rather than recurrence equations, are particularly effective in power systems with large dimensionality of the state vector. Chapter 2 analyzes subtleties in the mathematical structure of Hamiltonian algorithms, which may admit discrete process rates explicitly dependent on time intervals, θ. To date, however, the Hamiltonian algorithms used in power systems are limited to models with θ-independent discrete rates (Sieniutycz and Berry, 2000; Sieniutycz et al., 2011; Radulescu et al., 2006a, 2006b).

10.12. BIOLOGICAL FUEL CELLS AND BIOLOGICAL SOURCES OF HYDROGEN

BFCs have recently received considerable attention (Szewczyk, 2006; Li, 2006). Szewczyk (2006) introduces two categories of BFCs: enzymatic (EFCs) and microbial (MFCs). The former, due to the immobilization of biocatalysts, offer high current densities and possibilities of the device miniaturization. The latter, because of the presence of living organisms, use complex substrates and can do prolonged work. The substrate complexity property increases the range of reagents from which power can be produced. Bacteria are seen as self-reproducing catalysts that assure the continual production of power. Scholtz and Schroder (2003) regard an MFC as, basically, a system that harvests electrons generated in microbial metabolism and channels them for the electric current generation. Since the power generation and waste disposal are two key challenges in the quest for sustainable societies, MFC can, in principle, provide an elegant solution by linking both tasks. Long-lived FCs without diffusional electron mediators may convert simple and abundant sugars into electricity with efficiency exceeding 80% (Scholtz and Schroder, 2003).

According to Rabaey and Verstraete (2005) MFCs present novel biotechnology for energy generation and provide new opportunities for the sustainable power production from biodegradable, reduced compounds. Like "traditional" FCs, MFCs convert energy, available in biosubstrates, directly into electricity. The concept of microorganisms as catalysts in FCs was explored already in the 1970s, and MFCs treating domestic wastewater were presented in 1991 (Rabaey and Verstraete, 2005). However, it is only recently that practical MFCs with an enhanced power output have been developed. MFCs function on carbohydrates but also on complex substrates contained in wastewaters. Since available information about the energy metabolism and bacteria using the anode as electron acceptor is still limited, few electron transfer mechanisms have been proposed. Different metabolic pathways are used by the bacteria depending on the operational parameters of a MFC. This determines the selection and performance of specific organisms. Rabaey and Verstraete (2005) discuss how bacteria use the anode as an electron acceptor and to what extent they generate electrical output. They also evaluate the MFC technology relatively to current alternatives for energy generation.

Logan (2008) and Lovley (2008) consider microbe-sustained conversion of organic matter to electricity. Broad application of MFCs requires a substantial increase in current density; a better understanding of the system microbiology may help. Numerous investigations have greatly expanded the range of microorganisms functioning either as electrode-reducing microorganisms at the anode or as electrode-oxidizing microorganisms at the cathode. Microorganisms that can completely oxidize organic compounds with an electrode serving as the sole electron acceptor are expected to be primary contributors to power generation. Lovley (2008) specifies several proposed mechanisms for electron transfer to anodes including: direct electron transfer via cytochromes, long-range electron transfer via microbial nanowires, electron flow through a conductive biofilm matrix containing cytochromes, and soluble electron shuttles. Which mechanisms are most important depends on the microorganisms and the thickness of the anode biofilm (Lovley, 2008). In a recent report of Lovley group, an FC is described in which a microorganism oxidizes glucose to carbon dioxide at neutral pH. The reaction occurs via direct bioelectrocatalysis: the microorganism uses the anode itself as terminal oxidant to which liberated electrons are transferred, and does so with 83% efficiency (Palmore, 2004). These findings are significant because they demonstrate a substantial progress for harvesting energy from the environment by using microorganisms.

Gregoire and Becker (2012) consider design and performance of a MFC for the conversion of a lignocellulosic crop residue to electricity. Agricultural crop residues contain high biochemical energy as cellulose and lignin; a portion of this biomass could be harvested for conversion to biopower. The authors explore the potential for converting this biomass directly to electricity in a MFC. Their integrated approach aims to develop a solid-substrate MFC in which cellulose hydrolysis, fermentation, and anode respiration occur in a single chamber. Electricity is produced continuously from corncob pellets. Addition of rumen fluid increases power yield, presumably by providing growth factors to anode-respiring bacteria. Periodic exposure to oxygen also increases power yield. Under working conditions the maximum power density is $230 \, mW/m^3$. Problems of power limits in active biosystems still need more investigation.

Bioethanol as a source of hydrogen can be a useful medium fueling SOFCs (Hernandez and Kafarov, 2009). In recent years, SOFCs running on pure hydrogen or synthesis gas have drawn considerable attention, due to their application in distributed power systems. Arteaga-Pereza et al. (2009) focus on process simulation and heat exchanger network synthesis in autosustainable SOFC system fueled by bioethanol.

The catalytic steam reforming of bio-based alcohols, mainly bioethanol, is an interesting new technology based on the environmental compatibility of hydrogen energy when compared with other feedstocks. Ethanol has many advantages as a source of hydrogen, since it is easy to store, handle, and transport in a safe way due to its lower toxicity and volatility. In addition, that alcohol can be bioproduced from a wide variety of biomass sources, including sugarcane molasses, lignocelluloses, and waste materials from agroindustries (Arteaga-Pereza et al.,

2009). On the other hand, if the fermentation of biomass is carried out to obtain bioethanol, the total net emissions of CO_2 are essentially lower than those of fossils. Bioethanol can be converted into a hydrogen-rich gas stream using the reforming technologies (steam reforming, partial oxidation, and autothermal reforming). Ethanol steam reforming is the most studied technology because of the high yield and selectivity level obtained when carried out with a proper catalyst.

Arteaga-Pereza et al. (2009) carry out a simulation and heat integration of a SOFC integrated with an ethanol steam reforming system. Ethanol reaction is described by an original kinetic model. The operation is carried out by using synthesis gas produced in an ethanol steam reforming plant. High operating temperatures of SOFC (873–1273K) allow cogeneration and heat reuse. The system and FC efficiencies are studied under different conditions, temperature ($723\,K < T < 873\,K$), water to ethanol molar ratio ($3 < R < 6$), and fuel utilization coefficient ($0.7 < FUC < 0.9$). The SOFC off-gases are mixed and fed to an afterburner that provides heat to the process. Two heat exchanger networks are designed considering the influence of the fuel utilization coefficient at the cell electrodes. If the SOFC is operated at fuel utilization less than 0.8, a self-sufficient limit could be set; otherwise extra ethanol must be combusted with an overall efficiency penalty. A process flow diagram is proposed in order to obtain high efficiency and to avoid the use of any external source of energy.

10.13. ENERGY AND SIZE LIMITS FOR LIVING ORGANISMS IN BIOLOGICAL SYSTEMS

In the next section of this chapter, we consider energy and size limits of organisms in dynamic biosystems. First we analyze the significance of physical laws in establishing optimal forms of organisms. Next we point out the role of Bejan's (2000) constructal theory in thermodynamic optimization, and, in particular, in determining the number of bifurcations of human respiratory tree. We stress the common functioning of the physical laws in animate and inanimate systems. Finally, we briefly analyze some problems of biological evolution. This edition of the book omits many details of an extremum principle for an entropy-based complexity that governs the processes of biological development and evolution. The principle implies that some developmental processes progress in a relatively undisturbed manner, whereas others terminate rapidly due to inherent instabilities. Detailed treatments of these issues are available (Szwast, 1997; Dall and Cuthill, 1999; Sieniutycz, 2000a; Szwast et al., 2002; Sieniutycz and Jeżowski, 2009).

The structure of the last two sections is as follows. Energy and size limits are analyzed in the present section in the context of both traditional physical analyses and Bejan's constructal theory. Section 10.14 applies the meaning of complexity as a way to analyze evolutions of multiple organs or limbs in organisms with mutations (Dall and Cuthill, 1999). Classical thermodynamic quantities do not appear in complexity approaches, yet an associated statistical model implies, as in thermodynamics, extremum properties for a potential (Szwast et al., 2002). In effect, a great many features of living organisms can be predicted by describing their complex evolutions in terms of variational principles for shortest paths along with suitable transversality conditions. We terminate the chapter by stressing the impact of Darwinism and its view of evolution on economics and thermoeconomics.

Significant progress has recently been made toward explaining how the forms that are observed in nature are produced. Relatively simple differential systems were shown to produce bifurcations and singularities even without discontinuities in their coefficients or boundary conditions. A large number of researchers have shown bifurcations emerging out of the reaction–diffusion interactions. By applying physical principles we are now able to evaluate energy and dimension limits for living (e.g., aquatic) organisms and describe behavior of many organs functioning in these organisms (Bejan, 1997a, 2000; Bejan and Lorente, 2001). Size constraints can be attributed to the principles of thermodynamics that govern the energy budget and power generation of living organisms. While living systems have developed various strategies to manipulate their self-organization in order to satisfy the principles of both physics and biology, it is the physical laws that set limits to their size, functioning, and rate of development.

For example, the physical law of thermal conduction sets the size of warm-blooded aquatic animals that require a minimum diameter in order to survive in cold oceans. Species that survive in ecosystems are those that funnel energy into their own production and reproduction and contribute to autocatalytic processes in the ecosystem. There are data that show that poorly developed ecosystems degrade the incoming solar energy less effectively than more mature ecosystems. The present attitude is to view living systems as stable structures increasing the degradation of the incoming available energy, while surviving in a changing and sometimes unpredictable environment.

Ahlborn (1999) and Ahlborn and Blake (1999) have determined thermodynamic limits of body dimension of warm-blooded animals, and lower size limit of aquatic mammals. Bejan (1997a, 2000) has conducted analyses toward biological limits in the framework of his constructal theory, and obtained similar results. Warm-blooded animals are homothermal animals, high on the evolution tree, that can swim in oceans, roam the lands, and fly through the air. Still these species only occupy a relatively small range of body mass when they are compared to the 20 orders of magnitude of all creatures ($M = 10^{-15}$ kg for amoebae to 10^5 kg for a blue whale). Land animals that are able to run with endurance have body masses ranging from little mice of $M = 10^{-2}$ kg to big ungulates weighing about half ton (0.5×10^3 kg). Aquatic mammals span the range from $M = 10^1$ kg for baby seals to 10^5 kg for blue whales. Flying birds' size varies from 10^{-2} kg for humming birds to 15 kg for mute swans.

Ahlborn (1999) shows that these size limitations are a consequence of thermodynamics that sets the energy budget and power production of animals. It is the heat transfer rate through the body surface that imposes a lower size limit for warm-blooded animals in a cold environment. The available muscle power along with the minimum power necessary to stay aloft or to run with endurance determines upper limits for the body masses for birds and terrestrial animals. The basis of these limits is the metabolic power function G of animals.

The power of all animals in terms of their body mass can be described by a unique metabolic law that was first discovered by Kleiber (Ahlborn, 1999). His relation connects the metabolic energy production Γ_0 of the resting organism in Watts to its body mass M in kilograms:

$$G_0 = a_0 M^{0.73} \tag{10.53}$$

where the constant a_0 equals approximately 3.6. Animals expend energy Q to produce work W and power $P = dW/dt$ in their muscles. The metabolic rate of the active animal $G = dQ/dt$ is larger than G_0 by a factor b ($b = 10$–20 in some cases). The animals that generate mechanical power warm up. One can imagine a bird with body temperature T_i in a cold environment of temperature $T_e = T_i - \Delta T$. At all times the mechanical heat production $G = dQ/dt$ in the body volume must overcome the conductive losses of heat, lost through the body surface.

Assuming a spherical shape and the heat loss proportional to the surface, one can observe that the smaller the radius R of the body, the larger the surface to volume ratio A/V (equal to $3/R$ for a spherical shape). Clearly, the effect of heat losses increases when the radius R decreases. Thus, for any metabolic rate, there must be a certain minimal radius R_{min} below which elevated temperature of the body can no longer be sustained. With the idea that the metabolic heat production G must at least compensate losses, Ahlborn's (1999) analysis yields

$$R_{min} = \left(\frac{k \ \Delta T}{113} \right)^{0.84} \tag{10.54}$$

In this limiting situation, no amount of insulation is sufficient to maintain the body temperature at the homeostatic value T_i. The minimum of radius R (corresponding with an infinitely thick insulation) implies a minimum of the body mass

$$M_{min} = \left(\frac{4\pi}{3} \right) \rho R_{min}^3 = (k \ \Delta T)^{2.52} \tag{10.55}$$

For a temperature difference $\Delta T = 330$, $R_{min} = 1.6$ cm and $M_{min} = 17$ g. Both quantities R_{min} and M_{min} are sensitive functions of thermal conductivity k and temperature difference ΔT. A detailed consideration (Ahlborn, 1999) shows that warm-blooded animals of subcentimeter size can never live in a cold environment.

Aquatic mammals cannot exploit air in their feathers as an isolating material; they are insulated by incompressible body fat, with thermal conductivity of the order of magnitude $k = 0.2$ W/mK. Moreover, fast aquatic animals must have a form that minimizes their drag in order to keep their motion expenses small. They cannot have the spherical shape that reduces the surface area. In fact, they have adopted the ratio $2R/L$ in the range 0.2–0.3, where the sum of form drag and friction drag is a minimum (Ahlborn, 1999). These conditions can be used to determine a minimum mass of warm-blooded aquatic animals. For simplicity, these animals can be modeled as cylindrical objects. Again, the conductive heat losses must be offset by the metabolic heat production G_0. These heat losses become very large when the insulation thickness becomes small compared to R_i. This heat loss must be balanced by the metabolic heat production $G_0 = a_0 M^{0.73}$, where the body mass is proportional to R^3. With this reasoning, Ahlborn obtains an estimate for the insulation thickness $\Delta R = 3.5$ cm, as the quantity virtually independent of the body radius R. This finding is also confirmed by measurements.

With a similar reasoning it may be shown that a maximum size of birds can be estimated, because very big birds cannot generate enough power for flapping flight (Ahlborn, 1999). Also, as shown by Bejan (2000), astonishingly good agreement exists between theoretical formulae for flying speed and experimental data for animals. In an animal, the power needed to increase the kinetic energy necessary to fly is delivered by the muscles, and has its origin in metabolism. In a flying machine power is produced by the power plant installed on the board. In both cases power is extracted from the chemical exergy of the consumed food or fuel. The theory predicts a power formula in terms of the speed V and the body length scale D

$$P \sim \frac{\rho_b^2 g^2 D^4}{V} + \rho_{air} D^2 V^3 \tag{10.56}$$

This shows that the power necessary to fly is the sum of that for maintaining the body in the air (the first term) and that for overcoming the drag (the second term). Changes in the flying speed V induce changes of opposite signs in the two terms. For a fixed body length scale D, power has a minimum with respect to V that is obtained via minimization of the above power formula. The result is an optimal speed proportional to the body mass raised to the power 1/6:

$$V_{opt} \sim \frac{\rho_b^{1/3} g^{1/2} M^{1/6}}{\rho_{air}^{1/2}} \tag{10.57}$$

Cruising speeds of insects, birds, and airplanes correlate well with the speed for minimum power consumption (Equation (10.54)). Ahlborn's (1999) work implies that the metabolic rates of animals of different body masses M determine their heat production and their mechanical power output P. The metabolic heat production of small warm-blooded animals living in a cold environment sets a lower size limit for the body mass of birds and terrestrial mammals of $M_{t,b} > 0.01$ kg and $M_{aq} > 10$ kg for fast aquatic mammals. In particular, Ahlborn (1999) has shown that birds must travel at the speed $v_{fl} = 15 M^{1/6}$ in order to stay aloft, but their muscles only generate the metabolic velocity $V_\Gamma < 50 M^{-0.25}$. These constraints lead to a limit for an upper body size for flying birds: $M_{b,max} < 15$ kg. Walking speeds scale with body mass as $v_{walk} = c M^{1/6}$, while propagation speeds for endurance running are nearly independent of body mass. This assures that walking is faster than endurance running for animals with body masses M larger than 600 kg.

Each living organism can be regarded as a complex system with very many elementary objects, links, and couplings. Such complex system also has many inputs and outputs, some of which serve to supply the consumed energy and substances to the organism (system), whereas the others (e.g., muscles) are able to deliver produced mechanical energy out of the organism. In fact, there is an analogy between an energy system of inanimate nature and a living organism. There are also at least two essential differences between the living system and a machine: the first one is a much larger complexity of the living system than a machine and the second is the presence of a huge number of information streams out and within the living system. The essential property of the living system, the capability to work, is, however, represented by typical work generators, as shown in many diverse examples. In many thermal machines energy generators contain closed loops (e.g., Carnot cycles or Brayton cycles) corresponding with cyclic mode of the energy yield. In living systems this is unnecessary because chemical generators are most common.

A number of recent developments in thermodynamic optimization focus on the generation of optimal geometric form (shape, structure, topology) in animate and inanimate flow systems, where flow configurations are free to vary. The principle that generates geometric form is associated with the search for best global performance (e.g., minimum flow resistance, minimum irreversibility) subject to global finiteness constraints (volume, weight, time). The resulting structures obtained in this manner have been named constructal designs (Bejan, 2000; Bejan and Lorente, 2001). The idea that the same objective and constraints principle accounts for the optimally shaped flow paths that occur in natural systems (animate and inanimate) is the content of the constructal theory (Bejan, 1997a, 1997b, 1997c, 1997d, 2000). It holds that flow architecture arises from maximization of flow access in time and in flow configurations that are free to morph. Examples of large classes of applications are obtained from various sectors of engineering: the distribution of heat transfer area in power plants, optimal sizing and shaping of flow channels and fins, optimized ratios of heat exchanger core structures, aerodynamic and hydrodynamic shapes, tree-shaped assemblies of convective fins, tree-shaped networks for fluid flow and other currents, optimal configurations for streams that undergo bifurcation or pairing, insulated pipe networks for the distribution of hot water and exergy over a fixed territory, and distribution networks for virtually everything that moves in society (goods, currency, information). The principle-based generation of flow geometry unites the thermodynamic optimization developments known in mechanical engineering with lesser known applications in civil engineering and social organization. These results, obtained in A. Bejan's research group (Bejan, 1997a, 1997b, 1997c, 1997d, 2000; Bejan and Lorente, 2001), generalize methods of thermodynamics, because they show how thermodynamic principles of design optimization contribute to the development of optimal configurations in engineering and social organization. The meaning of social organization, essential therein, can be understood in a broad sense, in particular as one that refers to populations of living organisms (e.g., those in ecosystems).

In fact, Bejan has developed the theory by which it is possible to predict macroscopic shapes, originated by the spatial organization in nature, both in living and in engineering systems. His result is an optimization principle, which, however, still has no ultimately defined optimization criterion. As he himself has pointed out, the theoretical basis of the architecture of many living and nonliving systems remains an unknown design principle (Bejan, 2000). The principles of design and optimization of engineered systems describe a deterministic background for the generation of geometric forms in natural systems. Shape and structure follow from the concomitance for better performance in both engineering and nature. This notion leads to the idea that the objective and constraints principle used in engineering constitutes also the mechanism from which the geometry emerges in natural flow systems. The principle accounts not only for tree-shaped flows but also for other geometric forms encountered in engineering and nature (round ducts, regularly spaced internal channels, rivers with proportionality between width and depth, etc.). It can also be applied to estimate shapes of living organisms in oceans.

Quite generally, flow systems with geometric structure exhibit at least two flow regimes, one with high resistance (slow, diffusion, walking) that fills the volumes of smallest finite scale and one or more with low resistance (e.g., fast, streams, channels, streets). The optimal balancing of regions with different flow regimes means that the material and channels must be distributed in certain ways. Better global performance is achieved when the distribution is relatively uniform, this, in spite of the gaping differences between the high- and the low-resistivity domains (Bejan and Lorente, 2001). According to Bejan, it is the optimal distribution of imperfection that generates form: "The system is destined to remain imperfect." The system works best when its imperfection (its internal flow resistances) is spread around, so that more and more of the internal points are stressed as much as the hardest working points. One good form leads to the next, as the constructal principle demands: objective served better while under the grip of global and local constrains. As Bejan states: "There is a time arrow to all these forms, and it points toward the better."

The constructal principle of optimized tree-flow architecture allows to predict many important allometric laws (Bejan, 2000, 2001a, 2001b), for example, Kleiber's law (10.53), in the form describing the proportionality between metabolic rate q_0 and body size M raised to the power 3/4:

$$q_0 \sim M^{3/4}$$

the proportionality between breathing and heart beating times t and body size M raised to the power 1/4:

$$t \sim M^{1/4}$$

the link between mass transfer contact area A and body mass M:

$$A \sim M^{7/8}$$

the proportionality between the optimal cruising speed V_{opt} of flying bodies (insects, birds, airplanes) and body mass M in *kilograms* raised to the power $1/6$:

$$V_{opt} \sim 30M^{1/6} \text{ m/s}$$

Bejan's constructal theory also explains why humans have a bronchial tree with 23 levels of bifurcation (Reis et al., 2004). The constructal theory of the flow architecture for the lung predicts and offers an explanation for the dimensions of the alveolar sac, the total length of the airways, the total alveolar surface area, and the total resistance to oxygen transport in the respiratory tree. We describe briefly some of these results below.

Reis et al. (2004) first considered the flow between the entrance of the trachea and the alveolar surface where oxygen meets the blood and carbon dioxide is removed from the blood, including the resistances due to bifurcations. They showed that the global resistance (J s/kg^2) to oxygen transportation within the respiratory tree is of the form

$$R_{ox} = \frac{256 \nu L_0}{\pi D_0^4 [(\phi_{ox})_0 - \phi_{ox}] \rho} (N+1) + \frac{0.13 (R_g)_{ox} T 2^{-2N/3}}{\pi L_0 D_{ox} \phi_{ox} \rho} \qquad (10.58)$$

In Equation (10.58), N describes the number of bifurcations of the bronchial tree, ν is the kinematic viscosity of the air, L_0 and D_0 represent trachea length and diameter, respectively, T is temperature, ϕ_{ox} and $(\phi_{ox})_0$ represent the relative concentration of oxygen in the alveoli and in the outside air, respectively, D_{ox} is the diffusivity of the oxygen in the air, R_g is the air constant, and ρ designates the density of the air.

The first term on the right-hand side of Equation (10.58) represents the global channel flow resistance (bronchiolar tree) while the second term matches the global diffusive resistance to oxygen transport in the alveoli. An equation of the type of Equation (10.58) also holds for the resistance to the carbon dioxide flow in the respiratory tree. These resistances have been evaluated by assuming body temperature of 36 °C and taking all suitable values at this temperature (Reis et al., 2004). The average value of oxygen's relative concentration within the respiratory tree, ϕ_{ox}, is evaluated from the alveolar air equation in the form $[(\phi_{ox})_0 - \phi_{ox}]Q - S = 0$, where $(\phi_{ox})_0 \sim 1/2(\phi_{air} + \phi_{ox})$ and ϕ_{air} are the oxygen's relative concentrations at the entrance of the trachea and in the external air, respectively, Q is the tidal air flow, and S is the rate of oxygen consumption. With $(\phi_{ox})_{air} = 0.2095$, $Q \sim 6 \times 10^{-3}$ m^3/min, and $S \sim 0.3 \times 10^{-3}$ m^3/min, Reis and coworkers have found $\phi_{ox} \sim 0.1095$. The value of the average relative concentration of CO_2 in the respiratory tree, $\phi_{cd} = 0.04$. In this case $S = 0.24 \times 10^{-3}$ m^3/min since the respiratory coefficient is close to 0.8 and $(\phi_{cd})_{air} \sim 0.315 \times 10^{-3}$.

Anatomic treatises indicate that L_0 is typically 15 cm, while the trachea diameter, D_0, is approximately 1.5 cm.

To find an optimal number of bifurcations N_{opt} that assures the minimum resistance (to both oxygen access and carbon dioxide removal), global resistance to oxygen and carbon dioxide transportation in the respiratory tree is plotted against number of bifurcations. Equation (10.58) also allows to find N_{opt} analytically. The analysis yields $N_{opt} = 23.4$ and 23.2 for the oxygen and the carbon dioxide transport, respectively (Reis et al., 2004). As the number of bifurcations must be an integer, N_{opt} must be 23. The first conclusion is that the human respiratory tree that bifurcates 23 times between the trachea and the alveoli is optimized both for oxygen access and for carbon dioxide removal. The trade-off between resistance to flow in the bronchial tree and the resistance to oxygen and carbon dioxide diffusion in the alveoli is achieved by the human respiratory tree, which has been optimized by nature in time. The second conclusion is that the actual flow architecture of the respiratory tree can be anticipated theoretically based on constructal theory applied to the minimization of global resistances to fluid access under geometric constraints.

Constructal relationships of the respiratory tree (Reis et al., 2004) show that the number of bifurcations that matches minimal global resistance to oxygen access and carbon dioxide removal depends on several environmental variables such as normal body temperature, oxygen and carbon dioxide diffusivities and concentrations in the air, air kinematic viscosity, and a morphological parameter, which is the length $\lambda = D_0^2/L_0$. As every individual lives with the same average environmental parameters and since the respiratory tree with 23 bifurcations is a characteristic of humans, the number λ is also a characteristic of humans. Thus, the ratio of the square of trachea diameter to its length should be the same for every human being. This result still awaits more solid confirmation by anatomists.

Another constructal relationship involving the length λ, the area allocated for the respiratory process (the total area of the alveoli) A, the volume of the lungs V, and the length of the respiratory tree, between the entrance of the trachea and the surface of the alveolus, L, has also been derived by Reis et al. (2004):

$$\lambda = \frac{D_0^2}{L_0} = 8.63 \frac{AL}{V} \left\{ \frac{\nu D_{ox}\phi_{ox}}{(R_g)_{ox}T[(\phi_{ox})_0 - \phi_{ox}]} \right\}^{1/2} \tag{10.59}$$

This equation implies that the nondimensional number AL/V defines characteristic length $\lambda = D_0^2/L_0$, which determines the optimal number of bifurcations of the respiratory tree. This result may be summarized as follows: "The alveolar area required for gas exchange A, the volume allocated to the respiratory system V, and the length of the respiratory tree L determine univocally the structure of the lungs, that is, the bifurcation level of the bronchial tree" (Reis et al., 2004). Keep in mind that A, L, and V are constraints of the respiratory system. For some modification and refinement of Reis et al. (2004) approach, see Przekop (2009).

10.14. A BRIEF COMMENTARY ON DEVELOPMENT AND EVOLUTION OF SPECIES

Mechanical views are usually abandoned when development and evolution of species are described. In a number of evolution-related works nonequilibrium thermodynamics is regarded as the basis for an alternate evolutionary hypothesis. Since Darwin (1859) various researchers followed various ways to describe evolution and various theoretical approaches have been used (Shannon and Weaver, 1969; Mackey, 1989; Lasota and Mackey, 1985; Denbigh and Denbigh, 1985; Shiner et al., 1999; DePamphilis, 2002; Zotin and Zotin, 1996; Pierce, 2002; Wickey, 1979, 1980; Brooks and Wiley, 1986; Capra, 1996; Saunders and Ho, 1976, 1981; Szwast, 1997; Sieniutycz, 2000a; Szwast et al., 2002; Schneider and Kay, 1994; Sieniutycz and Jeżowski, 2009). It has been argued that evolution is "an axiomatic consequence of organismic information obeying the second law of thermodynamics and it is only secondarily connected to natural selection" (Pierce, 2002). As the time arrow acts, entropy is produced and the information within a biological system becomes more complex or variable. The information-related complexity is shaped or organized through historical, developmental, and environmental constraints of which only latter refer to natural selection. Whenever the information within the system becomes too complex and disorganized, living organisms diversify, extinct, or speciate at bifurcation points of the system. Speciation is evolutionary development of a new species usually through the division of a single species into two separate genetically different groups. Disorganization (disorder) is a consequence of the addition of new information (e.g., mutations, etc.) into the system. Speciation or extinction occurs when a species is unable to reorganize their information. Feedback loops in the system are capable of amplifying even very small amount of information into an uncontrolled chaos (Capra, 1996). At the points of informational bifurcations new states and new order are created. A leading place in theoretical ecology is occupied by thermodynamics. In Johansen work (Jørgensen, 2000, 2001) exergy and specific exergy have been found together with nine other ecological attributes for ca. a dozen coastal ecosystems. The correlation of a number (11) of attributes is examined, and the extent to which exergy and specific exergy can be applied as indicators to assess ecosystem health is discussed. These concepts cover a range of important properties of ecosystems, but other indicators are also required to provide a sufficiently comprehensive assessment of ecosystem health. With a health criterion, enzymological aspects of bioconversion of chemicals can be considered (Ebeling and Feistel, 1992).

After passing the bifurcation event the system reorganizes itself. According to thermodynamicists, self-organization is the spontaneous creation of order in open entropy-exporting systems operating beyond a critical distance from equilibrium. Self-organization occurs because each descendant branch (obtained by bifurcation) has less entropy than its ancestral branch. As both descendant branches have a combined entropy equal to or higher than the ancestral branch, biological systems are allowed to evolve to form a bifurcation tree of hierarchical relationships. In effect, the occurrence of speciations, extinctions, or diversifications is stimulated by informational disorganization and formed through external environment as the natural selection. In this picture the entropic "drive toward randomness" underlies both modification and speciation, thus constituting the ultimate cause to evolution. Quantitative thermodynamic models of these phenomena are available (Wickey, 1979, 1980; Brooks and Wiley, 1986; Saunders and Ho, 1976, 1981; Szwast, 1997; Sieniutycz, 2000a, 2000b, 2000c, 2000d, 2000e; Szwast et al., 2002). The appropriate state space is often the space of probabilities (Szwast, 1997). Some of the models (Szwast et al., 2002) incorporate a maximum change of the entropy at the constrained (to a constant) distance in the state space. Spontaneous increase of complexity is regarded as a basic feature of evolution (Saunders and Ho, 1976).

Saunders and Ho (1976, 1981) decisively argue that complexity, rather than fitness or organization, is the quantity whose increase fixes a direction to evolutionary processes. They claim that the complexity increase is a consequence of the process by which a self-organizing system optimizes somehow its organization with respect to a criterion and tends to permit the addition of components more readily than their removal. In effect of considering organization as a structure on a set of components (analogously as topology on a set of points), the complexity becomes the basic quantity in evolution. For biological systems a reasonable measure of complexity is the different number of components they contain, so as to be consistent with well-known Williston's law that predicts that very similar components will either merge or specialize (Saunders and Ho, 1976, 1981). The increase in organization seems a secondary effect that emerges because more complex system requires more organization in order to survive. An example is large, randomly connected linear systems, whose likelihood to be stable is very small. To ensure their survival components must be added to them in a nonrandom way, which increases their organization. Staying with the idea of increasing complexity, the system's organization acts as a force that prevents loss of components and allows complexity increase by the integration of new ones.

When speaking about complexity as a quantity related to entropy, one issue is particularly important: a nonequilibrium entropy has to be necessarily applied because it is the difference between the maximum entropy and the actual (observed) entropy that quantifies the organization in a complex system. This refers to the various forms of organization: with respect to characters over species, cells over all cell types, organisms over genotypes, etc. (Pierce, 2002). Yet, if the role of the environment is essential, the idea of maximizing a potential function, more general than complexity or nonequilibrium entropy, subject to given side conditions or constraints, needs to be used (Sieniutycz, 2000a, 2000b, 2000c, 2000d, 2000e). Still, in view of large difficulties in applying statistical mechanics to even relatively simple systems, some authors doubt that it will ever be possible to effectively use the information-theoretic expression for entropy for complex living systems especially because they are fairly more complex than thermodynamic ones. Also they doubt whether other measures of complexity, the information content of the genome or one based on genetic code (DePamphilis, 2002; Capra, 1996), which are complexities measured by a molecular biologist, are appropriate and/or feasible for complex-system modeling.

However, there may be two different formulations of the principle of increase in complexity, an issue that is not stressed in the current literature sufficiently well. If complexity is considered as a function of variables characterizing current state of the system or related parameters (e.g., entropy, as described in Szwast et al., 2002), then the complexity value will always be greater at the end of the process stage than that at any of the previous states. In this sense the final complexity is a maximum (initial is a minimum) with respect to variations of a state or an indicator of this state. An example of application of this idea is described by Szwast et al. (2002) in an analysis of complexity with respect to entropy. This approach is associated with seeking *maximum* of Γ at a given distance of states J. In a dual problem one seeks minimum of distance in the state (probability) space at a given change of complexity or entropy (min J at a given $\Delta\Gamma$).

On the contrary, the principle of *minimum* increase in complexity (Saunders and Ho, 1976, 1981) means that when process rates are taken as controls, then the complexity increase in the natural evolution will be minimal in comparison with that calculated for any of virtual evolutions subject to the same constraints.

Of these two formulations, the first involves the mathematical programming for static maximization of Γ, whereas the second uses variational calculus for dynamic minimization of Γ. The virtue of the first principle is its simplicity and determining a sole point of a static extremum, whereas that of the second the information about dynamics in terms of the extremum conditions of the problem (Euler–Lagrange equations, Hamiltonian equations, or geodesic equations).

Entropy-based models, quantifying critical phenomena with an increase and subsequent reduction of organs in multiorgan organisms, can be applied with the inclusion of nonclassical statistical entropies, for example, q-entropies of Tsallis or Renyi (Salinas and Tsallis, 1999; Kaniadakis et al., 2002) that may modify unstable regions in the space of process probabilities (Sieniutycz and Jeżowski, 2009). A rationale for an initial increase of organ number in evolution of animals' organs or limbs is substantiated with these models. A variational problem searches for the maximum entropy subject to geometric constraint of the constant thermodynamic distance in the nonEuclidean space of independent probabilities p_i plus possibly other constraints. Tensor form of dynamics is obtained. Some developmental processes are shown to progress in a relatively undisturbed way, whereas others may terminate rapidly due to inherent instabilities (Szwast, 1997). For processes with a variable number of states the extremum principle provides quantitative investigation of biological development. The results show that a discrete gradient dynamics (governed by the entropy) can be predicted from variational principles for shortest paths and suitable transversality conditions (Szwast et al., 2002).

Systems capable of increasing their size and/or number of states (growing systems) can exhibit critical behavior when their size increases beyond a certain value. In effect, some developmental processes may progress in a relatively undisturbed way, whereas others may terminate rapidly due to strong instabilities. In evolution literature Williston's law is frequently quoted (Saunders and Ho, 1976, 1981) that subsumes the results of observation and comparative analysis. This law states that if an organism possesses many of the same or similar elements, a tendency appears to reduce the number of these elements along with the simultaneous modification (speciation) of these elements that are saved by the organism. In Szwast's examples (Szwast, 1997; Szwast et al., 2002), the evolution submanifolds describe organisms possessing $n - 1$ of identical organs or limbs (e.g., pairs of legs) and one organ (limb) being modified, specialized, or subjected to mutation. The results lead to a principle stating that an organism with larger number of similar organs or limbs (e.g., pairs of legs) is more susceptible to evolution toward an increase in the number of these organs. Yet, during reversible specialization of organs, the state of an organism can fall into the region of a catastrophic decrease in the number of these organs. These catastrophes constitute the price of specialization. The likelihood of falling in the catastrophe region increases with the number of organs (limbs). This explains why organisms possessing a large number of similar organs (limbs) ultimately reduce this number, despite the fact that they are more susceptible to evolutionary increase in organ number. This also agrees with the Williston's law of evolution (Saunders and Ho, 1976, 1981) that is confirmed by excavation experiments.

In the dynamic description of this problem an extremum principle provides a quantitative picture of biological development (Szwast et al., 2002). It follows that a discrete gradient dynamics, governed by the entropy potential, can be predicted from variational principles for shortest paths and suitable transversality conditions (Elsgolc, 1960; Lyusternik, 1983). General gradient dynamics, which governs the evolution in curvilinear spaces, is of Onsager's structure and is consistent with the entropy principle of extremality as the driving factor in the discrete dynamics of living systems, postulated by Gontar (2000).

Speciation is an evolutionary development of a new species usually through the division of a single species into two separate genetically different groups. For living organisms regarded as multiorgan and multilimb systems, speciation can be treated by a complexity criterion, Γ, usually originating from the classical statistical entropy of Shannon–Boltzmann. In order to penetrate a vaster spectrum of system stability (instability) properties, Tsallis entropy S may also be used. In speciation analyses classical thermodynamic quantities do not appear, yet the model used satisfies an extremum principle that, similarly as in thermodynamics, implies a maximum of entropy subject to the geometric constraint of a given thermodynamic distance.

An original optimization problem for a system with n states is that of maximum change of entropy or entropy-related complexity Γ between a point and the surface of constant distance from the point (Figure 10.7). Dual forms of this principle may also be considered, where one minimizes the thermodynamic length subject to a fixed change of the system's complexity or entropy (Figure 10.8). In this formulation, evolution paths are obtained by minimization of a length

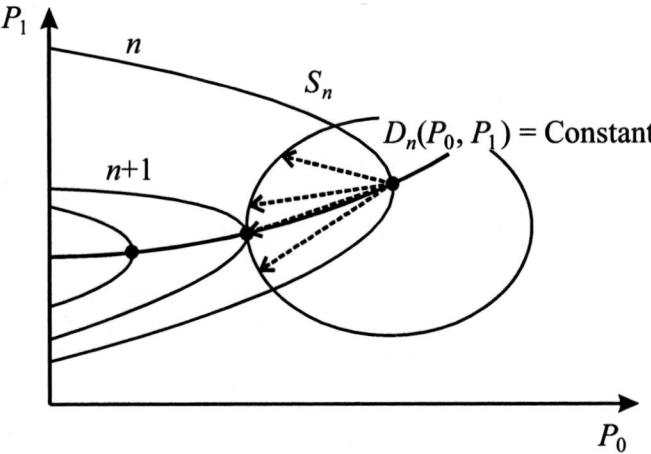

Figure 10.7 An original optimization problem for a system with n states is that of maximum change of entropy or complexity Γ between a point and the surface of constant distance D from the point.

functional in which independent probabilities p_i are constrained to reside on the constant-entropy manifold (Szwast et al., 2002).

The success of Darwinism and its view of evolution have induced economists who are interested in an evolutionary approach to economics to borrow concepts and tools from Darwinian theory. Particularly pronounced are constructions based on analogies to the theory of natural selection. Bioeconomics – the merging of views from biology and economics – accepts the idea that human economic activity and its evolution, meaningful over the past few centuries, may be considered an instance of fruitfully applying ideas from evolutionary biology and Darwinian theory (Witt, 1999). Thus, bioeconomics can be regarded as

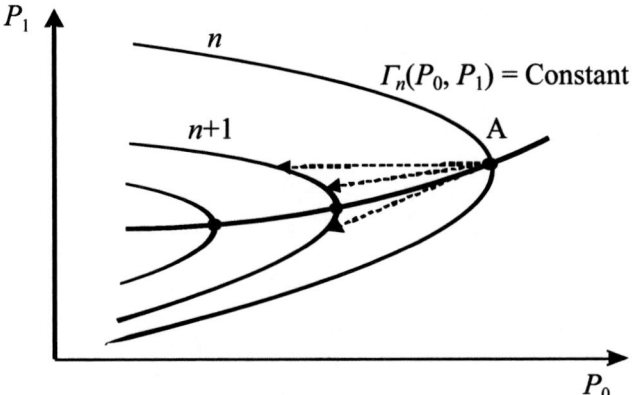

Figure 10.8 Dual optimization problem for a system with n states is that of minimum length between a point and the surface of constant entropy for the system with $n + 1$ states.

economics from a Darwinian perspective, and the Darwinian revolution can be a model for introducing a new paradigm in economic theory. In particular, the question of what a Darwinian world view might mean for assessing long-term economic evolution can be considered. Similarly, one could postulate the existence of biothermoeconomics as the thermoeconomics from a Darwinian perspective. This is consistent with the observation that understanding complex biological systems requires the integration of experimental and computational research—in other words, a systems biology approach. Computational biology and biothermoeconomics, through systems modeling and theoretical exploration, address questions fundamental to our understanding of life, and the progress in these disciplines will lead to practical innovations in medicine, drug discovery, and engineering (Kitano, 2002).

11 Systems theory in thermal and chemical engineering[*]

11.1. INTRODUCTION

A functional integrity accomplishing a certain goal is called a "system". In thermal and chemical engineering the term is usually used to refer to existing plants in which individual units are interconnected by suitable links. A system is usually constructed by the design of various units (subsystems) and the subsequent integration of these units. The links between the units are generally nonlinear, hence the properties of the system are not determined by additive combination of system outputs. Also, a performance criterion of the system is essentially different from a performance index of an individual unit.

The idea of the system approach is as old as European civilization. It has generally been accepted that the Aristotelean statement, *the whole is more than the sum of its parts*, was a first attempt to define the system problem (Klir, 1972; Ziębik, 1996). Yet, in past research the mechanistic approach prevailed, in which a problem was divided into parts analyzed separately. For many years no changes to this approach seemed to be necessary. It was only shortly before the Second World War that the system approach was discovered anew by Ludvig van Bertallanffy, philosopher and biologist, who stated that investigations of the parts must not only be supplemented by recognition of the role of the whole but that there exists a separate discipline, the theory of systems, dealing with investigations of the whole (Bertallanffy, 1973).

When the system structure is known, we can ask about its performance; this problem is known as "system analysis". Inversely, one can know the system's performance and ask about its structure; this problem is known as "system synthesis". Whereas analysis usually leads to a unique solution, synthesis does not generally provide a unique solution. In fact, the same performance can be shown by systems of various structures. Therefore, finding a structure that is most relevant to a given performance is a very important problem in system theory. The method of maximum entropy applied, for example, in thermodynamics, physics and urban modeling (Martyushev and Seleznev, 2006) is an example of the synthesis-type approach that leads to the most relevant structure of systems described in probability spaces.

In engineering, two basic approaches to investigations can be distinguished—process analysis and system analysis. Results of the process analysis are often

---

[*] An essential part of this chapter including figures and tables exploits Ziębik (1996) paper, with permission.

applied as input data for the system analysis. Ziębik's (1996) paper, discussed below, and other works by this author (Ziębik, 1986, 1990a, 1990b, 1991a, 1991b, 1995; Ziębik and Presz, 1993; Ziębik et al., 1994) deal with the methods of system analysis in the energy management of industrial plants. Ziębik's (1996) paper reviews the construction principles for mathematical models of industrial energy management (linear and nonlinear) and their applications, and provides an optimizing mathematical model for the preliminary design of industrial energy management. It is interesting because, to date, process analyses have been predominant.

The scope of this chapter is not only to review and classify the main methods and results obtained in the field but also to consider common objections caused by misunderstanding of these methods and results. In fact, the second part of the chapter contains a critical comparison of various methods applied for the systems of energy generation, such as second-law analyses, entropy generation minimization, approaches coming from ecology, and finite-time thermodynamics. Systems considered are those with transfer or rate processes that occur in a finite time and in equipment of finite dimensions. These processes include heat and separation operations and are found in heat and mass exchangers, thermal networks, energy convertors, energy recovery units, storage systems, chemical reactors, and chemical plants.

11.2. SYSTEM ENERGY ANALYSES

In process analysis the assessment of energy impact is constrained to the interior of the balance shield of the considered unit; this is how direct energy consumption is determined. Yet it should be realized that no real energy unit can operate as a separate system but is a component of a global energy system. In fact, the production and consumption of energy carriers always takes place within the network of interconnected thermal units. Therefore, both the direct consumption of energy carriers in the unit investigated and the indirect consumption occurring in other units must be analyzed. For this purpose the system approach should be applied (Bertallanffy, 1973; Klir, 1972; Leontief, 1951; Mielentilew, 1982). Application of the system approach to investigate problems of industrial thermal engineering started in the 1960s. The elaboration of a mathematical model for material and energy balance of an industrial plant using the example of ironworks may be considered (Szargut and Ziębik, 1972).

11.3. MATHEMATICAL MODELING OF INDUSTRIAL ENERGY MANAGEMENT

To proceed in a systematic way, a total system is often divided into subsystems. In an industrial plant the technological subsystem (consisting of technological pro-

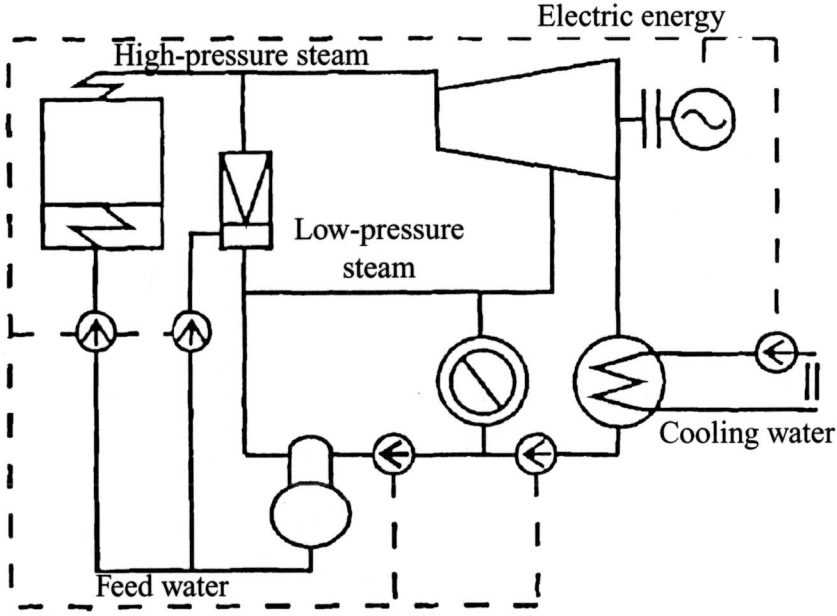

Figure 11.1 Heat and power generating plant—interbranch flows (Ziębik, 1996, with permission).

cesses) and the energy subsystem (energy management) are distinguished. The production of the energy branches is intended to cover the needs of the technological branches and, partially, also the plant's own consumption. The complexity of connections between energy management and technological branches, as well as the interdependences between the energy branches, causes the total energy management of an industrial plant to be more than the sum of energy processes considered separately. Some of these relations are of a feedback character. Therefore, all balance equations of energy carriers should be investigated as a whole. Thus, the energy management of an industrial plant is a system defined as a set of energy equipment and engines, as well as the inner relations between them and the external relation between energy management and the environment, the aim of which is the production, conversion, transmission and distribution of energy carriers consumed in industrial plants. Because of these relations energy management, treated as a complex, has attributes which its parts (the particular energy branches considered separately) do not possess.

As a simple example of an energy system a combined heat-and-power generating plant (Figure 11.1) can be considered. Another form of presenting such an energy system, besides a schematic diagram, is the binary input–output matrix (Table 11.1). Some relations situated under the main diagonal have a feedback character. The existence of feedback relations is responsible for the fact that the partial balances of energy carriers lead to an agreement of the balance by

Table 11.1 Input–output binary matrix (Ziębik, 1996, with permission)

	Electric energy	Feed water	High-pressure steam	Low-pressure steam	Cooling water
Electric energy	1	1	1	1	1
Feed water	0	0	1	1	1
High-pressure steam	1	0	0	1	0
Low-pressure steam	0	1	0	0	0
Cooling water	1	0	0	0	0

means of subsequent approximations. If, for example, the production of electric energy is increased, the production of high-pressure steam grows, too. But this increased production of high-pressure steam leads to another increase in electric energy production due to its own consumption of electric energy in the boiler house. And again, the demand for high-pressure steam will grow, again causing an increase in the demand for electric energy in the boiler house and so on, due to the existence of the feedback relation between the turbine and the boilers. Therefore, a mathematical model of the balance of energy systems of industrial plants has been prepared.

This model is a development of the *input–output analysis* (Ziębik, 1996), applied to energy management of industrial plants. The productive branch of an industrial plant is a technological and energy process producing a given major product, as well as optional by-products. If there is more than one source of energy carrier produced as the major product, the production must be divided into its basic part and peak part (e.g. the steam extraction nozzle of the turbine and the steam from the pressure-reducing valve). If a given energy carrier is the major product in one branch and a by-product in another, it should be considered as a whole in the balance equations of the major product (e.g. steam from the waste-heat boiler). In another case an energy carrier produced as a by-product can be treated as an individual fuel (e.g. blast-furnace gas. In some cases the production of energy carriers must be supplemented by external supplies (e.g. electric energy). Some energy carriers are only provided from outside (mainly fuels). Sometimes, a part of the production of energy carriers is sold to external consumers (e.g. heat and hot water). The possibility of accumulating energy carriers (e.g. steam, hot water, fuel and technical gases) has been taken into account (Ziębik, 1990a).

The set of balance equations of energy carriers is presented in Table 11.2, in which the particular symbols denote:

G_i, P_i	peak and basic part of the main production of energy carriers
$U_{ij}, \bar{U}_{ik}$	by-production of energy carriers in energy and technological branch, respectively
D_i	supply of the energy carrier
Z_{ij}, Z_{ik}	consumption of the energy carrier in energy and technological branch, respectively

Y_i	consumption of the energy carrier for the general needs of an industrial plant
V_i	increase of the energy carrier in the energy storage system
K_i	sale of the energy carrier
$i, j = 1, 2, \ldots, n$	number of the energy branch
$k = 1, 2, \ldots, p$	number of the technological branch

This model can be considered as a simulation model, but it can also be used in optimizations. In the simulation model values of the following variables can be calculated: peak part of the production of energy carriers – G_i, some of the external supplies or sales of energy carriers unknown a priori – D_i or K_i, increase of energy carriers in the energy storage system – V_i. This model may also be considered as a linear or nonlinear mathematical model.

11.4. LINEAR MODEL OF THE ENERGY BALANCE FOR AN INDUSTRIAL PLANT AND ITS APPLICATIONS

11.4.1. Model Equations

The general principle of balance equations is the assumption of linearity for the relations between consumption, by-production and main production. It means that the dependences $U_{ij}(G_j)$, $U_{ij}(P_j)$, $\bar{U}_{ik}(\bar{G}_k)$, $Z_{ij}(G_j)$, $Z_{ij}(P_j)$, $Z_{ik}(\bar{G}_k)$ occurring in Table 11.2 have a linear form. This assumption limits the application of this model to cases in which the coefficients of consumption and by-production can be assumed constant in the period of time considered. The increase of the energy

Table 11.2 Balance of energy carriers (Ziębik, 1996, with permission)

Energy carrier	Main production		By-production		External supply
	Peak part	Basic part	Energy subsystem	Technological subsystem	
Input part					
i	G_i	P_i	$\sum_{j=1}^{n}[U_{ij}(G_j) + U_{ij}(P_j)]$	$\sum_{j=1}^{p}[\bar{U}_{ik}(\bar{G}_k)]$	D_i

Energy carrier	Interbranch flows		General needs	Storage	Sale
	Energy subsystem	Technological subsystem			
Output part					
i	$\sum_{j=1}^{n}[Z_{ij}(G_j)] + Z_{ij}(P_j)$	$\sum_{j=1}^{p}[\bar{Z}_{ik}(\bar{G}_k)]$	Y_i	V_i	K_i

carrier in the energy storage system V_i can be neglected in this case. The set of balance equations of energy carriers in matrix notation looks like:

$$G + F_G G + D = A_G G + K + T \tag{11.1}$$

$$T = (\bar{A} - \bar{F})\bar{G} + Y - (E - A_P + F_P)P \tag{11.2}$$

where:

G, P	vectors of the peak and basic part of the production of energy carriers
$\bar{G}$	vector of the production of technological branches
$F_G, F_P, \bar{F}$	matrices of coefficients of energy carrier by-production
$A_G, A_P, \bar{A}$	matrices of coefficients of energy carrier consumption
D	vector of external supplies of energy carriers
K	vector of sale of energy carriers
Y	vector of consumption of energy carriers for general needs of a plant
E	unit matrix

The vector T contains quantities known a priori. After solving Equation (11.1), the element s of the vectors G or D, sometimes K, is usually obtained (the number of equations equals the number of unknown values). This is a simulation model. If the vectors G, P, and D are calculated simultaneously (surplus of unknown values), this problem is solved by means of linear programming.

11.4.2. Simulation of a Long-term Balance of the Energy System of an Industrial Plant

Due to the complexity of thermal processes in the energy subsystem a change in the production of one branch affects the production processes in all other branches because of interbranch dependences. Particularly, due to a change of production of the technological subsystem, a new variant of the energy balance must be prepared. In this case Equation (11.1), after transformation, takes the following form:

$$G = (E - A_G + F_G)^{-1}(T + K - D) \tag{11.3}$$

The elements of the inverse matrix take into account the direct, as well as indirect, relations between energy processes. The element s of the inverse matrix can be called the coefficient of the cumulative energy consumption concerning the energy management of the industrial plant considered.

Many variants of forecasting the energy balance plan may be calculated making use of matrix equations (11.3). In particular, the influence of changes of the selected element in the production of the technological subsystem (vector T) on the energy balance can be investigated.

11.4.3. Analysis of Exergy Balances

All values in this model can be expressed by exergy units (Szargut, 1983; Szargut and Ziębik, 1972; Ziębik, 1995). Then the losses of exergy may be calculated by means of a linear mathematical model of the energy balance (Table 11.2). The amount of supplied exergy is obtained by summing the element s of the jth column for the energy carriers passing to the energy subsystem. The useful effects of operation of the respective energy branch are: the exergy of the main product and the exergy of the by-products. The difference between the supplied exergy and the useful effects expresses the losses of exergy. The vector δB expressing the exergy losses of all energy branches takes the following form:

$$\delta B = [(A_G^T - F_G^T - E)G^D + (A_P^T - F_P^T - E)P^D + Y^T]b_e \qquad (11.4)$$

where b_e denotes the vector of specific exergy of energy carriers; the upper index D denotes the operation of creating a diagonal matrix from a vector.

11.4.4. System Analysis for the Rationalization of the Energy Management of Industrial Plants

The algorithm of the evaluation of system effects is based on Equation (11.3). In the system method the effects of rationalization of energy management are calculated at the boundary of the balance shield of an industrial plant. In this way the system of interior relations between energy processes has been taken into account.

The rationalization of industrial energy management directly influences, first of all, the coefficients of consumption and by-production of energy carriers (element s of matrices A_G, A_{Ps}, $\bar{A}$, F_G, F_P, $\bar{F}$). New values of these coefficients can be determined by means of process analysis (thermodynamic analysis).

If the rationalization has taken place in the kth technological branch of an industrial plant, the change in the consumption of energy carriers can be calculated from the equation:

$$\Delta \bar{Z}_k = [(\bar{A} - \bar{F})''_k - (\bar{A} - \bar{F})'_k]\bar{G}_k \qquad (11.5)$$

where:

$\Delta \tilde{Z}_k$	vector of changes of energy consumption in kth technological branch due to energy rationalization in this branch
$(\bar{A} - \bar{F})''_k$	column vector k belongs to matrix $(\bar{A} - \bar{F})$ after rationalization
$(\bar{A} - \bar{F})'_k$	the same as above but before rationalization
$\bar{G}_k$	amount of production of the kth technological branch

Substituting Equation (11.5) into (11.3) we obtain:

Table 11.3 Coefficients of consumption and by-production of energy carriers before and after the installation of evaporative cooling (Ziębik, 1996, with permission)

Coefficients of consumption and by-production of energy carriers	Before rationalization	After rationalization
Coefficient of industrial water consumption ($\Delta T_w = 10\,\text{K}$) (Mg/Mg r.p.)	0	0.098
Coefficients of soft water consumption (Mg/Mg r.p.)	0	0.098
Coefficient of by-production of medium-pressure steam (Mg/Mg r.p.)	0	0.0833

$$\Delta G_k = (E - A_G + F_G)^{-1}[(\bar{A} - \bar{F})''_k - (\bar{A} - \bar{F})'_k]\bar{G}_k \qquad (11.6)$$

where ΔG_k denotes a change of the vector G due to rationalization in the technological branch k.

In Equation (11.6) the interior relations between energy processes are taken into account by means of the inverse matrix $(E - A_G + F_G)^{-1}$. In this way the direct and indirect connections between energy carriers have been taken into account. The application of the inverse input–output matrix eliminates the laborious war of successive approximations in investigations of the influence of rationalization of the energy and technological process upon the industrial energy system as a whole.

As an example, system analysis of evaporative cooling in a heating furnace has been considered. The change from traditional water-cooling without evaporation to evaporative cooling brings direct energy benefits due to the decrease in industrial water consumption and the production of steam. The increase in energy consumption for the preparation of soft water is compensated by the useful effects of evaporative cooling. Table 11.3 contains values of coefficients of direct consumption and by-production of energy carriers before and after the installation of evaporative cooling.

The by-production of steam by the installation in evaporative cooling in a heating furnace substitutes the main production of medium-pressure steam in the heat-and-power generating plant. Due to mutual connections existing in the energy subsystem of an industrial plant (among other connections of feedback character) the by-production of medium-pressure steam influences the whole energy balance of the combined heat-and-power generating plant. Figure 11.2 presents the relations between pressurized industrial water and other energy branches.

The example of calculations and denotations in Figure 11.2 complies with the data basis of the examples of calculations contained in Ziębik (1990a, 1990b). The numbers correspond to the following energy carriers: l – low-pressure steam; m – medium-pressure steam; h – high-pressure steam; d – demineralized water; c – compressed air; p – pressurized industrial water; n – nonpressurized industrial water; e – electrical energy. The first index of the coefficients of specific

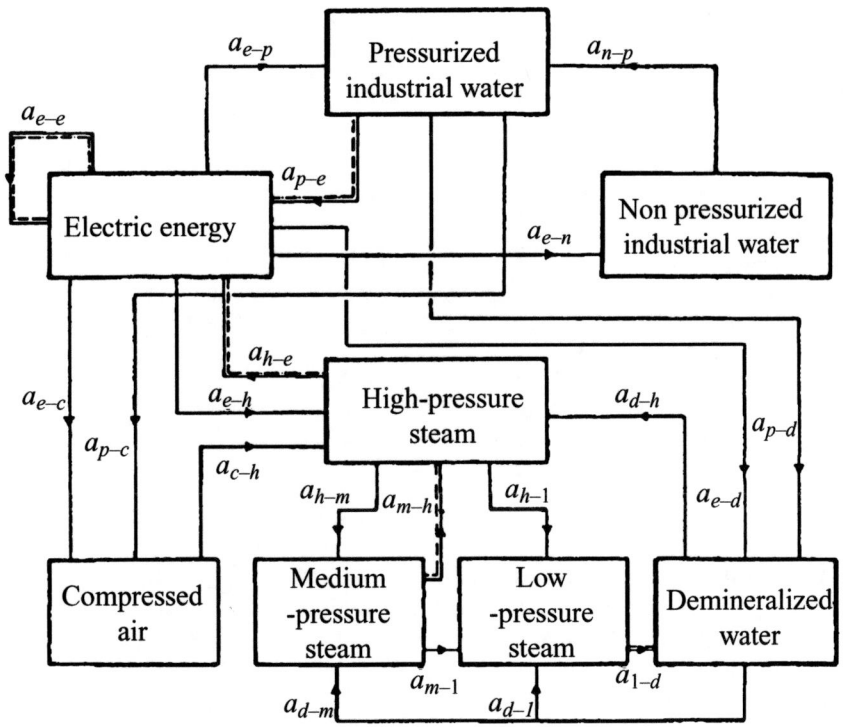

Figure 11.2 Schematic diagram of direct and indirect connections between pressurized industrial water and other energy branches (Ziębik, 1996, with permission).

consumption of energy a_{ij} denotes the consumed energy carrier, the second one concerns the consumption of energy branch.

Table 11.4 shows the results of Ziębik's system analysis obtained by means of a linear mathematical model of the energy balance of an industrial plant. Direct changes due to the rationalization of the cooling system of the heating furnace have been underlined. The other changes presented in Table 11.4 result from the interdependences existing in the energy system of the industrial plant considered. Due to some dependences of a feedback character, accurate results can be obtained only by means of a computer-aided mathematical model of energy management. The final result of the system analysis is a decrease in external supplies of fundamental fuels (mainly coal).

11.5. NONLINEAR MATHEMATICAL MODEL OF SHORT-TERM BALANCE OF INDUSTRIAL ENERGY SYSTEM

In order to achieve efficient control of the energy management of an industrial plant it is necessary to set up the energy balance for a shift and for 24 hours. In

Table 11.4 Results of the system analysis of evaporative cooling (Ziębik, 1996, with permission)

Energy carrier	Changes of major production, by-production and external supplies due to evaporative cooling			
	Unit	Major production	By-production	External supply
Soft water	kg/Mg r.p.	+86.3	+13.2	–
Demineralized water	kg/Mg r.p.	−41.8	−22.7	–
Low-pressure steam	kg/Mg r.p.	−9.8	0	–
Medium-pressure steam	kg/Mg r.p.	−98.9	+86.3	–
High-pressure steam	kg/Mg r.p.	−114.4	–	–
Compressed air	kmol/Mg r.p.	−0.021	–	–
Industrial water	Mg/Mg r.p.	−7.1	–	–
Electric energy	kWh/Mg r.p	−6.4	0	0
Power coal	MJ/Mg r.p.	–	–	−380.0
Natural gas	MJ/Mg r.p.	–	–	−4.0

this case the assumption about the linearity of relations between production and consumption is a far-fetched simplification. In the balance equations of the mathematical model for 24 hours, the energy characteristics of the particular engines and a complex of engines are applied. These energy characteristics are mostly nonlinear or piece-wise linear functions. The mathematical model is therefore a nonlinear one.

The following assumptions have been made in the nonlinear mathematical model:

- the balances of energy carriers are set up for time intervals of one hour; the energy balances for a shift and 24 hours are assembled by means of one-hour balances;
- the timetables of work and repair idle-time for energy and energy technological equipment are known; the plan of repair based on a long-term plan of energy balance is determined; it results from the connection of the model of long-term energy balance with that of short-term balance;
- from forecasts of hour-diagrams of the demand for energy carriers for a technological subsystem, the general needs of the plant and external consumers are known; the hour-diagrams show the average demands for energy carriers at particular hours of the considered shift or 24 hours;
- the characteristics of engines or a complex of these are given; these may be nonlinear piece-wise linear functions and sometimes linear dependences;
- the dependences of the consumption of energy carriers on the parameters of energy-technological processes are taken into account (e.g. the influence of the blast parameters and the injection of auxiliary fuels on the energy characteristics of the blast-furnace);
- the storage volume of energy carriers (gas holders, steam-storage cells and hot-water accumulators) has been taken into account;
- short-time fluctuations between the production and consumption of energy carriers existing in time intervals of one hour are covered by the ability to accumulate the heat and gas distribution network.

This model can be considered a simulation model but it can also be used for optimization purposes.

In the case of the simulation model the energy characteristics of the complexes of engines are assumed to be known. Also, the structure of the fuel-feeding system of an industrial plant and a part of the supplementary external supply are known.

The main aims of the model can be listed as follows:

- forecast of the energy balance of an industrial plant for a work-shift and 24 hours for the purpose of production control;
- hour-by-hour correction of the forecast of the energy balance;
- preparation of the energy balance of an industrial plant in case of failure.

The mathematical simulation model presents a set of balance equations of energy carriers for time intervals of one hour as described in Table 11.2.

In the case of the optimization model it has been assumed that the amount of production of a technological subsystem is known. The objective function is expressed by the following formula:

$$K_e = \sum_{i=1}^{n} [\kappa_{G_i} \dot{G}_i + \kappa_{P_i} \dot{P}_i + \kappa_{D_i} D_i + (\kappa_{Ds_i} - \kappa_{D_i}) + K_T \rightarrow \min] \qquad (11.7)$$

where K_e is the variable operating cost of energy management, κ_{G_i} is the variable operating unit cost of peak energy equipment (without the costs of external energy carriers), κ_{P_i} is the variable operating unit cost of basic energy equipment (without the costs of external energy carriers), κ_{D_i} is the unit cost of the basic part of the external supply of an energy carrier, κ_{Ds_i} is the unit cost of the peak part of the external supply of an energy carrier, D_{s_i} is the peak part of the external supply of an energy carrier, K_T are the losses in the technological system due to the deficiency of energy carriers.

The balance equation (Table 11.2) is the global constraint. The production capacity and limits of external supplies are local inequality constraints. This optimization problem is solved by means of the decomposition of the global optimization problem.

11.6. MATHEMATICAL OPTIMIZATION MODEL FOR THE PRELIMINARY DESIGN OF INDUSTRIAL ENERGY SYSTEMS

11.6.1. Aims

The aim of preliminary design is to choose the optimal variant of the structure of an industrial energy system from among numerous possible variants. This results from the variety of elements (engines and energy equipment) constituting the industrial energy system, the different variants of thermodynamic parameters characterizing these elements, as well as many possible combinations of mutual connections between the energy carriers themselves. Therefore, the application of the mathematical model of energy management balance presented in Section 11.3 is very expedient.

Based on a brief foredesign, the vectors $\bar{A}\bar{F}$ and $\bar{F}\bar{G}$ concerning the consumption and by-production of energy carriers in the technological subsystem and the vector of energy consumption for general needs (Y), as well as the vector of sale (K) are given. The matrices of the coefficients of consumption and by-production – A_G, A_P, F_G, and F_P – are formulated in the preliminary design. The vectors of the basic and peak part of main production (P and G), as well as the vector of supplied energy carriers (D) are calculated by means of optimization. The algorithm determining the industrial energy systems structure in the preliminary design contains solutions of the following partial problems:

(a) generation of a set of variants of the energy system based on the scenario of the energy management of a given industrial plant,
(b) determination of the structure of the binary input–output matrix and its structural analysis,
(c) determination of time-distribution functions of the demand for energy carriers,
(d) determination of the elements of the input–output matrix,
(e) determination of the optimal power rating and capacity of engines and other energy equipment.

Problems (c), (d), and (e) are closely interconnected. In order to solve these partial problems, the decomposition method of the global optimization problem must be applied.

11.6.2. Choice of Structure of the Input–Output Matrix and its Structural Analysis

First of all, a scenario of energy management has to be formulated based on which a general list of energy carriers is set up, providing a set of data concerning the engines and energy equipment used to produce the major product enumerated therein. Each energy carrier, being a major product, has a corresponding project. Connected with this is a set of designs comprising all technically possible and economically justified methods of production of a given energy carrier. A set of designs is created making use of information about engines and energy equipment. Each design is described by its binary input–output matrix characterizing the structure of the consumption and by-production of energy carriers. From the designs a set of all possible variants of structure of industrial energy systems is formed, choosing one design from each project. For each variant, matrices A_G^b, A_P^b, F_G^b and F_P^b are created by means of Boolean algebra. Next the input–output matrix ($A^b + F^b$) characterizing the interdependences in an industrial energy system is set up:

$$A^b + F^b = A_P^b + A_G^b + F_P^b A_G^b \tag{11.8}$$

where b denotes the binary matrix. This matrix is subjected to structural analysis in order to obtain a structure close to a triangular matrix, which is more effective for further calculations. First the energy carriers are divided into three groups: input-type (supplies from outside), center-type (energy carriers produced and consumed inside the energy systems), and output-type (for outside consumers). Next, inside the center group of energy carriers, strongly coherent subsystems are to be distinguished. Relations of the feedback type exist only among energy carriers belonging to strongly coherent subsystems. The separation of such subsystems transforms the *center* matrix to a matrix with diagonally arranged blocks (Popyrin, 1978).

11.6.3. Objective Function and Constraints

For each variant of the designed industrial energy system considered the power rating and capacity of engines and energy equipment, as well as the nominal amount of external supplies, are determined by optimization. The objective function to be minimized takes the following form:

$$K_R = (\rho_P + \beta_P)I_P + (\rho_P + \beta_G)I_G + (\rho_D + \beta_D)I_D$$
$$+ \alpha_P\dot{P}_n + \alpha_G\dot{G}_n + a_D\dot{D}_n + \kappa_P P + \kappa_G G + \kappa_D D + K_T \tag{11.9}$$

where:

K_R	annual costs of the industrial energy system
ρ	row vector of the discount rates
β	row vector of the rates of constant costs of repairs and maintenance
I	column vector of capital expenditure
α	row vector of the rates of prime costs
$\dot{P}_n, \dot{G}_n, \dot{D}_n$	power ratings or nominal capacity and supply
κ	row vector of the rates of operating costs
P, G, D	column vector of production and supply
K_T	annual costs of losses in the technological subsystem due to a deficiency of energy carriers

Equation (11.1) is the global constraint. Local inequality constraints result from the maximum demands for particular energy carriers and from the limitations of external supplies and capital expenditures:

$$\dot{P}_{ni} + \dot{G}_{ni} + \dot{D}_{ni} \le \Omega_{i\max} \tag{11.10}$$

$$\dot{D}_{ni} \le v_{i_{i\max}} \tag{11.11}$$

$$D_i \le v_i \tag{11.12}$$

$$I_{P_i} + I_{G_i} + I_{D_i} \le I_i \tag{11.13}$$

where $\dot{\Omega}_{i\max}$ is the maximum demand for energy carriers, $\dot{\nu}_{i_{\max}}$ is the maximum flux of external supply, ν_i is the annual limit of external supply, and I_i is the capital expenditure.

In order to solve the optimization problem (Equation (11.9)), the time-distribution function of the demand for energy carriers and the values of element s of the input–output matrix must be known. The technical coefficients of the consumption and by-production of energy carriers (that means element s of the input–output matrix) depend on the power rating and load of the engines and energy equipment. But in order to obtain the time-distribution function we must know the technical coefficients. Therefore, the setting up of time-distribution functions and the calculation of technical coefficients, as well as the determination of the power rating of engines and energy equipment, are related. In order to solve these problems the global optimization must be decomposed.

11.6.4. Decomposition of the Global Optimization Problem

Lagrange's method of decomposing the global optimization problem has been applied. Based on Equations (11.9) and (11.1) and neglecting the term in Equation (11.1) without influencing the results of optimization, the Lagrangian function takes the form:

$$L = K_R + \lambda[(A_P - F_P - E)P + (A_G - F_G - E)G - D] \rightarrow \min \tag{11.14}$$

where λ denotes the row vector of Lagrange's multipliers.

In order to determine Lagrange's multipliers, the procedure of coordination must be known. It has been proved that the matrix method of calculating the unit costs of energy carriers can be used as a coordination procedure.

Lagrange's decomposition method leads to an iterative procedure (Figure 11.3). In successive iterations, Lagrange's multipliers (unit costs of energy carriers) are determined by means of the matrix method on a higher level (complex of energy management). Next, optimization problems (optimal power rating, capacity of engines and energy equipment) are solved on lower level (particular energy carriers). Problems of the optimization of the particular energy carriers are solved according to their sequence in the upper triangular input–output matrix. In strongly coherent subsystems (i.e. subsystems with feedback) the inner iterative loops are solved. The determination of the optimum values of all decision variables in successive iterations is followed by a return to the level of coordination. Then a corrected balance of energy carriers is set up and the corrected values of the unit costs of energy carriers are calculated by means of the matrix method. Next, the corrected vector of the unit costs of energy carriers is applied to the level of optimization of the particular energy carriers (Figure 11.3).

Summing up, process analysis in thermal engineering is not a sufficient tool for the evaluation of the energy effects of operations aimed at improving ther-

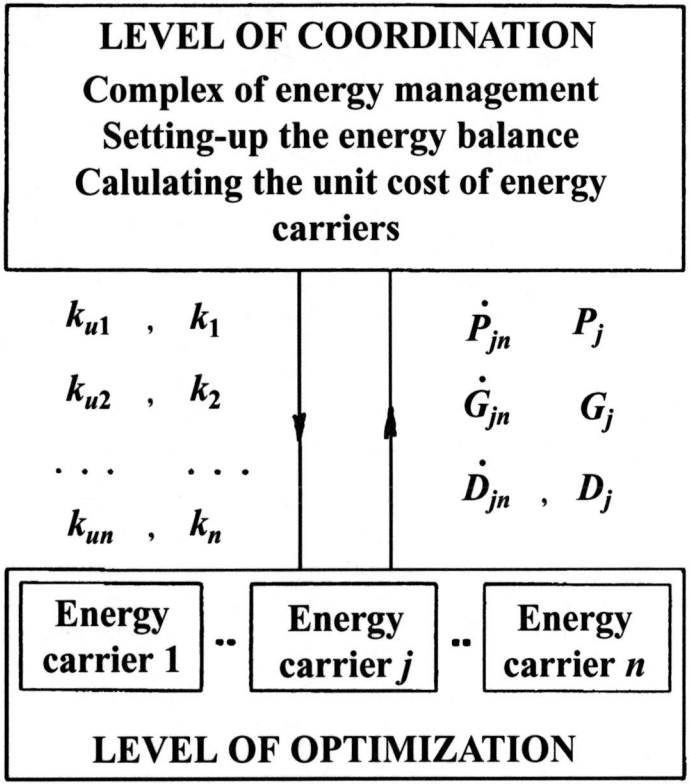

Figure 11.3 Schematic diagram of the decomposition method; k, k_u—average unit costs of energy carrier and unit cost of by-production of energy carriers, $\dot{P}_{jn}$, $\dot{G}_{jn}$, $\dot{D}_{jn}$—power rating and nominal supply, P_j, G_j, D_j—annual production and supply. Evaluation of the energy effects of operations aiming at an improvement of thermal processes (Ziębik, 1996, with permission).

mal processes (Ziębik, 1996). As energy carriers are produced and consumed in the network of interconnected thermal processes, the thermodynamic rationalization realized in one thermal process influences the energy balances of other processes. Industrial thermal engineering constitutes one whole, displaying properties which the respective thermal processes, treated separately, do not possess. Therefore, besides process analysis, system analysis ought to be applied in investigations concerning heat engineering.

System analysis in industrial thermal engineering is based on a mathematical model of the energy balance. The linear input–output model serves to analyze the heat management in the case of long-term balances. The nonlinear model is applied in short-term balances. Both models can be used as simulative or optimizing ones. For the purpose of preliminary design of heat management the optimization model of selecting the structure of heat engineering is used.

11.7. REMARKS ON DIVERSE METHODOLOGIES AND LINK WITH ECOLOGICAL CRITERIA

In this book, we investigate "nonclassical" thermodynamic limits which include kinetic effects. In particular we are interested in "dynamic" limits for systems evolving in time. The limits usually depend on operational constraints established under the condition that, in any circumstances, the key process will run with the required mean intensity, yet yield the desired product. The "nonclassicity" requirement usually yields bounds that may be orders of magnitude higher than classical ones known from textbooks. Consider, for example, heat consumption in a distillation column and evaluate a realistic limit that corresponds with the lower limit of heat associated with the use of theoretical plates instead of real plates of certain efficiency (lower than unity). This bound is usually 2–3 times larger than another (lower) bound; the heat consumed at minimum reflux conditions (in the case with an infinite number of theoretical stages). Next, the heat at minimum reflux is usually several times larger than the reversible evaporation heat. In effect, a design engineer must expect that the heat consumption should be at least an order of magnitude higher than the evaporation heat. In complex separation processes such as those with cycles, losses, and nonlinearities such evaluations are nontrivial. Complex optimization techniques must be used to obtain (possibly dynamic) limits for various operations. The union of applied thermodynamics and optimal control theory that derives from these limits has recently found applications in the design of solar engines, solar cells, semiconductor devices, photosynthesis engines and other sophisticated devices (see, e.g. de Vos, 1992).

Early papers on FTT stressed that its dynamical models can be optimized with respect to their external controls subject to optimization criteria (Andresen et al., 1977; Andresen, 1983). Thus not only power or entropy source but also economic criteria, such as profit or the sum of investment and operational costs, can be used in FTT. Examples of such optimization, that, of course, should not be confused with those leading to energy limits, are available (Berry et al., 2000). While it is true that the economic costs are influenced by irreversible properties of the system, the irreversibility does not follow from economic costs since no one-to-one relation exists between the two quantities. No prior recourse to economics to consider the irreversibility impact on the investment and operational costs is mandatory to get energy limits; physical considerations are sufficient as for other physiochemical properties. In particular, all irreversible extensions of exergy can be obtained without any prior consideration of process economies. This substantiates the relevance of energy limits and irreversible exergies obtained in a number of papers (Andresen et al., 1977, 1983; Andresen, 1983; Berry et al., 2000; Bejan, 1982; Tsirlin and Kazakov, 2002a, 2002b; Bejan and Errera, 1998; Sieniutycz, 1999d, 2001b).

Endoreversible modeling (imperfect reservoirs and perfect power generators, Rubin, 1979a, 1979b; Rubin and Andresen, 1979) is an anathema to adversaries of FTT or EGM but the magic words in the language of their followers. With the endoreversibility assumption classical thermodynamic equations for reversible engines and chillers may easily be combined with irreversible expressions describing energy and mass transports in imperfect reservoirs (those containing lossy elements in the form of various "conductances," "resistances," "boundary layers," "penetration depths," etc.). This results in relatively simple synthesizing models that are often susceptible to analytic treatments (Rubin, 1979a, 1979b; Rubin and Andresen, 1979; Hoffmann et al., 1997; Sieniutycz and Kubiak, 2002). The reason for this simplicity is the linearity of mathematical equations describing lossy parts of the system. When the linearity does not hold, as, for example, in the case of systems with radiative transport, then the description of the overall system is more complicated, yet it can still be obtained in a workable form (de Vos, 1992; Sieniutycz and Berry, 2000; Sieniutycz and de Vos, 2000; Sieniutycz, 2000c). When internal irrversibilities are included, the simplicity of endoreversible description is generally lost unless a simple but rough description of the internal entropy generation, based on the so-called internal factor, Φ, is accepted (Salah El-Din, 2001a, 2001b; Hoffmann et al., 1997; Sieniutycz and Szwast, 2003). Yet, models treated in contemporary thermodynamics of real thermal machines too often give up internal irreversibilities on account of external ones (Ng et al., 1997, 1998; Gordon and Ng, 2000; Heywood, 1975). Thus, all sorts of irreversibilities should be treated in FTT and EGM approaches although often at the expense of replacing analytical solutions with numerical ones.

Recent works on second-law analyses often deal with complex thermal systems composed of many objects and links; for their optimization the reader is referred to a review (Tsatsaronis, 1993). Advanced applications of thermodynamic limits, which include separation processes, may be regarded as prolongation of earlier ideas developed for second-law analyses of thermal systems (Gaggioli, 1980). Other works using second-law analyses include ecological applications of exergy; they are of interest here in view of their link with the theory of energy limits. A basic notion therein, supposedly of value in thermal technology, is the so-called cumulative exergy cost (CEC) defined as total consumption of exergy of natural resources necessary to yield (the unit of) a final product (Szargut, 1986). Also introduced is the notion of cumulative exergy loss, as the difference between the cumulative exergy cost and exergy of the considered product. In ecological research, various analogs and other criteria are introduced. There are general criteria (Odum, 1971; Ulanowicz, 1997; Jørgensen, 1988, 1997, 2000, 2001; Kay and Schneider, 1992) and special ones, related to thermal technology (Angulo-Brown, 1991; Yan, 1993).

The ecological cost (Szargut, 1986) measures the cumulative consumption of exergy of unrestorable resources burdening a definite product. Resulting technical indicators are used to forecast changes in demand for heat agents caused by changes in production level and technology of product yield and in costs of heat agents. This provides information about diverse exergy-consuming technologies, and is also suggested as a way to compare the technologies. Also, a so-called pro-ecological tax is proposed as the penalty for the negative effects of actions causing exhaustion of natural resources and contamination of natural environments (Szargut, 2001a, 2001b). All these applications involve nonequilibrium systems in which the sole use of the classical exergy is insufficient without including the associated notion of minimal (residual) dissipation of this exergy. This is, in fact, the realm into which we are driven with many analyses that lead to nonequilibrium applications of the exergy thermodyamics. They emerge since engineering processes must be limited by some irreversible processes allowing a minimum entropy production rather than by purely reversible processes. Limits following from reversible processes are usually too far from reality to be useful.

However, the method of cumulative exergy costs has its own imperfections and difficulties. Its definition of the sequential process is vague. The total consumption of exergy of natural resources, necessary to yield a product which defines the cumulative exergy cost, is, in fact, a dynamical notion burdened by sorts, locations and dates of various technologies, all affecting process efficiencies, semiproducts, controls, etc., and thus, in effect, influencing the cost definition. One way to improve this definition would be to include statistical measures of the process and its exergy consumption. Yet, no statistical procedure leading to an averaged sequence process that would add rigor to the definition of cumulative exergy costs was defined in the original work (Szargut, 1986, 2001a, 2001b, 2002a, 2002b). Moreover, as its definition shows, the cumulative exergy cost (CEC) is merely the exergy change of the whole (sequential) production process. Last but not least, the CEC method underestimates the significance of kinetic terms in treating practical processes. Thus, despite some suggestions implying its economic ramification (Szargut, 1986, 2001a, 2001b, 2002a, 2002b), the cumulative exergy cost is not a perfect measure of the economic costs of unrestorable resources consumed. In fact, this quantity is a rough representation of thermodynamic limit (work limit) of the same sort that is applied in the methods of EGM and FTT when they deal with limits of sequential operations. According to the Gouy–Stodola law (Gouy 1889; Stodola, 1905; Bejan, 1982) the exergy change represented by the CEC is proportional to the associated entropy production; this quantity should be minimized, yet no precise definition of the minimization constraints and operational quantities is given in the theory of cumulative exergy costs. While applying the CEC method to replace the minimization of the entropy production in both methods, EGM (Bejan, 1996a, 1996b) and FTT (Berry et al., 2000), its originators seem unaware that the origin of Szargut's approach is in fact similar. To define exergy costs as unique and usable quantities, effects of external controls and disturbances should be eliminated from their definition. This requires either statistical averaging or extremizing cumulative costs with respect to controls and disturbances. But, unlike FTT and EGM, the method of CEC does not attribute to its merit criteria potential functions and does not use techniques providing a unique result. Thus in the current definitions of the cumulative exergy cost and ecological cost, the mathematical structures of cost functions remain largely unknown. In fact, cumulative costs are not functions but rather functionals of controls, disturbances and path coordinates. To ensure potential properties in optimal costs their definition should imply a method that would eliminate the effect of controls; its absence makes the definition inexact. Without optimization, unique (potential) properties of costs are lost.

A solution to the above difficulties lies in finite-time thermodynamics (Berry et al., 2000) where extremal, potential cost functions are generated via optimization of costs with respect to controls (but not states). Mathematical structures of these costs as continuous and discrete functionals are recognized (Sieniutycz, 1973b, 1978, 1997a, 1997b, 1997c; Rubin and Andresen, 1979; Ondrechen et al., 1981, 1983; Andresen, 1983; Andresen et al., 1977, 1983, 1984; Berry et al., 2000). They are often called finite-time potentials.

The problem of finite-time potentials originated from knowledge of the fact that principal functions of extremal solutions to variational problems are potentials depending on process time and state and are nearly as old as analytical mechanics. Functions of this sort were first obtained in physics as extremum actions. Yet, for an arbitrary variational problem, where an analytical solution does not exist and only numerical solutions are possible, an effective numerical method of finding potential functions was established only in 1957, with the advent of Bellman's method of dynamic programming (Aris, 1964; Bellman, 1961, 1967). Classical thermodynamic potentials, which are time-independent quantities, are special cases of such generalized (time-dependent) potentials in the case of identically vanishing Hamiltonians. In particular, the purpose of some FTT analyses is the exposition of how the classical thermodynamic potential, entropy $S(x)$, emerges for an irreversible deterministic dynamics (Berry et al., 2000). Furthermore, these analyses can also demonstrate the emergence of "nonclassical thermodynamic potentials" from the same general approach. One of the nonclassical potentials is finite-time availability (Andresen et al., 1977; Andresen, 1983; Berry et al., 2000; Tsirlin and Kazakov, 2002a, 2002b).

The basic problem of extracting the most exergy from a continuous hot stream was formulated in the reversible limit by Bejan (1982). While its endoreversible generalization was posed soon after (Ondrechen et al., 1981; Andresen et al., 1983), the details were not completed until a decade and a half later (Bejan and Errera, 1998; Sieniutycz, 1999d, 2001b). Sieniutycz's paper (1999d) deals explicitly with cascades as genuine discrete systems with a small number of stages (described by difference equations) and his work (2001b) discusses the continuous limit and abandoning endoreversibility. Criticizing endoreversible limits (Sieniutycz, 2001b) and general methodologies of FTT and EGM, Szargut (2001b) mixes the problem of energy limits (a physical problem) with a different problem of economic optimum for the system (an economical problem).

Whereas a general analysis of the system's limiting possibilities (Sieniutycz, 2001b) is focused on the physical problem, much of Szargut's discussion (2001b) is centered on the economic one. This confusion is, perhaps, motivated by the observation that minimum energy data are sometimes used in the evaluation of economic optimum. A well-known example in the theory of Linde operation shows that the real work supplied to the compressor at economically optimal conditions may be a dozen times larger than the reversible (exergy) limit for the production of condensed air (Bosniakovic, 1965; Ciborowski, 1976). It is just this large difference between economically optimal work and reversible work limit (lower bound) that makes irreversible exergies useful. They provide higher lower limits for work consumption (lowered upper limits for work production) than the corresponding reversible results, and thus they are closer to real work.

The hierarchy of energy limits is an important issue (Andresen et al., 1984; Sieniutycz, 2001b). It begins with perfect (reversible) limits for reversible models and stretches by more and more exact models of imperfect processes, thus including more and more detailed contributions to the entropy production. The limits of zero rank are reversible ones; they stem from the classical exergy, they are the weakest and thus the worst. The limits of first rank are endoreversible limits; by assumption they correspond with the situation when the device that consumes (generates) work is perfect, as in the case of a Carnot machine, whereas the reservoirs can be imperfect. An endoreversible model is more general than the classical Carnot model, where no dissipation is possible in either of two reservoirs. The endoreversible modeling yields a locus of limiting irreversible states with minimum dissipation in bulks of reservoirs. Still the work generators or consumers remain ideal. When the work generator becomes imperfect, as in Figure 3.1 (Chapter 3, page, 91), limits of higher rank are defined (Berry et al., 2000; Sieniutycz and Kubiak, 2002; Sieniutycz and Szwast, 2003). The concept of such enhanced limits is motivated by the fact that only imperfect machines are encountered in thermal technology.

Entropy source minimization has no relevance to an economic optimum of a product yield, where economic cost criteria are attributed to a valuable final product (Tsatsaronis, 1993; El-Sayed, 1999; Clark, 1986). On the other hand, these economic criteria have little in common with physical limits on energy consumption or production.

Yet economic optimization of an energy network may require different methods than currently used in thermodynamics. Network or system thermodynamics (Peusner, 1986) can help in economic optimization of energy networks. Thermodynamics, especially as the subject of the thermal networks theory, is currently being extended to combine technical and economic approaches; the extended discipline is referred to as thermoeconomics (Section 11.9); it links the principles of economics with thermodynamic analyses of energy or resources (El-Sayed, 1999). Extra variables related to equipment and capital costs are introduced simultaneously along with purely thermodynamic variables. By including still more variables related to environmental effects ecological impact is treated.

Clark (1986) has stressed that economic cost analysis introduces a new and different set of variables which must be considered simultaneously with thermodynamic ones to obtain the best technically acceptable design. He has developed evaluation methods which bridge the interface between thermodynamic and economic considerations for the purpose of optimum design. An optimum configuration will vary depending on economic factors such as equipment and capital costs, sources of capital, period of investment, cost of displaced fuel, resale value, and so on. Other constraints follow from ecology when one takes into account not only the exergy consumption but also the exergy losses resulting from the deleterious impact of wastes on human activity and health, crops, forestry, natural resources, and so on. Energy and resource analysis has been defined as a particular set of procedures for evaluating the total energy (resource) requirements for the supply of a service or product (Berry, 1989). The quantification of economic analyses in technological systems is described in the literature on engineering economy. Thermodynamics has nonetheless helped correlate and predict selling prices of chemicals. The old, incorrect concept of prices based on enthalpy analysis has been replaced by that of prices based on analyses of free energy and exergy (Szargut and Petela, 1965). Exergy optimization, economic optimization and ecological criteria have been compared and considerable differences in various optima have been shown (Clark, 1986; El-Sayed, 1999).

Consideration of the impact of irreversibility on investment and operational costs is not only possible in the FTT method but it is more effective in FTT than in the classical thermal analyses due to the optionality of the optimization criterion in FTT (Andresen et al., 1984). In fact, FTT can effectively treat both categories of problems: those leading to limits and those evaluating economic decisions. Opinions that only endoreversible models can be treated in FTT are disproved by current analyses of real thermal machines with internal entropy generation, describing either imperfect engine cycles (Chen et al., 1996; Cheng et al., 1999; Salah El-Din, 2001a) or imperfect refrigeration cycles (El Haj Assad, 1999; Chen et al., 2000; Salah El-Din, 2001b). Experiments in the realm of chillers confirm the essential role of thermodynamics in describing and organizing diagnostic procedures for practical devices (Edera and Kojima, 2002; Chua et al., 1996; Ng et al., 1997, 1998). *Internal* entropy production is a basic parameter of analyses contained in recent books (Berry et al., 2000; Gordon and Ng, 2000). The alternative method of cumulative exergy costs (Szargut, 1986, 2001a, 2001b, 2002a, 2002b) does not ensure objective energy limits (Sieniutycz and Kubiak, 2002).

The definition of the limits on energy generation or consumption might be the subject of the attitude that *limits* should be attributed to systems with extreme efficiencies of energy converters. Since Carnot efficiency is the upper limit of all thermal generators a proposal was made (Sieniutycz, 1998b) that, as the perfect cycle, the Carnot cycle is a limiting abstract system that works with superconducting circulating fluid to eliminate the internal entropy production and to secure the closed loop of the cycle without cooling the circulating fluid. Moreover, following Landau and Lifshitz (1974), it was assumed that the power generating device (machine) is both perfect and very small. In the case of a CNCA system this assumption avoids the quantitative effect of the energy generator properties on those of the two-reservoir system. The assumption leads to the notion of a small Carnot machine, infinitesimal in relation to reservoirs. In cascade systems each such Carnot machine is a mechanical energy generator that only transforms the small amount of internal energy of the two-reservoir (nonequilibrium) system into the mechanical energy but otherwise is negligible and does not change the limiting work. Circulating Carnot fluid is, in case, a medium that does not change after a cycle, acting as a sort of catalyst. In fact, it also substantiates the use of "endoreversible" models that admit arbitrary dissipative properties within fluids of upper and lower reservoirs (resistances or boundary layers), but otherwise involve small Carnot units as ideal energy generators. Such an attitude is assumed by the researcher who wants to define energy limits in a two-reservoir system without consideration of a related three-body network composed of two reservoirs and a real thermal machine of a finite size.

This attitude has generated both followers and adversaries. The former claim that the system considered is actually the two-body (two-reservoir) system whereas the (small) Carnot machine is just an addition enabling the conversion of a small amount of energy. The representatives of this idea (Landau and Lifshitz, 1974; Sieniutycz, 1998b; Orlov and Rudenko, 1984; Szwast, 1994) claim that, only in this case, the adiabatic availability or reversible work from the two-body system, $\Delta(E_1 + E_2)_{(S_1 + S_2)}$, involves the properties of two bodies (both reservoirs). In fact, the associated exergy function does involve only the reservoir properties (Ondrechen et al., 1981). Yet, its adversaries (Gyftopoulos, 1999, 2002; Moran, 1998; Lior, 2002; Szargut, 1998, 2001b) supposedly want to see the CNCA system as a simple network (implying a three-body problem) in which there are two reservoirs and a thermal machine of a finite size. When the latter viewpoint is accepted, the size and other properties of the machine influence even reversible work, thus the universality of the results decreases. In conclusion, the introduction of thermal machines of finite sizes into the analyses, however valuable for real network descriptions, decreases the simplicity and universality of the modeling and results describing the limiting work.

11.8. CONTROL THERMODYNAMICS FOR EXPLICITLY DYNAMICAL SYSTEMS

Engineering systems deal most often with processes far from equilibrium, that is, those outside the realm of linear irreversible thermodynamics. These processes must go to the desired extent of completion within a finite time, and must produce at least a minimum amount of product. Furthermore, efficiency based solely on physical grounds is more often than not an insufficient criterion for the performance of a chemical engineering process. In view of the scarcity of resources, economic and ecological considerations also play important roles. Thus classical thermodynamics can at most place lowest limits on chemical engineering processes, and one must turn to irreversible thermodynamics for more information.

When one (say, upper) reservoir is finite a finite resource problem emerges in which power yield is necessarily a dynamical operation and the reservoir's temperature changes in time. Various dynamical systems have been analyzed in detail in Chapters 4–10; here we shall make only a few introductory remarks. Even for purely thermodynamic criteria, such as integral of exergy dissipated, a finite time constraint ensures the optimal performance function as a state function which is not classical from the viewpoint of equilibrium thermodynamics. In approaches involving optimal control (Salamon et al., 2001), optimization of a cost expression (or an associated entropy production) automatically eliminates controls from optimal costs expressions, thus generating a potential (R or R_σ). A potential depends only on initial and final states, duration and (in multistage processes) total number of stages. The terms "control thermodynamics" and "optimization thermodynamics" introduced in the Russian literature are particularly suitable for dynamical cases.

Suitable averaging procedures are proposed along with methods that use averaged criteria and models in optimization (Berry et al., 2000). It follows that an optimal sequence has a quasi-Hamiltonian structure that becomes Hamiltonian in the special cases of processes with optimal sizes of stages or a continuous limit (Sieniutycz, 1999c). Thus the well-known machineries of Pontryagin's maximum principle (Fan, 1966) and dynamic programming (Aris, 1964) can be included to generate functions of optimal cost. These theoretical achievements also enter the realm of economic criteria (Berry et al., 2000). In fact, problems in the method of cumulative exergy costs (Szargut, 1986) also belong to the group of dynamical operations as it involves sequences of operations and finite resources.

Control thermodynamics, which is the union of nonequilibrium thermodynamics and optimal control (Salamon et al., 2001), investigates the effect of constraints on time and rate on the optimal performance of generic processes through integral or sum expressions such as internal entropy generation $\sum \Delta S_k^i$, total exergy input $\sum \Delta B_k^{in}$, system's exergy change B_k, work W, and so on. (k is the stream index and superscripts in and i refer, respectively, to input and internal production.) Usually the goal of thermodynamic analysis is: (1) to find the paths of minimum $\sum \Delta B_k^{in}$ or $\sum \Delta S_k^i$ and realistic bounds on consumption of energy and resources in thermal, separation and chemical processes incorporating the minimal irreversibility, (2) to find the optimal strategies or controls for such processes, and (3) to refer these bounds to an actual process in order to verify its possible improvement. The bounds constructed on the basis of thermodynamic criteria, in particular exergy, are both relevant and useful. (Note that these bounds are in general functions of state and duration rather than numbers.) They generalize the well-known thermostatic bounds for finite rates and/or finite time. In this book they define thermodynamic limits rather than the economical consumption of exergy or resources for various generic processes.

Optimization techniques play a central role in obtaining the majority of bounds in control thermodynamics. The methods of linear programming (Dantzig, 1968) and nonlinear programming (Zangwill, 1974) are as a rule insufficient in those situations where functional extrema are sought. Instead, the application of optimal control techniques is necessary (Pontryagin et al., 1962; Leitman, 1966; Sieniutycz, 1991). Control thermodynamics often retain the philosophy of model idealization known from reversible thermodynamics (the Carnot cycle) but use somewhat more realistic models which have basic irreversibilities incorporated. The notion of thermodynamic metric and its consequence for establishing bounds as well as providing optimal paths deserves a mention (Hoffmann et al., 1989, and literature therein). These authors show that when a part of the dissipated energy remains within the system not all of the availability is necessarily lost. The bound defined by the thermodynamic length no longer limits the availability losses but rather a so-called work deficiency, W_d (where usually $W_d > -\Delta A$), or the total loss of availability that would have resulted if all the available work were lost to the environment.

11.9. INTERFACE OF ENERGY LIMITS, STRUCTURE DESIGN, THERMOECONOMICS AND ECOLOGY

Bejan (1982, 1987a, 1987b, 1996a, 1996b, 1988a) has synthesized a great deal of thermodynamic analyses of heat and mass transfer, fluid flow, and related design, on the basis of the minimization of entropy production σ. The minimum of the entropy produced, with terms containing temperature and pressure gradients, has been used in optimizations of fins and fin arrays of various geometries (Poulikakos and Bejan, 1982).

Complex energy systems, which exchange with their surroundings mass in the form of many cold and heat streams and may be chemically active, appear in almost every industry. The recognition that a common principle is the basis for the generation of the shape and structure of systems, whether living or inanimate, leads to the constructal theory, whose developments are systematized in Bejan (2000). Starting with the idea of competition between mechanical and thermal losses of exergy in flow systems Bejan developed his constructal theory of organization in nature (Bejan, 1997a, 1997b, 1997c, 2000) and has shown its applicability in many diverse fields including urban growth, economics and physiological processes. This theory provides an understanding of how naturally organized systems emerge and evolve. In addition to engineering, the theory has shown its inquiring and predictive potential in other areas, such as biology and medicine (allometric laws, structure of the respiratory and circulatory systems, bodily rhythms, organ and tissue structures) and earth sciences (circulation of planetary fluids, structures of river basins, etc.). For example, the theory explains why we have a bronchial tree with 23 levels of bifurcation (Reis et al., 2004, and Chapter 10). It also shows promising developments in social sciences.

The constructal theory describes a deterministic principle for the generation of geometric form in natural systems. Shape and structure emerge from the endeavor for better performance in both engineering and nature. The constructal theory applies the idea that the objective and constraints principle used in engineering is also the mechanism from which the geometry of natural flows emerges. The principle accounts for tree-shaped flows and other geometric forms found in engineering and nature: round ducts, regularly spaced internal channels, the proportionality between width and depth in rivers. Flow systems with geometric structure exhibit at least two flow regimes. One regime is slow (diffusion, walking), with high resistivity that fills the volumes of smallest finite scale, and one or more regimes are fast, with low resistivity (streams, channels, streets). The balancing of the regions with different regimes assures that material and channels are distributed optimally. Despite the differences between the regions of high and low resistance, better global performance is assured when the distribution is relatively uniform. The system works best when its internal flow resistances are spread around, so that more and more of the internal points are stressed as much as the hardest working points. One good form leads to another, the forms being constantly improved with time. Bejan (2000) synthesizes a vast spectrum of publications (Bejan and Ledezma, 1998; Bejan, 1997a, 1997b, 1997c, and many others).

The contemporary formulation of the constructal principle is as follows: For a finite-size system to persist it must evolve and organize in such a way that it provides easier access to the imposed currents that flow through it (Bejan, 2000). This formulation recognizes the natural tendency of imposed currents to construct paths of optimal access through a constrained open system. In the constructal theory the process of construction and shape optimization proceeds stage-wise from smaller to larger scales until a given volume is fully covered. High-conductivity paths emerge, that form a tree, and low-conductivity paths appear that fill the infinity of points of the given volume. Tree networks abound in nature in both animate and inanimate flow systems. They can be found in plants, roots, lungs, leaves, vascular tissues, river drainage basins, and dendritic clusters. In fluid trees, the smallest scale volumetric flow is by slow diffusion, while the larger scale flow is organized into faster channeled streams forming a tree-like structure. Channels of lower conductivity are tributaries of a channel of higher conductivity. An optimal geometry composed of low- and high-conductivity flow regimes coalesce all the volume-to-point flows. Artificial constructs and natural structures, such as internal computer arrangements or the arterial–vascular system in animals, exhibit the same cooperation between slow and fast transfer mechanisms, with the slow mode operating at the smallest scale. Tree networks are also observed in living communities; they can be explained in terms of minimization of the travel time between one point and a finite-size area of infinite destinations. The theory predicts a system's growth, in particular urban growth, from alleys to streets, avenues, and highways and drives multimodal transport systems that are set up to minimize travel time over long distances (Bejan, 2000; Bejan and Lorente, 2003, 2004). Systems with porous structures and fuel cells can be analyzed by the constructal theory (see, respectively, the literature discussions in Chapters 8 and 9). Applications which determine flow architectures in living systems are described in Chapter 10.

Reis and Bejan (2006) and Bejan (2006) examined the largest flow system on earth from the point of view of the constructal theory (Bejan, 1997a, 1997b, 1997c, 2000), which now constitutes the thermodynamics of nonequilibrium (flow) systems with configuration (Bejan and Lorente, 2004, 2005). They believe that there are two time arrows in physics. The old one is the time arrow of the second law of thermodynamics, the arrow of irreversibility. The new one is the time arrow of the constructal law, the arrow of how everything that flows acquires architecture or configuration. The constructal theory explains that existing configurations assure their survival by morphing in time toward easier flow configurations. Its time arrow unites physics with biology and engineering. A suggested vision (Bejan, 1997a, 1997b, 1997c; Reis and Bejan, 2006) involves the earth with its solar heat input, heat rejection, and wheels of atmospheric and oceanic circulation. Such an overall system constitutes a heat engine without a shaft whose maximized (not ideal) mechanical power output cannot be delivered to an extraterrestrial system. Instead, the earth engine dissipates all the mechanical power it produces (through air and water friction and other irreversibilities e.g. heat leaks). It does so by "spinning in its brake" the fastest that it can, hence winds and ocean currents, which proceed along the easiest routes. Because the flowing earth is a constructal heat engine, its flow configuration has evolved in such a way that it is the least imperfect that it can be. It produces maximum power, which it then dissipates at maximum rate. The heat engines of engineering and biology (power plants, animal motors) have shafts, rods, legs and wings that deliver the mechanical power to external entities that use the power (e.g. vehicles and animal bodies needing propulsion). Because the engines of engineering and biology are constructal, they morph in time toward flow configurations that make them the least imperfect that they can be. Therefore, they evolve toward producing maximum power, which means a time evolution toward minimum entropy generation rate (Reis and Bejan, 2006). Yet the standard system design includes economics (Moran and Shapiro, 2003, that is, an optimum is not associated with the maximum power. And a critical analysis of the constructal approach shows that it does not necessarily improve the flow performance if the internal branching of the flow is increased (Kuddusi and Egrican, 2007).

The standard system design involves principles of thermodynamics, hydrodynamics, materials engineering, economics, and mechanical design. The term thermoeconomics may be used to describe this complex field of application, although it is frequently applied in a narrow sense to methodologies combining economics and exergy theory to optimize the design and operation of energy systems.

Approximate optimization of the trade-off between investment costs and exergy-based exploitation costs offers useful estimates, as summarized by various researchers (Bejan, 1882; Bejan, 1987a, 1987b; Szargut and Petela, 1965; Szargut, 1970). On the other hand, approaches to "optimal design" that use the entropy (or exergy) generated as their optimization criteria make little sense from the standpoint of economic or thermoeconomic design.

The solution of a thermoeconomic problem is, in general, not equivalent to that of the corresponding thermodynamic problem. It does, however, reduce to thermodynamic optimizations in two special cases. The first case appears when the price of certain thermodynamic quantities such as the power produced becomes much larger than the prices of other participating quantities (Berry et al.,

1978). This limit represents an energy theory of value, that is, a value system in which one considers energy as the single valuable commodity (Berry et al., 1978). In the second case the economic value of the exergy unit is the same for all forms of matter and energy taking part in the process. Then the thermodynamic problem of the minimum exergy loss is equivalent to that of the minimum of the economic costs. This case is however quite special since the prices of the exergy units generally differ (Szargut and Petela, 1965). Nevertheless, a number of complex economic and ecological analyses have been born as generalizations of thermodynamic irreversibility analyses. While various performance criteria serve in various problems (this variety has already been admitted by control thermodynamics and extended versions of FTT), physical constraints are the product of thermodynamic analyses. Even if one replaces thermodynamic criteria by economic ones, the same optimum search method can often be used in both cases. Most of the methodological experience gained during a model formulation is preserved when passing from thermodynamic to economic optimization.

Systems with nontraditional energy sources have recently begun to become economically realistic. Examples of such sources are solar energy (photothermal and photovoltaic converters), wind energy, biomass energy of waterfalls, waves and tides, geothermal energy and convective-hydrothermal resources (de Vos, 1992; Bodvarsson and Witherspoon, 1989; Neill and Jensen, 1976; Gustavson, 1979). While the economic cost of renewables approaches zero, limits on their exergy consumption and thermodynamic efficiencies are finite and should be evaluated because the results of such evaluations may influence the equipment dimensions and hence its investment cost. Photothermal and photovoltaic conversions have been treated by Landsberg (1990) and Jarzebski (1990), and, in the framework of FTT, by de Vos (1992). The solar driven convection known as winds has been modeled in terms of the FTT of heat engines by treating the Earth's atmosphere as the working fluid (de Vos, 1992; Gordon, 1990, and references therein). Upper and lower limits for the performance coefficient of solar absorption cooling cycles followed from the first and second laws (Mansoori and Patel, 1979). These limits depend not only on the environmental temperatures of the cycle components but also on the thermodynamic properties of refrigerants, absorbents and mixtures thereof. Comparative studies of refrigerant–absorbent combinations are now possible.

Attempts at systematizing process synthesis by the energy limits theory and second-law analyses are known; an example is that of a methane liquefaction plant with the irreversibility of a unit as the selection criterion (King et al., 1972). As a rule the irreversibility of the plant decreases monotonically with each set of iterations during evolutionary synthesis; however, the final criteria are economic and operative (Gundersen and Naess, 1988). Thermodynamics is a valuable tool when dealing with new processes (Ranger et al., 1990; Le Goff et al., 1990a, 1990b, 1990c, 1990d; Grant, 1979). Performance studies of single units can help in the design of systems. System approaches contribute to the optimization of whole chemical or petrochemical industries and constitute a new trend in thermodynamic analyses.

For chemical plants the irreversibility rate $I = T_0\sigma$ provides a means for assessing the thermodynamic performance of individual plant components (I_k) as well as that of the whole plant (I_{tot}). This enables one to compare the exergy losses of the different plant components and to search for their relative changes caused by a selected parameter ξ_i. These relative changes are in the form of the coefficient of structural bounds, C_{sb}:

$$C_{sb} = \frac{\partial I_{tot}/\partial \xi_i}{\partial I_k/\partial \xi_i} \qquad (11.15)$$

(Brodyanskii, 1973). C_{sb} is a thermodynamic indicator of the system structure which offers a basis for systematic study and improvement of system structures (Gundersen and Naess, 1988). Second-law analyses help in establishing operating conditions of many industrial chemical processes such as coal gasification, coal combustion, ammonia synthesis, nitric acid production, production of methanol from naphtha and natural gas, and formaldehyde from methanol, and in reacting plasma systems (Denbigh, 1956; Bosniakovic, 1965; Riekert, 1974, 1979). Szargut et al. (1988) have indicated where a potential for improvement in fuel exists in chemical and metallurgical plants with large fossil fuel consumption.

As the theory of energy limits applies minimal values of the entropy production or exergy dissipation, the issue of exergy savings is related to economic optima and deserves consideration. At the interface of energy limits theory and (thermo)economics is the research of Tsatsaronis, Valero and associates and that of von Spakovsky and co-workers. The first group has developed a general theory of exergy saving and applied it in problems of exergetic cost and thermoeconomic cost, for the purpose of energy savings and thermoeconomics (Tsatsaronis, 1984; Valero et al., 1986a, 1986b, 1986c). A system for energetic/exergetic optimization of power plants (GAUDEAMO) was worked out (Valero et al., 1986d). Aspects of unification of thermoeconomic theories were considered (Valero et al., 1989). Thermoeconomics was applied in structural analysis and relative free energy function (Valero et al., 1992, 1993). Thermoeconomic analyses have shown their effectiveness in modeling gas turbine cogeneration systems and for calculating exergy in chemical processes (Lozano and Valero, 1993b). Tsatsaronis, Lozano and Valero have proposed the methodology of exergoeconomics and applied it to the exergetic cost theory (Tsatsaronis, 1984; Tsatsaronis and Winhold, 1984; Lozano and Valero, 1988, 1993a). The exergetic cost theory was applied, among others, to a stream boiler in a thermal generating station (Lozano and Valero, 1987). Other applications of the exergetic cost theory are also known (Valero et al., 1994a, 1994b, CGAM Problem). These treatments have contributed to practical and industrial applications of thermodynamic analyses in which the classical function of exergy and theory of exergetic cost are tools in thermoeconomic diagnoses of energy systems (Lozano et al., 1993, 1994).

The term exergoeconomics was coined by Tsatsaronis (1984), to indicate an exergy analysis with an economic analysis, in which the exergy costing principle is applied. For any other combination of a thermodynamic analysis with an economic analysis the broader term thermoeconomics should be used. Traditional exergy efficiencies (Sciubba and Wall, 2007) could mislead because they use the concept of "exergy input" instead of the more appropriate concept of the "exergy of fuel." The general concepts of "fuel" and "product" (in conjunction with exergy analysis and exergoeconomics), "cost per unit of exergy of fuel and product," "cost of exergy destruction" as well as the general formulation of exergy balances and cost balances were developed by Tsatsaronis' group (Tsatsaronis, 1984; Tsatsaronis and Winhold, 1984, 1985a, 1985b; Lazzaretto and Tsatsaronis, 1999, 2006). Energy conversion plants have been treated by exergoeconomic analyses (Tsatsaronis and Winhold, 1984, 1985a; Tsatsaronis et al., 1986, 1991, 1993, 1994). Examples have shown that the exergy approach leads engineers to develop new concepts of energy conversion systems based on exergetic evaluations or to significantly change the system design (Tsatsaronis et al., 1991, 1992; Tsatsaronis, 1999). Valero and Lozano exploited some of these developments, in which the contributions of investment costs and operating and maintenance expenses are neglected in the cost balances, to propose the "exergetic cost theory" (Valero et al., 1986a, 1986b, 1986c, 1986d; Lozano and Valero, 1993a, 1993b). In this theory, the useful additions were (a) the division of the variables by the cost of the fuel to the total system, and (b) the matrix formulation. The approaches used by Valero and co-workers mainly apply to existing systems, whereas the approaches used by Tsatsaronis' group focus on the design optimization of new systems. Significant contributions also came from the Padova group of researchers (Lazzaretto and Andreatta, 1995; Lazzaretto and Macor, 1995; Reini, 1994; Toffolo and Lazzaretto, 2003).

The field of thermoeconomics benefited by discussions of aspects of economic activity from an energy perspective and applications of statistical–mechanical theory to economics (Samuelson, 1947; Georgescu-Roegen, 1971; Gong and Wall, 1997; El-Sayed, 2003; Valero et al., 2006). The idea of linking thermodynamics and costing considerations was first explored by Lotka (1921), Keenan (1932), Benedict (1949, published by Benedict and Gyftopoulos in 1980), and Gilbert (1956): the essential idea that emerged from their very general papers was that entropic considerations should somehow be accounted for in monetary cost calculations. Beckmann (1953), Henatsch (1957) and Szargut (1957) explicitly addressed the cost allocation between co-generated steam and power in energy systems.

In the natural sciences, however, thermoeconomics is defined as the statistical physics of economic value (Georgescu-Roegen, 1971; Chen, 2005). According to Corning (Corning and Kline, 2000; Corning, 2002), thermoeconomics is based on the position that the role of energy in human development should be defined and understood through the second law of thermodynamics but in terms of such economic criteria as productivity, efficiency, and the costs and benefits of the various mechanisms for capturing and utilizing available energy to build biomass and do work. Thermoeconomists claim that human economic systems can be modeled as thermodynamic systems then, based on this premise, attempt to develop theoretical economic analogs of the first and second laws of thermodynamics (Burley and Foster, 1994). With this attitude, the exergy, that is, the measure of the useful work energy of a system, is the most important measure of value. In thermodynamics, thermal systems exchange heat, work, and/or mass with their surroundings; in this way, relations between the energy associated with the production, distribution, and consumption of goods and services can be determined. As the aim of many economic activities is to achieve a certain structure, thermoeconomics attempts to apply nonequilibrium thermodynamics, in which structures are dissipative in form, and information theory, in which information entropy is a central entity, to the modeling of economic activities in which the flows of energy and materials act to create scarce resources (Sieniutycz and Salamon, 1990; Baumgarter, 2004). Thermoeconomic activity is regarded as a dissipative structure sustained by transforming and exchanging resources, goods, and services. In engineering and industrial design, thermoeconomics applies to methodologies combining exergy and economics for optimizing the design and operation of thermal systems, in particular power generation units. Cost estimates accompany the design studies (Moran and Shapiro, 2003; Szargut et al., 1988).

An isomorphism between problems in thermodynamics and economics (Baumgarter, 2004) results in the laws of thermodynamics being inspiring for economists. Boulding (1966), Ayres and Kneese (1969), and Georgescu-Roegen (1971) applied thermodynamics when analyzing economy–environment interactions, and formally imbedded the economy in its biogeophysical context. In a first step, they formulated the materials balance principle, based on the thermodynamic law of mass conservation (Boulding, 1966; Ayres and Kneese, 1969; Ayres, 1978; Kneese et al., 1972), thus concluding that all resource inputs that enter a production eventually become waste. Up to now, this has been accepted theorem of resource, environmental and ecological economics. Simultaneously, Georgescu-Roegen (1971) developed an extensive critique of economics based on the laws of thermodynamics, and in particular the entropy law, which he considered to be "the most economic of all physical laws" (Georgescu-Roegen, 1971, p. 280). This initiated a lively discussion on whether thermodynamic laws are relevant to economics (Burness et al., 1980; Daly, 1992; Kåberger and Månsson, 2001; Khalil, 1990; Lozada, 1991, 1995; Norgaard, 1986; Townsend, 1992; Williamson, 1993; Young, 1991. While Georgescu-Roegen had drawn a principally correct picture of the irreversible transformations of energy and matter in economies, his analysis suffered from some flaws (Ayres, 1999; Baumgarter, 2004; Lozada, 1991, 1995). But as Georgescu-Roegen's research and many later studies have shown (Baumgärtner et al., 1996), the entropy law, if properly applied, yields insights into the irreversible nature of economy–environment interactions that are not otherwise available. Both the first and the second laws of thermodynamics therefore need to be combined in the study of how natural resources are extracted, used in production, and give rise to emissions and waste, thus leading to integrated models of ecological–economic systems (Odum,

1971; Ayres and Martinás, 1995; Ayres, 1998, 1999; Ayres et al., 1998, 2003; Baumgärtner, 2000; Faber et al., 1995; Faber, 1985; Perrings, 1987; Ruth, 1993, 1999).

The interconnection of the exergy concept with the "environmental issues" (taken in their broad sense) is of importance in the system context. Exergy *per se* is not a measure of environmental impact (Sciubba and Ulgiati, 2005; Sciubba and Wall, 2007), but, in essence, at the end of the life cycle of a device, plant or product, the exergy balance of the extraction–transformation–production–distribution–use–disposal cycle shows how many primary exergy resources have actually been consumed. In fact, the use of the entropy concept in ecology and ecological economics has been the subject of many works (Baumgärtner et al., 1996; Faber et al., 1996; Valero, 1998). Corning and Kline (2000) discuss what they call the thermoeconomics of living systems – a cybernetic and economic approach to analyzing the role of available energy in biological evolution – and relate this paradigm to a distinction they draw between various statistical or structural definitions of information and what they call "control information." In addition, with "global warming" in mind, some papers discuss exergy analysis of "CO_2 zero emission" high-efficiency plants (Calabrò et al., 2004).

All chemical processes consume unrestorable natural resources; the faster civilization develops, the sooner these are exhausted. Ecological components have been added to modify the formulations of energy limits theory and traditional problems of thermal technology (Angulo-Brown, 1991; Yan, 1993). Exergy has been used as a measure of the quality of resources (Szargut, 1986, 1987, 1989, 1990; Szargut and Morris, 1987; Szargut and Majza, 1989; Szargut et al., 1988; Månsson, 1990). It is important to calculate the rate at which industrial processes consume exergy resources. The cumulative consumption of exergy from unrestorable natural resources appearing in the chain of processes leading from natural raw materials to product expresses the ecological cost of the product (Szargut, 1978, 1986, 1990; Szargut and Majza, 1989). Exergetic ecological costs are applied to the optimization of production processes from the viewpoint of minimization of the consumption of natural resources. Indices of ecological costs determining the extent to which technological processes exhaust natural resources have been summarized (Szargut, 1989); related analyses of chemical processes are available in Szargut et al. (1988). One example treats the ecological second-law analysis of heat delivery from a complex heat–power station; the minimization of the consumption of unrestorable natural materials is achieved by using exergy (Szargut, 1990). Cumulative exergy cost seems to be suitable for industrial chemistry; in fact, it is a minimal work, as every exergy change is. However, in view of its nonuniqueness (dependence on accepted technology) and absence of statistical averaging in its definition, it is not an exact counterpart of an energy limit that must be a state-dependent physical quantity. As stated earlier, the method of cumulative exergy costs has serious imperfections. These are (a) vague sequential process, (b) date- and location-affected exergy consumption of resources defining cumulative cost, (c) undefined mathematical structure of costs. As pointed out earlier, cumulative costs are not functions but rather functionals of controls and state coordinates and, as such, they depend not only on process paths but also on the external controls and disturbances. Only *optimal* costs generated by optimization acquire properties of functions (potentials).

Starting with some earlier works (Beckmann, 1953; Szargut, 1957, 1970; Sieniutycz, 1973b; Szwast, 1990) the possibility of establishing a univocal and direct correlation between monetary price and physical value is debated, associated with various aspects of thermoeconomic analysis and methods of exergy accounting (Szargut, 1986, 1987, 1989, 1990; Tsatsaronis, 1987, 1993, 2007; Valero, 1999, 2003; Sciubba, 2000, 2001a, 2001b; Möller et al., 2006; Yantovski, 2007a, 2007b). Discussion of advanced zero emissions plant (AZEP) is included (Möller et al., 2006). Also, Szargut introduces the concepts of cumulative exergy consumption (CEC; Szargut, 1987) and thermoecological cost (TEC; Szargut, 2005), and postulates the application of exergy for the determination of the pro-ecological tax replacing actual personal taxes (Szargut, 2002b); see a discussion of Szargut's TEC definition and criticism of his exergy cost of human work based on nonrenewable natural resources (Yantovski, 2007b). As, in his opinion, the involvement of monetary costs is not compatible with exergy analysis, economic analysis, which is crucial for decision-making, should be done separately, after the thermodynamic one. Yet the data of cumulative exergy will be valid for years, as they are sensitive to technology changes only. Also, according to Sciubba and Wall (2007), the method of cumulative exergy consumption (CEC) provides a very clear picture of the "resources" incorporated in the production of goods. The method can be extended to include immaterial services, and, as in extended exergy accounting (EEA; Sciubba, 2000, 2001a, 2001b) it can account for labor and capital as well, thus paving the way to the calculation of an "exergy cost" of commodities measured in kJ/unit (common for CEC and EEA) instead of €/kJ (as in thermoeconomics). A problem for CEC is its neglect of labor and of all immaterial production factors in general. Finally, there are topological limitations for systems treated by the CEC method, which are, in principle, sequential systems (Sieniutycz, 2003b).

Sustainable energy systems engineering and, in particular, sustainable energy design can be promoted in the complex thermoeconomic context (Geworkian, 2007). In the energy conversion field, there is today practically no process or cycle analysis that does not include exergy considerations (Sciubba and Wall, 2007). Monographs on identification, modeling and simulation of energy technology systems by methods of systems theory involve graphs theory and matrix description of topological structures (Radwański et al., 1993; Portacha, 2002). The latter source also gives many examples of topological structures of electric power stations, and discusses costs of heat and power generation. Analyses of vapor power systems and case studies of exergy accounting of power plants are available (Bejan et al., 1996; Moran and Shapiro, 2003). Various cost-effective methods are used to improve energy resource utilization. One method is cogeneration, which sequentially yields power and heat (or steam).

Its aim is to accomplish the work and heat fluxes in an integrated system of the total expenditure less than that attributed to the sum of the process components. Two other methods are power recovery (by inserting a turbine into a pressurized gas or a liquid stream to capture a residual work) and waste heat recovery (to capture a residual exergy of heat, as in the case of exhaust gases of combustion engines). System-related optimization criteria include thermodynamic, economic, ecological, reliability and other terms. Multilevel optimization is the characteristic feature of models describing energy systems and industrial plants (Ziębik, 1990a). Exploitation of waste energy contained in biomass, biofuels, recycling of wastes, and various aspects of pro-ecological policy accompany the consideration of renewable energy sources (Lewandowski, 2001).

Heat-pumps may be components of optimum operation of complex thermal plants with cogeneration (Dentice d'Accadia et al., 2000). A plant may contain many gas-fueled engines with heat recovery. Each engine simultaneously drives an electric generator and the compressor of a heat-pump. The problem consists of selecting the operation mode with the lowest cost subject to process constraints. In another problem a method for determining the optimized dimensions of a ground source heat-pump system (GSHPS) heat exchanger are given (Marzbanrad et al., 2007). Optimum length and diameter for the heat exchanger is found for different mass flows by using the exergy form of the second law. The controls minimize entropy generation and result in increased efficiency of the heat-pump.

The book by Hirs (2000) contains in its Part 3 a number of examples on process integration, discussed in the context of: process systems, process units, processes with combined heat and power, and optimal design. An intelligent functional approach (Frangopoulos, 1983, 1990) is postulated to be a method for analysis and optimal synthesis–design–operation of complex systems including those with cogeneration (Dentice d'Accadia et al., 2000). Reviews of exergoeconomic methodologies are available (Tsatsaronis, 1987, 1993, 2007; Lozano et al., 1993; El-Sayed, 1999, 2003, 2007). A considerable amount of research deals with various aspects of thermoeconomic monitoring and diagnostics (Toffolo and Lazaretto, 2004; Valero et al., 2004; Zaleta-Aguilar et al., 2004; Verda, 2004; Sciubba, 2004) and the methodologies to generate fingerprints for malfunctioning devices (El-Sayed, 2007; Reini and Taccani, 2004; Toffolo and Lazaretto, 2004; Correas, 2004).

While there are still some unsolved issues as to the inclusion of environmental considerations into thermoeconomics, it is clear that thermoeconomic approaches will be more and more extensively adopted in the assessment of industrial processes and production cycles (Sciubba and Wall, 2007). The thermoeconomic mode of system synthesis is most likely the "next frontier": where a system's topological structure is sought for a required performance, and an optimization problem involves decisions not only related to process variables but also to the system's configuration. The objective functions can be formulated in the context of the methods of thermoeconomics, cumulative exergy content and extended exergy accounting, and thermoeconomics seems to be the one that is most likely to be used in the applications in the near future (Sciubba and Wall, 2007). The problem can be solved by mixed-integer linear programming (MILP) techniques (Kelahan and Gaddy, 1977; Salcedo, 1992; Floudas, 1995), genetic algorithms (Goldberg, 1989; Androulakis and Venkatasubramanian, 1991; Michalewicz, 1996; Garrad and Fraga, 1998; Upreti and Deb, 1997; Tayal and Fu, 1999; Yu et al., 2000; Toffolo and Lazzaretto, 2002; Ravagnani et al., 2005; Selbaş et al., 2006; Shopova and Vaklieva-Bancheva, 2006), and artificial intelligence techniques (Melli et al., 1992; Sciubba and Melli, 1998; Munos and Moore, 1999b).

A need is postulated for the analysis of energy systems not only on the basis of their operational life, but also on the basis of considerations of the manufacturing and recycling phases (Curti et al., 1993). Environmental and resource scarcity factors are taken into consideration. The main idea is to minimize the cumulated exergetic losses of operation, manufacture and recycling, distributing the exergy of manufacture (the so-called gray exergy) and recycling over the lifetime of the system. Additional levels of analysis consist of introducing pollution and resource scarcity factors and using these to artificially penalize the above-mentioned exergy terms. In this approach, the first level of analysis (exergy analysis with gray exergy) is applied to domestic heat-pumps operating with different evaporators. The pinch analysis (Linnhoff, 1979, and Chapter 12), extended to include exergy factors, falls within the framework of a global multidisciplinary analysis which considers economic, energetic, exergetic, and environmental factors in the light of sustainable development (Staine and Favrat, 1995). One of the essential themes of such a vast framework is the system's life cycle analysis that involves the manufacture, the exploitation and the recycling of components. This is more than in the original pinch method, which focuses primarily on economic and heat transfer issues. The extension of the pinch analysis to include exergy factors involves a global exergy balance and irreversibility considerations due to heat transfer, dissipation and the manufacture of components. The thermodynamic optimization of heat exchangers based on an optimal distribution of exergy losses is performed, and the gray exergy associated with the manufacture of shell and tube heat exchangers is calculated. Inclusion of electrical energy balance is also possible. Such a balance is particularly useful when introducing heat-pumps or power units. The extended composite curves offer a graphic representation of all the exergy losses by using a Carnot factor–heat rate diagram and an electric power–Carnot factor diagram. With such diagrams, the choice of the optimal pinch value (T_{min}) is determined for the minimum total exergy loss of the process. A methodology for the heat integration of industrial batch processes based on pinch analysis principles is also proposed (Krummenacher and Favrat, 2000). This methodology resorts to intermediate heat storage to prevent adverse effects on the operating flexibility. A minimum number of storage units and their range of feasible operation follows in terms of the amount of heat recovery. Heuristic rules emerge, related to minimum cost solutions. Heat-pump-surrounding (HPS) systems are modeled as gray systems, where the gray model serves to assess the system exergy efficiency (Ao et al., 2007). Gray system theory and its applications are summarized in Luo et al. (2001) and Liu and Yi (2005).

11.10. TOWARDS THE THERMOECONOMICS AND INTEGRATION OF HEAT ENERGY

As approaches to complex system design based on second-law analyses have been developed the emphasis has changed from exergy (energy) minimal to cost optimal units and networks (Gundersen and Naess, 1988; Tsatsaronis, 1993; Ziębik, 1996; Valero et al., 1994a, 1994b; Frangopoulos et al., 2002; Sciubba, 2001a, 2001b). While this reorientation does not constitute a real problem in FTT, where performance criteria can be arbitrary, analyses restricted to the classical availability and exergy criteria have a more limited chance of proving their full usefulness in such extended schemes.

Apparently, the optimal design point of a heat exchanger network can be calculated only by taking into proper account entropic losses, that is, exergy destruction (Sciubba and Wall, 2007). However, as stressed by Tsatsaronis (2007) such an approach has a number of severe flaws and no optimal design can be obtained by it for the following reasons:

(a) An optimal design point can, in general, be obtained by considering the economic and the environmental performance together with the thermodynamic one, but not just the thermodynamic performance.
(b) As the economic values of the exergy unit differ for various forms of exergy, the thermodynamic optimum is also questionable because the exergy destruction caused by friction is covered by mechanical (or electric) energy, whereas the exergy destruction caused by heat transfer is usually covered by the exergy of fossil fuels (or renewable energy). These exergy forms not only have different costs per unit of exergy but also different thermodynamic values, if we refer them to the same form of primary energy.
(c) The optimal design point of a heat exchanger cannot be obtained in isolation, because this depends on the relative position of the heat exchanger within the overall energy conversion system being studied.

In fact, the contemporary design of heat exchanger networks involves real-life objectives that contain both a quantitative part (cost of equipment, energy and resources) and a qualitative part (safety, operability, controllability, flexibility, etc.). The industrial problem is very complex and involves a combinatorial approach to the match between hot and cold streams, flow configuration choice, equality and inequality constraints on the temperature-dependent properties, materials, pressure drop limitations, and so on. Many of these issues are discussed in the second part of this book. All this leads to the trend towards complex thermoeconomies and requires reduction or abandonment of thermodynamic concepts. Optimization of energy systems works at various levels: synthesis (configuration), design (component characteristics), and operation (Frangopoulos et al., 2002), and possibly including human work (Sciubba, 2001a, 2001b). As one goes from operation optimization to design and synthesis, the problem is difficult from the methodological and computational viewpoints.

When physical criteria of performance are abandoned, methodologies developed for the analysis, improvement and optimization of energy systems must take into account not only energy consumption and financial resources expended but the scarcity of all resources as well as any possible thermal and/or chemical pollution and degradation of the environment. The current trend is to combine all these aspects into a global methodology (Frangopoulos and von Spakovsky, 1993; Von Spakovsky and Frangopoulos, 1993). Its essence is to mathematically combine in a single model second-law analysis with economics, subject process constraints including pollution, resource availability and the so-called gray exergy associated with the manufacture and recycling of the capital equipment used by an energy system. The benefit of such a global approach is a systematic search for the optimum system configuration, performance and cost (Von Spakovsky, 1994). Industry has the capability of finding such optima for individual components taking into account solely thermodynamic and economic considerations. On a system level, however, such a capability does not presently exist in industry, since the component optimums found on an individual basis do not necessarily correspond to an overall system optimum (Von Spakovsky, 1993, 1994; Von Spakovsky and Evans, 1990). An approach to exergy costing in exergoeconomics improves the fairness of the costing by taking a closer look at the cost formation and the monetary value of processes (Tsatsaronis et al., 1993).

In the following chapters of the book we shall apply system approaches to develop techniques of heat integration, a collection of approaches designed to reduce the cost of utilities and environmental impacts. The basic idea of heat integration is to apply the heat energy of hot streams to heat up cold streams; this causes cooling of hot streams and heating of cold streams along with decreased consumption of external utilities. An example is the use of the hot bottom product from distillation to increase the temperature of the feed stream directed to the distillation tower and to achieve a significant reduction of an external heat. To proceed in a systematic way, the total system is decomposed into three subsystems: (i) a basic process subsystem (BPS) composed of processes and apparatuses typical for the industry we are working in, (ii) a utility subsystem (US) that produces heating and cooling utilities and shaft work, and, (iii) a heat exchanger network (HEN), which is a network where process streams from BPS and utilities exchange heat. Interconnections among subsystems and the main features of heat integration are discussed in the next chapter.

12 Heat integration within process integration

We begin with heat energy integration since energy optimization is the main topic of the book. Investigation of heat integration started first and developments in this area have given rise to many achievements in designing various subsystems as well as total system integration.

In many industrial processing systems there are material process streams and processes that have to be cooled down and/or condensed and heated up and/or vaporized. The heating and cooling duties are mainly performed in an indirect way in recuperative heat exchangers. Direct heat exchange is possible but is a rather secondary means of heat exchange, particularly in chemical and relative industries.

In this chapter we present a simplified flow sheet of a typical chemical processing system, which is also common in other branches of industry such as petroleum refineries, food processing, the pharmaceutical industry and so on (Figure 12.1).

Raw materials, often after the pre-processing stage, are first fed into chemical and/or biochemical reactors. Then, the mixture from the reactors is separated into components that are final or intermediate products, and also waste. Reagents often require heating or cooling before reaction. Reactions can be endothermic or exothermic and thus need to be heated up or cooled down. The post-reaction mixture can be in various thermal states depending on reaction conditions. In order to separate it in downstream processes its thermal state often has to be changed. For instance, vapor mixture is usually condensed to reduce the cost

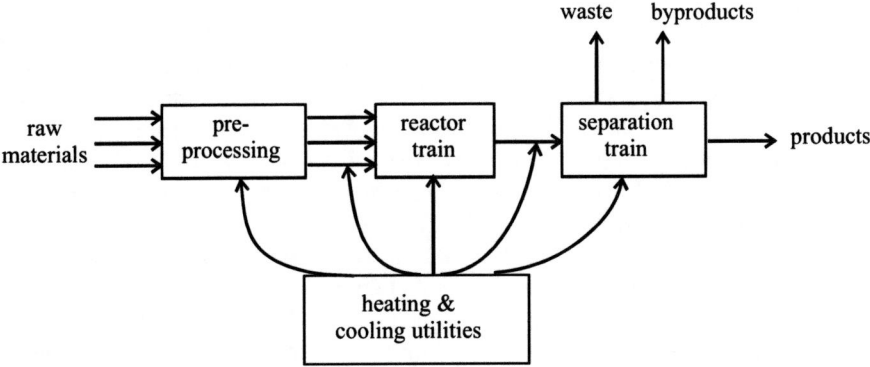

Figure 12.1 Simplified flow sheet of a typical process system in chemical industries.

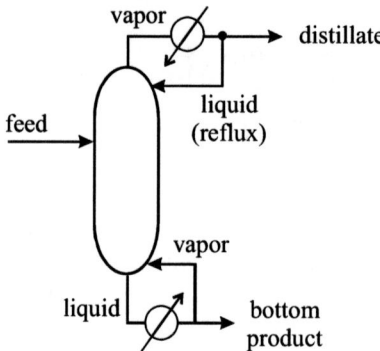

Figure 12.2 Scheme of a typical simple distillation column.

of transportation and, also, of separation processes. Additionally, many separation processes require heat energy input and/or output—energy separation agent (ESA) type processes. A good example is distillation, where heat input is necessary, usually to the reboiler, in order to vaporize the mixture from a column. At the same time, in a total or partial condenser, heat has to be removed to condense the vapor phase from the column—Figure 12.2.

To sum up, in a process system there are process streams that need cooling—they will be referred to as hot process streams—and streams which require heating—called cold process streams.

We can supply heat to each cold stream using heating utilities such as steam at various pressure levels, flue gas from furnaces and so on. Similarly, heat can be removed from each hot stream by cooling utilities such as water, air and refrigerants. Obviously, such a nonintegrated system is inferior in terms of money and environmental impact. Heating utilities are expensive since we have to buy energy sources and process them into heat energy. Most often energy sources applied in industry are not renewable—oil, coal, and so on. During conversion into heat energy waste products are produced that have to be treated, and thus increase further the cost of energy. Cooling water is also costly and has a strong environmental impact. We use fresh water from the environment, treat it, apply it to heat-exchange processes, then cool it down in cooling towers. Thus, cooling water is usually applied in a closed loop. However, due to inevitable losses in pipes and cooling towers, it is necessary to supplement it. Additional fresh water intake is also needed to prevent concentration of chemical substances. A similar procedure has to be applied for boiler feed water.

In order to diminish the cost of utilities and environmental impacts it is advantageous to integrate heat energy. The idea behind heat integration is simple: use heat energy of hot process streams to heat up cold process streams. In consequence, hot streams are cooled down and cold streams are heated up with a decreased consumption of utilities.

Heat integration has been applied in industry for many years even before the era of systematic approaches to process integration that began in the 1960s.

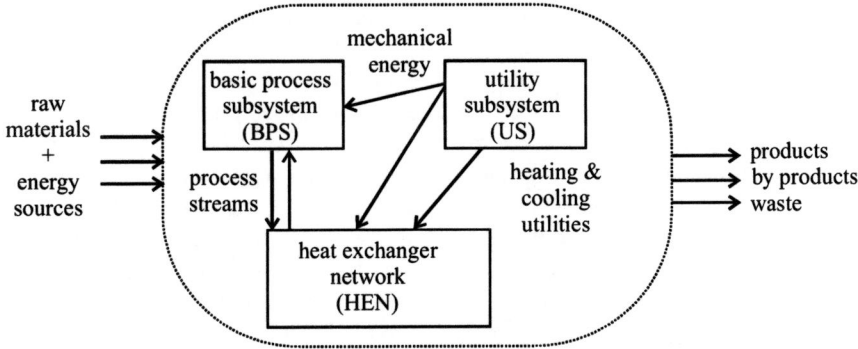

Figure 12.3 Block representation of a total heat-integrated process system.

However, heat integration was applied "locally", to one or a few processes. For instance, hot bottom products from distillation were often used to heat up feed to a distillation tower. There were no systematic approaches able to cope with larger problems featuring many hot and cold streams. Also, there was a lack of insight. From the late 1960s to the present, a huge number of papers more or less directly connected with heat integration have been published. For instance, Furman and Sahinidis (2002) listed almost 500 papers, published in the previous century, on heat-exchanger network design and related topics. Although many approaches to heat-exchanger network design and heat integration in general are available, there is still a gap between academic research and their industrial applications.

In order to analyze systematic methods for heat integration, we should proceed in a systematic manner. Let us consider the total process system as composed of three main elements:

- Basic process subsystems (BPSs) consisting of processes and apparatus typical for the branch of industry we are interested in
- Utility subsystems (USs) that produce shaft work together with heating and cooling utilities
- Heat-exchanger networks (HENs), that is, a network where process streams from BPSs and utilities exchange heat.

The representation of the system (often called block representation) shown in Figure 12.3 illustrates interconnections among the subsystems. The total site consumes raw materials, energy sources and also utilities such as water from the environment and solvents.

Heat integration is performed in a central HEN. Hot and cold process streams from the BPS are sent to the HEN, where they should reach their required thermal final states and are then sent back to the BPS or outside the system as products, by-products, or waste. Heating and cooling utilities are also transported to the HEN from the United States. Often these utilities come back to the US as, for instance, cooling water.

In a total heat-integrated system there are strong interactions: BPS ↔ HEN and US ↔ HEN as well as BPS ↔ US. The block representation does not show some other relations, such as the use of heat energy of process streams in power generation cycles that traditionally belong to USs. In fact, the traditional division into distinct subsystems such as in the block representation almost vanishes in a totally integrated system. A good example is a water network consisting of water-using and wastewater-treatment operations. Fresh water from the environment is applied in various water processes from BPSs (e.g. extraction) and also from USs (e.g. water intake to cooling water cycle). The water is used to wash tanks, pipes, and apparatus and finally as "hygienic" water. Some of these water-using operations are not included explicitly in the block representation. Next, wastewater from water-using processes is sent to the central water treatment station where it is treated before being discharged to the environment or to additional treatment outside the site. It is possible to distinguish two networks: water-using processes and water-treatment processes. However, redistribution of treatment processes reduces both water usage and cost of wastewater treatment. Thus, we should integrate both networks into a single integrated water network. This problem will be considered in depth in Chapter 20.

Despite some limitations of block representation we will use it in the following since it is sufficient to discuss major features of the heat-integration problem. Also, it is convenient for the explanation of developments to date in the field of process integration.

Principally there are two levels of heat integration:

- Level I (lower)—HEN design
- Level II (upper)—design of total heat-integrated site.

At the lower level we assume that parameters of process streams flowing into the HEN and exiting it are fixed. We also assume that we apply in the HEN the utilities that are produced in the US (though this assumption is often relaxed). The first assumption dictates that we do not make any changes in the BPS and the second that we keep the US intact. Thus, we focus only on the HEN. Our objective is to design an optimal HEN for given inlet and outlet parameters of streams exchanging heat in the HEN.

At the upper level we treat integration in a more holistic manner. All subsystems are considered simultaneously. The objective is to optimize the total heat-integrated system. We have to consider parameters and topology of the US and BPS as decision variables. Parameters of streams to the HEN are not fixed but depend on processes in both the BPS and US, as well as on heat exchange in the HEN. It is clear that solution of the upper level heat-integration problem requires solution of the lower level one. We can consider an iterative scheme, though it does not seem an efficient technique, and, more importantly, leads to inferior solutions—see for example, Biegler et al. (1997).

In the following chapters we limit the presentation mainly to the lower level problem. However, some explanations of basic ideas of total heat integration will also be given. There are many papers on simultaneous heat and process integration, including utility subsystems. Here we mention only some of them: Linnhoff and Townsend (1982), Papoulias and Grossmann (1983a), Townsend and Linnhoff (1983a, 1983b), Duran and Grossmann (1986), Linnhoff and Eastwood (1987), Colmanares and Seider (1989), Yee et al. (1990a), Dhole and Linnhoff (1994), Kovač Kralj and Glavič (1995, 2000), Makwana et al. (1998), Papalexandri et al. (1998), Kovač Kralj et al. (2000, 2005), Zhang and Zhu (2000), Bagajewicz and Ji (2001), Bagajewicz and Soto (2001), and Ji and Bagajewicz (2002).

To formalize the first level problem we have to know what data are available and what the objective is. Because BPS topology is fixed the process streams flowing into and exiting it are known. Hence, we know that hot process streams (NH) and cold process streams (NC) leave the apparatus in the BPS and that each stream is uniquely defined by flow rate, temperature, pressure and composition. These streams have to be fed back to apparatus in the BPS or sent outside the site and, thus, we know their required states, that is, temperature, pressure and composition.

The process streams leaving the apparatus enter the HEN and, thus, we will refer to them as inlet streams and will denote their parameters by the superscript "in". The streams leaving the HEN are the outlet streams and their parameters will be denoted by the superscript "out". It should be clear that "pure" heat-exchange processes are conceptually "removed" from the BPS and become part of the HEN. For instance, the distillation process is most often considered as the single unit consisting of distillation column, reboiler and condenser—see Figure 12.2. For heat-integration purposes we should divide it into column, reboiler and condenser. Both reboiler and condenser are now components of the HEN. Feed to column, vapor stream from the top of the column as well as liquid phase from the bottom are treated as process streams. The designer can exclude some streams from heat integration. For instance, he or she can decide that, due to control problems, the reboiler and condenser will not be subject to heat integration. The extent of heat integration is a problem, the solution of which depends on many factors. The seminal work is that by Ahmad and Hui (1991). The works of Hui and Ahmad (1994a, 1994b) are also important contributions to this problem. An example of more recent work on integration across several plants is the work of Rodera and Bagajewicz (1999), and Bagajewicz and Rodera (2000).

For a given utility subsystem we know the utilities that are available. Also, we know their parameters at the outlet from the US, that is, at the inlet to the HEN. For instance, in the case of steam we know the temperature and/or pressure. For cooling water its temperature leaving the cooling tower can be estimated with sufficient accuracy. The parameters of utilities leaving the HEN are more uncertain. However, in most cases the designer is able to estimate them or, at least, to assess their maximal or minimal admissible values. The problem of estimation of utility outlet parameters will be given deeper consideration in Section 13.5. Hence, we can acquire the knowledge of the inlet and outlet states for utilities but not their flow rates since they are optimization variables.

The aim is to design a HEN that ensures the minimum or maximum value of the adapted goal function. The approaches developed to date often apply the total annual cost (TAC) of the HEN. Also, simpler criteria such as cost of utilities are common, in earlier works in particular. More recently, multicriteria performance indices including ecological measures (life cycle analysis) have been addressed, for example, Li and Hua (2000), Ya and Shonnard (2000), and Mata et al. (2003).

The problem is complex, and to date no efficient, robust tool has been developed that is able to deal with industrial-size problems with a detailed model of apparatus that ensures at least good local optimum solutions. Nevertheless, several approaches exist. In fact they are numerous, and it is difficult for non-experts to assess their performance and applicability to a specific case. One of the main aims of the following chapters is to highlight their merits and limitations.

Heat and power (electricity) are not the only costly utilities consumed by various users within a total site. Water is also widely applied. Its cost continuously soars. Water networks were briefly characterized in the preceding paragraphs. The concept of water networks also stems from the process integration paradigm. Traditionally, water-using processes of various subsystems within a site were supplied from fresh water sources independently. Hence, they were not thought of as interlinked processes. The configuration of such traditional water supplies was as follows. Processes requiring water were supplied from a single or a few sources of fresh water. Wastewater streams from processes contaminated by various species were then mixed and sent to a central treatment station. Due to environmental regulations resulting in the soaring cost of fresh water and the increasing cost of water treatment, there is a strong need for minimizing both fresh water usage and wastewater generation. Integration is the simple, direct mean of reaching both aims. Some wastewater from processes can be reapplied—this is known as water reuse. Additionally, treatment processes can be redistributed and applied as regenerators, thus allowing for further increase of water reuse.

In doing so we can approach a zero water discharge system (closed circuit of process water). The water-using processes and regenerators create a total water network. The problem of designing optimal water networks is addressed in Chapter 20.

It is interesting to note that water network (WN) design problems can be considered as a somewhat simplified case of designing mass exchanger networks (MENs). The MEN consists of mass separating processes (MSA processes), that is, the separation processes that require additional substances or phases such as solvents or adsorbents. In MEN design problems a minimization of solvents and adsorbent usage is the main performance index (El-Halwagi, 1997, 2006). Thus, there exists a close resemblance between MEN and WN design problems. In both cases the networks are modeled as systems of simple mass exchangers (though there are some exceptions in WN) and the aim is to minimize utility usage (or WN total cost). The WN design problem seems the easier since it features only the single utility in water-using processes—fresh water. Hence, solutions developed for MEN should be applicable to WN as well. Nevertheless, design methods are often tailored for WN only.

Furthermore, the MEN/WN design problem bears a close resemblance to HEN design. In the former some substances are transferred from "reach" streams (having a higher concentration of transferred substances) to "lean" streams (of lower concentrations). In the latter, heat is transferred from hot streams to cold streams. The shortcut models (balances in particular) of the processes are similar. With utilities, solvents in MENs, fresh water in WNs and heating and cooling media in HENs have to be applied to reach the required outlet conditions of process streams. The main aim in both cases is to minimize utility consumption. The major difference between both problems is that in a HEN a single commodity (heat energy) is transferred, while in a MEN/WN it is often more than one. Nevertheless, due to this close resemblance many concepts and tools developed for heat integration and HEN design have been adapted for mass integration and MEN/WN design—see for example, El-Halwagi (1997, 2006). Streams in WNs often require heating and/or cooling due to conditions of water-using and wastewater-treatment operations. Simultaneous mass and heat integration is profitable because it reduces costs of both water and energy (Savulescu et al., 2002, 2005a, 2005b; Bagajewicz et al., 2002; Zhelev, 2005). A cooling water circuit is a special case of a water network. Optimization of cooling water usage requires both mass and heat integration. The problem is not addressed in this book: the reader is referred to textbooks by Smith (1995) and also to papers by Kim and Smith (2003, 2004a, 2004b).

MSA-driven processes (such as in MENs/WNs) are inherent elements of a BPS in many industries. There are also energy-driven processes (ESA processes) such as distillation, flash and evaporation. They can also be integrated to reduce the cost of energy. Heat integration in such processes is an important issue. To explain, let us consider distillation of an N-component mixture into pure components. Here, we assume the so-called sharp separator. Each separator has one feed stream and two outputs—see Figure 12.4.

Additionally, it is assumed that one of the product streams is an almost pure substance. To perform separation of an N-component mixture a train is required consisting of $N - 1$ sharp separators. This creates a network of separators. Decisions on what component to separate in each separator as pure product are key points since they largely influence the cost of separation. This gives rise to the problem of designing an optimal separation train; see for instance more recent works, such as Kovacs et al. (2000c) and Zhang and Linninger (2006). Additionally, in order to reduce the energy needed for separation, heat integration is applied. The columns can be heat-integrated via condensers and reboilers as illustrated in Figure 12.5.

The problem of designing an optimal heat-integrated distillation train requires concepts and tools from heat integration. There are numerous papers describing this problem. Here we list only a few of them: Aggarwal and Floudas (1990, 1992) and Petros et al. (2005).

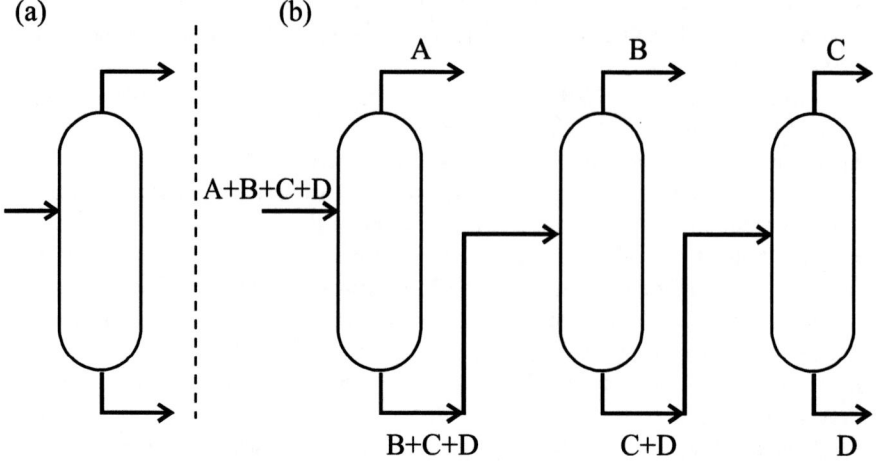

Figure 12.4 Scheme of sharp separator and separator sequence: (a) sharp separator; (b) example of sequence of sharp separators.

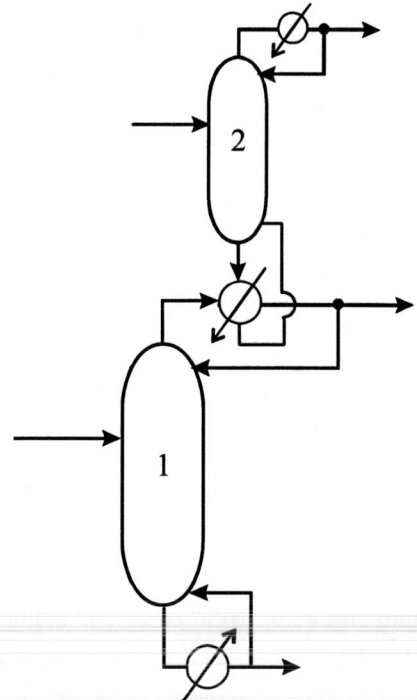

Figure 12.5 Example of two heat-integrated columns.

In process systems there are processes (subsystems) that require or produce heat and power energy as, for instance, cryogenic and refrigeration subsystems. The processes in the systems are thermodynamic cycles. Also, heat-pumps and power generators are thermodynamic cycles modeled by the same or similar thermodynamic rules. Notice that heat-pumps allow upgrading heat and hence are often applied in distillation trains. These thermodynamic cycles use or produce heat. Thus, heat-integration principles should apply to them, too. It is, however, necessary to account additionally for mechanical power (shaft energy). Investigations focusing on utility system design can be found in Varbanov et al. (2004a, 2004b), Shang and Kokossis (2004), and Oliveira and Matos (2004), to mention a few. Basics of utility system design are also explained in textbooks by Biegler et al. (1997) and Smith (1995, 2005).

The problems mentioned so far do not exhaust all the topics of process integration. For instance, investigations have been carried out on the synthesis of reactor networks, often with simultaneous design of subsequent separation trains, for example, Mehta and Kokossis (2000), Ismail et al. (2001), and Linke and Kokossis (2003). Also, approaches to designing total heat-integrated process systems have been suggested, for example, Manninen and Zhu (2001), Bedenik et al. (2004), Kovač Kralj et al. (2005), and Zhang and Xhu (2006).

Integration of processes may cause additional control problems. To avoid them it is necessary simultaneously to account for integration and control issues. The next level of integration arises within a complex design process—simultaneous integration of processes and control scheme. From among numerous contributions we mention a few: Gollapalli et al. (2000), Kookos and Perkins (2001), Blanco and Bandoni (2003) and Ulas and Diwekar (2006).

Similarly, aspects of flexibility and resiliency should be considered simultaneously with integration—see for instance Biegler et al. (1997). This also significantly increases the complexity of the problem. Among others, Konukman and Akman (2002) and Chen and Hung (2004) addressed heat-exchanger networks while for instance Bansal et al. (2000), Raspanti et al. (2000), Banerjee and Ierapetritou (2003) and Srinivasan et al. (2005) attacked the general process system.

Shrinking margins for chemical commodities and worldwide competition gave rise to the most recent paradigm—product design. This can be seen as a holistic integrated approach for the entire design process from the micro-/nano-molecular level to management problems at the level of enterprises. Environmental (ecological) aspects have also been taken into account. Papers by Hostrup et al. (1999), Wibowo and Ng (2001), Eden et al. (2004) and Papadopoulos and Linke (2006) are examples of works on simultaneous product and process design. Similarly, the works of Sighvi and Shenoy (2002), Hugo and Pistikopoulos (2005) and Guillen et al. (2006) are examples of more recent achievements on integration at the level of enterprises. A review can be found in Grossmann (2005). It is interesting to note that Sighvi and Shenoy (2002) proposed the application of the pinch concept from heat integration to cope with planning in the supply chain.

The reader is also referred to recent excellent reviews on various aspects of process and product integration: Wintermantel (1999), Westerberg and Subrahmanian (2000), Barnicki and Siirola (2004), Shapiro (2004), Li and Kraslawski (2004), Charpentier and McKenna (2004), Westerberg (2004), Gani (2004) and Sargent (2005).

Gundersen and Naess (1988) and Jeżowski (1994a, 1994b) gave a thorough review of heat integration and HENs in particular. Furman and Sahinidis (2002) cited almost 500 papers from the topic, together with short abstracts. Pinch technology methods for HEN design as well as certain optimization models for HEN design are addressed in detail in a monograph by Shenoy (1995). Books by Smith (1995, 2005), Seider et al. (1999, 2004), Floudas (1995), Biegler et al. (1997) and El-Halwagi (1997, 2006), and a monograph on heat integration and heat transfer in general by Linnhoff et al. (1982) are also excellent sources of information.

In the following chapters we limit ourselves to heat and water integration problems. Even within this limited scope we will limit the discussion to a brief presentation of some aspects of these two problems. In particular, we will not analyze heat and water integration in batch processes. Also, flexibility and resiliency will not be addressed. Thus, the material in the following chapters deals mainly with steady-state (continuous) processes with the assumption of certain data. The main emphasis is placed on heat-integration and heat-exchanger network design issues. First, we will address the basis of heat recovery developed mainly in the framework of pinch technology. Then a few chapters focus on HEN design. Heat-exchanger network targeting techniques will be addressed in depth since available methods are relatively simple and, most importantly, are able to solve large-scale industrial problems. Both insight- and optimization-based approaches in sequential and simultaneous frameworks will be addressed. HEN synthesis and retrofit are presented in two chapters. Finally, water network targeting and design is addressed in the last chapter.

13 Maximum heat recovery and its consequences for process system design

13.1. INTRODUCTION AND PROBLEM FORMULATION

Calculation of maximum heat (energy) recovery (commonly abbreviated to MER) is the core of the heat integration problem. The designer has to know how much heat can be recovered, or equivalently, what are the minimum loads of heating and cooling utilities. On this basis she/he can then decide whether to proceed to further stages of the design of a heat integrated system. Furthermore, techniques developed to calculate the MER give deep insights into the heat integration problem. These observations are a breakthrough in traditional designing approaches. Recognizing the pinch phenomenon as well as its effects on process integration allowed the development of techniques for designing total plants, utility subsystems, distillation trains and refrigeration cascades to name a few.

In this chapter we will present simple, usually graphical, methods for MER calculation belonging to the very popular pinch technology (PT). They have some limitations, which can be relaxed by mathematical approaches addressed ' in Chapter 15. However, they provide valuable insights and, usually, results that are of sufficient accuracy at an early stage of process system synthesis.

To begin the explanation we should first formulate the MER problem based on the preliminary description in Chapter 12.

Given are NH hot process streams (h_i) and NC cold process ones (c_j). For each process stream the inlet temperature (T^{in}) or the inlet state as well as the required outlet temperature (T^{out}) or the outlet state are known. Also, the mass flow rate of each process stream (F) is known.

Notice that in certain applications (e.g. formulating optimization models) it is convenient to apply index sets. Hence, we also define the sets:

$$H = \{i \,|\, i \text{ is hot process stream }\}, \quad C = \{j \,|\, j \text{ is cold process stream }\}$$

The aim is to find the minimum hot utilities, heat load (Qhu^{min}) and the minimum cold utilities, heat load (Qcu^{min}), which have to be supplied to reach final temperatures/states.

In fact the MER problem is a simplified version of the more general minimum utility cost (MUC) problem. In industry several heating utilities as well as cooling utilities are usually available. Most common are steam of differing pressure, flue gas, cooling water, air and various refrigerants. Thus, the designer has to choose the most appropriate (usually cheapest) ones and, also, has to know their heat loads and/or flow rates.

In the MUC problem, additionally to process stream data, we should know

- the number (NHU) of hot utilities available (hu_m; $m = 1, \ldots, NHU$), their inlet and outlet temperatures (states),
- the number (NCU) of cold utilities available (cu_n; $n = 1, \ldots, NCU$), their inlet and outlet temperatures (states).

Similarly to process streams the alternative is to define index sets:

$$HU = \{m \,|\, m \text{ is heating utility }\}, \quad CU = \{n \,|\, n \text{ is cooling utility }\}$$

The aim is to calculate the minimum cost of utilities to be used. Usually, the cost is defined using unit prices based on heat, that is, in (USD kWh^{-1}) and these parameters have to be known. The annual utility cost (UC) is given by:

$$UC = \left(\sum_{m=1}^{NHU} phu_m Qhu_m + \sum_{n=1}^{NCU} pcu_n Qcu_n \right) \theta \qquad (13.1)$$

where θ denotes operation time per year, in, for example, (h year^{-1}).

Solution of the MUC problem provides the designer with the optimal heat loads on utilities (Qhu_m, Qcu_n), which are to be applied.

In order to calculate heat loads in both MER and MUC problems the thermodynamic model for enthalpy change has to be chosen. The majority of approaches for heat integration developed to date assume very simple models:

- for no phase change

$$Q = F \, cp(T^{\text{in}} - T^{\text{out}}) \qquad (13.2)$$

- for phase change (vaporization or condensation)

$$Q = F \, Lp \qquad (13.3)$$

In order to unify both equations we will apply, following the literature, the single formula:

$$Q = CP \, \Delta T \qquad (13.4)$$

where

$$CP = F \, cp; \quad \Delta T = (T^{\text{in}} - T^{\text{out}}) \text{ for the case of no phase change} \qquad (13.4a)$$

$$CP = F \, Lp; \quad \Delta T = 1.0 \text{ for phase change only} \qquad (13.4b)$$

Notice that cp values of utilities are the results of MER/MUC problem solution.

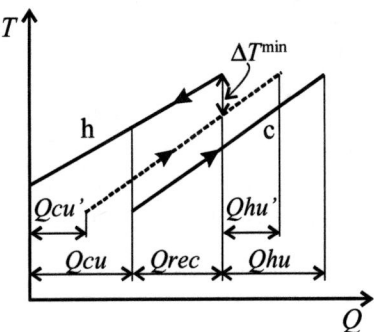

Figure 13.1 $T–Q$ plot for heat exchange between two streams.

Though Equation (13.4) seems oversimplified, the use of it is usually sufficient in the early design stage. Also, there are some means to account for cp values dependent on temperature. They will be discussed in the following.

13.2. COMPOSITE CURVE (CC) PLOT

The first valid approach for MER calculation was that by Hohmann in a PhD thesis (Hohmann, 1971). However, the achievement was not recognized for a long time. The method is now widely known as the CC plot—the name given by Linnhoff and co-workers from UMIST (University of Manchester, Institute of Science and Technology).

Here we present the approach with emphasis on the fact that it has roots in a simple, well-known heat exchange process between two streams. The process is often presented in basic textbooks and monographs in a $T–Q$ plot as shown in Figure 13.1 (by solid lines). There is one hot stream curve (*h*) and one cold stream curve (*c*). Because cp values are constant the curves are straight lines (more precisely we use here Equation (13.4a)).

The range where both curves cover each other represents the heat recovered ($Qrec$ in Figure 13.1). The range where the hot stream does not cover the cold stream represents the cold utility load (Qcu in Figure 13.1) that has to be supplied. Similarly, the uncovered segment of the cold stream represents the load on heating utility (Qhu in Figure 13.1). Figure 13.2 shows the serial

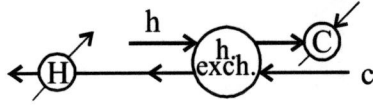

Figure 13.2 Typical arrangement of heat exchangers for heat exchange shown in Figure 13.1: H denotes heater; C denotes cooler; h. exch. denotes process heat exchanger.

arrangement: heat exchanger, heater, cooler that performs the heat exchange process from Figure 13.1.

It is also well known that by pushing the cold curve horizontally toward the hot one (or conversely, the hot toward the cold) we increase recovered heat and, in consequence, decrease both loads on utilities. The dashed line in Figure 13.1 represents the final position of the cold stream curve pushed against the hot stream. Such movement is limited by the thermodynamic condition:

$$T_h - T_c > 0.0 \tag{13.5}$$

In well-established industrial practice this thermodynamic condition is replaced by

$$T_h - T_c \geq \Delta T^{\min} \tag{13.6}$$

Usually practitioners apply $\Delta T^{\min}$ of the order 11.1 K (20 °F). This is, of course, a rule of thumb since $\Delta T^{\min}$ should be subject to optimization as the decision variable influencing the total cost of heat exchange.

Thus, to maximize heat recovery, the smallest vertical distance in temperature scale should be equal to given $\Delta T^{\min}$—see also Figure 13.1. One can see that this is the bottleneck for further increase of heat recovery. The temperature (usually that of hot stream) for which the distance is equal to $\Delta T^{\min}$ will be called the pinch following the nomenclature introduced by Linnhoff for the multistream case.

If cp values are constant, inlet or outlet temperatures of the hot or cold stream cause the pinch as shown in Figure 13.1. Hence, $\Delta T^{\min}$ can occur either at the hot or at the cold end of the heat exchanger. However, if cp values are temperature-dependent, the temperature profiles become curves, not straight lines, and an internal pinch can occur, as shown schematically in Figure 13.3. Notice that temperature approaches at both hot and cold ends of the heat exchanger are

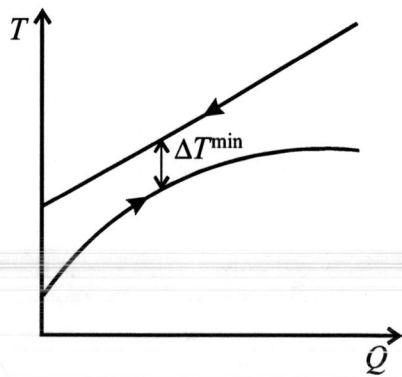

Figure 13.3 Internal pinch in two-stream heat exchange.

larger than $\Delta T^{\min}$. In order to account for internal pinch when rating a single heat exchanger segmentation should be performed.

To deal with multistream heat exchange the CC approach uses one hot composite curve (HCC) and one cold composite curve (CCC). The composite curve has a heat load equal to the sum of heat loads of the streams it is composed of. The curves are piece-wise straight lines since even though parameters cp are constant, the total heat load changes with temperature. The sharp changes at HCC or CCC will occur at the inlet temperature of streams since the heat load of the composite curve increases—a stream "enters" a curve. Also, a sharp change occurs at the outlet temperature of each stream because the heat load decreases—a stream "leaves" a curve. One can foresee that such changes can cause internal pinches similar to the effect of cp temperature-dependent values in the two-stream case. Thus, points of sharp changes of composite curves have to be treated with special care as potential pinches.

This reasoning forms the basis of the CC approach to the MER problem. Let us summarize it into a systematic procedure.

1. Calculation of HCC
   - Divide the total temperature range of NH hot process streams (from the maximal temperature to the minimal one) into temperature intervals ($k = 1$, ...) at inlet and outlet temperatures of each hot stream.
   - Calculate the heat loads in each interval k from Equation (13.7):

$$QH_k = \Delta T_k \sum_{i \in H_k} CP_i \qquad (13.7)$$

   where H_k denotes a set of hot streams in interval k, and ΔT_k is the temperature range of interval k.

2. Calculation of CCC—follow the procedure of point 1 for NC cold process streams
   - Divide the total temperature range of NC hot process streams (from the maximal temperature to the minimal one) into temperature intervals ($k = 1$, ...) at inlet and outlet temperatures of each hot stream.
   - Calculate the heat loads in each interval k from Equation (13.8):

$$QC_k = \Delta T_k \sum_{i \in C_k} CP_i \qquad (13.8)$$

   where C_k denotes a set of cold streams in interval k, and ΔT_k is the temperature range of interval k.

3. Draw the composite curves as a T–Q plot and move the CCC horizontally till the vertical distance becomes equal to fixed $\Delta T^{\min}$ (notice that the Q scale is relative). Similarly to the two-stream case, the point where the vertical distance is equal to $\Delta T^{\min}$ is the pinch.

Alternatively, the composite curves can be drawn directly into the plot without calculation of heat loads from Equations (13.7) and (13.8). First, the curves for

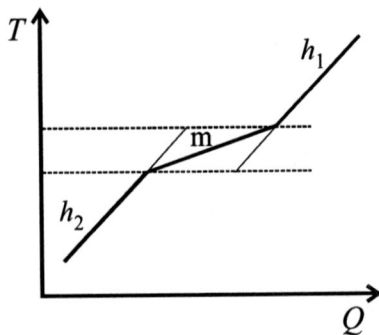

Figure 13.4 Illustration of stream merging into composite curve.

Table 13.1 Data for Example 13.1 ($\Delta T^{min} = 20\,K$)

Stream	T^{in} (°C)	T^{out} (°C)	CP (kW K^{-1})
h_1	150	60	2.0
h_2	90	60	8.0
c_1	20	125	2.5
c_2	25	100	3.0
hu	180	180	—

individual hot and cold streams are drawn separately in the plot. Then, the composite curves are constructed by merging hot or cold streams within the same temperature intervals. This merging procedure is illustrated in Figure 13.4 for two hot streams h_1 and h_2. The straight line denoted by "m" is the merged curve from parts of h_1 and h_2 in the same temperature interval. The entire composite curve is shown in bold. This graphical procedure was first proposed by Hohmann (1971).

The procedure described in points 1 to 3 above is illustrated by Example 13.1. Data are gathered in Table 13.1.

Construction of HCC

There are two temperature intervals: no. 1 from 150 to 90 °C and no. 2 from 90 to 60 °C. Heat loads of intervals are 120 and 300 kW, respectively. Segments 1, 2 of HCC are shown in Figure 13.5.

Construction of CCC

There are three temperature intervals: no. 1 from 125 to 100 °C, no. 2 from 100 to 25 °C and no. 3 from 25 to 20 °C. Heat loads of intervals are 62.5, 412.5, and 12.5 kW, respectively. Segments 1, 2, 3 of CCC are also shown in Figure 13.5.

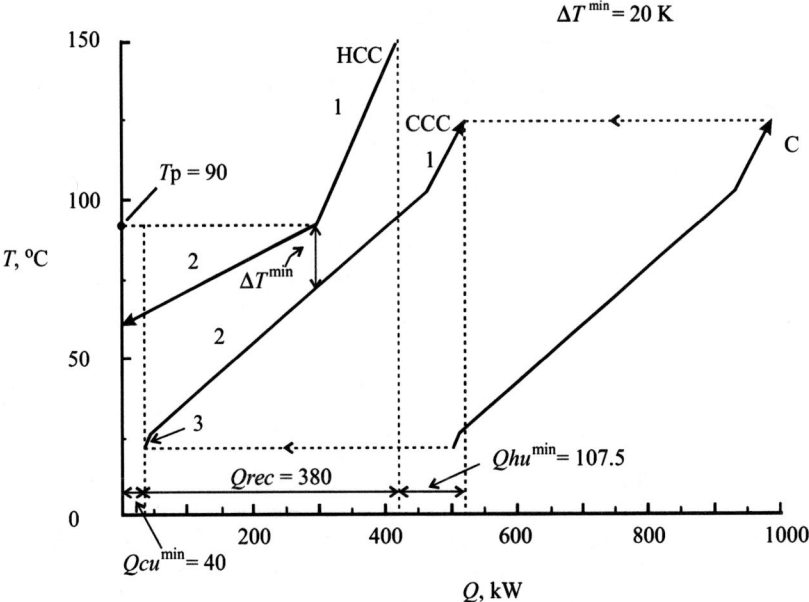

Figure 13.5 The CC plot for Example 13.1.

Plotting and pushing the curves

Next we draw the curves. Due to relative Q scale the cold composite is first drawn as line C in Figure 13.5 and then pushed horizontally to the position marked as CCC. Further shifting is not possible since the vertical distance becomes equal to $\Delta T^{min} = 20$ K. The pinch is at 90 °C, that is, at the inlet temperature of stream h_2. The minimum load of the hot utility (Qhu^{min}) is 107.5 kW and that of the cooling utility (Qcu^{min}) 40 kW.

The CC plot clearly illustrates the influence of ΔT^{min} on heat recovery. Similarly to the two-stream heat exchange case this parameter has a strong effect on the total cost of the heat exchanger network—this will be considered in depth in Chapters 15 and 16.

Some remarks should be made here with reference to the literature on heat integration. Often, the CC plot is presented using the so-called shifted temperature scale ($\bar{T}$). There are various options for re-scaling temperatures. Most often, temperatures of the hot streams are reduced by ΔT^{min} keeping the real (unchanged) temperatures of the cold streams. Also, it is possible to add ΔT^{min} values to the cold stream temperatures without changing the temperatures of the hot streams. Lastly, temperatures of the hot streams are decreased by $0.5 \, \Delta T^{min}$ and those of the cold streams increased by this value.

With a shifted temperature scale the composite curves touch each other at the pinch. The pinch value is read in the shifted scale. To recalculate it into real scale temperatures the rules of temperature shifting have to be known. Alternatively,

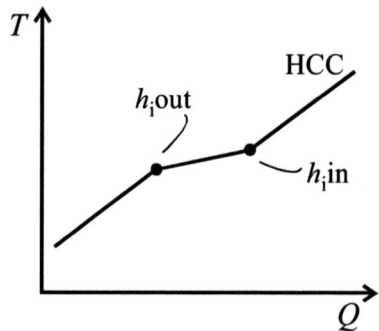

Figure 13.6 Illustration of noses at HCC.

the pinch temperature is presented using the notation: $T1$ (temperature of hot streams in real scale)/$T2$ (temperature of cold streams in real scale), for instance 90/70 in the example.

The use of shifted temperatures is not obligatory when constructing the CC plot. However, the shifted scale is very useful in heat integration calculations since it causes the ΔT^{min} value to be implicitly embedded into the temperature scale. Hot and cold streams within the same shifted temperature range can exchange heat since condition (13.6) is fulfilled. This makes formulation of the models and presentation of other approaches to the heat recovery problems easier. From this point we will apply shifted temperatures and use symbol T for them if not stated otherwise.

Analysis of the CC plot reveals that the pinch can occur only at the inlet temperature of one hot or cold stream. For the HCC, the downward "nose" occurs at the inlet temperature because its heat load increases while at the outlet temperature the nose is upwards since heat load decreases—see Figure 13.6. The opposite effects occur for the CCC. The pinch can occur at the downward nose of the HCC or upward nose of the CCC, hence only at the inlet temperature of a hot stream or of a cold one.

Recalling the procedure for the CC method we can now modify it by creating temperature intervals only at inlet temperatures. However, the decrease of computation burden is rather small.

The observation that the pinch can occur at inlet temperature only is important since it reduces the number of potential pinches. Duran and Grossmann (1986) made use of this fact when developing a fast procedure for MER. To explain it let us note that standard heat balances cannot be used to solve MER problems since they do not account for thermodynamic conditions (13.6) or (13.7). Let QH denote the total heat load of all NH hot process streams calculated from Equation (13.9) and QC the total heat load of all NC streams from Equation (13.10):

$$QH = \sum_{i=1}^{NH} CP_i \, \Delta T_i \qquad (13.9)$$

$$QC = \sum_{j=1}^{NC} CP_j \, \Delta T_j \qquad (13.10)$$

Seemingly, for QH higher than QC the difference $(QH - QC)$ shows the load of cold utility. In the opposite case the difference $(QC - QH)$ would be the hot utility load. This is only true for certain problems that have no (internal) pinch and are called "threshold" problems. Such threshold problems are created if we divide a pinched problem at the pinch (if only a single pinch exists). Both above-pinch and below-pinch subproblems are threshold ones. Thus, the MER problem for them can be solved using heat balances such as Equations (13.9) and (13.10). This is the basis of the Duran and Grossmann (1986) procedure, which can be presented as follows:

1. Assume the first (the highest or the lowest) inlet temperature of the process stream to be the potential pinch.
2. Divide the streams at the potential pinch and calculate heat loads on utilities as for the threshold problem, that is, from Equations (13.9) and (13.10).
3. Repeat points 2 and 3 for each inlet temperature.
4. Find the highest values of QH and QC calculated so far; the inlet temperature for these values is the real pinch, and the values (QH, QC) are the minimal utility loads for the problem.

To the authors' best knowledge this is the simplest procedure for calculating the MER. However, the disadvantage is that the method does not account for multiple pinches.

It is of importance that the rule "pinch at inlet temperature" is valid only if simplified Equation (13.4) can be applied. In other cases the pinch can appear in other temperatures similarly to "internal" pinches in the two-stream case—see Figure 13.3. Thus, the division at inlet (and outlet) temperatures is insufficient. To account for temperature-dependent cp in MER calculations, as well as in HEN design in general, stream segmentation is commonly advised. The streams are segmented into temperature ranges such that the use of condition (13.4) for the segments does ensure sufficiently accurate results. The stream segmentation mechanism has a great advantage in MER calculations since it does not call for changes in standard solution approaches to MER (or MUC) problems. It is sufficient to treat segments as original streams. There is also an alternative way of dealing with nonlinear enthalpy dependence on temperature by applying rigorous thermodynamic models. Castier and Querioz (2002) suggested a procedure, which works in frames of the problem table method. We will describe it in the following.

Finally, let us direct the reader's attention to a very important observation that can be drawn from a CC plot: the HCC above the pinch should not exchange heat with the CCC below the pinch.

In the opposite case the loads on utilities would increase. Assume that a certain amount of heat in the hot streams above the pinch has been used to heat up cold

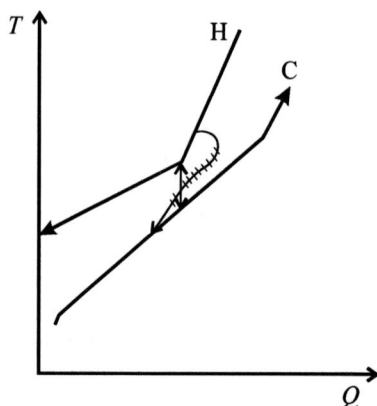

Figure 13.7 Illustration of the rule "no heat exchange via pinch" at the CC plot.

streams below the pinch. Due to thermodynamic constraint (13.6) this load cannot be compensated by hot streams below the pinch, since below-pinch hot streams cannot be used to heat up cold streams above the pinch. Hence, we would be forced to use an additional heating and cooling utility. This observation, illustrated in Figure 13.7 and called "no heat exchange across pinch" is probably one of the most important rules in pinch technology. We will also present an alternative (and more common) explanation of the rule when presenting the second approach for MER—the problem table (Pr-T) method.

13.3. PROBLEM TABLE (PR-T) METHOD

The Pr-T approach does not require drawing a plot. Similarly to the CC method it is limited to MER calculations, that is, the MUC problem cannot be solved with the original version of the Pr-T approach first presented in Linnhoff and Flower (1978). Here, we will explain the procedure using heat cascade representation. This representation is very useful for illustrating optimization approaches, particularly the transshipment model, addressed in Chapter 15. Also, we will use nomenclature relevant for the transshipment model.

To illustrate the Pr-T method we will employ Example 13.1. In the Pr-T method the streams should be in shifted scale. Here, we apply the most common shifting rules that is, temperature decrease of hot streams by ΔT^{min}—see Figure 13.8. The first step in the Pr-T method (data preparation) is division into temperature intervals (TI) at inlet temperatures of the streams (Figure 13.8). Notice that, due to the shifted temperature scale, the intervals are common for both hot and cold streams in contrast to the CC procedure.

Such division is sufficient to ensure that a pinch does not occur inside any TI. Obviously, the division at both inlet and outlet temperatures can be performed too. Figure 13.8 illustrates division into temperature intervals and, also, in

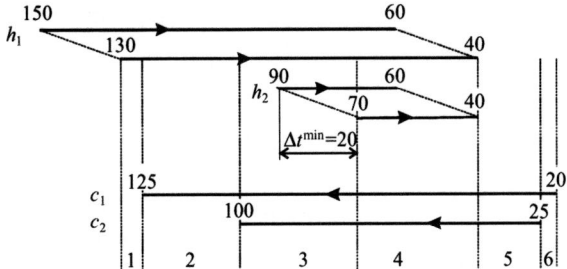

Figure 13.8 Division into temperature intervals and shifts of temperatures for Example 13.1.

Table 13.2 Problem table (Pr-T) method for Example 13.1

Interval k	Stage 1	Stage 2
1	$QH_1 = 10$; $QC_1 = 0$; $\Delta Q_1 = 10$	$Qhu^{min} = R_0 = R\ (1) = 107.5$
130–125	$R_1 = 0 + 10 = 10$	$R_1 = 107.5 + 10 = 117.5$
2	$QH_2 = 50$; $QC_2 = 62.5$; $\Delta Q_2 = -12.5$	
125–100	$R_2 = 10 - 12.5 = -2.5$	$R_2 = 117.5 - 12.5 = 105.0$
3	$QH_3 = 60$; $QC_3 = 165$; $\Delta Q_3 = -105$	
100–70	$R_3 = -105 - 2.5 = -107.5$	$R_3 = 105 - 105 = 0.0$ (pinch)
4	$QH_4 = 300$; $QC_4 = 165$; $\Delta Q_4 = 135$	
70–40	$R_4 = -107.5 + 135 = 27.5$	$R_4 = 0.0 + 135 = 135.0$
5	$QH_5 = 0$; $QC_5 = 82.5$; $\Delta Q_5 = -82.5$	
40–25	$R_5 = 27.5 - 82.5 = -55$	$R_5 = 135 - 82.5 = 52.5$
6	$QH_6 = 0$; $QC_6 = 12.5$; $\Delta Q_6 = -12.5$	
25–20	$R_6 = -55 - 12.5 = -67.5$	$Qcu^{min} = R_6 = 52.5 - 12.5 = 40.0$

Table 13.2 each row corresponds to one interval (here we divide at both inlet and outlet temperatures). The intervals are arranged in descending order, from the highest temperature interval (top interval) to the lowest one (bottom interval), and numbered from 1 to N ($N = 6$ in the example).

The calculations in the Pr-T approach are performed in two stages. In the first stage, starting from the first interval we calculate for each interval k:

- Total heat load of the hot streams in the interval

$$QH_k = \sum_{i \in H_k} CP_i\, \Delta T_i \qquad (13.11)$$

where H_k denotes set of hot streams in interval k and ΔT_i denotes temperature range of stream i in interval k
- Total heat load of the cold streams in the interval

$$QC_k = \sum_{j \in C_k} CP_j\, \Delta T_j \qquad (13.12)$$

where C_k denotes set of cold streams in interval k and ΔT_j denotes temperature range of stream j in interval k
• The difference of the loads

$$\Delta Q_k = QH_k - QC_k \qquad (13.13)$$

• The residual, which is the sum of the difference and residual from the interval $(k-1)$—notice that in the first stage residual R_0 is equal to 0.0

$$R_k = R_{k-1} + \Delta Q_k \qquad (13.14)$$

The results of the calculations for the example are shown in the second column of Table 13.2. It is important to observe that some residuals are negative. In the example they are negative in intervals 2, 3, 5, and 6. The residuals are interpreted as heat loads that are in excess in higher temperature intervals, and hence, can be used in lower temperature intervals. Thus, negative values have no meaning. They have to be nonnegative. To make the solution thermodynamically feasible let us pick up the lowest negative value of the residual. For the example at hand it is $R_3 = -107.5$ kW—see Table 13.2, interval no. 3. Then we add positive heat load of this residual (107.5 kW) to the top interval as residual R_0. Note that this heat load is equivalent to the load of external heating utility Qhu.

The calculations for correcting the residuals are performed in the second stage. They are similar to those in the first stage and require only minor change: residual R_0 should be equal to the absolute of the lowest negative value of residual from the first stage (107.5 kW in the example). Since values of QH_k, QC_k, and ΔQ_k are identical to those in the first stage we do not need to calculate them again. Only values of residuals R_k need recalculation. The third column of Table 13.2 contains the results for the example. One can observe that there is a nonzero residual from the last interval ($R_6 = 40$ kW). This residual is equivalent to the heat load of cooling utility—Qcu. Also, it is of importance that there is at least one zero residual (R_3 in the example). It means that no heat from a higher temperature interval is available in lower temperature intervals. For the example, no heat from interval no. 3 is available in interval no. 4. The lower temperature border of the interval with zero residual outflows is the pinch (in the example, interval no. 3 with shifted temperature of 70 °C).

More recently some modifications of the Pr-T method that slightly simplify calculations were developed by Salama (2005).

The results from the third column of Table 13.2 can be presented as a heat cascade shown in Figure 13.9.

To summarize, the major results from the Pr-T approach are the minimum hot utility load $Qhu^{min} = R_0$, the minimum cold utility load $Qcu^{min} = R_N$ and the pinch temperature (in shifted scale).

These utility loads are the minimal ones since if smaller (positive) R_0 would be used then at least one negative residual has appeared. Consider the use of heating a utility load larger than the minimum by some value, for instance

$$\overline{T}, \degree C \qquad Qhu^{\min} = 107.5 \text{ kW}$$

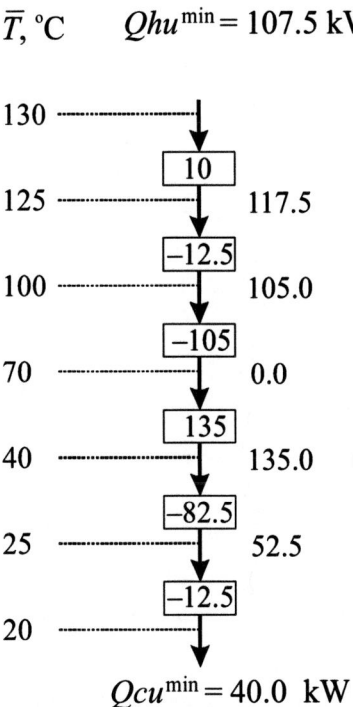

$$Qcu^{\min} = 40.0 \text{ kW}$$

Figure 13.9 Heat cascade representation for Example 13.1.

$Qhu = Qhu^{\min} + D$. D units of heat are in excess and have to "cascade" down across the pinch increasing $Qcu^{\min}$ by D units. Linnhoff et al. (1979) called this effect "double penalty" since it yields an increase of both utility loads.

From this observation Linnhoff et al. (1979) formulated the rule called here "no heat exchange across pinch": *The exchange of heat across pinch results in excess use of both heating and cooling utilities.*

Notice that it is an alternative version of the rule drawn out from the CC plot (Figure 13.7). The rule can be visualized in the so-called "bubble" plot, which is, in fact, a simplified version of the heat cascade representation from Figure 13.9. The bubble plot in Figure 13.10 shows only $Qhu^{\min}$ and $Qcu^{\min}$. Also, one can easily observe that there is "heat surplus" below the pinch and "heat deficit" above the pinch. Notice also that subproblems above and below the pinch need only one type of utility, not both. Hence, they are threshold problems. Notice that the method by Duran and Grossmann (1986) has made use of this fact.

Needless to say, application of cooling utilities above the pinch would increase hot utility usage. Similarly, application of hot utilities below the pinch increases cold utility consumption.

It is important to note that the rule "no heat exchange across pinch" applies to both direct and indirect heat exchange. Thus, when extracting data from an

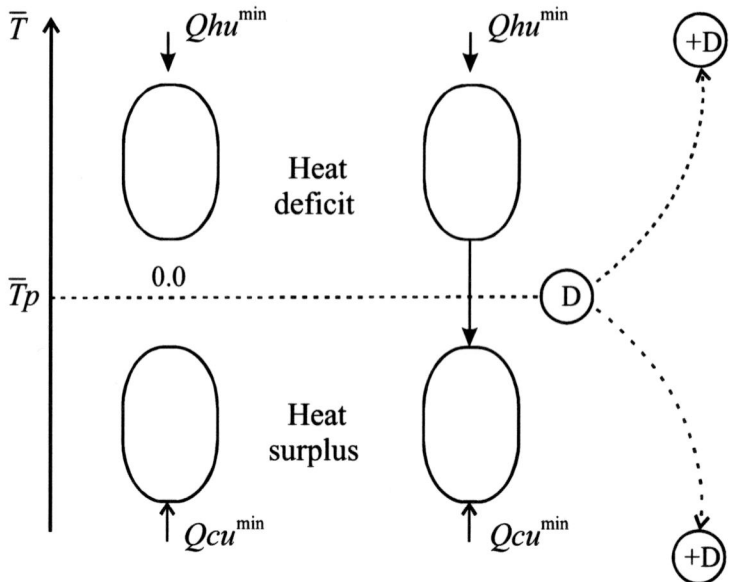

Figure 13.10 Bubble plot explaining the "no heat exchange across pinch" rule.

existing plant to heat recovery calculations the designer should not take into account existing mixers and splitters since they may not obey the rule. Only outlet and inlet stream parameters from processes have to be applied.

The question arises on how to ensure that the rule is met in process design. It can be directly achieved by dividing the streams at the pinch as shown in the bubble plot. The streams (parts of streams) above the pinch are separated from those below the pinch. Two independent, in terms of heat exchange, sets of streams are created. Designs are performed for them, and only at the final step are the two designs merged—see Chapter 18. Such division is treated as the rule in pinch technology though it can be circumvented, as we will show in Chapter 15.

13.4. GRAND COMPOSITE CURVE (GCC) PLOT

Though the CC and Pr-T methods are sufficient to solve MER problem, we present here a third approach, usually referred to as the grand composite curve (GCC plot). This is the graphical method that is often explained on the basis of the CC curve or the cascade representation from Pr-T calculations. The advantage of GCC over CC and Pr-T is that under some assumptions it is more convenient to use to solve the MUC problem. Also, the GCC is often applied to analyze heat and power integration in utility subsystems and in a total site—see Biegler et al. (1997) and Smith (2005) for more details. The seminal contributions

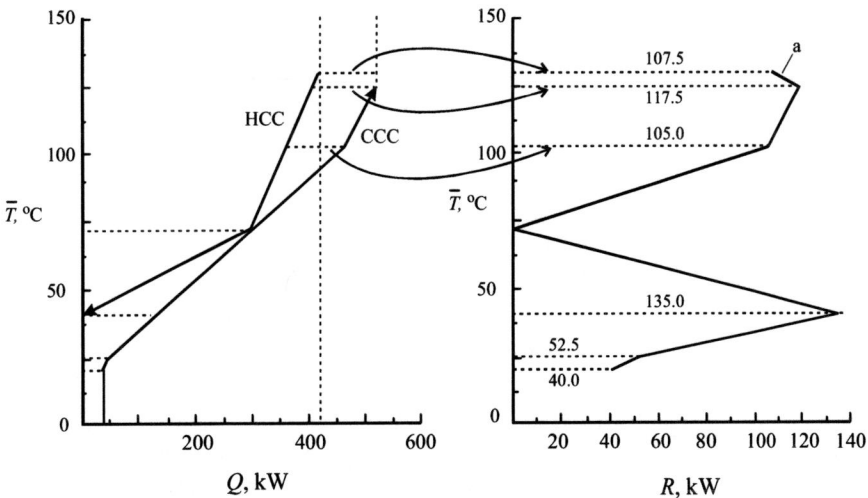

Figure 13.11 Construction of the GCC plot from the CC plot for Example 13.1.

are those by Townsend and Linnhoff (1983a, 1983b) on appropriate process placement in the context of total site. The "process placement" denotes process temperature range that allows heat integration.

As we have mentioned, the GCC plot can be constructed from the CC plot—this is illustrated in Figure 13.11 for data of Example 13.1. The procedure is explained in the following. Here it is important to note that the GCC presents the dependence of residual heat flow on temperature.

Comparing the residual values in column 3 of Table 13.2 and the values at the abscissa of the GCC plot at the same shifted temperatures (shown by dashed lines at the GCC plot) one notices that they are identical. Thus, the GCC plot can be interpreted as the curve showing heat cascading vs. shifted temperature. To draw the curve one uses residuals from Pr-T calculations, for instance from Figure 13.9, and connects the points: residual–shifted temperature by straight lines. The piece-wise straight lines obtained show heat cascading over the temperature range.

The construction shown in Figure 13.11 proves that the residuals can be found from the CC plot. They are equal to differences between the heat load of CCC and HCC at the same temperature. Hence, all the MER approaches presented up to this point—CC plot, Pr-T method and GCC plot—can be applied to extract the same information including the residuals. However, some information is better visualized in certain approaches and not so clearly visible in the others.

The GCC plot has some interesting features that help the designer to choose the best utilities from among available ones, that is, in solving the MUC problem. A single straight-line segment of the GCC from one temperature to the adjacent one in heat cascade has a load equal to the difference of the residuals for these temperatures. For instance segment "a" in Figure 13.11 has a load of 10 kW, that

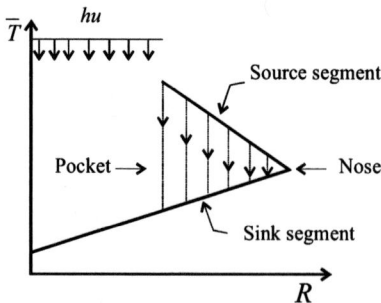

Figure 13.12 Internal heat recovery at GCC plot.

is, $R_1 - R_0 = 117.5 - 107.5$. If the difference is positive the segment is directed to the right hand side. In the opposite case, the segment is directed to the left hand side. An increase of residuals in some subsequent intervals means that there is an excess of heat—the intervals are heat donors (sources). A decrease indicates that the intervals are heat sinks. Source segments can supply heat to sink segments lying just below them. This indicates internal heat recovery between process streams. In GCC plots such source segments and sink segments create right side noses or pockets—see Figure 13.12. We can find the pockets and, in consequence, all internal heat recovery options. The rest of the heat to sinks above the pinch has to be added by heating utilities and the rest of the heat from sources below the pinch has to be removed by cooling utilities.

Note that up to this point we have used one heating utility of the highest temperature and one cooling utility with the lowest temperature.

Let us solve the MUC problem with the GCC plot. Formally, we can formulate this as an optimization task as follows.

Minimize UC defined by Equation (13.11) subject to conditions (13.15) and (13.16)

$$Qhu^{\min} = \sum_{m=1}^{NHU} Qhu_m \qquad (13.15)$$

$$Qcu^{\min} = \sum_{n=1}^{NCU} Qcu_n \qquad (13.16)$$

Parameters $Qhu^{\min}$ and $Qcu^{\min}$ are the minimal loads from the solution of the MER problem. To solve the problem with the GCC plot we have to assume the following costing rules:

(a) unit prices of hot utilities decrease with descending inlet temperature, that is,

$$phu_1 > phu_2 > \cdots \text{ if } T_{hu,1}^{\text{in}} > T_{hu,2}^{\text{in}} >$$

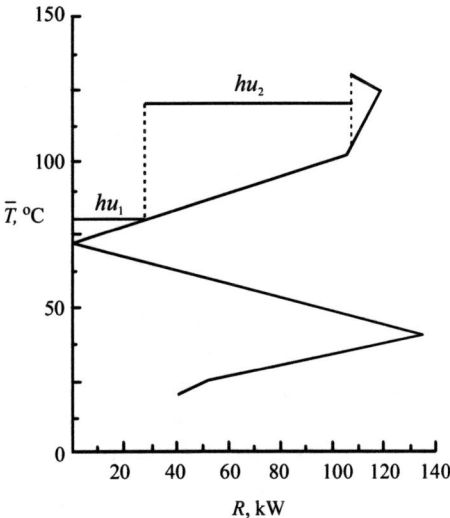

Figure 13.13 Solution of MUC problem for Example 13.1 at GCC plot.

(b) unit prices of cold utilities increase with descending inlet temperatures, that is,

$$pcu_1 > pcu_2 > \cdots \text{if } T_{cu,1}^{in} < T_{cu,2}^{in} <$$

The costing rules dictate that minimization of utilities cost can be achieved by maximizing in sequence the use of a less costly utility. Hence, for hot utilities we begin with the coldest one and maximize its load. We continue the procedure with the next utility from the sequence checking condition (13.15). For cold utilities the procedure is similar except for the fact that we should start with the hottest (cheapest) utility and take care of condition (13.16).

To illustrate the method we employ Example 13.1 once more. Assume that we have two kinds of steam available: hu_1 with inlet temperature of 80 °C and hu_2 with inlet temperature of 100 °C, both in shifted scale.

In the GCC plot we cut off pockets above the pinch—see Figure 13.13. Then, we draw the line (utility profile) for hu_1 at $\bar{T}_{hu,1}^{in}$. The maximum load we can assign to the cheaper utility hu_1 is 35 kW. Further increase is limited by the point where the utility profile line touches the GCC. This point is called the "utility pinch" to differentiate it from the process pinch, that is, the pinch caused by process streams. We cannot go further in order to ensure thermodynamic conditions (13.6). Since in this example there are two hot utilities, the rest of the heat, that is, $Qhu^{min} - 35 = 72.5$ kW, has to be supplied by hu_2.

13.5. SPECIAL TOPICS IN MER/MUC CALCULATIONS

13.5.1. *Nonpoint Utilities—Flow Rate Minimization*

In the example above we employed steam as the hot utility. This is the utility that changes phase but its temperature does not change. It is called the point utility since according to Equation (13.4b) its load is calculated with ΔT set at 1.0. Generally, the point utility is such that its temperature change is smaller than the temperature interval range within which the utility is applied. Nonpoint utilities have temperature changes larger than the appropriate temperature interval range. Thus, the designer does not know before solving the MUC problem whether some utilities are of the point type or not.

The nonpoint utilities cause some difficulties even when some of the optimization methods presented in Chapter 15 are used to solve the MUC problem. They are the result of the condition that their outlet temperatures should be fixed in data. Often, outlet temperatures are fixed by the designer at lower (for hot utilities) or higher (for cold utilities) values within an admissible range. Although the cost of utilities is most often based on their heat loads their flow rates should also be considered by the designer since they also influence the cost of transportation. For given minimal heat loads, the flow rates can be minimized by assuming the lowest (for hot utilities) or the highest (for hot utilities) admissible outlet temperature. This, however, increases the probability of fixing an inappropriate nonpoint utility outlet temperature. The explanation is as follows.

It clearly results from the rule "no heat exchange across pinch" that the outlet temperature of a hot utility should be above the process pinch and that of a cold utility below the pinch. This is a necessary condition for all kinds of utilities. Consider flue gas, which features a very broad temperature range from the flame temperature at the inlet to the stack temperature at the outlet. Flue gas is often applied to heat up high-grade streams in, for example, refineries. Thus, it is often possible that the process pinch will be above the outlet stack temperature. By forcing this outlet temperature the designer will increase the heat load on the flue gas above its minimal value. This effect is illustrated in Figure 13.14. The excess heat of gas denoted by ΔQ in the figure depends on pinch temperature and gas outlet temperature. To diminish or eliminate surplus heat load the outlet temperature has to be increased up to pinch temperature or to an even higher value. The gas at such a temperature can be applied to other duties. The problem was addressed in depth by Smith and Delaby (1991).

However, the above obvious condition on the outlet temperature of a nonpoint utility is insufficient though necessary. There are problems where even though the outlet temperature meets the necessary condition it can cause an excess use of utility heat if the designer decides to keep this temperature. To show the effect, consider the above pinch part of the GCC in Figure 13.15.

Let the hot utility hu have inlet temperature $\bar{T}^{in}$ and the outlet $\bar{T}^{out}$. Notice that $\bar{T}^{out}$ is higher than the pinch temperature. Line "hu" in Figure 13.15 shows the utility profile, that is, its temperature change. The line crosses the GCC

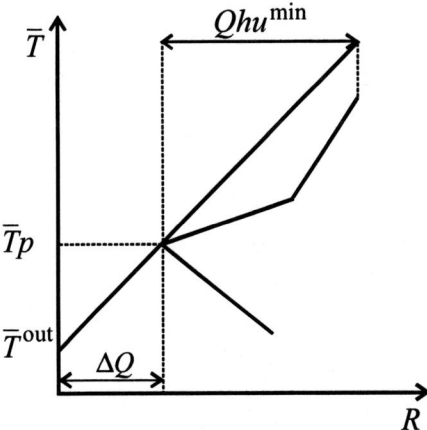

Figure 13.14 Illustration of the effect of the outlet temperature of a hot utility lower than the process pinch temperature.

and lies beneath it within a certain temperature range. This indicates that hot utility temperatures are lower in this range than temperatures of process streams, thus violating condition (13.6). To satisfy the condition and keep the outlet temperature at a given $\bar{T}^{out}$ value one has to increase the hot utility load above the minimum. The GCC curve has to be pushed horizontally right till condition (13.6) is met—dashed line in Figure 13.15. The minimal excess of heat load is achieved if utility line "*hu*" just touches the shifted GCC curve. The point where both lines touch each other is the utility pinch since the temperature difference between the utility and process streams equals ΔT^{min}. By following such a procedure the designer accepts a surplus heat utility load but keeps outlet temperature at the value fixed in the data.

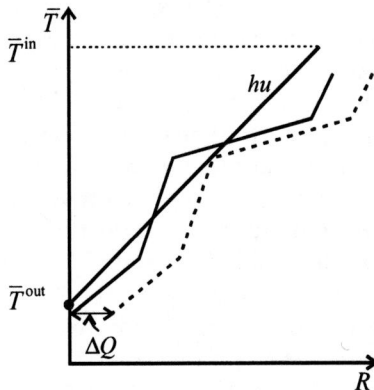

Figure 13.15 Illustration of excess use of a utility for the outlet temperature of a hot utility above process pinch.

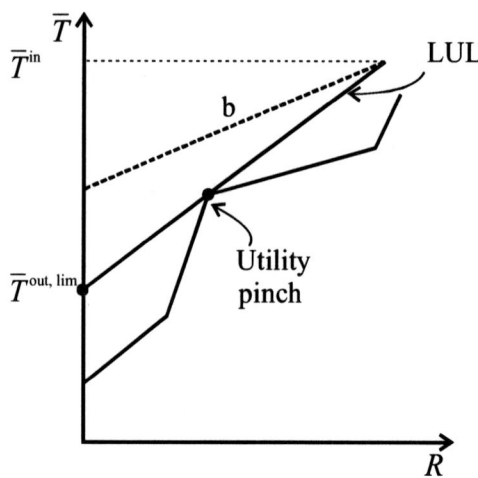

Figure 13.16 Limiting utility line in a GCC plot.

In order to preserve the minimal heat load of the utility it is necessary to increase the outlet temperature. Notice that the flow rate of the utility is proportional to the reciprocal of the tangent of the angle created by utility line—Figure 13.16. Thus, to minimize the flow rate we should maximize the angle. This is achieved if the utility line touches the GCC curve at at least one point. We can draw the utility line which starts at the utility inlet temperature and touches the GCC without crossing it—see line LUL in Figure 13.16. This is the limiting utility line (LUL) that minimizes the utility flow rate and ensures the minimal load on the nonpoint utility. The limiting utility line crosses the $\bar{T}$ axis at some limiting outlet temperature $\bar{T}^{out,lim}$. It is the highest thermodynamically feasible temperature for heating utilities and the lowest for cooling utilities. Jeżowski and Jeżowska (2002) showed that the concept behind the LUL is identical to that applied for calculating the minimal freshwater flow rate within the water pinch method (see Chapter 20).

Minimization of flow rate causes at least one utility pinch. Such pinches have some negative influence on the investment cost of heat exchanger network since they can increase heat transfer surface area and the number of heat exchangers. Hence, to eliminate utility pinches the designer can further increase (for hot utilities) or decrease (for cold utilities) the outlet temperature by increasing the flow rate, for instance to the position marked "b" in Figure 13.16.

Summing up, for nonpoint utilities the designer has some alternatives: to increase utility load fixing the assumed outlet temperature, to minimize flow rate keeping the minimal load but changing the outlet temperature, and to accept flow rate increase with minimal heat load but without utility pinches. All options should be considered at the targeting stage where the total cost of the heat exchanger network is minimized.

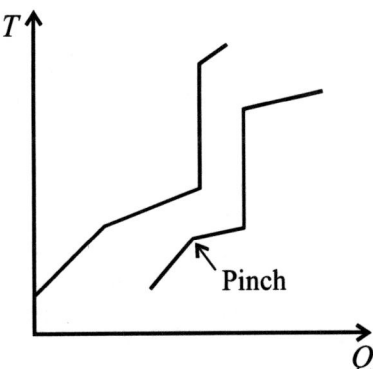

Figure 13.17 CC plot for the "empty" temperature interval.

13.5.2. Discontinuous Composite Curves

Lakshmanan and Fraga (2002) found a specific case where all the methods for MER calculations presented up to this point could yield mistaken results. The case occurs when there exists a discontinuity of composite curves in the CC plot caused by the "empty" temperature interval, that is, the interval where there are no process streams at all. Then, both composite curves feature vertical segments at this interval as shown in Figure 13.17.

While decreasing $\Delta T^{\min}$ by moving the curves horizontally we meet the point where vertical segments "clash" and we are not able to continue the movement. The $\Delta T^{\min}$ at the clash is called "critical $\Delta T^{\min}$". Any decrease of $\Delta T^{\min}$ below the critical value will not produce a decrease of utility heat loads. This observation is helpful in determining the optimal $\Delta T^{\min}$ value since critical $\Delta T^{\min}$ should be set as the lower bound on $\Delta T^{\min}$. Moreover, the Pr-T approach and optimization models, which do not provide visualization, will give mistaken results for $\Delta T^{\min}$ values lower than the critical one. More recently Salama (2006) developed a modified graphical approach based on the CC method that directly accounts for discontinuous composite curves.

13.5.3. The Use of Rigorous Thermodynamic Models for Calculating Heat Loads

The use of rigorous thermodynamic models is the alternative to stream segmentation in an attempt to ensure accurate calculations of heat loads. This alternative relies on using rigorous thermodynamic models instead of the simple Equation (13.4) for each segment. Castier and Querioz (2002) developed the procedure for MER calculation, which embeds rigorous thermodynamic models in the framework of the PT approach. The method requires iterations and also solution of optimization problems.

First, temperature intervals are created at inlet stream temperatures, similarly to the original Pr-T method. In each TI the following optimization problem has to be solved:

$$\min Q(T) = \sum_{i \in H} Q_i - \sum_{j \in C} Q_j \qquad (13.17)$$

where H and C are sets of hot and cold process streams in the interval; Q is the heat load of streams calculated by a rigorous model, and T is the temperature in shifted scale bounded by the highest and the lowest temperature of the interval.

Solution of this simple task with single optimization variable (T) is necessary to check whether a potential pinch exists within the interval. According to Castier and Querioz (2002) there is no such pinch in the interval if "$Q(T)$ is a monotically decreasing or increasing function of T in the interval, or if its only stationary point in the interval is a maximum or a saddle". In the opposite case $Q(T)$ has at least one minimum in the interval. Hence, it is necessary to divide the interval further at this minimum temperature.

The approach requires a lot of calculations including solution of the optimization problem with a robust optimizer. Segmentation seems a more efficient and straightforward solution.

13.6. SUMMARY AND FURTHER READING

In this chapter we have addressed in depth three basic methods for maximum energy recovery calculations: the CC plot, the Pr-T approach together with heat cascade and the GCC plot. All these methods have identical limitations. They do not account for restricted matches, or for forbidden ones in particular. By forbidden match between hot stream "h" and cold stream "c" we mean that both streams cannot exchange heat because of, for example, safety reasons. Such cases can be accounted for in the optimization-based methods addressed in Chapter 15.

Also, insight-based methods are not able to solve MER problems with the so-called "dual" matches. This term is applied to matches between two hot streams or between two cold streams. Such seemingly "strange" heat exchange can, however, decrease utility usage if there are forbidden matches which cause a decrease of heat recovery (in comparison with the case of no forbidden matches). Dual matches can also be accounted for by optimization approaches (see Chapter 15).

The original CC and Pr-T methods are the ones aimed at solving the MER problem. The GCC plot can be applied to solve a more general MUC problem though the solution is restricted to cases where costing rules based on temperatures apply. Mathematical approaches are able to solve MUC problems without this assumption. However, the importance of the GCC plot lies in its wide application to heat and power integration issues, particularly to the design of utility subsystems. The concept has been modified and extended in several works. For instance, Gorsek et al. (2006) applied site sink source profiles while

Anantharaman et al. (2006) developed energy level composite curves. Both are based on the GCC plot.

In fact, all the approaches presented in this chapter are based on the same concepts and give similar information. The choice of any one of them depends on the ease of its use in specific applications.

The reason why we have explained them in depth is because they provide deep insights into heat integration problems. In fact, models of optimization approaches have made use of the basic observations from them. Hence, it is of importance to understand the concepts behind the CC plot, the Pr-T approach and the GCC plot. This is necessary to understand heat integration issues in the following and, also, to make explanation of optimization-based approaches easier.

There exists also the nonoptimization approach, which can to some extent deal with multiple utilities, and forbidden and dual matches presented, for example, in Jeżowski and Friedler (1992). However, the approach is not general. Forbidden matches also can be dealt with in the PT-based approach by O'Young et al. (1988) although calculations are involved and the optimization approaches explained in Chapter 15 are easier to use.

As we have mentioned above, heat integration in batch processes is omitted from this book. The reader is referred to the following papers: Obeng and Ashton (1988), Kemp and Deakin (1989a, 1989b, 1989c), Kemp (1990, 1991), Corominas et al. (1994), Zhao et al. (1998a, 1998b), Adonyi et al. (2003), Bieler et al. (2004), Majozi (2006), Pourali et al. (2006).

14 Targeting and supertargeting in heat exchanger network design

14.1. TARGETING STAGE IN OVERALL DESIGN PROCESS

Targeting is a relatively new stage in a complex, usually iterative, process of designing process systems. The names "targeting" and "targets" (and also super-targeting) appeared in the literature on heat integration in the 1970s and were introduced by Prof. B. Linnhoff and his co-workers (UMIST, UK). The works addressed pinch technology approaches and thus targets are often thought of as specific features of these methods. However, targets can be also calculated by other, usually more systematic approaches. Also, targets are useful in designing other process systems, for example, mass exchanger networks, see El-Halwagi (1997, 2006) and reactor networks, see Mehta and Kokossis (2000), Ismail et al. (2001), Linke and Kokossis (2003). A special case of MEN, that is, water networks, will be addressed in Chapter 20.

The major objective of the targeting stage is to calculate the most important features (commonly called targets) such as total cost, number of apparatus and so on, of an optimal solution of a process system, subsystem or even a single process or apparatus, ahead of strictly "design stage" where the structure and parameters of the solution are finally determined. Calculation of targets is easier than the design. Solution techniques to determine targets are relatively simple and time-efficient even for large-scale problems. At the moment, designing methods for large process systems and for subsystems are time-consuming, require some assumptions, and do not ensure a global optimum or even a good local optimum. Due to these difficulties, calculating many alternative designs is not possible with available resources and in a realistic time period. However, very often we are able to calculate targets for various alternatives using available targeting algorithms. Since the results of the targeting stage provide the designer with sufficient information on important performance measures of the final design, she/he can then screen alternatives.

The targeting stage is thus the "pre-design" stage, which should be performed before the design step. Notice that the targeting stage has various applications:

- It can be applied to estimate the economic performance of a final design. The designer can stop the overall design process after the targeting step if the targets are not satisfactory. For instance, if expenditure for retrofitting a HEN in an existing plant is high, the economical benefits of revamping the system can be viewed as marginal.
- Often targets provide information for the design stage. In many sequential design methods, the targeting stage is necessary since it provides data for the subsequent design stage. Even in simultaneous design approaches the targets can be used to reduce the number of possible solutions, for instance to reduce a superstructure.

- In some cases, particularly in small "easy" problems, results from targeting are sufficient to perform the design by "hand," without any systematic method or computer aid.

Similarly to design approaches, targeting methods can be divided into two wide classes: sequential and simultaneous. Though the latter are more general and ensure more accurate results, sequential approaches are usually simpler and give the designer a deeper understanding of the problem. Pinch technology targeting is mature and also the most popular sequential approach for calculating targets for HEN design and for heat integration. Though there exist other systematic methods we will start with a description of targeting methods from pinch technology. The reason is that it gives problem understanding, and many systematic approaches have made use of this. Also, pinch technology imposes certain "rules and standards" that have been followed by many researchers though they are not always rigorous and may even be unnecessary since they may prevent better solutions being reached.

14.2. BASIS OF SEQUENTIAL APPROACHES FOR HEN TARGETING

The main objective of targeting is to estimate the goal function value of an optimal solution as well as some of its crucial parameters. Thus, first we have to adopt performance indices of HEN. Most often this is the total annual cost (TAC). TAC is calculated as the sum of cost of utilities and the investment cost of apparatus:

$$TAC = UC + r\,CAP \tag{14.1}$$

Basic pinch technology methods for calculating MER were presented in Chapter 13. Under some assumptions the cost of utilities (UC) is equal or proportional to MER. It is important to note that both MER and the cost of utilities depend on ΔT^{min}. Hence ΔT^{min} is the decision variable in calculating heat energy targets. It affects capital cost as well.

The capital cost of heat exchangers is usually taken into account when estimating HEN capital cost (CAP). It means that the cost of piping, splitters and mixers is not included. There are some approaches that try additionally to include the cost of pumps and pumping fluids, for example, Ahmad and Polley (1990) and Polley and Shahi (1991). Also, environmental issues have been included by Li and Hua (2000), Ya and Shonnard (2000) and Mata et al. (2003). Here, we limit ourselves to the capital cost of heat exchangers only since the majority of targeting methods developed to date have such limitations. The cost of a single heat exchanger is approximated by the following simple formula:

$$CAP = \alpha + \beta A^{\gamma} \tag{14.2}$$

The first term in Equation (14.2) is the fixed charge while the second correlates the capital cost of apparatus and its surface area. Parameters α, β, γ depend on the type of heat exchanger, construction materials and operating pressure. Notice that, these parameters are not treated as decision variables and designing approaches in most HEN targeting. Works by Hall et al. (1990) and Jegede and Polley (1992) are the exceptions, although the contributions do not allow for full optimization.

The investment cost of HEN consisting of NU heat exchangers ($e = 1, \ldots, NU$) is given by

$$CAP = \sum_{e=1}^{NU} (\alpha_e + \beta_e A_e^{\gamma_e}) \qquad (14.3)$$

In order to simplify the problem, the pinch technology approach imposes the conditions that:

1. All heat exchangers are of identical type, that is, all units feature the same parameters α, β, γ—this limitation was relaxed to a certain extent in Hall et al. (1990) and Jegede and Polley (1992).
2. The total heat transfer surface area of HEN is evenly distributed among heat exchangers, that is, each unit has the same area.

Under these assumptions the capital cost of HEN can be calculated from:

$$CAP = NU \left(\alpha + \beta \left[\frac{AT}{NU} \right]^{\gamma} \right) \qquad (14.4)$$

By considering Equations (14.1)–(14.4), it is easy to see that to minimize TAC one has to minimize the following parameters: utility cost (UC), total heat transfer surface area (AT) and the number of heat exchangers (NU). The minimal values of these parameters (for fixed ΔT^{min} value) are commonly called "targets" and the process of calculating their values is referred to as "targeting". We will also refer to them as minimal number of units (MNU) and minimal total surface area (MTA). All the targets depend on the ΔT^{min} value. This parameter is also the decision variable in design step. By "supertargeting" we refer (following Linnhoff) to calculation of the optimal ΔT^{min} value ($\Delta T^{min,opt}$). To find $\Delta T^{min,opt}$ we should solve the simple optimization problem:

$$\min_{\Delta T^{min}} (TAC) \qquad (14.5)$$

Various optimization techniques can be employed. Often (in pinch technology in particular) an illustrative graphical procedure is applied. The targets, including TAC, are calculated and then plotted for various values of ΔT^{min} from a fixed range. The optimal value or a small region around the optimum can be found visually from the plot: costs (targets) versus ΔT^{min}. Figure 14.1 shows typical

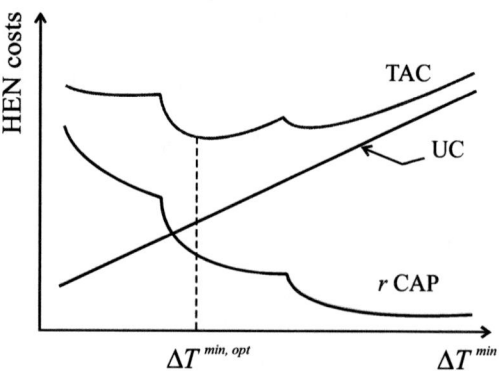

Figure 14.1 Typical plot: HEN targets vs. ΔT^{min}.

profiles. Heat recovery depends linearly on ΔT^{min}. Also, the cost of utilities is a linear function of ΔT^{min} provided that the same utilities are used in the total range. However, even under this assumption, there is a specific "threshold" value of ΔT^{min} at which a new type of utility appears with an increase of ΔT^{min}. Below threshold ΔT^{min} only one type of utility is necessary while above it both types—cooling and heating media—have to be used. Most often the threshold ΔT^{min} value is quite small and thus practical problems usually require both types of utilities. Even under simplifying assumptions 1 and 2 the total area target and hence capital cost is a nonlinear function of ΔT^{min} because of the nonlinear second term in Equation (14.4). Additional sharp "jumps" (step-like changes) are introduced into the CAP curve by unit changes of match number. The profiles of curves will become more complicated if the assumptions are removed. Notice that the $\Delta T^{min,opt}$ value depends on the goal function adopted. Though TAC is most often used it does not account for either cash flow or life cycles. Ya and Shonnard (2003) presented a few case studies of supertargeting using both economic and environmental evaluations based on life-cycle impact assessment. The results showed that the optimal $\Delta T^{min,opt}$ values for such criteria differ from that for TAC—they are smaller. However, the conclusion should be generalized with care due to the limited number of case studies calculated by Ya and Shonnard (2003).

There are some important provisos about sequential HEN targeting. First, the assumption in the preceding paragraph on single ΔT^{min} applied to calculate all targets, although inherent in pinch technology, can and even should be removed since it often prevents reaching a better TAC target. Thus, the designer should keep in mind that pinch technology targets provide rather conservative estimates. Note that the reason for forcing a single ΔT^{min} value is the rule of "no heat exchange via pinch"—see Chapter 13.

Secondly, the targeting philosophy discussed is valid for the HEN synthesis mode and should not be extended to retrofit without major changes. To date there is no rigorous insight-based method for HEN targeting under a retrofit scenario.

There exist several techniques from PT that shed some light on the problem but fail to produce good estimates, particularly in regard to investment cost. The works of Tjoe and Linnhoff (1986, 1988) are valuable but provide only general insights, highlighting substantial differences between targeting for synthesis and retrofit modes. As we will be explained in Chapter 19, the HEN retrofit problem is highly nonlinear, may exhibit discontinuous goal function and, moreover, is highly combinatorial. Hence, linearization and elimination of binary variables, inherent features of PT, cannot provide accurate targets, particularly for the investment cost of structural changes. More accurate results are achievable with the optimization approach by Yee and Grossmann (1987, 1988, 1991) or Ciric and Floudas (1989, 1990c). However, due to linear (MILP) optimization models they only show what structural changes are recommended but do not produce exact targets such as in synthesis. These approaches are mainly applied at the first stage of retrofitting HEN to screen structural changes, thus they reduce the combinatorial difficulty of the problem. Briones and Kokossis (1999a, b, c) developed more rigorous and more complex models. In the following chapters we will concentrate on calculating targets for HEN synthesis.

Thirdly, the notion of matches (single tube pass heat exchangers) and shells (multipass heat exchangers) requires further explanation. MER and MUC targets do not depend on heat exchanger type and construction. Solution methods do not apply design equations—heat balances and thermodynamic constraints are sufficient. However, the number of units in HEN as well as its total heat transfer surface area depends on heat exchanger type and construction. In order to make the problem manageable all existing approaches for MNU and MTA targets apply the assumption that the heat transfer coefficient of streams or, equivalently, overall heat transfer coefficients for possible matches, are given. Due to this, targeting methods have not dealt rigorously with various construction types of heat exchangers. However, both parameters—number of units and total area—depend on the mean temperature difference (MTD) which in turn is affected by the type of heat exchanger. For shell-and-tube heat exchangers the MTD value depends mainly on the number of tube passes and number of shells (though the former have a minor influence). Practically, with regard to the dependence of MTD on exchanger type, two cases are accounted for by targeting procedures developed to date:

(a) All heat exchangers are of shell-and-tube type, with single tube pass and strictly countercurrent flow. Notice that for such apparatus the mean temperature difference is equal to the logarithmic mean temperature difference (LMTD).
(b) Heat exchangers in HEN can feature a multipass tube arrangement.

A heat exchanger of type (a) is often referred to as a "match" (match between a hot and a cold stream) or 1-1 apparatus (i.e. heat exchanger with a single shell and single tube pass). We will follow this nomenclature. Hence, the minimum number of units target will be called the "minimum number of matches" target (MNM). The minimum total area for such apparatus will be abbreviated to "MTA-m".

In chemical and related industries such 1-1 heat exchangers are rarely used. More common are multipass heat exchangers, that is, those having more than one tube pass. They are often arranged in serial connections. A train of "m" heat exchangers each featuring "n" tube passes is called an "m-n" heat exchanger. Alternatively, we can say that such trains consist of "n" shells. Capital cost depends on the number of shells. Hence, targeting approaches have to calculate the number of shells. The minimum number of the unit target will be called the "minimum number of shells" target (MNS). Accordingly, the minimum total area target for multishell heat exchangers will be denoted by the acronym MTA-s.

Targeting approaches for multishell apparatus are more general. Apparently, they should encompass the 1-1 apparatus since the correction factor MTD for 1-1 apparatus reduces to 1.0. However, this is not always the case in existing targeting methods. It is the main reason for presenting, in Chapter 16, approaches for both matches and multishell heat exchangers. Also, the methods relevant only to the 1-1 apparatus are very often the foundation for targeting approaches accounting for multishell units. It is almost impossible to explain and understand the latter without knowledge of the former.

Notice that all targeting methods assume that heat exchangers in HEN are nonstandard ones and there are no limits on lower (zero is obviously the natural limit) and upper values of the heat transfer surface area (nonstandard heat exchangers have a heat transfer area that can take any value within reasonable limits, i.e. its area is not a discrete parameter). An approach to retrofitting HEN with standard heat exchangers will be addressed in Chapter 19. However, it does not employ standard targeting.

Last but not least, it is of great importance in sequential targeting methods that all the targets must be consistent. Which means that, independent of the sequence of calculations, all targets have to be achievable simultaneously. For instance, if the calculation sequence begins with the utility cost, as is the case in pinch technology and many other methods, the calculation methods for other targets have to ensure that this target will be also reached. Not all targeting approaches have this feature. For instance, some give the "absolute" minimum value of the target, such that it does not account for other targets. Even well-established pinch technology tools are not strictly accurate in this matter. This is caused by approximate solution techniques since the single ΔT^{min} concept should provide consistency of the general framework.

It is of importance that targeting approaches vary, not only as for solution techniques, but also they provide more or less useful results. For instance, a method for simultaneous targeting of utility cost and number of units gives the so-called "optimal heat load distribution", that is, values of heat loads of heat exchangers. The distribution also contains information on the structure of matches though usually insufficient. Other approaches also offer some information on match structure though it is less clear.

Pinch technology methods are based on the philosophy outlined in Chapter 13. Generally, it consists of transforming a single heat exchanger case into a multistream case. The techniques are simple, do not require optimization tools and use a lot of graphics. HEN design approaches from PT are, in practice, extensions of targeting procedures. There are also numerical or optimization-based techniques for calculating targets sequentially. They are often more rigorous and allow relaxation or removal of assumptions imposed in pinch technology methods.

14.3. BASIS OF SIMULTANEOUS APPROACHES FOR HEN TARGETING

Simultaneous methods provide possibilities for removing or relaxing some of the drawbacks of sequential methods. First, they are able to capture rigorously all or some targets simultaneously. Secondly, it is possible to eliminate some assumptions such as the even distribution of total heat transfer surface area among heat exchangers. The major outcome of simultaneous targeting is a heat load distribution in an optimal HEN.

However, care must be taken to ensure the simplicity and efficiency of simultaneous approaches since these features are of high significance at the targeting stage.

Mathematical optimization is usually the best choice to reach accurate results in a simultaneous approach. Minimization of match number and fixed charges of heat exchangers are necessary in the application of binary variables. Hence mixed-integer problems have to be solved. Moreover, nonlinearities in an optimization model cause nonlinear (MINLP) problems. They are very difficult to solve with available solvers to global optimum. Hence, efforts have been concentrated on developing linear models (MILP) since they are considered solvable to global optimum with widely available solvers. However, problem scale can also be a serious limitation for MILP as shown by Furman and Sahinidis (2001a, 2001b, 2004). The main difficulty in linearizing simultaneous HEN targeting optimization models is nonlinear capital cost—the term with the cost depending on surface area in Equation (14.2). Hence, the first simultaneous approach by Papoulias and Grossmann (1983b) accounted only for two targets: minimum cost of utilities and minimum number of units. Notice that more recently some authors, for instance Barbaro and Bagajewicz (2002), have claimed that initial linearization of capital provides sufficiently accurate results, though this is doubtful for the retrofit scenario in particular. In Chapter 17 we will address a targeting approach that accounts for TAC and does not need such linearization.

15 Minimum utility cost (MUC) target by optimization approaches

15.1. INTRODUCTION AND MER PROBLEM SOLUTION BY MATHEMATICAL PROGRAMMING

Despite the obvious advantages of PT approaches for MER/MUC calculations they all have some limitations, that is, do not account for forbidden, restricted, must-be and dual matches. Also, to solve the MUC problem by the GCC approach certain assumptions have to be imposed. Optimization approaches do not feature such limitations.

Probably, the most often-used optimization method for solving MUC problems with forbidden, restricted and must-be matches is that by Papoulias and Grossmann (1983b), referred to here as the standard MUC model. They developed a transshipment model formulation for modeling feasible heat exchange. To a certain extent the formulation is similar to heat cascade representation and, also, features a resemblance to Pr-T calculations. To show these features and, also, to begin with a simple case we will first explain a simple transshipment model for a solution of the MER problem using data from Example 13.1 (Table 13.1). Similarly to the methods from Chapter 13 we assume here that a single heating utility is supplied at the highest temperature and a single cooling one at the lowest temperature.

The typical transshipment formulation from operations research uses sources from which a commodity is sent to destinations via warehouses. The objective is to minimize the cost of transportation. To translate these terms into a heat recovery problem let us assume that heat is the commodity which is transferred from hot process streams and heating utilities (sources) to cold process streams and cooling utilities (destinations) via temperature intervals (warehouses). Also, heat can be transferred via residuals connecting intervals. The aim is to minimize the cost of heat transfer. In fact, it suffices to minimize utility usage/cost since heat transfer between process streams is free.

Figure 15.1 illustrates this concept for data of Example 13.1 using symbols applied to explain the Pr-T approach in Table 13.2. Notice, however, that in the following parameter R_0 is equivalent to Qhu and R_6 equals Qcu.

For each interval k heat balances are written as follows:

$$QH_k + R_{k-1} = QC_k + R_k \qquad (15.1)$$

Energy Optimization in Process Systems
©2013 Elsevier Ltd. All rights reserved.

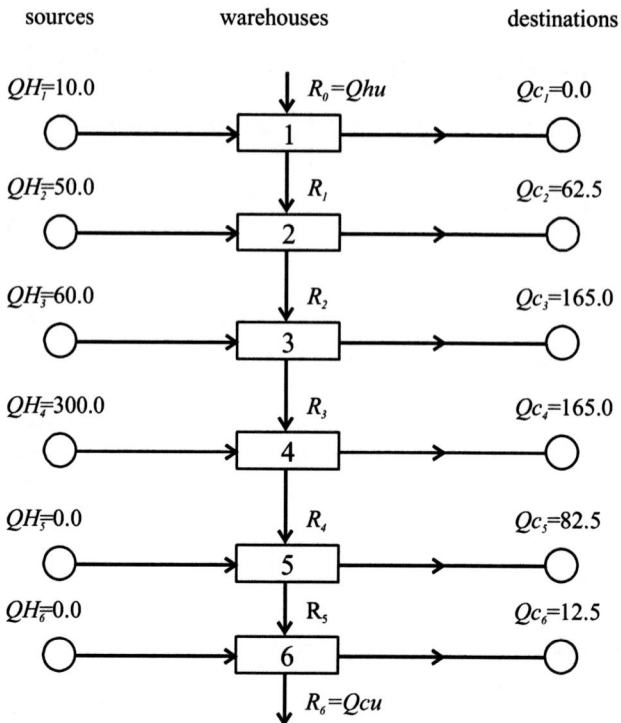

sources   warehouses   destinations

$QH_1=10.0$ $R_0=Qhu$ $Qc_1=0.0$

$QH_2=50.0$ R_1 $Qc_2=62.5$

$QH_3=60.0$ R_2 $Qc_3=165.0$

$QH_4=300.0$ R_3 $Qc_4=165.0$

$QH_5=0.0$ R_4 $Qc_5=82.5$

$QH_6=0.0$ R_5 $Qc_6=12.5$

$R_6=Qcu$

Figure 15.1 Transshipment representation for the MER problem and for data of Example 13.1.

The MER optimization model is:

$$\min R_0 \tag{15.2}$$

s.t.

$$[QH_k - QC_k] + [R_{k-1} - R_k] = 0, \quad k = 1, \dots, K$$
$$R_k \geq 0, \quad k = 0, \dots, K \tag{15.3}$$
$$QH_k \geq 0, \ QC_k \geq 0, \quad k = 1, \dots, K$$

In this model only R_0 (Qhu) is minimized, since due to the overall heat balance, R_K (Qcu) will be minimized, too. However, we can minimize the sum of R_0 and R_K as well.

It is interesting to notice that in the first stage of the Pr-T method we had to solve the same balances as in the MER optimization model. Look at the second column of Table 13.2 where calculations for the first stage of the Pr-T method are summarized. We have calculated first the difference $(QH_k - QC_k)$, the same as in the first square bracket in heat balances of constraints (15.3). They were denoted in Table 13.2 by ΔQ. Then, in the Pr-T approach, we have added to each ΔQ_k residual R_{k-1} flowing into interval k to calculate residual R_k exiting this interval, that is:

$$R_k = R_{k-1} + (QH_k - QC_k) \qquad (15.4)$$

Let us rearrange balances (15.3) in the optimization model in the same way. Using data from Table 13.2 or from Figure 15.1 we obtain:

$$R_1 = R_0 + 10 \qquad (15.5)$$

$$R_2 = R_1 - 12.5 \qquad (15.6)$$

$$R_3 = R_2 - 105.0 \qquad (15.7)$$

$$R_4 = R_3 + 135.0 \qquad (15.8)$$

$$R_5 = R_4 - 82.5 \qquad (15.9)$$

$$R_6 = R_5 - 12.5 \qquad (15.10)$$

Thus, we have six linear equations with seven unknowns: R_k; $k = 0, \ldots, 6$. The system has one degree of freedom. We can express all the equations in terms of one unknown. Since R_0, equivalent to Qhu, is to be minimized this variable is chosen as the unknown. Then, one will obtain, by substitution, the following equation set:

$$R_1 = R_0 + 10 \qquad (15.11)$$

$$R_2 = R_0 - 2.5 \qquad (15.12)$$

$$R_3 = R_0 - 107.5 \qquad (15.13)$$

$$R_4 = R_0 + 27.5 \qquad (15.14)$$

$$R_5 = R_0 - 55.0 \qquad (15.15)$$

$$R_6 = R_0 - 67.5 \qquad (15.16)$$

Notice that the second term of the right-hand side in each equation from (15.11) – (15.16) is identical to the values in the second column of Table 13.2 calculated in the first stage of the Pr-T approach and denoted there by R_1 to R_6. Hence, setting R_0 at zero in Equations (15.11) – (15.16) will produce the same values of R_1 to R_6 as in Table 13.2.

In the MER optimization model all the residuals have to be nonnegative according to Equation (15.3): $R_k \geq 0$, $k = 0, \ldots, 6$. To minimize R_0 it suffices to find the smallest negative (largest absolute) value of R_0 that will also satisfy the nonnegativity conditions for all R_k. For the above equation set the minimal value of R_0, which meets the requirements, is 107.5. Residual R_3 becomes, then, zero with all other residuals being positive. Thus, residual R_6, equivalent to Qcu, is equal to 40 (107.5 – 67.5 = 40).

The same solution has been found by applying two-stage calculations in the Pr-T approach. We can conclude that the MER problem formulated as the LP model (15.2) and (15.3) with one degree of freedom can be solved by a single-stage procedure. Equivalently, the Pr-T approach is the two-stage procedure of solving the LP model for MER with one degree of freedom.

15.2. MUC PROBLEM SOLUTION METHODS

In order to solve the MUC problem we have to include available utilities in the model, that is, to insert them into the intervals that exist according to their inlet and outlet temperature. The model will remain linear but will have more degrees of freedom. However, in order to account also for restricted matches it is necessary to expand the representation so as to capture matches between individual streams. Such a model developed first by Papoulias and Grossmann (1983b) is called the expanded transshipment model in Biegler et al. (1997) but we will refer to it as the standard model. Notice that it differs slightly from the original model of Papoulias and Grossmann (1983b) since they did not apply residuals for utilities that decrease model rigorousness.

The expanded transshipment model differs largely from the simple one in Figure 15.1. Its general illustration is shown in Figure 15.2. Each hot process stream and each heating utility can exchange heat in many temperature intervals provided that it is allowed by thermodynamic conditions (13.5). Residuals have to be defined for each hot process stream and each heating utility as well (this is not shown in Figure 15.2). Note that the residual of each hot process stream exiting the cascade has to be equal to zero since cold utilities are explicitly included in the model. Similarly, the residuals of process streams entering the cascade are equal to zero.

Given are hot process streams ($i \in H$), cold process streams ($j \in C$), available cooling utilities ($n \in CU$) and available heating utilities ($m \in HU$).

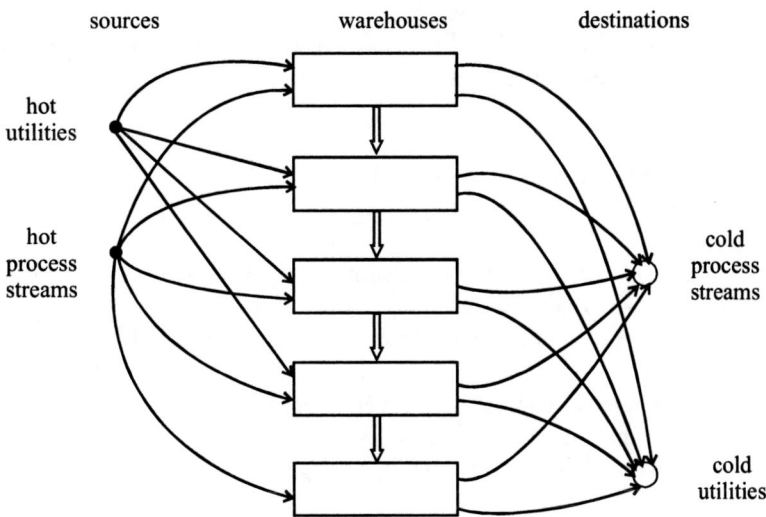

sources   warehouses   destinations

hot
utilities

hot
process
streams

cold
process
streams

cold
utilities

Figure 15.2 Overview of expanded transshipment formulation for MUC problem.

They are divided into temperature intervals following the rules for the Pr-T approach. The division at inlet temperatures of streams in shifted scale is sufficient. The intervals are numbered according to decreasing temperature from 1 to K and denoted by symbol k. Then the following index sets are created:

$$H_k = \{i \mid i \text{ is present in interval } k \text{ or can exchange}$$
$$\text{heat with } j \in C_k \text{ and } n \in CU_k\}$$

$$C_k = \{j \mid j \text{ is present in interval } k\ \}$$

$$HU_k = \{m \mid m \text{ is in interval } k \text{ or can exchange heat with } j \in C_k\ \}$$

$$CU_k = \{n \mid n \text{ is present in interval } k\ \}$$

Notice that hot stream $i \in H_k$ can exchange heat with both $j \in C_k$ and $n \in CU_k$ if the hot stream is present also in intervals higher than k. Hence, definitions of sets H_k and HU_k can be alternatively written as follows:

$$H_k = \{i \mid i \text{ is present in interval } k \text{ or is present in higher interval }\}$$

$$HU_k = \{m \mid m \text{ is in interval } k \text{ or is present in higher interval }\}$$

For each interval $k = 1, \ldots, K$ we have to calculate the heat load of each hot process stream and of each cold process stream. The loads are determined for known temperatures of intervals.

Let:

ΔQ_{ik}^{H} denote the heat load of $i \in H_k$
ΔQ_{jk}^{C} denote the heat load of $j \in C_k$.

Total heat load of cooling utility n is denoted by Qcu_n and that of heating utility m by Qhu_m. These loads are variables in the model.

Variables for heat loads exchanged in intervals are denoted by

q_{ijk}—for heat exchange between $i \in H_k$ and $j \in C_k$
q_{ink}—for heat exchange between $i \in H_k$ and $n \in CU_k$
q_{mjk}—for heat exchange between $m \in HU_k$ and $j \in C_k$

Notice that heat exchange between cold utilities and hot utilities is explicitly excluded. Also, due to the definitions of sets H_k and HU_k heat exchange is always feasible since constraints (13.5) and (13.6) are satisfied.

$R_{i,k}$ stands for residual of stream $i \in H_k$ exiting interval k and R_{mk} stands for residual of hot utility $m \in HU_k$ exiting interval k. It is important that residuals R_{i0}, R_{m0} entering the first interval and R_{iK}, R_{mK} exiting the last interval must be set at zero.

To complete the list of variables and parameters it is necessary to include as parameters unit prices of utilities phu_m ($m \in HU_k$) and pcu_n ($n \in CU_k$).

In order to facilitate understanding of the model we present here also another representation based on popular grid representation—Figure 15.3. It seems that it is easier to understand particularly for those familiar with the basic of pinch

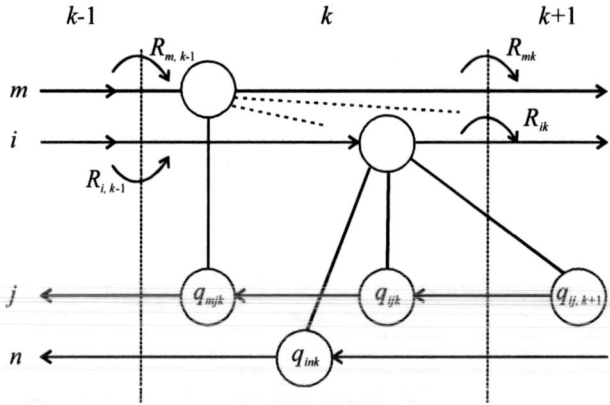

Figure 15.3 Grid representation of the expanded transshipment model for MUC.

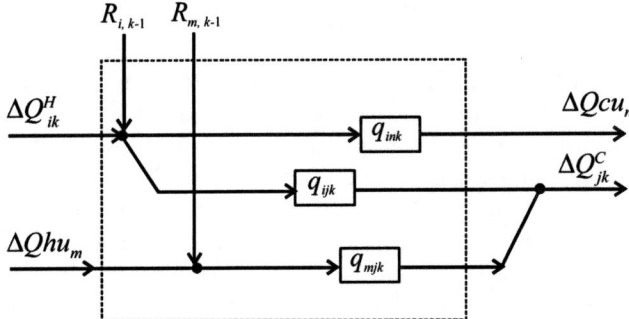

Figure 15.4 Heat exchange pattern in interval k of the transshipment formulation.

technology where grid representation is commonly applied in heat exchanger network design.

The basic equations of the transshipment model are heat balances. Heat exchange pattern for kth interval is shown in Figure 15.4.

Finally, we present below the standard transshipment model for MUC problem.

Standard transshipment model—S-MUC-Transs

$$\min UC = \sum_{m \in HU} (Qhu_m phu_m) + \sum_{n \in CU} (Qcu_n pcu_n) \tag{15.17}$$

s.t.

$$R_{ik} - R_{i,k-1} + \sum_{j \in C_k} q_{ijk} + \sum_{n \in CU_k} q_{ink} = \Delta Q^H_{ik}; \ i \in H_k; \ k = 1, \ldots, K \tag{15.18}$$

$$R_{mk} - R_{m,k-1} + \sum_{j \in C_k} q_{mjk} = Qhu_m; \ m \in HU_k; k = 1, \ldots, K \tag{15.19}$$

$$\sum_{i \in H_k} qi_{jk} + \sum_{m \in HU_k} q_{mjk} = \Delta Q^C_{jk}; \ j \in C_k; \ k = 1, \ldots, K \tag{15.20}$$

$$\sum_{i \in H_k} q_{ink} = Qcu_n; \ n \in CU_k; \ k = 1, \ldots, K \tag{15.21}$$

$$R_{i0} = R_{iK} = R_{mK} = 0; \ i \in H; \ m \in HU \tag{15.22}$$

$$R_{ik}, R_{mk}, q_{ijk}, q_{mjk}, q_{ink} \geq 0 \qquad (15.23)$$

Equations (15.18) and (15.20) are heat balances ensuring also that given values of heat loads of streams $i \in H_k$ and $j \in C_k$ are satisfied. Equations (15.19) and (15.21) define loads on utilities. Equation (15.22) sets the residuals of hot process streams entering the cascade and the residuals of all hot streams exiting the cascade at zero. Condition (15.23) ensures nonnegative values of all variables.

Forbidden matches can be simply accounted for by setting variables q_{ijk}, q_{ink} and q_{mjk} at zero for streams that are not allowed to match with each other. Alternatively, we can exclude such variables explicitly from the model by defining sets of forbidden matches. Also, it is possible to include other constraints on matches (restricted matches). Assume that in a match between $i \in H$ and $j \in C$ heat load should not be higher than the given value Q_{ij}. The condition in the model should be

$$\sum_{k=1}^{K} \sum_{i \in H_k} \sum_{j \in C_k} q_{ijk} \leq Q_{ij} \qquad (15.24)$$

There are also cases that some matches are imposed—must-be matches. Such cases are common in HEN retrofit scenarios. Must-be matches are inserted into the model by simply forcing appropriate variables to be higher than zero. Let a match between $i \in H$ and $j \in C$ be a must-be one. This condition can be written as follows:

$$\sum_{k=1}^{K} \sum_{i \in H_k} \sum_{j \in C_k} q_{ijk} > 0 \qquad (15.25)$$

However, accounting for dual matches is not simple since this requires binary variables and proper logical conditions. We will explain this in the following.

Here, we will show the alternative optimization model for MUC with restricted matches. It is based on a conceptually simpler transportation model that is also well known from operations research. The transportation model is, in fact, a reduced transshipment one since it does not involve warehouses. Thus, heat is transferred directly from sources (hot process streams and heating utilities) to destinations (cold process streams and cooling utilities). Division for temperature intervals is necessary to maintain feasible heat exchange and to account for the pinch. An illustration of the transportation model in grid representation is given in Figure 15.5.

Notice that there are no residuals in the formulation. This simplifies conceptually the model but increases the number of variables since far more variables defining heat exchange have to be used. The transportation MUC model was first presented in Cerda et al. (1983). We formulate here the model in a slightly different way using index sets and variables similar to those in the transshipment

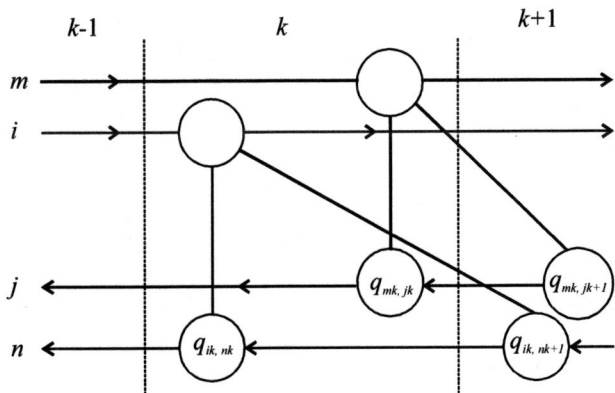

Figure 15.5 Transportation model in grid representation.

model. This is also caused by the fact that the transportation model for calculating the total area target that will be presented in Chapter 16 requires such a formulation.

We need here the following index sets:

$$H_k = \{i \,|\, i \text{ is present in interval } k \}$$

$$C_k = \{j \,|\, j \text{ is present in interval } k \}$$

$$HU_k = \{m \,|\, m \text{ is in interval } k \}$$

$$CU_k = \{n \,|\, n \text{ is present in interval } k \}$$

To formulate heat balances the following variables defining heat loads are necessary:

$q_{ik,jl}$—for heat exchange between $i \in H_k$ and $j \in C_l$
$q_{ik,nl}$—for heat exchange between $i \in H_k$ and $n \in CU_l$
$q_{mk,jl}$—for heat exchange between $m \in HU_k$ and $j \in C_l$

Notice that we have to use two indices for intervals (k and l) to define variables for heat exchange while in the transshipment model the index of one interval suffices. Hence, the number of variables in the transportation model is higher than that for the transshipment one even though the former has no residuals. Papoulias and Grossmann (1983b) gave a comparison of the number of variables in both models. However, with the present power of computers and taking into account linearity of the models this is not a serious drawback.

Notice that heat exchange between hot utilities $m \in HU$ and cold utilities $n \in CU$ is explicitly excluded. In order to maintain feasible heat exchange we should exclude from the formulation such matches that do not meet thermodynamic conditions (13.5) and (13.6). This is achieved by setting certain heat load variables at zero as defined by Equation (15.26). The conditions eliminate heat flow from intervals of lower temperature to those with higher temperature.

$$q_{ik,jl} = 0; \quad q_{mk,jl} = 0; \quad q_{ik,nl} = 0; \quad k < l; \quad k, l = 1, \ldots, K \qquad (15.26)$$

The data needed are the same as for the transshipment model S-MUC-Transs.

The transportation model for the MUC problem is as follows.

Transportation model for MUC: MUC–Transp model

$$\min UC = \sum_{m \in HU} (Qhu_m phu_m) + \sum_{n \in CU} (Qcu_n pcu_n) \qquad (15.27)$$

s.t.

$$\sum_{j \in C_l} q_{ik,jl} + \sum_{n \in CU_l} q_{ik,nl} = \Delta Q_{ik}^H; \quad i \in H_k; \quad k, l = 1, \ldots, K \qquad (15.28a)$$

$$\sum_{i \in H_k} q_{ik,jl} + \sum_{m \in H_k} q_{mk,jl} = \Delta Q_{jk}^C; \quad j \in C_l; \quad k, l = 1, \ldots, K \qquad (15.28b)$$

$$Qhu_m = \sum_{k,l=1}^{K} \sum_{m \in HU_m} \sum_{j \in C_l} q_{mk,jl}; \quad m \in HU \qquad (15.29)$$

$$Qcu_n = \sum_{k,l=1}^{K} \sum_{n \in CU_l} \sum_{i \in H_k} q_{mk,jl}; \quad n \in CU \qquad (15.30)$$

$$q_{ik,jl}, q_{mk,jl}, q_{ik,nl} \geq 0 \qquad (15.31)$$

Equations (15.28a) and (15.28b) are heat balances ensuring that given heat loads of process streams in intervals are met. Equations (15.29) and (15.30) are simply definitions of loads on hot and cold utilities. Finally, Equation (15.31) ensures nonnegative values of all model variables.

The inclusion of forbidden, must-be and restricted matches is identical to the transshipment model.

It is of importance that the standard transshipment model as well as the transportation formulation, even though widely used and presented in the basic monographs, for example, Biegler et al. (1997), Floudas (1995), have some drawbacks when applied to nonpoint utilities. This problem has also been discussed in Section 13.5 using the GCC approach. It was shown that if the outlet temperature of a nonpoint utility is fixed, one should treat the minimum (or the optimum) heat utility load with great care when calculating its flow rate. This is caused by potential pitfalls of heat exchange via process and utility pinches. In general the problem relies on accounting for the minimization of nonpoint utility flow rates within the MUC optimization model.

Both optimization models are even more "dangerous" in this respect than graphical approaches such as the GCC method since they do not provide visualization. Note that the final temperatures of utilities are not used in either optimization model. Hence, each utility can be supplied to each interval lower than its inlet interval irrespective of its outlet temperature. Furthermore, the flow rate of each utility can vary from interval to interval, as we will show in Example 15.2. In fact, a nonpoint utility is treated in both models as a "set" of point utilities, which are not correlated to each other by a continuity condition on flow rate. Thus, all the pitfalls that a designer can be trapped into when calculating flow rates are the same as in the case of pinch technology approaches. Finally, note that both models have no solutions if the temperature level of utilities provided in the data (i.e. their inlet temperatures) is not valid for a problem, for instance the heating utilities available have inlet temperatures too low.

These drawbacks were removed in the so-called generalized transshipment model of Shethna et al. (2000a) and Jeżowski et al. (2000a). This model accounts also for segmented nonpoint utilities.

The changes to formulation of the standard transshipment model are as follows:

1. To provide constant CP value of a segment of nonpoint utility in its entire temperature range; continuity relations for CP are added for each segment of the utility:

    CP = const. for all temperature intervals utility segment enters or can cascade heat

2. To provide constant flow rate for all segments of a particular utility continuity relations are added:

    utility flow rate = const. within intervals where utility exists according to its given temperature range

3. Slack heaters are placed on each cold process stream and each cooling utility in every interval the streams exist. The loads of slack heaters on cold nonpoint utilities encapsulate information on excess heat of cold utilities that is not useful in a process. The

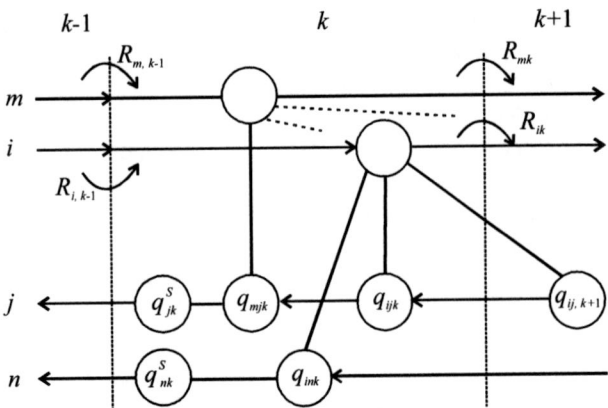

Figure 15.6 Illustration of the generalized transshipment model from Shethna et al. (2000a).

loads of slack heaters on cold process streams are necessary to determine whether inlet temperatures of hot utilities given in data are sufficient in regard to their temperature level.

4. Nonzero residual heat flow from last intervals for hot utilities and for hot process streams as well is allowed. This is necessary to get information on excess (not useful in a process) heat of hot utilities and insufficient (as for temperatures) cold utilities.

5. In the goal function weights are applied for all matches instead of unit prices of utilities in matches with utilities. Also, residual flows from the last temperature intervals of hot streams multiplied by proper weights have to be included in the goal function. This is because residuals act as a fictitious cold utility and have the unit price of these utilities.

An illustration of the generalized transshipment model in grid representation is shown in Figure 15.6.

The index sets are the same as in the standard transshipment model. In comparison with the latter, illustrated in Figure 15.3, the generalized model contains slack heaters at each cold process and utility stream in intervals where the streams exist. The loads on slack heaters are variables in optimization and are denoted by: $q_{jk}^S, j \in C_k$; $q_{nk}^S, n \in CU_k$.

To account for novel features of this model it is convenient to use symbols $K_m^{(1)}, K_n^{(1)}$ and $K_m^{(2)}, K_n^{(2)}$. Superscript (1) denotes the interval where utility m or n enters a heat cascade and superscript (2) denotes that interval where a utility exits. In order to account for flow rate continuity conditions parameters cp_{nk}, cp_{mk} are applied denoting specific heat of utilities m, n in interval k.

To formulate the generalized transshipment model we apply compact notation from Shethna et al. (2000a) that differs slightly from that of Biegler et al. (1997) but is better suited to this case. For the sake of clarity the model in the following does not account for CP continuity conditions of utility segments. The inclusion of such conditions requires additional symbols for intervals where the segments exist.

Generalized transshipment model for the MUC problem—(G-MUC-Transs model)

$$\min\left\{\sum_{k=1}^{K}\left(\sum_{i\in H_k}\sum_{j\in C_k}c_{ij}q_{ijk}+\sum_{m\in HU_k}\sum_{j\in C_k}phu_mq_{mjk}+\sum_{i\in H_k}\sum_{n\in CU_k}pcu_nq_{ink}\right)\right.$$

$$+\sum_{k=1}^{K}\left(\sum_{j\in C_k}c_j^Sq_{jk}^S+\sum_{n\in CU_k}c_n^Sq_{nk}^S\right)\right\}$$

$$+\left(\sum_{i\in H_k}c_i^RR_{iK_i^{(2)}}+\sum_{m\in HU_k}c_m^RR_{mK_i^{(2)}}\right)\tag{15.32}$$

s.t.

$$R_{ik}-R_{i,k-1}+\sum_{j\in C_k}q_{ijk}+\sum_{n\in CU_k}q_{njk}=\Delta Q_{ik}^H;\ i\in H_k,k=1,\dots,K\tag{15.33}$$

$$\sum_{i\in H_k}q_{ijk}+\sum_{m\in HU_k}q_{mjk}+q_{jk}^S=\Delta Q_{jk}^C;\ j\in C_k,k=1,\dots,K\tag{15.34}$$

$$\frac{\sum_{j\in C_k}q_{mj,k-1}}{\varphi_{m,k-1}\Delta T_{k-1}}=\frac{\sum_{j\in C_k}q_{mjk}}{\varphi_{mk}\Delta T_k};\ m\in HU,\ k=(K_m^{(1)}+1),\dots,K_m^{(2)}\tag{15.35}$$

$$\frac{\sum_{i\in H_k}q_{in,k-1}}{\varphi_{n,k-1}\Delta T_{k-1}}=\frac{\sum_{i\in H_k}q_{ink}}{\varphi_{nk}\Delta T_k};\ n\in CU,\ k=(K_n^{(1)}+1),\dots,K_n^{(2)}\tag{15.36}$$

$$R_{iK_i^{(1)}},R_{mK_m^{(1)}}=0\tag{15.37}$$

$$q_{ijk},q_{mjk},q_{ink}\geq0;\ q_{ik}^S;\ q_{nk}^S\geq0;\ R_{ik},R_{mk}\geq0\tag{15.38}$$

Equations (15.33) and (15.34) are heat balances similar to those in the standard transshipment model S-MUC-Transs. They differ in that heat loads on utilities $n\in CU$ and $m\in HU$ are not directly defined in the model. Next, loads of fictitious heaters have to be included. Equations (15.35) and (15.36) are continuity relations for flow rates of utilities. Notice that if a utility is of point type they are disabled since intervals $K^{(1)}$ and $K^{(2)}$ are identical for such a utility. Equality (15.37) sets residuals entering first intervals at zero and constraint (15.38) forces all variables to be nonnegative.

The goal function (15.32) requires detailed explanation. It consists of three items. The first item contains relative prices of matches: process–process stream and process stream–utility. Notice that matches of process–process streams are also charged with relative prices/weights c_{ij} to account for pinch crossing heat exchange. Such weights were not applied in MUC models in the preceding. The second item represents relative prices/weights on fictitious heater loads. The loads account for pinch crossing heat exchange of cold utilities. Lastly, the third term represents the penalty on excess heat loads of a hot process or hot utility that cannot be transferred to cold streams due to insufficient temperature levels of the utilities available. The relative prices/weights of all terms have to be balanced by appropriate choice of weights c_{ij}, $c_n{}^S$, $c_j{}^S$, $c_n{}^R$, and $c_i{}^R$ in Equation (15.32). Shethna et al. (2000a) derived the following conditions and formulae for the weights:

$$c_n^S = pcu_n(1 + \delta) \tag{15.39}$$

$$c_m^R = phu_m(1 + \delta) \tag{15.40}$$

$$c_i^R = \Delta_i; \; \Delta_i > \max_{m \in HU} (phu_m) \tag{15.41}$$

$$c_j^S = \Delta_j; \; \Delta_j > \max_{n \in CU} (pcu_n) \tag{15.42}$$

$$c_{ij} < \delta \left(\max_{m \in HU} (phu_m) + \max_{n \in CU} (pcu_n) \right) \tag{15.43}$$

Shethna et al. (2000a) proved that parameter σ in Equations (15.39), (15.40), and (15.43) has to satisfy conditions (15.44)–(15.47). However, it suffices to set this parameter at a very small positive value.

$$\delta < \min_{m \in HU} (c_m) \tag{15.44}$$

$$\delta < \frac{\min_{m \in HU}(c_m)}{\min_{n \in CU}(c_n)} \tag{15.45}$$

$$\delta < \min_{n \in CU} (c_n) \tag{15.46}$$

$$\delta < \frac{\min_{n \in CU}(c_n)}{\min_{m \in HU}(c_m)} \tag{15.47}$$

Rigorous proof for the conditions for weights is given in the original paper. The procedure of calculating the weights seems complex. However, Shethna et al. (2000a) noted that the weights could be chosen *ad hoc* such as for instance relative prices associated with temperatures in the GCC method (Chapter 13).

Certain parameters of interest can be calculated from the results of the G-MUC-Transs model *a posteriori*:

- loads on utilities

$$Qhu_m = \sum_{k=1}^{K} \sum_{j \in C_k} q_{mjk} + R_{mK_m^{(2)}}; \ m \in HU \tag{15.48}$$

$$Qcu_n = \sum_{k=1}^{K} \left(\sum_{i \in H_k} q_{imk} + q_{mk}^S + q_{jk}^S \right); \ n \in CU \tag{15.49}$$

- insufficient loads on utilities due to insufficient temperature level

$$Qhu^{IN} = \sum_{i \in H_k} R_{iK_i^{(2)}} \tag{15.50}$$

$$Qcu^{IN} = \sum_{j \in C} \sum_{k=1}^{K} q_{jk}^S \tag{15.51}$$

- utility heat load transferred via pinch

$$Qhu_m^{pinch} = R_{mK_m^{(2)}}; \ m \in HU \tag{15.53}$$

$$Qcu_n^{pinch} = \sum_{k=1}^{K} q_{nk}^s; \ n \in CU \tag{15.54}$$

These quantities are useful to construct a GCC plot with utility profiles. Notice that in Section 13.5 (Figure 13.6) we did this using the simple reasoning based on the water pinch concept. However, a solution of this optimization model provides

Table 15.1 Data for Example 15.1 ($\Delta T^{min} = 10$ K)

Stream	T^{in} (°C)	T^{out} (°C)	CP (kW K^{-1})
H4	249	138	10.55
H2	160	93	8.79
C1	60	160	7.62
C3	116	260	10.08
HU	270	140	–
CU	38	82	–

all the information in an "automated" manner and, also, we do not need specific costing rules. Also, restricted and forbidden matches can be accounted for. To show the importance of accounting for nonpoint utilities we present here two examples, 15.1 and 15.2. The data for the first one are gathered in Table 15.1. This problem has also been considered by Viswanathan and Evans (1987), who were the first to show that the solution of the standard transhipment model causes problems when calculating utility flow rates on its basis. In this example the results from the standard model are $Qhu^{min} = 453.5$ kW and $Qcu^{min} = 0.0$, that is, the example is the threshold task.

Viswanathan and Evans (1987) calculated changes of CP of heating utility HU in intervals and found that they vary from 0.386 to 10.375 kW K^{-1}. Hence, to ensure sufficient and the same flow rate in all intervals the designer should apply the highest value, that is, $CP = 10.375$ kW K^{-1}. This causes the excess heat load of 856.5 kW of the utility. Thus, the solution from the standard transshipment model is unrealistic since it requires a variable flow rate of HU in temperature intervals to reach MER—CP of the utility has to change from interval to interval. The solution of the G-MUC-Transs model also shows such an excess heat load and additionally indicates that to eliminate this excess load the designer has to increase the outlet temperature of HU to 225 °C. Jeżowski and Jeżowska (2002) achieved identical results using the GCC plot and water pinch concept. However, such a simple method can be insufficient for more complex problems, such as Example 15.2, taken also from Shethna et al. (2000a). Required data are in Table 15.2. There are nine process streams. Two hot utilities are available: flue gas of large temperature range and MP steam segmented into slightly superheated steam and saturated steam at 10 bars. Boiler feed water and cooling water are available as cooling utilities. Additionally, LP steam rising from the heat of process streams is included as an option. Formally, this steam is treated in the model as point cooling utility.

Notice that the problem goes beyond the standard MUC problem formulation since it includes steam generation. As to the weights in the model, Shethna et al. (2000a) showed that *ad hoc* relative prices that follow the common industrial logic satisfy conditions (15.39) to (15.47). The solution of the model is shown in Table 15.3. Notice that pinch temperature is at 160 °C.

The heat losses for flue gas and rising LP steam result from pinch crossing. The remedy is to adjust their outlet temperatures to pinch temperature. However,

Table 15.2 Stream data for Example 15.3

Stream	T^{in} (°C)	CP (kW K^{-1})
H1	327	100
H2	220	160
H3	220	60
H4	160	400
C5	100	100
C6	35	70
C7	85	350
C8	60	60
C9	140	200
Flue gas	1500	
MP steam	200	
	181	
Cooling water	10	
Boiler feed water	40	
LP steam	140	
Raising	141	

Table 15.3 Results for Example 15.2

Utility	T^{out}	Heat load	Inevitable loss	Status
Flue gas	160	17064.1	1782.7	Cross pinch
MP steam	180	215.9	0.0	Above pinch
Cooling water	15	23764.7	0.0	Below pinch
LP steam raising	160	1235.2	8.8	Cross pinch

this causes, in the case of flue gas, heat loss to the atmosphere and for rising steam additional expenses in the utility subsystem. Nevertheless, the designer is provided with clear options that he/she can further analyze in depth in the context of the total system.

15.3. DUAL MATCHES

Finally, we will explain how to include dual matches in both transshipment and transportation models. The use of the dual match can increase energy recovery if there is a forbidden match, which causes a bottleneck for heat recovery. Such a forbidden match increases utility usage to the value $Qu^{forbid,min}$, higher in comparison with the case without forbidden match Qu^{min}, that is, $Qu^{forbid,min} > Qu^{min}$ (notice that not each forbidden match necessarily increases utility usage). Under such circumstances an appropriate choice of dual match can decrease utility consumption from $Qu^{forbid,min}$ even to the Qu^{min}.

Let us consider the illustrative example in Figure 15.7. Assume that match of stream h1 with stream c1 is forbidden. Also, let the residual from interval $(k-1)$

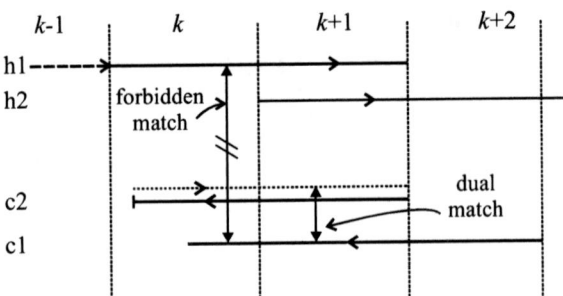

Figure 15.7 Illustration of dual match application.

and heat load of h1 be insufficient to heat up streams c1 and c2 in interval k. As a result, an additional heat load on the utility would be necessary. However, we can supply the heat that it lacks due to the forbidden match by matching cold streams c1 and c2. In this dual match c2 acts as a hot stream supplying heat to c1.

To explain the concept of inserting dual matches into the transshipment formulation we follow here an approach developed by Floudas (1995), which, to our knowledge is the only one presented in the literature. Let us consider the case of the dual match between two cold streams: c1 and c2 from Figure 15.7. Stream c2 has a higher inlet temperature and, thus, acts in the dual match as the hot one. Streams c1 and c2 exchange a certain heat load. To account for the fact that stream c2 can be cooled to below its inlet temperature we have to introduce an additional hot stream "h" which was not present in the original problem. This stream is shown in Figure 15.8 by the dashed line.

Notice that the temperatures in Figure 15.8 are in real scale. However, we have to remember that shifted scale should be applied to create intervals. The shifting is performed by diminishing the temperatures of hot streams, including also stream "h" that acts as a hot stream, by ΔT^{min}. One should be aware of the shifts when reading the following explanation.

Stream "h" has the inlet temperature T_h^{in} identical (in real scale) to the inlet temperature of cold stream c2 from problem data, denoted here by $T_{c2}^{in,data}$. Also, the CP value of stream "h" is the same as the CP value of stream c2. However, its outlet temperature T_h^{out} is unknown since the heat load in the dual match is

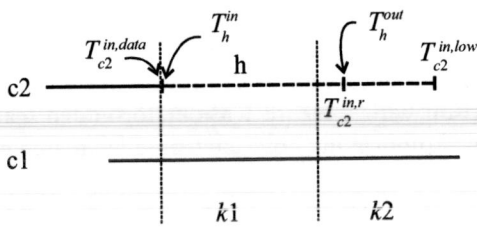

Figure 15.8 Illustration for dual match c1–c2.

unknown. This outlet temperature must be equal to the real inlet temperature of stream c2 ($T_{c2}^{in,r}$). Notice that prior division into intervals did not account for this real inlet temperature. The only information we are able to get on the possible value of $T_{c2}^{in,r}$ is its lower limit $T_{c2}^{in,low}$. This can be calculated assuming maximum heat load on match c1 with c2. Summing up, we have: new stream "h" with $CP_h = CP_{c2}$, $T_h^{in} = T_{c2}^{in,data}$ and unknown $T_h^{out} = T_{c2}^{in,r}$ limited by lower value $T_{c2}^{in,low}$.

In order to retain linearity of the heat exchange model it is necessary to keep intervals created on the basis of original inlet temperatures, that is, the real inlet temperature of c2 cannot be variable. Notice that the inlet temperature $T_{c2}^{in,r}$ can fall into any interval from $T_{c2}^{in,low}$ up to $T_{c2}^{in,data}$. To make further formulae simpler let us assume that for stream "h" its outlet temperature must fall in interval $k1$—see Figure 15.8. Thus, the heat load of stream "h" in interval $k1$ is given by the linear function:

$$\Delta Q_{h,k1}^H = CP_h(T_h^{in,data} - T_{c2}^{in,r}) \tag{15.55}$$

The heat load of c2 is the function of $T_{c2}^{in,r}$ but this temperature, due to the temperature shift, can be in the interval $k1$ or $k2$. To account for this fact we have to introduce two binaries y_k, for $k = k1, k2$ such that:

$$y_k = 1 \text{ if } T_{c2}^{in,r} \text{ is interval } k; \ y_k = 0 \text{ if not} \tag{15.56}$$

The obvious condition is that $T_{c2}^{in,r}$ can exist in one interval. This is ensured by the logical condition:

$$y_{k1} + y_{k2} = 1 \tag{15.57}$$

To maintain linearity of heat balance for stream "h" we have to add continuous variables $T_{c2}^{in,k}$ ($k = k1, k2$) to satisfy the condition:

$$T_{c2}^{in,k1} + T_{c2}^{in,k2} = T_{c2}^{in,r} \tag{15.58}$$

Finally, the model has to include limits on variables $T_{c2}^{in,k}$. For each interval $k = k1, k2$ there are the following temperature limits: lower limit T_k^{low} and upper limit T_k^{up}. Thus, the conditions are

$$T_k^{low} \le T_{c2}^{in,k} \le T_k^{up}, \quad k = k1, k2 \tag{15.59}$$

Notice that T_k^{low} is equal to $T_{c2}^{in,low}$.

Heat loads of c2 in intervals $k = k1$, $k2$ are both functions of $T_{c2}^{in,r}$. Since this variable is dependent on $T_{c2}^{in,k}$ according to Equation (15.58), we can formulate k linear equations ($k = k1$, $k2$):

$$\Delta Q_{hk}^{H} = a_k y_k + b_k T_{c2}^{in,k}, \quad k = k1, k2 \tag{15.60}$$

Notice that Equation (15.60) is, in fact, a piecewise linear function of $T_{c2}^{in,r}$. Parameters a_k and b_k in Equation (15.60) can be calculated from:

$$b_k = \frac{\Delta Q_{c2,k}^{C}(T_k^{up}) - \Delta Q_{c2,k}^{C}(T_k^{low})}{T_k^{up} - T_k^{low}}, \quad k = k1, k2 \tag{15.61a}$$

$$a_k = \Delta Q_{c2,k}^{C}(T_k^{up}) - b_k T_k^{low}, \quad k = k1, \ k2 \tag{15.61b}$$

All the equations from (15.55) to (15.6115.61a and 15.61bb) have to be added to the MUC optimization model that deals with the dual match.

Similar reasoning is applicable to the dual match of two originally hot process streams. In such a match a hot stream of lower inlet temperature acts as a cold stream and, thus, an additional cold stream has to be added to the heat recovery problem.

To account for a dual match we have to add binaries to an optimization model. Hence, the LP model becomes a MILP one. The model also contains new constraints. One can see that it becomes more complex. Thus, inserting all possible dual matches into the optimization model seems overcomplicated. The designer should, on the basis of problem analysis, chose an appropriate dual match and then incorporate proper constraints and variables for this match into the model. It is of importance that each dual match requires an additional heat exchanger and an additional stream in comparison with the case of no dual match. In consequence, the number of matches and, most likely, heat transfer area will increase. This can reduce or limit the profit from a decrease of utility cost.

15.4. MINIMUM UTILITY COST UNDER DISTURBANCES

In industry heat exchanger networks are subjected to disturbances of flow rates and inlet temperatures. The designer faces the problem of designing HEN with uncertain data. This problem is referred to in the literature as designing a flexible (resilient) network. Issues of controllability and operability are also considered in some works, for instance in Glemmestad et al. (1999) and Tellez et al. (2006). As mentioned in Chapter 12, these aspects are not addressed in full in this book. The reader is referred to numerous papers. The basic, early papers on flexible/resilient heat HENs are for instance: Saboo and Morari (1984, 1986a, 1986b), Kotjabasakis and Linnhoff (1986, 1988), Linnhoff and Kotjabasakis (1986), Grossmann and Floudas (1987), Colberg et al. (1989), Ratnam and Patwardhan (1991), Picon-Nunez and Polley (1995a, 1995b). Systematic solution methods have been proposed, among others, in Floudas and Grossmann (1986, 1987a, 1987b), Floudas and Ciric (1989, 1990), Cerda et al. (1990), Cerda and Galli (1990, 1991), Papalexandri and Pistikopoulos (1993a, 1993b, 1994), Aguilera and Nasini (1995, 1996). Jeżowski and Jeżowska (1999) presented a brief review and analysis of some basic approaches published up to 1998. Finally, more recent papers on this topic are, for instance, Novak and Kravanja (1999), Tantimurata et al. (2001), Aaltola (2002), Konukman and Akman (2002), Giovanini and Marchietti (2003), Chen and Hung (2004), Pintarič and Kravanja (2004), Verheyen and Zhang (2006) to name a few.

Here we will explain only some targeting issues. The objective is to provide insights into complex behavior of pinch changes under disturbances. Generally, for so-called nonconvex problems, pinches can jump within the whole temperature range along with changes of parameters. Multiple simultaneous pinches are common in contrast to cases with no disturbances. Utility usage (cost) often varies in wide ranges. If flow rates are only subject to disturbances with fixed inlet temperatures, pinches can only occur at temperature interval bounds (inlet temperatures). This significantly simplifies the problem though pinch jumps and multiple simultaneous pinches can still exist. Here we will present the method by Jeżowski et al. (2000b), which accounts only for CP disturbances.

We will assume, similarly to many other works, that CP disturbances are purely random and can take any value from the fixed ranges. Limiting upper and lower values are known ($CP^{\min}$, $CP^{\max}$).

The objective of the method is to find all the pinches that can occur under CP disturbances on condition that the cost of utilities is kept at a minimum. This requires a recursive procedure where at each step a MILP model has to be solved. The model uses the transshipment formulation. Thus, temperature intervals have to be created ($k = 1, \ldots, K$) with temperature differences ΔT_k. Recall that if residuals flowing via a temperature bound of interval k (T_k) are all zero than T_k is the pinch.

The objective of the MILP model is to determine the maximum number of pinches at the minimum utility cost (UC), that is:

$$\{\max (\text{no. of pinches})\} \text{ and } \{\min UC\} \qquad (15.62)$$

To eliminate multicriteria optimization for the max–min optimization, problem (15.62), is converted into minimization of the single criterion as follows:

$$\min \left[\sum_{k=1}^{K} y_k + \Phi \left(\sum_{m \in HU} phu_m Qhu_m + \sum_{n \in CU} pcu_n Qcu_n \right) \right] \qquad (15.63)$$

The second term defines the cost of utilities and is similar to the goal function (15.17) in the S-MUC-Transs model in Section 15.2. The first term of the criterion is equivalent to the number of nonpinched intervals since binary variable y_k is defined as follows:

$$y_k \begin{cases} 1 - \text{if there is a heat flow via temperature } T_k, \text{ i.e. } T_k \text{ is not the pinch} \\ 0 - \text{otherwise (i.e. } T_k \text{ is the pinch)} \end{cases}$$

$$(15.64)$$

Hence, minimization of $\sum_{k=1}^{K} y_k$ is equivalent to maximization of the number of pinches that can occur. Parameter Φ is the heuristic parameter that should ensure reaching both goals: maximum number of pinches at minimum utility cost. Jeżowski et al. (2000b) stated that it is sufficient to apply a small value from the range 10^{-4} to 10^{-2}. Alternatively, to apply more reliable value one can solve the S-MUC-Transs model for a certain *ad hoc* chosen CP value to determine the minimum utility cost (MUC`) and the number of pinches (NP`). Parameter Φ should satisfy the condition:

$$(\Phi MUC') \leq (N') \tag{15.65}$$

The full model for determining the maximum number of pinches at the minimum utility cost is as follows.

Uncertain-Transs model

$$\min \left[\sum_{k=1}^{K} y_k + \Phi \left(\sum_{m \in HU} phu_m Qhu_m + \sum_{n \in CU} pcu_n Qcu_n \right) \right] \tag{15.66}$$

s.t.

$$R_{ik} - R_{i,k-1} + \sum_{j \in C_k} q_{ijk} + \sum_{n \in CU_k} q_{ink} = CP_i \Delta T_k; \ i \in H_k; \ k = 1, \ldots, K$$

$$(15.67)$$

$$R_{mk} - R_{m,k-1} + \sum_{j \in C_k} q_{mjk} = Qhu_m; \ m \in HU_k; \ k = 1, \ldots, K \tag{15.68}$$

$$\sum_{i \in H_k} q_{ijk} + \sum_{m \in HU_k} q_{mjk} = CP_j \Delta T_k; \ j \in C_k; k = 1, \ldots, K \tag{15.69}$$

$$\sum_{i \in H_k} q_{ink} = Qcu_n; \ n \in CU_k; \ k = 1, \ldots, K \tag{15.70}$$

$$Q_k = \sum_{i \in H} R_{ik}; \ k = 1, \ldots, K \tag{15.71a}$$

$$Q_k \leq y_k Q_k^{\max}; \ k = 1, \ldots, K \tag{15.71b}$$

$$CP_i \geq CP_i^{\min}, CP_i \leq CP_i^{\max}; \ i \in H \tag{15.72}$$

$$CP_j \geq CP_j^{\min}, CP_j \leq CP_j^{\max}; \ j \in C \tag{15.73}$$

$$R_{i0} = R_{iK} = R_{mK} = 0; \ i \in H; \ m \in HU \tag{15.74}$$

$$R_{ik}, R_{mk}, q_{ijk}, q_{mjk}, q_{ink} \geq 0 \tag{15.75}$$

The first four constraints (15.67) to (15.70) of the model are almost identical to Equations (15.18) to (15.21) modeling heat exchange in the S-MUC-Transs model. The only difference is in the right-hand sides of Equations (15.67) and (15.69). Instead of given heat loads of streams in intervals in the S-MUC-Transs model the products of CP and temperature range of intervals ΔT have to be applied since CPs are variables. In spite of this, the model is still linear since ΔT_k are parameters. Inequality (15.71b) ensures that total residual flow via the upper bound of interval k is set at zero if the corresponding binary y_k is zero. The sum of stream residual flows is defined by Equation (15.71a). In fact both constraints can be merged. Parameters $Q_k^{\max}$ have to be properly chosen so as to ensure the desired effect. Jeżowski et al. (2000b) stated that it is sufficient to use a single large number for all $Q_k^{\max}$. Alternatively, $Q_k^{\max}$ (or a single $Q^{\max}$) can be more accurately estimated by solving the S-MUC-Transs model for some values of CP.

A solution of the Uncertain-Transs model gives the maximum number of pinches for admissible ranges of disturbances as a condition of the minimum cost of utilities. These pinches exist at certain CP values calculated by the model. However, the designer should know the locations of all possible simultaneous pinches (double, triple and so on) as well as single ones that can occur under admissible disturbances. A recursive procedure is to be applied with a solution of the model in each iteration. To eliminate identical results integer cuts have to be inserted into the model such that they cut off the solutions from the previous runs. Let define for each iteration p of the procedure index set P^p containing indices l of pinched intervals:

$P^p = \{l | y_l$ is 1 in run p of the procedure; that is, temperature T_l is not the pinch$\}$

Notice that P^1 is an empty set. Then, in each consecutive runs cutting conditions (15.76) has to be added to Uncertain-Transs model.

$$\sum_{l \in P^1} y_l \geq 1 \qquad (15.76)$$

The iterations continue until the model yields no solution (no new set of pinches can be determined).

Jeżowski et al. (2000b) found that the conditions are not sufficient since ill-defined solutions can be generated. An ill-defined solution is by definition such that $q_{k'}$ for some $k' \in K$ are zero but corresponding $y_{k'}$ are equal to 1, that is, there are pinches at $T_{k'}$ though binaries $y_{k'}$ do not show them as they should do. This effect results from the fact that logical condition (15.71) in the model does not eliminate a possibility of calculating such an ill-defined solution in the recursive procedure. Ill-defined solution can occur at further runs j ($j > 1$) if, because of integer cuts, such set P^r is computed such that it is a subset of previously calculated set P^p ($p < r$). In fact, such ill-defined solution P^r is equivalent to P^p. To eliminate ill-defined solutions Jeżowski et al. (2000b) added, together with integer cuts (15.76), additional constraints (15.77) in each run p.

$$Q \geq Q_l^{\min} \cdot y_l; \ l \in P^i \qquad (15.77)$$

Values of $Q_l^{\min}$ have to be given in data. Basically, it is sufficient to use sufficiently small values. However, $Q_l^{\min}$ corresponds to minimal value of heat "leakage" across a temperature T_l and a designer can assign a value that is meaningful for practical HEN design.

It is necessary to note that, in spite of additional constraints (15.77), the recursive procedure can produce solutions that are not of interest if one is looking only for pinches since some solutions are simply subsets of solutions from the previous iterations. However, they have different values of CP and different cost of utilities. Hence, they are important for control purposes. It is also worthwhile to note that differences in utility consumption are not simply caused by heat leakage $Q_l^{\min}$ via some pinches as in case of fixed data since there is an additional degree of freedom because CPs are variables. In the following we present Example 15.3 taken from Jeżowski et al. (2000b). It was also solved as Example 2 in Cerda et al. (1990), however with some mistakes—see the discussion in Jeżowski and Jeżowska (1999), Jeżowski et al. (2000b). Data for process streams

Table 15.4 Data for Example 15.3

Stream	T^{in} (°C)	T^{out} (°C)	CP^{min} (kW K^{-1})	CP^{max} (kW K^{-1})
H1	390	270	**1.4**	2.0
H2	360	180	1.5	1.5
H3	400	378	0.5	0.5
C4	200	370	**1.0**	1.5
C5	250	400	1.8	1.8

Table 15.5 Possible pinches for Example 15.2 according to Jeżowski et al. (2000b)

No	Pinches	Qhu	Qcu	Qu	CP_{C4}	CP_{H1}
1	360/370–250/260	14.364	38.182	52.546	1.336	2.0
2	250/260–200/210	34	30	64	1.5	2.0
3	390/400–360/370	13	60	73	1.2	2.0
4	250/260	14.473	38.136	52.609	1.337	2.0
5	360/370	14.355	38.321	52.676	1.335	2.0
6	390/400	13.000	61.700	74.700	1.130	2.0
7	No pinches	14.464	38.283	52.747	1.336	2.0

are shown in Table 15.4. A single heating utility and single cooling utility are applied and, thus, utility prices are not necessary.

$\Delta T^{min} = 10$ K, temperatures of streams in shifted scale, that is, temperatures of hot streams reduced by ΔT^{min}; ranges for CP disturbances in bold.

The solution to Uncertain–Transs model is (first iteration):

$Qhu = 14.364$, $Qcu = 38.182$, $Qu = 52.546$ with simultaneous pinches at 360/370 °C and 250/260 °C with $CP_{C4} = 1.336$ and $CP_{H1} = 2.0$.

All possible pinches calculated by the recursive procedure are listed in Table 15.5.

The example is simple with four processes only and, additionally, parameters CP of two streams are fixed. In spite of this the minimum load on utilities varies in wide range. Pinches can occur at practically all permissible positions with three sets of two simultaneous pinches. Jeżowski et al. (2000b) gave a second example showing a more complex picture of maximum heat recovery conditions. It shows a complexity of design and control measures needed to achieve full heat recovery under disturbances.

16 Minimum number of units (MNU) and minimum total surface area (MTA) targets

16.1. INTRODUCTION

As we have mentioned in Chapter 14, targeting approaches for minimum number of units and minimum total surface area have to account for the type of heat exchangers applied. Most often they are limited to shell-and-tube apparatus with pure counter-current flow. Such apparatus are called matches or 1-1 units (1 shell–1 tube pass). Thus, such methods are able to determine the minimum number of matches (MNM target) and the minimum total surface for matches (MTA-m target). Some methods are available for calculating targets for m-n heat exchangers featuring m shells each with n tube passes. One can employ them to determine minimum number of shells (MNS target) and minimum total surface area with shells (MTA-s target).

Targeting approaches for multishell apparatus are general. They should encompass 1-1 apparatus since the correction factor to MTD for 1-1 apparatus equals 1. However, this is not always the case in existing targeting methods. It is the main reason we will present here approaches for matches and multishell heat exchangers as well. Also, the methods relevant only for 1-1 apparatus are very often the starting points for targeting approaches accounting for multishell units. It is almost impossible to explain and understand the latter without knowledge of the former.

In order to calculate the MNU and MTA targets one should know the total number of hot streams—that is, hot process streams and heating utilities—and cold ones—that is, cold process streams and cooling utilities—that are considered for exchanging heat in a heat exchanger network. Thus, further division into utilities and process streams is not necessary and we will not differentiate between them in the following unless stated otherwise.

The necessary data for process streams are the same as those required for calculating the MUC target. Here, we need the same data for all utilities: it means that CP, T^{in} and T^{out} values are necessary. Additionally, to calculate minimum area target we have to know individual heat transfer coefficients (u) for all streams or, alternatively, overall heat transfer coefficients (U) for all pairs of hot and cold streams.

Energy Optimization in Process Systems

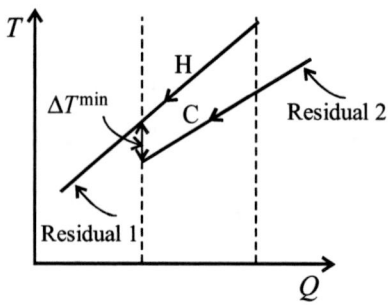

Figure 16.1 A match with a single residual stream.

16.2. MINIMUM NUMBER OF MATCHES (MNM) TARGET

Many researchers have sought a simple but rigorous formula for the minimum number of matches but to date it has not been found. However, there exist some approximations, which produce sufficiently accurate results in most cases.

Historically, the first equation for minimum number of units was given by Hohmann (1971):

$$N^{\min} = N - 1 \tag{16.1}$$

It is often applied to date and referred to as the "$N - 1$ rule." The equation is only the approximation but provides a good upper bound on the attainable minimum number of 1-1 units. Notice that the target from Equation (16.1) is the "global" minimum since it does not account for the MER/MUC target. However, both targets (MNM from Equation (16.1) and MER/MUC) are reachable in many cases.

A simple reasoning can develop the formula. Let us consider a set S of N streams: NH hot streams and NC cold ones. Then, choose hot stream H and cold stream C from the set S for heat exchange in a match. Assume that heat exchange between H and C can be performed in such a way that only a single residual stream is created. A residual stream is a segment (part) of a hot or cold stream that requires further cooling down or heating up to reach target conditions. The condition of a single residual stream created in the match H–C means that one stream (H or C) reaches the given temperatures while the second one does not, but only one part of it requires further cooling or heating. This is explained in Figure 16.1.

Notice that thermodynamic conditions (13.5), (13.6) can prevent reaching the single residual illustrated in Figure 16.2.

The residual stream is returned to the set of streams requiring further heat exchange. Thus, after a single match this set contains $(N - 1)$ streams. Following this scheme one can easily conclude that after $N - 1$ matches there will be no streams in set S. Hence, the final network will have $N - 1$ matches.

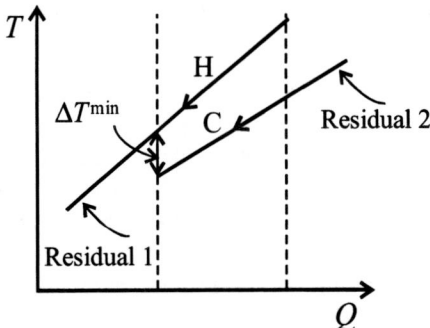

Figure 16.2 A match with two residual streams.

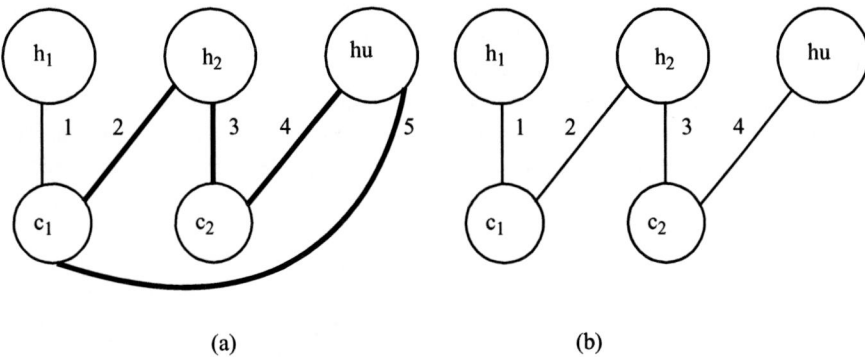

(a)   (b)

Figure 16.3 Graph representation of HEN: (a) general graph with loop 2–3–4–5, (b) tree obtained from the graph in (a) by eliminating the loop.

The reasoning is not rigorous since it requires additional proof that the "single residual match" is always attainable. Moreover, it does not account for the MER or MUC target.

The $N - 1$ rule was improved by Linnhoff et al. (1979) on the basis of graph theory. A HEN is represented by a planar graph in the following way. A stream is a node while an arc represents a match. Since heat exchange is performed between a hot stream/node and a cold stream/node the arcs can connect only hot nodes and cold nodes (notice that this assumption excludes dual matches). Hence the HEN graph is a bi-partite graph. An example is shown in Figure 16.3a.

It is known from basic graph theory that a tree, that is, graph with no loops, has the minimal number of arcs. Hence, to minimize the number of arcs/matches, the graph representing the HEN should be reduced to a tree. To reach the tree all loops in a graph have to be eliminated (often called in HEN literature "loop breaking"). Figure 16.3b shows the tree obtained by eliminating (breaking) the loop 2–3–4–5 from the graph of Figure 16.3a. The minimum number of arcs in a tree with fixed number of nodes (N) is defined by the Euler theorem:

$$N^{min} = N - N^{sub} \tag{16.2}$$

where N^{sub} denotes a number of independent subtrees.

Hence, providing that HEN in the form of a tree can be reached, Equation (16.2) with N equal to the number of streams gives the (global) minimum number of matches.

For a connected tree having no independent subtrees N^{sub} parameter is equal to 1, and then formula (16.2) becomes identical to Equation (16.1). Nevertheless, Equation (16.2) is more general. It also gives important advice for designing HEN since it shows a way to reduce the number of matches by forming independent subsets of streams that are in heat balance. The problem will be given deeper consideration in the following. It is, however, necessary to note here that there is no simple formula to calculate the N^{sub} value directly from the data—solution of an optimization model is necessary.

Formula (16.1), similar to Equation (16.2), does not account for the MER and MUC target. Therefore, the targets (N^{min} and MER/MUC) can be inconsistent. To reach consistency it is sufficient, according to pinch technology concepts, to take into account the basic rule ensuring the maximum energy recovery, that is, the rule of "no heat exchange across pinch"—see Chapter 13. Let the pinch for a problem be at temperature Tp. To follow the rule of "no heat exchange via pinch" one can simply divide the streams at temperature Tp into two subsets. In regard to heat exchange the subsets are independent. Additionally, each is of threshold type and requires only one type of utility. It is sufficient to apply formula (16.2) for both subsets, calculate N^{min} for them and, finally, add them together. In the general case of P pinches dividing the set of streams into $(P+1)$ subsets the equation for the MNM target consistent with the MER/MUC target is:

$$N^{min} = \sum_{p=1}^{P+1} (N_p - N_p^{sub}) \tag{16.3}$$

where N_p denotes the number of streams in subset p, and N_p^{sub} is the number of independent subtrees in subset p.

Using Example 13.1 (Table 13.1) we will show here the application of the formula and an increase of match number caused by the pinch division. The pinch temperature in Example 13.1 is at $Tp = 70\,°C$ and one hot and one cold utility is required. Since there are four process streams the minimum number of matches from Equation (16.1) is: $N^{min} = (4+2) - 1 = 5$.

The division at Tp produces two subsets of streams shown in Figure 16.4. The above pinch subset has three process streams and below the pinch four process streams. One heating utility is necessary above the pinch and one cooling

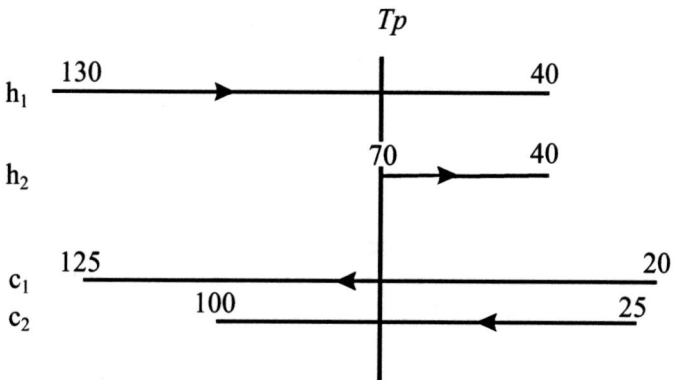

Figure 16.4 Division at pinch in Example 1.

medium below it. From formula (16.3) we obtain: $N^{min} = (3 - 1) + (4 - 1) = 7$. The number of matches increases by two.

For a long time Equations (16.2) and (16.3) have been considered as rigorous. To determine the N^{sub} parameter we should apply the MILP transshipment model abbreviated in the following to "N^{min} – Transs". The easiest and most commonly applied solution is to assume that N^{sub} equals 1 for each subset of streams created by pinch division. However, more recently Furman and Sahinidis (2004) showed that N^{min} according to Equation (16.2) could not always be reached even for threshold problems having no subsets. This was shown with an illustrative example and thus no conclusions were drawn on what conditions are necessary to reach N^{min} from Equation (16.2). In the authors' opinion formula (16.2) is, however, a quite good approximation of the minimum number of matches. For pinched problems Equation (16.3) overestimates the results and should be considered the upper bound. The reason is that in streams division at pinch temperature Tp calculated for a fixed value of the minimum temperature approaches (ΔT^{min}). In fact there is another reason—the rule deeply rooted in pinch technology concepts is that the same ΔT^{min} value should be applied to calculate all the targets. Hence in pinch technology approaches, the MER or MUC targets are calculated for a fixed ΔT^{min} value, often referred to as the HRAT (from heat recovery approach temperature). The same value is used to calculate other targets and also to design HEN. There are some arguments for the assumption, such as unification of decision variable in all sequentially calculated targets, simplification of calculating procedures and minimization of total surface area, but evidently the assumption can limit achieving optimal solutions in regard to total annual cost.

An alternative concept is to use the so-called "double temperature approach" (DTA). One ΔT^{min} value (i.e. HRAT) is applied for calculating the MER or MUC and, another, but not greater, to calculate other targets and also to design HEN. This second temperature approach is called EMAT (exchanger minimum approach temperature). It is of utmost importance that in DTA concept relation (16.4) must hold:

$$EMAT \leq HRAT \qquad (16.4)$$

It should be recognized that EMAT is only the lower limit on the temperature approach in matches. Hence, temperature approaches in exchangers can vary. Notice also that for EMAT fulfilling inequality (16.4) the minimum utility loads are not higher (they are lower for strict inequality in inequality (16.4)) than for HRAT. Hence the DTA concept provides an additional degree of freedom, since there is an additional chance of distributing heat loads among matches so as to reduce their number without increasing the MER/MUC above the minimum determined for HRAT. The reasoning is also valid for the MTA target. However, to make use of this degree of freedom, optimization is necessary though, in some simple cases, a reduction of matches is possible by inspection (Jeżowski, 1991). The papers by Trivedi et al. (1989a, 1989b) are some of the early contributions to HEN design with the DTA concept.

It is important to note that there is also another concept called the "diverse pinch approach" developed by Rev and Fonyo (1991), which features an apparent resemblance to the DTA notion. The diverse pinch method also applies different temperature approaches. However, the main objective is to minimize the total area of HEN. There also exist some other conceptual differences, for instance approaches that are match dependent. This topic will be considered in depth in Section 16.3 on minimum total surface area target.

Since the use of various concepts of temperature approach may cause misunderstandings, it is convenient to think in terms of "heat recovery level reachable for fixed HRAT and exchanger temperature approach EMAT to calculate heat loads". Fraser (1989) advocated the use of "heat flux" instead of the temperature approach as the decision variable in targeting. However, this suggestion, though interesting, has not found application in HEN targeting and design.

Finally, it should be noted that there is also another means to reduce the number of matches in comparison to the results from Equation (16.3). This technique can be applied with the single minimum temperature approach concept. The idea is to construct the matches in specific arrangements first suggested by Wood et al. (1986) and called here the "parallel pinch crossing arrangement". In order to explain the arrangement let us consider the illustrative example from Figure 16.5.

It is required that there exists at least one hot stream across the pinch (h_1 in Figure 16.5) and at least one hot stream with inlet temperature equal to pinch temperature (at-pinch stream, h_2 in Figure 16.5). Also, one cold pinch crossing stream is necessary (c in Figure 16.5) such as can be split into two branches c' and c''—see Figure 16.5. Branch c' has to feature the same CP value as pinch crossing hot stream h_1. This branch is matched with hot stream h_1 (match no. 1). The second branch exchanges heat with at-pinch hot stream h_2. It is required that the outlet temperature of branch c'' from match no. 2 is equal to the pinch

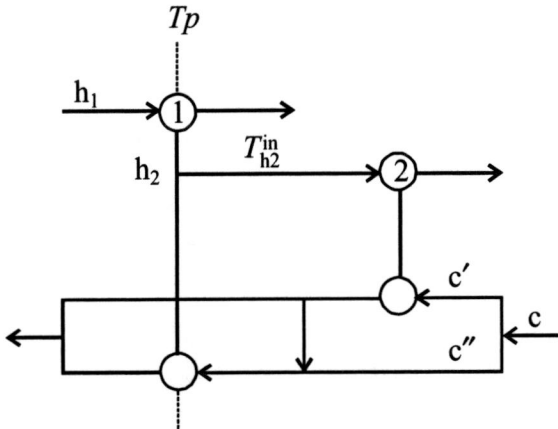

Figure 16.5 Illustration for parallel pinch crossing arrangement.

temperature (in shifted temperature scale). The required final temperature of stream c is reached by mixing two branches. For such conditions there is no heat exchange via pinch in both matches. Thus, the basic rule of pinch technology is met and the arrangement allows reduction of the number of matches from three (for pinch division) to two (the global minimum according to Equation (16.2)) in this example. There is no certainty that the parallel pinch arrangement always allows reaching of the MNM target from Equation (16.1) or (16.2)—a deeper analysis is also given in Jeżowski (1990a). However, the point is that such an arrangement features some disadvantages. First, it requires an additional splitter, which incurs additional cost and problems with HEN control and operation. Second, the pinch-crossing match (such as match 2 in Figure 16.5) often has a large area, particularly for multishell apparatus, since CP values of streams are identical. This may result in a small mean temperature difference in a large area and also a high number of shells. Summing up, the parallel pinch crossing arrangement seems to be of limited use in industry.

Let us proceed to calculations of the minimum match number with the DTA concept. As stated, this requires solution of an optimization model. In order to ensure the minimum area target will be consistent with the MER or MUC target the utility loads satisfying the MER/MUC must be known. It is of importance that they are calculated for a fixed HRAT. The first optimization model for N^{min} value was developed by Papoulias and Grossmann (1983b) and is still widely used. The model is based on the standard transshipment model for MUC (S-MUC-Transs) presented in Section 15.2. The transshipment formulation is applied to ensure feasible heat exchange. It is important that the solution to the model of Papoulias and Grossmann (1983b) provides not only an N^{min} value but also gives an optimal heat load distribution. By heat load distribution we mean heat loads of matches (q_{ij}) for hot ($i \in H$) and cold streams ($j \in C$) in the problem. The model solution yields such matches that are optimal in regard to both utility cost and number of matches, that is, they ensure meeting both the MER/MUC and MNM targets. Knowledge of heat distribution is a great advantage at in the HEN design stage though it does not fully define the structure since the splitting and mixing points are not known.

Here we start the presentation of the transshipment model for minimizing number of matches, abbreviated to MNM-Transs. The symbols applied are the same as in the S-MUC-Transs model (Section 15.2) apart from for some differences. First, it is not necessary to define separate index sets of hot and cold utilities. Instead, only two index sets are necessary:

$H = \{i \mid i$ is a hot stream$\}$; $C = \{j \mid j$ is a cold stream$\}$
After the division into intervals we create the following index sets:
$H_k = \{i \mid i$ is present in interval k or can exchange heat with $j \in C_k\}$
$C_k = \{j \mid j$ is present in interval $k\}$

Heat loads of all streams in intervals have to be given: ΔQ_{ik}^H for $i \in H_k$ and ΔQ_{jk}^C for $j \in C_k$.

To account for matches between streams $i \in H$ and $j \in C$ we have to introduce binary variable y_{ij}, such that:

$$y_{ij} = 1 \text{ if } i \in H \text{ exchanges heat with } j \in C$$
$$y_{ij} = 0 \text{ if not} \tag{16.5}$$

Due to the binary variable the model is of the MILP type.

The transshipment model for minimizing both utility cost and match number is as follows.

MNM—Transs model

$$\min \sum_{i \in H} \sum_{j \in C} y_{ij} \tag{16.6}$$

s.t.

$$R_{ik} - R_{i,k-1} + \sum_{j \in C_k} q_{ijk} = \Delta Q_{ik}^H; \quad i \in H_k; \quad k = 1, ..., K \tag{16.7}$$

$$\sum_{i \in H_k} q_{ijk} = \Delta Q_{jk}^C; \quad j \in C_k; \quad k = 1, ..., K \tag{16.8}$$

$$q_{ij} = \sum_{k=1}^{K} q_{ijk}; \quad i \in H; \quad j \in C \tag{16.9}$$

$$q_{ij} - y_{ij} q_{ij}^{\max} \leq 0; \quad i \in H; \quad j \in C \tag{16.10}$$

$$R_{i0}, R_{iK} = 0 \tag{16.11}$$

$$R_{ik}, q_{ijk} \geq 0; \quad i \in H; \quad j \in C; \quad k = 1, \ldots, K \tag{16.12}$$

Similarly to the S-MUC-Transs model, Equations (16.7) and (16.8) are heat balances ensuring that given heat loads of streams will be satisfied. Equation (16.9) is the definition of heat loads on matches. Of interest is inequality (16.10) since it is a typical example of a logical condition common in constructing models for designing optimal systems. Parameters $q_{ij}^{\max}$ are large numbers given in data. For $y_{ij} = 0$, that is, for the case of no heat exchange between streams i and j, condition (16.10) forces the heat load of the match to zero (notice that q_{ij} must be nonnegative according to Equation (16.12)). For $y_{ij} = 1$ inequality (16.10) has to be met, too—in fact it will be inactive since $q_{ij}^{\max}$ should be a sufficiently large number. Often a single parameter $q^{\max}$ is used for all matches, that is,

$$q_{ij}^{\max} = q^{\max}; \quad i \in H; j \in C \tag{16.13}$$

However, this may produce large "gaps" that can cause difficulties in solving the MILP model. To avoid problems we can use tighter approximations $q_{ij}^{\max}$ by calculating the maximum possible heat loads of matches for $i \in H$ and $j \in C$.

It is an easy task to include forbidden, must-be and restricted matches into the model.

The MNM-Transs model can be solved using a single HRAT as well as HRAT and EMAT. In the first case the temperature intervals are created for the HRAT value and the same as applied to calculate the optimum utility loads. Additionally, some authors divide the streams at pinches and solve the models for each subset of streams. This is, however, mainly for the purpose of comparing the results on the same basis as that applied in PT approaches. In frames of DTA concept utility loads are also calculated for a given HRAT. Then, temperature intervals are created using an EMAT that meets condition (16.4). It has been shown in several papers that such a scheme allows a lower number of matches than that with a single HRAT. In fact, in most cases $N^{\min}$ values from Equation (16.2) have been achieved for examples quoted in the literature since optimization accounts for independent subtrees. This is not possible with a single HRAT.

The question arises as to what EMAT value to apply. Too small values can result in such heat load distribution, which require large heat transfer area, and, in consequence, large TAC. Inferior results are possible without simultaneously accounting for all targets. An approach that optimizes HEN total annual cost will be addressed in Chapter 17. Notice that one can apply match dependent EMAT values provided one has the means to assess them sensibly. For instance, minimum temperature approaches in heaters are usually much higher than in heat exchangers in refrigeration cycles.

The standard formulation of HEN targeting and design problems requires that outlet temperatures of all streams have to be met as points, that is, the conditions on outlet temperatures are equalities (we will call them "hard conditions" in the following). However, in an industrial scenario some process streams are often allowed to vary outlet temperature within a certain range. Hence, Shethna and Jeżowski (2006) stated that the requirement of reaching fixed outlet temperatures by all streams is superfluous. The idea was also advocated by Johns (2001). Relaxation of hard conditions into weak ones, in the form of inequalities, should give additional freedom in targeting and designing procedures and, also, should result in better solutions. Shethna and Jeżowski (2006) developed a method for calculating the HLD optimal in respect of both utility cost and number of matches. The approach, referred to as the "near-balanced subnetworks" method, has the same objective of the MILP model of Papoulias and Grossmann (1983a, 1983b) but accounts for relaxed conditions on outlet temperatures. Mocsny and Govind (1984) proposed the first approach to the problem of targeting for a number of matches with relaxed temperatures. However, their method has some drawbacks and requires a large computation load—for details see Shethna and Jeżowski (2006).

Assume that a designer specifies a nominal outlet temperature of all streams and also provides ranges of the temperature variation for all streams or some of them. Such tolerances can be estimated on the basis of the knowledge of downstream processes. Tolerances will give additional freedom to create more subnetworks in some cases, particularly in large-scale industrial problems. For fixed nominal outlet temperatures such subsets/subnetworks are not in heat balance and hence they were referred to in Shethna and Jeżowski (2006) as near-balanced or near-independent subnetworks.

The possibility of designing a HEN with several near-independent subnetworks has the following advantages:

1. It reduces the total number of heat exchangers in a HEN according to Equation (16.2), and in consequence can decrease its total annual cost (TAC).
2. It makes operability and controllability problems easier to manage.
3. It makes the detailed design stage easier since subnetworks have a small number of streams.

The solution from Shethna and Jeżowski's (2006) approach meets the following conditions:

C1. Number of independent subnetworks making up the overall HEN is maximal (to minimize number of heat exchangers in overall HEN).
C2. Given tolerances on outlet temperatures or tolerances on heat loads of individual streams are satisfied.
C3. Heat loads of utilities are minimized (i.e. the loads necessary for compensating deviations from fixed nominal outlet temperatures).

Similar to the method of Papoulias and Grossmann (1983b) the MUC problem has to be solved first for a given HRAT value. As a result we have an initial set S of streams, that is, process streams and utilities defined by index sets:

$$H = \{i|i \text{ is a hot stream}\}; \quad C = \{j|j \text{ is a cold stream}\}$$

The approach concludes in a recursive procedure where at each step $(t = 1, \ldots, T)$ two near-independent subsets are calculated from the current set (S) of streams. One subset (S_1) is minimal while the second (S_2) is its complement. The minimal subset includes the minimal number of streams. Both subsets meet conditions C2 and C3 and also fulfill conditions:

$$S_1 \cup S_2 = S \tag{16.14}$$

$$S_1 \cap S_2 = \Theta \tag{16.15}$$

First, the initial set is divided into two subsets. In successive steps of the procedure the complement set S_2 from the previous step is further divided into two subsets: the minimal subset and its complement. The procedure ends at step T when no further division is possible, that is, set S_2 at step $(T + 1)$ cannot be divided into the minimal subset and the complement subset $(S_1$ at step $T + 1$ is an empty set). The tree shown in Figure 16.6 illustrates this recursive procedure. The set of minimal subsets from steps 1, …, T and the complement from step T form an overall HEN consisting of $T + 1$ near-independent subnetworks (see also Figure 16.6). Because this HEN consists of minimal subsets (plus the complement) it contains the maximum number of near-independent subnetworks, that is, the HEN meets condition C1. Additionally, since all the generated subsets meet conditions C2 and C3 the overall HEN also satisfies them.

For the given set S both subsets are calculated by solving the MILP optimization model formulated in the framework of the transshipment formulation with division into temperature intervals (TIs).

For each stream s in set S a binary variable is defined such that:

$$y_s = 1 \text{ if } s \text{ is in } S_1; \quad y_s = 0 \text{ if } s \text{ is in } S_2 \tag{16.16}$$

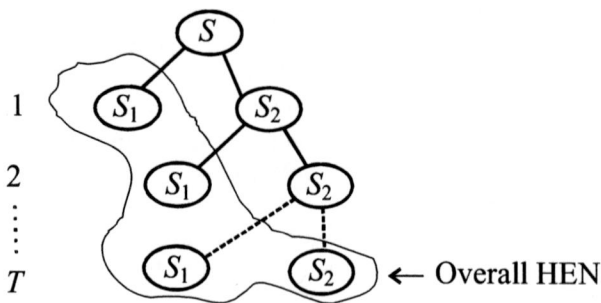

Figure 16.6 Illustration of the recursive procedure for generation of near-balanced subsets.

The minimal number of streams in the minimal subset is achieved by minimizing $\sum y_s$ for $s \in S$.

To ensure feasible heat exchange in both subsets two independent transshipment formulations have to be applied, one for the minimal subset and the second for its complement. To discriminate streams in the subsets the given enthalpy change of stream s is multiplied by y_s for each $s \in S_1$ and by $(1 - y_s)$ for each $s \in S_2$.

In order to account for tolerances and to control and minimize deviations from the given outlet temperatures, two types of variables are used:

1. residual heat flow from the last temperature interval of each hot process stream,
2. loads on slack heaters placed in the last temperature interval of each cold process stream.

The residuals control the temperature deviations for hot process streams while the loads of slack heaters control the deviations for cold process streams. In order to account for temperature tolerances a nonzero residual from the last interval has to be allowed in contrast to the transshipment models for fixed temperatures (MNM-Transs model). Also, heat balances for the last intervals of hot and cold process streams have to be relaxed into inequalities. Minimal deviations from fixed temperatures are achieved by minimizing total heat loads of slack heaters and total residual flows from the last intervals of hot streams.

Following Shethna and Jeżowski (2006) we apply here index set K for intervals since this allows compact presentation of the model.

$K = \{k \mid k$ is a number of the TI$\}$
Index sets H_k, C_k are the same as in the MNM-Transs model:
$H_k = \{i \mid i \in H$ and i is present in interval k or can cascade heat to interval $k\}$,
$C_k = \{j \mid j \in C$ and j is present in interval $k\}$

Application of HRAT used to calculate heat loads of utilities is not required to create intervals. Instead, an EMAT value such that $0.0 < \text{EMAT} \leq \text{HRAT}$ should be used similar to the transshipment models for heat recovery or HLD calculation.

It is convenient to distinguish the last temperature interval for each stream by defining the sets:

$K_i = \{K_i, i \in H$ is present in k_i but is not present in any $l > k_i$—that is, k_i is the last interval for $i\}$
$K_j = \{k_j, j \in C$, is present in k_j but is not present in any $l < k_j$—that is, k_j is the last interval for $j\}$

The data for the MILP optimization model of creating the subsets are as follows:

- minimum and maximum values of outlet temperatures (i.e. tolerances in terms of temperatures) or tolerances (*tol*) in terms of heat loads of streams
  $[TT_i^{\max}, TT_i^{\min}, TT_j^{\max}, TT_j^{\min}]$ or $[tol_i^{+/-}, tol_j^{+/-}]; i \in H, j \in C$

- enthalpy changes for streams in intervals (except the last ones) for temperature range of TIs
  ΔQ_k^H—enthalpy change of $i \in H_k$; $k \in K, k \notin K_i$
  ΔQ_k^C—enthalpy change of $j \in C_k$; $k \in K, k \notin K_j$

- minimal and maximal permissible enthalpy changes of streams in last intervals (here in terms of temperature tolerances)

$$(\Delta Q_k^H)^{\max} = CP_i(-TT_i^{\min} + T_i^{\text{last}}) \tag{16.17}$$

$$(\Delta Q_k^H)^{\min} = CP_i(-TT_i^{\max} + T_i^{\text{last}}) \tag{16.18}$$

$$(\Delta Q_k^H)^{\min} = CP_j(TT_j^{\min} - T_j^{\text{last}}) \tag{16.19}$$

$$(\Delta Q_k^H)^{\max} = CP_j(TT_j^{\max} - T_j^{\text{last}}) \tag{16.20}$$

where T_i^{last} is the highest temperature of the last TI for $i \in H$ and T_j^{last} is the lowest temperature of the last TI for $j \in C$.

Figure 16.7 illustrates the symbols and heat exchange in the last interval of hot stream (Figure 16.7a) and cold stream (Figure 16.7b).

The variables applied in the model are:

q_{ijk}—heat loads of matches for $i \in H, j \in C$ in intervals $k \in K$
R_{ik}—residual heat of $i \in H_k$ that leaves interval k (and enters interval $k + 1$)

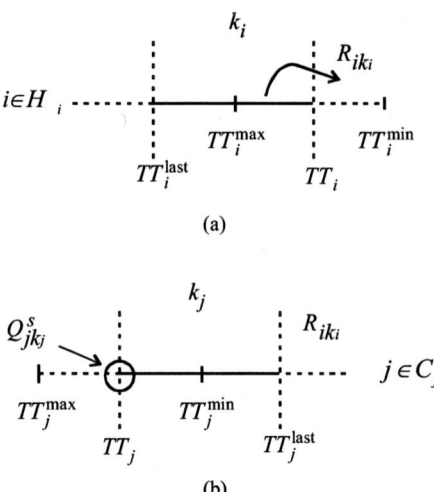

Figure 16.7 Illustration of heat exchange pattern and symbols for MNM-Transs-Relax model: (a) for hot streams, (b) for cold streams.

q_{jk}^s—heat loads of slack heaters; $j \in C_k$, $k \in K_j$
y_i—binary variables for $i \in H$; $y_i = 1$ if stream i is in subset S_1 and 0 if not
y_j—binary variables for $j \in C$; $y_j = 1$ if stream j is in subset S_1 and 0 if not.

The linear optimization problem for dividing a given stream set S into its minimal subset and its complement is as follows.

MNM-Transs-Relax model

$$\min\left\{ \left(\sum_{j \in C} y_j + \sum_{i \in H} y_i \right) + W \left(\sum_{j \in C_k, k \in K_j} q_{jk}^s + \sum_{i \in H_k, k \in K_i} R_{ik} \right) \right\} \quad (16.21)$$

s.t.

$$R_{ik} - R_{ik-1} + \sum_{j \in C_k} q_{ijk} = y_i \Delta Q_k^H; \quad i \in H, k \in K, k \notin K_i \quad (16.22)$$

$$\sum_{i \in H} q_{ijk} = y_j \Delta Q_k^C; \quad j \in C, k \in K, k \notin K_j \quad (16.23)$$

$$y_i(\Delta Q_k^H)^{\min} \leq R_{ik} - R_{ik-1} + \sum_{j \in Ck} q_{ijk} \leq y_i(\Delta Q_k^H)^{\max};$$
$$i \in H_k, k \in K_i \tag{16.24}$$

$$y_j(\Delta Q_k^C)^{\min} \leq q_{jk}^s + \sum_{i \in H_k} q_{ijk} \leq y_j(\Delta Q_k^C)^{\max};$$
$$j \in C_k, k \in K_j \tag{16.25}$$

$$R_{ik} - R_{ik-1} + \sum_{j \in C_k} q_{ijk} = (1 - y_i)\Delta Q_k^H; \quad i \in H, k \in K, k \notin K_i \tag{16.26}$$

$$\sum_{i \in Hk} q_{ijk} = (1 - y_j)\Delta Q_k^C; \quad j \in C, k \in K, k \notin K_j \tag{16.27}$$

$$(1 - y_i)(\Delta Q_k^H)^{\min} \leq -R_{ik-1} + \sum_{j \in C_k} q_{ijk} + R_{ik} \leq (1 - y_i)(\Delta Q_k^H)^{\max};$$
$$i \in H_k, k \in K_i \tag{16.28}$$

$$(1 - y_j)(\Delta Q_k^C)^{\min} \leq q_{jk}^s + \sum_{i \in H_k} q_{ijk} \leq (1 - y_j)(\Delta Q_k^C)^{\max}; \quad j \in C_k, k \in K_j$$
$$\tag{16.29}$$

$$\sum_{i \in H} y_i \geq 1 \tag{16.30a}$$

$$\sum_{j \in C} y_j \geq 1 \tag{16.30b}$$

$$R_{ik} \geq 0; \quad i \in H_k, k \in K, k \neq 1 \tag{16.31}$$

$$q_{ijk} \geq 0; \quad i \in H_k, j \in C_k \tag{16.32}$$

$$q_{jk}^s \geq 0; \quad j \in C_k \tag{16.33}$$

$$y_i = \{0, 1\}; \quad i \in H \tag{16.34}$$

$$y_j = \{0, 1\}; \quad j \in C \tag{16.35}$$

The objective function (16.21) minimizes the number of matches in the minimal subset. Also, it minimizes loads on slack heaters, that is, temperature relaxation of cold streams and residuals from last intervals of hot streams, that is, temperature relaxation of hot streams. Notice that these two terms in the goal function are competing. Hence, Shethna and Jeżowski (2006) applied weighting factor W to scale the heat load on the residuals and slack heaters in the objective function. The factor should ensure that an optimization procedure minimizes the number of streams in the minimal subset. They have chosen the scale to be the inverse of the net heat load on all hot streams including the hot utilities. They stated that such a scale ensures that the net value of the sum of all residual and slack heat loads is dimensionless and less than unity, thus guaranteeing the uniqueness of the objective function value.

Equality constraints (16.22) and (16.23) are heat balance equations for streams in subset S_1 for temperature intervals except for the last ones. In order to account for the tolerances on the outlet temperatures of streams in S_1, heat balances for the last intervals of streams are modeled as inequalities (16.24) and (16.25). Constraints (16.26) – (16.29) are identical to those mentioned above (see pairs: (16.22) and (16.26), (16.23) and (16.27), (16.24) and (16.28), and (16.25) and (16.29)) but they are active only for streams in the complement subset S_2. Constraints (16.22) – (16.29) altogether form two separate, relaxed transshipment models for both subsets.

Inequalities (16.30a) and (16.30b) have to be included to ensure that the minimization of the number of streams in subset S_1 will not generate an empty set. Shethna and Jeżowski (2006) set lower bound equals two (2) on the number of streams in the minimal subset (i.e. at least one hot stream and one cold stream).

HLD in both subsets can be easily calculated from the results. Alternatively, one can slightly modify the model to have the heat load of matches directly in the results. It is sufficient to introduce new variables q_{ij} such as those defined by Equation (16.9) in the MNM-Transs model. Notice that the left-hand sides of constraints (16.23) and (16.27) can be applied to define the variables.

Though the original model did not involve restricted matches this can be simply included. Shethna and Jeżowski (2006) also mentioned that segmented streams had been added to their approach and presented a proper example—Example (16.1) in the following.

The method did not account for total area and thus did not ensure a TAC optimum network. So, they proposed a method for generating various solutions to the problem. The networks can differ in respect to the number of near-independent subsets and also for total area. Notice that some solutions can feature the same number of subsets because the linearity of the optimization problem results in multiple global optima. Also, there are solutions that feature a

Table 16.1 Data for Example (16.1)

Stream	T^{in} (°C)	T^{out} (°C)	CP (kW·°C^{-1})	ΔH (kW)
C1	323	423	7.62	762.00
C2	389	495	6.08	644.48
C3	311	494	8.44	1544.5
C4	355	450	17.28	1641.60
C5	366	478	13.90	813.90
C6	311	490	8.44	1510.8
C7	350	450	17.28	1728.00
C8	352	468	13.90	1612.40
C9	366	478	8.44	945.28
C10	311	489	8.44	489.00
C11	422	478	21.78	1219.7
C12	339	411	13.84	996.48
H1	433	366	8.79	588.93
H2	522	411	10.55	1171.10
H3	510	349	14.77	2378.00
H4	544	422	12.56	1532.3
H5	472	339	17.73	2358.10
H6	505	339	14.77	2451.80
H7	544	420	12.56	1557.44
H8	475	339	17.73	2411.30
H9	583	478	12.53	1315.70
H10	517	366	8.32	1256.30
H11	511	339	6.96	1197.1

smaller number of subsets than the maximum. To generate all possible solutions one can apply the procedure outlined above. To prevent generation of identical solutions integer cuts have to be added to the optimization problem in the second and all successive runs.

It is of importance that the approach uses a small number of binaries, less than the MNM-Transs formulation. Furman and Sahinidis (2004) were not able to solve the MNM-Transs model in reasonable CPU time for certain large-scale tasks. Hence, they suggested some approximation solution techniques to circumvent the difficulty. The method of Shethna and Jeżowski (2006) is expected to work even for large-scale tasks featuring a large number of streams since the number of binaries is equal only to the total number of streams ($N = NH + NC$) while the MNM-Transs formulation requires many as $NH·NC$ binaries.

Here we reproduce two examples from Shethna and Jeżowski (2006). The first example (16.1) was introduced by Mocsny and Govind (1984). The problem involves 12 cold streams and 11 hot ones—see Table 16.1.

The approach by Mocsny and Govind (1984) was able to reach four nearbalanced subsets (see Figure 1 in Mocsny and Govind (1984)) using maximal tolerances of 1.0% in terms of temperature:

S1 = {H5, H8, H11, C4, C7}, S2 = {H4, H6, H10, C5, C6, C11, C12}
S3 = {H1, H2, H3, H9, C1, C3, C8, C10}, S4 = {H7, C2, C9}

Hence, the overall HEN will consist of 20 matches according to Equation (16.1). In order to compare violations of utility heat loads for the case of fixed nominal final temperatures in both methods, Shethna and Jeżowski (2006) calculated heat loads of utilities for the subsets that it would be necessary to add to reach such temperatures. The results for subsets S1 to S4 are

S1: $Qhu = 0.0\,kW$, $Qcu = 43.22\,kW$
S2: $Qhu = 43.22\,kW$, $Qcu = 0.0\,kW$
S3: $Qhu = 0.0\,kW$, $Qcu = 32.3\,kW$
S4: $Qhu = 32.3\,kW$, $Qcu = 0.0\,kW$.

Total additional loads are $Qhu = 75.5\,kW$ and $Qcu = 75.5\,kW$.

Shethna and Jeżowski (2006) applied the same tolerances as Mocsny and Govind (1984) as well as EMAT = HRAT = 10 K. They obtained 7 subsets:

{H7, C5}, {H4, C3}, {H3, C8, C1}, {H2, H1, C7}, {H6, C12, C10}, {H11, H10, C6, C9}, and {H8, H5, H9, C11, C4, C2, CU}

Thus, a total HEN will consist of 17 units in 7 separate subnetworks (vs. 20 units in 4 subnetworks in the solution from Mocsny and Govind, 1984) and requires additional loads of $Qhu = 61.7\,kW$ (vs. 75.5 kW in Mocsny and Govind, 1984) and $Qcu = 61.7\,kW$ (vs. 75.5 kW in Mocsny and Govind, 1984) if nominal outlet temperatures are to be met.

The next example (16.2) from Shethna and Jeżowski (2006) is an industrial problem from a crude oil distillation plant featuring 17 process streams and four utilities—Table 16.2. Notice that some streams were segmented to account for the temperature dependence of CP. Furthermore, streams AGO_steam, Die_steam, Bot_steam cannot match with any process streams. Using HRAT = EMAT = 7 K and 1% tolerances for temperatures three near-independent subnetworks were generated as shown in Table 16.3 by solid lines dividing the streams.

Treating the final temperatures in Table 16.2 as fixed values Shethna and Jeżowski (2006) calculated the heat loads necessary to heat balance the subsets. They were as follows: 898.9 kW for the first subset, −83.7 kW for the second one and 1207.0 kW for the third one. Then, after analysis of the total plant, they suggested the placement of these loads on streams and calculated the changes of final temperatures—shown in Table 16.3 in bold. One can see that the temperature violations are minor.

Table 16.2 Data for Example (16.2) taken from Shethna and Jeżowski (2006)

Stream name	T^{in} (°C)	T^{out} (°C)	CP (kW °C^{-1})	ΔQ (kW)
PA_3_Draw_COL1_TO_PA_3_Return_COL1	319.4	244.1	136.2	2451.7
WasteH2O_Main_TO_Cooled_WasteH2O_ Main	73.2	30.0	68.3	70.5
Lowtemp_crude_Main_TO_Preheat_ Crude_Main	30.0	85.8	327.0	18038.9
	85.8	142.8	327.0	
	142.8	213.8	391.6	
	213.8	232.2	482.9	
PA_2_Draw_COL1_TO_PA_2_Return_COL1	263.5	180.1	443.0	2451.7
PreFlashLiq_Main_TO_HotCrude_Main	232.2	287.9	469.0	12902.8
	287.9	343.3	503.0	
Residue_Main_TO_Cooled_Residue_Main	347.3	191.9	215.8	14272.2
	191.9	45.0	178.2	
AGO_Main_TO_Cooled_AGO_Main	297.3	195.2	21.9	928.1
	195.2	110.0	19.3	
Diesel_Main_TO_Cooled_Diesel_Main	248.0	158.0	68.3	2988.9
	158.0	50.0	58.9	
Naphtha_Main_TO_Cooled_Naphtha_Main	73.2	40.0	57.7	458.1
Kerosene_Main_TO_Cooled_Kerosene_Main	231.8	181.0	50.9	1296.1
	181.0	120.0	46.5	
PA_1_Draw_COL1_TO_PA_1_Return_COL1	167.1	120.7	173.1	3852.8
	120.7	69.6	158.2	
To_Condenser_COL1_TO_Naphtha_COL1	146.7	126.6	225.7	4366.7
	126.6	99.9	176.3	
	99.9	74.6	351.8	
	74.6	73.2	91.6	
KeroSS_ToReb_COL1_TO_KeroSS_ BoilUp_COL1	226.2	228.7	352.3	525.3
	228.7	231.8	424.9	
BottomSteam_Main	20.0	181.4	4.0	613.9
	181.4	181.9	3799.7	
	181.9	190.6	2.4	

Table 16.2 (*Continued*)

Stream name	T^{in} (°C)	T^{out} (°C)	CP (kW°C^{-1})	ΔQ (kW)
DieselSteam_Main	20.0	138.4	1.6	241.4
	138.4	138.9	1626.3	
	138.9	148.9	0.8	
AGOSteam_Main	20.0	138.4	1.3	201.2
	138.4	138.9	1355.3	
	138.9	148.9	0.7	
TrimDuty_COL1	345.6	351.5	1563.0	2218.6
Cooling water	20.0	25.0		19283.7
HP steam	250.0	249.0		1977.2
MP steam	175.0	174.0		224.7
Fired heat (1000)	1000.0	400.0		23803.9

Table 16.3 Solution to Example (16.2) according to Shethna and Jeżowski (2006)

Stream name	T^{in}, °C	T^{out}, °C	Calc. T^{out}, °C
Naphtha_Main_TO_Cooled_Naphtha_Main	73.2	40.0	40.0
To_Condenser_COL1_TO_Naphtha_COL1	146.7	73.2	72.5
Cooling Water	20.0	25.0	25.0
HP Steam	250.0	249.0	249.0
AGO_Main_TO_Cooled_AGO_Main	297.3	110.0	110.3
BottomSteam_Main	20.0	190.6	191.6
DieselSteam_Main	20.0	148.9	148.9
KeroSS_ToReb_COL1_TO_KeroSS_BoilUp_COL1	226.2	231.8	231.8
PA_2_Draw_COL1_TO_PA_2_Return_COL1	263.5	180.1	180.1
Residue_Main_TO_Cooled_Residue_Main	347.3	45.0	45.0
PA_3_Draw_COL1_TO_PA_3_Return_COL1	319.4	244.1	244.1
Diesel_Main_TO_Cooled_Diesel_Main	248.0	50.0	50.0
WasteH$_2$O_Main_TO_Cooled_WasteH$_2$O_Main	73.2	30.0	30.0
PA_1_Draw_COL1_TO_PA_1_Return_COL1	167.1	69.6	69.6
Kerosene_Main_TO_Cooled_Kerosene_Main	231.8	120.0	119.6
Fired Heat (1000)	1000.0	400.0	400.0
AGOSteam_Main	20.0	148.9	148.9
Lowtemp_crude_Main_TO_Preheat_Crude_Main	30.0	232.2	234.2
TrimDuty_COL1	345.6	351.5	351.5
PreFlashLiq_Main_TO_HotCrude_Main	232.2	343.3	343.3

16.3.   MINIMUM TOTAL AREA FOR MATCHES (MTA-M) TARGET

Historically, the first method for reaching the MTA-m target was developed by Townsend and Linnhoff (1984). Since it was presented at a conference held in Bath (UK) the approach has been commonly referred to as the "Bath formula". It is a simple, graphical and easy to understand method but requires strong assumptions. The key idea is to apply the composite curves to simulate overall counter-current heat exchange in a HEN. Similar to a single heat exchanger the counter-current pattern for heat exchange is aimed at minimizing the heat transfer surface area (it is somewhat surprising to note that the strict proof that counter-current heat transfer minimizes area was only recently presented by Maniousiouthakis and Martin (2004)). Obviously, the composite curves have to involve utilities with CP values calculated by any adequate MER/MUC method for a given HRAT. Hence, on a temperature–heat load plot they are balanced in contrast to the CC plot for heat recovery calculations in Chapter 13. The CC plot for balance curves is shown in Figure 16.8 for stream data of Example 13.1. Steam is applied as the heating utility and cooling water as the cold medium.

Counter-current flow imposes, in the balanced composite curve plot, the rule of "vertical heat transfer." It says that to minimize the area heat should be exchanged vertically as illustrated in Figure 16.8. To achieve a simple equation for area, enthalpy intervals ($k = 1, \ldots, K$) are created at each kink of both hot composite curves and cold ones (it is important to differentiate between temperature intervals and enthalpy intervals). More extensive division can be performed but it does not lead to more accurate results. The division into enthalpy intervals is also shown in Figure 16.8. Notice that for the enthalpy intervals created the corresponding temperatures are also known. Because composite streams are treated as "normal" streams the heat transfer coefficients for them have to be known. In the Bath approach the crude assumption of identical overall heat transfer coefficients in all matches is imposed. This condition allows the use of a simple formula (16.36), that is, the Bath formula, to calculate the total minimum surface area.

$$AT^{\min} = \frac{1}{U} \sum_{k=1}^{K} \left(\frac{q_k}{\mathrm{LMTD}_k} \right) \tag{16.36}$$

where U denotes the overall heat transfer coefficient, q_k is the heat load of enthalpy interval k and LMTD_k is the logarithmic mean temperature difference of enthalpy interval k.

It is clear, without any deeper analysis, that the approach is approximate, mainly due to the use of a single, fictitious overall heat transfer coefficient. Additionally, the original Bath formula does not account for forbidden, must-be and restricted matches. The method also has another drawback: there is no evidence

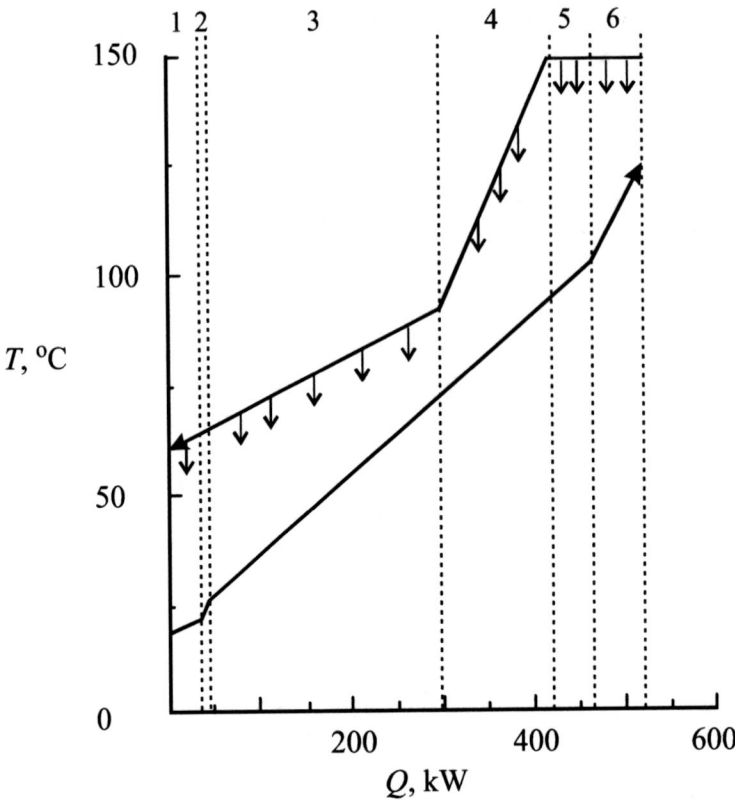

Figure 16.8 Illustration of a balanced composite curve plot.

of its consistency with the MNM target, though the use of a single HRAT value should maintain some consistency of targeting.

The Bath approach was significantly improved by Ahmad (1986) in his PhD thesis and then presented in Linnhoff and Ahmad (1990). The objective was to account for various stream heat transfer coefficients retaining the vertical heat transfer rule. To approach the aim, the so-called spaghetti structure concept has been suggested. In each enthalpy interval composite curves are decomposed into original streams and these streams (streams segments) are matched with each other. Each match has to be accomplished in the enthalpy interval it exists in. Temperatures of matches are fixed by the enthalpy intervals and heat loads of all matches have to be equal to the enthalpy change of an interval. This has to be achieved by adjusting heat capacity flow rates of streams by splits. Additionally, the number of hot streams has to be equal to the number of cold streams because each stream has to be matched within the interval it exists in to maintain vertical matching. Thus, streams have to be split and the split ratio has to be adjusted so as to reach the requirements. In a match in enthalpy interval k the ratio of heat capacity flow-rate of hot stream or a branch of hot stream to

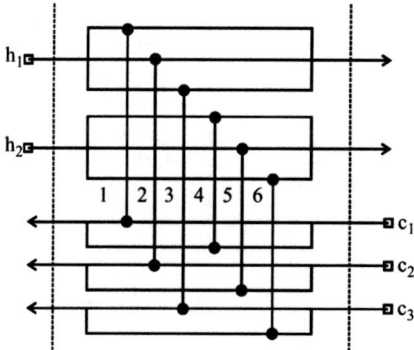

Figure 16.9 Illustration of a spaghetti structure in enthalpy interval k.

heat capacity flow-rate of a cold stream or a cold branch must be equal to the heat capacity ratio for the composite hot stream to the composite cold stream in this interval. Figure 16.9 illustrates the matches created in the enthalpy interval k. It is assumed in this figure that the hot composite curve in the interval consists of two original hot streams (h_1 and h_2) and three original cold streams (c_1, c_2 and c_3).

For such a spaghetti structure created in all enthalpy intervals Ahmad (1986) developed a simple formula (16.37) for the minimum total surface area target:

$$AT^{\min} = \sum_{k=1}^{K} \left[\left(\frac{q_k}{\text{LMTD}_k} \right) \sum_{s \in S_k} \left(\frac{1}{u_s} \right) \right] \qquad (16.37)$$

where S_k denotes a set of streams in enthalpy interval k and u_s is the heat transfer coefficient of streams $s \in S_k$.

Ahmad's method gives more accurate results than the Bath formula and does not require a large computation load. According to Linnhoff and Ahmad (1990) it produces reasonable results if condition (16.40) is satisfied:

$$\frac{u^{\max}}{u^{\min}} \leq 10 \qquad (16.38)$$

where $u^{\max}/u^{\min}$ denotes maximum/minimum value, respectively, of coefficients u_s of streams in a total problem.

However, Shenoy (1995) as well as Jeżowski et al. (2003e) showed that the deviations from accurate results could be large even if condition (16.40) is met and also that deviations depend on the problem. Thus, it is impossible to predict rigorously the range of applicability for Ahmad's approach, though it evidently depends on a range of stream heat transfer coefficient variations.

The Bath method modified by Ahmad is not rigorous since it is based on the perfectly counter-current heat exchange rule with a single $\Delta T^{\min}$ concept, that is,

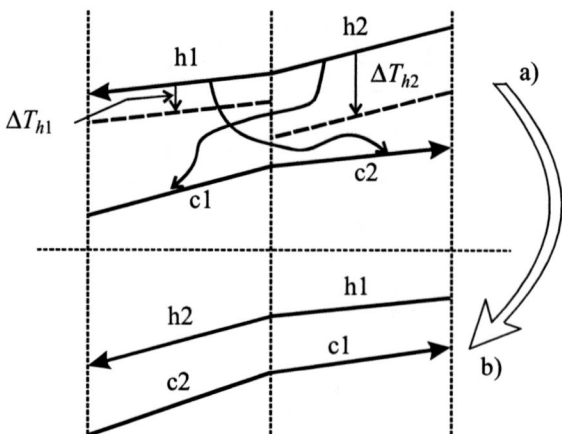

Figure 16.10 Illustration of criss-crossing due to various heat transfer coefficients: (a) criss-crossing necessary with single ΔT^{min}; (b) vertical heat transfer valid with stream contributions ΔT_s.

it employs a temperature approach independent of a match. The rule is rigorous for a single heat exchanger where there are two streams—see Manousiouthakis and Martin (2004). If more streams having various heat transfer coefficients exchange heat the rule of perfectly counter-current heat exchange does not ensure the minimum surface area. The vertical heat transfer rule does not capture the influence of heat transfer coefficients on the surface area. In order to account for this effect, deliberate criss-crossing, that is, nonvertical heat transfer, can be necessary. Let us consider the illustrative example shown first in Linnhoff and Ahmad (1990). We have two hot streams: liquid h1 and gas h2. There are also two cold streams: liquid c1 and gas c2. Liquid streams have high heat transfer coefficients while cold streams have low ones. The streams are located in intervals on the temperature–heat load plot as shown in Figure 16.10a.

Hence, according to the vertical heat transfer rule we should match hot liquid with cold gaseous stream in both intervals. Clearly, this can result in a high heat transfer area. In contrast, matches of both liquid streams and both gaseous streams will probably feature a lower area. This is shown in the plot of Figure 16.10a as criss-crossing. The question arises as to how to do such criss-crossing in a systematic way to reach the minimum surface area. Nishimura (1980) using the Pontryagin maximum principle proved that in order to minimize surface area one should apply various temperature approaches that depend on heat transfer coefficients of streams exchanging heat. Each stream s has a different contribution to the temperature approach of a match and this contribution should be equal to:

$$\Delta T_s = \frac{\kappa}{(u_s)^\beta} \tag{16.39}$$

where κ, β are parameters discussed in the following.

Hence, the temperature approach in a match of streams $i \in H$ and $j \in C$ is defined by Equation (16.40):

$$\Delta T_{ij} = \Delta T_i + \Delta T_j \tag{16.40}$$

The contributions ΔT_s shift streams vertically in the temperature–heat load plot as shown in Figure 16.10a. In consequence, the streams are moved from one interval to another as illustrated in Figure 16.10b. Finally, with stream contributions to the temperature approach we can follow the rule of vertical heat transfer—Figure 16.10b. According to Nishimura, parameter κ in Equation (16.39) is problem-dependent and β equals 0.5. Ahmad (1986) and other authors proposed a simple though iterative procedure for determining κ. It can be formulated as an optimization problem (16.41) with parameter κ as the unknown:

$$\min_{\kappa} \quad \text{abs} \left(Qhu^{\min}[HRAT] - Qhu^{\min} \left[\frac{\kappa}{(u_s)^{0.5}} \right] \right) \tag{16.41}$$

Symbol $Qhu^{\min}$ [·] in Equation (16.41) denotes the minimum load of heating utilities. The quantity depends on the temperature approach. Thus, $Qhu^{\min}$ [HRAT] denotes the minimal value of hot utility load calculated for fixed HRAT and $Qhu^{\min} \left[\frac{\kappa}{(u_s)^{0.5}} \right]$ the value of the load for temperature contributions. For the fixed HRAT optimization problem, Equation (16.41) features only one unknown—parameter κ. Any numerical technique such as, for example, the secant method can be applied to solve it. The solution is the value of κ for which stream contributions to temperature approaches in matches provide the same heat recovery level, that is, the same $Qu^{\min}$ value, as the use of the HRAT value.

The point is that the rigorous proof of Nishimura (1980) for temperature contributions is valid only for two specific cases of heat exchange: one cold stream with several hot ones or one hot stream with several cold ones. Rev and Fonyo (1991) extended this concept for the general case of several hot streams and several cold streams. Also, they applied the temperature contributions to the general problem of heat exchanger network design (not only to the MTA target). The approach has been named the "diverse pinch concept" since, with various stream contributions to temperature approaches, that is, with stream-dependent temperature shifts, there is no longer any single pinch point but diverse or dispersed pinches as illustrated in Figure 16.11.

The diverse pinch approach was adopted in many methods for targeting and designing HEN. The concept of using match-dependent temperature approaches is aimed at area minimization but it can also allow some reduction of unit number in comparison to single HRAT in pinch technology (though rather by chance).

The user of the diverse pinch approach may face some difficulties. They are caused mainly by uncertainties in the determination of parameters κ and β, in

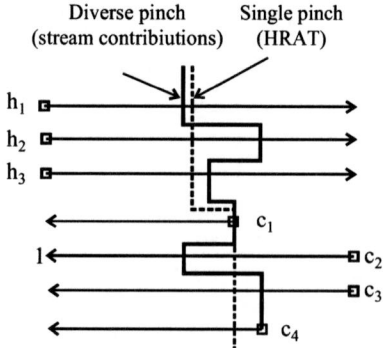

Figure 16.11 Illustration of the diverse pinch concept.

particular, in Equation (16.39). There are contradictory opinions on how to calculate them. Some authors did not find any substantial influence of the parameters; others have found the influence but did not suggest any general formula to estimate them or suggested application of optimization for calculating their values—e.g. Zhu et al. (1995a, 1995b). Furthermore, with regard to calculating only the MTA target the use of the diverse pinch method in the version of Rev and Fonyo does not systematically produce better results than the modified Bath formula (16.35)—see Rev and Fonyo (1993), Shenoy (1995) and Jeżowski et al. (2003e). More recently Serna and Jimenez (2004) reported that much better results could be achieved with a slightly modified version of the targeting procedure. Namely, they followed the division into intervals with shifts by stream contributions as suggested by Rev and Fonyo (1991) but applied real temperatures to calculate LMTD values in contrast to the original method, which uses temperatures with contributions.

The effect of match-dependent temperature approaches can be simply achieved by applying the DTA concept. Similar to calculations of the MNM target one can set EMAT at a very small value (even close to zero) and apply utilities calculated for fixed HRAT. Due to condition (16.4) there is a possibility of criss-crossing and of heat exchange via pinch. Hence there exists the possibility of rigorously accounting for the influence of various heat transfer coefficients of streams on total surface area. It is important that application of the DTA concept allows circumventing the difficulties of calculating stream contributions to temperature approaches.

There are several approaches for calculating the MTA target with DTA. All employ optimization. Colberg and Morari (1990) applied the transshipment model formulation to ensure feasible heat exchange within the spaghetti structure of Ahmad. This led to a complex optimization model with both temperature and enthalpy intervals. Additionally, the optimization model is of the NLP type with respect to both goal function and constraints. A slightly less rigorous approach was developed by Yee et al. (1990a, 1990b). They applied a special stage-wise structure, which is equivalent neither to temperature interval structure nor to enthalpy interval structure. Their model is also of the NLP type but only the goal function is nonlinear while constraints are linear. They then extended the approach to designing HEN which we will explain in Chapter 18.

Here, we will present in detail the linear optimization approach by Jeżowski et al. (2003e). It includes some heuristic elements, the validity of which were proved in the original paper. Its main advantage is in the linear optimization model, which can be solved with widely available solvers of the global optimum for very large problems. Also, it is of importance that the method was extended to simultaneous targeting of all the major targets—this approach will be addressed in Chapter 17. The LP method can also be used to calculate the MTA target for multishell heat exchangers (MTA-s). Thus, we will present it in the next section. Besides this LP method there are also others. They will be discussed briefly in the following.

16.4.   MINIMUM NUMBER OF SHELLS (MNS) TARGET

The most common approach to model multipass heat exchangers is to replace LMTD by mean temperature difference MTD defined by:

$$MTD = Ft \cdot \text{LMTD} \tag{16.42}$$

Correction factor Ft depends on the number of tube passes Nt, number of shells Ns and also parameters P and R. The latter parameters depend on temperatures at hot and cold sides of the heat exchanger—see Figure 16.12. R is the ratio of heat capacity flow rates (CP) and P is the thermal effectiveness of the heat exchanger:

$$R = \frac{t_h^1 - t_h^2}{t_c^2 - t_c^1} \tag{16.43}$$

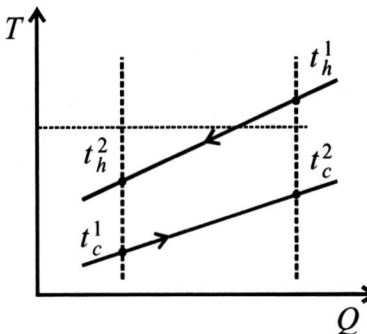

Figure 16.12 Explanation of symbols in Equations (16.43) and (16.44).

$$P = \frac{t_c^2 - t_c^1}{t_h^1 - t_c^1} \tag{16.44}$$

Correction factor Ft is usually determined from plots $Ft = f\ (P, R)$ for various numbers of shells and tube passes, or can also be calculated from complex equations available in textbooks.

It is important to note that:

- the influence of N_t on Ft is usually negligible and thus it is commonly assumed in HEN design that Nt equals 2,
- Ft factors can be very small or even indeterminate for some temperatures. The heat transfer surface area can then be very large or even indeterminate. To avoid such cases the condition is most often imposed that Ft factors should be not less than Ft^{min}. The lower limit is set at about 0.75–0.80. If Ft is smaller than the minimum number of shells is increased unless the Ft value meets the condition. Notice that this is one of the reasons for applying a multishell arrangement.

Due to the Ft correction factor to LMTD, the multishell apparatus can require more shells and a larger area to exchange the same amount of heat as countercurrent 1-1 heat exchangers. This results in calculating procedures for MNM and MTA-m targets being modified.

We start the description with the minimum number of shells (MNS) target. Similarly to targets for matches we will first explain methods from pinch technology. It should, however, be noticed that the first approach was suggested by Trivedi et al. (1987) but it did not produce reliable results. The important achievement that allowed the development of pinch technology based targets for MNS was due to changes in the traditional approach for determining the Ft factor. Ahmad et al. (1988, 1990) and Ahmad and Smith (1989) developed a method that was not only important for HEN targeting but also increased the reliability of rating single heat exchangers. Because of its advantages over traditional graphical technique we present it here in more detail.

Plots $Ft = f\ (P, R)$ given in basic textbooks to determine Ft values feature "asymptotic" profiles of curves at values below Ft of about 0.75 (notice that this is the lower bound on Ft). Hence, heat exchangers with such Ft values operate under unstable conditions since a small change of P or R value results in a drastically large change of Ft. The reason lies in small temperature approaches within a single heat exchanger, which can even become negative—observe the internal pinch illustrated in Figure 13.3. Ahmad et al. (1988) defined parameter P^{max} for each R to estimate the upper limit of the asymptotic region. More rigorously, the definition is:

$$\text{for } P \Rightarrow P^{max}$$

$$\text{abs} \left(\frac{d(Ft)}{d(P)} \right) \to \infty \quad \text{for each } R \tag{16.45}$$

They also developed the following equation for P^{max} for a single apparatus with two tube passes:

$$P^{max} = \frac{2}{R + 1 + \sqrt{R^2 + 1}} \tag{16.46}$$

From the definition of P^{max}, safe operation of a heat exchanger, that is, with $Ft > Ft^{min}$, is ensured if parameter P is greater than P^{max}. Alternatively, this can also be defined using parameter Xp:

$$P = Xp \quad P^{max}; \quad 0 < Xp < 1 \tag{16.47}$$

In general, Xp is the decision parameter in rating and designing heat exchangers—see, for instance, the paper by Galli and Cerda (2000) on HEN synthesis.

Here, we list the minimal values of Ft calculated for various values of Xp.

$$Xp = 0.75 \rightarrow Ft^{min} = 0.89$$

$$Xp = 0.80 \rightarrow Ft^{min} = 0.86$$

$$Xp = 0.85 \rightarrow Ft^{min} = 0.81$$

$$Xp = 0.90 \rightarrow Ft^{min} = 0.74$$

$$Xp = 0.95 \rightarrow Ft^{min} = 0.64$$

Ahmad et al. (1988) suggested using $Xp = 0.90$ since it gives a Ft^{min} value close to that applied in industrial practice. Based on Equations (16.46) and (16.47) it is possible to develop an equation for the number of shells Ns in a single heat exchanger:

- for $R = 1$

$$Ns = \frac{\ln((1 - R\,P)/1 - P)}{\ln W}$$

$$W = \frac{(R + 1) + \sqrt{R^2 + 1} - 2R\,Xp}{(R + 1) + \sqrt{R^2 + 1} - 2\,Xp} \tag{16.48a}$$

- for $R \neq 1$

$$Ns = \frac{P/1 - P(1 + 0.5\sqrt{2} - Xp))}{Xp} \tag{16.48b}$$

Equations (16.48a), (16.48b) have an important feature in respect to HEN targeting. Namely, they are additive in the sense that the sum of Ns values calculated for heat balanced segments of a heat exchanger gives the Ns value for the whole apparatus. Note that in order to retain physical meaning Ns has to be finally rounded off to an integer number after summation. The additiveness allowed a simple method to be developed for calculating the minimum number of shells on the basis of balanced composite curves in the enthalpy interval plot, the same as applied in area targeting. Also, the spaghetti structure notion was included. The approach was presented in full in Ahmad and Smith (1989) and Ahmad et al. (1990).

Since enthalpy intervals correspond to segments of a "composite" heat exchanger we can make use of additiveness of a shell number. In enthalpy interval k the minimum number of matches is $(N_k - 1)$ according to Equation (16.1), where N_k denotes the number of streams in this interval. For given Xp value and Nt set at 2 (two) each match in interval k requires Ns_k shells. The identical number of shells for each match results from the fact that streams in each match have the same temperatures equal to bounding temperatures of interval k. Hence, parameters P and R are identical as well as the number of shells—see Equation (16.48a). Thus, the number of shells in interval k equals the product of match number and number of shells per match:

$$(N_k - 1)Ns_k \tag{16.49}$$

For K enthalpy intervals the total minimum number of shells is given by:

$$Ns^{min} = \text{ent}\left(\sum_{k=1}^{K} (N_k - 1)Ns_k\right) \tag{16.50}$$

where ent (a) denotes the operation which transforms the real number a into integer one not less than a; N_k is the number of streams in interval k and Ns_k is the shell number calculated from Equations (16.48a) and (16.48b) for known temperatures of intervals.

To reach consistency with the MER/MUC target the number of shells should be determined independently for subproblems created by pinches similar to the calculation of the minimum match number by Equation (16.3).

There exists also an alternative version of Equation (16.50), see, for example, Ahmad and Smith (1990) and Hall et al. (1990). It is based on the so-called "stream contributions" concept used also to account for the influence of pressure drop and heat exchanger specifications on targets in Hall et al. (1990). According to this concept a contribution to a target is associated with a stream, that is, each stream brings some contribution into, for example, the total shell number in HEN. Hence, to calculate Ns^{min} we should add contributions to the total number of shells brought in by each stream.

Let us define index sets:

$I = \{i | i \text{ is stream in the problem}\}$
$K_i = \{k \mid \text{stream } i \text{ exists in enthalpy interval } k\}$

Contribution Ns_i brought in by stream i to the total number of shells is equal to:

$$Ns_i = \text{ent} \left(\sum_{k \in K_i} Ns_k \right) \tag{16.51}$$

Notice that contributions are rounded-off to integer numbers for each stream. The total number of shells can then be calculated from:

$$Ns^{\min} = \sum_{i \in I} Ns_i - \sum_{k=1}^{K} Ns_k \tag{16.52}$$

The equation was presented by Ahmad and Smith (1989) without rigorous derivation. Proof that Equation (16.52) can be derived from Equation (16.50) is presented in Jeżowski and Jeżowska (2002). However, this proof is only valid if we do not account for rounding-off operations. The rounding-off operations are more frequent in the case of the method based on stream contributions since rounding-off is performed for each stream contribution. One can expect that the results from both methods can be different and that those from Equation (16.52) will be higher. This was shown by Jeżowski and Jeżowska (2002) for some examples. In all of them the results from the stream contribution based approach were systematically higher. Moreover, those authors observed that Equation (16.50) could yield unreliable results, such that the number of shells is smaller than the number of counter-current 1-1 heat exchangers, that is, matches. Jeżowski and Jeżowska (2002) noticed that such results could be expected for high HRAT values.

Concluding, it can be stated that only Equation (16.52) applying stream contributions should be applied. It is expected to give rather conservative results. Moreover, it is difficult to estimate its accuracy since results from more rigorous approaches are not available. The Nmin-Transs model could be applied to determine the MNS target after inserting *MTD* parameters. However, there is no such extension in the literature.

16.5. MINIMUM TOTAL AREA FOR SHELLS (MTA-S) TARGET

Here, we explain approaches for the minimum surface area target for a multi-shell apparatus (MTA-s). We begin with the approach proposed by Jeżowski et al. (2003e). They focused on developing a linear model. To reach the goal they proposed creating sufficiently small temperature intervals such that temperatures in "mini-matches" among the mini-streams in intervals can be approximated by temperatures of intervals. Thus, for known temperatures of "mini-matches" the logarithmic mean temperature differences and *Ft* correction factors are also known. The result is that only heat loads of "mini-matches" are variables and the problem becomes linear. To model heat transfer Jeżowski et al. (2003e) applied the transportation model. Though it requires more variables than the transshipment one it has a distinct advantage over the transshipment formulation with regard to area calculation. This results from the lack of residuals that are in fact aggregate heat flows of various streams and hence do not feature a unique heat transfer coefficient.

Let us assume that sufficiently small, with regard to model requirements, temperature intervals have been created. They are numbered from $k = 1, \ldots, K$ as in the MUC transportation model—see Chapter 15. The index sets and variables for heat transfer are also identical to those in the MUC-Transp model in Chapter 15—Equations (15.27) to (15.31). Refer also to Figure 15.5 for model representation.

The log-mean temperature difference in a match (LMTD) of load $q_{ik,jl}$ is calculated from temperature differences determined for the intervals k and l. Hence, all matches between $i \in H_k$ and $j \in C_l$ have identical temperature differences (at both sides) as defined by Equation (16.53):

$$\left. \begin{array}{l} \Delta T^{hot}_{ik,jl} \equiv \Delta T^{hot}_{k,l} = T_{k-1} - T_{l-1} \\[2mm] \Delta T^{cold}_{ik,jl} \equiv \Delta T^{cold}_{k,l} = T_k - T_l \end{array} \right\} i \in H_k; \quad j \in C_{jl}; \quad k, l = 1, \ldots, K \quad (16.53)$$

For instance, in Figure 16.13a match $\{i1k, jl\}$ and match $\{i2k, jl\}$ have the same temperature approaches at both sides though they feature different heat loads. This can be totally achieved in the split spaghetti structure illustrated by Figure 16.13b.

In consequence, the approximations of LMTD and Ft parameters for matches are also calculated for intervals, according to:

$$\text{LMTD}_{ik,jl} \equiv \text{LMTD}_{k,l}; i \in H_k; j \in C_l; k, l = 1, \ldots, K \quad (16.54)$$

$$Ft_{ik,jl} \equiv Ft_{k,l}; i \in H_k; j \in C_l; k, l = 1, \ldots, K \quad (16.55)$$

Correction factors Ft can be calculated by any standard method.

The linear model for the minimum total surface area target, for both 1-1 and n-m apparatus, is as follows.

MTA-Transp model

$$\min \sum_{K=1}^{K} \sum_{l=1}^{K} \left(\frac{1}{\text{LMTD}_{kl} \quad Ft_{kl}} \right) \sum_{i \in H_k} \sum_{j \in C_l} \left(\frac{q_{ik,jl}}{U_{ij}} \right) \quad (16.56)$$

s.t.

$$\sum_{j \in C_l} q_{ik,jl} = \Delta Q^C_{ik}; \quad i \in H_k; \quad k, l = 1, \ldots, K \quad (16.57)$$

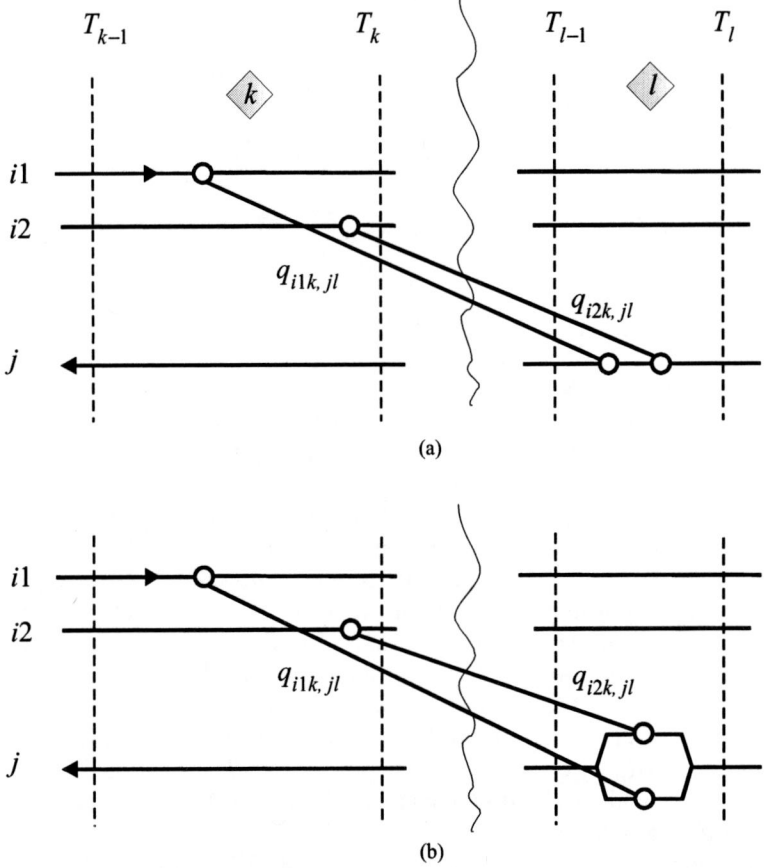

Figure 16.13 Illustration for approximations of temperature approaches: (a) approximations without splits, (b) approximations with splits.

$$\sum_{i \in H_k} q_{ik,jl} = \Delta Q_{jk}^C; \quad j \in C_l; \quad k, l = 1, ..., K \tag{16.58}$$

$$q_{ik,jl} \geq 0 \tag{16.59}$$

Constraints (16.57) and (16.58) are heat balances. Constraint (16.59) ensures that heat loads have to take only positive values. Conditions for must-be, forbidden and restricted matches have been omitted here but they can be simply added following for instance the MUC-Transs model in Section 15.2.

Proper division into temperature intervals is the crucial point of the method. Preliminary division is performed at inlet and outlet shifted temperatures of streams. As for choice of the approach temperature value for shifting temperatures, it is sufficient that a small number is used to meet thermodynamic feasibility constraints. Hence the preliminary division is similar to that in the MNM-Transs model, that also uses the DTA concept.

The mean size of the interval (dT_{mean}) is determined from Equation (16.60) where dT^{min} is the smallest interval after preliminary division:

$$dT_{mean} = \max [3 \cdot dT^{min}, 10] \qquad (16.60)$$

Each interval of size higher than dT_{mean} (e.g. dT) is divided into N_{ad} intervals where:

$$N_{ad} = \text{ent}[dT/dT_{mean}] \qquad (16.61)$$

where ent [a] means a nearest integer number not less than a.

The approximation of temperature approaches in matches applied in the model imposed a superstructure that is, in fact, the totally split spaghetti structure within temperature intervals similar to that from Colberg and Morari (1990). Element $i \in H_k/j \in C_l$ is split for as many branches of as many matches as it has in the solution of the model—see illustrative examples in Figure 16.13b and Figure 16.14. In some cases such a structure can cause an overestimation of area. As an example consider the totally split structure from the solution of the MTA-Transp model with two matches on adjacent intervals with the same streams as shown in Figure 16.14a. This can be represented in the more familiar fashion shown in Figure 16.14b. However, a serial connection can also be built from the solution of Figure 16.14a and matches can be merged as shown in Figure 16.14c. The parallel connection of Figure 16.14b has uneven distribution of driving forces while merged matches from Figure 16.14c give a more even one in the entire network.

Therefore, based on the solution from the MTA-Transp model the modified structure with merged matches such as shown in Figure 16.14c can be generated in the last step of the area targeting method. According to Jeżowski et al. (2003e) the modified structure usually has an area up to 10–15% less than the totally split spaghetti structure from the solution of the MTA-Transp model.

Here we reproduce two examples from Jeżowski et al. (2003e): Example 16.3 with data in Table 16.4 and Example 16.4 with data in Table 16.6. The aim is also to show the comparison of results from some popular MTA targeting

Table 16.4 Data for Example (16.3) (also problem 4S1t in Shenoy, 1995)

Stream	T^{in} (K)	T^{out} (K)	CP (kW K^{-1})	U (kW m^{-2} K^{-1})
H1	448	318	10	0.2
H2	398	338	15	2.0
C3	293	428	20	0.2
C4	313	385	15	2.0
HU	453	452	–	4.0
CU	288	298	–	2.0

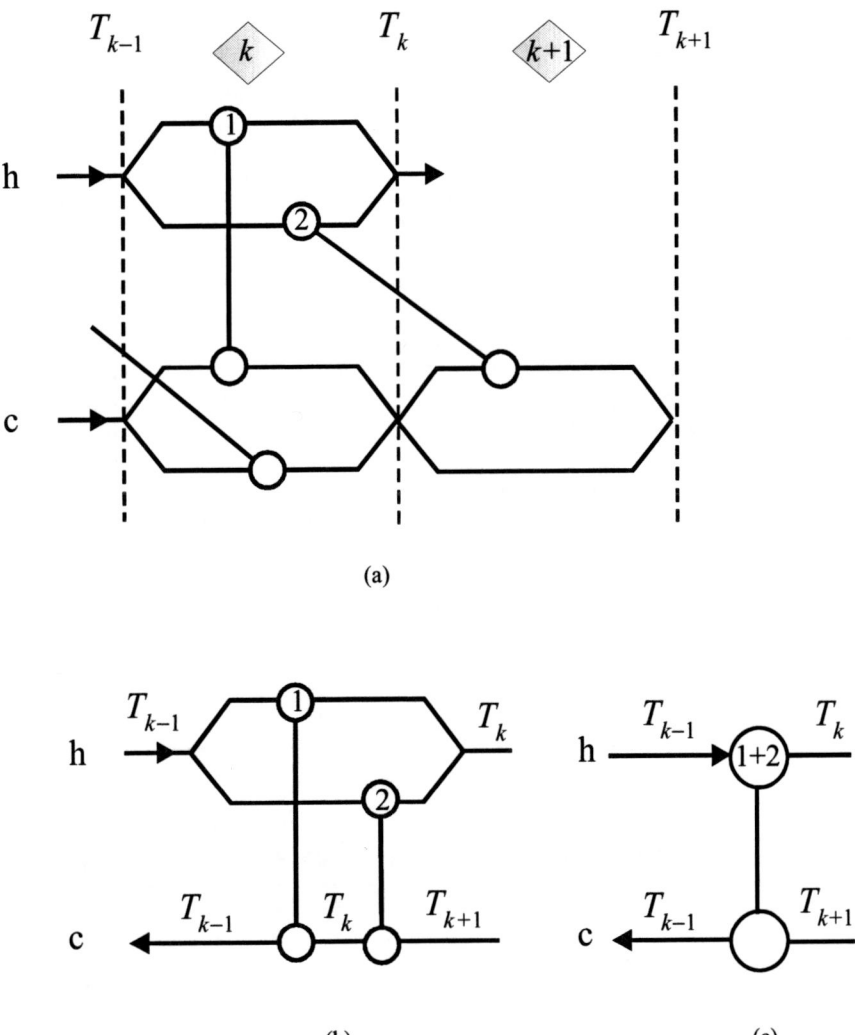

Figure 16.14 Illustration for spaghetti structure modification: (a) spaghetti structure of a solution from the MTA-Transp model, (b) standard presentation of the structure from 16.14a, (c) merged parallel matches in the modified structure.

approaches. Both problems feature various heat transfer coefficients, for instance the ratio of the maximal value to the minimal value in Example (16.3) is 20 and in Example (16.4) amounts to 40. Additionally, the latter example has forbidden matches and also multishell units.

The comparison of results is given in Table 16.5. In brackets there are relative errors calculated for the area target from Colberg and Morari (1990) treated as reference values (Table 16.7).

Table 16.5 Comparison of area targets for Example (16.3) (1-1 apparatus)

Utility usage	Bath formula[a]	Diverse pinch method of Rev and Fonyo (1991)[a]	Solution by method of Colberg and Morari (1990)[a]	Solution from Jeżowski et al. (2003e)
1580	484.53(11.52)	482.34(11.03)	434.42	432.43(0.002)
1713.33	442.76(12.40)	440.08(11.72)	393.90	394.06(0.0406)
1774.44	428.69(12.37)	426.44(11.72)	381.48	381.99(0.134)
1880	408.64(11.79)	408.04(11.63)	365.51	366.11(0.164)
1980	393.26(10.97)	393.71(11.09)	354.39	355.25(0.243)
2057.08	383.21(10.30)	384.92(10.79)	347.41	348.14(0.210)
2232.78	364.69(8.71)	368.229(9.76)	335.46	335.93(0.140)
2272.27	361.20(8.49)	365.09(9.65)	332.94	333.68(0.222)
2330	356.49(8.02)	360.84(9.33)	330.03	330.87(0.254)
2450	347.96(7.08)	353.14(8.68)	324.94	325.80(0.264)
2737.5	332.92(5.08)	340.11(7.35)	316.82	317.56(0.233)
2791.36	330.77(4.75)	339.05(7.37)	315.77	316.56(0.250)
2811.67	330.00(4.62)	336.07(6.55)	315.41	316.20(0.250)
2870.63	327.90(4.28)	330.31(5.04)	314.45	315.23(0.248)
2874.44	327.78(4.27)	329.93(4.96)	314.34	315.17(0.264)
2880	327.59(4.22)	329.40(4.80)	314.31	315.09(0.248)
3280	316.82(1.97)	312.78(0.67)	310.68	311.58(0.289)
3580	313.56(0.93)	311.05(0.12)	310.68	311.22(0.174)
3780	313.43(0.78)	313.90(0.93)	311.00	311.45(0.145)

[a] Values of area targets from Shenoy (1995).

Table 16.6 Data for Example (16.4) (HRAT = 10 K, Qhu = 620 kW, Qcu = 230 kW)

Stream	T^{in} (K)	T^{out} (K)	CP (kW K^{-1})	U (kW m^2 K^{-1})
H1	395	343	4	2.0
H2	405	288	6	0.2
C1	293	493	5	2.0
C2	353	383	10	0.2
Steam	520	519	–	2.0
Cooling water	278	288	–	2.0

Table 16.7 Comparison of area targets (MTA-m and MTA-s) for Example (16.4) (forbidden matches)

Match forbidden	Qhu/Qcu (kW)	Solution from Colberg and Morari (1990)		Solution from Jeżowski et al. (2003e)		Bath formula	
		MTA-m		MTA-m	MTA-s	MTA-m	MTA-s
Base case	620/230	258.8		260.60	349.4	295.70	295.7
H1–C2	620/230	259.7		260.60	349.4	295.78	385.9
H2–C1	840/450	–		241.00	322.4	147.68	180.7
H2–C2	662/272	–		244.90	326.6	240.59	385.9

Based on a broad sample of examples, Jeżowski et al. (2003e) concluded that:

(a) The Bath formula with spaghetti structure (16.37) gives targets substantially larger than those from rigorous approaches such as those of Colberg and Morari (1990). Only for cases with equal heat transfer coefficients are the results from the Bath method accurate. The errors increase with scale of heat transfer coefficient variation but also depend on minimum temperature approach (HRAT). Hence, this can result in mistakes in the supertargeting stage, which can produce nonoptimal values of the minimum temperature approach. As expected, the Bath formula is inadequate for problems with restricted matches.

(b) Diverse pinch method – version of Rev and Fonyo (1991) – does not produce reliable results and, in fact, its accuracy is similar to the Bath approach.

(c) The method of Jeżowski et al. (2003e) gives results that differ less than 1% from those of Colberg and Morari (1990) and have similar accuracy as the NLP approach of Yee and Grossmann (1990).

The core concept in the method by Jeżowski et al. (2003e) was to approach the situation where heat loads on hot and cold mini-segments exchanging heat are identical. At such a limit temperatures in "mini-matches" are equal to temperature interval temperatures. Hence, conditions (16.53) – (16.55) would be rigorously satisfied. To approach the equal load case extensive division into intervals is necessary which results in a huge number of variables in the optimization model. This observation was first noted by Colberg and Morari (1990). Jeżowski et al. (2003e) showed that it is sufficient to apply fewer divisions according to the heuristic procedure (Equations (16.60) and (16.61)) to reach a fairly accurate area target. However, the method includes the heuristic element. Manousiouthakis and Martin (2004) developed a more rigorous approach that also applies the concept of small equal heat load segments exchanging heat. However, they based their approach on the balanced CC plot and hence the condition of equal heat loads was satisfied. However, dense division into enthalpy intervals is still required to rigorously account for the influence of criss-crossing on the HEN area. Instead of a heuristic-based division, Manousiouthakis and Martin (2004) used an iterative procedure with a successive increase of number of intervals. The area targeting problem is posed as a two-staged linear optimization. First, the optimal assignment task is formulated to find optimal matching of intervals. The model calls for interval surface areas, which are parameters of the assignment task. Interval area is calculated as the sum of segment areas in the interval. These segment areas are determined by solving the second-stage optimization model where fractions of hot and cold streams exchanging heat are variables. The model is linear due to fixed enthalpy intervals. The target is calculated by an iterative procedure. In each step of the procedure the number of intervals is doubled. The iterations continue until the convergence of surface area values. The method by Manousiouthakis and Martin (2004) was presented for matches (1-1 heat exchangers), but its extension for multishell units is straightforward. It is important to note that it was applied for targeting both heat and mass exchangers by making use of the strict analogy between heat and mass integration (shown also in Chapter 12). However, only single contaminant cases in the MEN problem can be solved since this close analogy is limited to such a case. Manousiouthakis and Martin (2004) presented a solution to only a very simple example of the three stream problem. This requires 11 iterations. Hence, it is difficult to estimate the CPU time needed to solve large-scale tasks.

17 Simultaneous HEN targeting for total annual cost

Sequential targeting approaches are inferior in comparison with simultaneous methods since they are not able to rigorously account for trade-offs among targets. In the previous chapter we addressed the approach of aiming for two targets simultaneously: the MNM-Transs model that calculates such heat load distribution that is optimal in respect to utility cost (MUC target) and number of matches (MNM target). However, it does not account for cost incurred by area of heat exchangers. We noted that, for instance, the application of a small EMAT value may lead to large surface areas. As a consequence, the total annual cost may be high though the number of units and the cost of utilities are at a minimum. Also, due to the linearity of the MNM-Transs optimization model there exist multiple optimal solutions that can, however, differ significantly in total area and, as a result, in total annual cost (TAC)—see, for instance, Gundersen et al. (1996, 1997, 2000). Finally, it should be recalled that the MNM-Transs model requires a HRAT value as the parameter. Hence, to determine optimal HLD, supertargeting has to be performed to locate optimal HRAT, thus increasing significantly the computation load.

In order to account for TAC, Gundersen and Grossmann (1990) significantly modified the MNM-Transs model and developed a MILP formulation that accounts for cost of area, though indirectly. They embedded into this model a penalty mechanism for criss-crossing, that is, they not only imposed the rule of vertical heat transfer but also provided some allowance in order to leave freedom for some criss-crossing to account for various stream heat transfer coefficients. The approach of Gundersen and Grossmann (1990) did not systematically produce optimal solutions in regard to TAC. Gundersen and co-workers, in a series of papers (Gundersen et al., 1996, 1997, 2000) extended this concept, attempting to develop a method which produces a set of solutions that are ranked according to the values of TAC. Though they finally achieved results superior to those from Gundersen and Grossmann, the method was not able to systematically produce accurate results.

Here we explain the contribution addressed in Shethna et al. (2000b, 2002). The method requires recursive solution of the MILP model in a very small number of iterations or even converges in a one-step procedure. It allows dealing with forbidden, restricted, and must-be matches. Multishell apparatus are included, too. Moreover, it does not require a supertargeting stage to find the optimum value of the temperature approach. Finally, the influence of heat exchanger construction type can also be captured providing that cost coefficients

(α, β, γ) in Equation (14.2) can be estimated beforehand for each pair of streams (i, j).

The method is based on the transportation formulation, in particular on the transportation model for the MTA target (the MTA-Transp model in Section 16.5). Due to the linearity of the MTA-Transp model, the constraints accounting for area in the simultaneous targeting model are also linear. Modeling of heat exchange to minimize utility costs does not require nonlinear functions as we have shown in all models for MUC targeting (see MUC-Transs and MUC-Transp models in Section 15.2). Also, the minimization of a number of matches which is equivalent to minimizing fixed charge costs of heat exchangers can be modeled by linear function—see the MNM-Transs model in Section 16.2. Hence, all constraints in the simultaneous model of Shethna et al. (2000a, 2000b) are linear. The only nonlinearity is caused by the nonlinear term for area-dependent cost $(\beta(A)^\gamma)$ in Equation (14.2) for capital cost of heat exchangers. To eliminate the nonlinear term in capital cost depending on area they applied the Taylor series expansion around certain initial heat exchanger areas, truncated after the linear term. Hence, instead of using the original formula (14.3) for investment cost of heat exchangers, they applied linear dependence (17.1) with parameters β'_{ij} defined by Equation (17.2):

$$CAP_{ij} = \alpha_{ij} + \beta'_{ij}(A_{ij}) \tag{17.1}$$

$$\beta'_{ij} = \beta_{ij}\gamma_{ij}(A_{ij}^0)^{\gamma_{ij}-1} \tag{17.2}$$

where A_{ij}^0 denotes the initial value of A_{ij}; indices i, j state for streams i, j exchanging heat.

Due to the linearization, an iterative procedure is necessary. The convergence depends on values of initial areas A_{ij}^0. Shethna et al. noted that for A_{ij}^0 calculated from a rigorous approach to the MTA target, such as the MTA-Transp model, the convergence is very fast. If exponents γ_{ij} are not less than 0.6, which is usually the case, quite accurate results are reached in only a single step; that is, iterations are not needed. For smaller exponents γ_{ij} a second iteration can be necessary to obtain good results. Notice that, alternatively, linearization of capital cost, such as used in Barbaro and Bagajewicz (2002), can be applied, thus eliminating the application of iterations.

In order to calculate good initial values of A_{ij}^0 by the MTA-Transp model, division into small temperature intervals is necessary following the procedure for the MTA-Transp model in Section 16.5. Identical division is needed in the simultaneous targeting approach of Shethna et al. Also, the same parameters are applied in both models, such as logarithmic mean temperature differences LMTD, correction factors Ft and given heat loads on process streams in temperature intervals. Due to the fact that the simultaneous targeting model accounts for all targets, the choice of the minimum temperature approach EMAT in division into temperature intervals does not cause any problems. It is sufficient to apply any small value greater than zero—differentiating HRAT and EMAT is superfluous. Optimization will not allow for too small temperature approaches in matches, as it could in the MNM-Transs model, because investment expenses are included in the optimization criterion. Moreover, it is not necessary to first calculate heat loads of utilities for some fixed value of HRAT and then to search for the optimal HRAT in the supertargeting step. Optimal values of temperature approaches are determined by optimization. This greatly reduces the computation load.

All utilities that could be potentially applied or are available in the heat-integrated system should be inserted into temperature intervals according to their fixed inlet and outlet temperatures.

To formulate the simultaneous targeting model we need the following index sets for given sets of process and utility streams:

$$H = \{i|\, i \text{ is hot process stream}\}$$

$$C = \{j|\, j \text{ is cold process stream}\}$$

$$HU = \{m|\, m \text{ is heating utility}\}$$

$$CU = \{n|\, n \text{ is cooling utility}\}$$

The index sets in the following are created after the division into temperature intervals $k = 1, \ldots, K$. Temperature intervals are numbered in descending order from the highest temperature similar to optimization models for MUC, MNU and MTA targets in sequential approaches:

$$H_k = \{i|\, i \text{ is present in interval } k\}$$

$$C_k = \{j|\, j \text{ is present in interval } k\}$$

$$HU_k = \{m|\, m \text{ is present in interval } k\}$$

$$CU_k = \{n|\, n \text{ is present in interval } k\}$$

The sets above allow differentiation between process streams and utilities that is necessary in utility cost minimization. However, the distinction is unnecessary for modeling area and number of units' optimization. Thus, to make the model compact we follow the presentation from Shethna et al. (2000a, 2000b) and define also additional index sets for both the hot streams (hot process streams and heating utilities) and the cold streams (cold process streams and cooling utilities):

$$H' = \{i \mid i \text{ is hot stream}\}, \quad H'_k = \{i \mid i \text{ is present in interval } k\}$$

$$C' = \{j \mid j \text{ is cold stream}\}, \quad C'_k = \{j \mid j \text{ is present in interval } k\}$$

The parameters of the simultaneous targeting model are:

$\text{LMTD}_{k,l}; k, l = 1, \ldots, K$	logarithmic mean temperature differences
$Ft_{k,l}; k, l = 1, \ldots, K$	correction factors for LMTD of multipass heat exchangers
$\Delta Q_{ik}^H; i \in H_k, k = 1, \ldots, K$	heat loads (enthalpy changes) of hot process streams in intervals
$\Delta Q_{ik}^C; j \in C_l, k = 1, \ldots, K$	heat loads (enthalpy changes) of cold process streams in intervals
$\alpha_{ij}; i \in H'; j \in C'$	cost coefficients for capital cost of heat exchangers
$A_{ij}^0; i \in H'; j \in C'$	initial (starting) values of areas of heat exchangers calculated by a proper MTA targeting method
$\beta'_{ij}; i \in H'_k; j \in C'$	coefficients in linearized Equation (17.1) calculated from Equation (17.2)
$phu_m; m \in \text{HU}$	unit cost of heating utilities
$pcu_n; n \in \text{CU}$	unit cost of heating utilities

The following variables are used in the model:

$q_{ik,jl}$	load of heat exchange between $i \in H_k$ and $j \in C_l$
$q_{ik,nl}$	load of heat exchange between $i \in H_k$ and $n \in CU_l$
$q_{mk,jl}$	load of heat exchange between $m \in HU_k$ and $j \in C_l$
q_{ij}	load of heat exchange between $i \in H'$ and $j \in C'$
qhu_m	heat load of utility $m \in HU_k$
qcu_n	heat load of utility $n \in CU$
y_{ij}	binary variables defining existence of matches between $i \in H'$ and $j \in C'$

The binary variables are defined by

$$\begin{aligned} y_{ij} &= 1 \text{ if } i \in H' \text{ exchanges heat with } j \in C' \\ y_{ij} &= 0 \text{ if not} \end{aligned} \tag{17.3}$$

Notice that, similar to other optimization models presented so far, heat exchange among hot and cold utilities is explicitly excluded.

The MILP transportation model for calculating optimal heat load distribution in respect to TAC is as follows.

TAC-TRANSP MODEL

$$\min \left(\sum_{m \in HU} phu_m qhu_m + \sum_{n \in CU} pcu_n qcu_n \right)$$

$$+ \left(\sum_{k,l=1}^{K} \sum_{i \in H'} \sum_{j \in C'} \beta'_{ij} \frac{1}{(Ft_{k,l} \cdot LMTD_{k,l})} \frac{q_{i,j}}{(h_j + h_i)} \right) \quad (17.4)$$

$$+ \left(\sum_{j \in H'} \sum_{i \in C'} \alpha_{ij} y_{ij} \right)$$

s.t.

$$\sum_{j \in C'_l} q_{ik,jl} = \Delta Q_{ik}^H, \quad i \in H_k; \, k,l = 1, \ldots, K \quad (17.5)$$

$$\sum_{i \in H'_k} q_{ik,jl} = \Delta Q_{jl}^C, \quad j \in C_l; \, k,l = 1, \ldots, K \quad (17.6)$$

$$qhu_m = \sum_{k,l=1}^{K} \sum_{j \in C_k} q_{mk,jl}, \quad m \in HU \quad (17.7)$$

$$qcu_n = \sum_{k,l=1}^{K} \sum_{i \in H_k} q_{ik,nl}, \quad n \in CU \quad (17.8)$$

$$q_{ij} = \sum_{k,l=1}^{K} q_{ik,jl}, \quad i \in H'_k; \, j \in C'_l \quad (17.9)$$

$$q_{ij} \leq Q^{\max} y_{ij}, \quad i \in H'_k; \, j \in C'_k \quad (17.10)$$

$$q_{ik,jl} \geq 0; \, i \in H'_k; \, j \in C'_l; \, k,l = 1, \ldots, K; \, y_{ij} = 0, 1; \, i \in H'; \, j \in C' \quad (17.11)$$

Table 17.1 Data for Example (17.1)

Stream	T^{in} (°C)	T^{out} (°C)	CP (kW K^{-1})	U (kW m^{-2} K^{-1})	phu, pcu ($ kW year^{-1})
H1	175	45	10	0.2	–
H2	125	65	40	0.2	–
C1	20	155	20	0.2	–
C2	40	112	15	0.2	–
HU	180	179	–	0.2	100
CU	15	20	–	0.2	10

The total annual cost of the HEN is defined by Equation (17.4). The first term in the equation defines the cost of utilities, the second defines investment cost depending on area, and the third defines the fixed charge cost.

Equations (17.5) and (17.6) are heat balances of hot and cold process streams. Heat loads of utilities are defined by Equations (17.7) and (17.8), while heat loads on all matches are defined by Equation (17.9). Logical condition (17.10) ensures that loads of nonexisting matches are zero. Similar to other models, parameter Q^{max} can be replaced by values Q_{ij}^{max} calculated for all potential matches—see the description of the MNM-Transs model in Section 16.2. Finally, condition (17.11) ensures nonnegativity of all heat loads and also define binary variables y_{ij}.

Conditions on restricted matches as well as constraints on constant flow rate of utilities within temperature ranges can be simply inserted into the model.

Here, we will illustrate the application of the TAC-Transp model using some examples. The first Example (17.1) is simple. The data for this example (Table 17.1) were originally reported in Shenoy (1995). To calculate capital cost, identical cost parameters were applied for all heat exchangers as follows: $\alpha = 30\,000$, $\beta = 750$, $\gamma = 0.81$. Annualization factor r equals 0.322.

The problem has been solved for 1-1 and multishell heat exchangers. Optimal heat load distributions for both cases are shown in Table 17.2, for 1-1 heat exchangers in column (a) and for a multishell apparatus in column (b). It is important to note that they differ not only for values but also for matches. Cooling water exchanges heat with H1 in case (a) but with H2 in case (b).

Table 17.2 Optimal heat load distributions for Example (17.1)

(a) Model with 1-1 heat exchanger		(b) Model with multishell heat exchanger	
Match	Heat load (kW)	Match	Heat load (kW)
H2-C1	1320	H2-C1	1004
H1-C1	1020	H1-C1	1300
H2-C2	1080	H2-C2	1080
HU-C1	360	HU-C1	396
H1-CU	280	H2-CU	316

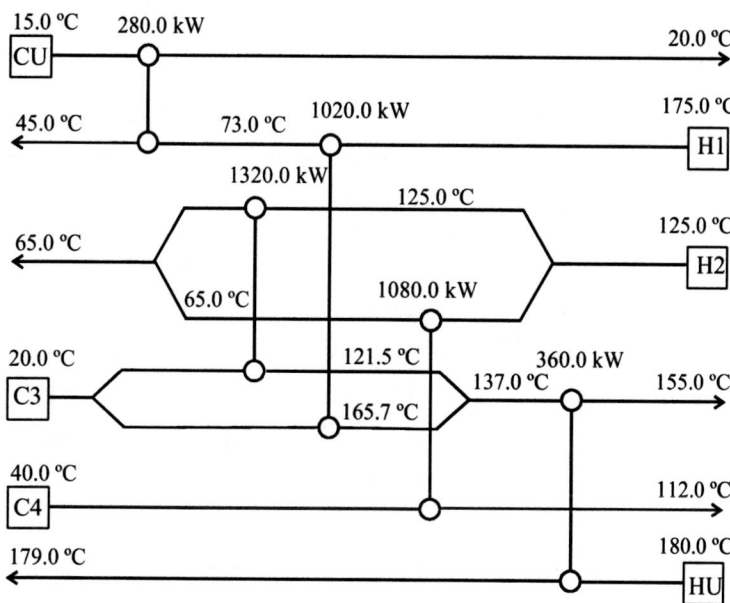

Total 1-2 exchanger area = 2365 m^2
Total capital cost = 830600 $
Total annual cost = 306340 $/yr

Figure 17.1 HEN for Example (17.1)—case (a) 1-1 heat exchangers.

Based on the heat load distributions HEN structures have been developed by inspection for both cases. They are shown in Figures 17.1 and 17.2, respectively. Notice that splitters had to be applied to achieve required loads. Information on HLD is insufficient to design the HEN directly. Moreover, it appears difficult to find a feasible HEN structure that strictly follows the HLD for the case of 1-2 heat exchangers. Hence, the solution in Figure 17.2 uses two units for matching H1-C1. Capital costs were calculated for both networks assuming 1-2 heat exchangers in both cases. Values of total surface area, capital cost and TAC are given in the figures. Notice that the TAC for solution (b) is lower by 8% than that for solution (a) even though the optimal HLD is not satisfied in solution (b). Hence, one can conclude that knowledge of optimal HLD calculated for matches to develop a HEN structure with multishell apparatus can lead to inferior results.

The next Example (17.2) taken from Shethna et al. (2002) is an industrial scale task. This problem was considered in Chapter 16 as Example 16.2 to illustrate the application of a MNM-Transs-sub model for generating near-independent networks. Table 16.2 gathers stream data. Cost parameters are: $\alpha = 10\,000$, $\beta = 500$, $\gamma = 0.80$; 1-2 heat exchangers were applied. Here, the problem was solved by a simultaneous targeting method for fixed outlet temperatures. The TAC optimal HLD determined by solving the TAC-Transp model in one step is presented

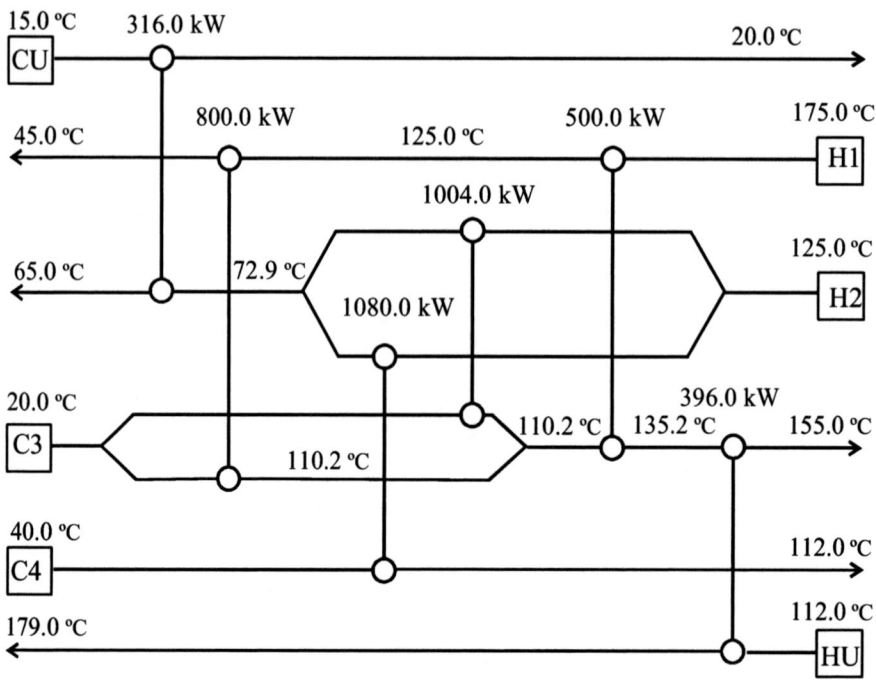

Total 1-2 exchanger area = 1911 m²
Total capital cost = 738300 $
Total annual cost = 280578 $/yr

Figure 17.2 HEN for Example (17.2)—case (b) multishell heat exchangers.

in Table 17.3. Notice that minimum temperature approaches calculated in the model are also available in the results.

Due to the linearity of the model it can be solved for fairly large industrial problems with widely available solvers. Here, we present as Example (17.3) the problem from Shethna et al. (2000b) containing 44 process streams and 10 potential utilities—stream data in Table 17.4. Notice that several streams are segmented. Cost parameters are: $\alpha = 30\,000$, $\beta = 2500$, $\gamma = 0.50$, $r = 0.336$; 1-2 heat exchangers were applied.

The results from the solution of the model (optimal heat load distribution) are gathered in Table 17.5. Shethna et al. (2000b) succeeded in developing the HEN on this basis (the graphical representation is not reproduced here). The network does not include splitters and this is only one possible structure for this HLD; likely this is not the TAC optimal HEN. Both Examples (17.1) and (17.2) show that the structure development stage depends not only on the problem scale but also on the possibility of reaching a feasible network without splitters. Table 17.5 contains information on heat exchangers from this network: temperatures of streams in matches and surface area of heat exchangers. It is interesting

Table 17.3 HLD from TAC-Transp model for Example (17.2)

Recommended matches		Load (MW)	EMAT (°C)
Hot stream	Cold stream		
PA2	Crude	6.6	4.0
PA2	Preflashlig	2.6	4.0
PA3	Preflashlig	10.2	4.0
Residue	Crude	30.7	4.0
Residue	Preflashlig	24.0	4.0
AGO	Crude	2.7	4.0
AGO	Preflashlig	1.2	4.0
Diesel	Crude	9.0	4.0
Kerosene	Crude	5.4	4.0
PA1	Crude	12.9	4.0
Condenser	Crude	7.5	8.3
PA2	KeroReb	1.0	4.0
Diesel	KeroReb	1.2	4.0
Naphta	CW	1.9	20.0
Condenser	CW	10.8	50.0
Residue	CW	5.0	25.0
Diesel	CW	2.3	30.0
Waste H$_2$O	CW	0.3	10.0
PA1	CW	3.2	50.0
HP steam	Bot_Steam	1.9	59.0
HP steam	Preflashlig	1.3	5.0
MP steam	AGO_Steam	0.8	26.0
MP steam	Bot_Steam	1.9	59.0
MP steam	Die_Steam	1.0	26.0
Fired heat	Trim duty	9.3	410.0
Fired heat	Preflashlig	14.8	96.0

Table 17.4 Stream data for Example (17.3) (from Shethna et al., 2000b)

Stream	T^{in} (°C)	T^{out} (°C)	CP (kW °C^{-1})	U (kW m^{-2} °C^{-1})	Cost ($ kW year^{-1})
S1	15.0	0.5	26.59	0.2	–
S2	−44.0	5.9	7.727	0.2	–
S3	−45.0	−44.0	1023	0.2	–
S4	−28.8	−33.0	428.9	0.2	–
S5	90.5	91.5	2539	0.2	–
S6	39.3	39.2	44 600	0.2	–
S7	157.0	158.9	4630	0.2	–
S8	81.1	76.9	533.8	0.2	–
	76.9	75.1	1222		
S9	102	103	3813.0	0.2	–

Table 17.4 (*Continued*)

Stream	T^{in} (°C)	T^{out} (°C)	CP (kW °C^{-1})	U (kW m^{-2} °C^{-1})	Cost ($ kW year^{-1})
S10	66.6	62.2	1259	0.2	–
	62.2	61.1	2505		
S11	61.1	52.6	16.52	0.2	–
	52.6	49.0	19.89		
S12	116.4	117.4	7115	0.2	–
S13	57.7	100.0	35.06	0.2	–
	100.0	160.0	35.14		
	160.0	198.0	37.84		
	198.0	200.0	2724		
S14	200.0	260.0	25.78	0.2	–
	260.0	520.0	67.30		
S15	162.5	125.0	70.48	0.2	–
S16	270.0	210.0	112.6	0.2	–
S17	125.0	119.0	123.4	0.2	–
	119.0	113.0	125.9		
	113.0	106.0	109.8		
	106.0	98.0	96.76		
	98.0	89.0	84.82		
	89.0	80.0	81.27		
S18	80.0	65.0	72.25	0.2	–
	65.0	51.0	66.56		
	51.0	40.0	119.4		
S19	150.0	208.0	5.905	0.2	–
S20	150.0	30.0	8.883	0.2	–
S21	6.7	210.0	3.243	0.2	–
S22	−28.8	30.0	4.133	0.2	–
	30.0	216.0	3.819		
S23	128.4	219.0	2.077	0.2	–
S24	215.0	205.0	1580.0	0.2	–
S25	260.0	116.1	6.988	0.2	–
S26	48.0	116.1	10.8	0.2	–
S27	105.0	102.0	302.3	0.2	–
	102.0	40.0	1.694		

Table 17.4 (*Continued*)

Stream	T^{in} (°C)	T^{out} (°C)	CP (kW °C^{-1})	U (kW m^{-2} °C^{-1})	Cost ($ kW year^{-1})
S28	113.2	95.0	137.8	0.2	–
	95.0	54.0	68.05		
	54.0	45.0	166.7		
S29	45.0	40.0	29.24	0.2	–
S30	54.6	135.0	23.23	0.2	–
S31	126.3	40.0	17.18	0.2	–
S32	97.7	87.0	61.85	0.2	–
	87.0	73.8	26.89		
	73.8	73.0	300.0		
	73.0	40.0	7.212		
S33	43.6	67.9	11.24	0.2	–
S34	101.5	102.5	992.5	0.2	–
S35	86.6	83.0	90.0	0.2	–
	83.0	73.0	23.9		
	73.0	45.0	6.464		
S36	115.0	112.7	6132.0	0.2	–
S37	87.8	84.6	678.8	0.2	–
	84.6	40.0	20.03		
S38	40.6	72.0	10.0	0.2	–
S39	97.6	98.6	535.6	0.2	–
S40	78.3	61.7	15.28	0.2	–
	61.7	40.0	3.356		
S41	97.6	71.8	242.4	0.2	–
Steam, 57 bar	405.0	404.0	–	0.2	272
Steam, 29 bar	250.0	249.0	–	0.2	244
Steam, 9 bar	175.0	174.0	–	0.2	125
Steam, 2.2 bar	123.0	122.0	–	0.2	25
Steam, 1.5 bar	111.0	110.0	–	0.2	20
Fuel gas	2050	280	–	0.2	105
Water	35.0	40.0	–	0.2	16
Air	35.0	36.0	–	0.2	10
Brine	−15.0	−14.0	–	0.2	200
Freon, 22 bar	−45	−44	–	0.2	350.0

Table 17.5 Heat exchanger network data and heat load distribution for Example (17.3)

Hot stream	Hot T^{in} (°C)	Hot T^{out} (°C)	Cold stream	Cold T^{in} (°C)	Cold T^{out} (°C)	Duty (kW)	1-2 Area (m²)
S1	15.0	0.5	S3	−45.0	−44.6	385.6	73.8
S6	39.3	39.3	S2	−44.0	5.9	385.6	70.6
S6	39.3	39.3	S3	−44.6	−44.0	637.3	76.2
S15	152.1	137.2	S13	75.9	105.7	1044.5	200.5
S16	231.3	210.0	S13	105.7	173.1	2404.3	318.7
S16	270.0	236.1	S7	158.1	158.9	3822.0	408.7
S17	120.2	115.1	S13	57.7	75.9	638.1	126.9
S16	236.1	233.0	S19	150.0	208.0	342.5	68.4
S20	70.5	30.0	S22	−28.8	60.6	360.0	141.2
S16	233.0	231.3	S23	128.4	219.0	188.2	42.9
S24	205.5	205.0	S7	157.9	158.1	807.5	170.8
S24	210.2	207.7	S9	102.0	103.0	3813.0	358.2
S24	207.1	205.5	S5	90.5	91.5	2539.4	220.2
S24	215.0	210.5	S12	116.4	117.4	7115.0	742.5
S15	162.5	152.1	S26	48.0	116.1	735.6	105.6
S25	260.0	116.1	S30	91.7	135.0	1005.6	176.7
S15	137.2	125.0	S30	54.6	91.7	862.1	154.8
S31	126.3	111.4	S21	6.7	85.6	255.9	39.6
S17	125.0	122.8	S33	43.6	67.9	273.2	40.5
S24	207.7	207.1	S34	101.5	102.5	992.5	94.1
S17	122.8	120.2	S38	40.6	72.0	314.1	49.1
S24	210.5	210.2	S39	97.6	98.6	535.6	47.7
S27	105.0	40.0	Air	35.3	35.3	1012.0	177.8
S40	78.3	40.0	Air	35.5	35.5	326.5	133.8
S37	87.8	40.0	Air	35.3	35.3	3066.0	898.9
S17	115.1	80.0	Air	35.9	36.0	3308.0	540.6
S35	86.6	45.0	Air	35.3	35.4	744.0	209.9
S8	81.1	75.1	Air	35.4	35.5	4442.0	1056.8
S36	115.0	112.7	Air	35.5	35.7	14 103.0	1802.5
S32	97.7	40.0	Air	35.2	35.3	1494.8	407.9
S18	80.0	40.0	Air	35.2	35.2	3328.9	2138.2
S10	66.6	61.1	Air	35.7	35.9	8293.0	3009.3
S31	111.4	40.0	Air	35.0	35.0	1227.0	468.9
S20	150.0	70.5	Air	36.0	36.0	706.0	106.1
S28	113.2	45.0	Air	35.0	35.2	6797.8	2226.7
S11	61.1	49.0	Air	35.4	35.4	212.0	112.8
S41	97.6	71.8	Air	35.9	35.9	242.4	50.9
S4	−28.8	−33.0	Freon 22	−45.0	−44.0	1801.2	1335.7
S6	39.3	39.2	Brine	−15.0	−14.0	3437.6	639.9
S29	45.0	40.0	Brine	−15.0	−15.0	146.2	25.5
Flue gas	2023.0	1595.0	S13	173.1	200.0	6391.0	39.9
Flue gas	319.7	280.0	S22	60.6	216.0	593.3	40.3
Flue gas	1595.0	319.7	S14	200.0	520.0	19 045.0	526.9
Flue gas	2050.0	2023.0	S21	85.6	210.0	403.4	2.1

to note here that the utility distribution from the model solution is close to that calculated by another method suggested in Shenoy et al. (1998)—for details see the original paper. Also, Shethna et al. (2000b) calculated surface area, number of shells and TAC for the HEN developed and compared them with corresponding targets determined by PT approaches (the targets were calculated assuming a heat recovery level identical to that from the optimal HLD). All of the parameters for this HEN were substantially lower even though the HEN is most likely only a local optimum. There is also evidence that PT targeting approaches may lead to conservative results, as we emphasized in Section 16.2. This example shows that HLD calculated from the TAC-Transp model can be applied to generate good solutions by hand even for large-scale problems though it does not guarantee the global optimum.

18 Heat exchanger network synthesis

18.1. INTRODUCTION

We will address the HEN design problem in two chapters: HEN synthesis in this chapter and HEN retrofit in the next. Separate treatment of these topics does not mean that they have to be solved by different techniques. In fact, there are some methods that can deal with both problems in a more-or-less common framework. For instance, simultaneous approaches addressed in this chapter require minor changes in order to be applicable to revamp design. The division into synthesis and retrofit results is for the following two reasons:

- The retrofit design problem is of significantly higher combinatorial character than synthesis. Also, the "nonlinearity" level is higher. Due to these features both insight-based and simultaneous methods have problems with efficient and robust solution, particularly if larger-scale problems are to be solved. Also, they are not able, at present, to account for such variables as for instance discrete parameters of standard heat exchangers. In the chapter on HEN retrofit (Chapter 19), we will describe the application of genetic algorithms to retrofitting HENs comprising standard heat exchangers.
- Existing general approaches to HEN retrofit are most often based on synthesis techniques. Hence, to explain the latter one should have knowledge of the former.

Even in the case of an HEN synthesis problem only numerous approaches have been suggested to date. Explanation of all of them is beyond the scope of this book. We have decided to present in more detail certain selected representative techniques from two wide classes: sequential and simultaneous. Sequential methods can be performed by insight-based techniques that do not use optimization at all or to a limited extent (as an auxiliary tool). Alternatively, all or some key stages can be performed with extensive use of optimization. Insight-based techniques usually are not reliable tools to attack larger-scale problems. Nevertheless, they are quite efficient for solving smaller ones. Also, they provide an understanding of optimization models. Hence we will address both insight-based and optimization sequential approaches for HEN grassroot design in the next two sections. Then selected contributions to simultaneous approaches will be addressed.

Energy Optimization in Process Systems

18.2. SEQUENTIAL APPROACHES

18.2.1. Pinch-Technology-Based Methods

Pinch-technology-based approaches to design heat exchanger networks are widely used in industry. They are simple, and employ graphical tools that visualize solution algorithms and thus give insight and greater understanding. Also, they give the designer control of the design procedure. However, the usefulness and simplicity of the calculating procedures are most often demonstrated by simple cases that do not show several potential pitfalls or the great computational load of the iterative, trial-and-error procedures needed for larger-scale problems.

PT approaches do not guarantee an optimal solution or even a near-optimal one. However, they lead to a better HEN than those designed by intuition and design experience. The user has to be aware that they are not systematical tools and a deep understanding of basic PT concepts is necessary to use them in practice.

The final pinch design method (PDM) developed by Ahmad (1986), also presented later in Linnhoff and Ahmad (1990) and Ahmad et al. (1990), evolved from the first version by Linnhoff and Hindmarsh (1983). The PDM allows calculation of network structure and basic parameters of heat exchangers and splitters. Because the PDM employs some heuristic elements, due to inherent simplifications hidden in PT concepts, a second design stage is often necessary. This is the evolution stage where the design is improved by further minimizing HEN total annual cost, particularly by improving trade-off: number of matches – cost of utilities. Also, NLP optimization of heat exchanger areas and split fractions can finally be performed for a network of fixed topology.

The targeting stage is of crucial importance in the PDM. It provides the targets and the optimum HRAT value (calculated by supertargeting) that is applied in the design. Also, targeting techniques are often the basis of design procedures and thus their knowledge is fundamental for performing the design stage.

Here, we limit the description of the PDM to a brief presentation of basic steps. Readers interested in a more thorough presentation are referred to Linnhoff et al. (1982), Shenoy (1995), Smith (2005) and papers by Linnhoff and Ahmad (1990) and Ahmad et al. (1990).

After the targeting stage the designer knows the optimal ΔT^{min} value ($\Delta T^{min, opt}$), pinch temperature/temperatures and values of all targets. To meet the "no heat exchange via pinch" rule the division at pinches must be performed. Most often a single pinch exists and we will explain the method for this case. If there are more pinches some subtasks are "between pinches" problems. They are very restricted as to structure and heat loads of matches. Though the basic rules of pinch technology are valid for between-pinches problems, special care should be paid to ensure that decisions on splits at one pinch are consistent with those valid for the adjacent one. Advice on design HEN with multiple pinches

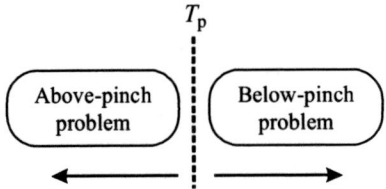

Figure 18.1 "Directions" of solution procedure in the PDM.

can be found in Jeżowski (1992a), who addressed an alternative approach to that suggested by Trivedi et al. (1989a).

The upper subtask (above the pinch) and lower one (below the pinch) are solved separately since they are treated as independent problems in the first stage of the PDM. The evolution stage is then applied to integrate them properly into one network, though simple merging is often treated as sufficient. The design, that is, structure development by matching streams, starts at the pinch and proceeds away from it as shown in Figure 18.1.

Thus, the streams at the pinch (at-pinch streams having pinch temperature) must be matched first. For the above-pinch task, all hot at-pinch process streams above the pinch must be matched with some cold at-pinch streams, in order to eliminate cold utility usage above the pinch. Likewise, below the pinch, all cold at-pinch streams must be matched with chosen hot at-pinch streams to avoid hot utility usage below the pinch. Hence, the number of branches of streams has to be properly adjusted at the pinch to engage all hot at-pinch streams (above the pinch) and all cold at-pinch streams (below the pinch) in matches. Notice that proper matching of at-pinch streams is a crucial point of the PDM since it decides on a possibility of reaching the MER target. This target is of the utmost importance since it greatly influences the total cost of the final solution.

To determine a match of at-pinch streams, that is, to find a proper branching scheme and to calculate heat load, the PDM offers the following rules:

1. Adjustment of CP values of streams to be matched (called CP rule in the following).
2. Adjustment of temperature difference profiles (called Δtm rule in the following).

Both rules are generalizations of earlier rules on stream branching and "tick-off" heuristic—see Linnhoff and Hindmarsh (1983). They are aimed at reaching all three targets. However, as we will show in the following, this requires tedious calculations and it is doubtful whether the goal can be achieved rigorously.

The rules are as follows:

CP rule: Streams chosen for heat exchange should have a ratio of CP values close to the ratio of CP values of appropriate segments of the composite curves, that is, the segments that contain the streams.

Δtm rule: Temperature profiles in a match should feature a close resemblance to temperature profiles of the appropriate segments of the composite curves.

Though the rules are simple, their application is often not. The designer usually faces the necessity of employing a trial-and-error procedure. She/he has to choose at-pinch streams for a match and check the rules. If they are not satisfied the designer has two options:

(a) change one stream in the match;
(b) split a hot or cold stream or both.

Additionally, when calculating the match, the designer has to view it in the context of all other possible at-pinch matches. Also, the rules are of a qualitative character.

The rules for branching streams h and c to be matched are:

A. For a choice of CP value for branches:
    B1. above pinch: $CP_h \leq CP_c$
    B2. below pinch: $CP_h \geq CP_c$

B. For the number of branches at the pinch (a branch can be an original stream or a branch of original stream)
    B1. above pinch: no. of hot branches $\leq$ no. of cold branches
    B2. below pinch: no. of hot branches $\geq$ no. of cold branches

After reaching a decision on branching, and choosing the split ratio, the designer has to determine the heat load of the match. To minimize the number of matches, use of the CP rule is advised. Also, the "tick off" rule, which says that a single residual stream should be created, is helpful (the heuristic results from the reasoning in Section 16.2 for achieving the MNM target by creating a single residual in each match).

The rules are tools. In the case of complex problems with many streams one may expect many trials in order to satisfy them. There is, however, a useful technique to control the structure development and match calculation. This is called the remaining problem analysis (RPA). The concept is based on extensive use and comparison of targets at steps of sequential choosing and rating of matches. To explain it let us assume that two streams (h_i and c_j) have been selected for the match at the current step of the sequential procedure. We assume also that the matches calculated upto this point are valid, that is, they allow the possibility of finally satisfying all targets. Heat load (Q_{ij}) and heat transfer area of the match (A_{ij}) can then be calculated.

Let:

AT^1, Qu^1, N^1 denote, respectively, the minimum total area, the maximum heat recovery and the minimum number of matches targets for an overall set of streams at the current design step, that is, streams unmatched and also streams h_i, c_j chosen for the match.

AT^2, Qu^2, N^2 denote the same targets for streams not matched and residual stream from the match h_i, c_j.

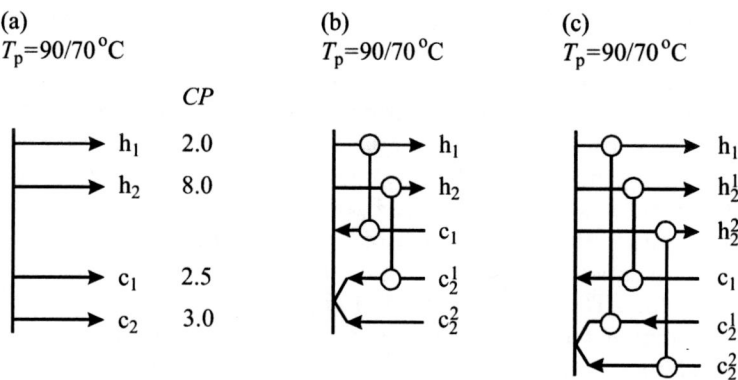

Figure 18.2 Illustration of the PDM for matching at-pinch streams in Example 13.1: (a) at pinch streams; (b) illustration for branching rule (B); (c) possible arrangement of at-pinch matches with many splits.

The match h_i with c_j is "good" (acceptable) if:

$$AT^1 - \left(AT^2 + A_{ij}\right) \le \varepsilon_A \tag{18.1}$$

$$N^1 - \left(N^2 + 1\right) \le \varepsilon_N \tag{18.2}$$

$$Qu^1 - \left(Qu^2 + Q_{ij}\right) \le \varepsilon_Q \tag{18.3}$$

where ε_A, ε_N, ε_Q are tolerances of the design.

To explain the PDM we present here an example of calculating the at-pinch matches for below the pinch subproblem in Example 13.1 (data in Table 13.1). The at-pinch streams are shown in Figure 18.2a.

First let us explain the logic behind branching rules B. Assume, for a moment, that stream c_2 has been split. Let the matches between both hot streams and two of the three cold streams be applied as shown in Figure 18.2b. In the result one cold at-pinch stream is left and the application of a heating utility would be necessary. It is important to notice that the split of stream c_2 does not spoil rule A. Hence both rules have been satisfied simultaneously.

We should start with the selection of streams using the CP rule. The ratio of CP for the composite curves just below the pinch is 1.82. Any possible match h_1–c_1, h_1–c_2, h_2–c_1, h_2–c_2 does not feature such a CP ratio. Hence we must proceed to branching, applying the CP rule and also following rules A and B. Stream h_2 is the candidate for splitting since it has the highest CP value. This gives three hot branches and two cold branches, which satisfies rule B and does not spoil rule A. However, there is the question of estimating CP values of the branches. The CP rule also gives us very crude advice since the CP ratio changes with temperature and composite curve segments are not equivalent to segments of streams in a match. The PDM also provides us with the Δtm rule. It requires drawing a plot of temperature difference between the composite curves and the cold streams. Then a similar plot for each potential match has to be drawn. The profiles of the two plots—for composite curves and for a match—should be similar in a proper temperature range. A close resemblance ensures vertical heat transfer in a match, and thus the minimum area (but only for identical heat transfer coefficients of streams, as we pointed out in Chapter 16). Notice that one can consider an additional split of stream c_2 since the number of hot and cold branches satisfies rule B. It is also valid as to rule A. The branches are shown in Figure 18.2c. The matches shown in Figure 18.2c, h_1–c_2^1, h_2^1–c_1 and h_2^2–c_2^2, give the minimum number according to the $N-1$ rule. The additional split gives more freedom in estimating CP values and thus has the potential to reach nearer to the MTA target. However, the designer should also follow the rule of thumb: try to keep the number of splitters to a minimum.

Even such a simple example illustrates the calculation load and the necessity of problem understanding with the PDM approach. Nevertheless, there are numerous applications of the PDM or other techniques from PT to industrial cases, for instance, the more recent contributions by Markowski (2000), Özkan and Dinčer (2001), Matijasevic and Otmaievic (2002), Noureldin and Hasan (2006) and Yoon et al. (2007).

Due to some limitations of the PDM, solutions obtained are rarely optimal and further improvements are possible. The most substantial effect on total annual cost can be achieved by minimizing the number of matches via eliminating loops, called "loop breaking." The general concept was explained in Section 16.2 within the discussion on the minimum number of matches—see for example, Figure 16.3. Because the final network is achieved in the PDM by simple merging of two (or more) subnetworks calculated independently, it usually has some loops across pinch/pinches. Also, a HEN can feature additional loops, not across pinch, because of the inadequacy of the PDM matching rules, particularly for matches of streams away from the pinch. The latter loops can be broken by a simple method developed by Su and Motard (1984). However, the technique should not be employed to break loops via pinch since it has no control mechanism for heat recovery decrease—see for example, Engel and Morari (1988). Notice that by breaking such a loop one increases utility usage if temperature approaches in matches are limited by the HRAT value, as is the case in the PDM. The increase is called energy penalty. Hence, an improvement of TAC can be achieved only if deleting a match causes a saving higher than the increase in utility cost. However, the use of the DTA concept can result in a smaller, or no, utility cost penalty compared to a single $\Delta T^{\min}$. This effect is identical to that explained in Section 16.2 on match number minimization. Relaxation of outlet temperature conditions can also increase the chances of minimizing the energy penalty.

The objective of loop breaking is to minimize the number of matches at minimum energy penalty. There are several more or less heuristic approaches to loop breaking, for instance contributions by Trivedi et al. (1990) and Zhu et al.

(1993, 1999). Works by Pethe et al. (1989), Gundersen et al. (1991) and Han et al. (1998) are also contributions to loop breaking identifications. Here we present a systematic method developed by Jeżowski et al. (2001a, 2001b) and briefly addressed in Jeżowski et al. (2000a).

The approach is based on a solution of the MILP optimization model for the HEN of a given structure, that is, the HEN from the structure development stage of the PDM. Notice that the structure is fixed but in the loop-breaking approach it can be subject to changes opposite to, for instance, parameter NLP optimization. To formulate the loop-breaking model we need to model a HEN of given topology. We will apply here also the model of Jeżowski et al. (2001a, 2001b) since it is compact, numerically efficient and easy to code. Notice that it can be employed also for other problems like HEN simulation, parameter optimization or HEN retrofit with the "network pinch" approach. The latter issue will be addressed in Chapter 19.

A HEN consists of the following elements: matches (i.e. process–process matches, heaters, coolers), splitters, mixers and substreams. A substream is a part of a stream (process stream and utility as well) between two apparatus, between the inlet to a HEN and an apparatus, and between an apparatus and the outlet from a HEN, where by apparatus we mean a match, a splitter or a mixer.

An index is assigned to each element. The following index sets are created:

$K = \{k \mid k$ is match (k^E—process–process stream match, k^H—heaters, k^C—coolers)\}

$L = \{l \mid l$ is substream (l^H—hot substream, l^C—cold substream)\}

$M = \{m \mid m$ is mixer\}

$N = \{n \mid n$ is splitter\}

The scheme of numbering substreams is explained in Figure 18.3. Also, matrices and vectors for this network (defined in the following) are inserted in this figure.

Substreams are numbered in ascending order from 1 starting with the upper-most stream in the grid representation. In the case of the hot stream the numbers increase from the left side and for cold streams in the opposite direction, that is, from the right to the left side of the grid representation. This convention has to be obeyed for mixers and splitters, too. It is important to note that a branch of a split is treated as a stream in the numbering scheme. Other elements, that is, apparatus, are sequentially numbered from 1, though this is not a strict requirement.

Similar to the models for MNU and MTA targets, index sets for streams are:

$H = \{i \mid i$ is hot stream\}

$C = \{j \mid j$ is cold stream\}

For the purposes of the model it is convenient to classify substreams into inlet and outlet substreams. Under the assumption that CP values of all streams do not

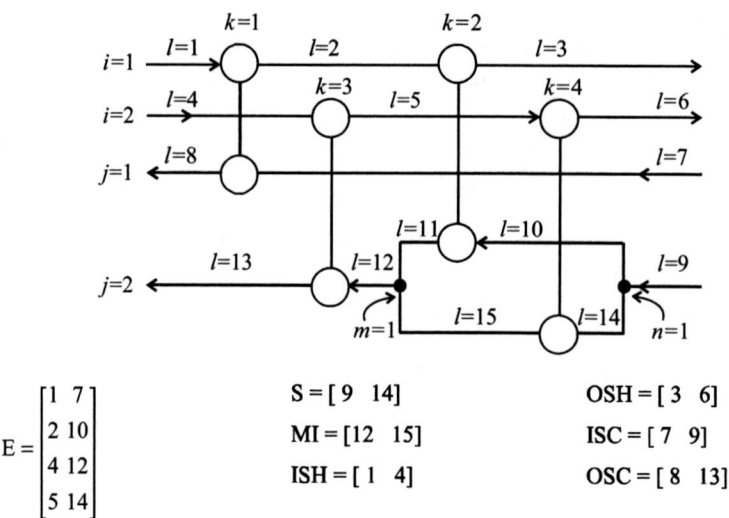

$$E = \begin{bmatrix} 1 & 7 \\ 2 & 10 \\ 4 & 12 \\ 5 & 14 \end{bmatrix}$$

$S = [9 \quad 14]$

$MI = [12 \quad 15]$

$ISH = [1 \quad 4]$

$OSH = [3 \quad 6]$

$ISC = [7 \quad 9]$

$OSC = [8 \quad 13]$

Figure 18.3 Illustration for the HEN mathematical model applied in the loop breaking method by Jeżowski et al. (2001a, 2001b).

depend on temperature, each substream has two uniquely assigned parameters, T and CP.

In order to simplify presentation of the model it is assumed in the following that a maximum of two branches can exist. This limitation can be easily removed since each branch can be further split into two and so on.

Let us define the following vectors and matrices based on the index sets and the numbering scheme defined above. The reader is also referred to Figure 18.3 for an explanation.

1. E—matrix for assigning substreams to matches. It has as many rows as matches exist in a HEN, and two columns. Elements of E have the following meaning:

   $e(k,1) = l^H$—number of inlet hot substream to kth match
   $e(k,2) = l^C$—number of inlet cold substream to kth match.

   Notice that from the convention of numbering substreams an outlet hot/cold substream has a number greater by one than the hot/cold inlet substream from match k. Hence matrix E provides sufficient information on assignments: match − substreams since numbers of outlet substreams are known from numbers of inlet substreams to the match.

2. S—matrix for assigning substreams to splitters. It has as many rows as splitters exist in a HEN, and two columns. Elements of S have the following meaning:

   $s(m,1) = l^S$—number of inlet substream to mth splitter
   $s(m,2) = l^b$—number of outlet substream to mth splitter such that $l^b \neq l^S + 1$ (it is called type "b" substream).

   By convention of numbering the second outlet substream always has number $l^S + 1$ and need not be included in S.

3. *MI*—matrix for assigning substreams to mixers. It has as many rows as mixers occur in a HEN and two columns. Elements of *MI* have the following meaning:

   $mi\ (n,1) = l^M$—number of outlet substream from mixer n

   $mi\ (n,2) = l^b$—number of inlet substream to nth mixer such that $l^b \neq l^M - 1$ (it is called type "b" substream).

   By convention of numbering the second inlet substream always has number $l^S - 1$ and need not be included in matrix *MI*.

4. *ISH*—vector for numbers of inlet hot substreams to a HEN, with elements:

   $ish\ (i) = l^H; i \in H$

5. *OSH*—vector for numbers of outlet hot substreams from a HEN, with elements:

   $osh\ (i) = l^H; i \in H$

6. *ISC*—vector for numbers of inlet cold substreams to a HEN, with elements:

   $isc\ (j) = l^C; j \in C$

7. *OSC*—vector for numbers of outlet cold substreams from a HEN, with elements:

   $osc\ (j) = l^C; j \in C$

8. *TIH*—vector of given inlet temperatures of hot streams, with elements:

   $tih\ (i) = T_i^{in}; i \in H$

9. *TIC*—vector of given inlet temperatures of cold streams, with elements:

   $tic\ (j) = T_j^{in}; j \in C$

10. *TOH*—vector of given outlet temperatures of hot streams, with elements:

    $toh\ (i) = T_i^{out}; i \in H$

11. *TOC*—vector of given outlet temperatures of cold streams, with elements:

    $toc\ (j) = T_j^{out}; j \in C$

12. *CP*—vector of given CPs of substreams in a HEN, with elements:

    $CP\ (l); l \in L$

    (since the assumption of constant *CP* of stream assignment of *CP* values to substreams from stream data is straightforward)

13. *Q*—vector of heat loads of matches in a HEN, with elements:

    $Q\ (k); k \in K$

14. *T*—vector of temperatures of substreams in a HEN, with elements:

    $T\ (l); l \in L$

    The vectors and matrices are sufficient to define a unique HEN. Though there are many matrices and vectors they are very compact.

   In order to define the existence or nonexistence of a match in a HEN it is necessary to use binary variables: $y(k); k \in K$, such that:

$$y(k) = \begin{cases} 1 & \text{if match } k \text{ is in HEN} \\ 0 & \text{if not} \end{cases} \tag{18.4}$$

In order to ensure linearity of the optimization model the following assumptions have to be imposed:

1. Heat transfer areas of heat exchangers will not be accounted for in optimization.
2. CP values of substreams have to be fixed. In consequence CP values for branches of split streams are fixed, too. Also, the number of branches cannot change. It is forbidden to eliminate an existing branch. Hence, one cannot eliminate all matches at a branch since Jeżowski et al. (2001a, 2001b) suggested that such an "empty" branch has no physical meaning except for control purposes as a bypass.

For the assumptions the model is of the MILP type with the following decision variables:

$$T(l), \quad Q(k) \text{ and } y(k); \quad l \in L, k \in K$$

Due to assumptions 1 and 2 the goal function cannot include area and split fractions. Hence, the goal function is given by Equation (18.5) and comprises utility cost and the term corresponding to the fixed cost of matches:

$$\min \left(\sum_{k \in K} y(k) + \eta \left[\sum_{k^H \in K} Q(k^H) + \sum_{k^C \in K} Q(k^C) \right] \right) \qquad (18.5)$$

Parameter η in Equation (18.5) is used to convert both terms: number of matches and utility load to common basis. The value of η has to be assumed by the user. Based on typical relations, investment cost of matches versus utility cost, Jeżowski et al. (2001a, 2001b) recommend a value of η of the order 0.01–0.001. It is important to note that the lower value is given for η and the higher preference is given for reducing the number of matches but less for saving energy. The designer can account for some specific features of an economic scenario by choosing the η value.

Before presenting a detailed optimization model we will first discuss its important features. In order to ensure the possibility of minimizing the energy penalty due to cross-pinch loop breaking Jeżowski et al. (2001a, 2001b) proposed to:

(a) use the DTA concept, that is, to apply EMAT $\leq$ HRAT,
(b) apply a "small" superstructure instead of the existing structure.

The notion of "small" superstructure is similar to that applied by Asante and Zhu (1996a) for the "network pinch" approach to retrofitting HEN (see Chapter 19). Such a superstructure involves some slack matches with zero heat load seeded by the designer into the existing structure. Slack matches can be activated by the optimization subroutine. Thus, some structural changes are possible additional to deletion of exchangers. Let us note that such structural changes are necessary to achieve an optimum since Sagli et al. (1990) and Gundersen et al. (1991) observed that it may be impossible to reach optimal HEN from a local optimum by the standard evolutionary approach only (i.e. without new matches and/or relocations of existing heat exchangers). There are no strict rules on how to built a superstructure. Jeżowski et al. (2001a, 2001b) advised adding empty heaters and coolers at the "end" of streams as well as at those locations which can result in independent subnetworks.

Constraints of the basic loop-breaking optimization model (called the LB model) are given in the following (the goal function is defined by Equation (18.5)).

Constraints of the LB Model

A. Assignments of inlet temperatures to substreams

$$T(ish(i)) = tih(i), \quad i \in H \tag{18.6}$$

$$T(isc(j)) = tic(j), \quad j \in J \tag{18.7}$$

B. Heat balances for matches (process exchangers, heaters, coolers)

$$\text{B1–hot streams} \quad T(e(k, 1)) - T(e(k, 1) + 1) = \frac{Q(k)}{CP(e(k, 1))}, \quad k \in K \tag{18.8}$$

$$\text{B2–cold streams} \quad T(e(k, 2) + 1) - T(e(k, 2)) = \frac{Q(k)}{CP(e(k, 2))}, \quad k \in K \tag{18.9}$$

Constraint (18.8) also ensures that outlet temperatures of hot substreams are not higher than inlet temperatures. The same effects have condition (18.9) for cold substreams.

C. Heat balances for splitters

$$\left. \begin{aligned} T(s(m, 1) + 1) &= T(s(m, 1)) \\ T(s(m, 2)) &= T(s(m, 1)) \end{aligned} \right\} m \in M \tag{18.10}$$

D. Heat balances for mixers

$$T(mi(n, 1) + 1) \\ = \frac{T(mi(n, 1)) \cdot CP(m((n, 1)) + T(mi(n, 2)) \cdot CP(mi(n, 2))}{CP(mi(n, 1) + 1)}, \quad n \in N \tag{18.11}$$

E. Thermodynamics constraints for matches

$$T(e(k, 1)) - T(e(k, 2) + 1) \geq \text{EMAT} + DT(y(k) - 1), \quad k \in K \tag{18.12}$$

$$T(e(k, 1) + 1) - T(e(k, 2)) \geq \text{EMAT} + DT(y(k) - 1), \quad k \in K \tag{18.13}$$

The above inequalities also deactivate thermodynamic conditions for nonexisting matches. Notice that for y_k equals 1 the left-hand sides become negative if parameter DT is given a sufficiently large number. Similar conditions will be applied in Chapters 19 and 20 in modeling HEN design optimization approaches.

F. Conditions on outlet temperatures from HEN

$$toh\,(i) = T(osh(i)), \quad i \in H \tag{18.14}$$

$$toc(j) = T(osc(j)), \quad j \in C \tag{18.15}$$

G. Logical constraints for heat loads of matches
G1. constraints ensuring that deleted matches feature zero heat loads

$$Q(k) \leq Q(k)^{\mathrm{max}} y(k), \quad k \in K \tag{18.16}$$

Parameter $Q(k)^{\mathrm{max}}$ can be replaced by large number Q^{max} identical for all $k \in K$.
G2. Logical conditions to prevent elimination of a branch

$$\sum_{k' \in K'} y(k') > 0 \tag{18.18}$$

where K'—set of matches placed at the same branch

H. Explicit constraints on variables

$$Q(k) \geq 0, \quad k \in K \tag{18.18a}$$

$$y(k) = 0, 1, \quad k \in K \tag{18.18b}$$

$$T(l) \geq 0, \quad l \in L \tag{18.18c}$$

Other problem-specific constraints can also be included. For instance, one can insert conditions (18.19) to prevent deleting a "must-be" match from a HEN:

$$y(kk) = 1, \quad kk \in KK \tag{18.19}$$

where KK is the subset of must-be matches.

Another restriction that can be helpful in optimization is to impose a lower bound on the final number of matches in a HEN. The constraint has the simple form:

$$\sum_{k \in K} y(k) \geq N^l \tag{18.20}$$

where N^l—lower bound on number of matches

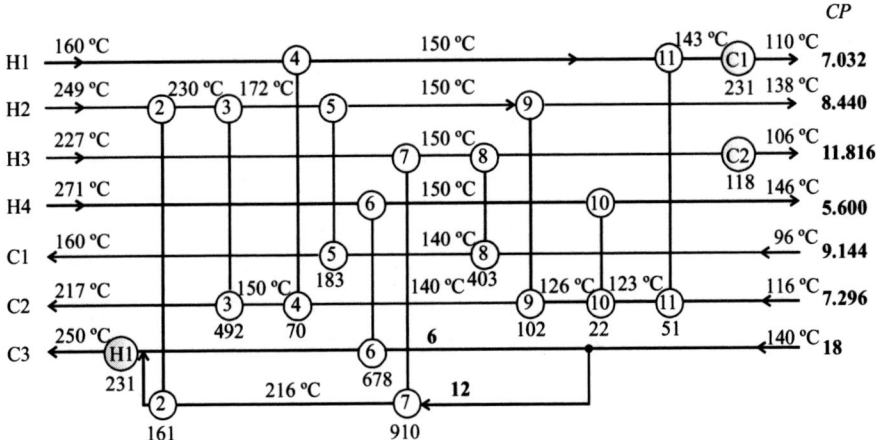

Figure 18.4 Initial HEN for Example 18.1: H1 – heater; C1, C2 – coolers; *CP* values in bold; heat load beneath heat exchanger.

Multiple global optima can exist for MILP optimization models. In order to generate all optimal solutions one has to add integer cuts to prevent solutions from previous runs having to be computed again.

Here we present Example 18.1 taken from Jeżowski et al. (2001a, 2001b) to show the possibilities for generating alternative solutions. The problem has been also investigated in Trivedi et al. (1990) and Zhu et al. (1993, 1999). The initial HEN is given in Figure 18.4. It contains 13 matches: 10 process heat exchangers, one heater and two coolers. The heat load of the heater is equal to 231 kW.

The HEN meets the basic "standards" of PT, that it divides at pinch ($Tp = 150/140$, HRAT = 10), consumes the minimum utility load and features the minimum number of matches according to Equation (16.2). As mentioned in the preceding, parameter η in goal function (18.5) can be employed to control trade-off: number of units–energy penalty. EMAT also influences this trade-off since the initial HEN had been designed with single ΔT^{min} equal to HRAT = 10. By varying parameters η and EMAT (HRAT is fixed at 10), Jeżowski et al. (2001a, 2001b) calculated certain alternative networks. Table 18.1 gathers the main parameters (number of matches and energy penalty) for the alternatives.

Table 18.1 Parameters of solutions to Example 18.1 according to Jeżowski et al. (2001a, 2001b)

Solution no. and control parameters	Number of matches	Energy penalty (kW)
1{$\eta > 0.001$; EMAT = 10 K}	8	522.07
2{$\eta > 0.001$; EMAT = 6 K}	8	375.00
3{$\eta > 0.001$; EMAT = 6 K}	8	32.80
4{$\eta < 0.001$; EMAT = 1 K}	7	749

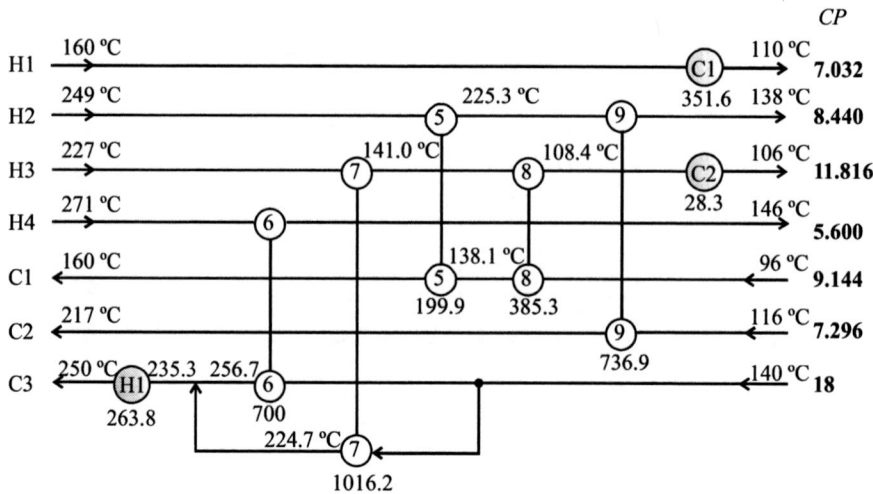

Figure 18.5 Solution no. 3 from Table 18.1 to Example 18.1 (symbols as in Figure 18.4).

Notice that the approach was able to reduce the number of units by five and even six in one case. Network no. 3 seems to be the best solution because of the small energy penalty. This is shown in Figure 18.5.

However, solution no. 4 from Table 18.1 can also be of interest if capital expenditure is high in relation to energy cost. It is interesting to observe the reason for achieving the number of units below the minimum from Equation (16.1), that is, below the $(N-1)$ rule—notice that the total number of streams (N) in this example is 9. The network is shown in Figure 18.6. Analysis reveals

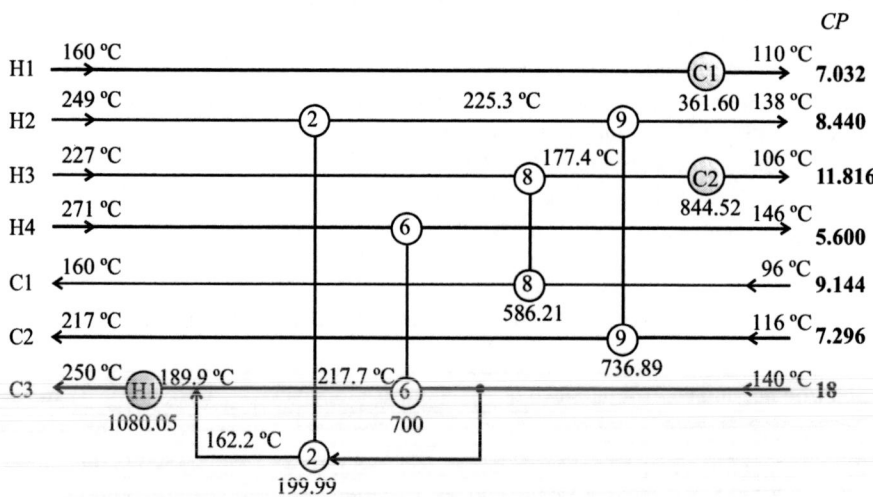

Figure 18.6 Solution no. 4 from Table 18.1 to Example 18.1 (symbols as in Figure 18.4).

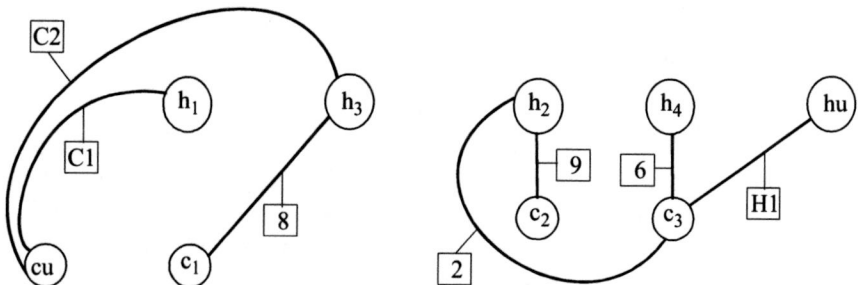

Figure 18.7 Two independent subnetworks in the network from Figure 18.6.

that seven matches result from the division into two subnetworks as Figure 18.7 illustrates.

In order to choose the best solution generated by the loop-breaking approach additional calculations are needed to determine the total cost of networks. Additionally, the solutions are not optimal in terms of TAC. This is the effect of model linearity. To account for TAC optimal solutions the assumption imposed by Jeżowski et al. (2001a, 2001b) as for fixed values of CP has to be removed. Also, it is necessary to include the capital cost of heat exchangers into the goal function and add design equations for the surface area of matches. The model will become more general but, probably, difficult to solve since it would be a MINLP.

18.2.2. Sequential Design with Optimization Approaches

Some optimization targeting approaches are able to give optimal heat load distribution (HLD) in regard to all or some targets, for example, the TAC-Transp model in Chapter 17. As we showed in that chapter, HLD does not uniquely define the HEN structure. The designer has to develop arrangements for all matches in the HLD. For small-scale problems, particularly those that do not require splitters and mixers, this can be performed by inspection with relatively few trials—see examples in Chapter 17. However, for more complex problems a systematic procedure is needed.

A HEN consists not only of heat exchangers but also of splitters and mixers. This fact had not been given sufficient consideration in the early stages of developing HEN design approaches. The paradigm of "homogenous networks" had dictated paying interest to heat exchange only. In consequence, splitters and mixers had not been included explicitly into conceptual models or into calculation procedures. For instance, even in insight-based methods, such as the PDM, stream branching causes serious difficulties. Also, there are no targeting approaches for determining the location and number of splitters and mixers since they are assumed to have no cost. Notice, however, that there are some more recent approaches that account for topological features—they will be mentioned in the next section.

To deal with a HEN synthesis problem in a systematic manner a superstructure concept is often employed as the most rigorous and general technique. The concept has found a wide application in simultaneous design methods for various systems. First a superstructure is created, and then an optimal structure is calculated by optimization. Simultaneous approaches to HEN synthesis are explained in the next section. However, the superstructure concept can be also applied to the design of a HEN with known HLD, that is, a sequential procedure for HEN synthesis. A superstructure, in general, must involve all possible structures of the HEN. Due to the fact that matches and their loads are fixed by HLD, the superstructure can be substantially reduced. This results in simplification of the optimization problem since it is sufficient to solve the NLP problem instead of the MINLP one.

There are two well-established propositions of superstructures for HEN synthesis: one by Floudas et al. (1986) and Ciric and Floudas (1990a, 1990b), and one by Yee et al. (1990a) and Yee and Grossmann (1990). The latter is described in full in the next section. Here, we will concentrate on the use of superstructure by Floudas (Ciric and Floudas), abbreviated to C-F superstructure in the following, to sequential HEN synthesis. For such a design mode the superstructure contains only those matches that correspond to given HLD. A C-F superstructure consists of superstructures for each match in the HLD. To allow for all possible arrangements, the match superstructure has the following features:

(a) There is a mixer at the inlet stream to the heat exchanger. The number of streams entering the mixer is equal to the number of heat exchangers associated with the stream.
(b) There is a splitter at the outlet stream from the heat exchanger. The number of streams from the splitter equals the number of heat exchangers associated with the stream.

Figure 18.8 shows an example of the superstructure for the illustrative case of two matches. This illustrative example corresponds to the HLD with two matches: h_1–c and h_2–c of loads $q_{h1,c}$ and $q_{h2,c}$.

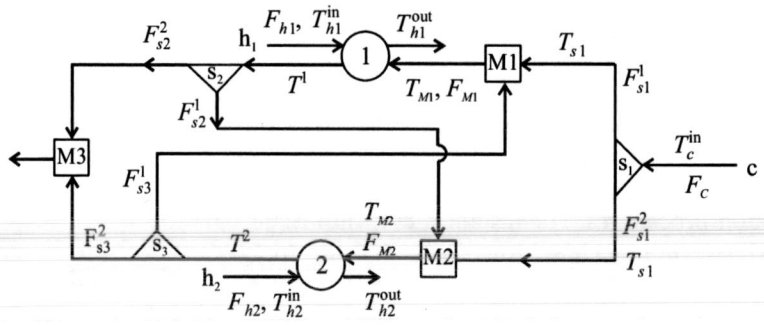

Figure 18.8 C-F superstructure for the illustrative example: S1, S2 – splitters; M1, M2 and M3 – mixers.

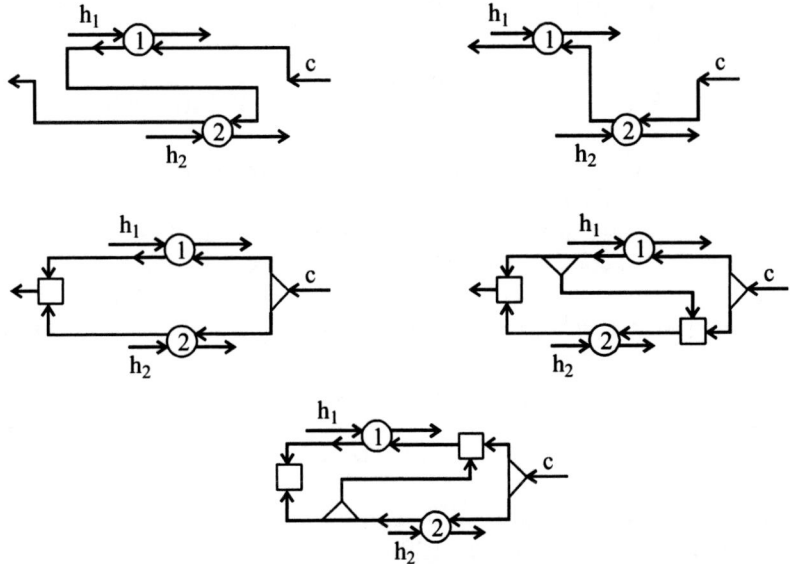

Figure 18.9 HEN structures embedded in the superstructure from Figure 18.8.

Note that the superstructure embeds all possible arrangements of two matches as shown in Figure 18.9, including two arrangements such as those discussed in Section 16.2 (see Figure 16.5) called "parallel pinch crossing arrangement". These connections have been criticized as being impractical.

In the sequential HEN synthesis utility loads are fixed. Also, the number of heat exchangers is given. Hence these parameters are not optimized. In consequence, no binary variables for heat exchangers are necessary. Due to the fact that no cost (including fixed charge) is given to splitters and mixers or to piping sections, binary variables are not needed for this equipment either. The optimization problem does not include discrete parameters, which are to be solved in simultaneous methods for HEN synthesis. To delete the splitter, mixer and the associated branches from the superstructure, continuous variables are sufficient. Namely, if the flow rate through connection is zero, the corresponding branches and apparatus are deleted.

Here, we begin the explanation of the optimization model for the C-F superstructure illustrated in Figure 18.8.

Parameters of the model:

F_c	flow rate of stream c
cp_c	heat capacity of stream c
F_{h1}, F_{h2}	flow rate of stream h_1, h_2 (respectively)
EMAT	minimum temperature approach in matches (notice that EMAT can be match dependent)

$q_{h1,c}$	heat load of match h_1-c in exchanger 1
$q_{h2,c}$	heat load of match h_2-c in exchanger 1
T_c^{in}, T_c^{out}	inlet, outlet temperature (respectively) of stream c
T_{h1}^{in}, T_{h1}^{in}	inlet, outlet temperature (respectively) of stream h_1 in exchanger 1
T_{h2}^{in}, T_{h2}^{in}	inlet, outlet temperature (respectively) of stream h_2 in exchanger 1
$U_{h1,c}$, $U_{h2,c}$	overall heat transfer coefficients in the matches
β, γ	cost coefficients for heat exchanger investment cost

The heat transfer surface area is calculated for 1-1 heat exchangers from the formula:

$$A = \frac{q}{\text{LMTD} \cdot U} \qquad (18.21)$$

Logarithmic mean temperature differences are most often calculated in HEN optimization models from Chen (1897) or Paterson (1984) approximations – Equations (18.22) and (18.23), respectively – to eliminate difficulties with the standard formula (18.24) for LMTD, particularly in cases where the temperature differences are identical. The Chen equation was employed in this model. It slightly overestimates the heat transfer area while Paterson's equation underestimates. Thus, Chen's approximation should provide a "safer" design. Also, this formula gives LMTD equal to zero if both temperature differences are zero, too. Generally, the influence of the approximation applied is minor and is usually discussed only when comparing results of synthesis methods. However, modeling a HEN using equations with LMTD parameters can result in some numerical difficulties. Hence, Akman et al. (2002) suggested applying a "LMTD-free" model. A similar heat exchanger model (though less general) will be presented in Chapter 19 as the retrofit approach by Bochenek (2003).

$$\text{LMTD} = [\Delta t1 \cdot \Delta t2 \cdot 0.5 \, (\Delta t1 + \Delta t2)]^{1/3} \qquad (18.22)$$

$$\text{LMTD} = \frac{2}{3} \left[(\Delta t1 + \Delta t2)^{1/2} \cdot \frac{1}{3} \left(\frac{\Delta t1 + \Delta t2}{2} \right) \right] \qquad (18.23)$$

$$\text{LMTD} = \frac{\Delta t1 - \Delta t2}{\ln (\Delta t1 / \Delta t2)} \qquad (18.24)$$

where $\Delta t1$, $\Delta t2$ are the temperature differences at the ends of a match.
The optimization model for HEN synthesis with known HLD is as follows.

Synthesis-HLD-C-F Model

A. Goal function (includes area-dependent cost of heat exchangers)

$$\min \left\{ \beta_{h1,c} \left(\frac{q_{h1,c}}{[\Delta t1_1 \Delta t2_1 0.5(\Delta t1_1 + \Delta t2_1)^{1/3}]U_{h1,c}} \right)^{\gamma_{h1,c}} \right.$$
$$\left. + \beta_{h2,c} \left(\frac{q_{h2,c}}{[\Delta t1_2 \Delta t2_2 0.5(\Delta t1_2 + \Delta t2_2)^{1/3}]U_{h2,c}} \right)^{\gamma_{h2,c}} \right\} \qquad (18.25)$$

B. Mass balances for splitters

$$F_c = F_{S1}^1 + F_{S1}^2 \qquad (18.26a)$$

$$F_{M1} = F_{S2}^1 + F_{S2}^2 \qquad (18.26b)$$

$$F_{M2} = F_{S3}^1 + F_{S3}^2 \qquad (18.26c)$$

C. Mass and heat balances for mixers

$$F_{M1} = F_{S1}^1 + F_{S3}^1 \qquad (18.27a)$$

$$F_{M1} T_{M1} = F_{S1}^1 T_c^{in} + F_{S3}^1 T^2 \qquad (18.27b)$$

$$F_{M2} = F_{S1}^2 + F_{S2}^1 \qquad (18.27c)$$

$$F_{M1} T_{M2} = F_{S2}^1 T_c^{in} + F_{S2}^1 T^1 \qquad (18.27d)$$

$$F_c = F_{S2}^2 + F_{S3}^2 \qquad (18.27e)$$

$$F_{M1} T_c^{out} = F_{S2}^2 T^1 + F_{S3}^1 T^2 \qquad (18.27f)$$

D. Heat balances for exchangers

$$q_{h1,c} = F_c \phi_c (T^1 - T_{M1}) \qquad (18.28a)$$

$$q_{h2,c} = F_c \phi_c (T^2 - T_{M2}) \qquad (18.28b)$$

E. Definitions of temperature differences in heat exchangers

$$\Delta t1_1 = T_{h1}^{in} - T^1 \qquad (18.29a)$$

$$\Delta t2_1 = T_{h1}^{out} - T_{M1} \qquad (18.29b)$$

$$\Delta t1_2 = T_{h2}^{in} - T^2 \tag{18.29c}$$

$$\Delta t2_2 = T_{h2}^{out} - T_{M2} \tag{18.29d}$$

F. Constraints on temperature differences

$$\Delta t1_1 \geq EMAT \tag{18.30a}$$

$$\Delta t2_1 \geq EMAT \tag{18.30b}$$

$$\Delta t2_1 \geq E \tag{18.30c}$$

$$\Delta t2_2 \geq EMAT \tag{18.30d}$$

G. Constraints on variables

$$F_{Sk}^1, F_{Sk}^2 \geq 0, \ k = 1, 2, 3$$
$$F_{m1}, F_{m2} \geq 0 \tag{18.31}$$

The above model is nonlinear due to the nonlinear goal function and heat balances of mixers. Biegler et al. (1997) suggested that it should not cause difficulties with calculating the global optimum. However, certain literature references show that NLP optimization of a HEN with fixed structure is not a trivial task, even for small problems. Some evidence for this is that the small optimization model for two exchangers has been included in the benchmark of test tasks for global optimizers. Quesada and Grossmann (1993) proposed an advanced global optimization algorithm with a spatial branch and bound procedure and convex under-estimators. Interestingly, the simple random search approach of Luus and Jaakola (see Chapter 1) was able to solve such NLP problems to global optima as reported in Luus (1993) though for small-scale tasks. A brief discussion of the optimization robustness issue will be given at the end of the chapter.

18.3. SIMULTANEOUS APPROACHES TO HEN SYNTHESIS

In this section we will mainly address the approaches applying the superstructure developed by Yee et al. (1990a, 1990b). The synthesis method of Ciric and Floudas (1990a, 1990b), with C-F superstructure is the alternative—in fact, it is even more rigorous. However, the former attracted greater attention and followers who proposed various extensions and modifications. Such wider application is, most likely, caused by the lower "level" of nonlinearity of the MINLP optimization model that should allow one to find at least good local optima with available solvers. The key feature of the model is that only the goal function is nonlinear while the model constraints are all linear.

The superstructure of Yee et al. (1990a, 1990b), called Y-G superstructure in the following, is based on the following key concepts:

- The superstructure is organized in a number of stages—stage-wise superstructure. Notice that temperatures of streams exiting stages are not fixed—they are decision variables in optimization in contrast to the division into temperature or enthalpy intervals.
- Streams entering each stage are split so as to achieve all possible matches between hot process and cold process streams. Then, the streams are mixed at the outlet of stages.

The following assumptions are imposed:

- The heater is placed at the highest temperature region of the cold process stream. One heater suffices for one process stream.
- The cooler is located at the lowest temperature region of the hot process stream. One cooler services one hot process stream.
- Mixing of streams is isothermal, that is, each branch of stream to a mixer has the same temperature.
- Heat exchangers are modeled as matches (1-1 units).

An illustration of the superstructure with two stages and for two hot (H1, H2) and two cold process streams (C1, C2) is shown in Figure 18.10.

Here, we present the optimization model following the explanation in Biegler et al. (1997). The formulation uses only one hot utility and one cold utility. Because steam is commonly applied, heating utility index s will be used instead of general index m. Likewise, index w (from cooling water) will denote cooling utility instead of general index n.

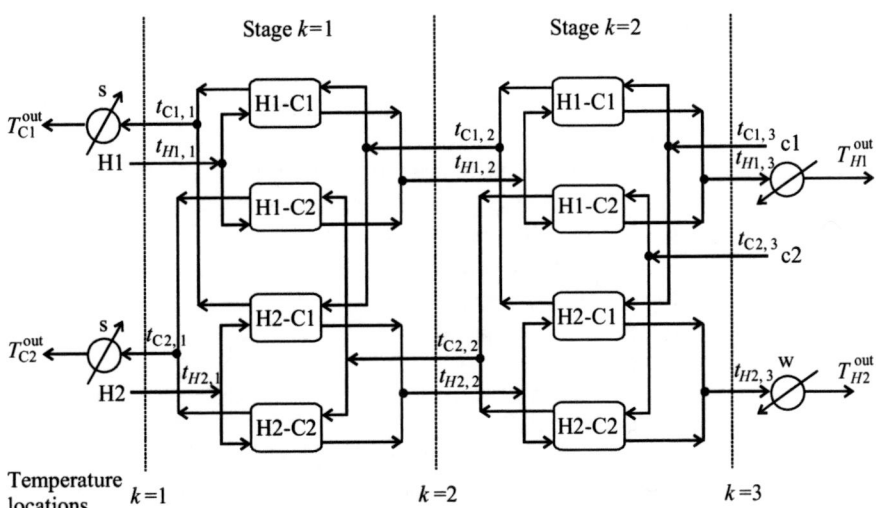

Figure 18.10 Illustration of the Y-G superstructure for HEN synthesis.

Indices and Index Sets

$H = \{i \mid i$ is hot process stream$\}$
$C = \{j \mid j$ is cold process stream$\}$
k—stage or temperature location; stages are numbered from 1 to K with descending temperature; for stage k there are two temperature locations, k at inlet and $k + 1$ at outlet
s—heating utility
w—cooling utility

Model Parameters

CP_i/CP_j	CP of stream i/j
EMAT	minimum temperature approach in matches
K	number of stages
phu_s	unit cost of heating utility s
pcu_w	unit cost of cooling utility w
$Q^{\max}, D$	sufficiently large numbers
$T_i^{\text{in}}/T_j^{\text{in}}$	inlet temperature of stream i/j to the superstructure
$T_i^{\text{out}}/T_j^{\text{out}}$	outlet temperature of i/j to the superstructure
$T_s^{\text{in}}/T_w^{\text{in}}$	inlet temperature of s/w
$T_s^{\text{out}}/T_w^{\text{out}}$	outlet temperature of s/w
U_{ij}, U_{sj}, U_{iw}	overall heat transfer coefficients in matches
α, β, γ	parameters for investment cost of a heat exchanger

Model Variables

LMTD_{ijk}	logarithmic mean temperature difference of match i with j in stage k
LMTD_{sj}	logarithmic mean temperature difference of match s with j
LMTD_{iw}	logarithmic mean temperature difference of match i with w
q_{ijk}	load of match i with j in stage k
q_{iw}	load of match i with w
q_{sj}	load of match s with j
t_{ik}	temperature of i at location k
t_{jk}	temperature of j at location k
y_{ijk}	binary variable for match i with j in stage k; $y_{ijk} = 1$ means that the match exists
y_{iw}^c	binary variable for match of i with w
y_{sj}^h	binary variable for match of s with j
Δt_{ijk}	temperature difference for match i with j at temperature location k

Δt_{iw} temperature difference in match i with w
Δt_{sj} temperature difference in match s with j

Notice that Biegler et al. (1997) defined temperature differences in matches with utilities (Δt_{iw}) and (Δt_{sj}) for outlet temperatures of utilities only. Thus, they assumed that inlet temperatures of utilities are sufficiently high for utility s and sufficiently low for utility w to ensure appropriate temperature differences. This assumption can be easily removed. Similar to the NLP model of Ciric and Floudas (Synthesis-HLD-C-F model) LMTD is approximated by Chen's equation (18.22).

Synthesis-Simult-Y-G Model

$$\min \left\{ \left[\sum_{j \in C} q_{sj} phu_s + \sum_{i \in H} q_{iw} pcu_w \right] \right\}$$

$$+ \left[\sum_{i \in H} \sum_{j \in C} \sum_{k=1}^{K} y_{ijk} \alpha_{ij} + \sum_{j \in C} y_{sj}^{h} \alpha_{sj} + \sum_{i \in H} y_{iw}^{c} \alpha_{iw} \right] \qquad (18.32)$$

$$+ \left[\sum_{i \in H} \sum_{j \in C} \sum_{k=1}^{K} \beta_{ij} \left(\frac{q_{ijk}}{U_{ij} \text{LMTD}_{ij}} \right)^{\gamma_{ij}} + \sum_{j \in C} \beta_{sj} \left(\frac{q_{sj}}{U_{sj} \text{LMTD}_{sj}} \right)^{\gamma_{sj}} \right.$$

$$\left. + \sum_{i \in H} \beta_{iw} \left(\frac{q_{iw}}{U_{iw} \text{LMTD}_{iw}} \right)^{\gamma_{iw}} \right]$$

The first term defines the total cost of both utilities, the second defines the fixed charge for heat exchangers, heaters and coolers and the last one defines the area-dependent cost of all heat exchangers. Here, for the sake of brevity, we inserted LMTD parameters into the goal function. It should be noticed, however, that this spoils the linearity of the model constraints. Hence, parameters LMTD should be replaced in Equation (18.32) by the right-hand sides of:

$$\text{LMTD}_{ijk} = [(\Delta t_{ijk} \Delta t_{ij,k+1}) 0.5 (\Delta t_{ijk} + \Delta t_{ij,k+1})]^{1/3}$$

$$\text{LMTD}_{iw} = [\Delta t_{iw} (T_i^{\text{out}} - T_w^{\text{in}}) 0.5 (\Delta t_{iw} + T_i^{\text{out}} - T_w^{\text{in}})]^{1/3}$$

$$\text{LMTD}_{sj} = [\Delta t_{sj} (T_s^{\text{in}} - T_j^{\text{out}}) 0.5 (\Delta t_{sj} + T_s^{\text{in}} - T_j^{\text{out}})]^{1/3}$$

The constraints of the model are as follows:

1. Overall heat balances of streams.
   The balances should also ensure that the required outlet temperatures of process streams will be reached. For hot process streams the required enthalpy changes must

be satisfied by matches with cold process streams and utility w. Likewise, enthalpy change of each cold stream must be equal to loads on matches with cold process streams and utility s.

$$CP_i \left(T_i^{in} - T_i^{out} \right) = \sum_{k=1}^{K} \sum_{j \in C} q_{ijk} + q_{iw}, \quad i \in H \tag{18.33a}$$

$$CP_j \left(T_j^{out} - T_j^{in} \right) = \sum_{k=1}^{K} \sum_{i \in H} q_{ijk} + q_{sj}, \quad j \in C \tag{18.33b}$$

2. Heat balances for stages.
   In addition to overall balances we have to include heat balances for each stage. Due to the assumption of isothermal mixing, heat balances of mixers are unnecessary. Also, CP parameters for branches are not required.

$$(t_{ik} - t_{i,k-1}) \, CP_i = \sum_{j \in C} q_{ijk}, \quad i \in H; \, k = 1, \ldots, K \tag{18.34a}$$

$$\left(t_{jk} - t_{j,k+1} \right) CP_j = \sum_{i \in H} q_{ijk}, \quad j \in C; \quad k = 1, \ldots, K \tag{18.34b}$$

3. Assignments of inlet temperatures to the superstructure.
   The following equalities are necessary to assign given inlet stream temperatures to hot streams entering the first stage and to cold streams entering the last stage:

$$t_{i1} = T_i^{in}, \quad i \in H \tag{18.35a}$$

$$t_{jK} = T_j^{in}, \quad j \in C \tag{18.35b}$$

4. Constraints ensuring feasibility of temperatures in the superstructure.
   The constraints enforce that the temperatures at outlets from successive stages decrease. Since heat load on a certain stream in a stage can be zero, weak inequalities should be used:

$$t_{ik} \geq t_{i,k+1}, \quad i \in H; \, k = 1, \ldots, K \tag{18.36a}$$

$$t_{jk} \geq t_{j,k+1}, \quad j \in C; \, k = 1, \ldots, K \tag{18.36b}$$

Also, it is necessary to impose a limit on the outlet temperature from each stage with respect to the fixed outlet temperature. The constraints have to account for the fact that some or all streams will be heated up in heaters and cooled down in coolers. Hence, weak inequalities have to be used:

$$t_{i,K+1} \geq T_i^{out}, \quad i \in H \tag{18.36c}$$

$$t_{j,1} \leq T_j^{out}, \quad j \in C \tag{18.36d}$$

5. Definitions of heat loads of heaters and coolers:

$$q_{sj} = CP_j(T_j^{out} - t_{j1}), \quad j \in C \tag{18.36c}$$

$$q_{iw} = CP_i(t_{i,K+1} - T_i^{out}), \quad i \in H \tag{18.36d}$$

6. Logical conditions on heat loads.
   The conditions ensure that matches which do not exist, that is, have binary variables equal zero, must also have zero heat load. For simplicity's sake, the single parameter Q^{max} is used in all constraints.

$$q_{ijk} - Q^{max}y_{ijk} \leq 0, \quad i \in H; j \in C; k = 1, \ldots, K \tag{18.37a}$$

$$q_{iw} - Q^{max}y_{iw}^c \leq 0, \quad i \in H; k = 1, \ldots, K \tag{18.37b}$$

$$q_{sj} - Q^{max}y_{sj}^h \leq 0, \quad j \in C; k = 1, \ldots, K \tag{18.37c}$$

7. Calculations of approach temperatures at temperature locations.
   For each match within the superstructure temperature approaches at both sides have to be calculated in the model. Note that they would be determined for nonexisting matches, too. Hence, logical conditions have to be applied to deactivate the thermodynamic conditions for nonexisting matches and to ensure that temperatures of matches yield positive temperature differences at both sides. Biegler et al. (1997) formulated the conditions as weak inequalities. Minimization of the HEN cost will cause maximization of temperature differences. Hence, constraints for active matches with binaries equal to 1 will be active.

$$\Delta t_{ijk} \leq \left(t_{ik} - t_{jk}\right) + D\left(1 - y_{ijk}\right), \quad i \in H; j \in C; k = 1, \ldots, K \tag{18.38a}$$

$$\Delta t_{ij,k+1} \leq \left(t_{i,k+1} - t_{j,k+1}\right) + D\left(1 - y_{ijk}\right), \quad i \in H; j \in C; k = 1, \ldots, K \tag{18.38b}$$

$$\Delta t_{sj} \leq \left(T_s^{out} - t_{j,1}\right) + D\left(1 - y_{sj}^h\right), \quad j \in C; k = 1, \ldots, K \tag{18.38c}$$

$$\Delta t_{iw} \leq \left(t_{i,K+1} - T_w^{out}\right) + D\left(1 - y_{iw}^c\right), \quad i \in H; k = 1, \ldots, K \tag{18.38d}$$

The above constraints ensure that temperature approaches are not negative for active matches. Usually, however, an additional condition is imposed that they should be larger than the given EMAT value. Hence, we should also add:

$$\Delta t_{ijk} \geq EMAT, \quad i \in H; j \in C; k = 1, \ldots, K+1 \tag{18.38e}$$

$$\Delta t_{iw} \geq EMAT, \quad i \in H \tag{18.38f}$$

$$\Delta t_{sj} \geq EMAT, \quad j \in C \tag{18.38g}$$

Notice that match-dependent EMAT can be applied.

8. Conditions on variables:

$$q_{ijk}, q_{iw}, q_{sj}, t_{ik}, t_{jk} \geq 0 \tag{18.39a}$$

$$y_{ijk}, y_{iw}^c, y_{sj}^h = 0, 1 \tag{18.39b}$$

The above model can be easily reduced to a targeting model for the minimum total surface area. Such a reduced model was referred to in Chapter 16 as the Yee and Grossmann method.

The Y-G superstructure and, in consequence, the model involves some simplifications. First, isothermal mixers are assumed at each stage. Second, all heaters and coolers are placed only at the ends of process streams, that is, at the lowest temperatures of hot streams and at the highest temperatures of cold streams. Note also that the superstructure does not allow placing more than one heat exchanger at one branch. Lastly, the number of stages is the heuristic parameter. Yee et al. (1990a, 1990b) suggested that setting it at the maximum of hot process and cold process streams will result in sufficiently accurate results. However, Daichendt and Grossmann (1994a, 1994b) observed that to achieve a high accuracy the number of stages has to be equal to the number of temperature intervals created for a fixed EMAT value. This, however, can lead to a huge number of variables and prevent an efficient solution. An excellent review on computational issues is given in Grossmann and Daichendt (1996).

Sorsak and Kravanja removed the limitation to 1-1 units—first in a short conference paper (Sorsak and Kravanja, 1999), and then in a journal paper (Sorsak and Kravanja, 2002). More importantly, they included in the latter work the possibility of heat exchanger type selection from among: double-pipe, shell-and-tube (U-tube) and plate-and-frame. The selection is performed on the basis of:

(a) capital cost via various cost coefficients,
(b) permissible surface area range,
(c) operational pressure and temperature limitations.

Table 18.2 shows the appropriate parameters suggested in Sorsak and Kravanja (2002).

The model of the heat exchanger does not include pressures and, hence, they are treated as fixed parameters. An initial pre-screening procedure suffices for

Table 18.2 Parameters for heat exchanger type selection according to Sorsak and Kravanja (2002)

Type	p^{max} (MPa)	T^{min}–T^{max} (°C)	A^{min}–A^{max} (m^2)	α ($/y)	β ($/y)
Double pipe	30.7	−100–600	0.25–200	1 937	201
Shell-and-tube	30.7	−200–600	10–1000	21 615	93
Plate-and-frame	1.6	−25–250	1–1200	17 034	61

Table 18.3 Parameters a, b in Equation (18.41) according to Sorsak and Kravanja (2002)

i	a_i	b_i
0	15.210 806	0.623 393
1	−1.603 737	−0.698 824
2	42.337 642	−0.250 187
3	–	0.434 342
4	–	0.301 423
5	–	0.171 389
6	–	0.141 410

elimination of heat exchanger types that do not satisfy pressure conditions. Permissible surface area ranges were accounted for by cost coefficients α and β. Sorsak and Kravanja (2002) applied the linearized investment cost function:

$$CAP = \alpha + \beta A \qquad (18.40)$$

Thus, the coefficients in Table 18.2 are not identical to those applied in the standard formula (14.2). The linearization also accounts for surface area ranges. Due to this, logical conditions for modeling area surface limitations were eliminated from the optimization model. However, conditions for operational temperatures have to be used.

In order to omit numerical difficulties with the standard formula (18.24) for correction factor Ft of shell-and-tube heat exchangers, Sorsak and Kravanja suggested the following approximation:

$$Ft = 1 - \frac{1}{\sum_{i=0}^{2} a_i [\lg(R)]^i} \tan \left(\frac{\pi P}{2 \sum_{j=0}^{6} b_j [\lg(R)]^j} \right) \qquad (18.41)$$

Parameters P and R are defined by Equations (16.43) and (16.44). Coefficients a, b from the original paper are listed in Table 18.3.

To present the model of Sorsak and Kravanja (2002) we have to define some new indices, parameters and variables with reference to the Synthesis-Simult-Y-G model. (Notice that the symbols, which will not appear in the following, have the same meaning as in the Yee and Grossman approach.)

In order to account for the heat exchanger type selection the Y-G superstructure has to be extended and an additional index set is necessary. Each match in the Y-G superstructure shown in Figure 18.10 is modeled as a heat exchanger type superstructure that comprises all types of interest. To account for the possibility that no available exchanger type satisfies all limitations of area, pressure and temperature ranges, the superstructure has to include bypass, too. Figure 18.11 illustrates this match type superstructure.

The index set of heat exchanger types is as follows:

HET = $\{l \mid l$ is heat exchanger type or bypass$\}$

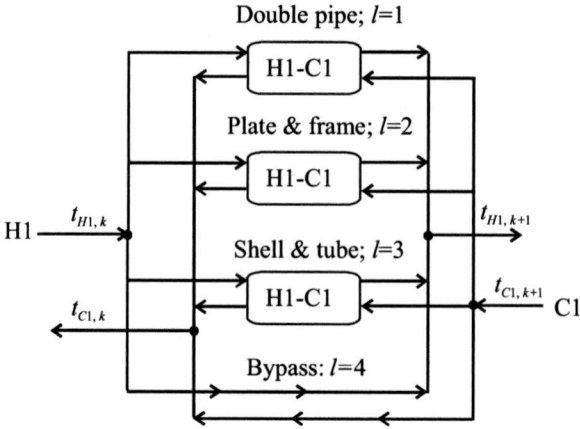

Figure 18.11 Illustration of match type superstructure according to Sorsak and Kravanja (2002).

Index 1 denotes double pipe heat exchanger, 2 denotes plate-and-frame, 3 denotes shell-and-tube and 4 denotes bypass.

Heaters and coolers are modeled as 1-1 units. The overall model keeps the assumption that single heating utility (s) and single cooling utility (w) are used. To shorten equations of the model we will use the index set for stages:

$$ST = \{k \mid k \text{ is stage of Y-G superstructure}\}$$

In addition to parameters of the Synthesis-Simult-Y-G model we need also the following:

DFt	sufficiently large number
Dlo	negative number (should be large as to absolute value)
$T_l^{\max}$	upper limit on permissible temperature for heat exchanger of type l $(l \neq 4)$
$T_l^{\min}$	lower limit on permissible temperature for heat exchanger of type l $(l \neq 4)$
α_s, β_s	parameters for investment cost of heater
α_w, β_w	parameters for investment cost of cooler
α_l, β_l	parameters for investment cost of process heat exchanger of type $l = 1, 2, 3$

Also, there are additional variables in the model. At first, additional index l has to be added to the variables associated with matches: $LMTD_{ijkl}$, q_{ijkl}, y_{ijkl}.

New variables are as follows:

ql_{ijkl}	heat load of heat exchanger
$ti_{ijkl}^A / ti_{ijkl}^B$	heat exchanger inlet/outlet temperature of stream i

t^A_{ijkl}/t^B_{ijkl} heat exchanger inlet/outlet temperature of stream j

$\Delta t^A_{ijkl}/\Delta t^A_{ijkl}$ temperature difference in heat exchanger at side A/B (hot/cold)

The model of Sorsak and Kravanja (2002) is presented in the following in a way similar to that of the Yee and Grossman model.

Synthesis-Simult-S-K Model

$$
\min \left\{ \left[\sum_{j \in C} q_{sj} phu_s + \sum_{i \in H} q_{iw} pcu_w \right] \right\}
$$

$$
+ \left[\sum_{i \in H} \sum_{j \in C} \sum_{k \in ST} \sum_{l \in HET, l \neq 4} y_{ijkl} \alpha_l + \sum_{j \in C} y^h_{sj} \alpha_{sj} + \sum_{i \in H} y^c_{iw} \alpha_{iw} \right]
$$

$$
+ \left[\sum_{i \in H} \sum_{j \in C} \sum_{k \in ST} \sum_{l \in HET, l \neq 4} \beta_l \left(\frac{q_{ijkl}}{U_{ij} Ft_{ijkl} LMTD_{ijkl}} \right) \right.
$$

$$
\left. + \sum_{j \in C} \beta_s \left(\frac{q_{sj}}{U_{sj} LMTD_{sj}} \right) + \sum_{i \in H} \beta_w \left(\frac{q_{iw}}{U_{iw} LMTD_{iw}} \right) \right]
$$

(18.42)

The goal function is similar to that from the Yee–Grossmann approach. Also, parameters LMTD should be replaced as follows:

$$
LMTD_{ijkl} = [(\Delta t^A_{ijkl} \Delta t^B_{ij,k+1,l}) 0.5 (\Delta t^A_{ijkl} + \Delta t^B_{ij,k+1,l})]^{1/3}
$$

$$
LMTD_{iw} = [\Delta t_{iw} (T^{out}_i - T^{in}_w) 0.5 (\Delta t_{iw} + T^{out}_i - T^{in}_w)]^{1/3}
$$

$$
LMTD_{sj} = [\Delta t_{sj} (T^{in}_s - T^{out}_j) 0.5 (\Delta t_{sj} + T^{in}_s - T^{out}_j)]^{1/3}
$$

The constraints of the model are listed in the following.

1. Overall heat balances of streams

$$
CP_i \left(T^{in}_i - T^{out}_i \right) = \sum_{k \in ST} \sum_{j \in C} q_{ijk} + q_{sj}, \quad i \in H
$$

(18.43a)

$$
CP_j \left(T^{out}_j - T^{in}_j \right) = \sum_{k \in ST} \sum_{i \in H} q_{ijk} + q_{iw}, \quad j \in C
$$

(18.43b)

2. Heat balances for stages

$$(t_{ik} - t_{i,k-1})CP_i = \sum_{j \in C} q_{ijk}, \quad i \in H; \ k \in ST \tag{18.44a}$$

$$(t_{jk} - t_{j,k+1})CP_j = \sum_{i \in H} q_{ijk}, \quad j \in C; \ k \in ST \tag{18.44b}$$

3. Assignments of inlet temperatures to the superstructure

$$t_{i1} = T_i^{in}, \quad i \in H \tag{18.45a}$$

$$t_{jK} = T_j^{in}, \quad j \in C \tag{18.45b}$$

4. Constraints ensuring monotonic decrease of temperatures and conditions on outlet temperatures

$$t_{ik} \geq t_{i,k+1}, \quad i \in H; \ k \in K \tag{18.46a}$$

$$t_{jk} \geq t_{j,k+1}, \quad j \in C; \ k \in K \tag{18.46b}$$

$$t_{i,K+1} \geq T_i^{out}, \quad i \in H \tag{18.46c}$$

$$t_{j,1} \leq T_j^{out}, \quad j \in C \tag{18.46d}$$

5. Definitions of heat loads

$$q_{sj} = CP_j(T_j^{out} - t_{j1}), \quad j \in C \tag{18.47a}$$

$$q_{iw} = CP_i(t_{i,K+1} - T_i^{out}), \quad i \in H \tag{18.47b}$$

$$q_{ijk} = \sum_{l \in HE, l \neq 4} ql_{ijkl}, \quad i \in H; \ j \in C; \ k \in ST \tag{18.47c}$$

6. Logical conditions on heat loads

$$q_{ijk} - Q^{max} \sum_{l \in HE, l \neq 4} y_{ijkl} \leq 0, \quad i \in H; \ j \in C; \ k \in ST \tag{18.48a}$$

$$ql_{ijkl} - Q^{max}y_{ijkl} \leq 0, \quad i \in H; \ j \in C; \ k \in S; \ l \in HE \tag{18.48b}$$

$$q_{iw} - Q^{max}y_{iw}^c \leq 0, \quad i \in H; \ k \in ST \tag{18.48c}$$

$$q_{sj} - Q^{max}y_{sj}^h \leq 0, \quad j \in C; \ k \in ST \tag{18.48d}$$

7. Calculations of approach temperatures at temperature locations with logical conditions

$$\Delta t_{ijk} \leq \left(t_{ik} - t_{jk}\right) + D\left(1 - y_{ijk}\right), \quad i \in H; j \in C; k = 1, \ldots, K \quad (18.49\text{a})$$

$$\Delta t_{ij,k+1} \leq \left(t_{i,k+1} - t_{j,k+1}\right) + D\left(1 - y_{ijk}\right), \quad i \in H; j \in C; k = 1, \ldots, K$$

$$(18.49\text{b})$$

$$\Delta t_{sj} \leq \left(T_s^{\text{out}} - t_{j,1}\right) + D\left(1 - y_{sj}^{\text{h}}\right), \quad j \in C; k = 1, \ldots, K \quad (18.49\text{c})$$

$$\Delta t_{iw} \leq \left(t_{i,K+1} - T_w^{\text{out}}\right) + D\left(1 - y_{iw}^{\text{c}}\right), \quad i \in H; k = 1, \ldots, K \quad (18.49\text{d})$$

Up to this point the constraints are almost identical to those in the Synthesis-Simult-Y-G model. The following conditions are necessary due to new features of the Sorsak and Kravanja model.

8. Conditions modeling feasible heat exchange for shell-and-tube heat exchangers ($l = 3$).

Approximation (18.41) requires additional conditions to ensure feasible heat exchange and, likely, numerical stability. For better understanding let us define once more parameters P and R using symbols of the model.

$$R = \frac{t_{ik} - t_{i,k+1}}{t_{j,k+1} - t_{jk}}$$

$$P = \frac{t_{j,k+1} - t_{jk}}{t_{ik} - t_{j,k+1}}$$

The constraints are

$$t_{jk} - t_{j,k+1} + D(1 - y_{ijk3}) \geq Dft \quad (18.50\text{a})$$

$$t_{ik} - t_{i,k+1} + D(1 - y_{ijk3}) \geq Dft \quad (18.50\text{b})$$

$$t_{ik} - t_{j,k+1} + D(1 - y_{ijk3}) \geq Dft \quad (18.50\text{c})$$

Notice that the additional condition necessary to fully satisfy feasible heat exchange is included also in point (7).

9. Constraints modeling influence of operating temperatures on heat exchanger type selection.

Hot streams:

$$t_{ik} = \sum_{l \in \text{HET}} ti_{ijkl}^{\text{A}}, \quad i \in H; j \in C; k \in ST \quad (18.51\text{a})$$

$$t_{i,k+1} = \sum_{l \in \text{HET}} ti^{\text{B}}_{ij,k+1,l}, \quad i \in H; j \in C; k \in ST \tag{18.51b}$$

$$ti^{\text{A}}_{ijkl} \leq y_{ijkl} T^{\max}_l, \quad i \in H; j \in C; k \in ST; l \in \text{HET}, l \neq 4 \tag{18.51c}$$

$$ti^{\text{B}}_{ij,k+1,l} \leq y_{ij,k+1,l} T^{\max}_l, \quad i \in H; j \in C; k \in ST; l \in \text{HET}, l \neq 4 \tag{18.51d}$$

$$ti^{\text{A}}_{ijkl} \geq y_{ijkl} T^{\min}_l, \quad i \in H; j \in C; k \in ST; l \in \text{HET}, l \neq 4 \tag{18.51e}$$

$$ti^{\text{B}}_{ij,k+1,l} \geq y_{ij,k+1,l} T^{\min}_l, \quad i \in H; j \in C; k \in ST; l \in \text{HET}, l \neq 4 \tag{18.51f}$$

Cold streams:

$$t_{jk} = \sum_{l \in \text{HET}} tj^{\text{A}}_{ijkl}, \quad i \in H; j \in C; k \in ST \tag{18.52a}$$

$$t_{j,k+1} = \sum_{l \in \text{HET}} tj^{\text{B}}_{ij,k+1,l}, \quad i \in H; j \in C; k \in ST \tag{18.52b}$$

$$tj^{\text{A}}_{ijkl} \leq y_{ijkl} T^{\max}_l, \quad i \in H; j \in C; k \in ST; l \in \text{HET}, l \neq 4 \tag{18.52c}$$

$$tj^{\text{B}}_{ij,k+1,l} \leq y_{ij,k+1,l} T^{\max}_l, \quad i \in H; j \in C; k \in ST; l \in \text{HET}, l \neq 4 \tag{18.52d}$$

$$tj^{\text{A}}_{ijkl} \geq y_{ijkl} T^{\min}_l, \quad i \in H; j \in C; k \in ST; l \in \text{HET}, l \neq 4 \tag{18.52e}$$

$$tj^{\text{B}}_{ij,k+1,l} \geq y_{ij,k+1,l} T^{\min}_l, \quad i \in H; j \in C; k \in ST; l \in \text{HET}, l \neq 4 \tag{18.52f}$$

10. Definitions and logical conditions for temperature differences.

$$\Delta t_{ijk} = \sum_{l \in \text{HET}} \Delta t^{\text{A}}_{ijkl}, \quad i \in H; j \in C; k \in ST \tag{18.53a}$$

$$\Delta t_{ij,k+1} = \sum_{l \in \text{HET}} \Delta t^{\text{B}}_{ijkl}, \quad i \in H; j \in C; k \in ST \tag{18.53b}$$

$$\sum_{l \in \text{HET}} \Delta t^{\text{A}}_{ijkl} \leq y_{ijkl} D, \quad i \in H; j \in C; k \in ST; l \in \text{HET} \tag{18.53c}$$

$$\sum_{l \in \text{HET}} \Delta t^{\text{B}}_{ijkl} \leq y_{ijkl} D, \quad i \in H; j \in C; k \in ST; l \in \text{HET} \tag{18.53d}$$

11. Logical conditions for bypasses.
The conditions allow for negative temperature differences for bypasses.

$$\Delta t_{ijk4} \geq y_{ijk4} Dlo, \quad i \in H; j \in C; k \in ST \tag{18.54a}$$

$$\Delta t_{ij,k+1,4} \geq y_{ijk4}Dlo, \quad i \in H; j \in C; k \in ST \tag{18.54b}$$

12. Logical conditions to ensure that a single heat exchanger type or bypass can be selected.

$$\sum_{l \in HET} y_{ijkl} = 1, \quad i \in H; j \in C; k \in ST \tag{18.55}$$

13. Conditions on variables.

$$q_{ijk}, q_{iw}, q_{sj}, t_{ik}, t_{jk}, ti_{ijkl}^A, ti_{ijkl}^B, tj_{ijkl}^A, tj_{ijkl}^B \geq 0 \tag{18.56a}$$

$$y_{ijkl}, y_{iw}^c, y_{sj}^h = 0, 1 \tag{18.56b}$$

To cope with complex MINLP, Sorsak and Kravanja applied an advanced optimization method—significantly modified outer approximation with equality relaxation (OA/ER) coded in software MIPSYN. A multilevel strategy has been developed where a space of integer variables increases from level to level. Additionally, several "tricks" have been applied, such as for example, integer-infeasible path optimization, to enhance optimization. The largest example given in Sorsak and Kravanja (2002) contained 13 hot streams and 7 cold streams. They applied 21 stages in the Y-G superstructure. The number of binaries amounts to 5753. To cope with such a huge number of binaries they had to divide the problem at pinch. Nevertheless, program execution was stopped at the second level since too long a CPU time (more than 47 hours) was required for the second level (with larger binary space).

The question arises as to whether large MINLP synthesis models are solvable for even medium scale industrial problems. Notice that the model by Sorsak and Kravanja retains many assumptions of the Y-G superstructure and the optimization model, for instance that of isothermal mixing. The elimination of the latter can significantly improve the results, as Björk and Westerlund (2002) have shown for a small example solved by a slightly modified Synthesis-Simult-Y-G model. However, they also noticed a higher CPU time in comparison with the case of isothermal mixing. It is also interesting to note that Björk and Westerlund claimed that solution of the Synthesis-Simult-Y-G model is easier if the goal function is linear but constraints are not (even with isothermal mixers). Namely, they proposed applying parameters A^* in the last term of the goal function (18.32). Hence, it has the form:

$$\sum_{i \in H}\sum_{j \in C}\sum_{k=1}^{K} \beta_{ijk}A_{ijk}^* + \sum_{j \in C}\beta_{sj}A_{sj}^* + \sum_{i \in H}\beta_{iw}A_{iw}^* \tag{18.57}$$

As a result, the area equations have to be added to the constraints, such as Equation (18.58) – here shown for matches of process streams only:

$$\frac{q_{ijk}}{(A^*)^{1/\gamma_{ij}}} - U_{ij}\mathrm{LMTD}_{ijk} \le 0, \quad i \in H; j \in C; k = 1, \ldots, K \qquad (18.58)$$

The goal function becomes linear at the cost of introducing nonlinear constraints into the MINLP model. Such a "trick" improves the solution if convexification of sigmonial terms is applied as in Björk and Westerlund (2002). The conclusion is that seemingly small changes in problem formulation can enhance the solution. However, they have to be tailored for the optimization subroutine.

There are still many limitations of the simultaneous approaches from the point of industrial needs. They suffer, for instance, from lack of: stream segmentation, application of standard heat exchangers, calculation of stream heat transfer coefficients within the optimization model and accounting for certain structural limitations. The latest issue was addressed in a series of papers by Galli and Cerda (1998a, 1998b, 1998c, 1998d, 2000).

The important features they included are for instance:

(a) inserting restricted matches,
(b) disallowing certain sequence of matches,
(c) disallowing certain locations of splitters over some streams,
(d) forcing locations of specified matches at inlets or outlets of specified streams,
(e) limitations on splitter number and number of branches over particular streams and/or within a network,
(f) restraining the set of matches with which a potential heat match can be arranged in parallel over a given process stream.

To model such conditions, Galli and Cerda suggested several logical conditions. The point is that they require additional binaries and, of course, additional constraints. The number of binary variables drastically increases.

Despite the generality and rigorousness of the superstructure concept for HEN synthesis the approach is unable, at present, to cope with large-scale problems or even smaller ones if all industrial needs are to be taken into consideration. The main reason is in existing optimization techniques for MINLP (the problem of superstructure completeness has not been discussed in HEN synthesis).

There are at least two ways to cope with these difficulties:

(a) development of robust efficient optimization techniques that can guarantee global minimum (though usually under certain conditions),
(b) application of other stochastic meta-heuristic strategies, problem decomposition and problem linearization.

Significant progress has been seen with respect to MINLP deterministic solvers. One example is the solution technique of the Synthesis-Simult-S-K model proposed by Sorsak and Kravanja (2002). The literature on this problem is ample, for instance: Floudas et al. (1986), Floudas and Ciric (1989), Floudas et al. (1989), Quesada and Grossmann (1993), Daichendt and Grossmann

(1994a, 1994b), Grossmann and Daichendt (1996), Zamorra and Grossmann (1998a, 1998b) and Lee and Grossmann (2001, 2003). However, convergence to the global optimum can be guaranteed only for a simple model of heat exchangers, together with other typical limitations. The problem of scale is still unresolved.

Stochastic/meta-heuristic approaches have been proposed in order to eliminate the necessity of solving the complex MINLP model in equation-oriented form (or in the form of logical notation in disjunctive programming). It is worth noting that even pure random generation results in "good" solutions. Chakraborty and Ghosh (1999) first showed this for HENs with no splits and then Pariyani et al. (2006) contributed with the extension on HENs with splitters. The random search optimization approach was tried to solve simple HEN problems as early as the 1970s—see Kelahan and Gaddy (1977). Simulated annealing was applied in Dolan et al. (1989, 1990). Genetic algorithms were employed by Lewin et al. (1998), Lewin (1998) and also Ravagnani et al. (2005). The more recent tabu search meta-heuristic was tried with success by Lin and Miller (2004). Hybrid two-level approaches have also found application to HEN design. Stochastic optimization is used at the first level for structure optimization while deterministic optimization is used at the second to solve the NLP problem of parameter optimization. Such a strategy, with SA at the first level, was proposed in Athier et al. (1997) for HEN synthesis and then by Athier et al. (1998) for retrofit. We will describe in more detail a similar two-level method in Chapter 19. Promising effects can be observed since some solutions reported in the works on stochastic programming are better than those calculated by solution of the MINLP model with deterministic solvers. However, the methods developed to date are able to cope, in reasonable CPU time, with at most medium-size problems with typical simplifications. It is important, however, that the SA-based approach by Dolan et al. (1989, 1990) was able to deal with standard heat exchangers where the deterministic method would, most likely, have failed. The retrofit approach for such an apparatus will be presented in Chapter 19.

HEN problem decomposition is the alternative, particularly for large-scale problems. Zhu et al. (1995a, 1995b) and Zhu (1997) proposed decomposition into blocks with some heuristic rules. This was not rigorous decomposition. Björk and Petersson (2003) developed a sequential two-stage method. The main idea was to divide the entire initial set of streams into independent subsets to reduce the problem scale for synthesizing subsets in the second stage. The grouping has been performed with genetic algorithms. Division into independent subsets has also been applied in a sequential approach by Petersson (2005). The steps of the procedure are:

- solution of LP with goal function which does not include fixed cost of heat exchangers
- solution of MILP model to reduce number of matches by inserting fixed charges into goal function
- solution of MILP to identify independent subsets of matches
- solution of MILP models for each subset.

Notice that investment costs have been linearized and that multishell apparatus are not considered. For medium-size problem with nine (9) process streams Petersson (2005) reached the best solution reported to date in the literature. This is the evidence that some relaxation of the sequential procedure is not crucial. CPU time depends largely on parameter setting (mainly number of intervals) and ranges, for the problem, from 1000 to about 2700 seconds for good solutions. Much larger problems with 39 process streams require CPU time up to 60 826 seconds though solutions of similar "quality" have been obtained in a very short time, of the order of minutes. However, the best network reported consumes more utilities than the minimum according to the MER target.

Finally, Barbaro and Bagajewicz (2002, 2005) developed an MILP model for both synthesis and retrofit (though retrofit has not been addressed in full). They applied a linearized goal function similar to the approaches of Sorsak and Kravanja, and Petersson. The MILP model constraints are based on transshipment/transportation formulation for temperature intervals. The model is very complex and requires numerous binary variables. Nevertheless, it is MILP and, hence, much easier to solve than MINLP. The CPU time required to solve medium size problems presented in this work is very short in comparison with other approaches. The results are at least close to optimal solutions in the literature. This simultaneous approach can deal with multishell heat exchangers. Hence, the concept of model linearization seems a very promising way forward for further work.

Hybrid approaches may help in the quest for solving large-scale problems. There are several examples of embedding "intelligent" heuristics which largely reduce solution space such as Zhu et al. (1995a, 1995b), Zhu (1997), Trivedi et al. (1989b), Jeżowski (1990b, 1992b) and Gundersen et al. (2000). Lastly, tools from artificial intelligence may also enhance the solution.

19 Heat exchanger network retrofit

19.1. INTRODUCTION

In contrast to HEN synthesis a standard formulation of the retrofit problem has not been precisely defined in the literature, or more accurately there is no standard problem formulation. There are difficulties with applying general and possibly accurate formulae for calculating the fixed cost of structural changes such as apparatus relocation since they are often case-dependent. Also, there are several reasons for HEN retrofit such as an improvement of heat recovery, adjusting an existing HEN for increased throughput of a total process system (de-bottlenecking) or both. HEN retrofit for de-bottlenecking requires consideration of the influence of a drop in pressure. We will not address this case in the following. Here we formulate the HEN retrofit problem as follows:

> *Given a HEN in existing plant. Given the required fixed outlet and inlet parameters of process streams and a set of available utilities. Determine HEN topology and parameters of apparatus (heat exchangers, splitters and mixers) such that optimize the performance index applied. Specified technological and structural constraints have to be met.*

Any reliable approach should allow for all possible structural changes that can be applied in industry. "Physically" HEN retrofit is performed by structural and parameter changes. The former concern the location of heat exchangers, splitters and mixers. Notice that the term "change of location" also means the addition and/or removal of apparatus. It is often assumed that an existing heat exchanger should not be removed since it is a "waste" of equipment. However, there may exist the possibility of employing it for another function. Notice also that splitters and mixers in HEN are assigned to each other and elimination of a splitter causes elimination of its associated mixer. This is only one example of strong assignments of structural changes in a HEN retrofit. Such correlations among structural changes cause serious difficulties when formulating and solving a HEN retrofit optimization model since logical conditions and binaries are required. The type of heat exchanger, its configuration and certain heat transfer enhancements can also be considered as structural (discrete) changes—see for example, Smith (2005). Parameter changes in HEN revamping concern key parameters of heat exchangers and splitters. For a heat exchanger we can change its surface area, that is, increase or decrease it. It is commonly assumed that no cost penalty is associated with area decrease. An increase of surface area is charged according to Equation (19.1), while the cost of adding a new heat exchanger is calculated from formula (14.2).

$$CAP = \beta \max\{0, (A^{\text{new}} - A^{\text{existing}})^{\gamma}\} \qquad (19.1)$$

Energy Optimization in Process Systems
©2013 Elsevier Ltd. All rights reserved.

where A^{new} denotes required surface area and $A^{existing}$ that of the existing heat exchanger.

Parameter change for a splitter relies on a change of split ratio or equivalently a change of flow rate via branches. Notice that if the flow rate through a branch becomes zero, the branch is eliminated. This is not charged similarly to elimination or addition of a new splitter or mixer. Typical cost functions in literature approaches to HEN retrofit result in most of them focusing mainly on parameter and structural changes of heat exchangers. Structural modifications of mixers and splitters are, in a few methods, accounted for by appropriate constraints or "structural complexity" indices. Hence, a typical economical model for HEN retrofit cost is rather crude. In industry all changes cost money, but it is likely that such expenses are case-dependent.

Structural changes to a heat exchanger differ as to cost. Here we list them all in ascending order of cost:

1. Heat exchanger shifting, that is, relocation without changing streams (also called re-sequencing, for example, Smith, 2005).
2. Change of one stream in heat exchanger.
3. Change of both streams in heat exchanger.
4. Addition of new heat exchanger.

Changes 2 and 3 are also referred to as repiping, for example, Smith (2005), and often considered to be identical in cost. Notice that in practice economic functions are nonlinear, and can also be discontinuous and nondifferentiable. For instance, correlation of the fixed cost of a heat exchanger may change at certain values of surface area. Among others, Gundersen (1989) pointed out the highly nonlinear profile of cost function for HEN retrofit.

Similar to HEN synthesis approaches, retrofitting methods can be divided into two wide classes: insight based and optimization based. Within the first class we can further specify two groups: methods that stem directly from pinch technology (PT) concepts and those that use HEN evolution techniques (though they are also closely related to PT). In the case of PT, first attempts at HEN retrofit were addressed in Tjoe and Linnhoff (1986, 1988). They gave insights and, often, some targets but did not account sufficiently for investment cost. Some improvements concerned with the cost of structural changes and heat exchanger area increase were proposed in, for instance: Carlson et al. (1993), Carlson and Berntsson (1995), Nordman and Berntsson (2001) and Bengtsson et al. (2002). However, they did not apply systematic approaches and, thus a trial-and-error procedure driven by the designer is needed to locate a good solution. Despite these drawbacks, PT-based approaches are user-friendly for designers and have found industrial applications, for instance: Rossiter et al. (1999), Markowski (2000), Urbaniec et al. (2000), Lababidi et al. (2000) and Matijasevic and Otmaievic (2002).

The network pinch method developed by Asante and Zhu (1996a, 1996b, 1997) and Zhu and Asante (1999) presents significant progress. It will be explained in Section 19.2.

Evolutionary approaches to HEN retrofit, that also made extensive use of PT concepts from the evolution stage of grassroot design (see Chapter 18), were presented in van Reisen et al. (1995, 1998), Nielsen et al. (1997), Amidpour and Polley (1997), Lakshmanan and Banares-Alcantara (1998), Abbas et al. (1999), Varbanov and Klemes (1999) and Varbanov et al. (2000). Heat energy reduction is the main aim, mainly by heat load shifting along paths between heater(s)–cooler(s) in existing HEN. Loop breaking, heat load shifting via loops and HEN decomposition have also been proposed. Similar to PT approaches, investment costs cannot be accounted for rigorously. However, the techniques are able to locate quite good local optima (though for small-scale tasks) and are user-friendly since they apply familiar logic and representation.

Optimization methods can also be divided into two groups: sequential and simultaneous.

The method of Yee and Grossmann was presented in a series of works (Yee and Grossmann, 1987, 1988, 1991). They started investigations with the two-stage approach. First, at the targeting stage, the MILP model is solved to find structural changes that should not be included in the superstructure. Then, the MINLP model of reduced superstructure is solved. A similar approach was developed by Ciric and Floudas (1989) and later, reformulated to a simultaneous one in Ciric and Floudas (1990c). Also, Yee and Grossmann (1991) finally developed a one-stage approach applying a complex but general superstructure specific for HEN retrofit while Ciric and Floudas (1990a, 1990b, 1990c) made use of the superstructure they had developed for HEN synthesis—see Section 18.3. Costs of structural changes as well as the cost of additional area were addressed rigorously, particularly in the method of Ciric and Floudas. Ma et al. (1998, 1999, 2000) also developed a simultaneous method. They adopted the superstructure developed by Yee and Grossman for HEN synthesis. The main focus in the work of Ma et al. was on linearization of the MINLP superstructure model. They achieved this by fixing the LMTD parameters of matches. An iterative procedure had to be employed to relax the assumption.

The approach by Briones and Kokossis (1999a, 1999b, 1999c) is sequential. First, targeting is performed in two stages by solving MILP problems. Notice that, in contrast to the early methods of Yee and Grossmann and Ciric and Floudas, the aim of targeting stages is not to reduce the superstructure by finding invalid structural changes but to calculate good structural changes. Therefore, the result of targeting is a set of "good" HENs of fixed structure. The networks are then subjected to NLP parameter optimization in the last step of the method. The approach by Briones and Kokossis can be applied for multishell heat exchangers and accounts for dependence of stream CP values on temperature by stream segmentation. Athier et al. (1998) developed a two-level simultaneous approach. They applied simulated annealing (SA) optimization at the first primary level to optimize the HEN structure. A population of structurally modified HENs is generated. Then, at the second (slave) level, each member of the HEN population is subjected to parameter optimization by solving the NLP problem for HEN with fixed topology. A similar concept was applied by Bochenek (2003) in the genetic algorithm-based method, which will be presented in Section 19.3. Athier et al. (1998) assumed that a HEN is divided into two parts: an external one with utility exchangers and an internal one with process heat exchangers—see also Athier et al. (1997). This means that utility heat exchangers are located at the ends of streams similarly to the superstructure of Yee and Grossmann addressed in Section 18.4. To code superstructure and structures Athier et al. (1998) employed concepts from Dolan et al. (1990, 1989) developed for HEN synthesis with the SA method. In a conference paper, Barbaro and Bagajewicz (2002) presented a unified framework for both HEN synthesis and retrofit. A detailed description of the synthesis method is given in Barbaro and Bagajewicz (2005). However, in both works there is insufficient information on the changes in the HEN synthesis optimization model that are required for the retrofit case.

Sorsak and Kravanja (2004) also developed a simultaneous HEN retrofit method by solving a MINLP model of superstructure. We will explain it in the following. They employed the synthesis superstructure of Yee and Grossmann. In order to reduce computation time for large-scale problems, Björk and Nordman (2005) proposed decomposing HEN into smaller independent subnetworks. The decomposition was performed by genetic algorithms (GA). The MINLP superstructure model of Yee and Grossmann has to be solved for each subnetwork. The model applied differs slightly from the original one since Björk and Nordman (2005) followed a modified version of Björk and Westerlund (2002) for HEN synthesis. GA-based decomposition required a large computation burden.

In this chapter we will describe three approaches in detail. First, the network pinch method by Asante and Zhu with some modifications, and then two simultaneous methods: that of Sorsak and Kravanja (2004) applying deterministic optimization, and the GA-based procedure of Bochenek (2003).

19.2. NETWORK PINCH METHOD

The method has been addressed in a series of journal and conference papers: Asante and Zhu (1996a, 1996b, 1997) and Zhu and Asante (1999). The contributions are based on the Ph.D. thesis of Asante and thus we will refer to the method as Asante's approach. Additionally, the presentation in the following makes use of remarks and extensions suggested in Wałczyk et al. (2004, 2005) and Jeżowski et al. (2005b).

Similar to process pinch, network pinch is such a temperature in the composite curve plot (in shifted temperature scale) where both curves touch each other. The important difference is that the plot is made for parameters (temperatures and heat loads) of an existing network. A CC plot with process pinch, that is for process streams data, shows the heat recovery that can be achieved by a proper grass-root design while the CC plot in the network pinch method depicts heat recovery in an existing network. The pinches (process and network ones) and heat recovery levels can differ. The heat recovery level in an existing HEN cannot be higher than that calculated with the MER/MUC approaches, providing that an identical minimum temperature approach is applied. To differentiate energy recovery levels for both cases we will refer to that for process streams (i.e. for grass-root design) as the "global MER" or by superscript "glob". Notice that it is possible to reach the global MER also in retrofit but it is usually too expensive due to high investment—see for example, Tjoe and Linnhoff (1986, 1988).

The important achievement of Asante and Zhu is to clearly prove that network pinch is most often caused by structural "design mistakes" since even with heat surface area of matches in existing HEN topology approaching infinity, the global MER could not be reached. Also, Asante's approach allows identification of structural inadequacies and provides tools for eliminating them by structural changes. Network pinches are caused by so-called pinching matches, which are structural obstacles for achieving a higher energy recovery level. Since pinching match (or matches) causes network pinch, pinching match/matches should be eliminated by applying available structural changes. An illustrative figure is often employed to convey this idea, see for example, the papers by Asante and Zhu and also Figure 18.34a in Smith (2005). However, the picture is only clear in specific cases of some hot/cold streams or a single cold/hot stream where a single network pinch exists. Some network pinches can appear with several pinching matches for them. It is also important to note that relocation of a pinching match is not always necessary to increase energy recovery. Relocating other matches and/or performing other available structural modifications can, in effect, produce the elimination of pinching matches, too.

Asante's method consists in performing a single structural change in sequence following a sequential procedure typical in PT (Tjoe and Linnhoff, 1986, 1988) and other insight-based methods. The approach is user-driven but the user is provided with more systematic and reliable techniques than only rules such as those in the PDM for HEN synthesis. Two optimization models are available to find an optimum (or a good) decision. First, the LP model allows determination of pinching matches. Second, the MILP model calculates beneficial structural changes. However, the objective of optimization is minimization of utility consumption/cost. To account for investment cost, Asante proposed following the hierarchy resulting from the cost of structural changes. The preferences, in descending order, are as follows:

1. Heat exchanger relocation (according to sequence of relocations given in the preceding, that is, re-sequencing and change of streams in matches).
2. Stream split.
3. Addition of a new heat exchanger (match).

Stream splitting is controlled by a heuristic. The MILP model is necessary to find structural modification for a selected type of structural change. Seemingly, if change type is fixed and the pinching match is known, a proper structural change such as heat exchanger relocation could be detected by inspection. However, there are usually alternative relocations of certain matches; the "quality" of them cannot be predicted precisely (notice that each relocation effects heat load re-distribution). Hence, optimization is necessary to locate the best alternative from among potential "moves".

Though the logic behind the network pinch method is clear and gives promise of finding at least a "good" solution, the realization is not easy and requires, in general, a large computation effort with extensive knowledge of the user who keeps control over the overall design process. Under these conditions good results can be achieved as shown in Fernadez-Polanco et al. (2000) for industrial-scale tasks.

The network pinch method is performed sequentially in three major steps:

1. Identification of pinching match/matches by solving the LP optimization model.
2. Calculation of structural and resulting heat load changes that increase energy recovery level at possibly small cost. This is a designer-driven procedure. The designer is facilitated by information on network pinches (and also some other parameters as has been shown in Wałczyk et al. (2005)). Structural changes are performed sequentially, following the cost preferences, by solving the MILP optimization model in each step. A set of "good" solutions is selected for further "refinements" in Step 3.
3. The networks obtained in the previous stage are subjected to parameter optimization, by solving NLP problems with heat load (or surface area) of matches and split ratio as variables. The goal is to minimize TAC. To account for some structural changes in regard to splitters, a concept of "small superstructure" from Asante and Zhu (1996a) can be employed.

Further explanation of Steps 1 and 2 is presented in the following.

Step 1: Identification of Pinching Matches

The aim of identification of pinching matches is to find those heat exchangers in a HEN of given structure which, by definition of pinching matches, feature temperature approaches at the hot or cold end equal to given $\Delta T^{\min}$ under the condition that the heat load distribution in matches is optimal, i.e. that it minimizes utility loads or cost of utilities. For a $\Delta T^{\min}$ parameter close to zero, the surface area of pinching heat exchangers can approach infinity. If energy recovery is larger than global MER or utility cost is larger than global MUC, then the only outcome is that changes of surface area are insufficient. Structural changes have to be applied, too.

The LP model is based on the mathematical model of network with fixed topology. The HEN model for loop breaking (Section 18.2) can be employed—see also Jeżowski et al. (2005b) and Wałczyk et al. (2005). Many symbols in the following are taken from this model. The objective is to minimize loads (or cost) of utilities, and also number of pinches:

$$\min \left(\sum_{k \in K^{\mathrm{H}}} Q(k) + \alpha \cdot \sum_{k \in K} [P^{\mathrm{H}}(k) + P^{\mathrm{C}}(k)] \right) \qquad (19.2)$$

K^H and K are index sets for heaters and heat exchangers, respectively. Parameter α is heuristic and should be from range (0.0; 0.1). $P^H(k)$ and $P^C(k)$ are variables defined by:

$$P^H(k)/P^C(k) = \begin{cases} 1 & \text{if temperature difference at hot/cold side of} \\ & \text{heat exchanger } k \text{ equals } \Delta T^{\min} \\ 0 & \text{if not} \end{cases} \qquad (19.3)$$

It is of importance that variables $P^H(k)$, $P^C(k)$ are not binaries but continuous variables bounded by 0.0 and 1.0. The optimization model requires, then, the following conditions:

$$t_h^1(k) - t_c^2(k) \geq \Delta T^{\min} + (1 - P^H(k)), \quad k \in K \qquad (19.4)$$

$$t_h^2(k) - t_c^1(k) \geq \Delta T^{\min} + (1 - P^C(k)), \quad k \in K \qquad (19.5)$$

$$0 \leq P^H(k) \leq 1, \quad k \in K \qquad (19.6)$$

$$0 \leq P^C(k) \leq 1, \quad k \in K \qquad (19.7)$$

Symbols t_h^1, t_c^2 denote temperatures at the hot side while t_c^2, t_h^1 denote temperatures at the cold side of the heat exchanger—see also Figure 16.2.

The final result of the LP model solution provides values of $P^H(k)$, $P^C(k)$ such that: k is pinching match if $P^H(k)$ or $P^C(k)$ is 1.0.

Step 2: MILP Model for Structural Changes

Recall that in the network pinch approach the structural change by stream split is performed by rule of thumb. The rule suggested by Asante is:

Insert splitter at the location where the network pinch covers with the process pinch.

Notice also that the split ratio has to be fixed. The optimization model then becomes linear since temperatures and heat loads are the only variables (see also the loop-breaking model in Section 18.2). Also, the concept of a small superstructure employed for the loop-breaking is well suited to the purpose of structure change calculations. Dummy matches are built into the existing structure at locations selected by the designer. Existing heat exchangers or new ones can be relocated or inserted into locations of dummy matches. Hence, explanation of this MILP formulation is reduced here in order to highlight its specific features.

The original approach of Asante employed cost (loads) of utilities as the goal function. Jeżowski et al. (2005b) suggested an extended criterion:

$$\min\left\{\sum_{k \in K^{H}} Q(k) + \alpha \sum_{k \in K^{N}} y(k) + \beta \sum_{k \in K^{R}} y(k)\right\} \qquad (19.8)$$

where K^{N} and K^{R} are index sets of new matches and relocated heat exchangers, respectively.

Heuristic parameters α and β are given in the data and they should reflect typical relations: cost of utilities–cost of heat exchanger relocation or cost of new matches. Appropriate but simple logical conditions have to be added to control structural changes. They depend on the type of change and on designer decisions on possible relocations. Consider, for instance, that a pinching match or a match that is the source of heat recovery bottleneck (k^{P}) has been identified as a candidate for relocation, and some dummy matches ($k^{d} \in K^{d}$) have been inserted into a small superstructure. Next, assume that the designer has decided that pinching match k^{P} has to be eliminated by removing it from the current location and inserting it into one from among dummy matches.

The conditions, then, are:

$$\sum_{k^{d} \in K^{d}} y(k^{d}) = 1 \qquad (19.9)$$

$$y(k^{P}) = 0 \qquad (19.10)$$

If, however, the designer is not certain that removing match k^{P} is obligatory the condition should be:

$$\sum_{k^{d} \in K^{d}} y(k^{d}) + y(k^{P}) = 1 \qquad (19.11)$$

Thus, the designer can control the retrofit by:

(a) creating a small superstructure, that is, fixing the number of dummy matches and their locations within HEN topology;
(b) formulating logical conditions in the MILP model.

Providing that the MILP optimization model really produces optimal results (though only for given superstructure and model conditions) the question of investment cost is still open. The hierarchy of structural changes and additional terms in goal function (19.8) are not sufficient means to account rigorously for capital cost. By a sequence of changes we can reach a solution that is cheap with regard to utilities but expensive in terms of investment. Asante suggested a tree search type procedure. The tree root corresponds to initial HEN. A branch is created by a sequence of structural changes (according to cost hierarchy). Each node is a network calculated by solution of the MILP model. Also, for each node investment cost is determined. It is important to notice that every structural change made at the higher level (parent node) is preserved in all lower level nodes of the parent (in descendant nodes). The procedure is not fully systematic in the sense that the definition of the branching operator is not precise. Asante suggested that the type of structural modification fully defines an operator. However, for a selected modification type, the designer can use various superstructures and logical conditions. Notice that this "not systematic" character is the consequence of user-driven philosophy and a tendency to reduce the complexity and size of the MILP model. Wałczyk et al. (2005) suggested another deficiency that results from preservation of selected changes in descendant nodes. Due to a sequential procedure and its simplifications, imposing selected modifications can be the wrong decision from the point of final solution quality. Hence, they suggested treating the changes performed as potential modifications that are not fixed. To illustrate the idea let us consider an illustrative example. Assume that, at a certain step of the procedure, heat exchanger k has been moved from location "a" into location "b". In Asante's approach position "a" is treated as forbidden for apparatus k in successive steps but this is not the case in the proposed method. Wałczyk et al. (2005) called the concept "expanding subspace" since superstructure gradually expands "around" the superstructure applied for the first structural modification. Hence, the solution space is not arbitrarily decomposed but the initial search region expands into neighboring promising search regions. The disadvantage is that the optimization model also increases.

In order to illustrate the network pinch method with the suggested search technique we will present one example solved by Wałczyk et al. (2005).

Example 19.1

This is a slightly modified problem from Fernadez-Polanco et al. (2000) of the revamp design for a large-scale HEN from a crude oil distillation plant. Parameters of 17 process streams are gathered in Table 19.1.

Steam of temperature $500\,°C$ and cooling water of temperature range $10\,°C$ (inlet) to $15\,°C$ (outlet) are used as utilities. Minimum temperature approach ΔT^{min} is set at $20.3\,°C$. Wałczyk et al. (2005) did not consider the investment cost of retrofit.

The aim was energy consumption reduction with user-controlled stopping criterion. Also, they did not use stream split to avoid control problems.

The topology of the existing network is shown in Figure 19.1 while heat loads of matches in the figure are from the solution of the LP model (identification of pinching matches).

The network comprises 29 heat exchangers including heaters and coolers. Altogether, it is a large-scale industrial problem. The global minimum loads on utilities calculated by Wałczyk et al. (2005) are: $Qhu^{min,glob} = 47\,394.23\,kW$, $Qcu^{min,glob} = 6644.21\,kW$, process pinch $T_P = 178.6\,°C/158.3\,°C$. The solution of

Table 19.1 Parameters of process streams in Example 19.1

Stream	Temperatures		CP (kW K^{-1})
	T^{in} (°C)	T^{out} (°C)	
H1	162.8	107.2	9.5
H2	158.1	60.0	204.1
H3	261.4	150.0	120.1
H4	170.8	50.0	25.7
H5	217.8	40.0	18.7
H6	262.9	70.0	27.2
H7	315.6	90.0	10.1
H8	240.1	90.0	29.1
H9	291.7	210.0	149.4
H10	291.7	90.0	46.4
H11	353.2	90.0	67.1
C1	20.0	220.0	277.8
C2	211.3	348.0	383.9
C3	158.3	162.8	1501.8
C4	315.3	370.0	271.9
C5	144.0	350.0	7.6
C6	63.2	110.0	21.2

the LP model (identification of pinching matches) for a given HEN topology and process stream data yields: $Qhu^{min,0} = 73\,855.34$ kW, $Qcu^{min,0} = 33\,024.83$ kW, and pinching matches are exchangers 5, 8, 11 and 13.

The difference between the global minimum utility loads and the minimum for given structure is high, allowing room for improvements by structural modifications. Wałczyk et al. (2005) presented in detail the calculations for six iterations of the retrofit procedure with the "expanding subspace" concept.

The conclusion that can be drawn from this description is that careful analysis is necessary to be able to create a small superstructure for each step. The analysis should be based not only on results from the LP model but also on targeting for synthesis. It is debatable whether one should undertake the task of reducing the superstructure instead of applying a larger one. Solution of a MILP problem with a number of binaries of the order one or two hundred should not create problems. The size of the problem has more importance if stream segmentation is necessary. With stream segmentation the LP model for pinching matches becomes a MILP and the MILP model for structural changes increases drastically with regard to the number of binaries and complex logical conditions. Such a model is presented in Zhu and Asante (1999) and an alternative version can be found in Wałczyk et al. (2004).

Here we present the solution to Example 19.1 shown in Figure 19.2 in compact form .

In order to explain the brief comments in the figure, let us decode the notation for Step 1, which is "h.exch 2: locations (a), (b), n.c.". The full meaning is: "heat

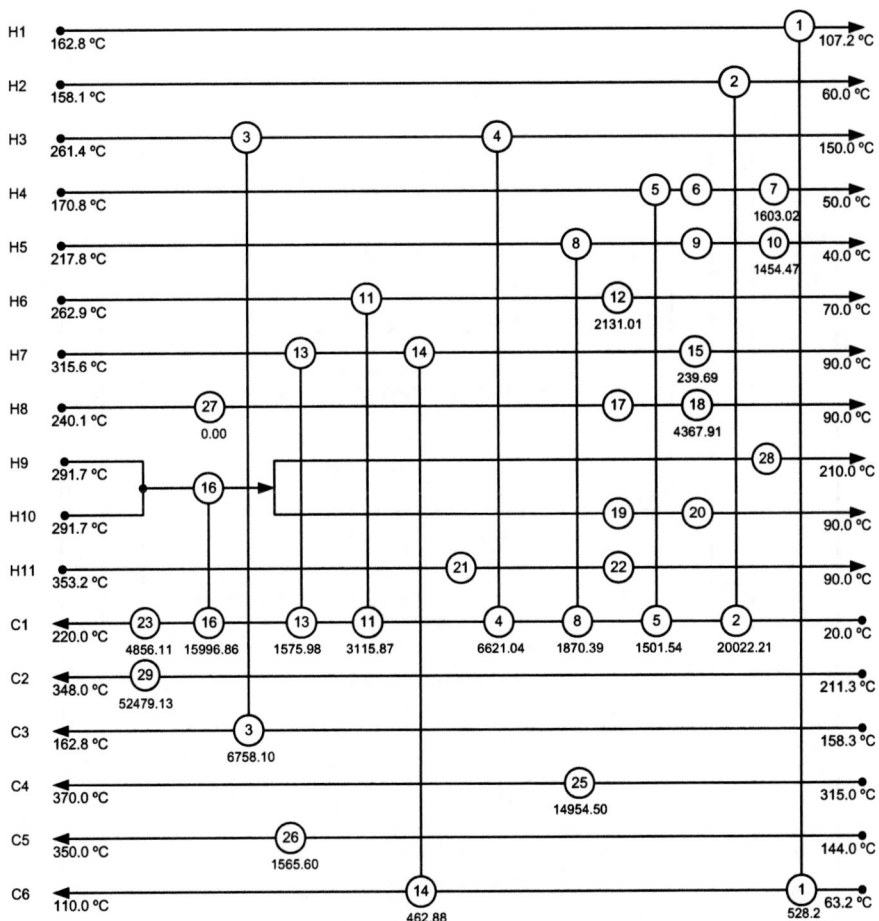

Figure 19.1 Topology of existing HEN in Example 19.1—adopted from Fernadez-Polanco et al. (2000); heat loads from the solution of the LP model.

exchanger no. 2 can be relocated into positions marked **a, b** or can be left in its existing position". Notice that the description defines logical conditions in the MILP model.

The figure provides important information on the decisions and the results in 6 iterations of the algorithm. To identify applied locations of heat exchangers and dummy matches, Figure 19.3 with initial superstructure should be consulted. It is important to notice that, for instance, heat exchanger 2 changes locations in some steps, which would be impossible in the approach suggested by Asante.

The final solution is shown in Figure 19.4.

The network still suffers from some structural deficiencies, as for instance heat exchange via process pinch in exchangers 4, 8, 11, 31 (exchanger 31

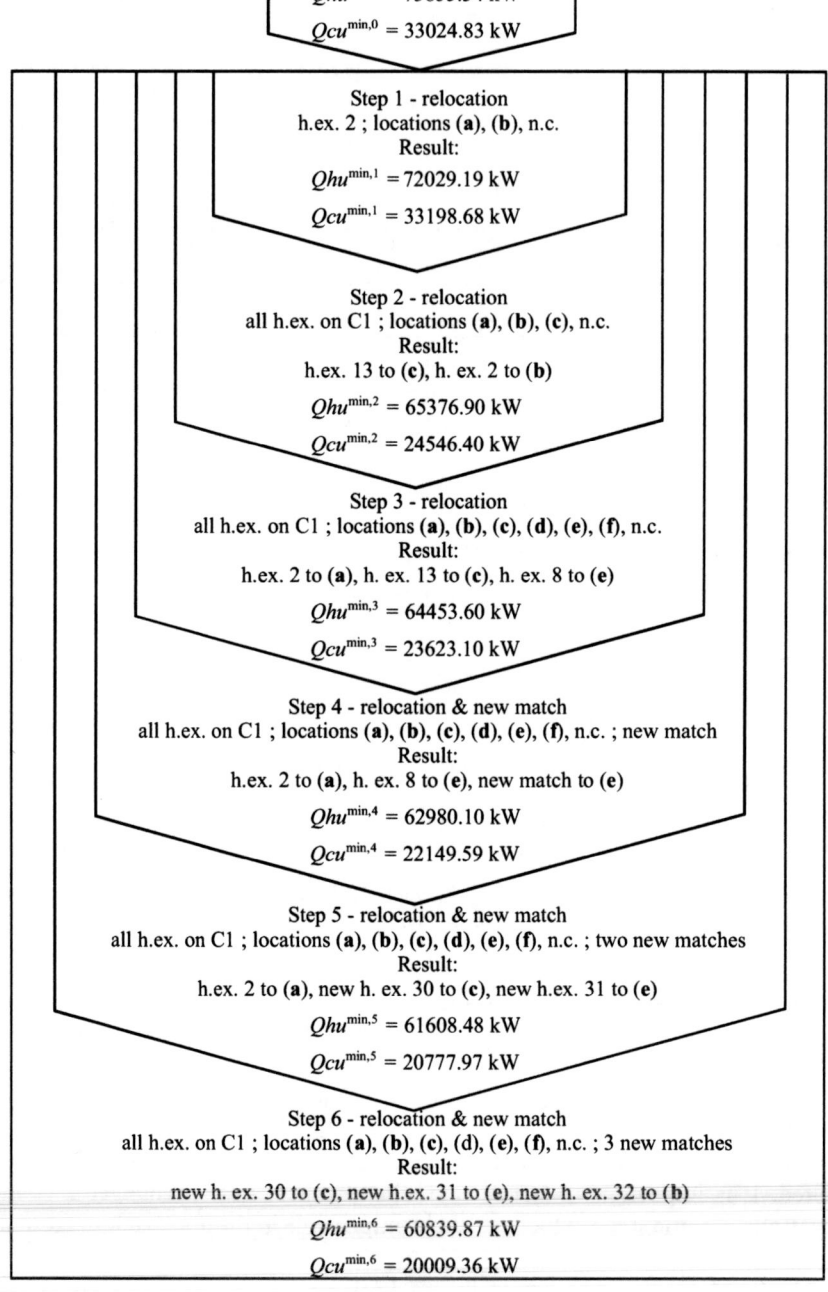

Figure 19.2 Summary of the solution procedure and results for Example 19.1 (from Wałczyk et al., 2005); n.c. denotes "no change".

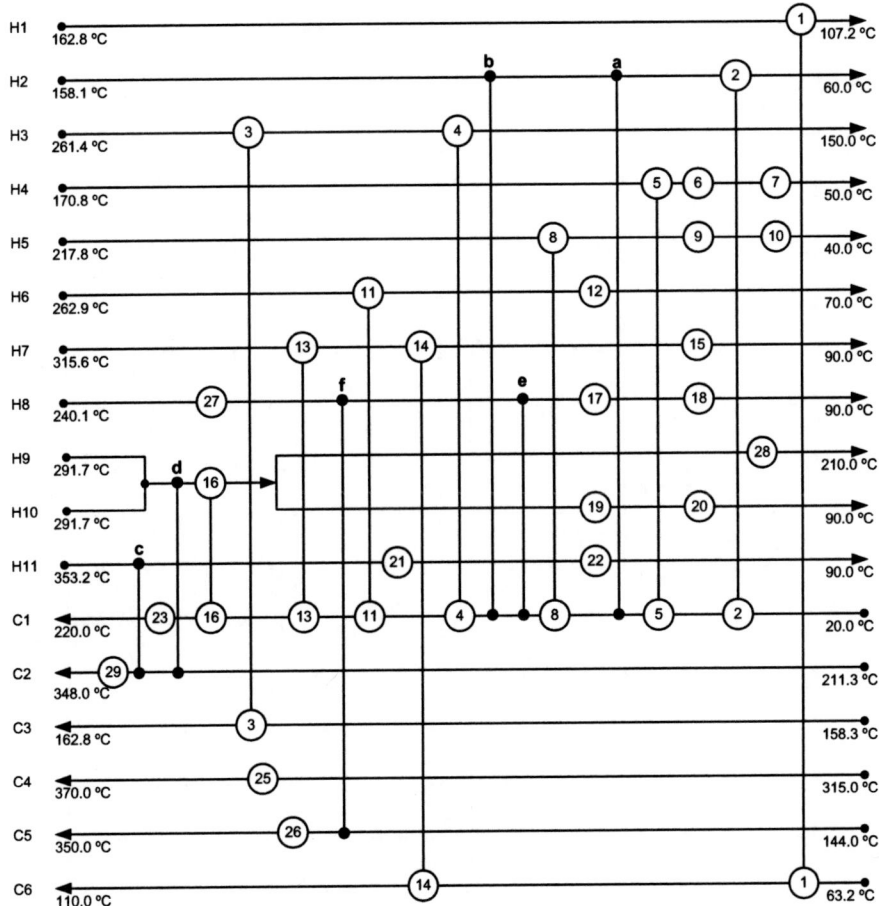

Figure 19.3 Illustration of the superstructure with possible locations of heat exchangers for Example 19.1.

in location **e**). In consequence, heat loads on utilities ($Qhu = 60\,839.87\,\text{kW}$, $Qcu = 20\,009.36\,\text{kW}$) are larger than the global optima. Nevertheless, relatively few structural changes (some relocations plus 3 new matches: (**b**), (**c**), (**e**)) allow a large reduction in the consumption of utilities even in comparison with the minimum for the existing topology ($Qhu^{\text{min},0} = 73\,855.34\,\text{kW}$, $Qcu^{\text{min},0} = 33\,024.83\,\text{kW}$).

To conclude, it is possible to state that the network pinch approach is a useful technique even for large-scale problems. It allows achievement of good networks. However, its application requires expertise and appropriate computer aids for interaction with the designer. The advantage is that no sophisticated optimization solvers are necessary.

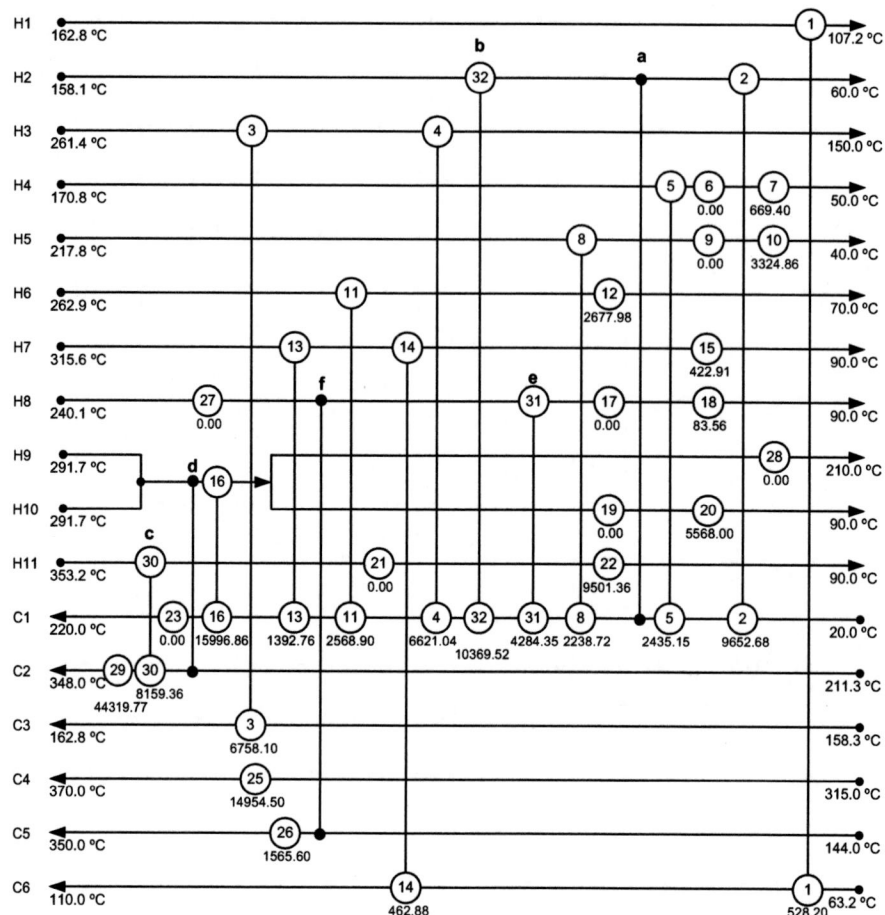

Figure 19.4 Final solution to Example 19.1 from Wałczyk et al. (2005).

19.3. SIMULTANEOUS APPROACHES FOR HEN RETROFIT

In this section we will address two simultaneous methods. First, the method of Sorsak and Kravanja (2004) applying deterministic MINLP optimization and then the proposition of Bochenek (2003) based on the genetic algorithm framework.

The former is very close to that for HEN synthesis—see the Synthesis-Simult-S-K model in Section 18.3. It applies the Yee-Grossmann superstructure and preserves the assumptions for the synthesis case. Also, linearized investment costs are used. Heat exchanger type selection is included.

The changes, with regard to the synthesis model, are explained in the following.

New Parameters (in Comparison with Synthesis-Simult-S-K Model)

A_{ijkl}	surface area of heat exchanger
POS_{ijkl}	parameter defining existence of heat exchanger within initial network
$POS_{ijkjl} = 0$	means that exchanger of type l in stage k for streams i, j does not exist
α^r_{ijkl}	fixed relocation cost for existing heat exchanger
α^d_{ijkl}	fixed de-installation cost of existing heat exchanger

It is also of importance that index set HET has more than four elements. Additionally, indices l higher than four ($l > 4$) denote existing heat exchangers. The goal function is the sum of items 1, 2, 3 and 4:

1. Cost of utilities

$$\left\{ \left[\sum_{j \in C} q_{sj} p h u_s + \sum_{i \in H} q_{iw} p c u_w \right] \right\} \tag{19.12a}$$

2. Annual investment cost of heat exchangers in the network

$$\sum_{i \in H} \sum_{j \in C} \sum_{k \in ST} \sum_{l \in HET_{l \neq 4}} \left[y_{ijkl} \alpha_l + \beta_l \max \left\langle 0, \frac{q_{ijkl}}{U_{ij} Ft_{ijkl} LMTD_{ijkl}} - A_l \right\rangle \right] \tag{19.12b}$$

3. Annual cost of relocating existing heat exchangers

$$\sum_{i \in H} \sum_{j \in C} \sum_{k \in ST} \sum_{l \in HET_{l>4}} \left[y_{ijkl} \alpha^r_{ijkl} + \beta_l \max \left\langle 0, \frac{q_{ijkl}}{U_{ij} Ft_{ijkl} LMTD_{ijkl}} - A_l \right\rangle \right]_{NOS_{ijkl}=0} \tag{19.12c}$$

4. Cost of de-installation of existing heat exchangers

$$\sum_{l \in HET_{l \neq 4}} \left[\alpha^d_{ijkl} \left(1 - \sum_{i \in H} \sum_{j \in C} \sum_{k \in ST} y_{ijkl} \right) \right] \tag{19.12d}$$

In addition to constraints (18.43) – (18.56) of the Synthesis–Simult-S-K model the following have to be added:

• Limit on the number of relocations

$$\sum_{i \in H} \sum_{j \in C} \sum_{k \in ST} \sum_{\substack{l \in HET \\ l>4}} (y_{ijkl})_{POS_{ijkl}=0} \leq REL \tag{19.13}$$

- Limit on the number of new heat exchangers

$$\sum_{i \in H} \sum_{j \in C} \sum_{k \in ST} \sum_{\substack{l \in HET \\ l < 4}} (y_{ijkl}) \leq NEW \qquad (19.14)$$

- Constraint to limit single placement of existing heat exchanger

$$\sum_{i \in H} \sum_{j \in C} \sum_{k \in ST} (y_{ijkl}) \leq 1, \quad l \in HET, \, l > 4 \qquad (19.15)$$

Parameters REL, NEW are given in the data. They can be used to control CPU time.

To solve the model the same software and multilevel procedure have been applied as for the Synthesis-Simult-S-K model. The largest example solved with the full procedure featured 11 process streams and required 11 000 CPUs on the computer with 566 MHz CPU and 256 MB of RAM. It is important to note that Sorsak and Kravanja (2004) stated that solution of the retrofit problem for a HEN consisting of standard heat exchangers is not possible with existing MINLP solvers.

Problems with the solution of complex MINLP models caused increasing interest in the application of meta-heuristic techniques. We have mentioned several such works on HEN synthesis in Chapter 18. Contributions to other systems were mentioned in Chapter 1. Here, we will address an approach for HEN retrofit developed by Bochenek (2003). It was also presented, though only partially, in Jeżowski et al. (2007) and briefly in Bochenek and Jeżowski (2004a, 2004b, 2006). It is of importance that the method is applicable for standard heat exchangers.

The approach consists in a two-level iterative procedure with the use of a genetic algorithm (GA) at both levels. Structural optimization is performed at the first (primary) level while in the second (slave) level parameters of heat exchangers and splitters are optimized. The second level is executed for each HEN structure generated at the primary level to find values of apparatus parameters and to calculate the full HEN performance index. The algorithm starts from the initial parent population, which is generated randomly from the existing HEN. Figure 19.5 illustrates the overall procedure—the single cycle of GA.

Though GAs are often applied as general-purpose optimizers for problems formulated in a standard equation-oriented manner, the best results are achieved if the GA applies solution space, that is, representation and codes, that are "natural" and tailored to the problem. The crucial point of the work by Bochenek has been the development of efficient problem representation and coding. He applied a chromosome that is a direct representation of network topology. The variable encapsulates all topological features of a HEN. It will be referred to as structural matrix SM. Also, feasible search space has been defined for all possible structural changes in the SM matrix. Due to such problem

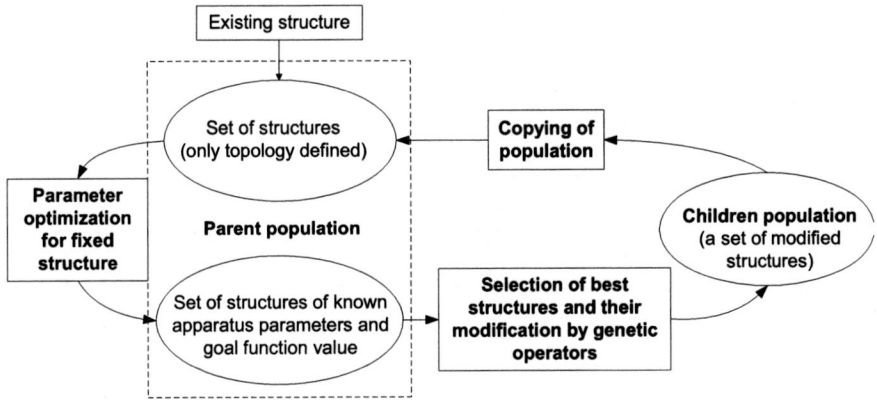

Figure 19.5 General flow sheet of Bochenek's approach—single cycle of GA.

representation it was possible to eliminate the use of a classical superstructure model applied in simultaneous approaches with the deterministic optimization solver addressed earlier.

To describe the method, we will first present the retrofit superstructure concept and coding system for both superstructure and structures. Then, genetic operators applied at the first optimization level will be given. A brief description of the optimization algorithm at the second level will follow. An example of an application will end the presentation.

Bochenek's approach does not apply a general superstructure that embeds all potentially feasible solutions. This can be used but, likely, would result in prohibitive CPU time. The proposed superstructure is based on the topology of an existing HEN. Conceptually, it bears some resemblance to the small superstructure. Also, it is visually very similar to it and, in consequence, to HEN grid representation. However, additional splitters, mixers and branches should be added by the designer in contrast to the small superstructure. The designer controls the locations and the number of these new elements. He/she should employ his/her problem understanding, case-specific conditions and, preferably, additional information on network pinches, process pinches and pinching matches. The next difference is that, in comparison to the small superstructure, the designer does not insert dummy matches but dummy nodes. The notion of nodes is applied to heat exchangers, splitters and mixers. There are hot and cold nodes for heat exchange. Connecting a hot node with a cold one creates a match. This gives numerous alternatives for new matches and, in consequence, for various relocations. This feature will also be shown in the following.

Due to the coding scheme and mechanism of genetic operators applied, it is convenient to explain HEN retrofit superstructure as composed of two building elements: a superstructure of stream flow configurations (stream superstructure) and a superstructure for heat exchangers (match superstructure). First, the stream superstructure is built by inserting nodes of new splitters and mixers. Then, hot

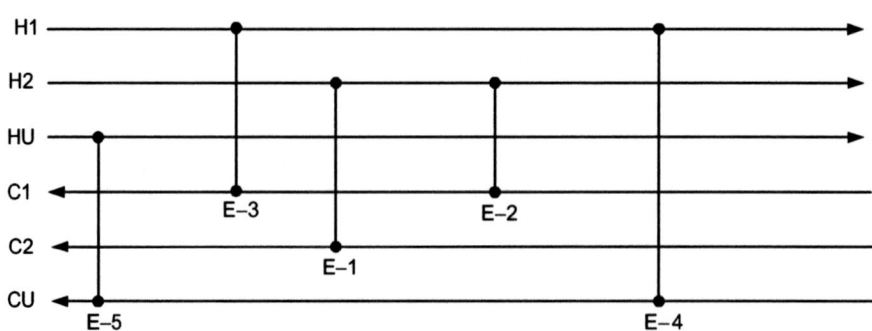

Figure 19.6 Topology of existing HEN for Example 19.2.

Table 19.2 Stream parameters for Example 19.2

Stream	T^{in} (K)	T^{out} (K)	CP (kW K^{-1})	Cost ($ kW^{-1} y^{-1})
H1	443	333	30	
H2	423	303	15	
HU	450	450		80
C1	293	408	20	
C2	353	413	40	
CU	293	313		20

and cold nodes for heat exchangers are added into this superstructure, creating a full retrofit superstructure.

To illustrate the superstructure development stage let us consider the example taken from Jeżowski et al. (2007) —here Example 19.2 (note that full description will be given in the following). The initial HEN topology is shown in Figure 19.6, and Table 19.2 provides stream information. A *HRAT* value of 10 K is assumed.

To develop stream configuration for the superstructure, MER calculations were done for given streams. Process pinch is at 153 K/163 K. Additionally, the solution of the LP model from the network pinch approach reveals that exchangers E1 and E2 are pinching matches and network pinch coincides with process pinch. Following the heuristic of Asante, Bochenek located the splitter at stream H2. An additional splitter has been added at stream C1 since its inlet temperature equals the network pinch. Stream superstructure applied by Bochenek is shown in Figure 19.7. Next, some cold and hot nodes were inserted based on problem analysis. The final superstructure is shown in Figure 19.7.

The important issue in GA (and similar optimization techniques such as simulated annealing) is how to code the superstructure and structures generated from it. The representation is of importance since it largely influences calculation load

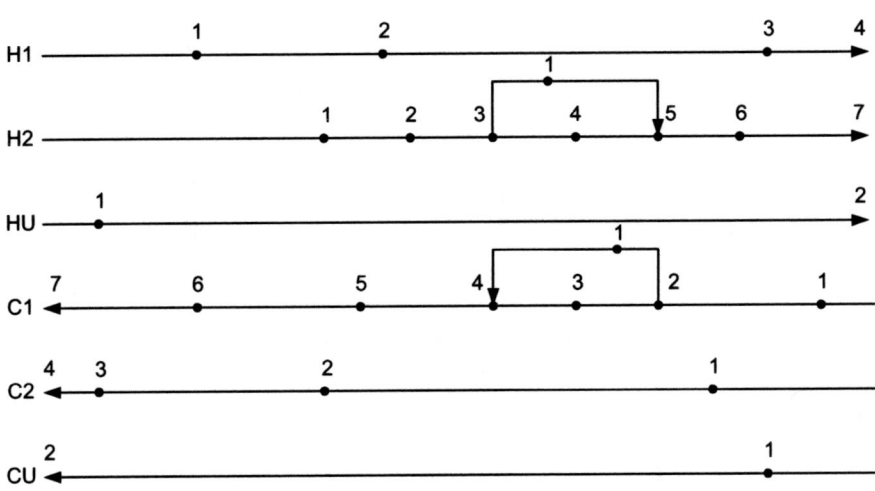

Figure 19.7 Retrofit superstructure by Bochenek for Example 19.2.

(see for instance Dolan et al., 1990, 1989). In the following we will explain the coding scheme proposed by Bochenek.

All branches (main branches equivalent to streams and side branches from splitters) and nodes in the superstructure are numbered, separately for hot and cold streams. The only rule for numbering branches is that main branches are given numbers first and then side branches. Nodes are given numbers separately for each branch, starting from the inlet of a stream to its outlet. Figure 19.8 illustrates the scheme of numbering. The stream superstructure from this figure embeds five different stream flow configurations shown in Figure 19.9. This shows that even for such a simple example there are many possible flow configurations.

Each node for the heat exchanger is given a unique address consisting of two items: *branch no.–node no.* Nodes with addresses: 1.1, 1.4, 1.7, 1.9, 1.11, 2.1, 2.2, 3.1, 3.2, 4.1, 4.2 are heat exchanger nodes in Figure 19.8.

It is worth noting that in contrast to the Y-G HEN synthesis superstructure the number of heat exchangers at side branches is not limited to one. It is also of importance that no assumptions are imposed on the location of utility exchangers in contrast to other retrofit approaches where such exchangers are inserted at

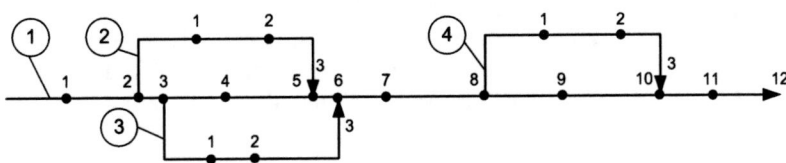

Figure 19.8 Illustration of numbering scheme and stream superstructure in the method by Bochenek: numbers of branches in circles.

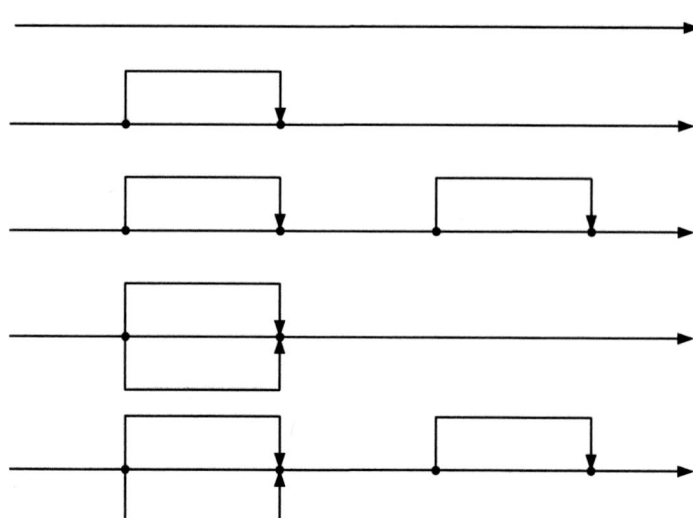

Figure 19.9 Stream flow configurations in the stream superstructure from Figure 19.8.

the "ends" of streams. Finally, the limitation of isothermal mixing does not exist.

The code for defining both a superstructure and a structure is based on matrix representation. The information on the stream flow configuration is stored in node vectors NOD^H, NOD^C and split matrices SPL^H, SPL^C that are created separately for hot and cold streams. Vectors NOD and matrices SPL have the following structure:

- $NOD = [n_1, \ldots, n_i, \ldots, n_{NB}]$; the vector has NB elements, where NB is equal to total number of branches; n_i denotes the number of nodes at the ith branch
- $SPL = [i = 1, \ldots, NS; j = 1, 2, 3]$; the matrix has NS rows (where NS is equal to total number of side branches) and three columns ($j = 1, 2, 3$). The number of the main branch from which a side branch is created is in the first column ($j = 1$), the split node number at the main branch in the second column ($j = 2$) and the mixing node number at the main branch is in the third column ($j = 3$).

Each structure embedded in the retrofit superstructure is created by assigning to each heat exchanger (active in this structure) unique addresses of hot and cold nodes from the superstructure. The addresses of each heat exchanger are coded as matrix SM such that $SM [i = 1, \ldots, NA; j = 1, \ldots, 4]$; the matrix has NA rows (where NA is equal to maximal number of heat exchangers which can exist in a HEN and has to be given in data) and four columns ($j = 1, \ldots, 4$), the first two columns ($j = 1, 2$) involve hot node addresses while columns 3 and 4 involve cold node addresses (two columns are used for one address since it consists of two items).

Figure 19.10 illustrates the general view of matrix SM. Figure 19.11a gives examples of SPL and NOD matrices for the illustrative superstructure and Figure 19.11b shows matrix SM for one structure within the superstructure.

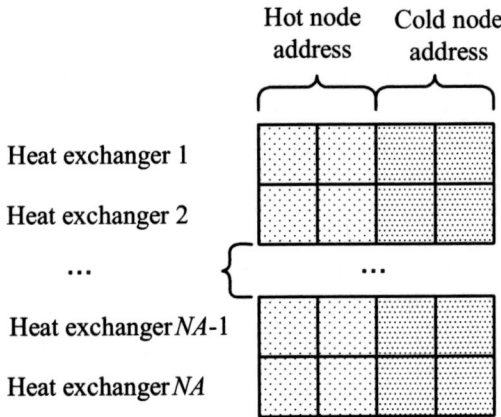

Figure 19.10 General view of matrix *SM*.

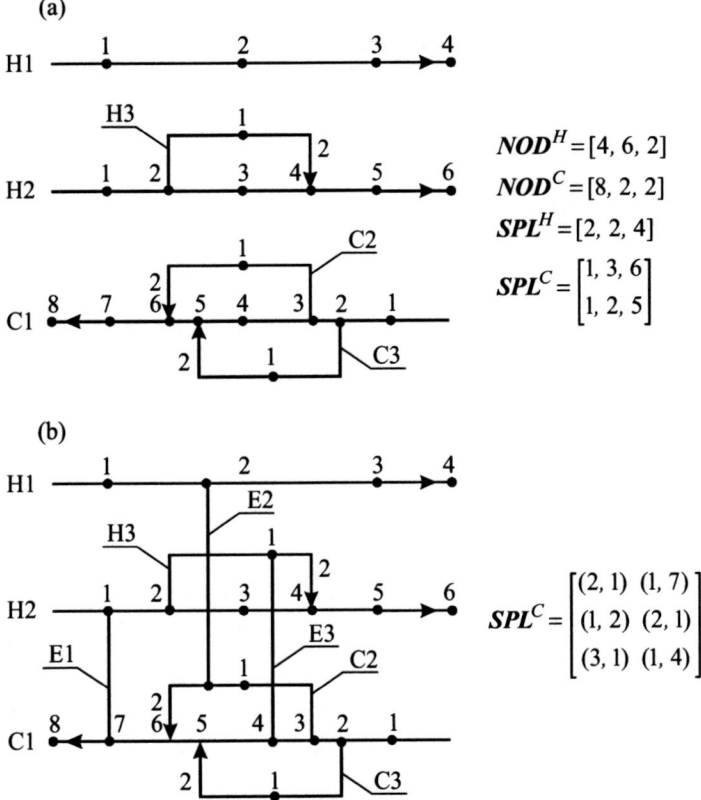

$$NOD^H = [4, 6, 2]$$

$$NOD^C = [8, 2, 2]$$

$$SPL^H = [2, 2, 4]$$

$$SPL^C = \begin{bmatrix} 1, 3, 6 \\ 1, 2, 5 \end{bmatrix}$$

$$SPL^C = \begin{bmatrix} (2, 1) & (1, 7) \\ (1, 2) & (2, 1) \\ (3, 1) & (1, 4) \end{bmatrix}$$

Figure 19.11 Examples of matrices coding the superstructure and structures: (a) matrices *SPL*, *NOD* for example superstructure; (b) matrix *SM* for the structure embedded in the superstructure from (a).

It is important to note that genetic operators act only on matrix *SM*. Thus, only this matrix is subjected to changes while vectors *SPL, NOD* are kept intact. The triple: *SM* and *SPL, NOD* is sufficient to fully define the generated HEN from the superstructure. Bochenek developed an efficient decoding algorithm that also generates all equations of the HEN optimization model at the second level and the cost of structural changes from the given triple.

In order to perform structural retrofit changes a set of genetic operators was developed by Bochenek. The operators modify the topology of HEN by making changes in structural matrix *SM*. Two crossover operators (single-point and multipoint crossover) and seven mutation operators were applied. The single-point crossover operation is illustrated in Figure 19.12, where the dashed horizontal line shows the crossover point. The locations of heat exchangers that are below crossover point are changed by this operator. It is important to note that the single-point crossover operator can produce solutions that are structurally infeasible. Such a case is shown in Figure 19.12 where two heat exchangers are assigned to the single node (hot node 2.1 in Children structure at the right-hand side). Hence, Bochenek developed a repairing mechanism to convert such solutions into structurally feasible ones. The repairing mechanism removes in the

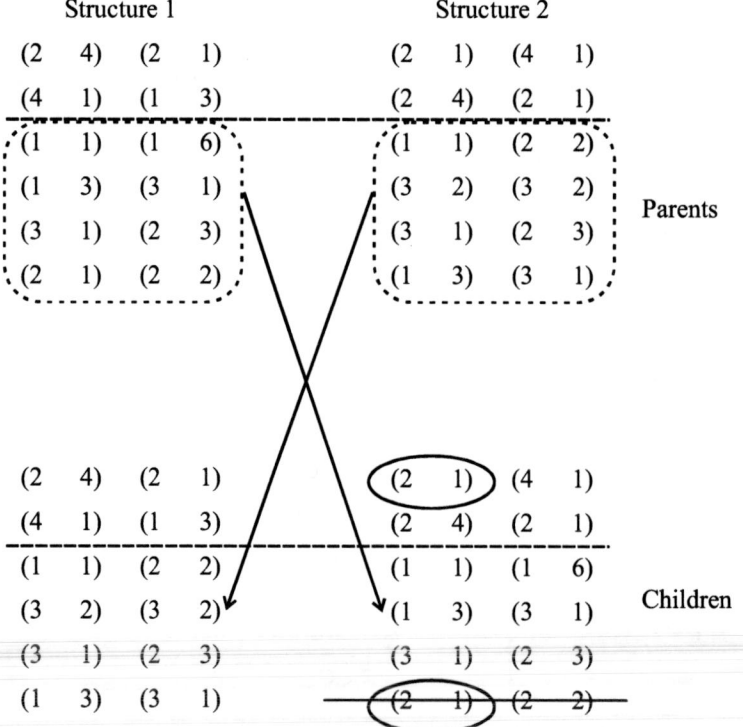

Figure 19.12 Illustration of single-point crossover operation on HEN structures with offspring reparation.

second (right-side) Children solution the "surplus" exchanger—see Figure 19.12. The multiple-point crossover operator consists in such exchange of randomly chosen heat exchangers that does not produce infeasible structures.

The following mutation operators were applied in the GA method:

1. Relocation of heat exchanger hot node
2. Relocation of heat exchanger cold node
3. Relocation of both nodes of a heat exchanger
4. Exchange of hot nodes between two heat exchangers
5. Exchange of cold nodes between two heat exchangers
6. Addition of a new heat exchanger
7. Removal of a heat exchanger.

The mutation operators are simple. They act on a hot/cold node or both by changing the number of branches or nodes in heat exchanger address (matrix SM). For instance, a change of node number causes a shift of heat exchanger.

It is worth noting that due to the coding system developed, the operators for stream superstructure are unnecessary. The decoding algorithm traces the influence of heat exchanger relocations on current stream flow configuration. To illustrate this, consider a simple example. Let a certain mutation operator have a heat exchanger relocated from position (a) to (b) as shown in Figure 19.13a and b. Before the relocation, the heat exchanger node at the side branch was inactive (Figure 13a), and thus the branch and the splitter node were inactive, too. After the relocation, both were de-activated. The decoding procedure keeps track of such structural changes.

The applied genetic operators allow all the structural changes that are possible in HEN revamp design to be performed. Hence, they can generate all the feasible structures within the retrofit superstructure.

To present the algorithm of the second level (parameter optimization) we will begin with its overview.

The problem, which is to be solved at the second level, is typical parameter optimization of HEN with fixed topology – the NLP problem – providing that nonstandard heat exchangers are to be used, or the MINLP if standard apparatus are to be applied. Here we limit ourselves to a brief description highlighting specific features of Bochenek's method, particularly the technique of dealing

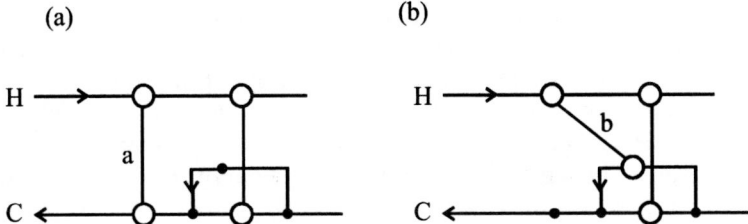

Figure 19.13 Example for the explanation of influence of match relocation on stream flow configuration.

with constraints of the HEN optimization model.

The constraints of the HEN mathematical model are as follows (see also the Synthesis-HLD-C-F-model in Section 18.2 or the loop-breaking model in Section 18.1):

1. Equality constraints of the HEN model, the comprise heat and mass balances of mixers and splitters, heat balances and design equations of heat exchangers.
2. Equality conditions on required values of outlet temperatures of streams leaving a HEN.
3. Inequality constraints on permissible temperature approaches in heat exchangers—thermodynamic constraints with *EMAT*.
4. Various other technological and economic constraints that are often problem-dependent.

Equality constraints from group 1 are nonlinear. In general, they cause serious difficulties for all classes of stochastic optimization approaches, including GA. Bochenek developed a scheme of solving them sequentially as a set of linear equations with regard to dependent variables. This will be shown in the following. Equality constraints from group 2 were relaxed to inequalities. This relaxation is similar to that applied in the generation of independent subnetworks (see Chapter 16) and also can mean the relaxing of impractical hard constraints in some cases. Relaxed constraints from group 2 as well as inequalities from groups 3 and 4 are accounted for in an augmented penalty function.

To solve this NLP/MINLP problem, Bochenek applied the GEN-COM procedure described in Chapter 1. The chromosome of the HEN is coded as the real number vector containing the surface area of all heat exchangers and split ratio of splitters since they are all decision variables. One crossover (single-point) operator and one mutation (uniform mutation) operator was applied. Though the number of operators is small it suffices for efficient solution of the NLP or MINLP model of the HEN.

The HEN model and its solution algorithm are presented below.

The assumptions imposed on the model, similar to those used in other works, are as follows:

- Counter-current flow scheme is applied in all heat exchangers (1-1 heat exchangers)
- *CP* values are constant (and given) in each heat exchanger.

The model applied by Bochenek is based on that from Kotjabasakis and Linnhoff (1986, 1988) and Linnhoff and Kotjabasakis (1986) who, in turn, have adapted it from textbooks on heat transfer, for instance Kays and London (1958). Bochenek and Jeżowski (1999) applied it to NLP optimization by an adaptive random search. The remarks and concerns about the original model addressed by Picon-Nunez and Polley (1995a, 1995b) have been accounted for in the solution algorithm. The model does not use LMTD parameters and, for each heat exchanger of area A within the structure, has the form:

$$T^{\text{out}} = \alpha T^{\text{in}} + (1 - \alpha)T^{\text{in*}} \qquad (19.16)$$

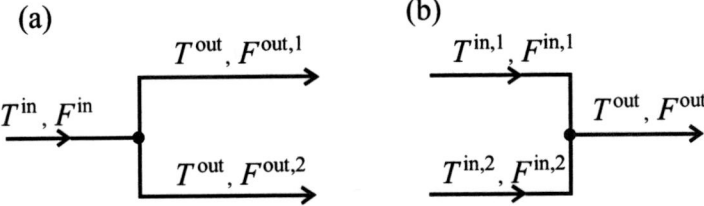

Figure 19.14 Explanations for symbols used in equations for splitters and mixers.

$$\alpha = \frac{(\gamma^* - \gamma) \exp(\gamma^*)}{\gamma^* \exp(\gamma^*) - \gamma \exp(\gamma)} \tag{19.16a}$$

$$\gamma = \frac{UA}{CP}; \quad \gamma^* = \frac{UA}{CP^*} \tag{19.16b}$$

Superscript (*) denotes one from the two streams exchanging heat while parameters of the second stream have no superscript. The advantages of the model are:

- It is linear in respect to temperatures (for fixed surface area and flow rates)
- It uses heat exchanger surface area directly instead of heat loads
- It does not require thermodynamic constraints on temperature approaches—for proof see Bochenek and Jeżowski (1999). Additionally, problems with calculation of LMTD are eliminated.

Notice that a similar though more general model was applied in Akman et al. (2002). Also, a similar model has been used in some works on HEN design with pressure drops as data instead of heat transfer coefficients. The reader interested in this problem is referred to, for instance: Ahmad and Polley (1990), Polley and Shahi (1991), Nie and Zhu (1999), Silva and Zemp (2000) and Frausto-Hernadez et al. (2003).

The basic model is limited to fixed values of split ratio. Since in this retrofit approach split ratios are variables it was extended by adding balances of splitters and mixers as follows (symbols applied in the equations are explained in Figure 19.14, number of branches is limited to two in the explanation):

(a) mass balances of splitters

$$F^{in} = F^{out,1} + F^{out,2} \tag{19.17}$$

(b) mass balances of mixers

$$F^{out} = F^{in,1} + F^{in,2} \tag{19.18}$$

(c) heat balances for splitters (isothermal splitters)

$$T^{\text{out},1} = T^{\text{in}}, \quad T^{\text{out},2} = T^{\text{in}} \tag{19.19}$$

(d) heat balances for mixers

$$\frac{T^{\text{in},1} CP^{\text{in},1}}{CP^{\text{in},1} + CP^{\text{in},2}} + \frac{T^{\text{in},2} CP^{\text{in},2}}{CP^{\text{in},1} + CP^{\text{in},2}} = T^{\text{out}} \tag{19.20}$$

(e) continuity conditions for inactive nodes

$$T^{\text{out}} = T^{\text{in}}; \quad G^{\text{out}} = G^{\text{in}} \tag{19.21}$$

(f) conditions for inlet nodes, that is, assignments of inlet temperature to inlet node of main branches.

The decoding algorithm generates the model equations for each HEN created at the first level of the approach.

The sources of nonlinearities in model constraints are Equations (19.16) and (19.20). Notice, however, that Equations (19.16a and b) become linear in respect to temperatures if surface areas and CPs are fixed. Likewise, Equation (19.20) become linear in respect to temperatures if heat capacity flow rates are fixed. By dividing the variables in the optimization model into two subsets, a subset of decision (independent) variables, and a subset of dependent (state) variables we can solve Equations (19.16) and (19.20) as a system of linear equations. The values of decision variables are generated by GA and thus fixed when solving equality constraints. To solve the HEN parameter optimization problem Bochenek has chosen heat transfer surface areas and CP values of side branches as decision variables. Hence, all the constraints are linear in respect to dependent variables—the temperatures calculated from model equations. This is a very useful technique in stochastic optimization (the general explanation is in Section 1.5).

As an example of method application we will present the solution to the problem referred to as Example 19.2 in the preceding. This problem was also solved in Yee and Grossmann (1987, 1991), Ciric and Floudas (1990c) and Lakshmanan and Banares-Alcantara (1998). Stream parameters are given in Table 19.2. The network consists of 5 heat exchangers as shown in Figure 19.6. Figure 19.7 depicts the superstructure applied by Bochenek. The construction of it has been explained in the preceding. Parameters of heat exchangers are gathered in Table 19.3. This HEN uses as much as 1500 kW of hot utility and 1900 kW of cold utility. Thus, the annual cost of utilities amounts to 158 000 $ y^{-1}. For minimum temperature approach (HRAT) equal to 10 K, the (global) minimum heat loads of hot and cold utilities are: 200 kW and 600 kW, respectively. It results in a minimum utility cost of 28 000 $ y^{-1}. Following Ciric and Floudas (1990c) the goal function is the annual investment cost of retrofitting the HEN on the condition that utility cost should not be greater than 29 000 $ y^{-1}.

Table 19.3 Heat exchanger parameters for Example 19.2

Heat exchanger	Heat exchanger area (m²)
1	46.74
2	68.72
3	38.31
4	40.23
5	5.40

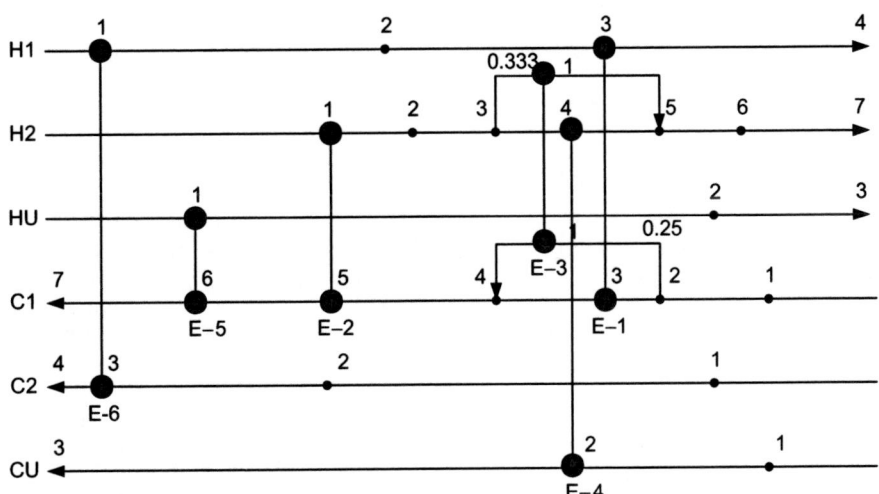

Figure 19.15 Optimal solution from Ciric and Floudas (1990c) for Example 19.2 (with Bochenek's superstructure at the background).

The costs of retrofit changes amount to:

- Cost of additional heat exchanger area, [$]: $1300 \, \Delta A^{0.6}$
- Cost of area for a new heat exchanger, [$]: $1300 \, A^{0.6}$
- Cost of new heat exchanger (match), [$]: 3000
- Cost of heat exchanger relocation, [$]: 300.

Notice that the cost of heat exchanger relocation does not depend on the type of relocation. Also, the investment cost of the new match does not comprise a fixed charge.

The best solution to the problem was obtained by Ciric and Floudas (1990c) using a simultaneous approach with the deterministic optimization method (General Benders Decomposition with Outer Approximation). This network is shown in Figure 19.15 and the parameters of heat exchangers are gathered in Table 19.4.

The problem was solved by Bochenek for both standard apparatus and non-standard. Discrete values of surface area applied for standard heat exchangers are shown in Table 19.5.

Table 19.4 Parameters of heat exchangers for the optimal solution from Ciric and Floudas (1990c) for Example 19.2

Heat exchanger	Heat exchanger area (m²)
1	51.92
2	68.71
3	37.5
4	30.11
5	5.4
6	164.71

Table 19.5 Values of discrete surface area applied by Bochenek (2003) for Example 19.2

Apparatus no.	Existing area (m²)	Set of standard areas (m²)						
1	46.74	38.00	40.00	42.00	46.74	49.00	51.92	54.00
2	68.72	54.00	57.00	0.00	62.00	65.00	68.72	70.00
3	38.31	30.00	32.00	35.00	34.00	36.00	37.50	38.31
4	40.23	22.00	24.00	26.00	28.00	30.11	35.00	40.23
5	5.40	3.00	5.40	10.00	15.00	20.00	30.00	35.00
6	–	145.00	150.00	155.00	160.00	162.00	164.71	167.00

The HEN with nonstandard apparatus calculated by Bochenek is shown in Figure 19.16. The parameters of heat exchangers are given in Table 19.6. A different solution, in Figure 19.17, was calculated for standard heat exchangers. The parameters of the solutions are gathered in Table 19.6.

The comparison of basic parameters characterizing the networks with that calculated by Ciric and Floudas (1990c) is shown in Table 19.7. The solution

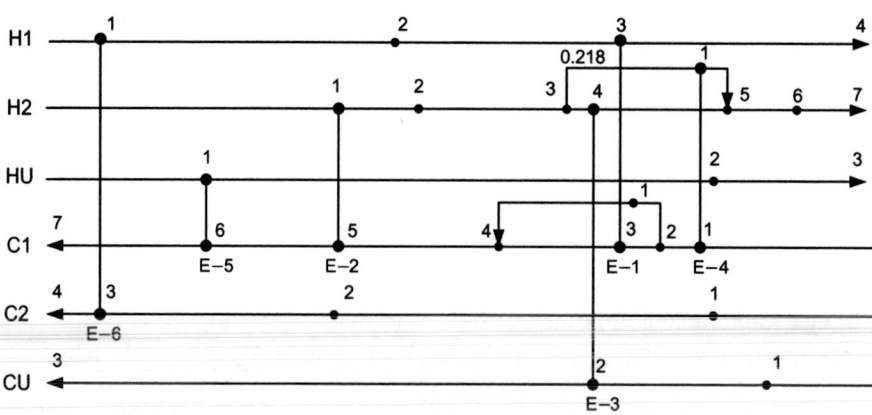

Figure 19.16 Example 19.2—optimal solution by Bochenek for nonstandard heat exchangers (superstructure shown in background).

Table 19.6 Parameters of heat exchangers in the optimal HENs from the method of Bochenek (2003)

Apparatus no.	Nonstandard area (m²)	Standard area (m²)
1	46.75	40.00
2	68.29	54.00
3	26.62	32.00
4	6.80	30.11
5	5.02	5.40
6	139.17	160.00

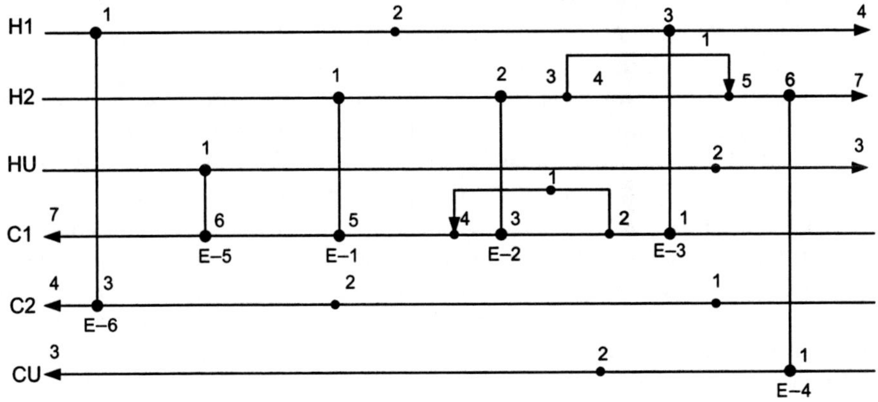

Figure 19.17 Example 19.2—optimal solution by Bochenek for standard heat exchangers (superstructure shown in background).

for nonstandard apparatus calculated by the GA-based approach is much better in terms of investment cost. Also, the utility cost is lower. For standard apparatus the method was also able to find a better solution than that by Ciric and Floudas (1990c). Notice, however, that it depends on parameters of standard heat exchangers.

Table 19.7 Comparison of the HEN calculated by Bochenek with the solution from Ciric and Floudas (1990c)

	Bochenek's solution for nonstandard areas	Bochenek's solution for standard areas	Solution from Ciric and Floudas (1990c)
Total investment cost ($)	29 923.17	32 115.89	35 784.00
Area increase of existing heat exchangers (m²)	0.00	0.00	5.18
Total area increase (m²)	139.17	160.00	169.90
Number of new branches	1	0	2
Annual utility cost ($ y⁻¹)	28 283.21	28 850.00	28 744.12

The following control parameters of the GA procedure have been applied in calculations for nonstandard apparatus: at the first level (structural optimization): population of 100 members and 200 generations; at the second level (parameter optimization): population of 50 members and 150 generations. For standard apparatus the parameters are: population of 100 members and 200 generations at the first level and population of 15 members and 20 generations at the second level.

Bochenek also reported typical probabilities of genetic operators: single-point crossover: 0.16, multipoint crossover: 0.34, relocation of heat exchanger—single node: 0.1, relocation of a heat exchanger—both nodes: 0.1, exchange of nodes between two heat exchangers: 0.1.

Bochenek found that optimization robustness only slightly depends on operators' probabilities but the effect of the number of members on population and the number of generations is substantial. Computation times, with an Athlon 1.5 GHz processor, were high, of the order of 12 hours for nonstandard apparatus and of 2 hours for standard ones. However, the main advantage of the approach is its ability to handle discrete parameters of heat exchangers. Notice that CPU time for such a heat exchangers is acceptable. Also, it is important to note that the quality of the solution is much higher. This also indicates that MINLP retrofit models are difficult to solve to global optima. Additionally, the method of Bochenek eliminates some limitations of MINLP models such as the placement of utility heat exchanger at the "ends" of streams and the restriction of a single heat exchanger at one branch. Finally, it can be concluded that the method is advantageous for industrial applications. However, it requires some enhancements to speed up calculations.

20 Approaches to water network design

20.1. INTRODUCTION

Water is applied in many processes from various branches of industry. Certain compounds, called contaminants in the following, are transferred to water and then have to be removed from wastewater in treatment processes before discharge to the environment.

Water is used in typical separation processes such as extraction, absorption or distillation with steam. These are mass exchange operations where water serves as a separating agent. Additionally, water is necessary to wash various types of equipment and in cyclones and filtrations. Also, a great amount of make-up water is consumed in steam boilers and in cooling water cycles.

Independent of treatment efficiency, water is degraded before exiting back to the environment. Water treatment processes are expensive. Hence, the minimization of freshwater usage as well as wastewater generation in process systems is of great environmental and economic importance. Significant effects can be achieved by optimization of water networks, that is, a network involving both water-using processes and regeneration/treatment units. Ultimately, zero discharge, that is, a closed water circuit, can be reached.

The means to reach substantial reduction of freshwater consumption by water network optimization are

- water reuse in water-using processes
- application of regeneration processes by using redistributed water treatment.

Often, in existing industrial complexes the water is fed to processes in a parallel arrangement, that is, freshwater (and only freshwater) is fed to each water-using process. Next, water streams after each process are mixed and sent to a central treatment station. Neither water reuse nor regeneration is applied. Such a traditional scheme of water network is shown schematically in Figure 20.1.

This network arrangement requires a large amount of freshwater and generates a high quantity of wastewater for treatment. Water reuse by applying water exiting certain processes in others substantially reduces freshwater consumption. This is often referred to as the water reuse network (WRN). An illustrative example of such a WRN is shown in Figure 20.2.

A lot of work has gone into the design and optimization of WRNs. They do not account for the treatment/regeneration processes—wastewater treatment is performed in a separate station. Hence, it is often called a "water-usage network" or a "water-using processes network." As a consequence, methods are needed that focus on designing a network of water treatment processes as a separate

Energy Optimization in Process Systems

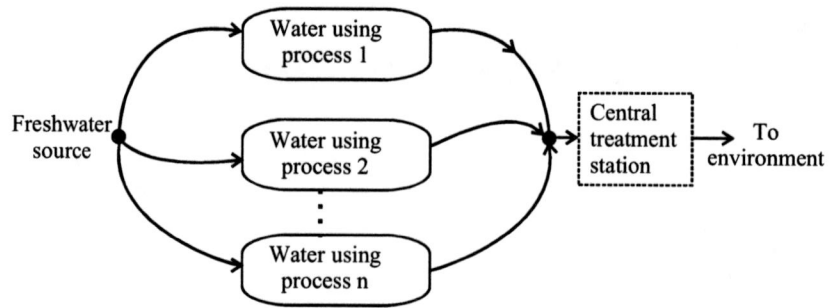

Figure 20.1 Traditional water network: freshwater supplied to processes in parallel with central water treatment station.

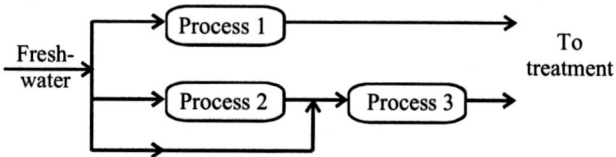

Figure 20.2 Example of water reuse network (WRN).

subsystem—a wastewater treatment network (WTN). This is not a centralized network where one common wastewater stream passes all treatment technologies but a distributed network with a mixing and bypassing scheme such as in Figure 20.3.

Advantages of the redistributed system result from the fact that the cost of treatment is proportional to wastewater flow rate. Thus, a centralized WTN has to be expensive in most cases since a total wastewater stream flows through all treatment units. In a distributed wastewater treatment process, streams are either treated separately or only partially mixed, which reduces the effluent flow rate to be processed. A proper design of the treatment system should segregate the streams for treatment and only combine them if it is appropriate. Minimization of wastewater flow rate through treatment processes is thus the main objective.

Further reduction of both freshwater intake and wastewater generation can be reached by additional application of wastewater regeneration processes within

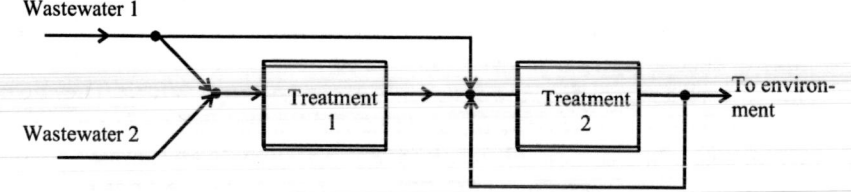

Figure 20.3 Example of distributed wastewater treatment network (WTN).

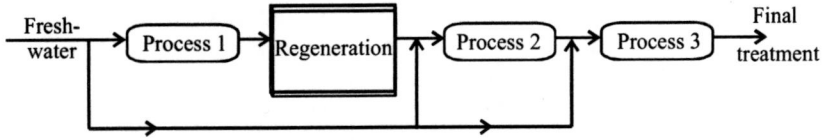

Figure 20.4 Regeneration without recycles.

the network of water-using processes. Figures 20.4 and 20.5 illustrate this. It is necessary to distinguish two cases of regeneration: regeneration without recycles (Figure 20.4) and regeneration with recycles (Figure 20.5).

In the former the water after regeneration is not to be applied in the processes where it was used before in order to prevent accumulation of harmful substances of very low concentration.

In numerous papers regeneration operations are applied solely to decrease freshwater consumption by allowing wider water reuse. There are no limitations imposed on final contaminant concentrations in effluents. Hence, additional final treatment is often necessary to comply with environmental regulations. Such contributions deal with water network with reuse and regeneration (WNRR). Then, design of a water treatment network is necessary. Interactions between two systems, WNR or WNRR and WTN, have to be properly accounted for, see for example, Kuo and Smith (1998b). However, if environmental conditions are imposed on final contaminant concentrations of effluent wastewater streams, the difference between regeneration and treatment processes becomes irrelevant. A result, total (integrated) water network (TWN) or water usage and treatment network (WUTN) is dealt with. The term "water allocation" problem is also used. In view of process integration such a concept is valid since it allows accounting for the whole system simultaneously. The division into a separate network of water-using processes (WNRR) and a network of treatment operations (WTN) becomes unnecessary. By designing TWN it is possible to reach a closed circuit of process water, that is, it is possible to eliminate water discharge to the environment (zero discharge). Freshwater is necessary as make-up water only. At present, it seems to be an expensive solution because of, for example, the accumulation of solid deposits (see e.g. Diepolder and Brown, 1992; Bagajewicz, 2000). However, zero discharge is the ideal solution with regard to sustainable industrial activity.

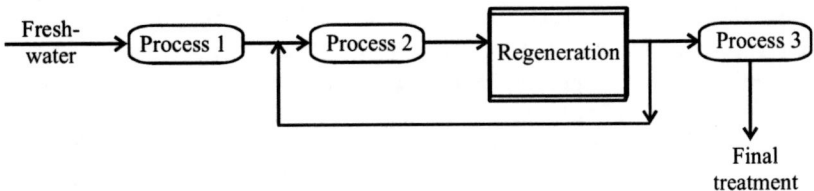

Figure 20.5 Regeneration with recycles.

Though minimization of freshwater consumption and wastewater treated is often assumed to be the only objective, other operating costs together with investment expenses should be taken into account. The cost of wastewater regeneration or treatment is often accounted for in designing methods for WRN, WNRR, and TWN. Some methods include the cost of pipelines or the number of connections. Pumping cost is included in the goal function in a few works. Finally, Ku-Pineda and Tan (2006) applied an "environmental" performance index—the sustainable process index developed by Krotschek and Narodoslavsky (1996).

Generally, a total water network consists of various freshwater/raw water sources, water-using operations, treatment/regeneration processes and sites for wastewater disposal. Mixers and splitters are also necessary to provide redistribution among sources, processes and sinks and a necessary means for reuse and regeneration. In the next section we will present models of processes and data that are commonly applied in approaches to water network design.

The concept of a water network in a single industrial site can be extended to industrial complexes of industrial (eco) parks and urban water networks, for example, Keckler and Allen (1999), Nobel and Allen (2000), Wenzel et al. (2002) and Foo et al. (2005). An industrial park is an area with a high concentration of plants that are water consumers and wastewater producers. Due to various limitations on feed water quality and various levels of wastewater contamination we can also make use of water reuse and regeneration. The plants are then treated as processes. Similarly, in public buildings such as hospitals and hotels, freshwater consumption and wastewater generation can also be reduced by means of reuse and regeneration (Foo et al., 2005; Alwi and Manan, 2006).

Notice that in systems operating in batch-wise mode tanks also are used in water networks. Some authors, for example, Feng and Seider (2001), Wang et al. (2003), Cao et al. (2004) and Feng and Chu (2004) claim that they should also be applied in continuous operation mode. Here, batch processes are not considered. The reader is referred to Almató et al. (1999), Jödicke et al. (2001), Foo et al. (2005), Majozi (2005a,b) and Chang and Li (2006). Additionally, a number of papers deal with simultaneous heat integration in water networks. Notice that water-using operations, treatment/regeneration processes and wastewater disposal sites have some requirements as for water temperature and phase state. To meet such conditions cooling and/or heating is necessary. Heat exchangers are used as well as heating and cooling utilities. To reduce the number of heaters and coolers as well as the cost of utilities heat integration should be applied. It is clear that heat integration needs to be carried out simultaneously with mass integration, that is, with optimization of the "basic" water network consisting of water-using and treatment processes. This topic is not addressed here: the reader is referred to Bagajewicz et al. (2002) and Savulescu et al. (2002, 2005a, 2005b). Similar to the HEN problem we will also omit analysis of WN design under uncertainty. This complex problem is addressed in the following papers: Huang and Wang (1999), Koppol and Bagajewicz (2003), Tan and Cruz (2004), and Al-Redhwan et al. (2005).

20.2. MATHEMATICAL MODELS AND DATA FOR WATER NETWORK PROBLEM

- Contaminants

All chemicals and other substances, such as solid phase suspensions that are transferred in water-using processes, treatment operations and/or subjected to environmental regulations, are contaminants. The number and types of contaminants are fixed.

- Freshwater/raw water sources

It is assumed that the number of available water sources is known. Also, water quality in terms of contaminant concentrations as well as unit costs of freshwater are given for each source. Upper limits on water availability in terms of maximum flow rates are also known.

- Water-using processes

The number of water-using processes is considered a fixed parameter in all contributions. However, models of processes differ. Wang and Smith (1994a) introduced mass transfer water-using processes, also called "quality controlled" operations. Many other authors have adopted the model. This is a simple counter-current mass exchanger (p) in which given loads of contaminants $i(L_p^i)$ are transferred to a water stream from a real or a fictitious process stream as shown in Figure 20.6.

Because contaminant concentrations are very small, the assumption of a constant flow rate of both streams is most often acceptable. Thus, the water stream mass balance equation of contaminant i in process p is:

$$L_p^i = F_p(C_p^{i,\text{out}} - C_p^{i,\text{in}}) \tag{20.1}$$

Concentrations of contaminants in process streams are known from process conditions. Hence, maximal allowable contaminant concentrations in water streams can be estimated from equilibrium conditions. In some cases such as equipment washing these concentrations depend on solubility, fouling or corrosion limits. To account for process kinetics the equilibrium concentrations are reduced by a small value. Let equilibrium concentration of specie i in water stream be C^{i^*}. Then, maximum allowable concentration is: $C^{i^*} - \varepsilon$. It is interesting to note that parameter ε plays a similar role to ΔT^{min} (HRAT) in heat integration. In the following we will apply concentrations in a shifted concentration scale, that is, reduced by parameter ε. Note also that an identical model has been applied by El-Halwagi (1997, 2005) for general mass exchanger networks.

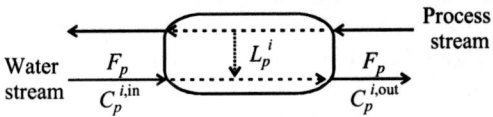

Figure 20.6 Scheme of water-using process p modeled as mass exchanger.

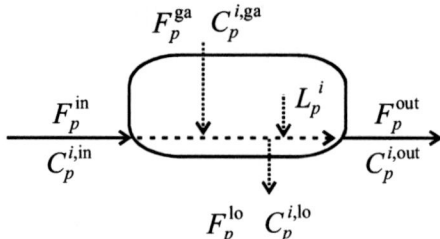

Figure 20.7 Water-using process with water gains and losses.

Thus, for mass transfer water-using processes the following data are required for each mass transfer water-using process p and each contaminant i:

- mass loads of contaminants L_p^i
- maximum permissible inlet concentration of contaminants $C_p^{i,\mathrm{in,max}}$
- maximum permissible outlet concentration of contaminants $C_p^{i,\mathrm{out,max}}$.

Finally, the model of mass transfer process consists of balance equation (20.1) and inequalities (20.2) and (20.3):

$$C_p^{i,\mathrm{in}} \leq C_p^{i,\mathrm{in,max}} \tag{20.2}$$

$$C_p^{i,\mathrm{out}} \leq C_p^{i,\mathrm{out,max}} \tag{20.3}$$

The basic model of Wang and Smith (1994a) has been extended in some works by inclusion of water gains and losses—see Figure 20.7. Most often either losses or gains have to be accounted for.

It is assumed that parameters F_p^{ga}, $C_p^{i,\mathrm{ga}}$, F_p^{lo}, $C_p^{i,\mathrm{lo}}$ are known and, thus, balance equation (20.1) becomes:

$$L_p^i = F_p^{\mathrm{in}} C_p^{i,\mathrm{in}} - F_p^{\mathrm{out}} C_p^{i,\mathrm{out}} + F_p^{\mathrm{ga}} C_p^{i,\mathrm{ga}} - F_p^{\mathrm{lo}} C_p^{i,\mathrm{lo}} \tag{20.4}$$

Inequality constraints (20.2) and (20.3) have to be added, too.

Dhole et al. (1996) noticed that not all water-using processes in a process system can be modeled as mass exchangers. Chemical reactions, which use water as reagents or generate water as product, are an example. Also, cooling water cycles and boiler cycles are only consumers of make-up water. There are no mass transfer processes in such equipment. These processes are often called "quantity controlled water-using processes". We will refer to them in the following as nonmass transfer processes. Generally, they can be modeled as sources and/or sinks (demands) of water streams. Notice also that chemical sites in eco-parks and various operations by urban water consumers are most often referred to as sources and demands.

Generally, water sources have a fixed water stream flow rate and fixed concentrations of contaminants. Water sinks also feature fixed flow rates but

(a)   (b)

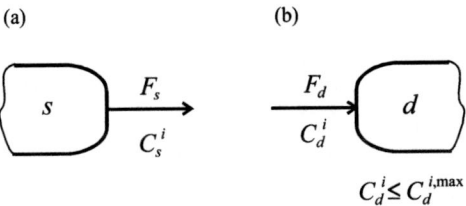

Figure 20.8 Representation of nonmass transfer processes as source and sink: (a) water source s; (b) water sink d.

contaminant concentrations are variables albeit limited by given maximal values. Water sources and sinks are illustrated by Figure 20.8a and b.

Note that if a process is water sink and water source at the same time it can also be represented by two items: single sink and single source.

Some other situations can exist, too. Assume that a water-using process is a mass transfer one but flow rate is fixed due to, for example, hydrodynamics restrictions. For instance, Dunn and Wenzel (2001), Dunn et al. (2001) and Wenzel et al. (2002) stated that this restriction is common in industry for the retrofit scenario. Wang and Smith (1995) have also considered such a condition. Thus, if contaminant mass loads and water stream flow rate are fixed, the only variables are contaminant concentrations. Such processes can also be modeled by a pair: sink and source. Prakash and Shenoy (2005a,b) and Agrawal and Shenoy (2006) considered processes named "fixed flow rate" that have fixed but different values of flow rate at inlet and outlet. They can also be treated as source and sink. It is worth noting that Manan and Foo (2003) and Agrawal and Shenoy (2006) employed the concepts and techniques for WN with sources and demands for designing hydrogen networks. Dunn and Wenzel (2001) claimed that in industrial retrofit conditions even inlet and outlet concentrations should be fixed. Representation of sink and source can also be applied to the last case, but there are no degrees of freedom since all parameters are fixed.

- Treatment/regeneration processes

Material balance of specie i for treatment/regenerator t is given by Equation (20.5)—see also Figure 20.9 for symbols:

$$L_t^i = F_t(C_t^{i,\text{in}} - C_t^{i,\text{out}}) \tag{20.5}$$

Note that constant flow rate is assumed similar to balance (20.1) of the water-using process. There are two simple design equations commonly used in the

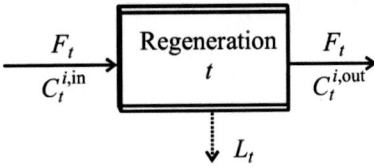

Figure 20.9 Scheme of regeneration/treatment process.

literature. Only Galan and Grossmann (1998) contributed with a detailed model of specific treatment operation in frames of WTN synthesis. Equation (20.6) fixes concentrations of some or all contaminants at the outlet. The outlet concentrations $C_t^{i,\text{out}*}$ are independent of flow rate and also of inlet conditions. These concentrations are parameters of the model:

$$C_t^{i,\text{out}} = C_t^{i,\text{out}*} \tag{20.6}$$

Equation (20.7) defines recovery ratio of some or all contaminants in a process. These ratios are assumed to be known:

$$\Psi_t^i = \frac{F_t C_t^{i,\text{in}} - F_t C_t^{i,\text{out}}}{F_t C_t^{i,\text{in}}} = \frac{L_t^i}{F_t C_t^{i,\text{in}}} \tag{20.7}$$

Some other conditions can be included. For instance, inequality (20.8) limits inlet concentration of selected or all substances. Also, total flow rate of wastewater stream through treatment/regeneration unit can be limited by Equation (20.9):

$$C_t^{i,\text{in}} \le C_t^{i,\text{in},\text{max}} \tag{20.8}$$

$$F_t \le F_t^{\text{max}} \tag{20.9}$$

It is most often assumed in approaches for WNRR, TWN and WTN that treatment/regeneration processes are known. Only Galan and Grossmann (1998) and Karuppiah and Grossmann (2006) took the choice of treatment technology into account. One unit is usually used for a single treatment/regeneration operation. Heuristic rules were developed in Chang and Li (2005) for estimating a number of units and their arrangement. Notice that neither number of units nor their arrangements were decision variables in the design method of Chang and Li (2005).

- Wastewater disposal sites

The number of wastewater disposal sites is known. For each site w, the limiting, environmental concentrations for each contaminant have to be given—Equation (20.10). Additional conditions on allowable maximum wastewater flow rate can be added—Equation (20.11). More complex conditions on water disposal to irrigation pools were applied only in Wenzel et al. (2002):

$$C^i \le C_w^{i,e} \tag{20.10}$$

$$F_w \le F_w^{\text{max}} \tag{20.11}$$

20.3. OVERVIEW OF APPROACHES IN THE LITERATURE

Takama et al. (1980) are the authors of the seminal paper on water network design. They were the first to formulate an optimal water allocation problem that includes both water-using and wastewater treatment processes in a petrochemical refinery. They were also the first to employ a superstructure optimization approach to cope with the problem. To the best of the authors' knowledge, no works on this problem were published before 1994. A substantial output of contributions started in 1994 beginning with the important work by Wang and Smith (1994a) on designing WNRR (see also Wang and Smith (1994b) on WTN). Bagajewicz (2000) presented the first review paper on achievements from the area. He focused mainly on mathematical methods showing their power in solving complex problems. Certain techniques, mainly those from pinch technology, were presented in a very detailed way in a monograph by Mann and Liu (1999). They are also addressed in the textbook by Smith (2005).

Here we classify the approaches into two broad groups:

1. Insight-based methods
2. Optimization-based methods

20.3.1. Insight-based approaches to water networks

Similar to HEN design methods, existing approaches to WN using the (water) pinch concept seem to be the most popular. They are organized in two stages: targeting and structure development. Wang and Smith (1994a) developed a graphical procedure for targeting a minimum freshwater flow rate for water networks consisting of mass exchange water-using processes. They considered water reuse and regeneration, that is, WNRR. Guidelines were presented on manual design of networks. In the same year Wang and Smith (1994b) described a conceptually similar technique for targeting and designing WTN with minimum wastewater flow rate.

In order to make the presentation systematic we will present developments on networks with water-usage processes and networks with treatment processes separately. We will focus on issues that provide deep insights into the problem.

The water pinch targeting method for water-using processes of mass transfer type will be described applying the simple Example 20.1 taken from Wang and Smith (1994a). Notice that we present the technique in a slightly different way than that in the original paper. In Example 20.1 there are four water-using processes—see Table 20.1.

For each process the mass loads of the contaminant and the maximal values of inlet and outlet concentrations in water streams are given in the table (notice that they are in shifted concentration scale). It is assumed that there is a single freshwater source and contaminant concentration in this freshwater is 0.0. Using the data we can simply calculate total freshwater usage for a parallel arrangement such as that in Figure 20.1. For each process $p = 1, \ldots, 4$ in the parallel network

Table 20.1 Data for Example 20.1 (taken from Wang and Smith, 1994a)

Process number	1	2	3	4
Contaminant mass load ($kg\,h^{-1}$)	2	5	30	4
C_{in}^{max}(ppm)	0	50	50	400
C_{out}^{max}(ppm)	100	100	800	800

the minimum flow rate of freshwater amounts to:

$$F_{fw,p} = \frac{L_p}{C_p^{in,max} - C^0} \tag{20.12}$$

Notice that index i for contaminant was dropped since there is one contaminant in the example.

For Example 20.1 the values of freshwater flow rate are 20, 50, 37.5, and $5\,t\,h^{-1}$, respectively. The total flow rate of freshwater that has to be applied in the parallel arrangement without reuse is thus $112.5\,t\,h^{-1}$.

In the water pinch approach, similar to heat pinch methods, the total range of contaminant concentration in the data is divided into intervals. The bounds of an interval correspond to the maximum inlet and the maximum outlet contaminant concentration of a process. The procedure for creating concentration intervals is illustrated in Figure 20.10 for Example 20.1.

We employ here a representation similar to that commonly used in heat integration problems though it was not applied in the original water pinch approach from Wang and Smith (1994a). For each interval ($j = 1, \ldots, J$), where intervals are numbered from that of the lowest concentration in ascending order, the total

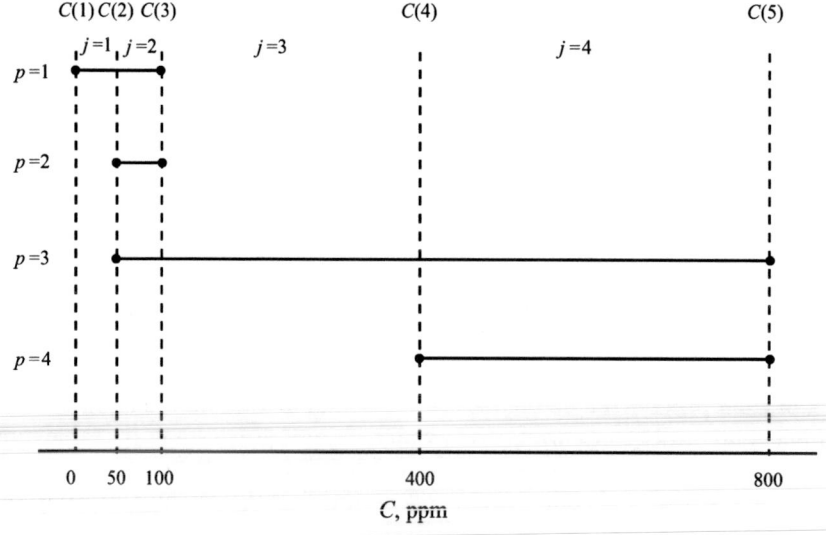

Figure 20.10 Illustration for creating concentration intervals for data of Example 20.1.

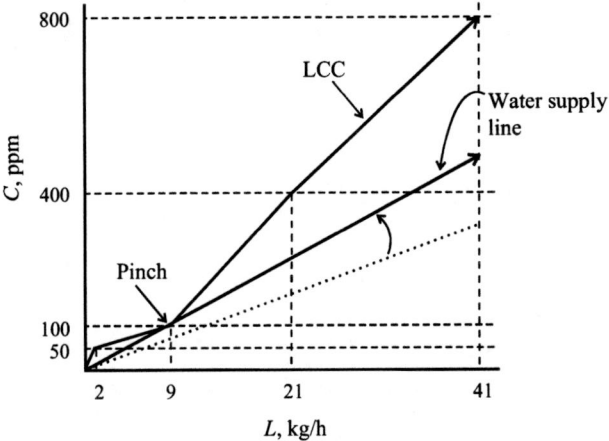

Figure 20.11 Water composite curve plot for Example 20.1.

contaminant load of all processes that fall into the interval is calculated. Mass load of contaminant in process p existing in interval j is calculated from:

$$L_p(j) = \frac{\Delta C(j)}{C_p^{out,max} - C_p^{in,max}} \qquad (20.13)$$

where $\Delta C(j)$ denotes concentration range of interval j.

Parameter $L_p(j)$ is equal to 0.0 if the concentration range in process p does not fall into the concentration range of interval j. Then, values of the total mass loads $L(j)$ are determined from:

$$L(j) = \sum_{p \in P} L_p(j) \qquad (20.14)$$

In Example 20.1, values of $L(j)$ are $L(1) = 1 \, kg \, h^{-1}$, $L(2) = 8 \, kg \, h^{-1}$, $L(3) = 12 \, kg \, h^{-1}$, $L(4) = 20 \, kg \, h^{-1}$, respectively. Cumulative loads are then calculated by summing up values of $L(j)$ in successive intervals. In Example 20.1 they are 1, 9, 21, and 41 $kg \, h^{-1}$. A plot of contaminant concentration (C) vs. cumulative mass load of contaminant (L), shown in Figure 20.11, is then prepared.

This is a piecewise straight-line curve, segments of which represent profiles of concentration changes in concentration intervals. Such a curve is called here a limiting composite curve (LCC).

There exists an alternative fully graphical way of constructing the WCC plot. Instead of calculating cumulative mass loads we first transfer directly the data for individual processes, that is, their mass loads and inlet and outlet concentrations, to a plot concentration vs. mass load. Notice that the mass load scale is relative. The plot is shown in Figure 20.12 for the data of Example 20.1.

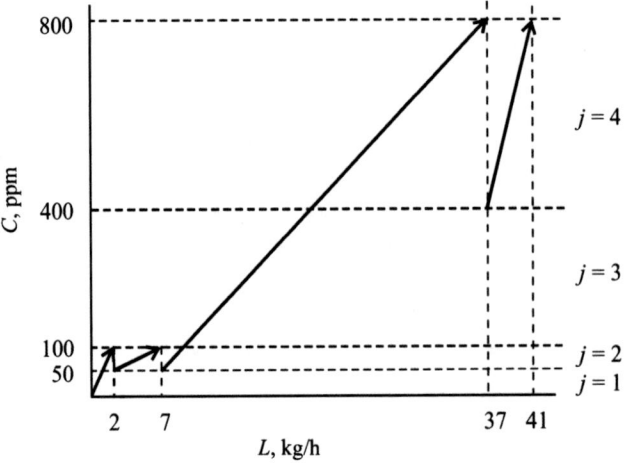

Figure 20.12 Illustration of graphical procedure for constructing LCC–concentration profiles of processes.

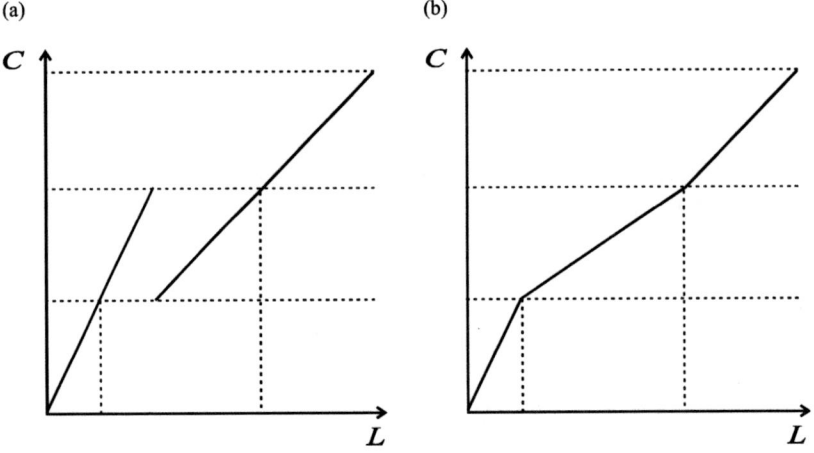

Figure 20.13 Illustration of graphical procedure for constructing WCC–construction of a segment of LCC: (a) profiles of two individual streams, (b) composite curve.

Concentration intervals are also imposed on the plot. Next, the composite curve is constructed for parts of processes in common intervals. For instance, in Example 20.1, we have to make the composite curve from three segments in interval $j = 2$ and two segments in interval $j = 4$. The composite curve is made by superimposing the segments onto each other according to the procedure illustrated in Figure 20.13. The final plot is identical to that presented in Figure 20.11.

Notice that the procedure is identical to that for constructing the CC plot for heat integration described in Chapter 13.

The reciprocal of tangent of a segment slope angle in an LCC plot defines water flow rate in concentration interval. The freshwater flow rate is defined by a straight line, called the water supply line (WSL) – see Figure 20.10 – drawn from the concentration equal to the contaminant concentration in the freshwater source (in Example 20.1 the concentration is zero). For instance, the dotted straight line in Figure 20.11 is the feasible water supply line. The reciprocal of tangent of the line slope angle is equal to the freshwater flow rate. In order to minimize the flow rate we should maximize the angle. However, to meet the condition that the contaminant concentration of the composite curve must not be lower than that for the water supply line the latter cannot cross the LCC. Therefore, to minimize the freshwater flow rate and to maintain the condition we should push the WSL to the position at which it only touches LCC—see the solid line in Figure 20.11.

This way of constructing the WSL ensures that

1. the condition that contaminant concentration in water streams of processes is not higher than given maximal values – see inequalities (20.2) and (20.3) – is always met,
2. the freshwater flow rate is minimal (since tangent of WSL slope angle is the permissible maximum).

The point where the WSL touches the composite curve is called the (water) pinch since it limits further the decrease of the freshwater flow rate. To define water pinch its concentration C^{pinch} will be used in the following.

In Example 20.1 the minimum freshwater flow rate $F_{fw}^{\min}$ determined from the construction described above of the WSL equals $90\,\text{t h}^{-1}$. Recall that for a parallel structure the freshwater flow rate is $112.5\,\text{t h}^{-1}$. The saving ($22.5\,\text{t h}^{-1}$) is due to water reuse.

Some extensions to the targeting approach for water reuse were added in Wang and Smith (1995) to allow for the inclusion of multiple water sources, fixed flow rate via processes and water losses. Jeżowski et al. (2006) developed a simple optimization linear model that accounts for the minimum cost of freshwater for multiple sources of various unit prices, availability limitations and contaminant concentrations. The approach is limited to a single contaminant case.

Only slight modifications are necessary to determine the minimum freshwater flow rate for regeneration with reuse (WNRR) from the LCC plot. The water supply line has to be changed into a composite curve (stepwise segmented straight line) since it represents water streams before and after regeneration. Wang and Smith (1994a) proved that to minimize freshwater flow rate, contaminant concentration at the inlet to regeneration process C^{reg} has to meet the condition:

$$C^{\text{reg}} \geq C^{\text{pinch}} \tag{20.15}$$

Notice that pinch concentration C^{pinch} corresponds to the concentration at the pinch calculated for water reuse only (WRN). The inequality should be taken as

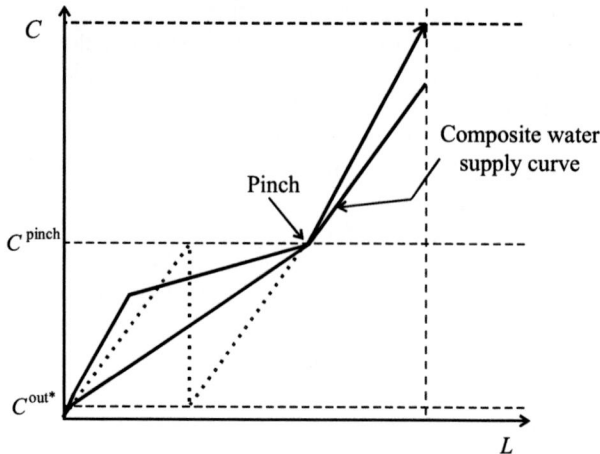

Figure 20.14 Construction of WCC plot for regeneration without recycling.

equality in order also to minimize wastewater flow rate:

$$C^{reg} = C^{pinch} \tag{20.16}$$

Figure 20.14 illustrates the construction of a composite water supply line for data of Example 20.1 for a regeneration process with fixed outlet concentration—Equation (20.6). At first, the dotted lines are drawn for water before and after regeneration and then the solid composite curve is made from them according to the rules for constructing the composite curve shown in Figure 20.13. Notice that both dotted lines have to lie in parallel since flow rate before and after regeneration is identical. This construction is valid for regeneration without recycling (but with reuse). Notice that with reuse and regeneration without recycling the freshwater flow rate drops to $46.2\,\mathrm{t\,h^{-1}}$, for Example 20.1. Regeneration with recycling further reduces the freshwater usage since the composite water supply curve has its starting point below the outlet concentration from regeneration C^{out*}. In Example 20.1 it is possible to reach F_{fw}^{min} equal to $20\,\mathrm{t\,h^{-1}}$—see Wang and Smith (1994a). However, recycling results in a higher flow rate via the regenerator.

The importance of the water pinch approach relies not only on targeting. It allows drawing out of important conditions that have to be met in water networks with minimum freshwater consumption. The conditions are applied as basic guidelines or subsidiary limitations in various approaches for designing WN. They are as follows:

- contaminant concentration in processes below the pinch has to be lower than C^{pinch}
- water with contaminant concentration lower than C^{pinch} should not be used in processes above the pinch
- water for regeneration should feature contaminant concentration equal to C^{pinch}—condition (20.16).

Note that they all stem from the fact that there exists a surplus of water above the pinch and a water shortage below it, similar to heat integration—see, for example, the "bubble plot" in Figure 13.10.

The water pinch targeting approach is valid only for one contaminant. Wang and Smith (1994a) also developed a shifting procedure to deal with multicontaminant problems. The procedure is quite complex and has not been applied in other approaches though the basic concept of a shifting concentration scale for contaminants was employed in Alva-Argaez et al. (1999). A simpler version of a shifting mechanism is given in the book by Mann and Liu (1999).

Based on the WCC plot and pinch consequences, Wang and Smith (1994a) proposed two design techniques but only one of them called "minimum number of water sources" seems valid—see Bagajewicz (2000). However, it is iterative and does not guarantee optimum results. A more rigorous approach was developed by Kuo and Smith (1998a,b) though it also requires tedious calculations. Large-scale industrial application of these techniques can be found in Thevendiraraj et al. (2003).

Certain other totally or partially insight-based methods for designing WN with mass transfer processes were suggested by Lin et al. (1997), Olesen and Polley (1997), Castro et al. (1999), Gómez et al. (2001), Parthasarathy and Gopal (2001), Hamad et al. (2003), and Gomes et al. (2007) to list a few. Insight-based methods for designing WN with nonmass transfer operations were addressed, among others, in El-Halwagi et al. (2003) and Prakash and Shenoy (2005a,b). Dhole et al. (1996) presented a general description of "water pinch" methodology marketed by Linnhoff-March.

They are all most often limited to freshwater minimization and usually to a single contaminant. Olesen and Polley (1996) criticized such approaches as impractical due to lack of piping cost or "geographical" restrictions. More recently, Alwi and Manan (2006) contributed with a detailed cost analysis for network retrofit. Iterative manual procedures have to be employed that are based on heuristics, logical reasoning and water pinch conditions. Nevertheless, such methods have been employed for industrial applications, for instance, those reported by Baetens and Tainsh (2000), Gautham and Gopal (2001), Wenzel et al. (2002), Thevendiraraj et al. (2003) and Hamad et al. (2003).

Feng and Seider (2001) suggested a simple systematic approach for WRN with a single contaminant. Then the approach was extended for the multicontaminant case by Wang et al. (2003). Finally, regenerators were added by Cao et al. (2004). In these contributions, water mains were employed similarly to Kuo and Smith (1998a,b). Notice, however, that water mains in the latter approach were used only as a calculation aid while in the former they were already included in the network.

The targeting and design methods described to this point are aimed at mass transfer processes. Dhole et al. (1996 Jan) proposed a graphical targeting method that deals with only nonmass transfer processes. Polley and Polley (2000) found that it could lead to incorrect results. Sorin and Bedard (1999) proposed an approach named "evolutionary table" which did not account for multiple

Table 20.2 Data for Example 20.2 (taken from Prakash and Shenoy, 2005a)

	Contaminant concentration (ppm)	Flow rate (t h^{-1})
Sinks		
D1	20	50
D2	50	100
D3	100	80
D4	200	70
Sources		
S1	50	50
S2	100	100
S3	150	70
S4	250	60

pinches. A graphical iterative method called "water surplus diagram" was developed by Hallale (2002). It accounts for both mass and nontransfer processes. A conceptually similar method was developed by El-Halwagi et al. (2003). The systematic approaches by Manan et al. (2004), Prakash and Shenoy (2005a) and Agrawal and Shenoy (2006) can deal with both types of process within a single framework. However, they require treating mass transfer processes as being composed of a pair: source–sink. Hence, an assumption on fixed outlet contaminant concentration has to be imposed on mass transfer operations. The methods are valid only for a single contaminant. Note that insight-based targeting methods for mass transfer processes employ the same condition: outlet concentration is set at the maximum. This seems valid if flow rate minimization is the only objective, and if a single contaminant exits. Here, we will explain the approach by Prakash and Shenoy (2005a), since it requires fewer calculations. The method does not need a graphic although the final solution subroutine results from a plot. This approach is primarily tailored for nonmass transfer processes and thus we will explain it using Example 20.2—for data see Table 20.2.

The sources and sinks, separately, are first arranged in increasing order of contaminant concentration. Notice that freshwater is included in the sources. It is assumed in the example that it has no contaminant. Mass loads of contaminant are calculated and inserted into the table. They are determined from mass balances (20.1) assuming that inlet concentration is zero and outlet concentration equals that in the data for sink or source, respectively. Then, cumulative flow rate and cumulative contaminant mass load are calculated separately for sources and sinks. The values are presented in Table 20.3. Let us assume for a moment that a plot—cumulative contaminant load vs. cumulative flow rate—is constructed from the values in Table 20.3. Two step-wise straight line curves for sources and sinks are shown in Figure 20.15a.

Prakash and Shenoy (2005a) proved that the contaminant load of the source should be less than or equal to the contaminant load of the sink at every point of the plot in order to meet constraints on contaminant concentration. Thus, to

Table 20.3 Results of calculations needed in the method of Prakash and Shenoy (2005a)

	Contaminant load (kg h^{-1})	Cumulative flow rate (t h^{-1})	Cumulative load (kg h^{-1})	Cumulative flow rate for minimum freshwater intake (t h^{-1})
Sinks				
D1	1	50	1	
D2	5	150	6	
D3	8	230	14	
D4	14	300	28	
Sources				
Freshwater	0	0	0	70
S1	2.5	50	2.5	120
S2	10.0	150	12.5	220
S3	10.5	220	23.0	290
S4	15.0	280	38.0	350

satisfy the condition, the source curve should be shifted horizontally until it is below the sink curve. Such a shift implies an increase of flow rate without any increase of contaminant load. This is possible only by the addition of fresh water to the sources (water of zero contaminant concentration). The shift distance of the source curve corresponds to freshwater intake. In order to minimize the intake it is sufficient for such a shift that the source curve touches the sink curve at least at one point—this is the pinch point. Figure 20.15b illustrates such a shift. As shown in the figure, the minimum flow rate can be found directly from the plot. However, drawing the plot and shifting the curve is not necessary. Prakash and Shenoy (2005a) proved that the pinch can occur only at the sink curve vertex. Therefore, they proposed calculating the horizontal distance at each sink vertex by simple interpolation of source curve data from Table 20.3. Consider, as an example, the first sink vertex with cumulative mass of 1 kg h^{-1}. The value lies in the first segment of the source curve, that is, between cumulative loads from 0 to 2.5 kg h^{-1} and cumulative flow rates from 0.0 and 50 t h^{-1}. Interpolated cumulative flow rate for cumulated load of 1 kg h^{-1} is 20 t h^{-1}. For three other vertices the loads are 85, 160 and 180 t h^{-1}, respectively. Next, we should calculate differences between interpolated flow rate and cumulative flow rate from Table 20.3 for each source. They are as follows: -30, -65, -70 and -60. Similar to the Pr-T approach for heat recovery (see Chapter 13) the most negative value, taken as a positive number, determines the minimum freshwater flow rate. Hence, the solution for Example 20.2 is freshwater intake equal to 70 t h^{-1}. Feasible cumulative flow rates are shown in the last column of Table 20.3. The pinch concentration is defined by the slope of the source curve at the pinch according to mass balance (20.1) with inlet concentration set at zero.

In order to deal with both nonmass transfer and mass transfer processes, the latter have to be transformed into a source–sink representation. We will illustrate this using the data of Example 20.1—see Table 20.1. Inlet streams are treated

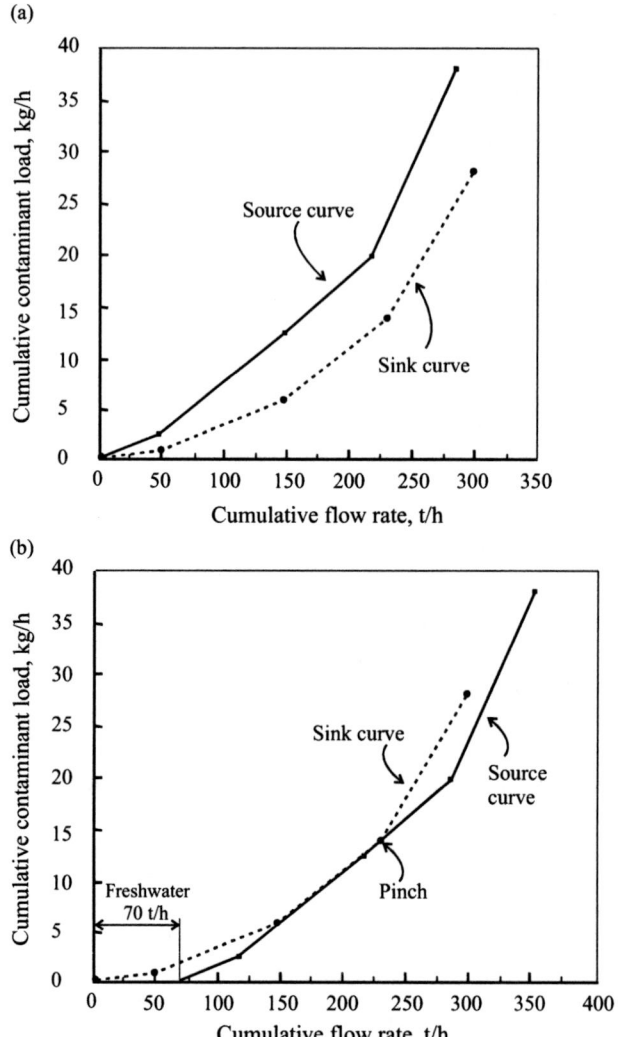

Figure 20.15 Plot for determining the minimum freshwater flow rate for Example 20.2: (a) source and sink curves before shifting; (b) source and sink curves after shifting.

as sources while outlet ones are treated as sinks. Then, flow rates are calculated for both from mass balance (20.1) with both inlet and outlet concentrations set at the maximum values. The results are shown in Table 20.4.

The method of Manan et al. (2004) is also similar to the Pr-T approach for heat recovery calculations. Moreover, the results can be directly inserted in water cascade representations identical to the heat cascade (Chapter 13). This allows visualization of possible changes of processes. Notice that both targeting approaches are limited to a single contaminant.

Table 20.4 Data of Example 20.1 transformed into source–sink representation

	Contaminant concentration (ppm)	Flow rate (t h^{-1})	Contaminant mass load (t h^{-1})
Sources			
1*a*	0	20	0
2*a*	50	100	5
3*a*	50	40	2
4*a*	400	10	4
Sinks			
1*b*	100	20	2
2*b*	100	100	10
3*b*	800	40	32
4*b*	800	10	8

Prakash and Shenoy (2005a) also developed the design method. Basically, this technique applies to both types of water-using processes. However, as noted by the authors, it produces solutions too complex in the case of mass transfer processes. Hence, Prakash and Shenoy (2005a) suggested three rules that aim at maximization of concentration gradient. It is interesting to note that one of the rules forces outlet concentration to maximum values. Hence, it is identical to the main necessary condition for freshwater minimization for the single contaminant case of Savelski and Bagajewicz (2000). In the case of nonmass transfer processes, Prakash and Shenoy (2005a) applied to the design water network the nearest neighbors' principle. It states that: "to satisfy a particular water sink, the source streams to be chosen are the nearest available neighbors to the sink in terms of contaminant concentration". The principle and the design approach were extended in Prakash and Shenoy (2005b).

Water treatment network—insight-based methods

Wang and Smith (1994b) developed a graphical targeting method for wastewater flow rate minimization that is based on an almost identical concept as the WCC plot. We will describe it for the simple example of a single contaminant and a single type of regeneration process. Let the environmental limit on contaminant concentration be C^e. Data has to provide also the number of wastewater streams of fixed initial concentrations and flow rates. Contaminant mass loads in streams can be directly calculated from the data for outlet concentration C^e. First, we draw a plot – contaminant mass loads vs. concentration – for individual streams following the procedure for the LCC curve in Figure 20.12. Then, segments of the composite curve are constructed following the procedure described for water-using operations and illustrated by Figure 20.13. Having constructed the composite curve for the wastewater streams, a treatment line is then drawn against it. The reciprocal of the slope of the treatment line

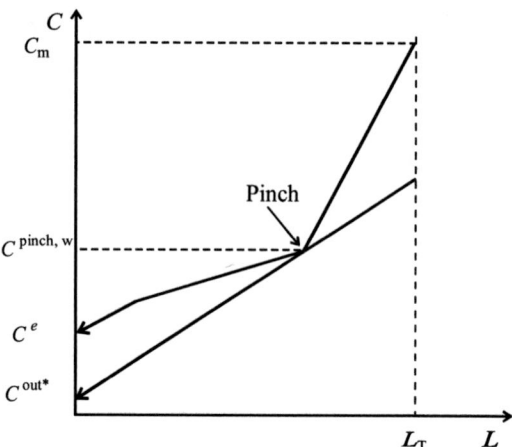

Figure 20.16 Wastewater composite curve plot.

is the wastewater flow rate treated. The slope should be maximized to minimize the treatment flow rate. However, the slope of the treatment line is limited by both the composite curve of the wastewater streams and the efficiency of the treatment process defined by Equation (20.6) or (20.7). Let us assume for simplicity's sake that the regeneration process is defined by a fixed outlet concentration C^{out*} according to Equation (20.6). The treatment flow rate is minimized by rotating the treatment line around the fixed point C^{out*} anticlockwise until it touches the composite curve of the wastewater streams. The procedure and the final placement are shown in Figure 20.16. The symbol L_T denotes total contaminant load while C_m stands for the highest contaminant concentration. The pinch is created at the point where the treatment line touches the composite curve. It is denoted by $C^{pinch,w}$. Wang and Smith (1994b) explained how to find the origin of the treatment line for treatment processes modeled by Equation (20.7).

The wastewater composite curve plot gives not only the minimum treatment flow rate and concentrations of contaminant; it also shows the designer which streams require treating and which streams do not. This information results from pinch location. The wastewater streams with a contaminant concentration level higher than the contaminant concentration level at the pinch point $C^{pinch,w}$ must pass through the treatment process. The wastewater streams with contaminant concentration level equal to $C^{pinch,w}$ should partially bypass the treatment process. Finally, the wastewater streams with contaminant concentration level lower than $C^{pinch,w}$ should bypass treatment.

Based on these observations, Wang and Smith (1994b) developed a manual, iterative design method that accounts also for more complex cases such as multiple treatment processes and multicontaminant wastewater streams. Kuo and Smith (1997) found that this method had some drawbacks, particularly if multiple processes and multicontaminant cases were to be solved. They proposed several modifications, which largely improved the rigor of the approach. Generally, this is a sequential procedure that consists of the following steps:

Step 1. Targeting subnetworks for all treatment processes and contaminants.
Step 2. Estimation of mixing exergy losses for contaminants and treatment processes operating in parallel.
Step 3. Selection of treatment process with minimum mixing exergy loss.
Step 4. Re-targeting subnetworks for remaining treatment processes.

Due to the use of exergy losses as the objective the approach does not guarantee an optimal cost network. Statyukha et al. (2008) employed a solution from Kuo and Smith's (1997) method to find an optimal cost network. The case studies in Statyukha et al. (2008) prove that even a simple direct optimization NLP algorithm is able to locate the optimum cost solution if it initializes from such a point.

20.3.2. Optimization-based approaches to water networks

The majority of optimization-based approaches apply superstructure optimization. They differ as models of processes, simultaneous or sequential procedures, optimization techniques, goal functions and technological constraints. Alva-Argaez et al. (1999) were the only group to employ a transshipment formulation in WN design. The main aim of the model was targeting freshwater usage but the results also provide certain structural features that can be applied to develop the final network. They developed an MILP optimization model that accounts for the multicontaminant case with concentration shifts.

In order to explain approaches using the superstructure concept we begin with the superstructure and its optimization model. The superstructure considered in the following is generic since it deals with TWN (or WUTN). Hence, it can also be applied to solve simpler cases such as WRN, WNRR or WTN. It embeds two types of water-using processes and accounts for the multicontaminant case. In order to simplify presentation we limit ourselves to a single freshwater source and a single wastewater disposal site. Extension for multiple freshwater sources and multiple disposal sites is straightforward. Also, we will assume that the number of treatment/regeneration processes is fixed and the treatment technology is known for each process. Each treatment process is performed in a single unit. Notice that the assumptions allow circumvention of optimising the arrangement of treatment units.

To build the superstructure, numerous splitters and mixers are necessary. One mixer is attached at the inlet and one splitter at the outlet of any mass transfer water-using process as well as the regeneration/treatment process. Each source of the nonmass transfer process has a splitter at the outlet and each sink of the nonmass transfer process has a mixer at the inlet. Freshwater from a source is redistributed by a splitter (freshwater splitter). A mixer is associated with the wastewater disposal site (wastewater mixer).

A freshwater splitter can redistribute a certain amount of freshwater to any of the mass transfer processes and any sink of the nonmass transfer processes. Connections between freshwater splitter and regenerators as well as between freshwater splitter and final wastewater mixer are explicitly excluded since they have no economical or technical sense.

Splitters of mass transfer water-using processes, and of sources of nonmass transfer, and of regeneration processes, are connected to mixers of mass transfer processes, sinks of nonmass transfer processes, regenerators and also to the final wastewater mixer. Self-recycles in mass transfer water-using processes and regenerators are included, too. They are commonly inserted in the regeneration processes. Here, we also embed them in mass transfer water-using processes to account for cases of fixed flow rate (retrofit scenario) following the remarks of Wenzel et al. (2002) and Dunn et al. (2001). Wang and Smith (1995) also applied such self-recycles to maintain fixed flow rates via processes. The items of TWN superstructure are shown in Figure 20.17.

Figure 20.18 illustrates the scheme of a TWN superstructure. The symbols are explained in the following.

To formulate the optimization model the following indices, parameters and variables are used.

Index sets:

$I = \{i|\ i$ is contaminant in TWN$\}$
$P = \{p|\ p$ is mass transfer water-using process$\}$
$S = \{s|\ s$ is source of nonmass transfer water-using process$\}$
$D = \{d|\ d$ is sink of nonmass transfer water-using process$\}$
$T = \{t|\ t$ is treatment/regeneration process$\}$

It is convenient to divide set T into two subsets. Subset $T1$ contains indices of treatment units modeled by Equation (20.6) while $T2$ contains those modeled by Equation (20.7). The subsets satisfy conditions (20.17) and (20.18).

$$T1 \cup T2 = T \tag{20.17}$$

$$T1 \cap T2 = \varnothing \tag{20.18}$$

Parameters:

AR—annualized factor for investment on treatment operations
C_{fw}^i—contaminant i concentration in freshwater source

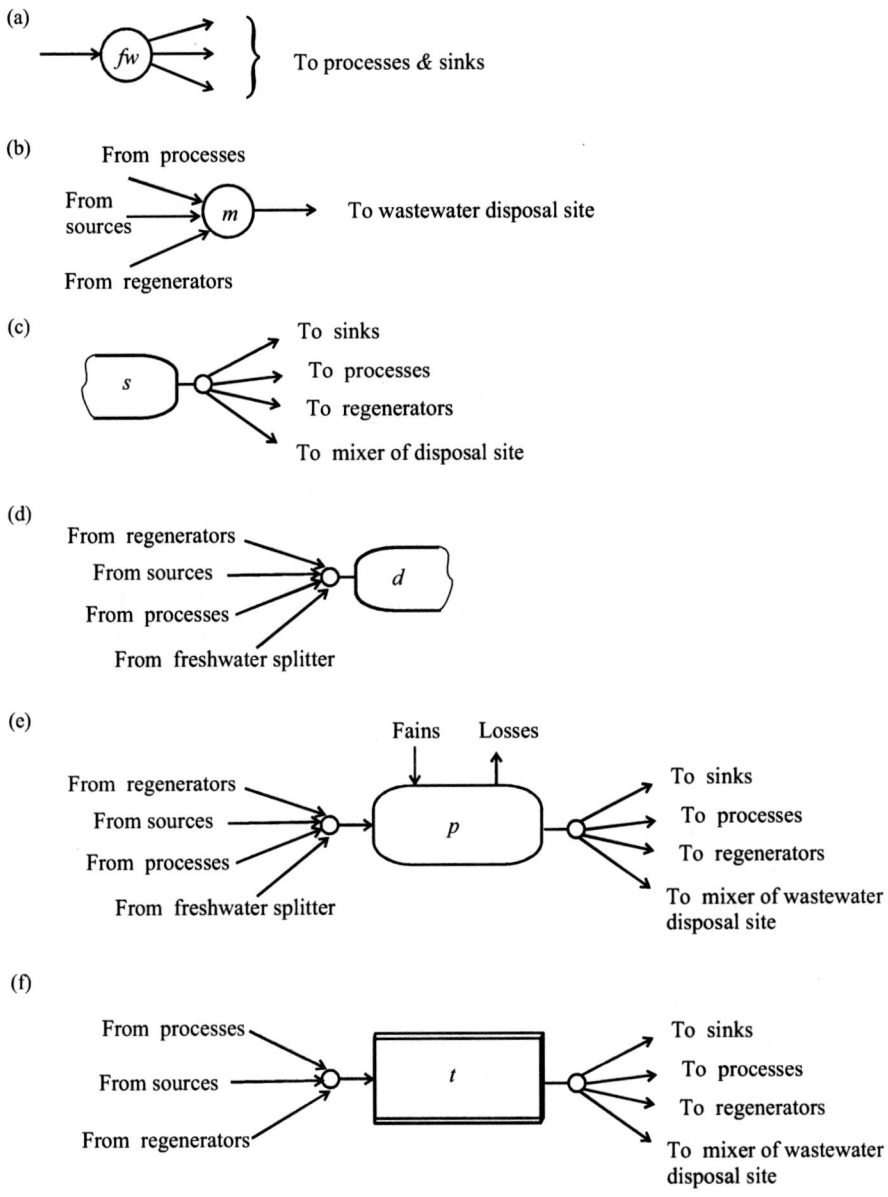

Figure 20.17 Building blocks of TWN superstructure: (a) freshwater splitter; (b) mixer of wastewater disposal site; (c) source of nonmass transfer process; (d) sink of nonmass transfer process; (e) mass transfer process; (f) treatment/regeneration process.

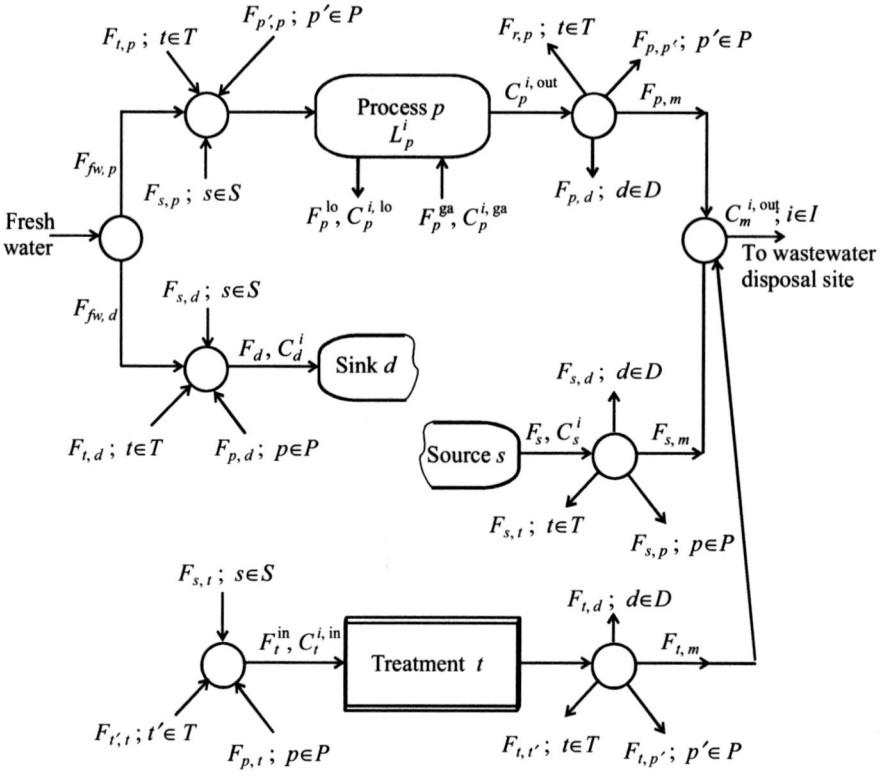

Figure 20.18 Scheme of TWN superstructure.

$C^{i,e}$—contaminant i limiting concentration in wastewater stream to environment

$C_p^{i,in,max}/C_p^{i,out,max}$—maximum value of inlet/outlet contaminant i concentration in mass transfer water-using processes p

$C_p^{i,ga}/C_p^{i,lo}$—contaminant i concentration in gains/losses in mass transfer water-using process p

$C_t^{i,in,max}$—maximum permissible inlet concentration of contaminant i at the inlet to treatment/regeneration processes t

C_s^i/F_s—contaminant i concentration/flow rate for source s

$C_d^{i,max}/F_d$—maximum contaminant i concentration/flow rate for sink d

$C_t^{i,out*}$—outlet concentration of contaminant i at the outlet from treatment/regeneration process t modeled by Equation (20.6)

F_p^{ga}/F_p^{lo}—flow rate of gains/losses in mass transfer water-using processes p

F_t^{max}—maximum permissible flow rate via treatment/regeneration process t

F^*—sufficiently large number, higher than any possible flow rate in the network

H—time of operation per year
L_p^i—mass load of contaminant i transferred to water in mass transfer water-using process p
α_{fw}—unit cost of freshwater
$\beta_{fw,p}/\beta_{fw,d}$—fixed cost of piping section from freshwater splitter fw to mass transfer water-using process p/sink d
$\beta_{t,t'}/\beta_{t,p}/\beta_{t,m}/\beta_{t,d}$—fixed cost of piping section from treatment process t to: treatment process t'/mass transfer water-using process p/mixer of disposal site m/sink d
$\beta_{p,p'}/\beta_{p,t}/\beta_{p,s}/\beta_{p,m}$—fixed cost of piping section from mass transfer water-using processes p to: mass transfer water-using process p'/treatment t/source d/mixer of wastewater disposal site m
$\beta_{s,p}/\beta_{s,t}/\beta_{s,d}/\beta_{s,m}$—fixed cost of piping section from source s to: mass transfer water-using processes p/treatment t/sink d/mixer m of wastewater disposal site.

Notice that the fixed charges listed above should reflect distances between equipment in TWN.

γ_t/η_t—cost parameters in operation/investment expenses on treatment
κ_t—parameter in Equation (20.21) for investment cost of treatment
ψ_t^i—removal ratio of contaminant i for treatment/regeneration processes modeled by Equation (20.8).

For each piping section, parameters β listed above are defined; minimum flow rate values are given, too. They are denoted by $F_{i,j}^{\min}$ with i,j state for indices of mixers and splitters at the operations: t, p, s, d, fw, m.
Variables:

$C_p^{i,\text{in}}/C_p^{i,\text{out}}$—inlet/outlet contaminant i concentration in mass transfer water-using processes p
C_d^i—contaminant i concentration for sink d
$C_m^{i,\text{out}}$—outlet concentration of contaminant i from mixer m of wastewater disposal site
$C_t^{i,\text{in}}$—inlet concentration of contaminant i to treatment/regeneration processes t
$C_t^{i,\text{out}}$—outlet concentration of contaminant i at the outlet from treatment/regeneration process t modeled by Equation (20.7)
F—sufficiently large number
$F_{fw,p}/F_{fw,d}$—flow rate from freshwater splitter fw to mass transfer water-using process p/sink d
$F_{t,t'}/F_t, _p/F_{t,m}/F_{t,d}$—flow rate from treatment process t to treatment process t'/mass transfer water-using process p/mixer m of disposal site/sink d

$F_{p,p'}/F_{p,t}/F_{p,s}/F_{p,m}$—flow rate from mass transfer water-using process p to mass transfer water-using process p'/treatment t/source s/mixer m of wastewater disposal site

$F_{s,p}/F_{s,t}/F_{s,d}/F_{s,m}$—flow rate from source s of nonmass transfer water-using processes to mass transfer water-using process p/treatment t/sink d/mixer m of wastewater disposal site.

Additionally, binary variables are defined. They are applied to control the fixed cost of piping sections and, in consequence, to simplify the structure and reduce the cost of TWN. For each connection a single binary has to be used. In order to make the description concise we will define it in a general way:

$$y_{i,j} = \begin{cases} 1 & \text{if the conection between } i \text{ and } j \text{ exists} \\ 0 & \text{if not} \end{cases} \tag{20.19}$$

with i, j state for indices p, t, s, d, fw, m similarly as in parameters β and $F^{\min}$.

The total cost of TWN consists of the following items: cost of freshwater, cost of treatment operations and cost of piping. Additionally, one can take into account the cost of stream transportation and cost of water disposal. Here, for the sake of simplicity we do not consider them.

Cost of freshwater is given by:

$$\Phi_{fw} = \alpha_{fw} \left(\sum_{p \in P} F_{fw,p} + \sum_{d \in D} F_{fw,d} \right) \tag{20.20}$$

Cost of treatment/regeneration processes consists of operation and investment expenses according to:

$$\Phi_{\text{treat}} = H \sum_{t \in T} \gamma_t F_t + AR \sum_{t \in T} \eta_t (F_t)^{\kappa_t} \tag{20.21}$$

where H denotes operating time per year and AR is return on capital.

Note that, in order to reduce the number of variables in the optimization model, variable flow rates F_t are calculated via basic variables from:

$$F_t = \sum_{s \in S} F_{s,t} + \sum_{p \in P} F_{p,t} + \sum_{t' \in T} F_{t',t}; \quad t \in T \tag{20.22}$$

Finally, the fixed cost of piping sections are calculated from:

$$
\Phi_{\text{pipe}} = AR
\begin{bmatrix}
\begin{cases}
\begin{aligned}
&\sum_{p \in P} \beta_{fw,p} y_{fw,p} + \sum_{p \in P} \sum_{p' \in P} \beta_{p,p'} y_{p,p'} + \\
&\sum_{p \in P} \beta_{p,m} y_{p,m} + \sum_{p \in P} \sum_{t \in T} \beta_{p,t} y_{p,t}
\end{aligned}
\end{cases} \\
\left. + \sum_{p \in P} \sum_{d \in D} \beta_{p,d} y_{p,d} \right\} + \left\{ \sum_{d \in D} \beta_{fw,d} y_{fw,d} \right\} + \\
+ \begin{cases}
\sum_{s \in S} \sum_{p \in P} \beta_{s,p} y_{s,p} + \sum_{s \in S} \sum_{t \in T} \beta_{s,t} y_{s,t} \\
+ \sum_{s \in S} \beta_{s,m} y_{s,m}
\end{cases} \\
+ \begin{cases}
\sum_{t \in T} \sum_{t' \in T} \beta_{t,t'} y_{t,t'} + \sum_{t \in T} \beta_{t,m} y_{t,m}
\begin{array}{l}
+ \sum_{t \in T} \sum_{p \in P} \beta_{t,p} y_{t,p} \\
+ \sum_{t \in T} \sum_{d \in D} \beta_{t,d} y_{t,d}
\end{array}
\end{cases}
\end{bmatrix}
$$

$$(20.23)$$

Note that, in addition to fixed charges, it is possible to account for the operating cost of piping, that is, cost of transportation through them.

The optimization model of the superstructure for the TWN problem is given in the following. To model processes, mixers and splitters we have adapted the formulations of Bagajewicz (2000). The equations are written for arrangements: mixer–process–splitters, mixer–treatment–splitter rather than for individual equipment. Also, flow rates are employed in modeling splitters instead of split ratios. This was also advised by Karuppiah and Grossmann (2006). Additionally, application of flow rates is advantageous in the stochastic optimization approach addressed in the following section.

20.3.2.1. Optimization Model for TWN: TWN–Superstr–Optim

$$\min \text{TAC} = \Phi_{fw} + \Phi_{\text{treat}} + \Phi_{\text{pipe}} \tag{20.24}$$

s.t.

$$
\begin{aligned}
& F_{fw,p} + \sum_{s \in S} F_{s,p} + \sum_{p' \in P} F_{p',p} + \sum_{t \in T} F_{t,p} + F_p^{\text{ga}} - F_p^{\text{los}} \\
& -F_{p,m} - \sum_{p' \in P} F_{p,p'} - \sum_{d \in D} F_{p,d} - \sum_{t \in T} F_{p,t} = 0; \quad p \in P
\end{aligned}
\tag{20.25}
$$

$$
F_{fw,p} \cdot C_{fw}^i + \sum_{s \in S} (F_{s,p} \cdot C_s^i) + \sum_{p' \in P} (F_{p',p} \cdot C_{p'}^{i,\text{out}})
$$

$$
+ \sum_{t \in T2} (F_{t,p} \cdot C_t^{i,\text{out}}) + \sum_{t' \in T1} F_{t',t} C_{t'}^{i,\text{out}*}
$$

$$
+ (F_p^{\text{ga}} \cdot C_p^{i,\text{ga}}) + L_p^i - \left(F_p^{\text{los}} \cdot C_p^{i,\text{los}} \right) \tag{20.26}
$$

$$
- C_p^{i,\text{out}} \cdot \left(F_{p,m} + \sum_{p' \in P} F_{p,p'} + \sum_{d \in D} F_{p,d} + \sum_{t \in T} F_{p,t} \right)
$$

$$
= 0; \quad i \in I, \ p \in P
$$

$$
\sum_{s \in S} \left(F_{s,p} \cdot C_s^i \right) + \sum_{p' \in P} \left(F_{p',p} \cdot C_p^{i,\text{out}} \right)
$$

$$
+ \sum_{t \in T2} \left(F_{t,p} \cdot C_t^{i,\text{out}} \right) + \sum_{t \in T1} \left(F_{t,p} \cdot C_t^{i,\text{out}*} \right)
$$

$$
- \left(\sum_{s \in S} F_{s,p} + \sum_{p' \in P} F_{p',p} + \sum_{t \in T} F_{t,p} + F_{fw,p} \right) \tag{20.27}
$$

$$
\cdot C_p^{i,\text{in}} = 0 \quad i \in I, \ p \in P
$$

$$
\sum_{s \in S} F_{s,t} + \sum_{p \in P} F_{p,t} + \sum_{t' \in T} F_{t',t} - F_{t,m}
$$

$$
- \sum_{p \in P} F_{t,p} - \sum_{d \in D} F_{t,d} - \sum_{t' \in T} F_{t,t'} = 0; \quad t \in T \tag{20.28}
$$

$$
\left(1 - \psi_t^i \right) \sum_{s \in S} \left(F_{s,t} \cdot C_s^i \right)
$$

$$
+ \sum_{t' \in T2} \left(F_{t',t} \cdot C_{t'}^{i,\text{out}} \right)
$$

$$
+ \sum_{t' \in T1} \left(F_{t',t} \cdot C_{t'}^{i,\text{out}*} \right)
$$

$$
+ \left(1 - \psi_t^i \right) \sum_{p \in P} \left(F_{p,t} \cdot C_p^{i,\text{out}} \right) \tag{20.29}
$$

$$
- C_t^{i,\text{out}} \cdot \left(F_{t,m} + \sum_{p \in P} F_{t,p} + \sum_{d \in D} F_{t,d} \right)
$$

$$
= 0; \quad i \in I, \ t \in T2
$$

$$\sum_{s \in S} \left(F_{s,t} \cdot C_s^i \right) + \sum_{p \in P} \left(F_{p,t} \cdot C_p^{i,\text{out}} \right)$$

$$+ \sum_{t' \in R2} \left(F_{t',t} \cdot C_{t'}^{i,\text{out}} \right) + \sum_{t' \in T21} \left(F_{t',t} \cdot C_{t'}^{i,\text{out}*} \right) \tag{20.30}$$

$$- \left(\sum_{s \in S} F_{s,t} + \sum_{p \in P} F_{p,t} + \sum_{t' \in T} F_{t',t} \right) \cdot C_t^{i,\text{in}} = 0; \quad i \in I, \, t \in T$$

$$F_s - \sum_{p \in P} F_{s,p} - \sum_{d \in D} F_{s,d} - \sum_{t \in T} F_{s,t} - F_{s,m} = 0; \quad s \in S \tag{20.31}$$

$$F_{fw,d} + \sum_{p \in P} F_{p,d} + \sum_{s \in S} F_{s,d} + \sum_{t \in T} F_{t,d} - F_d = 0; \quad d \in D \tag{20.32}$$

$$F_{fw,d} \cdot C_{fw}^i + \sum_{p \in P} (F_{p,d} C_p^{i,\text{out}}) +$$

$$\sum_{s \in S} (F_{s,d} C_s^i) + \sum_{t \in T2} (F_{t,d} C_t^{i,\text{out}}) + \sum_{t \in T1} (F_{t,d} C_t^{i,\text{out}*})$$

$$- \underbrace{\left(F_{fw,d} + \sum_{p \in P} F_{p,d} + \sum_{s \in S} F_{s,d} + \sum_{t \in T} F_{t,d} \right)}_{F_d} \tag{20.33}$$

$$\cdot C_d^i = 0; \quad i \in I, \, d \in D$$

$$\sum_{p \in P} (F_{p,m} C_p^{i,\text{out}}) + \sum_{t \in T1} (F_{t,m} C_t^{i,\text{out}*})$$

$$+ \sum_{t \in T2} (F_{t,m} C_t^{i,\text{out}}) + \sum_{s \in S} (F_{s,m} C_s^{i,\text{out}}) \tag{20.34}$$

$$= \left(\sum_{p \in P} F_{p,m} + \sum_{t \in T} F_{t,m} + \sum_{s \in S} F_{s,m} \right) C_m^{i,\text{out}}; \quad i \in I$$

$$F_{fw,p} + \sum_{s \in S} F_{s,p} + \sum_{p' \in P} F_{p',p} + \sum_{t \in T} F_{t,p} \leq F_p^{\max}; \quad p \in P \tag{20.35}$$

$$\sum_{s \in S} F_{s,t} + \sum_{p \in P} F_{p,t} + \sum_{t' \in T} F_{t',t} \leq F_t^{\text{in,max}}; \quad t \in T \tag{20.36}$$

$$F_{p,p'} \leq F^* y_{p,p'}; \quad F_{p,m} \leq F^* y_{p,m}; \quad F_{p,t} \leq F^* y_{p,t}; \quad F_{p,d} \leq F^* y_{p,d}; \quad p \in P \tag{20.37}$$

$$F_{t,t'} \leq F^* y_{t,t'}; \quad F_{t,m} \leq F^* y_{t,m};$$
$$F_{t,p} \leq F^* y_{t,p}; \quad F_{t,d} \leq F^* y_{t,d}; \quad t, t' \in T, \quad p \in P, \quad d \in D \tag{20.38}$$

$$F_{s,p} \leq F^* y_{s,p}; \quad F_{s,m} \leq F^* y_{s,m}; \quad F_{s,t} \leq F^* y_{s,t}; \quad p, p' \in P, \ s \in S, \ t \in T \tag{20.39}$$

$$F_{fw,p} \leq F^* y_{fw,p}; \quad p \in P \tag{20.40}$$

$$F_{fw,d} \leq F^* y_{fw,d}; \quad d \in D \tag{20.41}$$

$$F_{t,t'} \geq F_{t,t'}^{\min} y_{t,t'}; \quad F_{t,m} \geq F_{t,m}^{\min} y_{t,m};$$
$$F_{t,p} \geq F_{t,p}^{\min} y_{t,p}; \quad F_{t,d} \geq F_{t,d}^{\min} y_{t,d}; \quad t, t' \in T, \ p \in P, \ d \in D \tag{20.42}$$

$$F_{s,p} \geq F_{s,p}^{\min} y_{s,p}; \quad F_{s,m} \geq F_{s,m}^{\min} y_{s,m}; \quad F_{s,t} \geq F_{s,t}^{\min} y_{s,t}; \quad s \in S, \ p \in P, \ t \in T \tag{20.43}$$

$$F_{fw,p} \geq F_{fw,p}^{\min} y_{fw,p}; \quad p \in P \tag{20.44}$$

$$F_{fw,d} \geq F_{fw,d}^{\min} y_{fw,d}; \quad d \in D \tag{20.45}$$

$$C_p^{i,\text{in}} \leq C_p^{i,\text{in,max}}; \quad i \in I, p \in P \tag{20.46}$$

$$C_p^{i,\text{out}} \leq C_p^{i,\text{out,max}}; \quad i \in I, p \in P \tag{20.47}$$

$$C_d^i \leq C_d^{i,\text{max}}; \quad i \in I, d \in D \tag{20.48}$$

$$C_t^{i,\text{in}} \leq C_t^{i,\text{in,max}}; \quad t \in T \tag{20.49}$$

$$C_m^{i,\text{out}} \leq C^{i,e}; \quad i \in I \tag{20.50}$$

Equations (20.25), (20.26), and (20.27) are mass balances of mass transfer water-using processes and associated mixers and splitters. The first equation (20.25) is the general mass balance of the arrangement mixer–process–splitter, while the second (20.26) is the mass balance of contaminants. The third equation (20.27) defines mass balances of contaminants in the mixer. Balances for other processes are organized in the same way. The general mass balance of the sequences mixer–treatment–splitter is defined by Equation (20.28) while Equation (20.29) is the mass balance of contaminants. Notice that the latter is necessary only for those treatment processes which belong to subset T2. Contaminant mass balances of mixers attached to treatment processes are given by Equation (20.30). For sources of nonmass transfer processes, mass balances of their splitters are sufficient (Equation (20.31)). Equations (20.32) and (20.33) model the arrangements mixer-source of a nonmass transfer process. The last equality (20.34) is the mass balance of contaminants for the mixer of the wastewater disposal site.

Inequalities (20.35) and (20.36) impose the upper limits on flow rates via mass transfer water-using processes and treatment processes. Inequalities (20.37)–(20.41) are logical conditions that force flow rates via piping sections to zero if the associated binary variable is zero. They are inactive if binary variables are equal to zero. Inequalities (20.42)–(20.45) ensure that flow rates via piping sections will not be smaller than fixed lower limits. Conditions on maximum permissible contaminant concentrations in processes, sources and treatment processes are defined by inequalities (20.46)–(20.49). Finally, inequality (20.50) ensures that environmental constraints on contaminant concentrations in wastewater to disposal are met.

Additional conditions can also be included. For instance, conditions on forbidden and/or must-be connections have the simple form: $y_{ij} = 0$/$y_{ij} = 1$. Note that they are very useful in a retrofit scenario. The binaries are also helpful in simplifying network topology. The number of sources that can feed operations and the number of inlet connections for operations can be restricted as shown in, for example, Gunaratnam et al. (2005).

As we have mentioned in the preceding, the model is generic and can be directly applied to simpler problems as follows:

- WNRR—remove mixer of wastewater disposal site (note then that index m will denote parameters and variables of wastewater streams to final treatment, and index t will denote regenerators)
- WRN—remove mixer of wastewater disposal site and eliminate variables with index t ($T = \varnothing$)
- WTN—eliminate variables with indices p, s, d (P, S, $D = \varnothing$) except those which have m as the second index, also remove freshwater source and the associated splitter.

To design water networks with mass transfer water-using processes only, it is sufficient to put $S = \varnothing$ and $D = \varnothing$. Likewise, for nonmass transfer processes, set P is an empty set.

It is of importance that the model constraints become linear only for WRNs consisting of nonmass transfer water-using processes with a single contaminant. This results from the fact that bi-linear terms in mass balances for contaminants, that are the source of nonlinearity, vanish. In other cases the model is nonlinear. This is the reason for limiting insight-based approaches to a single contaminant. The model is, additionally, of the MINLP type if fixed costs of piping sections or restrictions on junctions are to be accounted for. Due to model complexity it is very difficult to solve it to global optimum with currently available optimization methods, particularly for larger problems.

Approaches applying the superstructure differ in many respects: goal function, scope of the superstructure (TWN, WRN, WNRR and WTN), one-stage or multistage procedure, type of technological constraints, solvers used and so on. Total water network problems have been tackled in relatively few papers, for instance Takama et al. (1980), Huang et al. (1999), Benkő et al. (1999), Gunaratnam et al. (2005), Chang and Li (2005) and Karuppiah and Grossmann (2006). Koppol et al. (2003) investigated the influence of final concentrations of effluents discharged to the environment on the cost of a total system in a WNRR problem. Many works have addressed WRN; fewer contributions have aimed at WNRR. Alva-Argaez et al. (1998, 1999), Bagajewicz et al. (2000), Yang et al. (2000), Bagajewicz and Savelski (2001), Tsai and Chang (2001), Savelski and Bagajewicz (2001), Koppol et al. (2003), Lee and Grossmann (2003), Shakhnovskij et al. (2004), Ullmer et al. (2005), and Gunaratnam et al. (2005) applied deterministic optimization solvers to solve the superstructure optimization model. However, several authors decided to apply meta-heuristic/stochastic approaches to cope with nonlinearities. A genetic algorithms framework was applied by Tsai and Chang (2001), Prakotpol and Srinophakun (2004), Shafiei et al. (2004) and Lavric et al. (2005). Poplewski (2004) decided to employ an adaptive random search technique (see Section 1.2)—this will be given great consideration in the following.

Only a few works have employed a cost function since it results in the MINLP problem. Even a NLP problem with freshwater minimization may cause problems for existing deterministic solvers. The most common practice to circumvent the difficulty is to fix contaminant concentration in the outlet stream at the maximum value. This works well in the case of a single contaminant. Notice that in fact this assumption is inherent for water pinch methodology and approaches using the source–sink concept. Savelski and Bagajewicz (2000, 2003) gave rigorous proof, first for a single contaminant and then for a multiple one, that the condition ensures reaching minimum freshwater intake. The problem is that in the latter case the maximum concentration condition concerns one from the existing contaminants. This limiting contaminant may vary from process to process and there is no information as to what it is. In a series of papers Bagajewicz and co-workers developed a tree search procedure that requires solution of a series of MILP problems. The method is rigorous and does not require global optimization of NLP solvers. However, it only guarantees networks with minimum freshwater consumption: not necessarily with minimum cost. Due to two-stage optimization the solutions also feature minimum cost of piping but on condition that freshwater consumption reaches its global minimum. This is not equivalent to minimization of the total cost since cost minimization may require an increase of freshwater intake above the global minimum. Although under current cost relations the cost of freshwater almost always prevails, there may exist industrial problems with high charges on piping and transportation. An example is shown in Lee and Grossmann (2003); as well as Example 20.4 in the following, even though it does not seem to be an industrial case study.

The condition of fixed contaminant concentration applied to regeneration processes was also used by Koppol et al. (2003) to attack the WNRR problem. A similar tree search procedure was applied for both water-using processes and regenerators. The approach seems to call for a large computation load in the case of large-scale industrial problems.

The optimization approaches to the WTN problem suggested to date are similar to those for TWN/WRN/WNRR; see Galan and Grossmann (1998), Lee and Grossmann (2002), Hernandez-Suarez et al. (2004) and Meyer and Floudas (2006). Their contributions employed deterministic optimization techniques.

Here, we will describe in more detail a method for solving the WNRR problem proposed by Poplewski (2004) in his PhD thesis. Basic ideas, particularly on WRN design, can also be found in Poplewski et al. (2002), Poplewski and Jeżowski (2003, 2005) and Jeżowski et al. (2003d, 2007). The approach applies the stochastic random search optimization LJ–FR algorithm described in Chapter 1.

Though the WNRR problem is slightly simpler than the TWN problem it is still an advantageous task, particularly for the ARS technique due to the existence of binaries. To cope with large-scale cases Poplewski developed solutions to the following subproblems on how to:

- cope with numerous binaries
- deal with constraints
- find a feasible starting point and "good" initial search sizes of variables.

The equality constraints were solved directly in a sequential manner according to the general scheme from Section 1.5 and explained in the following for the water network case.

To deal with the mixed integer problem, Poplewski treated binary variables as dependent ones. This was necessary since ARS techniques aren't efficient tools for discrete valued models on a large scale. For any flow rate, which is a decision variable given a value less than the fixed minimum by the algorithm, the value of zero is assigned to the associated binary variable. In the opposite case the binary is kept at 1. This was coded in the LJ–FR method by replacing constraints (20.43) to (20.45) by logical statements written here in pseudo-code as follows:

$$\text{IF } F_{i,j} < F_{i,j}^{\min} \text{ THEN}$$

$$y_{i,j} \wedge F_{i,j} \Rightarrow 0 \tag{20.51}$$

$$\text{ELSE } y_{i,j} \Rightarrow 1$$

with i, j state for indices of the decision parameters.

If flow rates $F_{i,j}$ are not the decision variables but dependent ones, such a scheme cannot be applied since it would result in invalid mass balances. Hence, for dependent variables, the logical statements have the form:

$$\text{IF } F_{i,j} < F_{i,j}^{\min} \text{ THEN } y_{i,j} \Rightarrow 0 \text{ ELSE } y_{i,j} \Rightarrow 1 \tag{20.52}$$

It is important to note that logical conditions (20.37)–(20.41) in the MINLP model are not necessary and are removed from problem formulation. Note that such logical conditions are a source of difficulties in the solution of MINLP or

MILP problems by deterministic techniques. They cause a "gap" effect since it is difficult to achieve close estimation of the F^* parameter and usually a very large number is employed. Poplewski found that application of $F^{\min}$ parameters greater than zero had an advantageous effect on optimization efficacy. He advised setting the parameters greater than zero, even for problems with goal functions that do not require binary variables. The lower bounds on flow rates can be estimated by economic and hydraulic considerations similar to those in Savelski and Bagajewicz (2001).

Next, Poplewski developed a way of dealing with equality constraints by sequentially solving them as linear equations and sets of simultaneous linear equations. All nonlinearities in the constraints of the WNRR problem are caused by bilinear terms—products of flow rates and concentrations. By choosing flow rates of some water streams as the decision variables and making use of sequential solution of constraints in the ARS approach it was possible to solve all equality constraints as linear ones with regard to concentrations.

The solution procedure for equalities is, hence, as follows.

Decision variables:

$$F_{fw,p}, \; F_{p,p'}, \; F_{p',p}, F_{p,d}, F_{p,t}, \; F_{s,p}, \; F_{s,d}, F_{s,t}, F_{t,d}, F_{t,p}, F_{t,t'}, F_{t',t}$$

Dependent variables calculated from equations and logical statements:

$$F_{p,m}, F_{s,t}, F_{t,m}, F_{fw,d}, C_p^{i,\text{out}}, C_p^{i,\text{in}}, C_d^i, C_t^{i,\text{in}}, C_t^{i,\text{out}} \;\; (t \in T1), \text{all binaries}$$

For each decision variable generated by the LJ–FR algorithm logical statement (20.51) is used to determine values of binary variables associated with them.

The dependent variables are calculated in the sequence shown below.

1. Calculate the values of flow rates from processes to final treatment ($F_{p,m}$) from Equations (20.53a) and determine binaries from Equation (20.53b). Equations (20.53a) are obtained from Equation (20.25). Equations (20.53a) and (20.53b) are solved in sequential fashion.

$$F_{p,m} = F_{fw,p} + \sum_{s \in S} F_{s,p} + \sum_{p' \in P} F_{p',p} + \sum_{t \in T} F_{t,p}$$

$$+ F_p^{\text{ga}} - F_p^{\text{los}} - \sum_{p' \in P} F_{p,p'} - \sum_{d \in D} F_{p,d} \tag{20.53a}$$

$$- \sum_{t \in T} F_{p,t}; \quad p \in P$$

$$y_{p,m} \Rightarrow 0 \quad \text{if} \quad F_{p,m} = 0.0 \quad \text{else} \quad y_{p,m} \Rightarrow 1; \quad p \in P \tag{20.53b}$$

2. Calculate the values of flow rates from regenerators to final treatment ($F_{t,m}$) from Equations (20.54a). Equations (20.54a) are obtained from Equation (20.28). Deter-

mine binary variables from Equation (20.54b). Flow rates and binaries are calculated in sequence.

$$F_{t,m} = \sum_{s \in S} F_{s,t} + \sum_{p \in P} F_{p,t} + \sum_{t' \in T} F_{t',t} - \sum_{p \in P} F_{t,p} - \sum_{d \in D} F_{t,d} - \sum_{t' \in T} F_{t,t'} \quad t \in T \quad (20.54a)$$

$$y_{t,m} \Rightarrow 0 \quad \text{if} \quad F_{t,m} = 0.0 \quad \text{else} \quad y_{t,m} \Rightarrow 1; \quad t \in T \quad (20.54b)$$

3. Calculate concentrations $C_t^{i,\text{out}}$, $C_t^{i,\text{out}}$ by solving the set of simultaneous equations (20.26) and (20.29).
4. Solve Equation (20.55) in sequence to determine concentrations C_p^{in}. The equations are obtained from balances (20.27).

$$C_p^{i,\text{in}} = \frac{\sum_{s \in S} \left(F_{s,p} \cdot C_s^i \right) + \sum_{p' \in P} \left(F_{p',p} \cdot C_{p'}^{i,\text{out}} \right) + \sum_{t \in T} \left(F_{t,p} \cdot C_t^{i,\text{out}} \right) + \sum_{t \in T1} \left(F_{t,p} \cdot C_t^{i,\text{out}*} \right)}{\left(\sum_{s \in S} F_{s,p} + \sum_{p' \in P} F_{p',p} + \sum_{t \in T} F_{t,p} + F_{fw,p} \right)};$$

$$i \in I, \, p \in P \quad (20.55)$$

5. Solve Equation (20.56) in sequence to determine concentrations $C_t^{i,\text{in}}$. The equations are obtained from balances (20.30).

$$C_t^{i,\text{in}} = \frac{\sum_{s \in S} \left(F_{s,t} \cdot C_s^i \right) + \sum_{p \in P} \left(F_{p,t} \cdot C_p^{i,\text{out}} \right) + \sum_{t' \in T} \left(F_{t',t} \cdot C_{t'}^{i,\text{out}} \right) + \sum_{t' \in T1} \left(F_{t',t} \cdot C_{t'}^{i,\text{out}*} \right)}{\left(\sum_{s \in S} F_{s,t} + \sum_{p \in P} F_{p,t} + \sum_{t' \in T} F_{t',t} \right)};$$

$$i \in I, \, t \in T$$

$$(20.56)$$

6. Solve Equations (20.57a), obtained from Equation (20.31), and determine binary variables from (20.57b) in sequence.

$$F_{s,m} = F_s - \sum_{p \in P} F_{s,p} - \sum_{d \in D} F_{s,d} - \sum_{t \in T} F_{s,t}; \quad s \in S \quad (20.57a)$$

$$y_{s,m} \Rightarrow 0 \quad \text{if} \quad F_{s,m} = 0.0 \quad \text{else} \quad y_{s,m} \Rightarrow 1; \quad s \in S \quad (20.57b)$$

7. Solve Equations (20.58a), obtained from Equation (20.32), and determine binary variables from Equation (20.58b) in sequence.

$$F_{fw,d} = F_d - \left(\sum_{p \in P} F_{p,d} + \sum_{s \in S} F_{s,d} + \sum_{r \in R} F_{r,d} \right); \quad d \in D \quad (20.58a)$$

$$y_{fw,d} \Rightarrow 0 \quad \text{if} \quad F_{fw,d} = 0.0 \quad \text{else} \quad y_{fw,d} \Rightarrow 1; \quad d \in D \qquad (20.58b)$$

8. Solve Equation (20.59) in sequence to determine concentrations C_d^i, Equations (20.59) are obtained from Equation (20.33).

$$C_d^i = \frac{\begin{aligned}&F_{fw,d} \cdot C_{fw}^i + \sum_{p \in P}\left(F_{p,d} \cdot C_p^{i,\text{out}}\right) \\ &+ \sum_{s \in S}\left(F_{s,d} \cdot C_s^i\right) + \sum_{t \in T}\left(F_{t,d} \cdot C_t^{i,\text{out}}\right) + \sum_{t \in T1}\left(F_{t,d} \cdot C_t^{i,\text{out}*}\right)\end{aligned}}{F_d};$$

$$i \in I,\, d \in D \qquad\qquad\qquad (20.59)$$

Inequality constraints are dealt with by the death penalty scheme. After each step of the sequential procedure outlined above inequality constraints in the model are checked, and, if not met, new values of decision variables are generated according to the LJ–FR algorithm. Next, Poplewski proposed a very simple initialization scheme for the LJ–FR algorithm. He applied a network with no reuse and zero flow rates to regenerations, that is, parallel feeding with freshwater of all water-using processes. The flow rates of freshwater streams to processes are then calculated from:

$$F_{fw,p/d}^{ps} = \max_{i \in I}\left(\frac{L_p^i}{C_p^{i,\text{out}}}\right); \quad i \in I,\, p \in P,\, s \in S \qquad (20.60)$$

The upper limits on flow rates are calculated from:

$$F_{fw,p}^u = \sum_{p \in P} F_{fw,p}^{ps} \qquad (20.61)$$

The final search regions for each decision variable were fixed at values of the order 10^{-3}–10^{-4}. These values were recommended in Jeżowski and Bochenek (2002) and Jeżowski and Jeżowska (2003). They also make physical sense since they correspond to flow rates of the order 1–$0.1\,\text{kg}\,\text{h}^{-1}$ in many industrial problems. Such small values have no practical meaning. The initializing procedure provides a poor feasible starting solution in terms of goal function value, and upper bounds on the decision variables are not tight. However, it is very simple and ensures that no feasible solution will be eliminated.

The initial starting point and the search region sizes provided satisfactory robustness and efficiency of optimization for WNRR and WRN design problems.

To illustrate the capabilities of the approach we will present some examples on WRN and WNRR problems from amongst several solved by Poplewski. It is worth noting that he was able to achieve the global optimum in most cases or at least a network very close to it.

Example 20.3, WRN in a crude oil refinery system, has three contaminants and a nonlinear goal function—a case rather seldom solved by other approaches.

Table 20.5 Data for Example 20.3

Process number	Contaminant	L_p^i(kgh^{-1})	$C_p^{i,in,max}$(ppm)	$C_p^{i,out,max}$(ppm)
1 Distillation (steam stripping)	Hydrocarbon	0.675	0	15
	H$_2$S	18.000	0	400
	Salt	1.575	0	35
2 Hydrodesulfurization	Hydrocarbon	3.400	20	120
	H$_2$S	414.800	300	12500
	Salt	4.590	45	180
3 Desalter	Hydrocarbon	5.600	120	220
	H$_2$S	1.400	20	45
	Salt	520.800	200	9500

The problem was first considered by Wang and Smith (1994a) and then in Alva-Argaez et al. (1998). Savelski and Bagajewicz (2003) and Doyle and Smith (1997) also solved it but with freshwater usage as the goal function. The data are given in Table 20.5. The goal function (20.62) consists of freshwater cost—the first term in Equation (20.62), and of investment cost (second term), as well as operation cost (third term) of the final water treatment.

$$\min \left(8600 \sum_{p \in P} F_{fw,p} + 34200 \left(\sum_{p \in P} F_{p,t} \right)^{0.7} + 860 \cdot 1.0067 \sum_{p \in P} F_{p,t} \right) \quad (20.62)$$

The best solution from Doyle and Smith (1997) shown in Figure 20.19a features a cost of 1.277×10^6 \$/y. Poplewski's solution, Figure 20.19b, reduced the cost by 676 \$/y.

Savelski and Bagajewicz (2003) reported the same network as the optimum with regard to freshwater usage. A much worse solution with a cost of 1.322×10^6 \$/y was achieved by Alva-Argaez et al. (1999). The network in Figure 20.19b has the connection with small flow rate (0.0671 t h^{-1}). In order to eliminate it, constraints (20.43) were added to the model and the method yielded the network in Figure 20.19a.

The data for Example 20.4 first presented in Olesen and Polley (1997) are given in Table 20.6. The WRN problem has six water-using processes, a single contaminant and a single freshwater source with contaminant concentration equal to zero. Olesen and Polley (1997), as well as Savelski and Bagajewicz (2001), applied freshwater consumption as the goal function. The optimal network obtained in both works was also reached by Poplewski's approach. It consumes 157.14 t h^{-1} of freshwater.

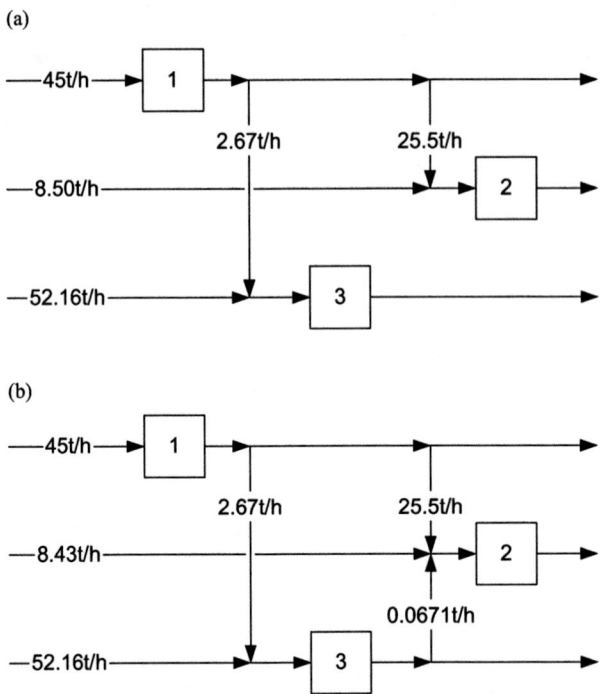

Figure 20.19 The solutions for Example 20.3: (a) from Doyle and Smith (1997), (b) from the method of Poplewski.

Lee and Grossmann (2003) applied the same process data but used cost function (20.63), which is the sum of freshwater cost and cost of all interconnections.

$$\min\left(\sum_{p \in P} F_{fw,p} + \sum_{p,r \in P} 10y_{r,p}\right) \qquad (20.63)$$

Applying advanced disjunctive programming, Lee and Grossmann (2003) reached the global optimum shown in Figure 20.20a: WRN with nine connections and freshwater usage of $160\,\mathrm{t\,h^{-1}}$, that is, with the cost of 250. Note that,

Table 20.6 Data for Example 20.4

Process number p	$L_p^i \,(\mathrm{kgh^{-1}})$	$C_p^{i,\mathrm{in,max}}$ (ppm)	$C_p^{i,\mathrm{out,max}}$ (ppm)
1	2.0	25	80
2	5.0	25	100
3	4.0	25	200
4	5.0	50	100
5	30.0	50	800
6	4.0	400	800

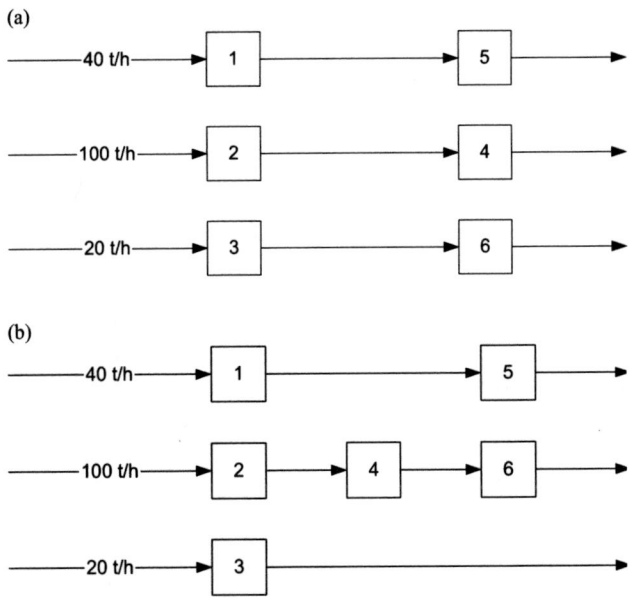

Figure 20.20 Cost optimal networks for Example (20.4): (a) solution from Lee and Grossmann (2003); (b) the second global optimum calculated by Poplewski.

due to the costly interconnections, the freshwater usage is larger than the minimum value ($157.14\,t\,h^{-1}$). The basic approach of Poplewski described in the preceding was not able to calculate the global optimum. The reason was that the flow rates via piping segments "processes–final treatment" are dependent variables. Hence the algorithm was not able to minimize the number of all connections (however, it did minimize the number of connections among processes). To cope with the difficulty Poplewski added the penalty term (20.64) to the original goal function (20.63), which penalizes connections to treatment with small flow rates.

$$\sum_{p \in P} \ln\left(1 + F_{p,t}\right) \tag{20.64}$$

With the augmented goal function it was possible to calculate the solution of Lee and Grossmann (2003) and also the second global optimum shown in Figure 20.20b. This network has the same goal function (freshwater consumption and number of pipes) but a different structure.

The next problem, Example 20.5, was first addressed in Wang and Smith (1995). It is an interesting industrial case study of specialty chemicals. Five processes, listed in Table 20.7 together with necessary data, require water. The goal is to minimize freshwater usage. The single pollutant, solid suspension, is taken into account in all the processes. Two processes feature water loss and one

Table 20.7 Data for Example 20.5 from Wang and Smith (1995)

Process/apparatus	Inlet water stream $(t\,h^{-1})$	Outlet water stream $(t\,h^{-1})$	C_{in}^{max} (ppm)	C_{out}^{max} (ppm)
Reactor	80	20	100	1000
Cyclone	50	50	200	700
Filtration	10	40	0	100
Steam boiler	10	10	0	10
Cooling system	15	5	10	100

process, filtration, water gain. To take into account the retrofit scenario, that is, to meet mass balances of existing processes, Wang and Smith (1995) fixed inlet and outlet flow rates of water streams in the processes.

Hence, in order to solve the example by Poplewski's approach, it is necessary to apply the procedure for a retrofit scenario.

The solution by Wang and Smith (1995) presented in Figure 20.21a features a freshwater flow rate of $90.7\,t\,h^{-1}$. They noticed that the total freshwater usage is likely to be larger by $5\,t\,h^{-1}$ since the application of wastewater from the cooling system, polluted by corrosion products, as the feed to the reaction can be forbidden. Thus, Poplewski included this restriction as the forbidden match in the optimization model. The network shown in Figure 20.21b was calculated. The solution has the same freshwater consumption $(90.69\,t\,h^{-1})$ but the match cooling system-reaction is eliminated. Average CPU time with a Pentium III 800 MHz processor was of the order of 14 s.

Example 20.6 concerns the design of a water network with reuse and regeneration for a paperboard mill. The problem was first addressed in Koppol et al. (2003). The mill consists of chests for pulping and diluting paper fibers, a paper machine, showers and added treatments. Paper fibers are the single contaminant of water, which is measured by total suspended solids (TSS). In the paper machine, the paper fibers are removed and the white-water of two different concentrations exit the process. This machine is considered as consisting of two processes (rich and lean). Though they are in fact regenerators in our approach they are modeled as water-using processes but with diverse mass transfer direction, that is, the contaminant load has a negative sign in the balances. The rich white-water has a high concentration of fibers and exits the paper machine into a silo. The lean white-water with a lower concentration of fibers comes from suction boxes. Note that to reach a throughput of 600 t/day the flow rates through rich and lean processes of the paper machine have to be kept fixed. The other processes with numbers 1, 4, 5 and 6 in Table 20.8 are typical water-using processes of the mass transfer type. The data for processes in the mill are gathered in Table 20.8.

Koppol et al. (2003a) considered typical treatment processes applied in the pulp and paper industry:

- dissolved air flotation (called physical treatment)
- combination of physical treatment and membrane cleaning up (ultrafiltration).

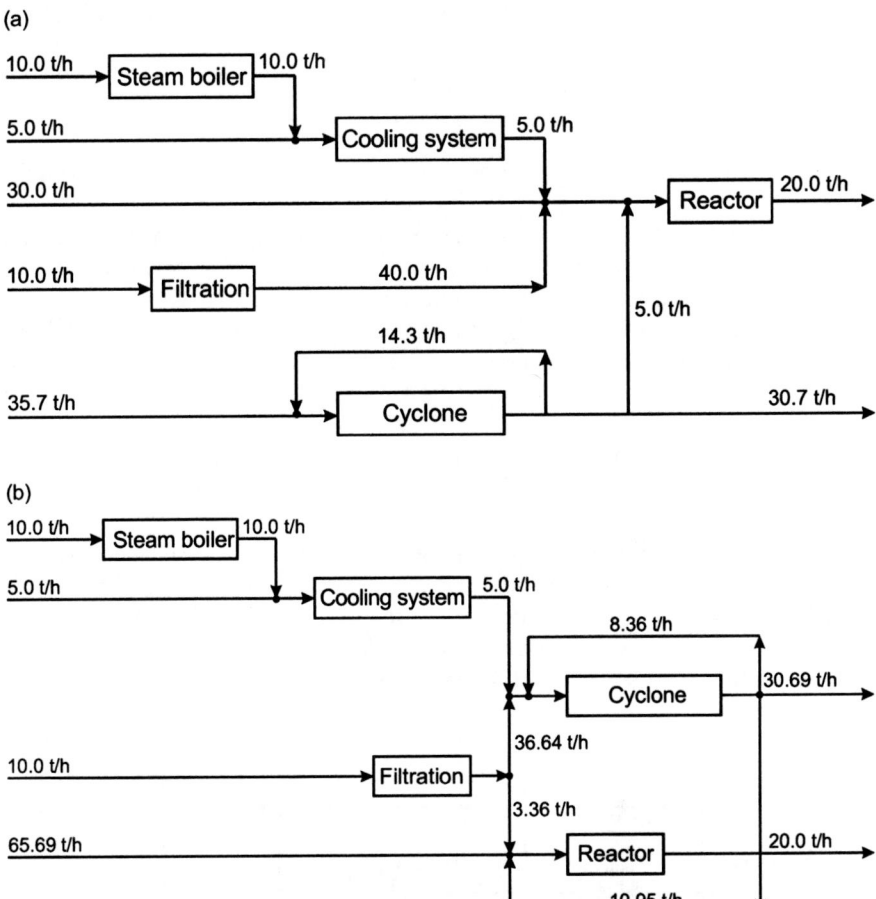

Figure 20.21 Solutions to Example 20.5: (a) the network from Wang and Smith (1995); (b) the solution by Poplewski with forbidden match cooling system-reaction.

Table 20.8 Data for processes in the paper mill—Example 20.6

Process	Load (kg h^{-1})	C_{in}^{max}(ppm)	C_{out}^{max}(ppm)
1. Pulping/dilution	24800	500	5000
2. Paper machine (rich)	−18225	5000	500
3. Paper machine (lean)	−6480	5000	200
4. Deckle showers	21.67	100	500
5. Cylinder showers	37.50	300	600
6. Felt showers	3.33	20	100

Treatment type	$C_{t,out}^{min}$(ppm)	Cost ($\$ \, t^{-1}$)
1. Physical	30	0.15
2. Physical and membrane	2	0.9

Both regenerators are modeled according to Equation (20.6) with limiting outlet concentrations as follows:

- 30 ppm for physical regeneration
- 2 ppm for combined physical and membrane treatment.

The final centralized treatment consists of a sequence of physical and biological treatments.

The aim of the project is to design an optimal WNRR with regard to operation cost, which consists of the cost of freshwater and the cost of treatment. The cost of treatment by a physical regenerator is $0.15\,\$\,t^{-1}$ and that by a physical and membrane regenerator is 0.9 $/t. The cost of final treatment of wastewater which passes through a regenerator is $0.35\,\$\,t^{-1}$. The cost of freshwater and final treatment (without regenerators) is $1.65\,\$\,t^{-1}$. Operating time amounts to 350 days per year.

The network with parallel splitting of freshwater stream supplied at processes consumes $1492\,t\,h^{-1}$ of freshwater and has an operating cost of 20.68 million $/y.

Following Koppol et al. (2003), Poplewski first solved the case with reuse only. The network obtained uses $363.9\,t\,h^{-1}$ of freshwater and has an operating cost of 5.04 million $/y. The solution is presented in Figure 20.22.

Next, physical treatment (dissolved air flotation) was included as a regenerator, that is, wastewater reuse was allowed but a combined physical and membrane

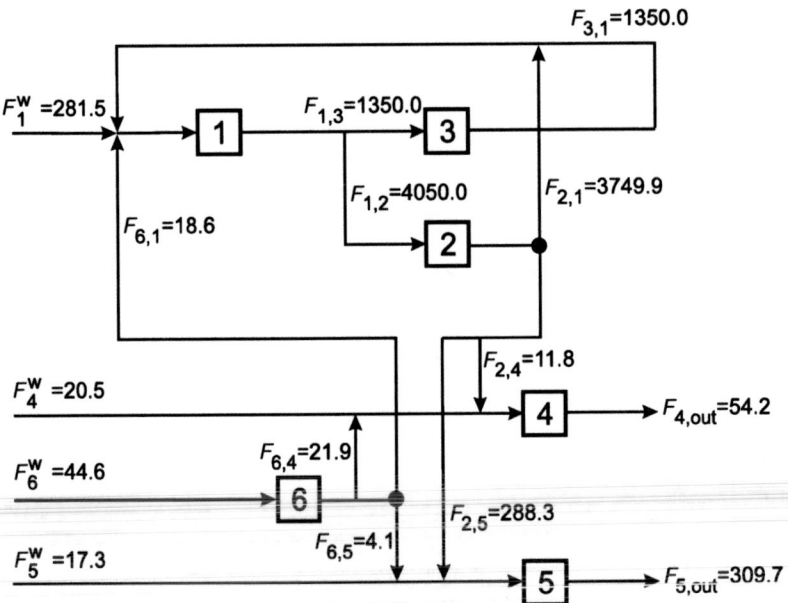

Figure 20.22 Water network with reuse for Example 20.6.

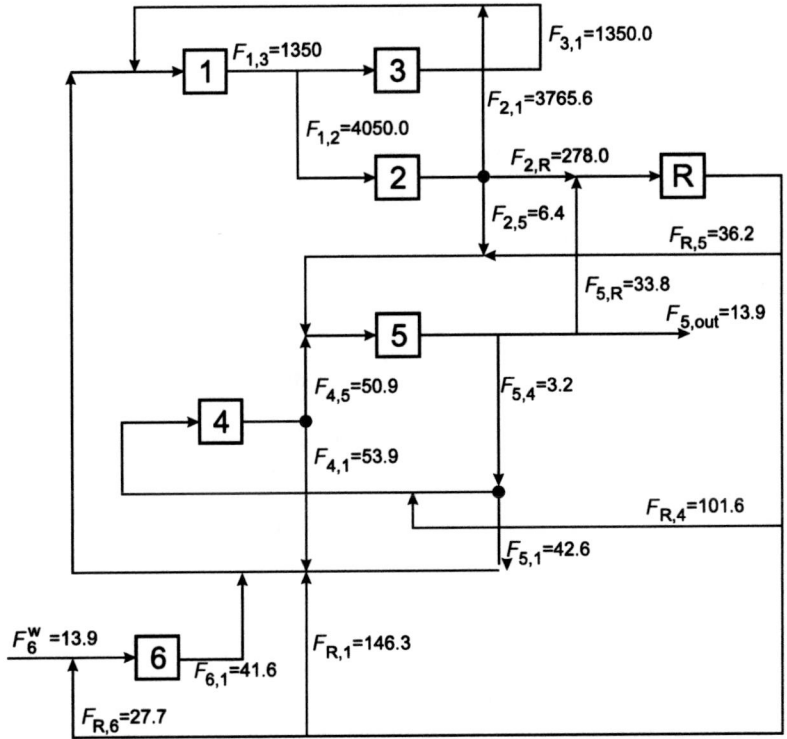

Figure 20.23 Water network with reuse and physical regeneration for Example 20.6.

regenerator was not considered. The optimal network for such assumptions features largely reduced freshwater consumption, that is, the freshwater needed amounts to $13.875\,\text{t}\,\text{h}^{-1}$. $311.8\,\text{t}\,\text{h}^{-1}$ of wastewater is processed by the physical regenerator and the discharge to final treatment amounts to $13.875\,\text{t}\,\text{h}^{-1}$. The network is shown in Figure 20.23.

Finally, the case with water reuse and combined physical and ultrafiltration treatment as regenerators was solved. The consumption of freshwater falls to zero, that is, a closed cycle of wastewater was reached. The network is shown in Figure 20.24. The stream of wastewater to physical and membrane treatment is $293.9\,\text{t}\,\text{h}^{-1}$. Due to the high cost of the membrane regeneration the operation cost of this networks is 2.22 million $/y. The same solution was reported in Koppol et al. (2003) as the optimum. It was achieved with the use of a time-consuming sequential procedure in the form of a decision tree.

Finally, we present here Example 20.7 to illustrate the solution to the WRN problem with both mass transfer and nonmass transfer processes. The problem was taken from Prakash and Shenoy (2005a) and comprises four sources, four sinks and four mass transfer water-using operations. The data are presented in Table 20.9. The goal is to minimize freshwater usage. The solution achieved by Prakash and Shenoy (2005a) consumes $155\,\text{t}\,\text{h}^{-1}$ of freshwater.

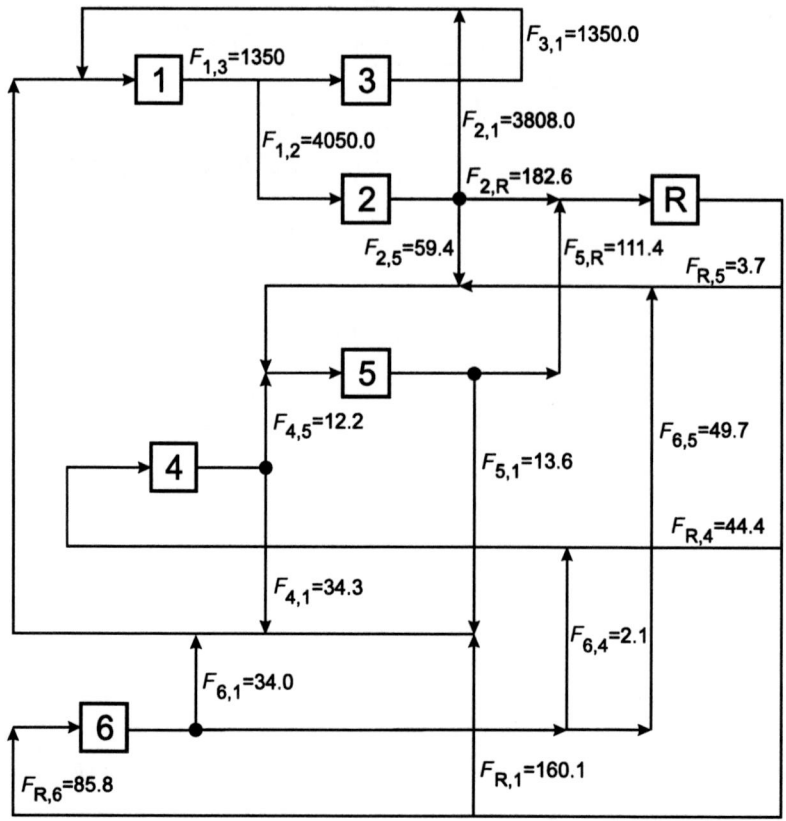

Figure 20.24 Water network with reuse and combined regeneration for Example 20.6.

Table 20.9 Example 20.7—data for sources and sinks

Source	Flow rate (t h^{-1})	C_{out} (ppm)
1	50	20
2	100	50
3	80	100
4	70	200

Sink	Flow rate (t h^{-1})	C_{in} (ppm)
1	50	50
2	100	100
3	70	150
4	60	250

Table 20.10 Example 20.7—data for mass-transfer processes

Process	Contaminant mass load (kg h^{-1})	C_{in}^{max} (ppm)	C_{out}^{max} (ppm)
1	2	0	100
2	5	50	100
3	30	50	800
4	4	400	800

Table 20.11 Optimal solution by Poplewski's approach to Example 20.7

$F_{i,j}$		Source					Process			
		F_{fw}	1	2	3	4	1	2	3	4
Sink	$F_{s,t}$				35.78	38.92			37.84	20.68
	1	30								
	2	48.63	2.74					26.66		
	3			57.75				22.25		
	4			7.59	3.98		8.42	13.75	2.16	13.02
Process	1	20						8.31	3.01	
	2	50		8.3					3.36	
	3	6.37	27.26							
	4			3.47	30.2					

The optimal solution calculated by Poplewski's method also reaches a fresh-water flow rate of 155 t h^{-1}. The flow rates in the network are shown in Table 20.11.

The cases presented show that the approach by Poplewski is able to cope with various goal functions, multiple contaminants and various technological constraints. It is also able to solve both synthesis and retrofit problems. Computation times for the problems are reasonable and do not exceed approximately a few minutes with a standard PC. This is, however, CPU time for a single run while the ARS optimization strategy requires several runs. Poplewski's experience was that usually up to 20 runs is sufficient even for difficult cases. To date, the method has not been tested to date for most general TWN problems.

References

Aaltola, J. (2002). Simultaneous synthesis of flexible heat exchanger network. *Appl. Therm. Eng.*, 22, 907–918.

Aarst, E. and Korst, J. (1989). *Simulated Annealing and Boltzmann Machines—A Stochastic Approach to Combinatorial Optimization and Neural Computers*. Wiley, New York.

Abadie, J. (1967). On the Kuhn–Tucker theorem. In: J. Abadie (Ed.), *Nonlinear Programming*. Noth-Holland, Amsterdam, pp. 21–36.

Abbas, H.A., Wiggins, G.A., Lakshmanan, R. and Morton, W. (1999). Heat exchanger network retrofit via constraint logic programming. *Comput. Chem. Eng.*, 21(Suppl.), S129–S132.

Adjiman, C.S., Androulakis, O.P. and Floudas, C.A. (2000). Global optimization of mixed-integer nonlinear problems. *AIChE J.*, 46(9), 1769–1797.

Adonyi, R., Romero, J., Puigjaner, L. and Friedler, F. (2003). Incorporating heat integration in batch process scheduling. *Appl. Therm. Eng.*, 23, 1743–1762.

Aggarwal, A. and Floudas, C.A. (1990). Synthesis of general distillation sequence—nonsharp separation. *Comput. Chem. Eng.*, 14, 631–653.

Aggarwal, A. and Floudas, C.A. (1992). Synthesis of heat integrated non-sharp distillation sequences. *Comput. Chem. Eng.*, 16, 89.

Aghbalou, F., Badia, F. and Illa, J. (2006). Exergetic optimization of solar collector and thermal energy storage system. *Intern. J. of Heat and Mass Transfer*, 49, 1255–1263.

Agrawal, V. and Shenoy, U.V. (2006). Unified conceptual approach to targeting and design of water and hydrogen networks. *AIChE J.*, 52(3), 1071–1082.

Aguilera, N. and Nasini, G. (1995). Flexibility test for heat exchanger networks with uncertain flowrates. *Comput. Chem. Eng.*, 19(9), 1007–1017.

Aguilera, N. and Nasini, G. (1996). Flexibility test for heat exchanger networks with non-overlapping inlet temperature variations. *Comput. Chem. Eng.*, 20(10), 1227–1240.

Ahlborn, B.K. (1999). Thermodynamic limits of body dimension of warm-blooded animals. *J. Non-Equilib. Thermodyn.*, 40, 407–504.

Ahlborn, B.K. and Blake, R.W. (1999). Lower size limit of aquatic mammals. *Am. J. Phys.*, 67, 1–3.

Ahmad, S. (1986). Heat exchanger networks: cost trade-offs in energy and capital. Ph.D. Thesis. University of Manchester, UK.

Ahmad, S. and Hui, D.C.W. (1991). Heat recovery between areas of integrity. *Comput. Chem. Eng.*, 15, 809–832.

Ahmad, S. and Polley, G.T. (1990). Debottlenecking of heat exchanger networks. *Heat Recovery Syst. CHP*, 10(4), 369–385.

Ahmad, S. and Smith, R. (1989). Targets and design for minimum number of shells in heat exchanger networks. *Chem. Eng. Res. Des.*, 67, 481–494.

Ahmad, S., Linnhoff, B. and Smith, R. (1988). Design of multipass heat exchangers: an alternative approach. *J. Heat Transfer (Trans. ASME)*, 110, 304–309.

Ahmad, S., Linnhoff, B. and Smith, R. (1990). Cost optimum heat exchanger networks. II. Targets and synthesis for detailed capital cost models. *Comput. Chem. Eng.*, 14, 751–767.

Energy Optimization in Process Systems

Akman, U., Uygun, K., Uzturk, D. and Konukman, A. (2002). HEN optimizations without using logarithmic-mean temperature difference. *AIChE J.*, 48(3), 596–606.

Alaphilippe, M., Bonnet, S. and Stouffs, P. (2007). Low power thermodynamic solar energy conversion: coupling of a parabolic trough concentrator and an Ericsson engine. *Int. J. Thermodyn.*, 10, 37–45.

Ali, M.M., Torn, A. and Vitanen, S. (1997). A numerical comparison of some modified controlled random search algorithms. *J. Global Optim.*, 11, 377–385.

Almató, M., Espuña, A. and Puigjaner, L. (1999). Optimisation of water use in batch-process industries. *Comput. Chem. Eng.*, 23, 1427–1437.

Al-Redhwan, S.A., Crittenden, B.D. and Lababidi, H.M.S. (2005). Wastewater minimization under uncertain operational conditions. *Comput. Chem. Eng.*, 29, 1009–10021.

Alva-Argaez, A., Kokossis, A.C. and Smith, R. (1998). Wastewater minimisation of industrial systems using an integrated approach. *Comput. Chem. Eng.*, 22(Suppl.), S741–S744.

Alva-Argaez, A., Vallianatos, A. and Kokossis, C.A. (1999). A multi-contaminant trans-shipment model for mass exchange networks and wastewater minimisation problems. *Comput. Chem. Eng.*, 23, 439–1453.

Alvarez, L.H.R. (1999). A class of solvable singular stochastic control problems. *Stochastics Stochastics Rep.*, 67, 83–122.

Alvarez, L.H.R. (2001). Singular stochastic control, linear diffusions, and optimal stopping: a class of solvable problems. *SIAM J. Control Optim.*, 39, 1697–1710.

Alwi, S.R.W. and Manan, Z.A. (2006). SHARPS: a new cost-screening technique to attain cost-effective minimum water network. *AIChE J.*, 52(11), 3981–3988.

Amarger, R.J., Biegler, L.T. and Grossmann, I.E. (1992). An automated modelling and reformulation system for design optimization. *Comput. Chem. Eng.*, 6(7), 623–636.

Amidpour, M. and Polley, G.T. (1997). Application of problem decomposition in process integration. *Trans. IChemE*, 75, 53–63.

Amundson, N. (1966). *Mathematical Methods in Chemical Engineering*. Prentice Hall, Englewood Cliffs.

Anantharaman, R., Abbas, O.S. and Gundersen, T. (2006). Energy Level Composite Curves—a new graphical methodology for the integration of energy intensive process. *Appl. Therm. Eng.*, 26, 1378–1384.

Ando, Y., Tanaka, T., Doi, T. and Takashima, T. (2001). A study on a thermally regenerative fuel cell utilizing low-temperature thermal energy. *Energy Convers. Manage.*, 42, 1807–1816.

Andre, J., Siary, P. and Dognon, T. (2001). An improvement of the standard genetic algorithm fighting premature convergence in continuous optimization. *Adv. Eng. Software*, 32, 49–60.

Andresen, B. (1983). *Finite-Time Thermodynamics*. University of Copenhagen, Englewood Cliffs.

Andresen, B. and Gordon, J. (1992a). Optimal paths for minimizing entropy production in a common class of finite-time heating and cooling processes. *Int. J. Heat Fluid Flow*, 13, 294–299.

Andresen, B. and Gordon, J. (1992b). Optimal heating and cooling strategies for heat exchanger design. *J. Appl. Phys.*, 71, 76–79.

Andresen, B. and Gordon, J. (1994). Constant thermodynamic speed for minimizing entropy production in thermodynamic processes and simulated annealing. *Phys. Rev.E*, 50, 4346–4351.

Andresen, B. and Salamon, P. (2000). Distillation by thermodynamic geometry. In: S. Sieniutyczand and A. de Vos (Eds.), *Thermodynamics of Energy Conversion and Transport*. Springer, New York, pp. 340–352 (Chapter 12).

Andresen, B., Salamon, P. and Berry, R.S. (1977). Thermodynamics in finite time:extremals for imperfect engines. *J. Chem. Phys.*, 66, 1571.

Andresen, B., Rubin, M.H. and Berry, R.S. (1983). Availability for finite time processes. General theory and model. *J. Phys. Chem*, 87, 2704–2713.

Andresen, B., Salamon, P. and Berry, R.S. (1984). Thermodynamics in finite time. *Phys. Today* (September issue), 62-70.

Androulakis, I.P. and Venkatasubramanian, V. (1991). A genetic algorithmic frame-work for process design and optimization. *Comput. Chem. Eng.*, 4(15), 217–228.

Angulo-Brown, F. (1991). An ecological optimization-criterion for finite-time heat engines. *J. Appl. Phys.*, 69, 7465–7469.

Ao, Y., Chen, J., Duan, M. and Shen, S. (2007). Gray modeling of heat-pump-surrounding system exergy efficiency, *Proceedings of 2007 IEEE International Conference on Grey Systems and Intelligent Services*, Nanjing, China, November 18–20, 2007.

Aoki, I. (1995). Entropy production in living systems: from organisms to ecosystems. *Thermochim. Acta*, 250, 359–370.

Aoki, I. (2001). Entropy and exergy principles in living systems. In: S.E. Jorgensen (Ed.), *Thermodynamics and Ecological Modelling*. Lewis (Publishers of CRC Press), Boca Raton, pp. 165–190.

Aris, R. (1961). *The Optimal Design of Chemical Reactors*. Academic Press, New York.

Aris, R. (1964). *Discrete Dynamic Programming*. Blaisdell, New York.

Arpaci, V. and Selamet, A. (1988). Entropy production in flames. *Combust. Flame*, 73, 251–259.

Arteaga-Pereza, L.E., Casas, Y., Peralta, L.M., Kafarov, V., Dewulfd, J. and Giunta, P. (2009). An auto-sustainable solid oxide fuel cell system fueled by bio-ethanol. Process simulation and heat exchanger network synthesis. *Chemical Engineering Journal*, 150, 242–251.

Asante, N.D.K. and Zhu, X.X. (1996a). An automated approach for heat exchanger network retrofit featuring minimal topology modifications. *Comput. Chem. Eng.*, 20(Suppl.), S7–S12.

Asante, N.D.K. and Zhu, X.X. (1996b). A new method for automated retrofit of heat exchanger networks, *Congress CHISA'96*, Prague, Czech Republic H7.1.

Asante, N.D.K. and Zhu, X.X. (1997). An automated and interactive approach for heat exchanger network retrofit. *Trans. IChemE*, 75, 349–360.

Asante, N.D.K., Zhu, X.X. and Wood, R.M. (1996). Simultaneous evolution and optimization of heat exchanger network stream split configurations, *IChemE Annual Research Meeting*, Leeds, UK.

Athans, M. and Falb, P.C. (1966). *Optimal Control*. McGraw-Hill, New York.

Atherton, R.W. and Homsy, G.M. (1975). On the existence and formulation of variational principles for nonlinear differential equations. *Stud. Appl. Math*, 54, 31–80.

Athier, G., Floquet, P., Pibouleau, L. and Domenech, S. (1997). Process optimization by simulated annealing and NLP procedures. Application to heat exchanger network synthesis. *Comput. Chem. Eng.*, 21(Suppl.), S475–S480.

Athier, G., Floquet, P., Pibouleau, L. and Domenech, S. (1998). A mixed method for retrofitting heat-exchanger networks. *Comput. Chem. Eng.*, 22(Suppl.), S505–S511.

Ayres, R.U. (1978). *Resources Environment and Economics—Applications of the Materials/Energy Balance Principle*. Wiley, New York.

Ayres, R.U. (1998). Eco-thermodynamics: economics and the second law. *Ecol. Econ*, 26, 189–209.

Ayres, R.U. (1999). The second law, the fourth law, recycling, and limits to growth. *Ecol. Econ*, 29, 473–483.

Ayres, R.U. and Kneese, A.V. (1969). Production, consumption, and externalities. *Am. Econ. Rev*, 59, 282–297.

Ayres, R.U. and Martinás, K. (1995). Waste potential entropy: the ultimate ecotoxic?. *Economie Appliquée*, 48, 95–120.

Ayres, R.U., Ayres, L.W. and Martinás, K. (1996). Eco-thermodynamics: exergy and life cycle analysis. Working Paper (96/./EPS). INSEAD, Fontainebleau, France, January 1996.

Ayres, R.U., Ayres, L.W. and Martinás, K. (1998). Exergy, waste accounting, and life-cycle analysis. *Energy*, 23, 355–363.

Ayres, R.U., Ayres, L.W. and Warr, B. (2003). Exergy, power and work in the US economy 1900–1998. *Energy Int. J.*, 28, 219–273.

Azzaro-Pantel, C., Bernal-Haro, L., Baudet, P., Domenech, S. and Pibouleau, L. (1998). A two stage methodology for short-term batch plant scheduling: discrete-event simulation and genetic algorithm. *Comput. Chem. Eng.*, 22(10), 1461–1481.

Babu, B.V. and Angira, R. (2006). Modified differential evolution (MDE) for optimization of non-linear chemical processes. *Comput. Chem. Eng.*, 30, 989–1002.

Badescu, V. (1998a). Accurate upper bounds for the conversion efficiency of black-body radiation energy into work. *Phys. Lett. A*, 244, 31–34.

Badescu, V. (1998b). Accurate upper bounds for the efficiency of converting solar energy into work. *J. Phys. D*, 31, 820–825.

Badescu, V. (1999). Simple upper bound efficiencies for endoreversible conversion of thermal radiation. *J. Non-Equilib. Thermodyn*, 24, 196–202.

Badescu, V. (2000). Accurate upper bound efficiency for solar thermal power generation. *Int. J. Solar Energy*, 20, 149–160.

Badescu, V. (2004). Letter to the. *Solar Energy*, 20, 149–160.

Badescu, V. (2008). Exact and approximate statistical approaches for the exergy of black-body radiation. *Central Eur. J. Phys.*, 6(2), 344–350.

Baetens, D. and Tainsh, R.A. (2000). Towards an efficient water management strategy: Vito's integrated approach, *Congress CHISA'2002*, Prague, Czech Republic.

Bagajewicz, M. (2000). A review of recent design procedures for water networks in refineries and process plants. *Comput. Chem. Eng.* (24), 2093–2113.

Bagajewicz, M.J. and Ji, S. (2001). Rigorous procedure for the design of conventional atmospheric crude fractionation units. Part I: Targeting. *Ind. Eng. Chem. Res.*, 40, 617–626.

Bagajewicz, M.J. and Rodera, R. (2000). Energy savings in the total site heat integration across many plants. *Comput. Chem. Eng.*, 24, 1237–1242.

Bagajewicz, M.J. and Savelski, M.J. (2001). On the use of linear models for the design of water utilization systems in process plants with single contaminants. *Trans. IChemE*, 79(A5), 600–610.

Bagajewicz, M.J. and Soto, J. (2001). Rigorous procedure for the design of conventional atmospheric crude fractionation units. Part II: Heat exchanger network. *Ind. Eng.Chem. Res.*, 40, 627–634.

Bagajewicz, M.J., Rivas, M. and Savelski, M.J. (2000). A robust method to obtain optimal and sub-optimal design and retrofit solutions of water utilization systems with multiple contaminants in process plants. *Comput. Chem. Eng.*, 24, 1461–1466.

Bagajewicz, M.J., Rodera, H. and Savelski, M.J. (2002). Energy-efficient water utilization systems in process plants. *Comput. Chem. Eng.*, 26(1), 59–79.

Baierlein, R. (2001). The elusive chemical potential. *Am. J. Phys.*, 69(4), 423–434.

Bampi, F. and Morro, A. (1982). The inverse problem of the calculus of variations applied to continuum physics. *J. Math. Phys.*, 23, 2312–2321.

Band, Y.B., Kafri, O. and Salamon, P. (1982). Optimization of a model external combustion engine. *J. Appl. Phys.*, 53, 29.

Banerjee, I. and Ierapetritou, M.G. (2003). Parametric process synthesis for general non-linear models. *Comput. Chem. Eng.*, 27, 1499–1512.

Banga, J.R. and Seider, W.D. (1996). Global optimization of chemical processes using stochastic algorithms. In: C.A. Floudas and P.M. Pardalos (Eds.), *State of Global Optimization*. Kluwer Academic Publishers, The Netherlands, pp. 563–583.

Banga, J.R., Irizarry-Rivera, R. and Seider, W.D. (1998). Stochastic optimization for optimal and model-predictive control. *Comput. Chem. Eng.*, 22(4–5), 603–612.

Bansal, V., Perkins, J.D. and Pistikopoulos, E.N. (2000). Flexibility analysis and design under linear systems by parametric programming. *AIChE J*, 46(2), 335–354.

Barbaro, A. and Bagajewicz, M.J. (2002). New one-stage MILP procedure for HEN design and retrofit, *Congress CHISA'2002*, Prague, Czech Republic.

Barbaro, A.B. and Bagajewicz, M.J. (2005). New rigorous one-step MILP formulation for heat exchanger network synthesis. *Comput. Chem. Eng.*, 29, 1945–1976.

Barbir, F. and Gomez, T. (1997). Efficiency and economics of proton exchange membrane (PEM) fuel cells. *Intern. J. of Hydrogen Energy*, 22, 1027–1037.

Bardi, M. and Capuzzo-Dolcetta, I. (1997). *Optimal Control and Viscosity Solutions of Hamilton–Jacobi–Bellman Equations (with appendices by Maurizio Falcone and Pierpaolo Soravia)*. Birkhauser, Boston.

Barles, G. (1990). An approach to deterministic control problems with unbounded data. *Annales de L'Inst Le Henri Poincaré. Analyse Non Linéaire*, 7, 235–258.

Barles, G. (1993). Discontinuous viscosity solutions of first-order Hamilton–Jacobi equations: a guided visit. *Nonlinear Anal*, 20(9), 1123–1134.

Barles, G. (1994). *Solutions de viscosite des equations de Hamilton–Jacobi. Collection Mathematiques et Applications de la SMAI 17*. Springer Verlag, Paris.

Barles, G. and Perthame, B. (1988). Exit time problems in optimal control and the vanishing viscosity method. *SIAM J. Control Optim*, 26, 1133–1148.

Barles, G. and Souganidis, P.M. (1991). Convergence of approximation schemes for fully nonlinear second order equations. *Asymptotic Anal*, 4, 271–283.

Barnicki, S.D. and Siirola, J.F. (2004). Process synthesis prospective. *Comput. Chem. Eng.*, 28, 441–446.

Barron, E.N. and Jensen, R. (1986). The Pontryagin maximum principle from dynamic programming and viscosity solutions to the first-order partial differential equations. *Trans. Am. Math. Soc*, 298, 635–641.

Barron, E.N. and Jensen, R. (1990). Semicontinuous viscosity solutions for Hamilton–Jacobi equations with convex Hamiltonians. *Comm. Partial Differ. Equations*, 15(12), 1713–1742.

Barsow, A.S. (1961). *What is Linear Programming*. Państwowe Wydawnictwa Naukowe, Warsaw.

Bauer, F., Grune, L. and Semmler, W. (2006). Adaptive spline interpolation for Hamilton–Jacobi–Bellman equations. *Appl. Numer. Math.*, 56, 1196–1210.

Baumgärtner, S. (2004). Thermodynamic models. In: J. Proops and P. Safonov (Eds.), *Modelling in Ecological Economics*. Edward Elgar, Cheltenham, pp. 102–129.

Baumgärtner, S. (2000). *Ambivalent Joint Production and the Natural Environment. An Economic and Thermodynamic Analysis.* Physica Verlag, Heidelberg.

Baumgärtner, S., Faber, M. and Proops, R. (1996). The use of the entropy concept in Ecological Economics. In: M. Faber, R. Manstetten and J. Proops (Eds.), *Ecological Economics—Concepts and Methods.* Edward Elgar, Cheltenham, pp. 115–135.

Bazykin, A.D., Khibnik, A.J. and Aponina, E.A. (1983). A model of evolutionary appearance of dissipative structure in ecosystems. *J. Math. Biol.*, 18, 13–23.

Beckel, D., Bieberle-Hütter, A., Harvey, A., Infortuna, A., Muecke, U.P., Prestat, M., Rupp, J.L.M. and Gauckler, L.J. (2007). Thin films for micro solid oxide fuel cells. *J. of Power Sources*, 173(1), 325–345.

Beckmann, H. (1953). Die Verteilung der Selbstkosten in Industrie und Heizkraftwerken auf Strom und Heizdampf. *BWK*, 5, 37–44.

Bedenik, N.I., Pahor, B. and Kravanja, Z. (2004). An integrated strategy for the hierarchical multivel MINLP synthesis of overall process flowsheets using the combined synthesis/analysis approach. *Comput. Chem. Eng.*, 28, 693–706.

Bednarek, G., Strumiłło, Cz. and Kudra, T. (1981). Mathematical modelling of dielectric drying. In: T. Stopinski (Ed.), *Materials of IV Symposium on Drying.* Instytut Chemii Przemysłowej, Warsaw.

Bedringas, K.W., Ertesvag, I.S., Byggstoyl, S. and Magnussen, B.F. (1997). Exergy Analysis of solid-oxide fuel cell (SOFC) systems. *Energy*, 22, 403–412.

Bejan, A. (1982). *Entropy Generation Through Heat and Fluid Flow.* Wiley, New York.

Bejan, A. (1987a). Unification of three different theories concerning the ideal conversion of enclosed radiation. *J. Sol. Energy Eng.*, 109, 46–51.

Bejan, A. (1987b). The thermodynamic design of heat and mass transfer processes and devices. *Heat Fluid Flow*, 8, 258–276.

Bejan, A. (1988a). *Advanced Engineering Thermodynamics.* Wiley-Interscience, New York.

Bejan, A. (1988b). Theory of heat transfer-irreversible power plants. *Int. J. Heat Mass Transfer*, 31, 1211.

Bejan, A. (1996a). *Entropy Generation Minimization, The Method of Thermodynamic Optimization of Finite-Size Systems and Finite-Time Processes.* CRC Press, Boca Raton.

Bejan, A. (1996b). Entropy generation minimization: the new thermodynamics of finite-size devices and finite time processes. *J. Appl. Phys.*, 79, 1191–1218.

Bejan, A. (1996c). Notes on the history of the method of entropy generation minimization (finite time thermodynamics). *J. Non-Equilib. Thermodyn.*, 21, 239–242.

Bejan, A. (1996d). Models of power plants that generate minimum entropy while operating at maximum power. *Am. J. Phys.*, 64, 1054–1059.

Bejan, A. (1996e). Street network theory of organization in nature. *J. Adv. Transport.*, 30, 85–107.

Bejan, A. (1997a). Theory of organization in Nature: pulsating physiological processes. *Int. J. Heat Mass Transfer*, 40, 2097–2104.

Bejan, A. (1997b). Constructal theory: from thermodynamic and geometric optimization to predicting shape in nature. *Energy Convers. Manage.*, 39, 1705–1718.

Bejan, A. (1997c). Constructal theory network of conducting paths for cooling a heat generating volume. *Int. J. Heat Mass Transfer*, 40, 799–816.

Bejan, A. (1997d). *Advanced Engineering Thermodynamics* (2nd edn.). John Wiley and Sons, New York.

Bejan, A. (2000). *Shape and Structure from Engineering to Nature.* Cambridge University Press, Cambridge, UK pp. 234–245.

Bejan, A. (2001a). The tree of convective heat streams: its thermal insulation function and the predicted 3/4-power relation between body heat loss and body size. *Int. J. Heat Mass Transfer,* 44, 699–704.

Bejan, A. (2001b). Thermodynamic optimization of geometry in engineering flow systems. *Exergy Int. J.,* 1(4), 269–277.

Bejan, A. (2003). Constructal comment on a Fermat-type principle for heat flow (Letter to Editor). *Int. J. Heat Mass Transfer,* 46, 1885–1886.

Bejan, A. (2005). The constructal law of organization in nature: tree-shaped flows and body size. *J. Exp. Biol.,* 208, 1677–1686.

Bejan, A. (2006). *Advanced Engineering Thermodynamics* (3rd edn.). Wiley, Hoboken.

Bejan, A. and Dan, N. (1996). Analogy between electrical machines and heat transfer— irreversible heat engines. *Int. J. Heat Mass Transfer,* 39, 3659–3666.

Bejan, A. and Dan, N. (1997). Maximum work from an electric battery model. *Energy,* 22, 93–102.

Bejan, A. and Errera, M.R. (1998). Maximum power from a hot stream. *Int. J. Heat Mass Transfer,* 41, 2025–2036.

Bejan, A. and Ledezma, G.A. (1998). Streets tree networks and urban growth: optimal geometry for quickest access between a finite-size volume and one point. *Phys. A,* 255, 211–217.

Bejan, A. and Lorente, S. (2001). Thermodynamic optimization of flow geometry in mechanical and civil engineering. *J. Non-Equilib. Thermodyn.,* 26, 305–354.

Bejan, A. and Lorente, S. (2003). Chapter 4. In: R.N. Rosa, A.H. Reis and A.F. Miguel (Eds.), *Bejan's Constructal Theory of Shape and Structure.* Evora Geophysics Center, Evora, Portugal.

Bejan, A. and Lorente, S. (2004). The constructal law and the thermodynamics of flow systems with configuration. *Int. J. Heat Mass Transfer,* 47, 3203–3214.

Bejan, A. and Lorente, S. (2005). *La Loi Constructale.* L'Harmattan, Paris.

Bejan, A. and Paynter, H.M. (1976). *Solved Problems in Thermodynamics. Problem VIID.* Massachusetts Institute of Technology, Cambridge MA.

Bejan, A. and Smith, J.L. (1974). Thermodynamic optimization of mechanical supports for cryogenic apparatus. *Cryogenics,* 14, 158.

Bejan, A., Tsatsaronis, G. and Moran, M. (1996). *Thermal Design and Optimization.* Wiley, New York.

Bellman, R.E. (1957). *Dynamic Programming.* Princeton University Press, Princeton.

Bellman, R.E. (1961). *Adaptive Control Processes: A Guided Tour.* Princeton University Press, Princeton pp. 1–35.

Bellman, R.E. (1967). *Introduction to Mathematical Theory of Control Processes.* Academic Press, New York.

Bellman, R. and Dreyfus, S. (1962). *Applied Dynamic Programming.* Princeton University Press, Princeton.

Bellman, R. and Dreyfus, S.E. (1967). *Dynamic Programming Applications (in Polish).* Państwowe Wydawnictwa Ekonomiczne, Warsaw.

Bellman, R.E. and Kalaba, R. (1965). *Dynamic Programming and Modern Control Theory.* Academic Press, New York.

Benedict, M. and Gyftopoulos, E. (1980). Economic selection of the components of an air separation process. In: R.A. Gaggioli (Ed.), *Thermo-dynamics: Second Law Analysis, ACS Symposium Series 122.* American Chemical Society, Washington, D.C.

Benelmir, R. and Feidt, M. (1997). Thermoeconomics and finite size thermodynamics for the optimization of a heat pump. *Intern. J. of Energy Environment and Economics*, 5(1), 129–133.

Benelmir, R. and Feidt, M. (1998). Energy cogeneration systems and energy management strategy. *Energy Convers. Manage*, 39(16/18), 1791–1802.

Bengtsson, C., Nordman, R. and Berntsson, T. (2002). Utilization of excess heat in the pulp and paper industry—a case study of technical and economic opportunities. *Appl. Therm. Eng.*, 22, 1069–1081.

Benkő, N., Rév, E., Szitkai, Z. and Fonyó, Z. (1999). Optimal water use and treatment allocation. *Comput. Chem. Eng.* (Suppl.), 157–160.

Bensoussan, A. (1981). Lectures on stochastic control. *Nonlinear Filtering and Stochastic Control, Lecture Notes in Math*. Springer-Verlag, Berlin, vol. 972 pp. 1–62.

Benton, S. (1977). *The Hamilton–Jacobi Equation: A Global Approach*. Academic Press, New York.

Benzinger, J.B., Satterfield, M.B., Hogarth, W.H.J., Nehlsen, J.P. and Kevrekidis, I.G. (2006). The power performance curve for engineering analysis of fuel cells. *J. of Power Sources*, 155, 272–285.

Berdicevskii, V.L. (1983). *Variational Principles in Mechanics of Continua*. Nauka, Moscow.

Bernal-Haro, L., Azzaro-Pantel, C., Domenech, S. and Pibouleau, L. (1998). Design of multipurpose batch chemical plants using a genetic algorithm. *Comput. Chem. Eng.*, 22(Suppl.), S777–S780.

Berry, R.S. (1989). Energy future: time horizons and instability. *Environment*, 31, 42.

Berry, R.S., Salamon, P. and Heal, G. (1978). On the relation between thermodynamic and economic optima. *Resour. Energy*, 1, 125–127.

Berry, R.S., Kazakov, V.A., Sieniutycz, S., Szwast, Z. and Tsirlin, A.M. (2000). *Thermodynamic Optimization of Finite Time Processes*. Wiley, Chichester p. 117.

Bertallanffy, L. (1973). *General System Theory. Foundations, Development, Applications* (revised edition). George Braziler, New York.

Bessette, N.F. (1994). Modeling and simulation for SOFC power systems. Ph.D. Dissertation. Georgia Institute of Technology, pp. 1–212.

Beveridge, G.S. and Schechter, S. (1970). *Optimization: Theory and Practice*. McGraw-Hill, New York.

Bhattacharyya, D., Rengaswamy, R. and Finnerty, C. (2009). Dynamic modeling and validation studies of a tubular solid oxide fuel cell. *Chem. Eng. Sci.*, 64, 2158–2172.

Biegler, L.T., Grossmann, I.E. and Westerberg, A.W. (1997). *Systematic Methods of Chemical Process Design*. Prentice Hall, Englewood Cliffs, New Jersey.

Bieler, P.S., Fischer, U. and Hungerbühler, K. (2004). Modeling the energy consumption of chemical batch plants: bottom-up approach. *Ind. Eng. Chem. Res.*, 43, 7785–7795.

Bismut, J.-M. (1978). An introductory approach to duality in optimal stochastic control. *SIAM Rev.*, 20, 62–78.

Björk, K.-M. and Nordmann, R. (2005). Solving large-scale retrofit heat exchanger network synthesis problems with mathematical optimization methods. *Chem. Eng. Process.*, 44, 869–876.

Björk, K.-M. and Petersson, F. (2003). Optimization of large-scale heat exchanger network synthesis problems, *International Conference Modeling and Simulation, IASTED*, Palm Springs, USA, pp. 313–318.

Björk, K.-M. and Westerlund, T. (2002). Global optimization of heat exchanger network synthesis problems with and without the isothermal mixing assumption. *Comput. Chem. Eng.*, 26, 1581–1593.

Blanco, A.M. and Bandoni, J.A. (2003). Interaction between process design and process operability of chemical processes: an eigenvalue optimization approach. *Comput. Chem. Eng.*, 27, 1291–1301.

Bochenek, R. (2003). Optymalna modyfikacja sieci wymienników ciepła przy zastosowaniu algorytmów genetycznych. Ph.D. Thesis. Rzeszów-Wrocław.

Bochenek, R. and Jeżowski, J. (1999). Adaptive random search approach for retrofitting flexible heat exchanger networks with fixed topology. *Hung. J. Ind. Chem.*, 27(2), 89–97.

Bochenek, R. and Jeżowski, J. (2000). Optymalizacja metodą adaptacyjnego przeszukiwania losowego w projektowaniu procesów. II. Przykłady zastosowań w inżynierii procesowej. *Inż. Chem. Proc.*, 21, 465–487.

Bochenek, R. and Jeżowski, J. (2004a). Kodowanie i optymalizacja struktury systemów technologicznych w strategiach stochastycznych. *Inż. Chem. Proc.*, 25(3/1), 715–720.

Bochenek, R. and Jeżowski, J. (2004b). Optymalizacja parametryczna sieci wymienników ciepła przy zastosowaniu algorytmów genetycznych. *Inż. Chem. Proc.*, 25(3/1), 721–726.

Bochenek, R. and Jeżowski, J. (2006). Genetic algorithms approach for retrofitting heat exchanger network with standard apparatus. In: W. Marquardt and C. Pantelides (Eds.), *Computer-Aided Chemical Engineering, 21A. 16th European Symposium on Computer Aided Process Engineering and 9th International Symposium on Process Systems Engineering*. Elsevier, Amsterdam, pp. 871–876.

Bochenek, R., Jeżowski, J., Poplewski, G. and Jeżowska, A. (1999). Studies in adaptive random search optimization for MINLP problems. *Comput. Chem. Eng., Suppl.*, S483–S486.

Bochenek, R., Jeżowski, J. and Słoma, R. (2001). Zastosowanie algorytmów genetycznych do optymalizacji systemów technologii chemicznej. *Inż. Chem. Proces.*, 22(3B), 247–252.

Bochenek, R., Jeżowski, J. and Jeżowska, A. (2005). Genetic algorithms optimization and its use in chemical engineering. Studia Universitatis Babes-Bolyai, Seria Chemia 2 Anul L: 73–83.

Bockris, J.O'M. (1980). *Energy Options*. Halsted Press, New York.

Bockris, J.O'M. (1984). Hydrogen: then and now. *Hydrogen Energy Progress*, V, 3–10.

Bockris, J.O'M. and Kainthla, R.C. (1986). The conversion of light and water to hydrogen and electric power. *Hydrogen Energy Process*, VI(2), 449.

Bockris, J.O'M., Bhardwaj, R.C. and Tennakoon, C.L.K. (1992). The electrochemistry of waste removal. *J. Serb. Chem. Soc.*, 57, 799–818.

Bodvarsson, G.S. and Witherspoon, P.A. (1989). Geothermal reservoir engineering, part1. *Geotherm. Sci. Technol.*, 1, 1–2.

Boetius, F. (2001a). Bounded variation singular stochastic control and associated Dynkin game. In: M. Kohlmann and S. Tang (Eds.), *Mathematical Finance, Trends in Mathematics*. Birkhauser, Basel, pp. 111–120.

Boetius, F. (2001b). Singular stochastic control and its relations to Dynkin game and entry—exit problems. Ph.D. Thesis. University of Konstanz, Konstanz, Germany.

Boetius, F. and Kohlmann, M. (1998). Connections between optimal stopping and singular stochastic control. *Stochastic Process. Appl.*, 77, 253–281.

Boltyanski, V.G. (1969). *Mathematical Methods of Optimal Control.* Nauka, Moscow.

Boltyanski, V.G. (1973). *Optimal Control of Discrete Systems.* Nauka, Moscow.

Bosniakovic, F. (1965). *Technische Thermodynamik.* I und II. Theodor Steinkopff, Dresden.

Boulding, K.E. (1966). The economics of the coming spaceship Earth. In: H. Jarrett (Ed.), *Environmental Quality in a Growing Economy.* Johns Hopkins University Press, Baltimore, pp. 3–14.

Box, M.J. (1965a). A new method of constrained optimization and comparison with other methods. *Comput. J.,* 8, 42.

Box, M.J. (1965b). A comparison of several current optimization methods and the use of transformations in constrained problems. *Comput. J.,* 9, 67.

Boyle, G. (2004). *Renewable Energy* (2nd edition). The Open University Bath Press, Glasgow.

Bracken, J. and McCormick, G.P. (1968). *Selected Applications of Nonlinear Programming.* Wiley, New York.

Briones, V. and Kokossis, A.C. (1999a). Hypertargets: a Conceptual Programming approach for the optimization of industrial heat exchanger networks—I. Grassroots design and network complexity. *Chem. Eng. Sci.,* 54, 519–539.

Briones, V. and Kokossis, A.C. (1999b). Hypertargets: a Conceptual Programming approach for the optimization of industrial heat exchanger networks—II. Retrofit design. *Chem. Eng. Sci.,* 54, 541–561.

Briones, V. and Kokossis, A.C. (1999c). Hypertargets: a Conceptual Programming approach for the optimization of industrial heat exchanger networks—III. Industrial applications. *Chem. Eng. Sci.,* 54, 685–706.

Brodowicz, K. and Dyakowski, T. (1990). *Heat Pumps (in Polish).* Państwowe Wydawnictwa Naukowe, Warsaw.

Brodyanskii, V.M. (1973). *Exergy Method of Thermodynamic Analysis (in Russian).* Energia, Moscow.

Brodyanskii, V.M., Fratsher, V. and Mihalek, K. (1988). *Exergy Method and its Applications (in Russian).* Energoatomizdat, Moscow.

Brooks, D.R. and Wiley, E.O. (1986). *Evolution as Entropy: Towards a Unified Theory of Biology.* University of Chicago Press, Chicago.

Brunin, O., Feidt, M. and Hivet, B. (1997). Comparaison des domaines de fonctionnement de certaines pompes à chaleur à compression et d'une pompe à compression-absorption. *Int. J. Refrigeration,* 20(5), 308–318.

Bryson, S. and Levy, D. (2006). Mapped WENO and weighted power ENO reconstructions in semi-discrete central schemes for Hamilton–Jacobi equations. *Appl. Numer.Math.,* 56, 1211–1224.

Burley, P. and Foster, J. (1994). *Economics and Thermodynamics—New Perspectives on Economic Analysis.* Kluwer Academic Publishers, Dordrecht.

Burness, H.S., Cummings, R.G., Morris, G. and Paik, I. (1980). Thermodynamic and economic concepts as related to resource-use policies. *Land Econ,* 56, 1–9.

Cadenillas, A. and Haussmann, U. (1994). The stochastic maximum principle for a singular control problem. *Stochastics Stochastics Rep.,* 49, 211–237.

Cadenillas, A. and Karatzas, I. (1995). The stochastic maximum principle for linear, convex optimal control with random coefficients. *SIAM J. Control Optim.,* 33, 590–624.

Calabro, A., Girardi, G., Fiorini, P. and Sciubba, E. (2004). Exergy analysis of a "CO_2 zero emission" high efficiency plant, *Proceedings of ECOS'04*, Guanajuato, Mexico, July 2004.

Calise, F., Dentice d'Accadia, M., Palombo, A. and Vanoli, L. (2006a). Simulation and exergy analysis of a hybrid Solid Oxide Fuel Cell (SOFC)–Gas Turbine System. *Energy*, 31, 3278–3299.

Calise, F., Dentice d'Accadia, M., Vanoli, L. and von Spakovsky, M.R. (2006b). Single-level optimization of a hybrid SOFC–GT power plant. *Journal of Power Sources*, 159, 1169–1185.

Calise, F., Dentice d'Accadia, M., Palombo, A. and Vanoli, L. (2007). A detailed one-dimensional finite-volume simulation model of a tubular SOFC and a pre-reformer. *Int. J. Thermodyn.*, 10, 87–96.

Camilli, F. and Prados, E. (2006). Shape-from-shading with discontinuous image brightness. *Appl. Numer. Math.*, 56, 1225–1237.

Campanari, S. (2001). Thermodynamic model and parametric analysis of a tubular SOFC module. *J. Power Sources*, 92, 26–34.

Campbell, J.R. and Gaddy, J.L. (1976). Methodology for simultaneous optimization with reliability: nuclear PWR example. *AIChE J.*, 6(22), 1050–1055.

Cannarsa, P. and Frankowska, H. (1991). Some characterizations of optimal trajectories in control theory. *SIAM J. Control Optim.*, 29, 1322–1347.

Canon, M.D., Cullun, C.D. and Polak, E. (1970). *Theory of Optimal Control and Mathematical Programming*. McGraw-Hill, New York.

Cao, C., Feng, X. and Duan, X. (2004). Design of water network with internal mains for multi-contaminant wastewater regeneration recycle. *Trans. IChemE*, 82(Part A), 1-6.

Caplan, S.R. and Essig, A. (1983). *Bioenergetics and Linear Nonequilibrium Thermodynamics*. Harvard University Press, Cambridge, MA.

Capra, F. (1996). *The Web of Life: A New Scientific Understanding of Living Systems*. Doubleday, New York.

Capuzzo Dolcetta, I.I. (1983). On a discrete approximation of the Hamilton–Jacobi equation of dynamic programming. *Appl. Math. Optim.*, 4, 367–377.

Capuzzo Dolcetta, I.I. and Falcone, M. (1989). Discrete dynamic programming and viscosity solutions of the Bellman equation. *Annales de L'Instit Le Henri Poincaré. Analyse Non Linéaire*, 6, 303–328.

Capuzzo Dolcetta, I.I. and Ishii, H. (1984). Approximate solutions of the Bellman equation of deterministic control theory. *Appl. Math. Optim.*, 2, 161–181.

Capuzzo-Dolcetta, I. and Perthame, P. (1996). On some analogy between different approaches to first order PDE's with nonsmooth coefficients. *Adv. Math. Sci. Appl.*, 6, 689–703.

Capuzzo Dolcetta, I., Fleming, W.H. and Zolezzi, T. (1985). Recent Mathematical Methods in Dynamic Programming. Di AutoriVari. *Lecture Notes In Mathematics*. Springer Verlag, Berlin, vol. 1220 pp. 1–202.

Caratheodory, C. (1935). *Variationsrechnung und partielle Differentialgleichungen erster Ordnung*. Teubner, Leipzig.

Cardoso, M.F., Salcedo, R.L. and de Azevedo, S.F. (1994). Nonequilibrium simulated annealing: a faster approach to combinatorial minimization. *Ind. Eng. Chem. Res.*, 33, 1908–1918.

Cardoso, M.F., Salcedo, R.L. and de Azevedo, S.F. (1996). The simplex-simulated annealing approach to continuous non-linear optimization. *Comput. Chem. Eng.*, 20(9), 1065–1080.

Cardoso, M.F., Salcedo, R.L., de Azevedo, S.F. and Barbosa, D. (1997). A simulated annealing approach to the solution of MINLP problems. *Comput. Chem. Eng.*, 21(12), 1349–1364.

Carlson, A. and Berntsson, T. (1995). Retrofit of heat exchanger networks based on detailed match calculations, *Proceedings of 30th Intersociety Energy Conversion Engineering Conference*. Orlando, Florida.

Carlson, A., Franck, P. and Berntsson, T. (1993). Design better heat exchanger network retrofits. *Chem. Eng. Prog.* (3), 87–96.

Carrasco, E.F. and Banga, J.R. (1997). Dynamic optimization of batch reactors using adaptive stochastic algorithms. *Ind. Eng. Chem. Res.*, 36, 2252–2261.

Carrington, C.G. and Sun, Z.F. (1992). Second Law Analysis of combined heat and mass transfer in internal and external flows. *Int. J. Heat Fluid Flow*, 13, 65–70.

Castell, C.M.L., Lakshmanan, R., Skilling, J.M. and Benares-Alcantra, R. (1998). Optimisation of process plant layout using genetic algorithms. *Comput. Chem. Eng.*, 22(Suppl.), S993–S996.

Castier, M. and Querioz, E.M. (2002). Energy targeting in heat exchanger network synthesis using rigorous physical property calculations. *Ind. Eng. Chem. Res.*, 42, 1511–1515.

Castro, P., Matos, H., Fernandes, M.C. and Nunes, C.P. (1999). Improvements for mass exchange networks design. *Chem. Eng. Sci.*, 54, 1649–1665.

Caviglia, G. (1988). Composite variational principles and the determination of conservation laws. *J. Math. Phys.*, 29, 812–816.

Caviglia, G. and Morro, A. (1987). Nether-type conservation laws for perfect fluid motions. *J. Math. Phys.*, 28, 1056–1060.

Cavin, L., Fischer, U., Glover, F. and Hungerbühler, K. (2004). Multi-objective process design in mult-purpose batch plants using Tabu Search optimization algorithm. *Comput. Chem. Eng.*, 28, 459–478.

Cavin, L., Fischer, U., Mošat, A. and Hungerbühler, K. (2005). Batch process optimization in a multipurpose plant using Tabu Search with a design-space diversification. *Comput. Chem. Eng.*, 29, 1770–1786.

Cenusa, V.E., Benelmir, R., Feidt, M. and Badea, A. (2003). On gas turbines, and combined cycles. *Int. J. of Energy, Environment and Economics*, 11(4), 236–258.

Cenusa, V.E., Badea, A., Feidt, M. and Benelmir, R. (2004). Exergetic optimization of the heat recovery steam generators by imposing the total heat transfer area. *Int. J. Thermodynamics*, 7(3), 149–156.

Cerda, J. and Galli, N. (1990). Synthesis of flexible heat exchanger networks—II. Nonconvex networks with large temperature variations. *Comput. Chem. Eng.*, 14(2), 213–225.

Cerda, J. and Galli, N. (1991). Synthesis of flexible heat exchanger networks—III. Temperature and flowrate variations. *Comput. Chem. Eng.*, 15(1), 7–24.

Cerda, J., Westerberg, A.W., Mason, D. and Linnhoff, B. (1983). Minimum utility usage in heat exchanger network synthesis. *Chem. Eng. Sci.*, 38(3), 373–387.

Cerda, J., Galli, N. and Isla, M.A. (1990). Synthesis of flexible heat exchanger networks—I. Convex networks. *Comput. Chem. Eng.*, 14(2), 197–211.

Chaisantikulwat, A., Diaz-Goano, C. and Meadows, E.S. (2008). Dynamic modelling and control of planar anode-supported solid oxide fuel cell. *Computers and Chemical Engineering*, 32, 2365–2381.

Chakraborty, S. and Ghosh, P. (1999). Heat exchanger network synthesis: the possibility of randomization. *Chem. Eng. J.*, 72, 209–216.

Chambadal, P. (1957). *Les Centrals Nuclearies*. Armand Colin, Paris pp 41–58.

Chambadal, P. (1963). *Evolution et Applications du Concept d'Entropie*. Dunod, Paris.

Chan, S.H., Ho, H.K. and Tian, Y. (2003). Multi-level modeling of SOFC-gas turbine hybrid system. *Int. J. Hydrogen Energy*, 28, 889–900.

Chang, C.-T. and Li, B.-H. (2005). Improved optimization strategies for generating practical water-usage and -treatment network structures. *Ind. Eng. Chem. Res.*, 44, 3607–3618.

Chang, C.-T. and Li, B.-H. (2006). Optimal design of wastewater equalization systems in batch processes. *Comput. Chem. Eng.*, 30, 797–806.

Charnes, A. and Cooper, W. (1961). *Management Models and Industrial Applications of Linear Programming*. Wiley, New York.

Charpentier, J.C. and McKenna, T.F. (2004). Managing complex systems: some trends for the future of chemical and process engineering. *Chem. Eng. Sci.*, 59, 1617–1646.

Chaudhuri, P.D. and Diwekar, U.M. (1997a). Synthesis under uncertainty with simulators. *Comput. Chem. Eng.*, 7(21), 733–738.

Chaudhuri, P.D. and Diwekar, U.M. (1997b). An automated approach for the optimal design of heat exchangers. *Ind. Eng. Chem. Res.*, 36, 3685–3693.

Chen, C.-L. and Hung, P.-S. (2004). Simultaneous synthesis of flexible heat-exchange networks with uncertain source-stream temperatures and flow rates. *Ind. Eng. Chem. Res.*, 42, 5916–5928.

Chen, G.-Q. and Su, B. (2003). Discontinuous solutions for Hamilton-Jacobi equations: uniqueness and regularity. *Discrete Continuous Dyn. Syst.*, 9, 167–192.

Chen, J.J.J. (1987). Letter to the Editor: Comments on improvement on a replacement for the logarithmic mean. *Chem. Eng. Sci.*, 42, 1489–2488.

Chen, J. (2005). *The Physical Foundation of Economics—An Analytical Thermodynamic Theory*. World Scientific, Singapore.

Chen, J. and Yan, Z. (1989). Unified description of endoreversible cycles. *Phys. Rev. A*, 39, 4140–4147.

Chen, J. and Zhao, Y. (2009). Modeling and optimization of a typical fuel cell–heat engine hybrid system and its parametric design criteria. *J. of Power Sources*, 186, 96–103.

Chen, J., Bi, Y. and Wu, Ch. (1999). A generalized model of a combined heat pump cycle and its performance. In: Ch. Wu, L. Chen and J. Chen (Eds.), *Advance in Recent Finite Time Thermodynamics*. Nova Science, New York, pp. 525–540.

Chen, J., Wu, Ch. and Kiang, R.J. (1996). Maximum specific power output of an irreversible radiant heat engine. *Energy Convers. Manage.*, 37, 17–22.

Chen, J., Yan, Z., Chen, L. and Andresen, B. (2000). On the performance of irreversible combined heat pump cycles. In: G.G. Hirs (Ed.), *ECOS 2000 Proceedings*, University Twente, The Netherlands, pp. 269–277.

Chen, J., Yan, Z., Lin, G. and Andresen, B. (2001). On the Curzon–Ahlborn efficiency and its connection with the efficiencies of real heat engines. *Energy Convers. Manage.*, 42, 173–181.

Chen, L., Sun, F. and Chen, W. (1990). Influence of heat transfer law on the performance of a Carnot engine. *J. Eng. Thermophys.*, 11, 241–243.

Chen, L., Sun, F. and Wu, C. (1999a). Effect of heat transfer law on the performance of a generalized irreversible Carnot engine. *J. Phys. D: Appl. Phys.*, 32, 99–105.

Chen, L., Sun, F. and Wu, C. (2006). Optimal configuration of a two-heat-reservoir heat-engine with heat leak and finite thermal capacity. *Appl. Energy*, 83(2), 71–81.

Chen, L., Wu, Ch. and Sun, F. (1996). Cooling load versus COP characteristics for an irreversible air refrigeration cycle. *Energy Convers. Manage.*, 37, 17–22.

Chen, L., Wu, C. and Sun, F. (1999b). Finite time thermodynamic optimization or entropy generation minimization of energy systems. *J. Non-Equilib. Thermodyn.*, 24, 327–359.

Chen, L., Xia, S. and Sun, F. (2009). Optimal paths for minimizing entropy generation during heat transfer processes with a generalized heat transfer law. *J. Appl. Phys.*, 105(4), 044907.

Chen, L., Xia, D., Sun, F., (2010). Ecological optimization of generalized irreversible chemical engines. Int. J. Chemical Reactor Engineering, 8(1), Article A121. ISSN (Online) 1542-6580, DOI: 10.2202/1542-6580.2361, July 2010.

Chen, L., Xia, D., Sun, F., (2011). Maximum power output of multistage continuous and discrete isothermal endoreversible chemical engine system with linear mass transfer law. *Intern. J. of Chemical Reactor Engineering*, 9(1), Article A10. ISSN (Online) 1542-6580, DOI: 10.1515/1542-6580.2470.

Chen, L., Zhoua, J., Sun, F. and Wu, Ch. (2004). Ecological optimization for generalized irreversible Carnot engines. *Applied Energy*, 77(3), 327–338.

Chen, T.-Y. and Su, J.-J. (2002). Efficiency improvement of simulated annealing in optimal structural design. *Adv. Eng. Software*, 33, 675–680.

Cheng, Ch.-Y., Lee, S.-H. and Chen, Ch.-K. (1999). Power optimizations of an irreversible Brayton heat engine with finite heat reservoirs. In: Ch. Wu, L. Chen and J. Chen (Eds.), *Advances in Recent Finite Time Thermodynamics*. Nova Science, New York, pp. 425–448.

Chew, E.P., Ong, C.J. and Lim, K.H. (2002). Variable period adaptive genetic algorithm. *Comput. Industrial Eng.*, 42, 353–360.

Chiang, A.C. (1992). *Elements of Dynamic Optimization*. Waveland Press, New York (Prospect Heights).

Choi, S.H., Ko, J.W. and Manousiouthakis, V. (1999). A stochastic approach to global optimization of chemical processes. *Comput. Chem. Eng.*, 23(9), 1351–1356.

Chow, P.-L., Menaldi, J.-L. and Robin, M. (1995). Additive control of stochastic linear systems with finite horizon. *SIAM J. Control Optim.*, 23, 858–899.

Chua, H.T., Ng, K.C. and Gordon, J.M. (1996). Experimental study of the fundamental properties of reciprocating chillers and its relation to thermodynamic modelling and chiller design. *Int. J. Heat Mass Transfer*, 39, 2195–2204.

Ciano, C., Cali, M. and Verda, V. (2009). Analysis of entropy generation for the performance improvement of a tubular Solid Oxide Fuel Cell stack. *Int. J. of Thermodynamics*, 12(1), 1–8.

Ciborowski, J. (1965). *Fundamentals of Chemical Engineering*. Wydawnictwa Naukowo Techniczne, Warsaw.

Ciborowski, J. (1976). *Basic Chemical Engineering (in Polish)*. Wydawnictwa Naukowo Techniczne, Warsaw.

Ciric, A.R. and Floudas, C.A. (1989). A retrofit approach for heat exchanger networks. *Comput. Chem. Eng.*, 13(6), 703–715.

Ciric, A.R. and Floudas, C.A. (1990a). Application of the simultaneous match-network optimization approach to the pseudo-pinch problem. *Comput. Chem. Eng.*, 14(3), 241–250.

Ciric, A.R. and Floudas, C.A. (1990b). Heat exchanger network synthesis without decomposition. *Comput. Chem. Eng.*, 15(6), 385–396.

Ciric, A.R. and Floudas, C.A. (1990c). A mixed integer nonlinear programming model for retrofitting heat-exchanger networks. *Ind. Eng. Chem. Res.*, 29, 239–251.

Clark, J.A. (1986). Thermodynamic optimization. Interface with economic analysis. *J. Non-Equilib. Thermodyn*, 10, 85–122.

Clarke, F.H. (1997). *Non-Smooth Analysis and Control Theory*. Springer-Verlag, New York.

Coclite, G.M. and Risebro, N.H. (2002). Conservation laws with a time dependent discontinuous coefficients, preprint, available at url: www.math.ntnu.no/conservation.

Coclite, G.M. and Risebro, N.H. (2003). Viscosity solutions of Hamilton–Jacobi equations with discontinuous coefficients. *Preprint*, 2003-2023 http://www.math.ntnu.no/conservation/2003/023.html.

Coclite, G.M. and Risebro, N.H. (2005). Conservation laws with a time dependent discontinuous coefficients. *SIAM J. Math. Anal.*, 36(4), 1293–1309.

Coclite, G.M. and Risebro, N.H., (2007). The Research Report funded by the BeMatA program of the Research Council of Norway and by the European network HYKE, funded by the EC as contract HPRN-CT-2002-00282. November 7, 2007.

Colberg, R.D. and Morari, M. (1990). Area and capital cost targets for heat exchanger network synthesis with constrained matches and unequal heat transfer coefficients. *Comput. Chem. Eng.*, 14(1), 1–22.

Colberg, R.D., Morari, M. and Townsend, D.W. (1989). A resilience target for heat exchanger network synthesis. *Comput. Chem. Eng.*, 13(7), 821–837.

Colmanares, T.R. and Seider, T.R. (1989). Synthesis of utility systems integrated with chemical processes. *Ind. Eng. Chem. Res.*, 28, 84–93.

D. Corne, M. Dorigo and F. Glover (Eds.) (1999). *New Ideas in Optimization*. McGraw-Hill, Berkshire.

Corning, P.A. (2002). Thermoeconomics—Beyond the Second Law. *J. Bioecon*, 4, 57–88 (also at: www.complexsystems.org).

Corning, P.A. and Kline, S.J. (2000). Thermodynamics, information and life revisited. Part II. Thermoeconomics and control information. *Syst. Res. Behav. Sci.*, 15, 453–482.

Corominas, J., Espuna, A. and Puigjaner, L. (1994). Method to incorporate energy integration considerations in multiproduct batch processes. *Comput. Chem. Eng.*, 18(11/12), 1043–1055.

Correas, L. (2004). On the thermoeconomic approach to the diagnosis of energy system malfunctions suitability to real time monitoring. *Int. J. Thermodyn.*, 7, 85–94 (the special issue on thermoeconomic diagnosis).

Costa, L. and Oliveira, P. (2005). Evolutionary algorithms approach to the solution of mixed-integer non-linear programming problems. *Comput. Chem. Eng.*, 25, 257–266.

Costea, M. and Feidt, M. (1999). The effect of the overall heat transfer coefficient variation on the optimal distribution of the heat transfer surface conductance on area in a Stirling engine. *Energy Convers. Manage*, 39(16/18), 1753–1761.

Costea, M., Feidt, M., Petrescu. (1999). Synthesis on Stirling engine optimization. Chapter in the book: *Thermodynamic optimization of complex energy systems*. Edited by A. Bejan and E. Mamut. Kluwer Academics, pp. 403–410.

Cownden, R., Nahon, M. and Rosen, M. (2001). Exergy analysis of fuel cells for transportation applications. *Exergy*, 1, 112–121.

Crandall, M.G. (1997). Viscosity solutions: a primer. In: Viscosity Solutions and Applications (Montecatini Terme, 1995). *Lecture Notes in Mathematics*. Springer, Berlin, 1660 pp. 1–43.

Crandall, M.G. and Lions, P.L. (1981). Condition d'unicité pour les solutions generalisé et des équations de Hamilton–Jacobi du premier ordre. *C. R. Acad. Sci. Paxris S'er. I Math.*, 292, 183–186.

Crandall, M.G. and Lions, P.L. (1983). Viscosity solutions of Hamilton–Jacobi equations. *Trans. Am. Math. Soc.*, 277, 1–42.

Crandall, M.G. and Lions, P.L. (1984). Two approximations of solutions of Hamilton–Jacobi equations. *Math. Comput.*, 43, 1–19.

Crandall, M.G. and Lions, P.L. (1986). On existence and uniqueness of solutions of Hamilton–Jacobi equations. *Nonlinear Anal. Theor. Methods Appl.*, 10, 353–370.

Crandall, M.G., Evans, L.C. and Lions, P.L. (1984). Some properties of viscosity solutions of Hamilton–Jacobi equations. *Trans. Am. Math. Soc.*, 282, 487–502.

Crandall, M.G., Ishii, H. and Lions, P.-L. (1992). User's guide to viscosity solutions of second order partial differential equations. *Bull. Am. Math. Soc.*, 27(1), 1–67.

Cross, A.D., Jones, P.L. and Lawton, J. (1982). Simultaneous energy and mass transfer in radiofrequency fields, I and II. Part I: Validation of the theoretical model. Part II: Characteristics of radiofrequency dryers. *Trans. IChemE*, 60 67 and 75.

Cullum, J. (1971). An explicit procedure for discretizing continuous optimal control problems. *J. Optim. Theor. Appl.*, 8.(1).

Curti, V., Favrat, D. and von Spakovsky, M. (1993). Contribution to a life cycle exergy analysis of heat pumps with various evaporator sources, *LENI-CONF-1993-027*, Lausanne.

Curzon, F.L. and Ahlborn, B. (1975). Efficiency of Carnot engine at maximum power output. *Am. J. Phys.*, 43, 22–24.

Daichendt, M.M. and Grossmann, I.E. (1994a). Preliminary screening procedure for the MINLP synthesis of process systems I. Aggregation and decomposition techniques. *Comput. Chem. Eng.*, 18(8), 663–677.

Daichendt, M.M. and Grossmann, I.E. (1994b). Preliminary screening procedure for the MINLP synthesis of process systems II. Heat exchanger networks. *Comput. Chem.Eng.*, 18(8), 679–709.

Daly, H.E. (1992). Is the entropy law relevant to the economics of natural resources? Yes, of course it is! *J. Environ. Econ. Manage.*, 23, 91–95.

Dall, S.R.X. and Cuthill, I.C. (1999). Mutation rates: does complexity matter?. *J. Theor. Biol.*, 198, 283–285.

Dal Maso, G. and Frankowska, H. (2000). Value functions for Bolza problems with discontinuous Lagrangians and Hamilton–Jacobi inequalities. *ESAIM Control Optim. Calc. Var.*, 5, 369–393 (electronic).

Dal Maso, G. and Frankowska, H. (2002). Autonomous integral functionals with discontinuous nonconvex integrands: Lipschitz regularity of minimizers, DuBois–Reymond necessary conditions, and Hamilton–Jacobi equations. Preprint, available at the url: http://cvgmt.sns.it/papers/dalfra02/.

Dantzig, G.B. (1968). *Linear Programming and Extensions*. Princeton University Press, Princeton.

Darwin, C. (1859). *The Origin of Species*. Penguin, London.

David, P. (1990). Thermodynamics and process design. *Chem. Engineer* (May), 31–33.

Davidon, W.C. (1959). Variable metric method of minimization. *A.E.C. Research and Development Report* ANL-5990.

Das, H., Cummings, P.T. and Le Van, M.D. (1990). Scheduling of serial multiproduct batch processes via simulated annealing. *Comput. Chem. Eng.*, 14(12), 1351–1362.

De Groot, S.R. and Mazur, P. (1984). *Non-Equilibrium Thermodynamics*. Dover, New York.

Delsman, E.R. Uju, C.U., de Croon, M.H.J.M., Schouten, J.C., Ptasinski, K.J. (2006). Exergy analysis of an integrated fuel processor and fuel cell (FP–FC) system. Energy, 31 (2006) 3300–3309.

Demirel, Y. (2002). Nonequilibrium Thermodynamics. Elsevier, Oxford.

Denbigh, K.G. (1956). The second law efficiency of chemical processes. Chem. Eng. Sci., 6, 1–9.

Denbigh, K.G. and Denbigh, J.S. (1985). Entropy in Relation to Incomplete Knowledge. Cambridge University Press, Cambridge.

Denn, M.M. and Aris, R. (1965). Second order variational equations and the strong maximum principle. Chem. Eng. Sci., 20, 373.

Dentice d'Accadia, M., Sasso, M. and Sibilio, S. (2000). Optimum operation of a thermal plant with cogeneration and heat pumps. In: G.G. Hirs (Ed.), ECOS 2000 Proceedings. Part 3: Process Integration. University Twente, The Netherlands, pp. 1411–1424.

Denton, J. (2002). Thermal cycles in classical thermodynamics and nonequilibrium thermodynamics in contrast with finite time thermodynamics. Energy Convers. Manage, 43, 1583–1618.

DePamphilis, M.L. (Ed.) (2002). Gene Expression at the Beginning of Animal Development. Elsevier, Oxford.

de Miguel, R. (2006a). On the nonequilibrium thermodynamics of large departure from Butler-Volmer behavior. The Journal of Physical Chemistry B Letters, 110, 8176–8178.

de Miguel, R. (2006b). Thermal effects in a modified law of mass action. Chemical Physics, 321, 62–68.

de Vos, A. (1991). Endoreversible thermodynamics and chemical reactions. J. Phys. Chem., 95, 4534–4540.

de Vos, A. (1992). Endoreversible Thermodynamics of Solar Energy Conversion. Clarendon Press, Oxford pp. 29–51.

de Vos, A. (1993a). The endoreversible theory of solar energy conversion: a tutorial. Solar Energy Mater. Solar Cells, 31, 75–93.

de Vos, A. (1993b). The endoreversible thermodynamics of hybrid photothermalphotovoltaic converter, Eleventh E.C. Photovoltaic Solar Energy Conference, Montreux, Switzerland, October 12-16, pp. 363–366.

de Vos, A. (1995). Thermodynamics of photochemical solar energy conversion. Solar Energy Mater. Solar Cells, 38, 11–22.

de Vos, A. (1997). Endoreversible economics. Energy Convers. Manage., 38, 311–317.

de Vos, A. and Pauwels, H. (1981). On the thermodynamic limit of photovoltaic solar energy conversion. J. Appl. Phys., 25, 119–123.

de Vos, A. and van der Wei, P. (1992). Endoreversible models for the conversion of solar energy into wind energy. J. Non-Equilib. Thermodyn., 17, 77–89.

de Vos, A. and van der Wei, P. (1993). The efficiency of the conversion of solar energy into wind energy by means of Hadley cells. Theor. Appl. Climatol., 46, 193–202.

de Vos, A., Landsberg, P., Baruch, P. and Parrott, J.E. (1993). Entropy fluxes, endoreversibility and solar energy conversion. J. Appl. Phys., 74, 3631–3637.

Dhole, V.D. and Linnhoff, B. (1994). Total site targets for fuel, co-generation, emissions, and cooling. Comput. Chem. Eng., 18(Suppl.), S101–S109.

Dhole, R.V., Ramchandani, N., Tainsh, R.A. and Wasilewski, M. (1996 Jan). Pinch technology can be harnessed to minimize raw-water demand and wastewater generation alike. Chem. Eng., 100–103.

Dincer, I. and Rosen, M.A. (2007). *Exergy: Energy, Environment and Sustainable Development*. Elsevier, Oxford.

Di Pietro, D.A., Lo Forte, S. and Pasolini, N. (2006). Mass preserving finite element implementations of the level set method. *Appl. Numer. Math.*, 56, 1179–1195.

Diepolder, P. and Brown, C. (1992). Is 'zero discharge' realistic?. *Hydrocarbon Process (October)*, 129–131.

Diwekar, U.M. and Xu, W. (2005). Improved genetic algorithms for deterministic optimization under uncertainty. Part I. Algorithms development. *Ind. Eng. Chem. Res.*, 44, 7132–7137.

Dolan, W.B., Cummings, P.T. and Le Van, M.D. (1989). Process optimization via simulated annealing: application to network design. *AIChE J.*, 35(5), 725–736.

Dolan, W.B., Cummings, P.T. and Le Van, M.D. (1990). Algorithmic efficiency of simulated annealing for heat exchanger network design. *Comput. Chem. Eng.*, 14(10), 1039–1050.

Domanski, R. (1990). *Storage of Thermal Energy (in Polish)*. PWN, Warsaw. Doroghov, N.N., Maikov, G.P. and Tsirlin, A.M. (1973). Conditions of optimality for various forms of control process (in Russian). Izv. AN SSSR, ser. Tech. Cybern., 1973(5),11–19.

Doyle, S.J. and Smith, R. (1997). Targeting water reuse with multiple contaminants. *Trans. IChemE Part B*, 75(3), 181–189.

I.G.C. Dryden (Ed.) (1975). *The Efficient Use of Energy*. IPC Science and Technology Press, Guildford.

Dueck, G. and Scheuer, T. (1990). Threshold accepting: a general purpose optimization algorithm appearing superior to simulated annealing. *J. Comput. Phys.*, 90, 161–175.

Duffie, J.A. and Beckman, W.A. (1974). *Solar Energy Thermal Processes*. Wiley, New York.

Dufour, F. and Miller, B. (2004). Singular stochastic control problems. *SIAM J. Control Optim.*, 43, 708–730.

Dufour, F. and Miller, B. (2006). Maximum principle for singular stochastic control problems. *SIAM J. Control Optim.*, 45, 668–698.

Duminil, M. (1976). Basic principles of thermodynamics as applied to heat pumps. In: E. Camatini and T. Kester (Eds.), *Heat Pumps and their Application to Energy Conservation*. Noordhoff, Leyden.

Dunn, R.F. and Wenzel, H. (2001). Process integration design methods for water conservation and wastewater reduction in industry. Part 1: Design for single contaminants. *Clean Products Process*, 3, 307–318.

Dunn, R.F., Wenzel, H. and Overcash, M.R. (2001). Process integration design methods for water conservation and wastewater reduction in industry. Part 2: Design for multiple contaminants. *Clean Products Process*, 3, 319–329.

Duran, M.A. and Grossmann, I.E. (1986). Simultaneous optimization and heat integration of chemical processes. *AIChE J.*, 32(1), 123–138.

Durmayaz, A., Sogut, O.S., Sahin, B. and Yavuz, H. (2004). Optimization of thermal systems based on finite-time thermodynamics and thermoeconomics. *Progress in Energy and Combustion Science*, 30(2), 175–217.

Ebeling, W. and Feistel, R. (1992). Theory of selforganization and evolution: the role of entropy, value and information. *J. Non-Equilib. Thermodyn.*, 17, 303–332.

Eden, M.R., Jorgensen, S.B., Gani, R. and El-Halwagi, M.M. (2004). A novel framework for simultaneous separation process and product design. *Chem. Eng. Process*, 43, 595–608.

Edera, M. and Kojima, H. (2002). Development of a new gas absorption chiller heater advanced utilization of waste heat from gas-driven co-generation systems for air conditioning. *Energy Convers. Manage.*, 43, 1493–1501.

Edgar, T.E., Himmelblau, D.M. and Lasdon, L.S. (2001). *Optimization of Chemical Processes* (International edition (2nd edn.)). McGraw Hill, Boston, MA.

Edwards, K., Edgar, T.F. and Manousiouthakis, V.I. (1998). Kinetic model reduction using genetic algorithms. *Comput. Chem. Eng.*, 22(1–2), 239–246.

Anon. EG&G Technical Services Inc. (2004). *Fuel Cell Handbook*, (Seventh Edition, Under Contract No. DE-AM26-99FT40575), U.S. Department of Energy, Office of Fossil Energy, National Energy Technology Laboratory, P.O. Box 880, Morgantown, West Virginia 26507-0880, November.

Ekman, A., Liukkonen, S. and Kontturi, K. (1978). Diffusion and electric conduction in multicomponent electrolyte systems. *Electrochemica Acta*, 23, 243–250.

El-Genk, M.S. and Tournier, J.M. (2000). Design optimization and integration of Nickel/Haynes-25 AMTEC cells into radioisotope power systems. *Energy Convers.Manage.*, 41, 1703–1728.

El Haj Assad, M. (1999). Performance characteristics of an irreversible refrigerator. In: Ch. Wu, L. Chen and J. Chen (Eds.), *Advance in Recent Finite Time Thermodynamics*. Nova Science, New York, pp. 181–188.

El-Halwagi, M.M. (1997). *Pollution Prevention Through Process Integration. Systematic design tools*. Academic Press, San Diego.

El-Halwagi, M.M. (2006). *Process Integration*. Elsevier, Amsterdam.

El-Halwagi, M.M., Gabriel, F. and Harell, D. (2003). Process design and control: rigorous graphical targeting for resource conservation via material recycle/reuse networks. *Ind. Eng. Chem. Res.*, 42, 4319–4328.

El Karoui, N. and Karatzas, I. (1988). Probabilistic aspects of finite-fuel, reflected follower problems. *Acta Appl. Math.*, 11, 223–258.

El Karoui, N. and Karatzas, I. (1991). A new approach to the Skorohod problem, and its applications. *Stochastics Stochastics Rep.*, 34, 57–82.

Elliott, R.J. (1990). The optimal control of diffusions. *Appl. Math. Optim.*, 22, 229–240.

Elliott, R.J. and Kohlmann, M. (1994). The second order minimum principle and adjoint process. *Stochastics Stochastics Rep.*, 46, 25–39.

El-Sayed, Y.M. (1999). Thermodynamics and thermoeconomics. *Int. J. Appl. Thermodyn.*, 2, 5–18.

El-Sayed, Y.M. (2003). *The Thermoeconomics of Energy Conversions*. Pergamon, Oxford p. 4.

El-Sayed, Y. (2007). Fingerprinting the malfunction of devices. *Int. J. Thermodyn.*, 10, 79–85.

Elsgolc, L.E. (1960). *Variational Calculus*. Państwowe Wydawnictwa Naukowe, Warsaw.

El-Wakil, M.M. (1962). *Nuclear Power Engineering*. McGraw-Hill, New York.

El-Wakil, M.M. (1971). *Nuclear Energy Conversion*. International Textbook Co., Scranton, PA.

Engel, P. and Morari, M. (1988). Limitations of the primary loop breaking method for synthesis of heat exchanger networks. *Comput. Chem. Eng.*, 12(4), 307–310.

Escher, C., Kloczkowski, A. and Ross, J. (1985). Increased power output and resonance effects in a thermal engine driven by a first or second order model reaction. *J. Chem. Phys.*, 82, 2457.

Evans, L. (1994). Some min-max methods for the Hamilton–Jacobi equation. *Indiana Univ. Math J.*, 33, 31–50.

Evans, L. and Souganidis, P.E. (1984). Differential games and representation formulas for solutions of Hamilton–Jacobi–Isaacs equation. *Indiana Univ. Math J.*, 33, 773–796.

Evans, L.C. and Souganidis, P.E. (1989). A PDE approach to geometric optics for certain reaction diffusion equations. *Indiana Univ. Math J.*, 38, 141–172.

Everdell, X. (1965). *Introduction to Chemical Thermodynamics*. The English Universities Press, London.

Faber, M. (1985). A biophysical approach to the economy: entropy, environment and resources. In: W. van Gool and J.J.C. Bruggink (Eds.), *Energy and Time in the Economic and Physical Sciences*. North Holland, Amsterdam, pp. 315–335.

Faber, M., Jöst, F. and Manstetten, R. (1995). Limits and perspectives on the concept of sustainable development. *Economie Appliquée*, 48, 233–251.

Faber, M., Manstetten, R. and Proops, J. (1996). *Ecological Economics—Concepts and Methods*. Edward Elgar, Cheltenham.

Falcone, M. (1987). A numerical approach to the infinite horizon problem of deterministic control theory. *Appl. Math. Optim.*, 15, 1–13.

Falcone, M. and Ferretti, R. (1994). Discrete time high-order schemes for viscosity solutions of Hamilton–Jacobi–Bellman equations. *Numer. Math.*, 67, 315–344.

Falcone, M. and Ferretti, R. (2006). Editorial. *Appl. Numer. Math.*, 56, 1135.

Falcone, M. and Makridakis, Ch. (2001). *Numerical Methods for Viscosity Solutions and Applications*. World Scientific, Singapore.

Fan, L.T. (1966). *The Continuous Maximum Principle. A Study of Complex System Optimization*. Wiley, New York pp. 329–342.

Fan, L.T. and Wang, C.S. (1964). *The Discrete Maximum Principle. A Study of Multistage System Optimization*. Wiley, New York.

Fan, L.T., Erickson, L.E. and Hwang, C.L. (1971). *Methods of Optimization, vol. 2. Equality Constraints and Optimization*. Kansas State University, Manhattan.

Feidt, M. (1996). Optimisation d'un cycle de Brayton moteur en contact avec des capacités thermiques finies. *Rev. Gén. Therm.*, 418/419, 662–666.

Feidt, M. (1997). Sur une systématique des cycles imparfaits. *Entropie*, (Numéro Spécial "De la thermotechnique à la thermodynamique"), 205, 53–61.

Feidt, M. (1999). Thermodynamics and optimization of reverse cycle machines. In: A. Bejan and E. Mamut (Eds.), *Thermodynamic Optimization of Complex Energy Systems*. Kluwer Academic Press, Dordrecht, pp. 385–401.

Feidt, M. (2001). Reconsideration of efficiency of processes and systems from a non equilibrium point of view. *Int. J. of Energy, Environment and Economics*, 11(1), 31–49.

Feidt, M. (2002). Depletion of ozone and the greenhouse effect: a new goal for the design of inverse cycle machines. *Int. J. Energy Research*, Special issue edited by A. Bejan, M. Feidt, and E. Mamut, 26(7), 653–574 (see also Foreword, p. 543-544).

Feidt, M. (2003). Advanced Thermodynamics of reverse cycle machine. Chap. I in *Low Temperature and Cryogenic Refrigeration*, Edited by Sadik Kakaç, M. R. Avelino, and H. F. Smirnov, Kluwer Academic Press, p. 39–82

Feidt, M. (2006). *Energétique: concepts et applications*. Dunod Editeur, Paris.

Feidt, M., Lesalos, K., Costea, M. and Petrescu, S. (2002). Optimal allocation of HEX inventory associated with fixed power output or fixed heat transfer rate input. *Int. J. Applied Thermodynamics*, 5(1), 25–36.

Feidt, M. and Lottin, O. (2004). Fuel cell implementation in transportation: state of the art. *Int. J. of Energy, Environment and Economics*, 12(2), 13–35.

Feidt, M., Costea, M., Petre, C. and Petrescu, S. (2007). Optimization of the direct Carnot cycle. *Applied Thermal Eng.*, 27, 829–839.

Fellner, Ch. and Newman, J. (2000). High-power batteries for use in hybrid vehicles. *J.Power Sources*, 85, 229–256.

Feng, X. and Chu, K.H. (2004). Cost optimization of industrial wastewater reuse systems. *Trans. IChemE*, 82(B3), 249–255.

Feng, X. and Seider, W.D. (2001). New structure and design methodology for water networks. *Ind. Eng. Chem. Res.*, 40, 4140–4146.

Fernadez-Polanco, D., Klemes, J. and Plesu, V. (2000). Process integration analysis and retrofit of the crude distillation unit heat exchanger network in an oil refinery, *Congress CHISA 2000*, Prague, Czech Republic.

Fewell, M.E., Reid, R.L., Murphy, L.M. and Ward, D.S. (1981). First and second law analysis of steam steadily flowing through constant-diameter pipes, *Proceedings of the 3rd Annual Conference on Systems Simulation, Economic Analysis—Solar Heating and Cooling Operational Results*, Reno, NV, April 1981, pp. 712–718.

Fiacco, A.V. and McCormick, G.P. (1968). *Nonlinear Programming: Sequential Unconstrained Minimization Techniques*. Wiley, New York.

Fialho, I.J. and Georgiu, T.T. (1999). Worst case analysis of nonlinear systems. *IEEE Trans. Automatic Control*, 44, 1180–1196.

Findeisen, W. (1974). *Multilevel Control Systems (in Polish)*. PWN, Warsaw.

Findeisen, W., Szymanowski, J. and Wierzbicki, A. (1974). *Computational Methods of Optimization (in Polish)*. Technological University Publications, Warsaw.

Findeisen, W., Szymanowski, J. and Wierzbicki, A. (1977). *Theory and Computational Methods of Optimization*. Państwowe Wydawnictwa Naukowe, Warsaw.

Findeisen, W., Szymanowski, J. and Wierzbicki, J. (1980). *Theory and Computational Methods of Optimization* (2nd edn.). Panstwowe Wydawnictwa Naukowe, Warsaw.

Finlayson, B.A. (1972). *The Method of Weighted Residuals and Variational Principles*. Academic Press, New York.

Fleming, W.H. and Rishel, R. (1975). *Deterministic and Stochastic Optimal Control*. Springer Verlag, Berlin.

Fleming, W. and Soner, H. (1993). *Controlled Markov Processes and Viscosity Solutions. Applications of Mathematics*. Springer Verlag, Berlin.

Fletcher, R. (1965). Function minimization without evaluating derivatives—a review. *Comput. J.*, 8, 33.

Fletcher, R. (1969). *Optimization*. Academic Press, New York.

Fletcher, R. and Powell, M.J.D. (1963). A rapidly convergent descent method for minimization. *Comput. J.*, 6, 163.

Fletcher, R. and Reeves, C.M. (1964). Function minimization by conjugate gradients. *Comput. J.*, 7, 149.

Floquet, P., Pibouleau, L. and Domenech, S. (1994). Separation sequence synthesis: how to use simulated annealing procedure. *Comput. Chem. Eng.*, 18(11/12), 1141–1148.

Floudas, C.A. (1995). *Nonlinear and Mixed-Integer Optimization. Fundamentals and Applications*. Oxford University Press, New York.

Floudas, C.A. and Ciric, A.R. (1989). Strategies for overcoming uncertainties in heat exchanger network synthesis. *Comput. Chem. Eng.*, 13(10), 1133–1152.

Floudas, C.A. and Ciric, A.R. (1990). Corrigendum—Strategies for overcoming uncertainties in heat exchanger network synthesis. *Comput. Chem. Eng.*, 14(8), 1.

Floudas, C.A. and Grossmann, I.E. (1986). Synthesis of flexible heat exchanger networks for multiperiod operation. *Comput. Chem. Eng.*, 10(2), 153–168.

Floudas, C.A. and Grossmann, I.E. (1987a). Automatic generation of multiperiod heat exchanger network configuration. *Comput. Chem. Eng.*, 11(2), 123–142.

Floudas, C.A. and Grossmann, I.E. (1987b). Synthesis of flexible heat exchanger networks with uncertain flowrates and temperatures. *Comput. Chem. Eng.*, 11(4), 319–336.

Floudas, C.A. and Pardalos, P.M. (1990). A collection of test problems for constrained global optimization algorithms, *Lecture Notes in Computation Science*. Springer, Berlin.

Floudas, C.A., Ciric, A.R. and Grossmann, I.E. (1986). Automatic synthesis of optimum heat exchanger configurations. *AIChE J.*, 32(2), 276–290.

Floudas, C.A., Aggarwal, A. and Ciric, A.R. (1989). Global optimum search for nonconvex NLP and MINLP problems. *Comput. Chem. Eng.*, 13(10), 1117–1132.

Fonyo, Z. and Rev, E. (1981). Paper 6.9.4, *Proceedings of the 2nd World Congress of Chemical Engineering, vol. II*, Montreal, Canada, October 1980.

Foo, D.C.Y., Manan, Z.A. and Tan, Y.L. (2005). Synthesis of maximum water recovery network for batch process systems. *J. Cleaner Prod.*, 13, 1381–1394.

Forland, K.S., Forland, T. and Ratkje, S.K. (1988). *Irreversible Thermodynamics. Theory and Applications*. Wiley, Chichester.

Fraga, E.S. and Senos Matias, T.R. (1996). Synthesis and optimization of nonideal distillation system using a parallel genetic algorithm. *Comput. Chem. Eng.*, 20(Suppl.), S79–S84.

Frangopoulos, Ch.A. (1983). Thermodynamic functional analysis: a method for optimal design or improvement of complex thermal systems. Ph.D. Thesis. Georgia State University, Atlanta, ISA.

Frangopoulos, Ch.A. (1990). Optimization of synthesis-design-operation of a cogeneration system by the Intelligent Functional Approach, *Proceedings of Florence World Energy Research Symposium*, Florence, Italy pp. 597–609 and 805-815.

Frangopoulos, Ch.A. and Von Spakovsky, M.R. (1993). A global environomic approach for energy systems analysis and optimization – I, *Proceedings of the International Conference Energy Systems and Ecology ENSEC' 93*, Cracow, Poland, July 5-9, 1993, pp. 123–132.

Frangopoulos, Ch.A., von Spakovsky, M.R. and Sciubba, E. (2002). A brief review of methods for the design and synthesis optimisation of energy systems, *Proceedings of ECOS 2002, vol. I*, Berlin, Germany, July 3-5, 2002 pp. 306–316.

Fraser, D.M. (1989). The use of minimum flux instead of minimum approach temperature as a design specification for heat exchanger networks. *Chem. Eng. Sci.*, 44(5), 1121–1127.

Fratcher, W. and Bayer, W. (1981). Exergy thermodynamics. *Chemische Technik*, 33H, 1.

Frausto-Hernandez, S., Rico-Ramirez, V., Jimenez-Gutierrez, A. and Hernandez-Castro, S. (2003). MINLP synthesis of heat exchanger networks considering pressure drops effects. *Comput. Chem. Eng.*, 27, 1143–1152.

Friese, T., Ulbig, P. and Schultz, S. (1998). Use of evolutionary algorithms for the calculation of group contribution parameters in order to predict thermodynamic properties. Part 1: Genetic algorithms. *Comput. Chem. Eng.*, 22(11), 1559–1572.

Fritzmoriss, R.E. and Mach, R.S.H. (1980). Improving distillation column using thermodynamic availability analysis. *AIChE J.*, 26, 265.

Furman, K.C. and Sahinidis, N.V. (2001a). Computational complexity of heat exchanger network synthesis. *Comput. Chem. Eng.*, 25(9–10), 1371–1390.

Furman, K.C. and Sahinidis, N.V. (2001b). Response to W. R. Johns' Commentary. *Comput. Chem. Eng.*, 25, 13911–21393.

Furman, K.C. and Sahinidis, N.V. (2002). A critical revue and annotated bibliography for heat exchanger network synthesis in the 20th century. *Ind. Eng. Chem. Res.*, 41, 2335–2370.

Furman, K.C. and Sahinidis, N.V. (2004). Approximations algorithms for the minimum number of matches problem in heat exchanger network synthesis. *Ind. Eng. Chem.Res.*, 43, 3554–3565.

Gadewar, S.B., Doherty, M.F. and Malone, M.F. (2001). A systematic method for reaction invariants and mole balances for complex chemistries. *Computers and Chemical Engineering*, 25, 1199–1217.

Gaggioli, R.A. (Ed.) (1980). *Thermodynamics: Second Law Analysis*. American Chemical Society, Washington, D.C..

Gaines, L.D. and Gaddy, J.L. (1976). Process optimization by flow sheet simulation. *Ind. Eng. Chem. Proc. Des. Dev.*, 1(15), 206–211.

Galan, B. and Grossmann, I.E. (1998). Optimal design of distributed wastewater treatment networks. *Ind. Eng. Chem. Res.*, 37, 4036–4048.

Galli, M.R. and Cerda, J. (1998a). A designer-controlled framework for the synthesis of heat exchanger networks involving non-isothermal mixers and multiple units over split streams. *Comput. Chem. Eng.*, 22(Suppl.), S813–S816.

Galli, M.R. and Cerda, J. (1998b). Synthesis of structural-constrained exchanger networks: I. Series networks. *Comput. Chem. Eng.*, 22(7–8), 819–839.

Galli, M.R. and Cerda, J. (1998c). Synthesis of structural-constrained exchanger networks: II. Split networks. *Comput. Chem. Eng.*, 22(7–8), 1017–1035.

Galli, M.R. and Cerda, J. (1998d). A customized MILP approach to the synthesis of heat recovery networks reaching specified topology targets. *Ind. Eng. Chem. Res.*, 37, 2479–2495.

Galli, M.R. and Cerda, J. (2000). Synthesis of heat exchanger networks featuring a minimum number of constrained-size shells of 1-2 type. *Appl. Therm. Eng.*, 20, 1443–1467.

Gani, R. (2004). Chemical product design: challenges and opportunities. *Comput. Chem. Eng.*, 28, 2441–2457.

Garcia-Palomares, U.M. and Rodriguez, J.F. (2002). New sequential and parallel derivative-free algorithms for unconstrained minimization. *SIAM J. Optim.*, 13(1), 79–96.

Garg, H.P., Mullick, S.C. and Bhargava, A.K. (1985). *Solar Thermal Energy Storage*. Reidel, Dordrecht.

Garrard, A. and Fraga, E.S. (1998). Mass exchange network synthesis using geneticalgorithms. *Comput. Chem. Eng.*, 22(12), 1837–1850.

Garrett, D.E. (1989). *Chemical Engineering Economics*. Van Nostrand Reinhold, Amsterdam.

Gass, S.I. (1958). *Linear Programming: Methods and Applications*. McGraw-Hill, New York.

Gautham, P. and Gopal, K. (2001). Systematic reallocation of aqueous resources using mass integration in a typical pulp mill. *Adv. Environ. Res.*, 5(1), 61–79.

Gelfand, J.M. and Fomin, S.W. (1963). *Variational Calculus. Revised English Edition*. Prentice Hall, Englewood Cliffs R.A. Silverman (Trans.).

Georgescu-Roegen, N. (1971). *The Entropy Law and the Economic Process*. Harvard University Press, Harvard (copryright by the President and Fellows of Harvard College; ISBN 1: 58348-600-3).

Geworkian, P. (2007). *Sustainable Energy Systems Engineering. The Complete Green Building Design Structure*. McGraw-Hill, New York.

Gibbs, W. (1948). *Collected Works*. Yale University Press, New Haven, vol. I, pp. 33–54.

Giga, Y. and Sato, M.-H. (2001). A level set approach to semicontinuous viscosity solutions for Cauchy problems. *Commun. Partial Differential Equations*, 26, 813–839.

Gilbert, R. (1956). La recherche des economiques de l'energie par l'analyse entropique. *Genie Chemique*, 75(4), 89–94.

Giovanini, L. and Marchietti, J.L. (2003). Low-level flexible-structure control applied to heat exchanger networks. *Comput. Chem. Eng.*, 27, 1129–1142.

Glansdorff, P. and Prigogine, I. (1971). Thermodynamic Theory of Structure, *Stability and Fluctuations*. John Wiley & Sons, New York.

Glazunov, Yu.T. (1983). The variational method for the solution of the combined heat and mass transfer problems. *Int. J. Heat Mass Transfer*, 26, 1815–1822.

Glemmestad, B., Skogestad, S. and Gundersen, T. (1999). Optimal operation of heat exchanger networks. *Comput. Chem. Eng.*, 23, 509–522.

Goldberg, D.E. (1989). *Genetic Algorithm in Search, Optimization and Machine Learning*. Adison-Wesley, Reading, MA.

Gollapalli, U., Dantus, M.M. and High, K.A. (2000). Environmental and control issues in design. *Comput. Chem. Eng.*, 24, 1709–1712.

Gołąb, S. (1956). *Tensor Calculus*. Państwowe Wydawnictwa Naukowe, Warsaw.

Gomes, J.F.S., Queiroz, E.M. and Pessoa, F.L.P. (2007). Design procedure for water/wastewater minimization: single contaminant. *J. Cleaner Prod.*, 15, 474–485.

Gómez, M.J., Savelski, M. and Bagajewicz, M.J. (2001). On a systematic design procedure for single component water utilization systems in process plants. *Chem. Eng. Commun.*, 186, 183–203.

Gong, M. and Wall, G., et al. (1997). On exergetics, economics and optimization of technical processes to meet environmental conditions. In: Ruixian Cai, (Ed.), *Presented at TAIES'97. Thermodynamic Analysis and Improvement of Energy Systems*, Beijing, China, June 10-13. BeijingWorld, Warsaw, pp. 403–413 ISBN: 7-5062-3264-Z, http://www.exergy.se/ftp/execopt.pdf.

Gontar, V. (2000). Entropy principle of extremality as a driving force in the discrete dynamics of complex and living systems. *Chaos Solitons Fractals*, 11, 231–236.

Gordon, J.M. (1988). On optimized solar-driven heat engines. *Solar Energy*, 40, 457.

Gordon, J.M. (1989). Maximum power point characteristics of heat engines as a general thermodynamic problem (proves that finite size is equivalent to finite time). *Am. J. Phys.*, 57, 1136–1142.

Gordon, J.M. (1990). Nonequilibrium thermodynamics for solar energy applications. In: S. Sieniutycz and P. Salamon (Eds.), *Finite-Time Thermodynamics and Thermoeconomics. Adv. Thermodyn.* 4. Taylor and Francis, New York, pp. 95–120.

Gordon, J.M. (1991). Generalized power versus efficiency characteristics of heat engines: the thermoelectric generator as an instructive illustration. *Am. J. Phys.*, 59, 551–555.

Gordon, J.M. and Huleihil, M. (1991). On optimizing maximum-power heat engines. *J. Appl. Phys.*, 69, 1–7.

Gordon, J.M. and Huleihil, M. (1992). General performance characteristics of real heat engines. *J. Appl. Phys.*, 72, 829–837.

Gordon, J.M. and Ng, K.C. (2000). *Cool Thermodynamics*. Cambridge International Science Publishing, UK.

Gorsek, G., Glavic, P. and Bogataj, M. (2006). Design of the optimal total site heat recovery system using SSSP approach. *Chem. Eng. Proc.*, 45, 372–382.

de Gouvea, T.M. and Odloak, D. (1998). A new treatment of inconsistent quadratic programs in a SQP-based algorithm. *Comput. Chem. Eng.*, 22(11), 1623–1651.

Gouy, G. (1989). Sur l'Energie utilisable. *J. Phys.*, 8, 501.

Grabert, H., Hänggi, P. and Oppenheim, I. (1983). Fluctuations in reversible chemical reactions. *Phys. A*, 117, 300–316.

Grant, C.D. (1979). *Energy Conservation in the Chemical and Process Industries*. George Goldwin, London.

Gregoire, K.P. and Becker, J.G. (2012). Design and characterization of a microbial fuel cell for the conversion of a lignocellulosic crop residue to electricity. *Bioresource Technology*, 119, 208–215.

Greig, D.M. (1980). *Optimisation*. Longman, New York.

Grossmann, I.E. (2005). Enterprise-wide optimization: a new frontier in process systems engineering. *AIChE J.*, 51(7), 1846–1857.

Grossmann, I.E. and Daichendt, M.M. (1996). New trends in optimization based approaches to process synthesis. *Comput. Chem. Eng.*, 20(6–7), 665–683.

Grossmann, I.E. and Floudas, C.A. (1987). Active constraint strategy for flexibility analysis in chemical processes. *Comput. Chem. Eng.*, 11(6), 675–693.

Grossmann, I.E. and Sargent, R.W.H. (1979). Optimum design of multipurpose chemical plants. *Ind. Eng. Chem. Proc. Des. Dev.*, 18(2), 343–348.

Grossmann, I.E., Westerberg, C.A. and Biegler, L.T. (1987). Retrofit design of processes, *Conference: Foundations of Computer-Aided Process Operations*, Park City, UT, USA.

Grosu, L., Benelmir, R. and Feidt, M. (1999). Technico-economic simulation and optimization of a compression refrigerating machine. *Energy Convers. Manage.*, 40, 1651–1660.

Guerra, M. (2005). Discontinuous Hamiltonian flows for nonlinear control systems. *Rend. Sem. Mat. Univ. Pol. Torino Control Theory and Stabil.*, I, 63, 4.

Guillen, G., Badell, M., Espuna, A. and Puigjaner, L. (2006). Simultaneous optimization of process operations and financial decisions to enhance the integrated planning/scheduling of chemical supply chains. *Comput. Chem. Eng.*, 30, 421–436.

Gumiński, K. (1982). *Thermodynamics* (4th edition). PWN, Warsaw p. 328.

Gunaratnam, M., Alva-Argáez, A., Kokossis, C.A., Kim, J.-K. and Smith, R. (2005). Automated design of total water systems. *Ind. Eng. Chem. Res.*, 44, 588–599.

Gundersen, T. (1989). Retrofit process design research and applications of systematic methods, *Conference: Foundations of Computer-Aided Process Design*, Snowmass Village, CO, USA.

Gundersen, T. (2001). Cost optimal heat exchanger networks: talk at Gordon Research Conference, *Modern Developments in Thermodnamics*. Ventura, CA.

Gundersen, T. and Grossmann, I.E. (1990). Improved optimization strategies for automated heat exchanger network synthesis through physical insights. *Comput. Chem. Eng.*, 14, 925–944.

Gundersen, T. and Naess, L. (1988). The synthesis of cost optimal heat exchanger networks. *Comput. Chem. Eng.*, 12(6), 503–530.

Gundersen, T., Sagli, B. and Kiste, K. (1991). Problem in sequential and simultaneous strategies for heat exchanger network synthesis. In: L. Puigjaner and A. Espuna (Eds.), *Computer-Oriented Process Engineering*. Elsevier Science Publishers, The Netherlands, pp. 105–116.

Gundersen, T., Douvold, S. and Hashemi-Ahmady, A. (1996). An extended vertical MILP model for heat exchanger network synthesis. *Comput. Chem. Eng.*, 20(Suppl.), S97–S102.

Gundersen, T., Traedal, P. and Hashemi-Ahmady, A. (1997). Improved sequential strategy for the synthesis of near-optimal heat exchanger networks. *Comput. Chem. Eng.*, 21(Suppl.), S59–S64.

Gundersen, T., Hashemi-Ahmady, A., Zamora, J.M. and Grossmann, I.E. (2000). A sequential framework for optimal synthesis of industrial size heat exchanger networks, *Congress CHISA 2000*, Prague, Czech Republic (paper H.8.1).

Gustavson, M. (1979). Limits to wind power utilization. *Science*, 204, 13.

Gyftopoulos, E. (1999). Infinite time (reversible) versus finite time (irreversible) thermodynamics: a misconceived distinction. *Energy Int. J.*, 24, 1035–1039.

Gyftopoulos, E. (2002). On the Courzon-Ahlborn efficiency and its lack of connection to power-producing processes. *Energy Convers. Manage.*, 42, 609–615.

Gyftopoulos, E.P. (2003). A tribute to energy systems scientists and engineers. *Int. J. of Thermodynamics*, 6, 49–57.

Hadley, G. (1962). *Linear Programming*. Addison-Wesley, Reading, MA.

Hadley, G. (1964). *Linear and Dynamic Programming*. Addison-Wesley, Reading, MA.

Halkin, H. (1966). A maximum principle of the Pontryagin type for systems described by nonlinear difference equations. *SIAM J. Control Ser. A*, 4, 528–547.

Hall, S.G., Ahmad, S. and Smith, R. (1990). Capital cost targets for heat exchange rnetworks comprising mixed materials of construction, pressure ratings and exchanger types. *Comput. Chem. Eng.*, 14(3), 319–335.

Hallale, N. (2002). A new graphical targeting method for water minimisation. *Adv. Environ. Res.*, 6, 377–390.

Hamad, A., Aidan, A. and Douboni, M. (2003). Cost-effective wastewater treatment and recycling in mini-plants using mass integration. *Clean Technol. Environ. Policy*, 4, 246–256.

Han, Z.J., Zhu, Z.J., Rao, M. and Chuang, K.T. (1998). Determination of independent loops in heat exchanger networks. *Chem. Eng. Commun.*, 164, 191–204.

Hanke, M. and Li, P. (2000). Simulated annealing for the optimisation of batch distillation processes. *Comput. Chem. Eng.*, 24, 1–8.

Hancock, H. (1960). *Theory of Maxima and Minima*. Dover, New York.

Harland, J.R. and Salamon, P. (1988). Simulated annealing: a review of the thermodynamic approach. *Nucl. Phys. B*, 5A(Proc. Suppl.), 109.

Haseli, Y., Dincer, I. and Naterer, G.F. (2008). Thermodynamic modeling of a gas turbine cycle combined with a solid oxide fuel cell. *Int. J. of Hydrogen Energy*, 33, 5811–5822.

Haussmann, U.G. (1986). A stochastic maximum principle for optimal control of diffusions, *Pitman Research Notes Math Series 151*. Longman Scientific and Technical/John Wiley Sons, Harlow, UK/New York.

Haynes, C. (2001). Clarifying reversible efficiency misconceptions of high temperature fuel cells in relation to reversible heat engines. *Journal of Power Sources*, 92, 199–203.

Haynes, C. and Wepfer, W.J. (2000). "Design for Power" of a commercial grade tubular solid oxide fuel cell. *Energy Convers. Manage.*, 41, 1123–1139.

Henatsch, A. (1957). Thermodynamische-energie wirtschaftliche Bewertung von Warmeaustauscher. *Wiss. Zeitsch. der Techn. Hochs. f. Verkehrswes. Dresden*, 14, 57–67.

Hendrix, E.M.T., Ortigosa, P.M. and Garcia, I. (2001). On success rates for controlled random search. *J. Global Optim.*, 21, 239–263.

Herivel, J.W. (1955). The derivation of the equations of motion of an ideal fluid by Hamilton's principle. *Proc. Cambridge Philos. Soc.*, 51, 344–349.

Hernandez, L. and Kafarov, V. (2009). Use of bioethanol for sustainable electrical energy production. *Int. J. of Hydrogen Energy*, 34(3), 7041–7050.

Hernandez-Suarez, R., Castellanos-Fernandez, J. and Zamorra, J.M. (2004). Superstructure decomposition and parametric optimization approach for the synthesis of distributed wastewater treatment networks. *Ind. Eng. Chem. Res.*, 42, 2175–2191.

Heuckroth, M.W., Gaddy, J.L. and Gaines, L.D. (1976). An examination of the adaptive random search technique. *AIChE J.*, 22(4), 744–750.

Haywood, R.W. (1975). *Analysis of Engineering Cycles*. Pergamon, Oxford.

G.G. Hirs (Ed.) (2000). *ECOS 2000 Proceedings. Part 3 Process Integration*. University Twente, The Netherlands.

Hirschenhofer, J.H., Stauffer, D.B., Engleman, R.R. and Klett, M.G. (1998). *Fuel Cell Handbook* (fourth edition). Parsons Corporation, Reading, PA.

Hjelmfelt, A., Schreiber, I. and Ross, J. (1991). Efficiency of power production in simple nonlinear electrochemical systems. *J. Phys. Chem.*, 95, 6048.

Hoffmann, K.-H. (1990). Optima and bounds for irreversible thermodynamicprocesses. In: S. Sieniutycz and P. Salamon (Eds.). *Finite-Time Thermodynamics and Thermoeconomics*. Adv. Thermodyn. Taylor and Francis, New York, Series 4, pp. 22–65.

Hoffmann, K.H., Andresen, B. and Salamon, P. (1989). Measures of dissipation. *Phys. Rev. A*, 39, 3618–3621.

Hoffmann, K.-H., Burzler, J.M. and Schubert, M. (1997). Endoreversible thermodynamics. *J. Non-Equilib. Thermodyn.*, 22, 311–355.

Hohmann, E.C. (1971). Optimum networks for heat exchange. Ph.D. Thesis. University of South California.

Hooke, R. and Jeeves, T.A. (1961). Direct search solution of numerical and statistical problems. *J. Assoc. Comp. Mach.*, 8, 212.

Hostrup, M., Harper, P.M. and Gani, R. (1999). Design environmentally benign processes: integration of solvent and separation process synthesis. *Comput. Chem. Eng.*, 23, 1395–1414.

Hotz, N., Lee, M.-T., Grigoropoulos, C.P., Senn, S.M. and Poulikakos, D. (2006). Exergetic analysis of fuel cell micropowerplants fed by methanol. *Intern. J. Heat and Mass Transfer*, 49, 2397–2411.

Hou, Y., Zhuang, M. and Wan, G. (2007). The analysis for the efficiency properties of the fuel cell engine. *Renewable Energy*, 32, 1175–1186.

Householder, A.S. (1953). *Principles of Numerical Analysis*. McGraw-Hill, New York.

Huang, C.-H., Chang, C.-T. and Ling, H.-C. (1999). A mathematical programming model for water usage and treatment network design. *Ind. Eng. Chem. Res.*, 38, 2666–2679.

Huang, Y.C. and Wang, X.Z. (1999). Application of fuzzy causal networks to waste water teatment plants. *Chem. Eng. Sci.*, 54, 2731–2738.

Huber, M.L. (1994). Structural optimization of vapour pressure correlations using simulated annealing and threshold accepting: application to R134a. *Comput. Chem. Eng.*, 18(10), 929–932.

Hugo, A. and Pistikopoulos, E.N. (2005). Environmentally conscious long-range planning and design of supply chain networks. *J. Cleaner Prod.*, 13, 1471–1491.

Hui, C.W. and Ahmad, S. (1994a). Minimum cost heat recovery between separate plant regions. *Comput. Chem. Eng.*, 18(8), 711–728.

Hui, C.W. and Ahmad, S. (1994b). Total site integration using the utility system. *Comput. Chem. Eng.*, 18(8), 729–742.

Hussain, M.M., Li, X. and Dincer, I. (2007). Mathematical modeling of transport phenomena in porous SOFC anodes. *Intern. J. of Thermal Sciences*, 46, 48–56.

Hwang, S.-F. and He, R.-S. (2006). Improving real-parameter genetic algorithm with simulated annealing for engineering problems. *Adv. Eng. Software*, 37, 406–418.

Imre, I. (Ed.) (1990), *Issue on solar drying. Drying Technol.*, 8 (No. 2).

Imre, I., Farkas, L., Fabri, L. and Hecker, G. (1986). *Economical Aspects of Solar Drying.* International Meeting on Energy Savings in Drying Processes, Liege.

Ishii, H. (1984). Uniqueness of unbounded viscosity solution of Hamilton–Jacobi equation. *Indiana Univ. Math J.*, 33, 721–749.

Ishii, H. (1985). Hamilton–Jacobi equations with discontinuous Hamiltonians on arbitrary open sets. *Bull. Fac. Sci. Eng. Chuo Univ.*, 28, 33–77.

Ishii, H. (1987). Perron's method for Hamilton–Jacobi equations. *Duke Math. J.*, 55, 369–384.

Ishii, H. and Ramaswamy, M. (1995). Uniqueness results for a class of Hamilton–Jacobi equations with singular coefficients. *Comm. Partial Differential Equations*, 20, 2187–2213.

Ismail, S.R., Proios, P. and Pistikopulos, E.N. (2001). Modular synthesis framework for combined separation/reaction systems. *AIChE J.*, 47(3), 629–649.

Itahara, S. and Stiel, L.I. (1966). Optimal design of multiple-effect evaporators by dynamic programming. *Ind. Eng. Chem. Proc. Design Dev.*, 5, 309.

Jaakola, T.H.I. and Luus, R. (1974). A note on the application of nonlinear programming to chemical-process optimization. *Oper. Res.*, 22(2), 415–417.

Jaluria, Y. (2008). *Design and Optimization of Thermal Systems*, Second Edition. Series: Dekker Mechanical Engineering Volume: 209. CRC Taylor and Francis Group, New York.

Jang, W.-K., Hahn, J. and Hall, K.R. (2005). Genetic/quadratic search algorithm for plant economic optimizations using a process simulator. *Comput. Chem. Eng.*, 30, 285–294.

Jarzebski, Z.M. (1990). *Solar Energy Photovoltaic Conversion*. PWN, Warsaw.

Jegede, F.O. and Polley, G.T. (1992). Capital cost targets for networks with nonuniform heat exchanger specifications. *Comput. Chem. Eng.*, 16(5), 477–495.

Jensen, R. (1988). The maximum principle for viscosity solutions of fully nonlinear second order partial differential equations. *Arch. Rat. Mech. Anal.*, 101, 1–27.

Jeter, J. (1981). Maximum conversion efficiency for the utilization of direct solar radiation. *Solar Energy*, 26, 231–236.

Jeżowski, J. (1990a). Simple synthesis method of heat exchanger network withminimum number of units. *Chem. Eng. Sci.*, 45, 1928–1932.

Jeżowski, J. (1990b). Linear programming based method of heat exchanger network synthesis. *Inż. Chem. Proc.*, 10, 299–312.

Jeżowski, J. (1991). A note on the use of dual temperature approach in heat exchanger network synthesis. *Comput. Chem. Eng.*, 15, 305–312.

Jeżowski, J. (1992a). The pinch design method for tasks with multiple pinches. *Comput. Chem. Eng.*, 16, 129–133.

Jeżowski, J. (1992b). SYNHEN: Micromputer oriented package of programs for heat exchanger network synthesis. *Comput. Chem. Eng.*, 16, 691–706.

Jeżowski, J. (1994a). Heat exchanger network grassroot and retrofit design. The review of the state-of-the art: Part I. Heat exchanger network targeting and insight based methods of synthesis. *Hung. J. Ind. Chem*, 22, 279–294.

Jeżowski, J. (1994b). Heat exchanger network grassroot and retrofit design. The review of the state-of-the art: Part II. Heat exchanger network synthesis by mathematical methods and approaches for retrofit design. *Hung. J. Ind. Chem*, 22, 295–308.

Jeżowski, J. and Bochenek, R. (2000). Optymalizacja metodą adaptacyjnego przeszukiwania losowego w projektowaniu procesów. I. Opis metody optymalizacji. *Inż. Chem. Proc.*, 21, 443–463.

Jeżowski, J. and Bochenek, R. (2002). Experiences with the use of the Luus–Jaakola algorithm and its modifications in optimization of process engineering problems. In: Rein. Luus (Ed.), *Recent Developments in Optimization and Optimal Control in Chemical Engineering*. Research Signpost, Trivandrum, India, pp. 89–114.

Jeżowski, J. and Friedler, F. (1992). A simple approach for maximum heat recovery calculations. *Chem. Eng. Sci.*, 47, 1481–1494.

Jeżowski, J. and Jeżowska, A. (1999). Some remarks on heat exchanger networks targeting under uncertainty. *Hung. J. Ind. Chem*, 27(1), 17–24.

Jeżowski, J. and Jeżowska, A. (2002). Some remarks on heat exchanger networks targeting. *Chem. Pap.*, 56(6), 362–368.

Jeżowski, J. and Jeżowska, A. (2003). Adaptive random search optimization procedures in solver Opti-Sto. *Inż. Chem. Proc.*, 24(2), 261–279.

Jeżowski, J., Shethna, H.K., Bochenek, R. and Castillo, F.J.L. (2000a). On extensions of approaches for heat recovery calculations in integrated chemical process systems. *Comput. Chem.*, 24, 595–601.

Jeżowski, J., Bochenek, R. and Jeżowska, A. (2000b). Pinch locations at heat capacity flow-rate disturbances of streams for minimum utility cost heat exchanger networks. *Appl. Therm. Eng.*, 20, 1481–1494.

Jeżowski, J., Bochenek, R. and Jeżowska, A. (2001a). Loop breaking in heat exchanger networks by mathematical programming. *Appl. Therm. Eng.*, 21, 1429–1448.

Jeżowski, J., Poplewski, G. and Słoma, R. (2001b). Zastosowanie symulowanego wyżarzania do optymalizacji aparatów i procesów. *Inż. Chem. Proc.*, 22(3C), 573–578.

Jeżowski, J., Poplewski, G. and Jeżowska, A. (2003a). Optymalizacja metodą symulowanego wyżarzania w inżynierii chemicznej i procesowej. I. Algorytm optymalizacji z zastosowaniem metody simpleksów i symulowanego wyżarzania. *Inż. Chem. Proc.*, 24, 47–62.

Jeżowski, J., Poplewski, G. and Jeżowska, A. (2003b). Optymalizacja metodą symulowanego wyżarzania w inżynierii chemicznej i procesowej. II. Przykłady zastosowania i dobór wartości parametrów. *Inż. Chem. Proc.*, 24, 63–79.

Jeżowski, J., Poplewski, G., Jeżowska, A. and Słoma, R. (2003c). Parameter settingsin a simulated annealing/simplex optimization method, *4th European Congress of Chemical Engineering*, Granada, Spain, 2003 (P-9.2-052).

Jeżowski, J., Poplewski, G. and Jeżowska, A. (2003d). Optimization of water usage in chemical industry. *Environ. Protect. Eng.*, 29(1), 97–117.

Jeżowski, J., Shethna, H.K. and Castillo, F.J.L. (2003e). Area target for heat exchanger networks using linear programming. *Ind. Eng. Chem. Res.*, 42(8), 1723–1730.

Jeżowski, J., Bochenek, R. and Ziomek, G. (2005a). Random search optimization approach for highly multi-modal nonlinear problems. *Adv. Eng. Software*, 36(8), 504–517.

Jeżowski, J., Bochenek, R., Wałczyk, K. and Jeżowska, A. (2005b). Metoda i wspomaganie komputerowe do optymalnej modyfikacji sieci wymienników ciepła. I. Opis metody modyfikacji i modeli optymalizacyjnych. *Inż. Chem. Proc.*, 26, 745–758.

Jeżowski, J., Wałczyk, K., Szachnowskij, A. and Jeżowska, A. (2006). Systematic methods for calculating minimum flow rate and cost of water in industrial plants. *Chem. Proc. Eng.*, 27(3/2), 1137–1154.

Jeżowski, J., Bochenek, R. and Poplewski, G. (2007). On application of stochastic optimization techniques to designing heat exchanger- and water networks. *Chem. Eng. Process. (Special Issue)*, 46(11), 1160–1174.

Ji, S. and Bagajewicz, M. (2002). Design of crude fractionation units with preflashing or prefractionation: energy targeting. *Ind. Eng. Chem. Res.*, 41, 3003–3011.

Jiang, W., Fang, R., Khana, J.A. and Dougal, R.A. (2006). Parameter setting and analysis of a dynamic tubular SOFC model. *Journal of Power Sources*, 162, 316–326.

Jin, S. and Wen, X. (2005). Hamiltonian-preserving schemes for the Liouville equation with discontinuous potentials. *Commun. Math. Sci.*, 3, 285–315.

Jin, S. and Liao, X. (2006). A Hamiltonian-preserving scheme for high frequency elastic waves in heterogeneous media. *J. Hyperbolic Differential Equations*, 3, 741–777.

Jin, S. and Wen, X. (2006a). A Hamiltonian-preserving scheme for the Liouville equation of geometrical optics with transmissions and reflections. *SIAM J. Numer. Anal.*, 44, 1801–1828.

Jin, S. and Wen, X. (2006b). Hamiltonian-preserving schemes for the Liouville equation of geometrical optics with discontinuous local wave speeds. *J. Comput. Phys.*, 214, 672–697.

Jin, S. and Wen, X. (2006c). Computation of transmissions and reflections in geometrical optics via the reduced Liouville equation. *Wave Motion*, 43(8), 667–688.

Jin, S., Wu, H. and Huang, Z. (2007). A Hybrid Phase-Flow Method for Hamiltonian Systems with Discontinuous Hamiltonians, private communication of S. Jin.

Jödicke, G., Fischer, U. and Hungerbühler, K. (2001). Wastewater reuse: a new approach to screen for designs with minimal total costs. *Comput. Chem. Eng.*, 25, 203–215.

Johannessen, E. and Kjelstrup, S. (2005). A highway in state space for reactors with minimum entropy production. *Chem. Eng. Sci.*, 60, 3347–3361.

Johns, W.R. (2001). [Commentary] Computational complexity of heat exchanger networks (Furman & Sahinidis). *Comput. Chem. Eng.*, 25, 1391–1393.

Johnson, L. (2001). Imperfect symmetry: action principles in ecology and evolution. In: S.E. Jorgensen (Ed.), *Thermodynamics and Ecological Modelling*. Lewis Publishers of CRC Press, Boca Raton, pp. 229–286.

Jordan, B.W. and Polak, E. (1964a). Theory of a class of discrete optimal control systems. *J. Electron. Control*, 17, 697–713.

Jordan, B.W. and Polak, E. (1964b). Optimal control of aperiodic discrete time systems. *SIAM J. Control Ser. A*, 2, 332–338.

Jørgensen, S.E. (1988). Use of models as an experimental tool to show that structural changes are accompanied by increased exergy. *Ecol. Modell.*, 41, 117–126.

Jørgensen, S.E. (1997). *Integration of Ecosystem Theories: A Pattern* (second edition). Kluwer Academic Publishers, Dordrecht.

Jørgensen, S.E. (2000). Application of exergy and specific exergy as ecological indicators of coastal areas. *Aquat. Ecosyst. Health Manage.*, 3(3), 419–430.

S.E. Jørgensen (Ed.) (2001). *Thermodynamics and Ecological Modelling*. Lewis Publishers of CRC Press, Boca Raton.

Jørgensen, S.E., Mejer, H., Nielsen, S.N. and Teuber, J. (1998). The evolution of the thermodynamic equilibrium in the expanding universe. *Phys. Scr.*, 58, 543–544.

Jung, J.H., Lee, C.H. and In-Beum, Lee (1998). A genetic algorithm for scheduling of multi-product batch processes (Short Note). *Comput. Chem. Eng.*, 22(11), 1725–1730.

Kåberger, T. and Månsson, B. (2001). Entropy and economic processes—physics perspectives. *Ecol. Econ.*, 36, 165–179.

Kamien, M.I. and Schwartz, N.L. (1992). *Dynamic Optimization. The Calculus of Variations and Optimal Control in Economics and Management*. Wiley, New York.

Kaniadakis, G., Lissia, M. and A. Rapisarda (Eds.) (2002), *Non-extensive thermodynamics and its applications. Proceedings of the International School and Worksshop, NEXT 2001*, Villiasimius, Italy, May 23-30, 2001. *Phys. A, vol.* 305 (1 + 2).

G.E. Kanieviec (Ed.) (1978). *Mathematical Modelling and System Analysis of Heat Equipment*. Naukova Dumka, Kijew.

Kakaç, S., Pramuanjaroenkij, A., Zhou, X.Y., (2007). A review of numerical modeling of solid oxide fuel cells. International Journal of Hydrogen Energy, 32, 761–786.

Kanieviec, G.E. (1982). *Heat Exchangers and Heat Exchange Systems*. Naukova Dumka, Kijew.

Kaplinskiy, A.M. and Propoy, A.I. (1970). About stochastic approach to non-linear programming problem. *Automation Remote Control*, 3, 34–41.

Karatzas, I. (1983). A class of singular stochastic control problems. *Adv. Appl. Probab.*, 15, 225–254.

Karatzas, I. (1985). Probabilistic aspects of finite-fuel stochastic control. *Proc. Natl. Acad. Sci. U. S. A.*, 82, 5579–5581.

Karatzas, I. and Shreve, S.E. (1984). Connections between optimal stopping and singular stochastic control I. Monotone follower problems. *SIAM J. Control Optim.*, 22, 856–877.

Karatzas, I. and Shreve, S.E. (1985). Connections between optimal stopping and singular stochastic control II. Reflected follower problems. *SIAM J. Control Optim.*, 23, 433–451.

Karatzas, I. and Shreve, S.E. (1986). Equivalent models for finite-fuel stochastic control. *Stochastics*, 18, 245–276.

Karlsen, K.H. and Risebro, N.H. (2002). A note on front tracking and the equivalence between viscosity solutions of Hamilton–Jacobi equations and entropy solutions of scalar conservation laws. *Nonlinear Anal. Theory Methods Appl.*, 50, 455–469.

Karlsen, K.H, Risebro, N.H. and Towers, J.D. (2003). L^1 stability for entropy solutions of nonlinear degenerate parabolic convection-diffusion equations with discontinuous coefficients. *Preprint*, available at the url: www.math.ntnu.no/conservation.

Karuppiah, R. and Grossmann, I.E. (2006). Global optimization for the synthesis of integrated water systems in chemical processes. *Comput. Chem. Eng.*, 30, 650–673.

Kasat, R.B., Kunzru, D., Saraf, D.N. and Gupta, S.K. (2002). Multiobjective optimization of industrial FCC units using elitist nondominated sorting genetic algorithm. *Ind. Eng. Chem. Res.*, 41, 4765–4776.

Katare, S., Bhan, A., Caruthers, J.M., Delgass, N.C. and Venkatasubramanian, V. (2004). A hybrid genetic algorithm for efficient parameter estimation of large kinetic model. *Comput. Chem. Eng.*, 28, 2569–2581.

Kato, K. and Wen, C.H. (1969). Bubble assemblage model for fluidized-bed catalytic reactors. *Chem. Eng. Sci.*, 24, 1351–1368.

Katz, S. (1962). Best operating points for staged systems. *Ind. Eng. Chem. Fundam.*, 1, 226.

Kaushik, S.G., Chandra, S. and Gadhi, S.M.B. (1985). Thermodynamic feasibility ofdouble effect generation absorption system using water–salt and alcohol–salt mixtures as working fluids. *Heat Recov. Syst.*, 5, 19.

Kaushik, S.G. and Kumar, R. (1987). A comparative study of an absorber heat recovery cycle for solar refrigeration using NH_3-refrigerant with liquid–solid absorbent. *Int. J. Energy Res.*, 4, 123.

Kay, J. and Schneider, E.D. (1992). Thermodynamics and measures of ecological integrity, *Proceedings of Ecological Indicators*. Elsevier, Amsterdam, pp. 159–182.

Kays, W.M. and London, A.L. (1958). *Compact Heat Exchangers*. McGraw-Hill, New York.

Keckler, S.E. and Allen, D.T. (1999). Material reuse modeling: a case study of water reuse in an industrial park. *J. Ind. Ecol.*, 2(4), 79–92.

Keenan, J.H. (1932). A steam chart for second law analysis. *Mech. Eng. ASME*, 54, 195–204.

Keenan, J.H. (1941). *Thermodynamics*. Wiley, New York.

Kelahan, R.C. and Gaddy, J.J. (1977). Synthesis of heat exchange networks by mixed integer optimization. *AIChE J.*, 23(6), 423–435.

Keller, J.U. (1982). Efficiency coefficients in energy conversions with cogeneration. *Energy*, 7, 637.

Kemp, I.C. (1990). Applications of the time dependent heat cascade analysis in process integration. *Heat Recov. Syst. CHP*, 10(4), 423–435.

Kemp, I.C. (1991). Some aspects of the practical application of Pinch Technology methods. *Chem. Eng. Res. Des.*, 69(6), 471–479.

Kemp, I.C. and Deakin, A.W. (1989a). The cascade analysis for energy and process integration of batch processes. Part 1: Calculation of energy targets. *Chem. Eng. Res.Des.*, 67(5), 495–509.

Kemp, I.C. and Deakin, A.W. (1989b). The cascade analysis for energy and process mintegration of batch processes. Part 3: A case study. *Chem. Eng. Res. Des.*, 67(5), 517–525.

Kemp, I.C. and Deakin, A.W. (1989c). The cascade analysis for energy and process integration of batch processes. Part 2: Calculation of energy targets. *Chem. Eng. Res. Des.*, 67(5), 495–509.

Khalil, E.L. (1990). Entropy law and exhaustion of natural resources: Is Nicholas Georgescu–Roegen's paradigm defensible?. *Ecol. Econ.*, 2, 163–178.

Kharton, V.V., Naumovich, E.N., Tikhonovich, V.N., Bashmakov, I.A., Boginsky, L.S. and Kovalevsky, A.V. (1999). Testing tubular solid oxide fuel cells in nonsteady-state conditions. *J. of Power Sources*, 79, 242–249.

Kim, J.-K. and Smith, R. (2003). Automated retrofit design of cooling-water systems. *AIChE J.*, 49(7), 1712–1730.

Kim, J.-K. and Smith, R. (2004a). Automated design of discontinuous water systems. *Trans. IChemE*, 82(B3), 238–248.

Kim, J.-K. and Smith, R. (2004b). Cooling system design for water and wastewater minimization. *Ind. Eng. Chem. Res.*, 43, 608–613.

Kimia, B., Tannenbaum, A. and Zucker, S. (1994). On optimal control methods in computer vision and image processing. In: B. ter Haar Romeny (Ed.), *Geometry Driven Diffusion in Computer Vision*. Kluwer, Dordrecht, pp. 307–338.

King, C.J., Gantz, D.W. and Barnes, F.J. (1972). Systematic evolutionary system synthesis. *Ind. Eng. Chem. Process Design Dev.*, 11, 271.

Kirkpatrick, S., Gellat, C.D. and Vecchi, M.P. (1983). Optimization by simulated annealing. *Science*, 200, 671–680.

Kirwan, A.D. (2004). Intrinsic photon entropy? The darkside of light?. *Int. J. Eng. Sci.*, 42, 725–734.

Kitano, H. (2002). Overview: computational systems biology. *Nature*, 420, 206–210.

Kjelstrup, S., Sauar, E., van der Koi, H. and Bedeaux, D. (1999). The driving force distribution for minimum lost work in a chemical reactor close to and far from equilibrium. I. Theory. Ind. Eng. Chem. Res., 38, 3046–3050.

Kjelstrup, S., Johannessen, E., Rosjorde, A., Nummendal, L. and Bedeaux, D. (2000). Minimizing the entropy production of the methanol producing reaction in a methanol reactor. Int. J. Appl. Thermodyn., 3, 147–153.

Klausen, R.A. and Risebro, N.H. (1999). Stability of conservation laws with discontinuous coefficients. J. Differential Equations, 157, 41–60.

Klingenberg, C. and Risebro, N.H. (1995). Convex conservation laws with discontinuous coefficients. Existence, uniqueness and asymptotic behavior. Comm. Partial Differential Equations, 20, 1959–1990.

Klingenberg, C. and Risebro, N.H. (2001). Stability of a resonant system of conservation laws modeling polymer flow with gravitation. J. Differential Equations, 170, 344–380.

Klir, G.J. (1972). Trends in General Systems Theory. John Wiley, New York.

Kneese, A.V., Ayres, R.U. and d'Arge, R.C. (1972). Economics and the Environment: A Materials Balance Approach. Resources for the Future, Washington.

Kobrinski, N. (1972). Fundamentals of Control in Economic Systems. Wydawnictwa Naukowo Techniczne, Warsaw.

Kocis, G.R. and Grossmann, I.E. (1988). Global optimization of nonconvex mixedinteger nonlinear programming (MINLP) problems in process synthesis. Ind. Eng. Chem. Res., 27, 1407–1421.

Kocis, G.R. and Grossmann, I.E. (1989a). Computational experience with DICOPT solving MINLP problems in process systems engineering. Comput. Chem. Eng., 13(3), 307–315.

Kocis, G.R. and Grossmann, I.E. (1989b). A modelling and decomposition strategy for the MINLP optimization of process flowsheets. Comput. Chem. Eng., 13(7), 797–819.

Kodal, A. (1999). Heating rate maximization for an irreversible heat pump with a general heat transfer law. In: Ch. Wu, L. Chen and J. Chen (Eds.), Advance in Recent Finite Time Thermodynamics. Nova Science, New York, pp. 299–306.

Konukman, A.E.S. and Akman, U. (2002). Simultaneous flexibility targeting and synthesis of minimum-utility heat-exchanger networks with superstructure-based simultaneous MILP formulation. Chem. Eng. Process, 41, 501–518.

Kooijman, S.A.L.M. and Nisbet, R.M. (2001). How light and nutrients affect life in a closed bottle. In: S.E. Jorgensen (Ed.), Thermodynamics and Ecological Modelling. Lewis Publishers of CRC Press, Boca Raton, pp. 17–60.

Kookos, I.O. and Perkins, J.D. (2001). An algorithm for simultaneous design and control. Ind. Eng. Chem. Res., 40, 4079–4088.

Koppol, A. and Bagajewicz, M. (2003). Financial risk management in the design of water utilization systems in process plants. Ind. Eng. Chem. Res., 42(21), 5249–5255.

Koppol, A.P.R., Bagajewicz, M.J., Dericks, B.J. and Savelski, M.J. (2003). On zero water discharge solutions in the process industry. Adv. Environ. Res., 8(2), 151–171.

Kotas, T.J. (1985). Exergy Method of Thermal Plant Analysis. Butterworths, Borough Green.

Kotas, T.J. (1986). Exergy method of thermal and chemical plant analysis. Chem. Eng. Res. Des., 64, 212.

Kotjabasakis, E. and Linnhoff, B. (1986). Sensitivity tables for the design of flexible processes (1)—How much contingency in heat exchanger networks is cost-effective?. Chem. Eng. Res. Des., 64, 197–211.

Kotjabasakis, E. and Linnhoff, B. (1988). Sensitivity tables for the design of flexible processes (2)—a case study, *Understanding Process Integration II. IChemE Symposium
Series 109*, Institution of Chemical Engineers, Rugby, England, pp. 181–203.

Kovač Kralj, A. and Glavič, P. (1995). Retrofit of complex and energy intensive
processes—I. *Comput. Chem. Eng.*, 19(12), 1255–1270.

Kovač Kralj, A. and Glavič, P. (2000). Simultaneous retrofit of complex and energy
intensive processes—III. *Comput. Chem. Eng.*, 24, 1229–1235.

Kovač Kralj, A., Glavič, P. and Kravanja, Z. (2000a). Retrofit of complex and energy
intensive processes—II. Simultaneous superstructural approach. *Comput. Chem. Eng.*,
24, 125–138.

Kovač Kralj, A., Glavic, P. and Kravanja, Z. (2005). Heat integration between processes:
integrated structure and MINLP model. *Comput. Chem. Eng.*, 29, 1699–1711.

Kovacs, Z., Ercsey, E., Friedler, F. and Fan, L.T. (2000b). Separation-network synthesis:
global optimum through rigorous superstructure. *Comput. Chem. Eng.*, 24, 1881–
1900.

Kowalik, J. and Osborne, M.R. (1968). *Methods for Unconstrained Optimization Problems*. Elsevier, New York.

Kramarz, J. and Wyczesany, A. (1989a). The exergy analysis of the mazout and coal
gasification processes. *Chem. Stos.*, 33, 621.

Kramarz, J. and Wyczesany, A. (1989b). Exergy analysis of three complex hydrogen plants
working for the purpose of the coal hydrogenation. *Koks, Smola, Gaz*, 3, 111.

Kramarz, J. and Wyczesany, A. (1992). Application of the second law analysis for the
energy efficiency studies of various coal to liquid fuels processes. Part 5. Which way
to coal liquefaction?. *Polish J. Appl. Chem.*, 36, 51–59.

Krane, R.J. (1987). A second law analysis of the optimum design and operation of thermal
energy storage systems. *Int. J. Heat Mass Transfer*, 30, 43.

Kreider, J.F. and Kreith, F. (1975). *Solar Heating and Cooling*. McGraw-Hill, New York.

Krotschek, C. and Narodoslavsky, M. (1996). The Sustainable Process Index—a new
dimension in ecological evolution. *Ecol. Eng.*, 6, 241–258.

Krummenacher, P. and Favrat, D. (2000). Indirect and mixed direct–indirect heat integration of batch processes based on Pinch analysis. In: G.G. Hirs (Ed.), *ECOS
2000 Proceedings. Part 3: Process Integration*. University Twente, The Netherlands,
pp. 1411–1424.

Kruzkov, S.N. (1970). First order quasi-linear equations in several independent variables.
Math. USSR Sbornik, 10, 217–243.

Krylov, N.V. (1980). *Controlled Diffusion Processes*. Springer-Verlag, Berlin.

Ku, H. and Karimi, I. (1991). An evaluation of simulated annealing for batch process
scheduling. *Ind. Eng. Chem. Res.*, 30, 163–169.

Kubiak, M. (2005). Thermodynamic limits for production and consumption of mechanical energy in theory of heat pumps and heat engines. Ph.D. Thesis (in Polish).
Wydawnictwa Politechniki Warszawskiej, Warsaw.

Kuchonthara, P., Bhattacharya, S. and Tsutsumi, A. (2003). Energy recuperation in solid
oxide fuel cell (SOFC) and gas turbine (GT) combined system. *J. of Power Sources*,
117, 7–13.

Kuddusi, L. and Egrican, N. (2008). A critical review of constructal theory. *Energy
Convers. Manage.*, 49, 1283–1294.

Kuhn, H.W. and Tucker, A.W. (1951). Nonlinear programming. In: J. Neyman (Ed.),
*Proceedings of the Second Berkeley Symposium on Mathematical Statistics and Prob-
ability*. University of California Press, Berkeley, pp. 481–493.

Kunii, D. and Levenspiel, O. (1969). *Fluidization Engineering*. Wiley, New York.

Kunii, D. and Levenspiel, O. (1991). *Fluidization Engineering*. Butterworth Heinemann, Newton.

Kushner, H.J. (1965). On the stochastic maximum principle: fixed time of control. *J. Math. Anal. Appl.*, 11, 78–92.

Kuo, W.C.J. and Smith, R. (1997). Effluent treatment system design. *Chem. Eng. Sci.*, 52, 4273–4290.

Kuo, W.C.J. and Smith, R. (1998a). Design of water-using systems involving regeneration. *Trans IChemE*, 76(Part A), 287–301.

Kuo, W.C.J. and Smith, R. (1998b). Designing for the interactions between water-use and effluent treatment. *Trans. IChemE*, 76(Part A), 287–301.

Kupershmidt, B.A. (1992). *The Variational Principles of Dynamics*. World Scientific, New York.

Ku-Pineda, V. and Tan, R.R. (2006). Environmental performance optimization using process water integration and Sustainable Process Index. *J. Cleaner Prod.*, 14, 1586–1592.

Kuran, P. (2006). Nonlinear models for production of mechanical energy in imperfect generators driven by thermal or solar energy (in Polish). Ph.D. Thesis. Wydawnictwa Politechniki Warszawskiej, Warsaw.

Kuran, P. and Sieniutycz, S. (2007). Pogląd na obliczanie egzergii i strumienia egzergii promieniowania słonecznego. *Inżynieria i Aparatura Chemiczna*, 6, 7–14.

Kuznecov, A.G., Rudenko, A.A. and Tsirlin, A.M. (1985). Optimal control of thermodynamics systems with finite heat capacity. *Automation Remote Control*, 6, 62–72.

Lababidi, H.M.S., Alatiqi, I.M. and Nayfeh, L.J. (2000). Energy retrofit study of an ammonia plant. *Appl. Therm. Eng.*, 20, 1495–1503.

Lakshmanan, R. and Banares-Alcantara, R. (1998). Retrofit by inspection using thermodynamic process visualisation. *Comput. Chem. Eng.*, 22(Suppl), S809–S812.

Lakshmanan, R. and Fraga, E.S. (2002). Pinch location and minimum temperature approach for discontinuous composite curves. *Comput. Chem. Eng.*, 26, 779–783.

Lanczos, C. (1949). *The Variational Principles of Mechanics*. Toronto University Press, Toronto.

Landau, L.D. and Lifshitz, E. (1971). *Mechanics*. Pergamon, Oxford.

Landau, L.D. and Lifshitz, E. (1974). *Statistical Physics*. Pergamon, Oxford.

Landsberg, P. (1990). Statistics and thermodynamics of energy conversion from radiation. In: Nonequilibrium theory and extremum principles. *Adv. Thermodyn. Ser*, 3, 482–536.

Landsberg, P.T. (1984). Can entropy and order increase together? *Phys. Lett.*, 102 A, 171.

Landsberg, P.T. and Mallinson, J.R. (1976). Thermodynamic constraints, effective temperatures and solar cells, *Coll. Int. sur l'Electricite Solaire*, CNES, Toulouse, pp. 27–35.

Larson, R.E. (1967). *State Increment Dynamic Programming*. Elsevier, New York.

Larminie, J. and Dinks, A. (2000). *Fuel Cell Systems Explained*. Wiley, New York http://en.wikipedia.org/wiki/Fuel_cell.

Lasdon, L.S., Mitter, S.K. and Waren, A.D. (1967). The conjugate gradient method for optimal control problems. *IEE Trans. Automatic Control*, AC-12, 132.

Lasota, A. and Mackey, M.C. (1985). *Probabilistic Properties of Deterministic Systems*. Cambridge University Press, Cambridge.

Lavenda, B.H. (1985). *Nonequilibrium Statistical Thermodynamics*. Wiley, Chichester.

Lax, P.D. (1973). Hyperbolic systems of conservation laws and the mathematical theory of shock waves, *Conference Board of the Mathematical Sciences Regional Conference Series in Applied Mathematics, No. 11.* Society for Industrial and Applied Mathematics, Philadelphia.

Lavric, V., Iancu, P. and Pleşu, V. (2005). Genetic algorithm optimisation of water consumption and wastewater network topology. *J. Cleaner Prod*, 13, 1405–1415.

Lazzaretto, A. and Andreatta, R. (1995). Algebraic formulation of a process-based exergycosting method. In: R.J. Krane (Ed.). *Symposium on Thermodynamics and the Design, Analysis, and Improvement of Energy Systems, AES-Vol. 35*, ASME, New York, pp. 395–403.

Lazzaretto, A. and Macor, A. (1995). Direct calculation of average and marginal costs from the productive structure of an energy system. *J. Energy Resour. Technol*, 117, 171–178.

Lazzaretto, A. and Tsatsaronis, G. (1999). On the calculation of efficiencies and costs in thermal systems. *ASME, AES*, 39, 421–430.

Lazzaretto, A. and Tsatsaronis, G. (2006). SPECO: A systematic and general methodology for calculating efficiencies and costs in thermal systems. *Energy Int. J*, 31, 1257–1289.

Ledezma, G.A., Bejan, A. and Errera, M.R. (1997). Constructal tree networks for heat transfer. *J. Appl. Phys.*, 82, 89–100.

Lee, S. and Grossmann, I.E. (2001). A global optimisation algorithm for nonconvex generalized disjunctive programming and applications to process systems. *Comput. Chem. Eng.*, 25, 1675–1697.

Lee, S. and Grossmann, I.E. (2003). Global optimization of nonlinear generalized disjunctive programming with bilinear equality constraints: applications to process networks. *Comput. Chem. Eng.*, 27, 1557–1575.

Lee, E.B. and Marcus, L. (1967). *Foundations of Optimal Control Theory*. Wiley, New York.

Lee, Y.P., Rangaiah, G.P. and Luus, R. (1999). Phase and chemical equilibrium calculations by direct search optimization. *Comput. Chem. Eng.*, 23, 1183–1191.

Le Goff, P. (1979). *Energetique Industrielle*. Lavoisier, vol. 1, Paris.

Le Goff, P. (1980). *Energetique Industrielle*. Laviosier, vol. 2, Paris.

Le Goff, P., Le Goff, H., Barkaoui, M., Rmadane, A., Dietrich, E. and Trap, J.C. (1986). Chemical storage of enthalpy, entropy or exergy? An inventory of the thermal multipole system, *World Congress III of Chemical Engineering*, Japan, Tokyo, September.

Le Goff, P., Aoufoussi, Z. and Louis, G. (1988). Le Concept de "Quadripole Exergetique" et ses Applications. *J. de Chimie Physique*, 85, 247.

Le Goff, P., Matsuda, H. and Rivero, R. (1990a). Advances in chemical plants and heat transformers, *The 3rd IEA Heat Pump Conference*, Tokyo.

Le Goff, P., Rivero, R., de Oliveira, S. and Cachot, T. (1990b). Application of the enthalpy–Carnot factor diagram to the exergy analysis of distillation process, *Symposium on Thermodynamics and the Design Analysis and Improvement of Energy Systems, ASME Winter Annual Meeting*, Dallas, TX, November 25-30.

Le Goff, P., Schwarzer, B., Rivero, R., de Oliveira, S., Liu, B. and Aoufoussi, Z. (1990c). Exergy effectiveness and exergo-economic efficiency of sorption heat pumps and heat transformers, *3rd International Workshop on Research Activities on Advanced Heat Pumps*, Graz, September 24-26.

Le Goff, P., de Oliveira, S., Matsuda, H., Ranger, P.M., Rivero, R. and Liu, B. (1990d). Heat transformers for upgrading waste heat from industrial processes, *Proceedings of the FLOWER'S 90 Conference*, Firenze, May.

Leitman, G. (1966). *An Introduction to Optimal Control*. McGraw-Hill, New York.

Leitman, G. (1981). *The Calculus of Variations and Optimal Control*. Plenum Press, New York.

Lemański, M. (2003). Detailed balances of enthalpy and chemical species in the tubular SOFC. *Reports of the Institute of Fluid Flow Machinery*. No. 3175/2003, 1-26 (in Polish).

Lemański, M. (2007). Analiza obiegów energetycznych z ogniwem paliwowym i turbiną gazowo – parową. *Ph.D. Thesis*. Institute of Fluid Flow Machinery, Gdańsk.

Lemański, M. and Badur, J. (2004). Parametrical analysis of a tubular pressurized SOFC. *Arch. Thermodyn.*, 25, 53–72.

Lemański, M. and Badur, J. (2005). Ogniwo paliwowe SOFC z wewnętrznym reformingiem. *Inżynieria Chemiczna i Procesowa*, 26, 157–172.

Lemański, M., Topolski, J., Badur J. (2004). Analysis strategies for gas turbine-solid oxide fuel cell hybrid cycles. In: *Technical, Economic, and Environmental Aspects of Combined Cycle Power Plants*, Editor: Z. Domachowski, Gdańsk TU Press, pp. 213–220.

Lems, S, van der Koi, H.J. and de Swaan Arons., H.J. (2003). Thermodynamic optimization of energy transfer in (bio)chemical reaction systems. *Chemical Engineering Science*, 58, 2001–2009.

Leonard, D.L. and Van Long, N. (1994). *Optimal Control Theory and Static Optimization in Economics*. Cambridge University Press, Cambridge.

Leontief, W.W. (1951). *The Structure of American Economy 1919-1939*. Oxford University Press, New York.

Levenspiel, O. (1988). Chemical engineering grand adventure. *Chem. Eng. Sci.*, 43, 1427–1435.

Lewandowski, W.M. (2001). *Proecological Sources of Renewable Energy (in Polish)*. Wydawnictwa Naukowo Techniczne, Warsaw.

Lewin, D.R. (1998). A generalized method for HEN synthesis using stochastic optimization—II. The synthesis of cost-optimal network. *Comput. Chem. Eng.*, 22, 1387–1405.

Lewin, D.R., Wang, H. and Shalev, O. (1998). A generalized method for HEN synthesis using stochastic optimization—I. General framework and MER optimal synthesis. *Comput. Chem. Eng.*, 22(10), 1503–1515.

Lewins, J. (2003). Bejan's constructal theory of equal potential distribution. *Int. J. Heat Mass Transfer*, 46, 1541–1543.

Li, P.-W. and Chyu, M.K. (2004). Simulation of the chemical/electrochemical reactions and heat/mass transfer for a tubular SOFC in a stack. *J. Power Sources*, 124, 487–498.

Li, P.-W. and Chyu, M.K. (2005). Electrochemical and transport phenomena in Solid Oxide Fuel Cells. *Transactions of the ASME*, 127, 1344–1362.

Li, P.-W. and Suzuki, K. (2004). Numerical Modeling and Performance Study of a Tubular SOFC. *Journal of The Electrochemical Society*, 151, A548–A557.

Li, P.-W., Tao, G. and Liu, H. (2011). Effect of the geometries of current collectors on the power density in a solid oxide fuel cell. *International Journal of Energy and Environmental Engineering*, 2(3), 1–11.

Li, Z.H. and Hua, B. (2000). Modeling and optimizing for heat exchanger networks synthesis based on expert system and exergo-economic objective function. *Comput. Chem. Eng.*, 24, 1223–1228.

Li, X. and Kraslawski, A. (2004). Conceptual process synthesis: past and current trends. *Chem. Eng. Process*, 4, 589–600.

Li, J. and Rhinehart, R. (1998). Heuristic random optimization. *Comput. Chem. Eng.*, 3(22), 427–444.

Li, P., Löwe, K., Arellano-Garcia, H. and Wozny, G. (2000). Integration of simulated annealing to a simulation tool for dynamic optimization of chemical processes. *Chem. Eng. Process.*, 39, 357–363.

Li, J., Chen, L., Sun, F. and Wu, C. (2008a). Power vs. efficiency characteristic of an endoreversible Carnot heat engine with heat transfer law $q \propto (\Delta T^n)^m$. *Int. J. Ambient Energy*, 29(3), 149–152.

Li, J., Chen, L. and Sun, F. (2008b). Performance optimization for an endoreversible Carnot refrigerator with complex heat transfer law. *J. Energy Inst.*, 81(3), 168–170.

Li, J., Chen, L. and Sun, F. (2008c). Heating load vs. COP characteristic of an endoreversible Carnot heat pump subjected to heat transfer law $q \propto (\Delta T^n)^m$. *Appl. Energy*, 85(2–3), 96–100.

Li, J., Chen, L. and Sun, F. (2009a). Optimum work in real systems with a class of finite thermal capacity reservoirs. *Math. Comput. Modell.*, 49, 542–547.

Li, J., Chen, L. and Sun, F. (2009b). Maximum work output of multistage continuous Carnot heat engine system with finite reservoirs of thermal capacity and radiation between heat source and working fluid. *Thermal Sci.*, 14, 1–9.

Li, J., Chen, L. and Sun, F. (2010). Fundamental optimal relation of a generalized irreversible Carnot heat pump with complex heat transfer law. *Pramana J. Phys.*, 74(2), 219–230.

Li, X. (2006). *Principles of Fuel Cells*. Taylor and Francis, New York.

Liao, B. and Luus, R. (2005). Comparison of the Luus–Jaakola optimization procedure and the genetic algorithm. *Eng. Optim.*, 37(4), 381–398.

Lima, R.M., Salcedo, R.L. and Barbosa, D. (2006). SIMOP: Efficient reactive distillation optimization using stochastic optimizers. *Chem. Eng. Sci.*, 61, 1718–1739.

Lin, B. and Miller, D.C. (2004). Solving heat exchanger network synthesis problems with Tabu Search. *Comput. Chem. Eng.*, 28, 1451–1464.

Lin, G., Chen, J. (Jincan), and Bruck, E. (2004). Irreversible chemical-engines and their optimal performance analysis, *Applied Energy*, 78, 123-136.

Lin, B.L., Wu, R.S. and Liaw, S.L. (1997). A heuristic approach algorithm for the optimization of water distribution networks. *Water Sci. Technol.*, 36(5), 219–226.

Lin, P.-H. and Hong, Ch.-W. (2006). On the start-up transient simulation of a turbo fuel cell system. *J. of Power Sources*, 160, 1230–1241.

Linke, P. and Kokossis, C.A. (2003). On the robust application of stochastic optimization technology for the synthesis of reaction/separation systems. *Comput. Chem. Eng.*, 27, 733–758.

Linnhoff, B. (1979). Thermodynamic analysis in the design of process networks. Ph.D. Dissertation. University of Leeds, UK.

Linnhoff, B. and Ahmad, S. (1990). Cost optimum heat exchanger networks. 1. Minimum energy and capital using simple models for capital cost. Comput. Chem. Eng., 14, 729–750.

Linnhoff, B. and Eastwood, A.R. (1987). Overall site optimisation by pinch technology. *Chem. Eng. Res. Des*, 65, 408–414.

Linnhoff, B. and Flower, J.R. (1978). Synthesis of heat exchanger networks: systematic generation of energy optimal networks. *AIChE J*, 24(4), 633–642.

Linnhoff, B. and Hindmarsh, E. (1983). The pinch design method for heat exchanger networks. *Chem. Eng. Sci.*, 38(5), 745–763.

Linnhoff, B. and Kotjabasakis, E. (1986). Downstream paths for operable process design. *Chem. Eng. Prog.* (May), 23-28.

Linnhoff, B. and Townsend, B. (1982). Designing total energy systems. *Chem. Eng. Prog.*, 78(7), 72–80.

Linnhoff, B., Mason, D.R. and Wardle, I. (1979). Understanding heat exchanger networks. *Comput. Chem. Eng.*, 3, 295–302.

Linnhoff, B., Townsend, D.W., Boland, D., Hewitt, G.F., Thomas, B.E.A., Guy, A.R. and Marsland, R.H. (1982). *A User Guide on Process Integration for the Efficient Use of Energy*. IChemE, Rugby, UK.

Linnhoff, B., Polley, G.T. and Sahdev, V. (1988). General process improvements through Pinch technology. *Chem. Eng. Progr.* (June), 51.

Lions, P.L. (1982). Generalized solutions of Hamilton–Jacobi equations. *Research Notes in Mathematics*. Pitman, Boston, Vol. 69.

Lions, P.L. (1983a). On the Hamilton–Jacobi–Bellman equations. *Acta Appl.*, 1, 17–41.

Lions, P.-L. (1983b). Optimal control of diffusion processes and Hamilton–Jacobi–Bellman equations. Part 1: The dynamic programming principle and applications. Part 2: Viscosity solutions and uniqueness. *Comm. Partial Differential Equations*, 8, 1101–1174 and 1229–1276.

Lions, P.L. (1985). Optimal control and viscosity solutions. In: I. Capuzzo Dolcetta, W.H. Fleming and T. Zolezzi (Eds.), *Recent Mathematical Methods in Dynamic Programming, Lecture Notes in Mathematics, 1119*. Springer, Berlin.

Lions, P.L. and Souganidis, P.E. (1988). Viscosity solutions of second-order equations, stochastic control and stochastic differential games. In: W.H. Fleming and P.L. Lions (Eds.), *Stochastic Differential Systems, Stochastic Control Theory and Applications, IMA Vol. Math. Appl.*, 10. Springer, Berlin.

Lior, N., Dunbar, W.R. (1991). Understanding combustion irreversibility (Winner of the ASME Advanced Energy Systems Edward F. Obert Best Paper Award), in AES-Vol.25/HTD-vol. 191, *Second Law Analysis-Industrial and Environmental Applications*, Ed. By G.M. Reistag, M.J. Moran, W.J. Wepfer and N.Lior, ASME Winter Annular Meeting, pp. 81–90.

Lior, N. (2002). Thoughts about future power generation systems and the role of exergy analysis in their development. *Energy Convers. Manage*, 43, 1187–1198.

Lior, N., Dunbar, W.R. and Gaggioli, R.A. (1991). Combining fuel cells with fuel-fired power plants for improved energy efficiency. *Energy*, 16(10), 1259–1274.

Liu, S. and Lin, Yi. (2005). *Grey Information: Theory and Practical Applications (Advanced Information and Knowledge Processing)*. Springer Verlag, Berlin.

Llewellyn, R.W. (1963). *Linear Programming*. Holt, Rinehalt and Wilson, New York.

Locatelli, A. (2001). *Optimal Control. An Introduction*. Birkhauser, Basel.

Locatelli, M. (2000). Convergence of a simulated annealing algorithm for continuous global optimization. *J. Global Optim.*, 18, 219–234.

Logan, B.E. (2008). *Microbial fuel cells*. Wiley, New York.

Logan, J.D. (1977). *Invariant Variational Principles*. Academic Press, New York.

Lorente, S.J. (2007). Constructal view of electrokinetic transfer through porous media. *Phys.D: Appl. Phys.*, 40, 2941–2947.

Lorente, S. and Bejan, A. (2006). Heterogeneous porous media as multiscale structures for maximum flow access. *J. Appl. Phys.*, 100 11490-9-1–114909-8.

Lotka, A.J. (1921). Note on economic conversion factors of energy. *Am. J. Proc. Natl. Acad. Sci.*, 7, 192–197.

Lovley, D.R. (2008). The microbe electric: conversion of organic matter to electricity. *Current Opinions in Biotechnology*, 19, 564–571.

Lozada, G.A. (1991). A defense of Nicholas Georgescu-Roegens's paradigm. *Ecol. Econ.*, 3, 157–160.

Lozada, G.A. (1995). Georgescu–Roegens's defence of classical thermodynamics revisited. *Ecol. Econ.*, 14, 31–44.

Lozano, M.A. and Valero, A. (1987). Application of the exergetic cost theory to a stream boiler in a thermal generating station. In: M.J. Moran, S.S. Stecco and G.M. Reistad, (Eds.). *Analysis and Design of Advanced Energy Systems: Applications, AES, ASME Book No. G0377B*. ASME, New York, vol. 3-2, pp. 41–51.

Lozano, M.A. and Valero, A. (1988). Methodology for calculating exergy in chemical process. In: A.J. Wepfer, G. Tsatsaronis and R.A. Bajura (Eds.), *Thermodynamic Analysis of Chemically Reactive Systems, ASME Book No. G00449*. ASME, New York, pp. 77–86.

Lozano, M.A. and Valero, A. (1993a). Theory of exergetic cost. *Energy*, 18, 939–960.

Lozano, M.A. and Valero, A. (1993b). Thermoeconomic analysis of gas turbine cogeneration systems. In: H.J. Richter (Ed.), *Thermodynamic and Design Analysis and Improvement of Energy Systems, ASME Book no. H00874*. ASME, New York, pp. 311–320.

Lozano, M.A., Valero, A. and Serra, L. (1993). Theory of exergetic cost and thermoeconomic optimization. In: J. Szargut, Z. Kolenda, G. Tsatsaronis and A. Ziebik (Eds.), *Proceedings of the International Conference Energy Systems and Ecology ENSEC'93*, Cracow, Poland, July 5-9, pp. 339–350.

Lozano, M.A., Bartolome, J.L., Valero, A. and Reini, M. (1994). Thermoeconomic diagnosis of energy systems, *FLOWERS'94, Proceedings of the Florence World Energy Research Symposium* , pp. 149–156.

Luenberger, D.G. (1974). *Theory of Optimization*. Wydawnictwa Naukowo Techniczne, Warsaw.

Lucia, U. and Grazzini, G. (1997). Global analysis of dissipations due to irreversibility. *Rev. Gén. Therm*, 36, 605.

Luikov, A.V. and Mikhailov, Y.A. (1963). *Theory of Energy and Mass Transfer (in Russian)*. Gosenergoizdat, Moscow (English transl. by L.A. Fenn, Pergamon Press, London, 1968).

Lund, K. (1986). General thermal analysis of parallel-flow flat-plate solar collector absorbers. *Solar Energy*, 36, 443.

Lund, K. (1989). General thermal analysis of serpentine-flow flat-plate solar collector absorbers. *Solar Energy*, 42, 133.

Lund, K. (1990). Application of finite-time thermodynamics to solar power conversion. In: S. Sieniutycz and P. Salamon (Eds.), *Finite-Time Thermodynamics and Thermoeconomics, Advances in Thermodynamics 4*. Taylor and Francis, New York.

Luo, L. and Feidt, M. (1992). Thermodynamics of adsorption cycles; a theoretical study. *Heat Transfer Engineering*, 13(4), 1–13.

Luo, Y., Zhang, L. and Li, M. (2001). *Application of the Gray System Theory in the Mechanical Engineering*. China National Defense Science and Technology University Publisher, Changsha.

Lutz, A.E., Larson, R.S. and Keller, J.O. (2002). Thermodynamic comparison of fuel cells to the Carnot cycle. *Intern. J. of Hydrogen Energy*, 27, 1103–1111.

Luus, R. (1973). A direct approach to optimization of a complex system. *AIChE J.*, 19(3), 645–646.

Luus, R. (1974). Two-pass method for handling difficult equality constraints in optimization. *AIChE J.*, 20(3), 608–610.

Luus, R. (1975). Optimization of multistage recycle systems by direct search. *Can. J. Chem. Eng.*, 53, 217–220.

Luus, R. (1993). Optimization of heat exchanger networks. *Ind. Eng. Chem. Res.*, 32(11), 2633–2635.

Luus, R. (1996). Numerical convergence properties of iterative dynamic programming when applied to high dimensional systems. *Trans. IChemE*, 74(Part A), 55–62.

Luus, R. (2002). Comparison of LJ optimization procedure and IDP in solving optimal control problems. In: Rein Luus (Ed.), *Recent Developments in Optimization and Optimal Control in Chemical Engineering*. Research Signpost, Trivandrum, India, pp. 166–253.

Luus, R. and Brenek, P. (1989). Incorporation of gradient into random search optimization. *Chem. Eng. Technol.*, 12, 309–318.

Luus, R. and Jaakola, T.H.I. (1973). Optimization by direct search and systematic reduction of the size of search region. *AIChE J.*, 19(4), 760–766.

Luus, R., Iyer, R.S. and Woo, S.S. (2002). Handling equality constraints in direct search. In: Luus. Rein (Ed.), *Recent Developments in Optimization and Optimal Control in Chemical Engineering*. Research Signpost, Trivandrum, India, pp. 120–178.

Lyusternik, L.A. (1983). *The Shortest Lines Variational Problems*. Mir Publishers, Moscow.

Ma, K.L., Yee, T.F. and Hui, C.W. (1998). MILP utility and structural modification model for the retrofit of heat exchanger networks, *Congress CHISA'98*, Prague, Czech Republic.

Ma, K.L., Hui, C.W. and Yee, T.F. (1999). Constant approach temperature model for the retrofit of heat exchanger networks—simultaneous optimisation for energy, structural modifications and new area cost, *Proceedings of PRES'99*, Budapest, Hungary.

Ma, K.L., Hui, C.W. and Yee, T.F. (2000). Constant approach temperature model for HEN retrofit. *Appl. Therm. Eng.*, 20, 1505–1533.

Mackey, M.C. (1989). The dynamic origin of increasing entropy. *Rev. Modern Phys.*, 61, 981–1015.

MacLean, H. and Lave, L.B. (2003). Evaluating automobile fuel/propulsion system technologies. *Progress in Energy & Combustion Science*, 29, 1–69 (Lifecycle problems).

Maia, L.O.A., de Carvalho, L.A. and Qassim, R.Y. (1995). Synthesis of utility systems by simulated annealing. *Comput. Chem. Eng.*, 19(4), 481–488.

Maia, L.O. and Qassim, R.Y. (1997). Synthesis of utility systems with variable demands using simulated annealing. *Comput. Chem. Eng.*, 21(9), 947–950.

Maier, R.W. and Whiting, W.B. (1998). The variation of parameter settings and their effects on performance for the simulated annealing algorithm. *Comput. Chem. Eng.*, 23(1), 47–62.

Maikov, G.P. and Tsirlin, A.M. (1973). Conditions of optimality for various forms of control process. *Izv. AN SSSR, Ser. Tech. Cybern.*, 5, 63–71.

Majozi, T. (2005a). Wastewater minimisation using central reusable water storage in batch plants. *Comput. Chem. Eng.*, 29, 1631–1646.

Majozi, T. (2005b). An effective technique for wastewater minimisation in batch processes. *J. Cleaner Prod.*, 13, 1374–1380.

Majozi, T. (2006). Heat integration of multipurpose batch plants using a continuous-time framework. *Appl. Therm. Eng.*, 26, 1369–1377.

Makwana, Y., Smith, R. and Zhu, X.X. (1998). A novel approach for retrofit and operations management of existing total sites. *Comput. Chem. Eng.*, 22(Suppl.), S793–S796.

Malih, V. (1987). Thermodynamic constraints and possibilities of isothermal separation processes. *Vsesojuznyi Institut Nautshnoj Informacji* (VINITI) 2020, B87, 12.

Manan, Z.A. and Foo, D.C.Y. (2003). Setting targets for water and hydrogen networks using cascade analysis, *AIChE Annual Meeting*, San Francisco, CA, pp. 16–21.

Manan, Z.A., Tan, Y.L. and Foo, D.C.Y. (2004). Targeting the minimum water flow rate using water cascade analysis technique. *AIChE J.*, 50(12), 3169–3183.

Mangasarian, O.L. (1969). *Nonlinear Programming.* McGraw-Hill, New York.

Mann, Y.G. and Liu, Y.A. (1999). *Industrial Water Reuse and Wastewater Minimization.* McGraw-Hill, New York.

Manninen, J. and Zhu, X.X. (2001). Level-by-level flowsheet synthesis methodology for thermal system design. *AIChE J.*, 47(1), 142–159.

Manousiouthakis, V. and Martin, L.L. (2004). A minimum area (MA) targeting scheme for single component MEN and HEN synthesis. *Comput. Chem. Eng.*, 28, 1237–1247.

Manousiouthakis, V. and Sourlas, D. (1992). A global optimization approach to rationally constrained rational programming. *Chem. Eng. Commun.*, 115, 127–147.

Mansoori, G.A. and Patel, V. (1979). Thermodynamic basis for the choice of working fluids for solar absorption cooling systems. *Solar Energy*, 22, 483–491.

Månsson, B.Å. (1990) Thermodynamics and economics. In: S. Sieniutycz and P. Salamon (Eds.), *Finite-Time Thermodynamics and Thermoeconomics, Advances in Thermodynamics* Taylor and Francis, New York, vol. 4, pp. 153–174.

Maranas, C.D. and Floudas, C.A. (1997). Global optimization in generalized geometric programming. *Comput. Chem. Eng.*, 21(4), 351–369.

Marcoulaki, E.C. and Kokossis, C.A. (2000a). On the development of novel chemicals using a systematic synthesis approach. Part I. Optimisation framework. *Chem. Eng. Sci.*, 55, 2529–2546.

Marcoulaki, E.C. and Kokossis, C.A. (2000b). On the development of novel chemicals using a systematic synthesis approach. Part II. Solvent design. *Chem. Eng. Sci.*, 55, 2547–2561.

Marcoulaki, E.C., Kokossis, A.C. and Batzias, F.A. (2000). Novel chemicals for clean and efficient processes using stochastic optimization. *Comput. Chem. Eng.*, 24, 705–710.

Markowski, M. (2000). Reconstruction of a heat exchanger network under industrial constraints—the case of a crude distillation unit. *Appl. Therm. Eng.*, 20, 1535–1544.

Marsden, J.E. (1992). *Lectures on Mechanics.* Cambridge University Press, Cambridge.

Martin, D.L. and Gaddy, J.L. (1982). Process optimization with the adaptive randomly directed search. *AIChE Symp. Ser.*, 78(214), 99–107.

Martyushev, L.M. and Seleznev, V.D. (2006). Maximum entropy production principle in physics, chemistry and biology. *Phys. Rep.*, 426, 1–45.

Marzbanrad, J., Sharifzadegan, A. and Kahrobaeian, A. (2007). Thermodynamic optimization of GSHPS heat exchangers. *Int. J. Thermodyn.*, 10, 107–112.

Massa, C. (1986). On the thermodynamics of Planck's radiation. *Am. J. Phys.*, 54, 754.

Massa, C. (1989). Gravitational constraints on blackbody radiation. *Am. J. Phys.*, 57, 91.

Mata, T.M., Smith, R.L., Young, D.M. and Costa, C.A.V. (2003). Evaluating the environmental friendliness, economics and energy efficiency of chemical processes: heat integration. *Clean Tech. Environ. Policy*, 5, 302–309.

Mathur, M., Karle, S.B., Priye, S., Jayaraman, V.K. and Kulkarni, B.D. (2002). Ant colony approach to continuous function optimization. *Ind. Eng. Chem. Res.*, 39, 3814–3822.

Matijasevic, L. and Otmaievic, H. (2002). Energy recovery by pinch technology. *Appl. Therm. Eng.*, 22, 477–484.

McLachlan, R.I. (1993). Explicit Lie–Poisson integration and the Euler equations. *Phys. Rev. Lett.*, 71, 3043.

McLachlan, R.I. and Quispel, G.R.W. (1998). Generating functions for dynamical systems with symmetries, integrals and differential invariants. *Phys. D*, 112, 298–309.

McLachlan, R.I., Quispel, G.R.W. and Robidoux, N. (1999). Geometric integration using discrete gradients. *Philos. Trans. R. Soc. Lond. A*, 357, 1021–1045.

Mehta, V.L. and Kokossis, C.A. (2000). Nonisothermal synthesis of homogeneous and multiphase reactor networks. *AIChE J.*, 46(11), 2256–2273.

Melli, R., Sciubba, E. and Paoletti, B. (1992). Second-Law based synthesis and optimization of thermal systems: a third-generation A.I. code, *Proceedings of ECOS'92* .

Mench, M.M. (2008). *Fuel Cell Engines*. Wiley & Sons, Hoboken (New Jersey).

Mench, M.M., Wang, Ch-Y. and Thynell, S.T. (2009). *An Introduction to Fuel Cells and Related Transport Phenomena*. Report of The Pennsylvania State University, University Park, Pa, USA.

Metropolis, N., Rosenbluth, A., Rosenbluth, M., Teller, A. and Teller, E. (1953). Equation of state calculations by fast computing machines. *J. Chem. Phys.*, 21, 1087–1092.

Meyer, C.A. and Floudas, C.A. (2006). Global optimization of a combinatorially complex generalized pooling problem. *AIChE J.*, 52(3), 1027–1037.

Michalewicz, Z. (1996). *Genetic Algorithms + Data Structures = Evolution Programs (third edition)*. Springer-Verlag, New York.

Michalewicz, Z. and Fogel, D.B. (2002). *How to Solve it: Modern Heuristics*. Springer Verlag, Berlin.

Michinev, I., Stoyanov, S. and Keil, F.J. (2000). Investigation and modification of the LUUS-JAAKOLA global optimisation algorithm. *Hung. J. Ind. Chem.*, 28, 231–239.

Mielentilew, L.A. (1982). *Optimization of the Development and Control of Large Scale Energy Systems (in Russian)*. High School, Moscow.

Mihail, R. and Maria, G. (1986). A modified Matyas algorithm (MMA) for random process optimization. *Comput. Chem. Eng.*, 10(6), 539–544.

Mikhailov, Yu.A. and Glazunov, Yu.T. (1985). *Variational Methods in Theory of Nonlinear Heat and Mass Transfer*. Zinatne, Riga.

Milewski, J. (2004). Badanie układów energetycznych z ogniwem paliwowym SOFC. Ph.D. Thesis. Warsaw University of Technology.

Milewski, J. and Miller, A. (2012). Triple-layer based control strategy for molten carbonate fuel cell-hybrid system, Chemical and Process Engineering, 33 (3), 445-461.

Milewski, J., Lewandowski, J. and Miller, A. (2009). Reducing CO_2 emissions from a coal fired power plant by using a molten carbonate fuel cell. *Chemical and Process Engineering - Inżynieria Chemiczna i Procesowa*, 30, 341–350.

Milewski, J., Miller, A. and Badyda, K. (2010). The control strategy for high temperature fuel cell hybrid systems. *The Online Journal on Electronics and Electrical Engineering*, 2(4), 331–335.

Mironova, V.A., Amelkin, S.A. and Tsirlin, A.M. (2000). *Mathematical Methods of Finite Time Thermodynamics*. Chimija, Moscow.

Mocsny, D. and Govind, R. (1984). Decomposition strategy for the synthesis of minimum-unit heat exchanger networks. *AIChE J.*, 30(5), 853–856.

Mohanty, S. (2006). Multiobjective optimization of synthesis gas production using non-dominated sorting genetic algorithm. *Comput. Chem. Eng.*, 30, 1019–1025.

Moiseiwitsch, B.L. (1966). *Variational Principles*. Wiley Interscience, New York.

Molenda, J. (2006). High-temperature solid-oxide fuel cells: New trends in materials research, *Materials Science - Poland*, 24 (No. 1), 1-7.

Möller, B.F., Torisson, T. and Assadi, M. (2006). AZEP gas turbine combined cycle power plants—thermo-economic analysis. *Int. J. Thermodyn.*, 9, 21–28.

Moran, M.J. (1998). On second-law analysis and the failed promise of finite-time thermodynamics. *Energy*, 6, 517–519.

Moran, M.J. and Shapiro, H.N. (2003). *Fundamentals of Engineering Thermodynamics*. Wiley, New York.

Mordukhovich, B.S. (1988). Approximate maximum principle for finite-difference control systems. *Comput. Math. Math. Phys.*, 28, 106–114.

Mordukhovich, B.S. (1995). Discrete approximations and refined Euler–Lagrange conditions for nonconvex differential inclusions. *SIAM J. Control Optim.*, 33, 882–915.

Mordukhovich, B.S. and Shvartsman, I. (2004). The approximate maximum principle for constrained control systems. *SIAM J. Control Optim.*, 43, 1037–1062.

Mori, S. and Wen, C.Y. (1975). Estimation of bubble diameter in fluidized beds. *AIChEJ.*, 21, 109.

Morosuk, T., Morosuk, C. and Feidt, M. (2004). New proposal in the thermodynamic analysis of complex regeneration system. *Energy*, 29, 2517–2535.

Moser, F. and Schnitzer, H. (1985). *Heat Pumps in Industry*. Elsevier, Amsterdam.

Mrugala, R., Nulton, J.D., Schön, J.C. and Salamon, P. (1990). A statistical approach to the geometric structure of thermodynamics. *Phys. Rev. A*, 41, 3156–3165.

Murakami, Y., Uchiyama, H., Hasebe, S. and Hashimoto, I. (1997). Application of repetitive SA method to scheduling problems of chemical processes. *Comput. Chem. Eng.*, 21(Suppl), S1087–S1092.

Mullins, O.C. and Berry, R.S. (1984). Minimization of entropy production in distillation. *J. Phys. Chem.*, 88, 723.

Munos, R. and Moore, A. (2001). Variable resolution discretization in optimal control. *Machine Learning*, 1, 1–31.

Munos, R. and Moore, A. (1999a). Influence and variance of a Markov chain: application to adaptive discretization in optimal control, *IEEE Conference on Decision and Control* http://www.cmap.polytechnique.fr/~munos/papers/cdc99.ps.gz.

Munos, R. and Moore, A. (1999b). Variable resolution discretization for high accuracy solutions of optimal control problems, *International Joint Conference on Artificial Intelligence* http://www.cmap.polytechnique.fr/~munos/papers/ijcai99.ps.gz.

Munos, R. and Moore, A. (2001). Variable resolution discretization in optimal control. *Machine Learning Journal*, 49, 291-323 http://www.cmap.polytechnique.fr/~munos/papers/ML J2001.ps.gz.

Naka, Y., Teraskita, M. and Takamatsu, T. (1982). A thermodynamic approach to multicomponent distillation system synthesis. *AIChE J.*, 28, 812.

Narayan, V. and Diwekar, U.M. (1996). Synthesizing optimal waste blends. *Ind. Eng. Chem. Res.*, 35, 3519–3527.

Natalini, G. and Sciubba, E. (1994). A new criterion for configuration optimization of aircooled gas turbine blades, based on the minimization of the local entropy production. In: ASME (Ed.). *ASME WAM. Thermodynamics and the Design, Analysis and Improvement of Energy Systems.* ASME, Amsterdam, AES-Vol. 33, pp. 81–91.

Natalini, G. and Sciubba, E. (1999). Minimization of the local rates of entropy production in the design of air-cooled gas turbine blades. *ASME J. Eng. Gas T. Power*, 121 (July (No. 3)), 135-145.

Neill, D.T. and Jensen, W.P. (1976). Geothermal powered heat pumps to produce process heat, *Proceedings of the 11th Intersoc. Energy Conversion Engineering Conference,* USA.

Nelder, J.A. and Mead, R. (1965). A simplex method for function minimization. *Comput. J.*, 7, 308.

Newman, J. (1973). *Electrochemical Systems.* Englewood Cliffs, Prentice Hall.

Ng, K.C., Chua, H.T., Ong, W., Lee, S.S. and Gordon, J.M. (1997). Diagnostics and optimization of reciprocating chillers: theory and experiment. *Appl. Therm. Eng.*, 17, 263-276 (Erratum ibid, 17 (1997), 601-602).

Ng, K.C., Chua, H.T., Tu, K., Gordon, J.M., Kashiwagi, T., Akisawa, A. and Saha, B.B. (1998). The role of internal dissipation and process average temperature in chiller performance and diagnostics. *J. Appl. Phys.*, 83, 1831–1836.

Nie, X.R. and Zhu, X.X. (1999). Heat exchanger network retrofit considering pressure drop and heat transfer enhancement. *AIChE J.*, 45(6), 1239–1254.

Nielsen, J.S., Hansen, M.W. and Kristensen, K.P. (1997). Retrofit and optimisation of industrial heat exchanger networks. *Comput. Chem. Eng.*, 21(Suppl.), S469–S474.

Nishimura, H. (1980). A theory for the optimal synthesis of heat-exchanger systems. *J. Opt. Theory Appl.*, 30(3), 423–450.

Nobel, C.E. and Allen, D.T. (2000). Using geographic information systems (GIS) in industrial water reuse modelling. *Trans. IChemE*, 78(Part B), 295–303.

Nordman, R. and Berntsson, T. (2001). New pinch technology based HEN analysis methodologies for cost-effective retrofitting. *Can. J. Chem. Eng.*, 79(4), 655–662.

Norgaard, R.B. (1986). Thermodynamic and economic concepts as related to resource-use policies: synthesis. *Land Econ.*, 62, 325–328.

Noton, A.R.M. (1972). *Modern Control Engineering.* Pergamon, Oxford.

Nougues, J.M., Grau, M.D. and Puigjaner, L. (2002). Parameter estimation with genetic algorithm in control of fed-batch reactors. *Chem. Eng. Proc*, 41, 303–309.

Noureldin, M.B. and Hasan, A.K. (2006). Global energy targets and optimal operating conditions for waste energy recovery in Bisphenol-A plant. *Appl. Therm. Eng.*, 26, 374–381.

Novak, Z. and Kravanja, Z. (1999). Mixed-integer nonlinear programming problem process synthesis under uncertainty by reduced dimensional stochastic optimization. *Ind. Eng. Chem. Res.*, 38, 2680–2698.

Novikov, I.I. (1958). The efficiency of atomic power stations (a review). *J. Nucl. Energy II*, 7, 125-128 (English translation from *At. Energ.* 3 (1957), 409-412).

Novikov, I.I. (1984). *Thermodynamics.* Mashinostroenije, Moscow.

Novikov, I.I. and Vosskresenskii, K.D. (1977). *Thermodynamics and Heat Transfer.* Atomizdat, Moscow.

Nöther, E. (1918). Invariante Variationsprobleme. *Nachr. Akad. Wiss. Gottingen. Math-Phys*, Kl. II, 235-257.

Nöther, E. (1971). Invariant variational problems. *Transport Theory Stat. Phys.*, 1, 186–207 (translation by M.A. Tavel of the original article).

Nulton, J. and Salamon, P. (1988). Statistical mechanics of combinatorial optimization. *Phys. Rev. A*, 37, 1351.

Nuwayhid, R.Y. and Moukalled, F. (2002). Effect of heat leak on cascaded heat engines. *Energy Convers. Manage.*, 43, 2067–2083.

Obara, S. (2007a). Dynamic operation plan of a combined fuel cell cogeneration, solar module, and geo-thermal heat pump system using Genetic Algorithm. *Int. J. Energy Res.*, 31, 1275–1291.

Obara, S. (2007b). Arrangement plan for distributed fuel cells installed in urban areas. *Int. J. Energy Res.*, 31, 1323–1336.

Obeng, E.D.A. and Ashton, G.J. (1988). On pinch technology based procedures for the design of batch processes. *Chem. Eng. Res. Des.*, 66, 255–259.

Odum, H.T. (1971). *Environment, Power and Society*. Wiley, New York.

Oláh, K. (1997). Electrode Thermokinetics. *Periodica Polytechnica Chem. Eng.*, 41, 97–114.

Olesen, S.G. and Polley, G.T. (1996). Shorter communication: Dealing with plant geography and piping constraints in water network design. *Trans. IChemE*, 74(B4), 273–276.

Olesen, S.G. and Polley, G.T. (1997). A simple methodology for the design of water networks handling single contaminants. *Trans. IChemE*, 75(A4), 420–426.

Oliveira, F.A.P. and Matos, A.H. (2004). Multiperiod synthesis and operational planning of utility systems with environmental concerns. *Comput. Chem. Eng.*, 28, 745–753.

Ołdak. M. 2011. *Thermodynamics and Mathematical Modeling Hybrid Energy Systems with Solid Oxides Fuel Cells* (SOFC). MsD Thesis, (in Polish). Faculty of Chemical and Process Engineering, Warsaw University of Technology.

Ondrechen, M.J. (1990). Non-Lorentz cycles in nonequilibrium thermodynamics (in Finite-Time Thermodynamics and Thermoeconomics). *Adv. Thermodyn.*, 4, 139–152.

Ondrechen, M.J., Berry, R.S. and Andresen, B. (1980a). Thermodynamics in finite time: a chemically driven engine. *J. Chem. Phys.*, 72, 5118–5124.

Ondrechen, M.J., Andresen, B. and Berry, R.S. (1980b). Thermodynamics in finite time: processes with temperature dependent chemical reactions. *J. Chem. Phys.*, 73, 5838–5843.

Ondrechen, M.J., Andresen, B., Mozurkiewicz, M. and Berry, R.S. (1981). Maximum work from a finite reservoir by sequential Carnot cycles. *Am. J. Phys.*, 49, 681–685.

Ondrechen, M.J., Rubin, M.H. and Band, Y.B. (1983). The generalized Carnot cycle: a working fluid operating in finite time between finite heat sources and sinks. *J. Chem. Phys.*, 78, 4721–4727.

Opman, J.S. (1967). Application of exergy indices for determining the efficiency of drying equipment. In: I.L. Lyuboshitz (Ed.), *Problems of Intensifying of Heat and Mass Transfer in Drying and Thermal Processes*. Nauka i Technika, Minsk.

Ordonez, J.C., Bejan, A. and Cherry, R.S. (2003). Designed porous media: optimally nonuniform flow structures connecting one point with more points. *Int. J. Therm. Sci.*, 42, 857–870.

Ordonez, J.C., Chen, S., Vargas, J.V.C., Dias, F.G., Gardolinski, J.E.F.C. and Vlassov, D. (2007). Constructal flow structure for a single SOFC. *Int. J. Energy Res.*, 31, 1337–1357.

Orlov, V.N. and Berry, R.S. (1990). Power output from an irreversible heat engine with a nonuniform working fluid. *Phys. Rev. A*, 42, 7230–7235.

Orlov, V.N. and Berry, R.S. (1991a). Thermodynamic bounds for performance of nonuniform systems under finite-time constraints, *Proceedings of Ninth Symposium on Energy Engineering Sciences*, Argonne National Laboratory, May 13–15.

Orlov, V.N. and Berry, R.S. (1991b). Estimation of minimal heat consumption for heat-driven separation processes via methods of finite-time thermodynamics. *J. Phys. Chem.*, 95, 5624–5628.

Orlov, V.N. and Berry, R.S. (1992). Analytical and numerical estimates of efficiency for an irreversible heat engine with distributed working fluid. *Phys. Rev. A*, 45, 7202–7206.

Orlov, V.N. and Rudenko, A.V. (1984). Optimal control in the problems involving limiting possibilities of irreversible thermodynamic processes—a review. *Automatika i Telemekhanika*, 5, 7–42.

Ostrov, D.N. (1999). Viscosity solutions and convergence of monotone schemes for synthetic aperture radar shape from-shading equations with discontinuous intensities. *SJAM*, 59, 2060–2085.

Ostrov, D.N. (2000). Extending viscosity solutions to Eikonal equations with discontinuous spatial dependence. *Nonlinear Anal.*, 42, 709–736.

Ourique, J.E. and Telles, A.S. (1998). Computer-aided molecular design with simulated annealing and molecular graphs. *Comput. Chem. Eng.*, 22(Suppl.), S615–S618.

O'Young, D.L., Jenkins, D.M. and Linnhoff, B. (1988). The constrained problem table for heat exchanger networks, *IChemE Symposium, Understanding Process Integration II*, UMIST, Manchester, UK.

Özcelik, Y. and Özcelik, Z. (2004). Solving mixed integer chemical engineering problems via simulated annealing approach. *Chem. Biochem. Eng. Q.*, 18(4), 329–335.

Ozekmekci, K., Ozturk, A. and Onbasioglu, S. (2000). Availability without entropy: an application to refrigerant R134a. In: G.G. Hirs (Ed.), *ECOS 2000 Proceedings, Part 1*, Universiteit Twente, pp. 415–421.

Özkan, S. and Dinčer, S. (2001). Application for pinch design of heat exchanger networks by use of a computer code employing an improved problem algorithm table. *Energy Convers. Manage.*, 42, 2043–2051.

Özturk, A. (1997). Some thermodynamic relations involving availability not entropy. In: G. Manfrida (Ed.), *Flowers' 97, Florence World Energy Research Symposium*. SGE, Padova, pp. 193–198.

Palmore, G.T.R. (2004). Bioelectric power generation. *Trends in Biotechnology*, 22, 99–100.

Papadopoulos, A.I. and Linke, P. (2006). Multiobjective molecular design for integrated process-solvent systems synthesis. *AIChE J.*, 52(3), 1057–1070.

Papalexandri, K.P. and Pistikopoulos, E.N. (1993a). An MINLP retrofit approach for improving the flexibility of heat exchanger network. *Ann. Oper. Res.*, 42, 119–168.

Papalexandri, K.P. and Pistikopoulos, E.N. (1993b). A multiperiod MINLP mode for improving the flexibility of heat exchanger networks. *Comput. Chem. Eng.*, 17S(Suppl.), S111–S116.

Papalexandri, K.P. and Pistikopoulos, E.N. (1994). Synthesis and retrofit design of operable heat exchanger networks. 1. Flexibility and structural controllability aspects. *Ind. Eng. Chem. Res.*, 33, 1718–1737.

Papalexandri, K.P., Patsiatzis, D.I., Pistikopoulos, E.N. and Ebbesen, L. (1998). Heat integration aspects in crude preheat refinery section. *Comput. Chem. Eng.*, 22(Suppl.), S141–S148.

Papoulias, S.A. and Grossmann, I.E. (1983a). A structural optimization approach in process synthesis—I. Utility systems. *Comput. Chem. Eng.*, 7(6), 695–706.

Papoulias, S.A. and Grossmann, I.E. (1983b). A structural optimization approach in process synthesis—II. Heat recovery networks. *Comput. Chem. Eng.*, 7(6), 707–721.

Pariyani, A., Gupta, A. and Ghosh, P. (2006). Design of heat exchanger networks using randomized algorithm. *Comput. Chem. Eng.*, 30, 1046–1053.

Parthasarathy, G. and Gopal, K. (2001). Systematic reallocation of aqueous resources using mass integration in a typical pulp mill. *Adv. Environ. Res.*, 5(1), 61–79.

Patel, A.N., Mah, R.S.H. and Karimi, I.A. (1991). Preliminary design of multiproduct noncontinuous plants using simulated annealing. *Comput. Chem. Eng.*, 15(7), 451–469.

Paterson, W.R. (1984). A replacement for the logarithmic mean. *Chem. Eng. Sci.*, 39, 1635–1636.

Pavani, E.O., Aguirre, P.A. and Irazoqui, H.A. (1990). Optimal synthesis of heat and power systems: the operating line method. *Int. J. Heat Mass Transfer*, 33, 2683.

Pearson, J.D. (1969). Variable metric methods of optimization. *Comput. J.*, 11, 171.

Penfield, P. and Haus, H.A. (1968). *Electrodynamics of Moving Media*. MIT Press, Cambridge.

Peng, S. (1990). A general stochastic maximum principle for optimal control problems. *SIAM J. Control Optim.*, 28, 966–979.

Perrings, C. (1987). *Economy and Environment: A Theoretical Essay on the Interdependence of Economic and Environmental Systems*. Cambridge University Press, Cambridge.

Petela, R. (1964). Exergy of heat radiation. *J. Heat Transfer*, 86, 187–192.

Petela, R. (2003). Exergy of undiluted thermal radiation. *Solar Energy*, 74, 469–488.

Peters, M.S., Timmerhaus, K.D. and West, R.E. (2003). *Plant Design and Economics for Chemical Engineers* (fifth edition). McGraw-Hill, New York.

Pethe, S., Singh, R. and Knopf, F.C. (1989). A simple technique for locating loops in heat exchanger networks. *Comput. Chem. Eng.*, 13(7), 859–860.

Petros, P., Goula, N.F. and Pistikopoulos, E.N. (2005). Generalized modular framework for the synthesis of heat integrated distillation column sequences. *Chem. Eng. Sci.*, 60, 4678–4701.

Pettersson, F. (2005). Synthesis of large-scale heat exchanger network synthesis problems using a sequential match reduction approach. *Comput. Chem. Eng.*, 29, 993–1007.

Peusner, L. (1986). *Studies in Network Thermodynamics*. Elsevier, Amsterdam.

Picon-Nunez, M. and Polley, G.T. (1995a). Determination of the steady state response of heat exchanger networks without simulation. *Trans. IChemE*, 73(January), 49–58.

Picon-Nunez, M. and Polley, G.T. (1995b). Applying basic understanding of heat exchanger network behaviour to the problem of plant flexibility. *Trans. IChemE*, 73(November), 941–952.

Pierce, S.E. (2002). Non-equilibrium thermodynamics: an alternate evolutionary hypothesis. *Crossing Boundaries*, 1, 49–59.

Pintarič, Z.N. and Kravanja, Z. (2004). A strategy for MINLP synthesis of flexible and operable processes. *Comput. Chem. Eng.*, 28, 1105–1119.

Polley, G.T. and Polley, H.L. (2000). Design better water networks. *Chem. Eng. Prog.*, 96, 47–52.

Polley, G.T. and Shahi, M.H.P. (1991). Interfacing heat exchanger network synthesis and detailed heat exchanger design. *Trans. IChemE*, 69, 445–457.

Pontryagin, L.S., Boltyanski, V.A., Gamkrelidze, R.V. and Mischenko, E.F. (1962). *The Mathematical Theory of the Optimal Processes*. Wiley, New York.

Popescu, G., Radcenco, V., Costea, M. and Feidt, M. (1996). Optimisation thermo-dynamique en temps fini du moteur de Stirling endo- et exo-irréversible. *Rev. Gén. Therm.*, 418/419 656-661.

Poplewski, G. (2004). Optymalizacja sieci wody procesowej (Water network optimisation).Ph.D. Thesis. Rzeszow University of Technology, Rzeszow, Poland (under the supervision of J. Jeżowski).

Poplewski, G. and Jeżowski, J. (2003). Random search based approach for designing optimal water networks, *4th European Congress of Chemical Engineering*, Granada, Spain P-9.2-052.

Poplewski, G. and Jeżowski, J. (2005). Stochastic optimization based approach for designing cost optimal water networks. In: L. Puigjaner and A. Espuna (Eds.), *Computer-Aided Chemical Engineering, 20A*. Elsevier, Amsterdam, pp. 727–732.

Poplewski, G., Jeżowski, J. and Jeżowska, A. (2002). Optimal wastewater reuse networks design by adaptive random search optimization, *15th International Congress of Chemical and Process Engineering CHISA'2002*, Praha H.1.6.

Popyrin, L.S. (1978). *Mathematical Modelling and Optimization of Thermoenergy Facilities (in Russian)*. Energy, Moscow.

Portacha, J. (2002). *Energy Research in Heat Systems of Power Stations (in Polish)*. Technological University Publications, ITC PW, Warsaw.

Poświata, A. (2005). Optimization of drying of fine solids in bubble fluidized bed. Ph.D. Thesis. Warsaw University of Technology.

Poświata, A. (2012). Optimal discrete processes, nonlinear in time intervals: Theory and selected applications. *Cybernetics and Physics*, 1(2), 120–127.

Poświata, A. and Szwast, Z. (2000). Optimization of mass and heat transport processes in fluidized bed with bubbled phase. In: P.J.A.M. Kerkhof, W.J. Coulmans and G.D. Mooiweer (Eds.), *Proceedings of the 12th International Drying Symposium IDS'2000*. Noordwijkerhout, The Netherlands, pp. 28–31 August 2000 (paper nr 22 on CD rom).

Poświata, A. and Szwast, Z. (2001). Optymalizacja Procesów Wymiany Ciepła i Masy w Złożu Fluidalnym z Fazą Pęcherzykową. *Inż. Chemiczna i Procesowa*, 22, 1187–1192.

Poświata, A. and Szwast, Z. (2003). Minimization of exergy consumption in fluidized drying processes. In: N. Houbak (Ed.), *Proceedings of The 16th International Conference on Efficiency, Cost, Optimization, Simulation and Enviromental Impact of Energy Systems, ECOS'2003*. Technical University of Denmark, Copenhagen 30 June–02 July, 785-792.

Poświata, A. and Szwast, Z. (2007). Optimization of fine solid drying in bubble fluidized bed. *Transport Porous Media*, 66, 219–231.

Poświata, A. and Szwast, Z. (2009). Egzergetyczna optymalizacja poziomego fluidalnego wymiennika ciepła z uwzględnieniem dwufazowego modelu złoża. *Inżynieria i Aparatura Chemiczna*, 6, 156–157.

Poświata, A. and Szwast, Z. (2010). Minimum of exergy consumption in a horizontal fluidized heat exchanger. *Heat Transfer Research*, 41, 265–282.

Poulikakos, D. and Bejan, A. (1982). Fin geometry for minimum entropy generation in forced convections. *Trans. ASME, J. Heat Transfer*, 104, 616–623.

Pourali, O., Amidpour, M. and Rashtchian, D. (2006). Time decomposition in batch process integration. *Chem. Eng. Process.*, 45, 14–21.

Powell, M.J.D. (1964). An efficient method for finding the minimum of a function of several variables without calculating derivatives. *Comput. J.*, 7, 155.

Prados, E. and Faugeras, O. (2003). A mathematical and algorithmic study of the lambertian SFS problem for orthographic and pinhole cameras. *Technical Report RR-5005*. INRIA, 2003.

Prados, E. and Faugeras, O. (2006). *Shape From Shading. Handbook of Mathematical Models in Computer Vision*. Springer, Berlin, pp. 375–388.

Prados, E., Camilli, F. and Faugeras, O. (2004). A viscosity method for Shape-From-Shading without boundary data. *Technical Report RR-5296*. INRIA 2004.

Prakash, R. and Shenoy, U.V. (2005a). Targeting and design of water networks for fixed flowrate and fixed contaminant load operations. *Chem. Eng. Sci.*, 60, 255–268.

Prakash, R. and Shenoy, U.V. (2005b). Design and evolution of water networks by source shifts. *Chem. Eng. Sci.*, 60, 2089–2093.

Prakotpol, D. and Srinophakun, T. (2004). GAPinch: genetic algorithm toolbox for water pinch technology. *Chem. Eng. Process*, 43, 203–217.

Press, W.H. (1976). Theoretical maximum for energy from direct and diffused sunlight. *Nature*, 264, 734–735.

Press, W.H. and Teukolsky, S.A. (1991). Simulated annealing optimization over continuous spaces. *Comput. Phys.* (July–August), 426-429.

Press, W.H., Teukolsky, S.A., Velterling, W.T. and Flannery, B.P. (1992). *Numerical Recipes in C. The Art of Scientific Computing* (2nd edition). Cambridge University Press, Cambridge.

Price, D.W. and Tsatsaronis, G. (1993). Design Analysis of Dow-Based IGCC Power Plants. In: H.J. Richter (Ed.), *Thermodynamics and the Design, Analysis, and Improvement of Energy Systems, AES-Vol. 30/HTD-Vol. 226*. ASME, New York, NY, pp. 355–369.

Price, W.L. (1978). Global optimization by controlled random search. *J. Optim. Theory Appl*, 24, 333–348.

Przekop, R. (2009). Flow and oxygen transport resistance in human lungs, *Proceedings of the 8th World Congress in Engineering*, Montreal, 23-27 August 2009.

Pukrushpan, J.T., Stefanopoulou, A.G. and Peng, H. (2004). *Control of Fuel Cell Power Systems. Principles, Modeling, Analysis and Feedback Design*. Springer Verlag London Limited, London.

Quesada, I. and Grossmann, I.E. (1993). Global optimization algorithm for heat exchanger networks. *Ind. Eng. Chem. Res.*, 32, 487–499.

Quispel, G.R.W. (1995). Volume preserving integrators. *Phys. Lett. A*, 206, 26–30.

Quispel, G.R.W. and Dyt, C.P. (1998). Volume preserving integrators have linear error growth. *Phys. Lett. A*, 242, 25-30 and references therein.

Rabaey, K. and Verstraete, W. (2005). Microbial fuel cells: novel biotechnology for energy generation. *Trends in Biotechnology*, 23(6), 291–298.

Radcenco, V., Vasilescu, E. and Feidt, M. (2001). The substantiation of the exergetic method of analysis based on finite time thermodynamics. *Entropie*, 232, 5–8.

Radwański, E., Skowroński, P. and Twarowski, A. (1993). *Modeling Problems of Energy Technology Systems (in Polish)*. Technological University Publications, ITC PW, Warsaw.

Radulescu, M., Lottin, O., Feidt, M., Lombard, C., Le Noc, D. and Ledoze, S. (2006a). Experimental results with a natural gas cogeneration system using a PEMFC. *Journal of Power Sources*, 159(2), 1142–1146.

Radulescu, M., Lottin, O., Feidt, M., Lombard, C., Le Noc, D. and Ledoze, S. (2006b). Experimental theoretical analysis of the operation of a natural gas cogeneration system using a polymer exchange membrane fuel cell. *Chemical Eng. Science*, 61, 743–752.

Rajesh, J., Jayaraman, V.K. and Kulkarni, B.D. (2000). Taboo search algorithm for continuous function optimisation. *Trans. IChemE*, 78(Part A), 845–848.

Rangaiah, G.P. (1985). Studies in constrained optimization of chemical process problems. *Comput. Chem. Eng.*, 9(4), 395–404.

Ranger, P.-M., Matsuda, H. and Le Goff, P. (1990). Modelling of a new type of absorption heat pump combining rectification and reverse rectification. *J. Chem. Eng. Japan*, 23, 530.

Rant, Z. (1956). Exergie – ein neues Wort fur "Technische Arbeitsfähigkeit". *Forsch. Eng. Wes.*, 22, 36–39.

Rao, A.D. and Samuelsen, G.S. (2002). Analysis strategies for tubular solid oxide fuel cell based hybrid systems. *ASME J. Eng. Gas Turbines Power*, 124, 503–509.

Rao, A, Maclay, J. and Samuelsen, S. (2004). Efficiency of electrochemical systems. J. of Power Sources, 134, 181–184.

Rao, A.D. and Samuelsen, G.S. (2002). Analysis strategies for tubular solid oxide fuel cell based hybrid systems. *ASME J. Eng. Gas Turbines Power*, 124, 503–509.

Raspanti, C.G., Bandoni, J.A. and Biegler, L.T. (2000). New strategies for flexibility analysis and design under uncertainty. *Comput. Chem. Eng.*, 24, 2193–2209.

Ratkje, S.K. and Moller-Holst, S. (1993). Exergy effeciency and local heat production in solid oxide fuel cells. *Electrochim. Acta*, 38(2–3), 447–453.

Ratnam, R. and Patwardhan, S.V. (1991). Sensitivity analysis for heat exchanger networks. *Chem. Eng. Sci.*, 46(2), 451–458.

Ravagnani, M.A.S.S., Silva, A.P., Arroyo, A.A. and Constantino, A.A. (2005). Heat exchanger network synthesis and optimisation using genetic algorithms. *Appl. Therm. Eng.*, 25, 1003–1017.

Ravanjak, D., Ilič, G. and Može, E. (2004). Designing water reuse in a paper mill by means of computer modeling. *Chem. Biochem. Eng. Q.*, 18(1), 13–19.

Ravi, G., Gupta, S.K. and Ray, M.B. (2000). Multiobjective optimisation of cyclone separators using genetic algorithm. *Ind. Eng. Chem. Res.*, 39, 4272–4286.

Ravi, G., Gupta, S.K., Viswanathan, S. and Ray, M.B. (2002). Optimisation of Venturi scrubbers using genetic algorithm. *Ind. Eng. Chem. Res.*, 41, 2988–3002.

Reay, D.A. (1977). *Industrial Energy Conservation*. Pergamon Press, Oxford.

Reay, D.A. and Mac Michael, D.B.A. (1979). *Heat Pumps Design and Application*. Pergamon Press, London.

Reini, M.(1994). Analisi e sviluppo dei metodi termoeconomici per lo studio degli impianti di conversione dell'energia. Ph.D. Thesis. Dipartimento di Ingegneria Meccanica,University of Padova.

Reini, M. and Taccani, R. (2004). On the thermoeconomic approach to the diagnosis of energy system malfunctions. The role of the fuel impact formula. *Int. J. Thermodyn.*, 7, 61–72 (the special issue on thermoeconomic diagnosis).

Reis, A.H. and Bejan, A. (2006). Constructal theory of global circulation and climate. *Int. J. Heat Mass Transfer*, 49, 1857–1875.

Reis, A.H. and Miguel, A.F. (2006). Constructal theory and flow architectures in living systems. *Therm. Sci.*, 1, 57–64.

Reis, A.H., Miguel, A.F. and Aydin, M. (2004). Constructal theory of flow architectures of the lungs. *Med. Phys.*, 31, 1135–1140.

Reis, A.H., Miguel, A.F. and Bejan, A. (2006). Constructal theory of particle agglomeration and design of air-cleaning devices. *J. Phys. D: Appl. Phys.*, 39, 2311–2318.

Rev, E. and Fonyo, Z. (1991). Diverse pinch concept for heat exchange network synthesis: the case of different heat transfer conditions. *Chem. Eng. Sci.*, 46(7), 1524–1634.

Rev, E. and Fonyo, Z. (1993). Letter to the Editors. *Chem. Eng. Sci.*, 48(3), 627–628.

Riekert, L. (1974). The efficiency of energy-utilization in chemical processes. *Chem. Eng. Sci.*, 29, 1613.

Riekert, L. (1979). Large chemical plants. In: G.F. Froment (Ed.), *Proceedings 4th International Symposium*. Antwerp, London October, 35-64.

Rivera-Alvarez, A. and Bejan, B. (2003). Constructal geometry and operation of adsorption processes. *Int. J. Therm. Sci.*, 42, 983–994.

Ro, S.T. and Sohn, J.L. (2007). Some issues on performance analysis of fuel cells in thermodynamic point of view. *Journal of Power Sources*, 167, 295–301.

Roco, J.M.M., Velasco, S., Medina, A. and Hernandez, A.C. (1997). Optimum performance of a regenerative Brayton thermal cycle. *J. Appl. Phys.*, 82, 2735–2741.

Rodera, H. and Bagajewicz, M.J. (1999). Targeting procedures for energy savings by heat integration across plants. *AIChE J.*, 45(8), 1721–1742.

Romanchuk, B.G. (1995). Input–output analysis of feedback loops with saturation nonlinearities.Ph.D. Dissertation. University of Cambridge.

Romeijn, H.E., Zabinsky, Z.B., Graesser, D. and Neogi, S. (1999). New reflection generator for simulated annealing in mixed-integer/continuous global optimization. *J. Opt. Theory Appl.*, 101(2), 403–427.

R.N. Rosa, A.H. Reis and A.F. Miguel (Eds.) (2004). A brief appraisal of Professor Adrian Bejan's work, *Proceedings of the Symposium "Bejan's Constructal Theory of Shape and Structure"*, Centro de Geofísica de Évora, Univ.d'Évora, Portugal.

Rosen, J.B. (1960). The gradient projection method for non-linear programming. Part I. Linear constraints. *J. Soc. Ind. Appl. Math.*, 8, 181–217.

Rosen, J.B. (1961). The gradient projection method for non-linear programming. Part II. Nonlinear constraints. *J. Soc. Ind. Appl. Math*, 9, 514–532.

Rosenbrock, H.H. and Storey, C. (1966). *Computational Techniques for Chemical Engineers*. Pergamon, Oxford.

Rossiter, A., King, F. and Salyzyn, B. (1999). Syncrude Canada Ltd. enhances efficiency of the upgrader expansion project with pinch analysis. Paper no. T3001a. Presented at AIChE 1999, Spring National Meeting, Houston, Texas.

Roy-Aikins, J. (2002). Thermoeconomic optimisation of a hybrid microturbine and fuel cell. *ECOS 02 Conference Proceedings*.

Rozonoer, L.I. and Tsirlin, A.M. (1983). Optimal control of thermodynamic systems. *Automation Remote Control*, 1, 70–79 (2: 88-101, 3: 50-65).

Rubi, J.M. and Kjelstrup, S. (2003). Mesoscopic nonequilibrium thermodynamics gives the same thermodynamic basis to Butler-Volmer and Nernst Equations. *J. Phys. Chem. B*, 107, 13471–13477.

Rubin, M.H. (1979a). Optimal configuration of a class of irreversible heat engines. I. *Phys. Rev. A*, 19(1979), 1272–1276.

Rubin, M.H. (1979b). Optimal configuration of a class of irreversible heat engines II. *Phys. Rev. A*, 19(1979), 1277–1289.

Rubin, M.H. and Andresen, B. (1979). Optimal staging of endoreversible heat engines. *J. Appl. Phys.*, 53, 1–7.

Rund, H. (1966). *The Hamilton–Jacobi Theory in the Calculus of Variations*. Van Nostrand, London 1–32.

Russel, D.L. (1970). *Optimization Theory*. Benjamin, New York.

Ruth, M. (1993). *Integrating Economics, Ecology and Thermodynamics*. Kluwer, Dordrecht.

Ruth, M. (1999). Physical principles in environmental economic analysis. In: J.C.J.M. van den Bergh (Ed.), *Handbook of Environmental and Resource Economics*. Edward Elgar, Cheltenham, pp. 855–866.

Rymarz, C. (1999). Chaos and self-organization in living systems. *J. Tech. Phys.*, 40, 407–504.

Ryoo, H.S. and Sahinidis, N.V. (1995). Global optimization of nonconvex NLPs and MINLPs with applications in process design. *Comput. Chem. Eng.*, 19(5), 551–556.

Saboo, A.K. and Morari, M. (1984). Design of resilient processing plants—VIII. A resilience index for heat exchanger networks. *Chem. Eng. Sci.*, 40(8), 1553–1565.

Saboo, A.K. and Morari, M. (1986a). Reshex: An interactive software package for the synthesis and analysis of resilient heat-exchanger networks—I. Program description and application. *Comput. Chem. Eng.*, 10(6), 577–589.

Saboo, A.K. and Morari, M. (1986b). Reshex: An interactive software package for the synthesis and analysis of resilient heat-exchanger networks—II. *Comput. Chem. Eng.*, 10(6), 591–599.

Sagli, B., Gundersen, T. and Yee, T.F. (1990). Topology traps in evolutionary strategies for heat exchanger network synthesis. In: H. Bussemaker and P. Iedema (Eds.), *Computer Applications in Chemical Engineering, Process Technology Proceedings, vol. 9*. Elsevier Science Publishers, Amsterdam, The Netherlands.

Sahin, A.Z. (2002). Finite-time thermodynamic analysis of a solar driven heat engine. *Exergy Int. J.*, 1, 122–126.

Sahin, K.H. and Ciric, A.R. (1998). A dual temperature simulated annealing approach for solving bilevel programming problems. *Comput. Chem. Eng.*, 23, 11–25.

Salah El-Din, M.M. (2001a). Second law analysis of irreversible heat engines with variable temperature heat reservoirs. *Energy Convers. Manage.*, 42, 189–200.

Salah El-Din, M.M. (2001b). Performance analysis of heat pumps and refrigerators with variable reservoir temperatures. *Energy Convers. Manage.*, 42, 201–216.

Salama, A.I.A. (2005). Numerical techniques for determining heat energy targets in pinch analysis. *Comput. Chem. Eng.*, 29, 1861–1866.

Salama, A.I.A. (2006). Determination of the optimal heat energy targets in heat pinch analysis using geometry-based approach. *Comput. Chem. Eng.*, 30, 758–764.

Salamon, P. and Berry, R.S. (1983). Thermodynamic length and dissipated availability. *Phys. Rev. Lett.*, 51, 1127–1130.

Salamon, P., Andresen, B. and Berry, R.S. (1977). Thermodynamics in finite time. II. Potentials for finite-time processes. *Phys. Rev. A*, 15, 2094.

Salamon, P., Nitzan, A., Andresen, B. and Berry, R.S. (1980). Minimum entropy production and the optimization of heat engines. *Phys. Rev. A*, 219, 2115–2129.

Salamon, P., Nulton, J.D., Siragusa, G., Andersen, T.R. and Limon, A. (2001). Principles of control thermodynamics. *Energy*, 26, 307–319.

Salcedo, R.L. (1992). Solving nonconvex nonlinear programming and mixed-integer nonlinear programming problems with adaptive random search. *Ind. Eng. Chem. Res.*, 31, 262–273.

Salcedo, R.L. and Lima, R.L. (1999). On the optimum choice of decision variables for equation-oriented global optimization. *Ind. Eng. Chem. Res.*, 38, 4742–4758.

Salcedo, R.L., Goncalves, M.J. and de Azevedo, S.F. (1990). An improved random search algorithm for non-linear optimization. *Comput. Chem. Eng.*, 14(10), 1111–1126.

Salinas, S.R.A. and Tsallis, C. (1999). Nonextensive statistical mechanics and thermodynamics. *Braz. J. Phys.*, 29(special issue), 1-214.

Samuelson, P.A. (1947). *Foundations of Economic Analysis*. Harvard University Press, Cambridge, MA.

San, J.Y., Worek, W.M. and Lavan, Z. (1987). Entropy generation in convective heat transfer and isothermal convective mass transfer. *J. Heat Transfer*, 109, 647.

San, J.Y., Worek, W.M. and Lavan, Z. (1988). Entropy generation in combined heat and mass transfer. *Int. J. Heat Mass Transfer*, 30, 1359.

Santilli, R.M. (1977). Necessary and sufficient conditions for the existence of a Lagrangian in field theory. I. Variational approach to self-adjointness for tensorial field equations. *Ann. Phys. (NY)*, 103, 354–408.

Sargent, R. (2005). Process systems engineering: a retrospective view with questions for the future. *Comput. Chem. Eng.*, 29, 1237–1241.

Sauar, E., Nummendal, L. and Kjelstrup, S. (1999). The principle of equipartition of forces in chemical reactor design: the ammonia synthesis. *Comput. Chem. Eng.*, 23(Suppl.), 499–502.

Saunders, P.T. and Ho, M.W. (1976). On the increase in complexity in evolution I. *J. Theor. Biol.*, 63, 375–384.

Saunders, P.T. and Ho, M.W. (1981). On the increase in complexity in evolution II. The relativity of complexity and principle of minimum increase. *J. Theor. Biol.*, 90, 515–530.

Savelski, M.J. and Bagajewicz, M.J. (2000). On the optimality conditions of water utilization systems in process plants with single contaminants. *Chem. Eng. Sci.*, 55(21), 5035–5048.

Savelski, M.J. and Bagajewicz, M.J. (2001). Algorithmic procedure to design water utilization systems featuring a single contaminant in process plants. *Chem. Eng. Sci.*, 56(5), 1897–1911.

Savelski, M.J. and Bagajewicz, M.J. (2003). On the necessary conditions of optimality of water utilization systems in process plants with multiple contaminants. *Chem. Eng. Sci.*, 58(23–24), 5349–5362.

Savulescu, L., Sorin, M. and Smith, R. (2002). Direct and indirect heat transfer in water network systems. *Appl. Therm. Eng.*, 22, 981–988.

Savulescu, L., Kim, J.-K. and Smith, R. (2005a). Studies on simultaneous energy and water minimisation—Part I: Systems with no water re-use. *Chem. Eng. Sci.*, 60, 3279–3290.

Savulescu, L., Kim, J.-K. and Smith, R. (2005b). Studies on simultaneous energy and water minimisation—Part II: Systems with maximum re-use of water. *Chem. Eng. Sci.*, 60, 3291–3308.

Sayigh, A.A.M. (Ed.) (1977). *Solar Energy Engineering*. Academic Press, New York.

Schechter, R. (1967). *The Variational Method in Engineering*. McGraw-Hill, New York.

Schmidt, W.F. and Willmott, A.J. (1981). *Thermal Energy Storage and Regeneration*. Hemisphere-McGraw-Hill, New York.

Schneider, E. and Kay, J. (1994). Complexity and thermodynamics: towards a new ecology. *Future*, 26, 626–647.

Scholtz, F. and Schroder, U. (2003). Bacterial batteries. *Nature Biotechnology*, 21(10), 1151–1152.

Schrodinger, E. (1967). *What is Life?* Cambridge University Press, Cambridge (first edn in 1944).

Sciubba, E. (2000). On the possibility of establishing a univocal and direct correlation between monetary price and physical value: the concept of Extended Exergy Accounting, *Proceedings of the 2nd International Workshop on Advanced Energy Studies*, Porto Venere, Italy.

Sciubba, E. (2001a). Extended energy accounting: a proposal for a new measure of value, a talk at the Gordon research conference. *Modern Developments in Thermodynamics*, 11-16 March, Ventura, California.

Sciubba, E. (2001b). Beyond thermoeconomics? The concept of Extended Exergy Accounting and its application to the analysis and design of thermal systems. *Exergy Int. J.*, 1, 68–84.

Sciubba, E. (2004). Hybrid semi-quantitative monitoring and diagnostics of energy conversion processes. *Int. J. Thermodyn*, 7, 95–106 (the special issue on thermoeconomic diagnosis).

Sciubba, E. and Melli, R. (1998). *Artificial Intelligence in Thermal Systems Design: Concepts and Applications*. Nova Science-Pergamon Press, New York.

Sciubba, E. and Ulgiati, S. (2005). Energy and Exergy analyses: complementary methods or irreducible ideological options?. *Energy*, 10(30).

Sciubba, E. and Wall, G. (2007). A brief commented history of exergy from the beginnings to 2004. *Int. J. Thermodyn*, 10, 1–26.

Seculic, D.P. (1998). A fallacious argument in the finite thermodynamics concept of irreversibility. *J. Appl. Phys.*, 83, 4561–4565.

Seider, W.D., Seader, J.D. and Lewin, D.R. (1999). *Process Design Principles: Synthesis, Analysis, and Evaluation*. Wiley, New York.

Seider, W.D., Seader, J.D. and Lewin, D.R. (2004). *Product and Process Design Principles: Synthesis, Analysis and Evaluation* (2nd edition). Wiley, New York.

Seidler, I., Badach, A. and Molisz, W. (1980). *Methods of Solving of Optimization Problems*. Wydawnictwa Naukowo Techniczne, Warsaw.

Selba, R., Kizilkan, Ő. and Reppich, M. (2006). A new design approach for shell and tube heat exchangers using genetic algorithms from economic point of view. *Chem. Eng. Process.*, 45, 268–275.

Seliger, R.L. and Whitham, G.B. (1968). Variational principles in continuum mechanics. *Proc. R. Soc.*, 302A, 1–25.

Serna, M. and Jimenez, A. (2004). An area targeting algorithm for the synthesis of heat exchanger networks. *Chem. Eng. Sci.*, 59, 2517–2520.

Shaal, L. and Feidt, M. (1999). Optimization of the dynamic behaviour of a heat exchanger subject to tooling. Comparison of three optimization models. *Int. J. Appl. Thermodynamics*, 2(2), 89–96.

Shafiei, S., Domenech, R., Koteles, R. and Paris, J. (2004). System closure in pulp and paper mills: network analysis by genetic algorithm. *J. Cleaner Prod*, 12, 131–135.

Shakhnovskij, A., Jeżowski, J., Kvitka, A., Jeżowska, A. and Statiukha, G. (2004). Badania nad optymalizacją sieci wody procesowej przy zastosowaniu programowania matematycznego. *Inż. Chem. Proc.*, 25(3-3), 1607–1612.

Shang, Z. and Kokossis, C.A. (2004). A transhipment model for the optimisation of steam levels of total utility system for multiperiod operation. *Comput. Chem. Eng.*, 28, 1673–1688.

Shannon, C.E. and Weaver, W. (1969). *The Mathematical Theory of Communication.* The University of Illinois Press, Urbana.

Shapiro, J.F. (2004). Challenges of strategic supply chain planning and modelling. *Comput. Chem. Eng.*, 28, 855–861.

Shenoy, U.V. (1995). *Heat Exchanger Network Synthesis.* Gulf Publishing, Houston, TX.

Shenoy, U.V., Sinha, A. and Bandyopadhyay, S. (1998). Multiple utilities targeting for heat exchanger networks. *Trans. IChemE. Chem. Eng. Res. Des.*, 76, 259–272.

Shethna, H.K. and Jeżowski, J. (2006). Near independent subnetworks in heat exchanger network design. *Ind. Eng. Chem. Res.*, 45, 4629–4636.

Shethna, H.K., Jeżowski, J. and Castillo, F.J.L. (2000a). Generalized transshipment model for targeting of multiple utilities in heat exchanger networks. *Inż. Chem. Proc.*, 21, 625–643.

Shethna, H.K., Jeżowski, J. and Castillo, F.J.L. (2000b). A new methodology for simultaneous optimization of capital and operating cost targets in heat exchanger network design. *Appl. Therm. Eng.*, 20, 1577–1587.

Shethna, H.K., Jeżowski, J.M. and Castillo, F. (2002). Targets in heat exchanger networks using optimization models. In: Luus. Rein (Ed.), *Recent Developments in Optimization and Optimal Control in Chemical Engineering.* Research Signpost, Trivandrum, India, pp. 307–322.

Shi, Z., Chen, J. and Wu, Ch. (2002). Maximum work output of an electric battery and its load matching. *Energy Convers. Manage.*, 43, 241–247.

Shiner, J.S. (1984). A dissipative Lagrangian formulation of the network thermodynamics of (pseudo) first-order reaction-diffusion systems. *J. Chem. Phys.*, 81, 1455–1465.

Shiner, J.S. (1992). A Lagrangian formulation of chemical reaction dynamics far from equilibrium. *Flow, Diffusion and Rate Processes, Advances in Thermodynamics.* Taylor and Francis, New York, vol. 6, p. 248.

Shiner, J.S., Lüscher, H.-R. and Sieniutycz, S. (1992). Electrodynamical dynamics of excitable membranes. *Experientia*, 48, A 91.

Shiner, J.S., Fassari, J. and Sieniutycz, S. (1996). Lagrangian and network formulations of nonlinear electrochemical systems. In: J.S. Shiner (Ed.), *Entropy and Entropy Generation: Fundamentals and Applications.* Kluwer, Dordrecht, pp. 221–239.

Shiner, J.S., Davison, M. and Landsberg, P.T. (1999). Simple measure for complexity. *Phys. Rev. E*, 59, 1459–1464.

Shinnar, R. (1988). Thermodynamics analysis in chemical process and reactor design. *Chem. Eng. Sci.*, 43, 2303.

Shopova, E.G. and Vaklieva-Bancheva, N.G. (2006). BASIC—A genetic algorithm for engineering problems solution. *Comput. Chem. Eng.*, 20, 1293–1309.

Sidwell, R.W. and Grover Coors, W. (2005). Large limits of electrical efficiency in hydrocarbons fueled SOFCs. *Journal of Power Sources*, 143, 166–172.

Sieniutycz, S. (1972). Optimization of fluidized drying and moistening in case of variable temperature of inlet gas. Reports of Institute of Chemical Engineering, Warsaw Technical University, vol. 1, p. 99.

Sieniutycz, S. (1973a). The constancy of a hamiltonian in a discrete optimal process. Reports of Institute of Chemical Engineering, Warsaw Technical University, vol. 2, pp. 399–429.

Sieniutycz, S. (1973b). The thermodynamic approach to fluidized drying and moistening optimization. *AIChE J.*, 19, 277–285.

Sieniutycz, S. (1973c). Computing of thermodynamic functions in gas-moisture-solid systems. Reports of Institute of Chemical Engineering, Warsaw Technical University, vol. 2, pp. 323–349.

Sieniutycz, S. (1973d). Thermodynamic approach to fluidized drying and moistening optimization problems. ScD Thesis (in Polish). *Reports of Institute of Chemical Engineering*, Warsaw Technical University, vol. 2, pp. 71–198.

Sieniutycz, S. (1974). The constant Hamiltonian problem and an introduction to the mechanics of optimal discrete systems. *Reports of Institute of Chemical Engineering*, Warsaw Technical University, vol. 3, no. 1–2, pp. 27–53.

Sieniutycz, S. (1976). Theory of thermoeconomic optimization of continuous and multistage fluidized drying. ScD Thesis. *Reports of Institute of Chemical Engineering*, Warsaw Technical University, vol. 5, no. 1, p. 5.

Sieniutycz, S. (1978). *Optimization in Process Engineering* (1st edn.). Wydawnictwa Naukowo Techniczne, Warsaw.

Sieniutycz, S. (1984). A general theory of optimal discrete drying processes with a constant Hamiltonian. *Drying*, 84, 62–75.

Sieniutycz, S. (1985). Optimizing in a class of dispersed material drying operation, Chapter 40. In: N. Cheremissinoff (Ed.), *Handbook for Heat and Mass Transfer*. Gulf Publishing, Houston.

Sieniutycz, S. (1987). From a least action principle to mass action law and extended affinity. *Chem. Eng. Sci.*, 11, 2697.

Sieniutycz, S. (1991). *Optimization in Process Engineering* (2nd edn.). Wydawnictwa Naukowo Techniczne, Warsaw, pp. 151–194.

Sieniutycz, S. (1994). *Conservation Laws in Variational Thermo-Hydrodynamics*. Kluwer, Dordrecht.

Sieniutycz, S. (1997a). Hamilton–Jacobi–Bellman theory of dissipative thermal availability. *Phys. Rev.*, 56, 5051–5064.

Sieniutycz, S. (1997b). Irreversible Carnot problem of maximum work in a finite time via Hamilton–Jacobi–Bellman Theory. *J. Non-Equilib. Thermodyn.*, 22, 260–284.

Sieniutycz, S. (1997c). Carnot problem of maximum work from a finite resource interacting with environment in a finite time. *Phys. A*, 264, 234–263.

Sieniutycz, S. (1998a). Nonlinear thermokinetics of maximum work in finite time. *Int. J. Eng. Sci.*, 36, 557–597.

Sieniutycz, S. (1998b). Generalized thermodynamics of maximum work in finite time. *Open Syst. Inform. Dyn.*, 5, 369–390.

Sieniutycz, S. (1998c). Nonlinear dynamics of generalized carnot problem of maximum work received in finite time from a system of two continua with different temperatures. *Periodica Polytech. Ser. Chem. Eng.*, 42, 33–54.

Sieniutycz, S. (1998d). Non-equilibrium and classical thermodynamics for practical systems: today closer together than ever before. *Int. J. Appl. Thermodyn.*, 1, 9–19.

Sieniutycz, S. (1998e). Hamilton–Jacobi–Bellman theory of irreversible thermal exergy. *Int. J. Heat Mass Transfer*, 41, 183–195.

Sieniutycz, S. (1999a). Optimal control framework for multistage engines with heat and mass transfer. *J. Non-Equilib. Thermodyn.*, 24, 40–74.

Sieniutycz, S. (1999b). Endoreversible modeling and optimization of thermal machines by dynamic programming, Chapter 11. In: Ch. Wu (Ed.), *Recent Advances in Finite Time Thermodynamics*. Nova Science, New York.

Sieniutycz, S. (1999c). Thermodynamic framework for discrete optimal control in multistage thermal systems. *Phys. Rev. E*, 60, 1520–1534.

Sieniutycz, S. (1999d). Optimal control for multistage endoreversible engines with heat and mass transfer. In: A. Bejan and E. Mamut (Eds.), *Thermodynamics and Optimization of Complex Energy Systems*. Kluwer, Dordrecht, pp. 363–384.

Sieniutycz, S. (1999e). Optimal control in multistage flow systems by Hamilton–Jacobi–Bellman. *Arch. Thermodyn.*, 20, 27–48.

Sieniutycz, S. (2000a). Some thermodynamic aspects of development and bistability in complex multistage systems. *Open Sys. Inform. Dyn.*, 7, 309–326.

Sieniutycz, S. (2000b). Dynamic programming approach to a Fermat-type principle for heat flow. *Int. J. Heat Mass Transfer*, 43, 3453–3468.

Sieniutycz, S. (2000c). Hamilton–Jacobi–Bellman framework for optimal control in multistage energy systems. *Phys. Rep.*, 326, 165–285.

Sieniutycz, S. (2000d). Thermodynamic optimization for work-assisted heating and drying operations. *Energy Convers. Manage.*, 41, 2009–2031.

Sieniutycz, S. (2000e). Optimal control of active (work-producing) systems. *Inz Chem i Procesowa*, 21, 29–55.

Sieniutycz, S. (2001a). Work optimization in continuous and discrete systems with complex fluids. *J. Non-Newtonian Fluid Mech.*, 96, 341–370.

Sieniutycz, S. (2001b). Thermodynamic optimization for work-assisted and solar-assisted heat and mass transfer operations. *Arch. Thermodyn.*, 22, 17–36.

Sieniutycz, S. (2002a). Steady interphase heat transfer in nonlinear media—a variational principle. *Arch. Thermodyn.*, 23, 79–90.

Sieniutycz, S. (2002b). A Fermat-like principle for chemical reactions in heterogeneous systems. *Open Sys. Inform. Dyn.*, 9, 257–272.

Sieniutycz, S. (2002c). Irreversible thermodynamics for finite-rate fuel cells systems. In: Ulf Bosel (Ed.), *Proceedings of Fifth European Solid Oxide Fuel Cell Forum*. Luzerne, Switzerland July 1-5, vol 2., pp. 1023–1030.

Sieniutycz, S. (2003a). A synthesis of thermodynamic models unifying traditional and work-driven operations with heat and mass exchange. *Open Systems Inform. Dyn.*, 10, 31–49.

Sieniutycz, S. (2003b). Thermodynamic limits on production or consumption of mechanical energy in practical and industrial systems. *Progr. Energy Combust. Sci.*, 29, 193–246.

Sieniutycz, S. (2003c). Carnot controls to unify traditional and work-assisted operations with heat and mass transfer. *Int. J. Appl. Thermodyn.*, 6, 59–67.

Sieniutycz, S. (2004a). Limiting power from imperfect systems with fluid flow. *Arch. Thermodyn.*, 25, 69–80.

Sieniutycz, S. (2004b). Nonlinear macrokinetics of heat and mass transfer and chemical or electrochemical reactions. *Int. J. Heat Mass Transfer*, 47, 515–526.

Sieniutycz, S. (2004c). Extremum properties of entropy to determine dynamics of growth and evolution in complex systems. *Phys. A*, 340, 356–363.

Sieniutycz, S. (2006a). Thermodynamic limits in applications of energy of solar radiation. *Drying Technol.*, 24, 1139–1146.

Sieniutycz, S. (2006b). State transformations and Hamiltonian structures for optimal control in discrete systems. *Rep. Math. Phys.*, 49, 289–317.

Sieniutycz, S. (2007a). Frictional passage of fluid through inhomogeneous porous system: a variational principle. *Transp. Porous Med.*, 69, 239–257.

Sieniutycz, S. (2007b). A simple chemical engine in steady and dynamic situations. *Arch.Thermodyn.*, 28, 57–84.

Sieniutycz, S. (2007c). Hamilton–Jacobi–Bellman equations and dynamic programming for power-maximizing relaxation of radiation. *Int. J. Heat Mass Transfer*, 50, 2714–2732.

Sieniutycz, S. (2007d). Dynamical converters with power-producing relaxation of solar radiation. *Int. J. Therm. Sci.*, 66, 219–231.

Sieniutycz, S. (2008). Analysis of power and entropy generation in a chemical engine. *Intern. J. of Heat and Mass Transfer*, 51, 5859–5871.

Sieniutycz, S. (2009a). Dynamic bounds for power and efficiency of non-ideal energy converters under nonlinear transfer laws. *Energy*, 34, 334–340.

Sieniutycz, S. (2009b). Thermodynamics of simultaneous drying and power production. *Drying Technology*, 27, 322–335.

Sieniutycz, S. (2009c). Complex chemical systems with power production driven by heat and mass transfer. *International J. of Heat and Mass Transfer*, 52, 2453–2465.

Sieniutycz, S. (2009d). Dynamic programming and Lagrange multipliers for active relaxation of resources in non-equilibrium systems. *Applied Mathematical Modeling*, 33, 1457–1478.

Sieniutycz, S. (2010a). Thermodynamics of power production in fuel cells. *Chemical and Process Engineering*, 31, 81–105.

Sieniutycz, S. (2010b). Upper limits for power yield in thermal, chemical and electrochemical systems. Chapter in the book: *Current Themes in Engineering Science 2009*, pp. 181–196, Edited by. A. Korsunsky, American Institute of Physics,. (CP 1220, Melville, New York).

Sieniutycz, S. (2010c). Thermodynamic aspects of power generation in imperfect fuel cells I. *Intern J. of Ambient Energy*, 31(No4), 195–202.

Sieniutycz, S. (2010d). Finite-rate thermodynamics of power production in thermal, chemical and electrochemical systems. *Int. J. of Heat and Mass Transfer*, 53, 2864–2876.

Sieniutycz, S. (2011a). Thermodynamic aspects of power generation in imperfect fuel cells II. *Intern J. of Ambient Energy*, 32(No1), 46–56.

Sieniutycz, S. (2011b). Fuel cells as energy systems: efficiency, power limits and thermodynamic behavior. *Journal of Energy and Power Engineering (USA)*, 5(1), 17–28.

Sieniutycz, S. (2011c). Thermodynamic aspects of power production in energy systems. *Mathematics in Industry* 15, Edited by A. Chiru, Springer, Berlin, Vol.1, Chap.1, pages 3-30.

Sieniutycz, S. (2011d). Modeling and simulation of power yield in thermal, chemical and electrochemical systems: Fuel cell case. In: *Computer Aided Systems Theory*, Edited by R. Moreno-Diaz, F. Pilcher and A. Arencibia. Las Palmas, Gran Canaria, LNCS, Springer 2011.

Sieniutycz, S. (2011e). Thermodynamic basis of fuel cell systems. *Cybernetics and Physics*, 1, 1–6 2011. http://coms.physcon.ru/conf/11/editpaper.html.

Sieniutycz, S. (2011f). Identification and selection of unconstrained controls in power systems propelled by heat and mass transfer. *Int. J. of Heat and Mass Transfer*, 54, 938–948.

Sieniutycz, S. (2012a). Maximizing power yield in energy systems - A thermodynamic synthesis. *Applied Mathematical Modeling*, 36 (2012), 2197-2212.

Sieniutycz, S. (2012b). General properties of approaches maximizing power yield in thermo-chemical systems. *Renewable Energy & Power Quality Journal (RE&PQJ*, 10, 1-6. (paper RE&PQJ-10-1).

Sieniutycz, S. (2012c). Synthesizing models of power yield in thermo-electro-chemical systems. *Proceedings of the World Congress of Engineering 2012: The International Conference of Applied and Engineering Mathematics*, London, 3-7 July 2012, pp. 52–57, Edited by S.I. Ao, L. Gellman, D.W.L. Hukins, A. Hunter and A.M. Korsunski (ISBN: 978-988-19251-3-8).

Sieniutycz, S. and Berry, R.S. (2000). Discrete Hamiltonian analysis of endoreversible thermal cascades, Chapter 6. In: S. Sieniutycz and A. de Vos (Eds.), *Thermodynamics of Energy Conversion and Transport*. Springer, New York, pp. 143–172.

Sieniutycz, S. and de Vos, A. (2000). *Thermodynamics of Energy Conversion and Transport*. Springer, New York, pp. 143–172 (Chap.6).

Sieniutycz, S. and Farkas, H. (2005). *Variational and Extremum Principles in Macroscopic Systems*. Elsevier, Oxford, pp. 497–522.

Sieniutycz, S. and Jeżowski, J. (2009). Optimal decisions for chemical and electrochemical reactors, Chapter 9 of the first edition of the book: *Energy Optimization in Process Systems*. Elsevier, Oxford, pp. 321–366.

Sieniutycz, S. and Kubiak, M. (2002). Dynamical energy limits in traditional and work driven operations I. Heat–mechanical systems. *Int. J. Heat Mass Transfer*, 21, 2115–2129.

Sieniutycz, S. and Kuran, P. (2005). Nonlinear models for mechanical energy production in imperfect generators driven by thermal or solar energy. *Int. J. Heat Mass Transfer*, 48, 719–730.

Sieniutycz, S. and Kuran, P. (2006). Modeling thermal behavior and work flux in finite rate systems with radiation. *Int. J. Heat Mass Transfer*, 49, 3264–3283.

Sieniutycz, S. and Kuran, P. (2011). Egzergia, strumień egzergii i sprawność wykorzystania energii promieniowania elektromagnetycznego. *Prace Instytutu Inżynierii Chemicznej PW*, 35, 31–56.

S. Sieniutycz and P. Salamon (Eds.) (1990). *Advances in Thermodynamics, Vol. 4: Finite Time Thermodynamics and Thermoeconomics*. Taylor & Francis, New York.

Sieniutycz, S. and Poświata, A. (2011). Basic thermodynamic properties of fuel cell systems. *Chem. Eng. Transactions*, 24, 547–552.

Sieniutycz, S. and Poświata, A. (2012). Thermodynamic aspects of power production in thermal, chemical and electrochemical systems. *Energy*, 52, 1–9.

Sieniutycz, S. and Shiner, J.S. (1994). Thermodynamics of irreversible processes and its relation to chemical engineering: second law analyses and finite time thermodynamics. *J. Non-Equilib. Thermodyn.*, 19, 303–348.

Sieniutycz, S. and Szwast, Z. (1982a). *Practice in Optimization*. Wydawnictwa Naukowo Techniczne, Warszawa.

Sieniutycz, S. and Szwast, Z. (1982b). Optimization of multistage crosscurrent fluidized drying by a special algorithm of the discrete maximum principle with a constant Hamiltonian. *Chem. Eng. J.*, 25, 63–75.

Sieniutycz, S. and Szwast, Z. (1983). A discrete algorithm for optimization with a constant hamiltonian and its application to chemical engineering. *Int. J. Chem. Eng.*, 23, 155–166.

Sieniutycz, S. and Szwast, Z. (1999). Optimization of multistage endoreversible machines by a Pontryagin's-like Discrete Maximum Principle, Chapter 12. In: Ch. Wu (Ed.), *Advance in Recent Finite Time Thermodynamics*. Nova Science, New York, pp. 221–237.

Sieniutycz, S. and Szwast, Z. (2003). Work limits in imperfect sequential systems with heat and fluid flow. *J. Nonequilib. Thermodyn.*, 28, 85–114.

Sieniutycz, S. and von Spakovsky, M. (1998). Finite time extension of thermal exergy. *Energy Convers. Manage.*, 39, 1423–1447.

Sieniutycz, S., Szwast, Z., Kuran, P., Poświata, A., Zalewski, M., Przekop, R., Błesznowski, M. (2011). *Thermodynamics and Optimization of Chemical and Electrochemical Energy Generators with Applications to Fuel Cells* (in Polish). Research Report N N208019434 for the period 16.05.2008-15.05.2011. Warsaw TU, Faculty of Chemical and Process Engineering.

Sieniutycz, S., Błesznowski, M., Zieleniak, A. and Jewulski, J. (2012). Power generation in thermochemical and electrochemical systems – A thermodynamic theory. *Intern. J. Heat and Mass Transfer*, 55, 1197–1213.

Sighvi, A. and Shenoy, U.V. (2002). Aggregate planning in supply chains by pinch analysis. *IChemE Chem. Eng. Res. Des.*, 80(A6), 597–605.

Silva, M.L. and Zemp, R.J. (2000). Retrofit of pressure drop constrained heat exchanger networks. *Appl. Therm. Eng.*, 20, 1469–1480.

Singh, D., Lu, D.M. and Djilali, N.A. (1999). Two-dimensional analysis of mass transport in proton exchange membrane fuel cells. *Int. J. Eng. Sci.*, 37, 431–452.

Singh, M.K., Banerjee, T. and Khanna, A. (2005). Genetic algorithm to estimate interaction parameters of multicomponent systems for liquid–liquid equilibria. *Comput. Chem. Eng.*, 29, 1712–1719.

Smidth, D.R. (1974). *Variational Methods in Optimization*. Prentice Hall, Englewood Cliffs.

Smith, E.M.B. and Pantelides, C.C. (1997). Global optimisation of nonconvex MINLPs. *Comput. Chem. Eng.*, 21(Suppl.), S791–S796.

Smith, R. (2005). *Chemical Process Design*. Wiley, Chichester, UK.

Smith, R. (2000). State of the art in process integration. *Appl. Therm. Eng.*, 20(15–16), 1337–1345.

Smith, R. (2005). *Chemical Process Design*. Wiley, Chichester, UK.

Smith, R. and Delaby, O. (1991). Targeting glue gas emissions. *Trans. IChemE*, 69(Part A), 492–505.

Soner, M.H. (1986). Optimal control with state space constraint. *SIAM J. Control Optim.*, 24, 552–561.

Soravia, P. (1996). *H* ∞ control of nonlinear systems: differential games and viscosity solutions. *SIAM J. Control Optim.*, 34, 1071–1097.

Sorin, M. and Bedard, S. (1999). The global pinch point in water re-use networks. *Trans. IChemE*, 77(Part B), 305–308.

Sorsak, A. and Kravanja, Z. (1999). Simultaneous MINLP synthesis of heat and power integrated hat exchanger networks. *Comput. Chem. Eng.* (Suppl.), S143-S147.

Sorsak, A. and Kravanja, Z. (2002). Simultaneous MINLP synthesis of heat exchanger networks comprising different exchanger types. *Comput. Chem. Eng.*, 26, 599–615.

Sorsak, A. and Kravanja, Z. (2004). MINLP retrofit of heat exchanger networks comprising different exchanger types. *Comput. Chem. Eng.*, 28, 235–251.

Souganidis, P.E. (1985a). Existence of viscosity solutions of Hamilton–Jacobi equations. *J. Diff. Eq.*, 56, 345–390.

Souganidis, P.E. (1985b). Approximation schemes for viscosity solutions of Hamilton–Jacobi Equations. *J. Diff. Eq.*, 57, 1–43.

Spall, J.C. (2003). *Introduction to Stochastic Search and Optimization*. Wiley, Hoboken, New Jersey.

Spanner, D.C. (1964). *Introduction to Thermodynamics*. Acadamic Press, London.

Sparrow, R.E., Forder, G.J. and Rippin, D.W.T. (1975). The choice of equipment sizes for multiproduct batch plants. *Ind. Eng. Chem. Proc. Des. Dev.*, 14(3), 197–203.

Spirkl, W. and Ries, H. (1995). Optimal finite-time endoreversible processes. *Phys. Rev. E*, 52, 3485–3489.

Srinivas, M. and Rangaiah, G.P. (2006). Implementation and evaluation of random tunneling algorithm for chemical engineering applications. *Comput. Chem. Eng.*, 30, 1400–1415.

Srinivasan, R., Viswanathan, P., Vedam, H. and Nochur, A. (2005). A framework for managing transitions in chemical plants. *Comput. Chem. Eng.*, 29, 305–322.

Staine, F. and Favrat, D. (1995). Intégration énergétique de procédés industriels par la méthode du pincement étendue aux facteurs exergétiques. *LENI-REPORT-1995-022*. Lausanne.

Stambouli, A.B.E. and Traversa, E. (2002). Solid oxide fuel cells (SOFCs): a review of an environmentally clean and efficient source of energy. *Renewable Sustainable Energy Rev.*, 6, 433–455.

Stanciu, D., Marinescu, M. and Dobrovicescu, A. (2007). The influence of swirl angle on the irreversibilities in turbulent diffusion flames (and literature therein). *Int. J. Thermodyn.*, 10(No. 4), 143–153.

Standaert, F., Hemmes, K. and Woudstra, N. (1996). Analytical fuel cell modeling. *Journal of Power Sources*, 63, 221–234.

Statyukha, G., Kvitka, O., Dzhygyrey, I. and Jeżowski, J. (2008). A simple sequential approach for designing industrial wastewater treatment networks. *J. Cleaner Prod.*, 16, 215–224.

Stephan, E.A. and Chase, G.G. (2003). Use of genetic algorithms as an aid in modeling deep bed filtration. *Comput. Chem Eng.*, 27, 281–292.

Stephens, J.J. (1967). Alternate forms of the Herrivel–Lin variational principle. *Phys. Fluids*, 10, 76–77.

Stodola, A. (1905). *Steam Turbines* (with an Appendix on Gas Turbines and the Future of Heat Engines) (L.C. Loewenstein (Trans.)),Van Nostrand, New York.

Stromberg, T. (2003). On viscosity solutions of irregular Hamilton–Jacobi equations. *Archiv der Mathematik*, 81(6), 678–688.

Strumillo, Cz. (1983). *Fundamentals of Theory and Technology of Drying* (2nd edition). Wydawnictwa Naukowo Techniczne, Warsaw.

Strumillo, C. and Kudra, T. (1987). *Drying: Principles, Applications and Design*. Gordon and Breach, New York.

Strumillo, Cz. and Lopez-Cacicedo, C. (1984). Energy aspects in drying. In: A.S. Mujumdar (Ed.), *Handbook of Industrial Drying*. Marcel Dekker, New York.

Su, J.L. and Motard, R.L. (1984). Evolutionary synthesis of heat exchanger networks. *Comput. Chem. Eng.*, 8, 67–80.

Subbotina, N.N. (1989). The maximum principle and the superdifferential of the value function. *Problems Control Inform. Theory*, 18, 151–160.

Summanwar, V.S., Jayaraman, V.K., Kulkarni, B.D., Kusumakar, H.S., Gupta, K. and Rajesh, J. (2002). Solution of constrained optimization problems by multi-objective genetic algorithm. *Comput. Chem Eng.*, 26, 1481–1492.

Sundheim, B.R. (1964). Transport properties of liquid electrolytes. In: Sundheim, B.R. (Ed.), *Fused Salts*. Mc Graw Hill, New York, pp. 165–254.

Svirezhev, Yu.M. (2001a). Application of thermodynamic concepts to real ecosystems. In: S.E. Jorgensen (Ed.), *Thermodynamics and Ecological Modelling*. Lewis Publishers of CRC Press, Boca Raton, pp. 133–152.

Svirezhev, Yu.M. (2001b). Thermodynamics and ecology: far from thermodynamic equilibrium. In: S.E. Jorgensen (Ed.), *Thermodynamics and Ecological Modelling*. Lewis Publishers of CRC Press, Boca Raton, pp. 211–228.

Swift, G.W. (1988). Thermoacoustic engines. *J. Acoust. Soc. Am.*, 84, 1145.

Szargut, J. (1957). Towards a rational evaluation of steam prices (in Polish). *Gospodarka Cieplna*, 5, 104–106.

Szargut, J. (1970). Application of exergy in establishing of generalized technicaleconomical interdependences. *Arch. Bud. Maszyn*, 17, 105–108.

Szargut, J. (1978). Minimization of the consumption of natural resources. *Bull. Acad. Pol. Tech.*, 26, 41–45.

Szargut, J. (1983). *Thermodynamic and Economic Analysis of Industrial Energetics (in Polish)*. Wydawnictwa Naukowo Techniczne, Warsaw.

Szargut, J. (1986). Application of exergy for the calculation of ecological cost. *Bull. Polish Acad. Sci. Tech. Sci.*, 34(1986), 475–480.

Szargut, J. (1987). Analysis of cumulative exergy consumption. *Int. J. Energy Res.*, 11, 541–547.

Szargut, J. (1989). Indices of the cumulative consumption of energy and exergy. *Zeszyty Nauk. Polit. Śląskiej: Ser. Energetyka*, 106, 39–55.

Szargut, J. (1990). Analysis of cumulative exergy consumption and cumulative exergy losses. In: S. Sieniutycz and P. Salamon (Eds.), *Finite-Time Thermodynamics and Thermoeconomics, Advances in Thermodynamics vol. 4*. Taylor and Francis, New York, pp. 278–302.

Szargut, J. (1998). Problems of thermodynamic optimization. *Arch. Thermodyn.*, 19, 85–94.

Szargut J. (2001a). Exergy analysis in thermal technology, Ekspertyza KTiSp PAN: Współczesne Kierunki w Termodynamice, In: Bilicki, Z., Mikielewicz, J., Sieniutycz, S., (Eds.), Wydawnictwa Politechniki Gdańskiej.

Szargut, J. (2001b). Letter to the Editor. *Arch. Thermodyn.*, 22, 36.

Szargut, J. (2002a). Minimization of the depletion of non-renewable resources by means of the optimization of the design parameters, *Proceedings of ECOS 2002, Vol. I*, Berlin, Germany, July 3-5, pp. 326–333.

Szargut, J. (2002b). Application of exergy for the determination of the proecological tax replacing the actual personal taxes. *Energy*, 27, 379–389.

Szargut, J. (2005). *Exergy Method: Technical and Ecological Applications*. WIT Press, Southampton.

Szargut, J. and Morris, D.R. (1987). Cumulative exergy consumption and cumulative degree of perfection of chemical processes. *Int. J. Energy Res.*, 11, 245–261.

Szargut, J. and Majza, E. (1989). Thermodynamic imperfection and exergy losses at production of pig iron and steel. *Arch. Metall.*, 34, 197–216.

Szargut, J. and Petela, R. (1965). *Exergy*. Wydawnictwa Naukowo Techniczne, Warsaw.

Szargut, J. and Ziębik, A. (1972). Linear mathematical model of the material and energy balance of ironworks (in Polish). *Archiwum Energetyki*, 2, 1–18.

Szargut, J., Morris, D.R. and Steward, F. (1988). *Exergy Analysis of Thermal Chemical and Metallurgical Processes*. Hemisphere, New York.

Szewczyk, K. (2002). Balances of microbial growth (in Polish). *Biotechnology*, 57, 15–32.

Szewczyk, K. (2006). Biological fuel cells - Biologiczne ogniwa paliwowe (in Polish). *Na Pograniczu Chemii i Biologii*, 15, 179–208.

Szwast, Z. (1979). *Discrete Algorithms of Maximum Principle with a Constant Hamiltonian and their Selected Applications in Chemical Engineering*. Ph.D. Thesis. Institute of Chemical Engineering, Warsaw University of Technology.

Szwast, Z. (1988). Enhanced version of a discrete algorithm for optimization with a constant Hamiltonian. *Inz. Chem. Proc.*, 3, 529–545.

Szwast, Z. (1990). Exergy optimization in a class of drying systems with granular solids. In: S. Sieniutycz and P. Salamon (Eds.), *Finite-Time Thermodynamics and Thermoeconomics and Advances in Thermodynamics vol. 4*. Taylor and Francis, New York, pp. 209–248.

Szwast, Z. (1994). Discrete optimal control thermodynamic processes with a constant Hamiltonian. *Periodica Polytechnica: Physics and Nuclear Sci. Budapest T U*, 2, 85–109.

Szwast, Z. (1997). An approach to the evolution of selected living organisms by information-theoretic entropy (in Polish). *Reports of Faculty of Chemical and Process Engineering at Warsaw TU*, 24, 123–143.

Szwast, Z., Sieniutycz, S. and Shiner, J. (2002). Complexity principle of extremality in evolution of living organisms by information-theoretic entropy. *Chaos, Solitons Fractals*, 13, 1871–1888.

Szymanowski, J. (1971). Przegląd metod poszukiwania ekstremum z ograniczeniami. *Arch. Aut. i Telemech.*, 16(2), 205–213.

Szymanowski, J. and Brzostek, J. (1971). Porównanie bezgradientowych metod optymalizacji statycznej. *Arch. Aut. i Telemech.*, 16(1), 104–115.

Szymanowski, J. and Jastrzębski, S. (1971). Porównanie gradientowych metod optymalizacji statycznej. *Arch. Aut. i Telemech*, 16(2), 155–168.

Takama, N., Kuriyama, T., Shiroko, K. and Umeda, T. (1980). Optimal water allocation in a petroleum refinery. *Comput. Chem. Eng.*, 4, 251–258.

Tan, A. and Holland, L.R. (1990). Tangent law of refraction for linear conduction through an interface and related variational principle. *Am. J. Phys.*, 58, 991–998.

Tan, R.R. and Cruz, D.E. (2004). Synthesis of robust water reuse networks for singlecomponent retrofit problems using symmetric fuzzy linear programming. *Comput. Chem. Eng.*, 28, 2547–2551.

Tantimurata, L., Kokossis, A.C. and Muller, F.U. (2000). The heat exchanger network design as a paradigm of technology integration. *Appl. Therm. Eng.*, 20, 1589–1605.

Tantimurata, L., Asteris, G., Antonopoulos, D.K. and Kokossis, C.A. (2001). A conceptual programming approach for the design of flexible HENs. *Comput. Chem. Eng.*, 25, 887–892.

Tayal, M.C. and Fu, Y. (1999). Optimal design of heat exchangers: a genetic algorithm framework. *Ind. Eng. Chem. Res.*, 38, 456–467.

Teh, Y.S. and Rangaiah, G.P. (2002). A study of equation-solving and Gibbs free energy minimization methods for phase equilibrium calculations. *Trans. IChemE*, 80(Part A), 745–749.

Tellez, R., Svrcek, W.Y., Ross, T.J. and Young, B.R. (2006). Heat exchanger network process modifications for controllability using design reliability theory. *Comput. Chem. Eng.*, 30, 730–743.

Temple, B. (1982). Global solution of the Cauchy problem for a class of 2×2 non-strictly hyperbolic conservation laws. *Adv. Appl. Math*, 3, 335–375.

Tessitore, M.E. (1995). Optimality conditions for infinite horizon control problems. *Boll Un. Mater. Ital.*, 9B(7), 785–814.

Thevendiraraj, S., Klemeš, J., Paz, D., Aso, G. and Cardenas, G.J. (2003). Water and wastewater minimisation study of a citrus plant. *Resour. Conserv. Recycl.*, 27, 227–250.

Tjoe, T.N. and Linnhoff, B. (1986). Using Pinch technology for process retrofit. *Chem.Eng.* (April 28), 47-54.

Tjoe, T.N. and Linnhoff, B. (1988). Achieving the best energy targets. *AIChE Ann. Manage.*, Houston, TX, USA, paper 17d.

Toffolo, A. and Lazzaretto, A. (2002). Evolutionary algorithms for multi-objective energetic and economic optimization in thermal system design. *Energy*, 27, 549–567.

Toffolo, A. and Lazzaretto, A. (2003). A new thermoeconomic method for the location of causes of malfunctions in energy systems, *Proceedings of IMECE 2003, ASME Int. Mech. Eng. Congress and R&D Exposition*, Washington D.C., USA, November 15–21, 2003, file 2003-42689.

Toffolo, A. and Lazaretto, A. (2004). On the thermoeconomic approach to the diagnosis of energy system malfunctions indicators to diagnose malfunctions: application of anew indicator for the location of causes. *Int. J. Thermodyn.*, 7, 41–49 (the special issue on thermoeconomic diagnosis).

Tolman, R.C. (1949). *Relativity, Thermodynamics and Cosmology*. Clarendon Press, Oxford pp. 72, 269.

Tondeur, D. and Kvaalen, E. (1987). Equipartition of entropy production: an optimality criterion for transfer and separation processes. *Ind. Eng. Chem. Res.*, 26, 50–56.

Tourin, A. (1992). A comparison theorem for a piecewise Lipschitz continuous Hamiltonian and application to shape from shading problems. *Numer. Math.*, 62, 75–85.

Towers, J.D. (2000). Convergence of a difference scheme for conservation laws with a discontinuous flux. *SIAM J. Numer. Anal.*, 38, 681–698.

Towers, J.D. (2001). A difference scheme for conservation laws with a discontinuous flux—the nonconvex case. *SIAM J. Numer. Anal.*, 39, 1197–1218.

Townsend, D.W. and Linnhoff, B. (1983a). Heat and power networks in process design—Part I: Criteria for placement of heat engines and heat pumps in process networks. *AIChE J.*, 29(5), 742–748.

Townsend, D.W. and Linnhoff, B. (1983b). Heat and power networks in process design—Part II: Design procedure for equipment selection and process matching. *AIChE J.*, 29(5), 748–771.

Townsend, D.W. and Linnhoff, B. (1984). Surface area targets for heat exchanger networks. *IChemE Ann. Res. Manage* Bath, UK.

Townsend, K.N. (1992). Is the entropy law relevant to the economics of natural resource scarcity?. *J. Environ. Econ. Manage.*, 23, 96–100.

Trivedi, K.K., Roach, J.R. and O'Neill, B.K. (1987). Shell targeting in heat exchanger networks. *AIChE J.*, 33, 2097–2099.

Trivedi, K.K., O'Neill, B.K. and Roach, J.R. (1989a). Synthesis of heat exchanger networks featuring multiple pinch points. *Comput. Chem. Eng.*, 13(3), 291–294.

Trivedi, K.K., O'Neill, B.K. and Roach, J.R. (1989b). A new dual-temperature design method for the synthesis of heat exchanger networks. *Comput. Chem. Eng.*, 13(6), 667–685.

Trivedi, K.K., O'Neill, B.K., Roach, J.R. and Wood, R.M. (1990). Systematic energy relaxation in MER heat exchanger networks. *Comput. Chem. Eng.*, 14(6), 601–611.

Tsai, M.-J. and Chang, C.-T. (2001). Water usage and treatment network design using genetic algorithms. *Ind. Eng. Chem. Res.*, 40, 4874–4888.

Tsai, Y.-H., R., Giga, Y. and Osher, S. (2001). A level set approach for computing discontinuous solutions of a class of Hamilton–Jacobi Equations. http://citeseer.comp.nus.edu.sg/429204.html (see also this work in Math Comp. 2001).

Tsatsaronis, G. (1984). Combination of exergetic and economic analysis in energy conversion processes, *Energy Economics and Management in Industry, Proceedings of the European Congress*, Algarve, Portugal, April 2-5. Pergamon Press, Oxford, England vol. 1, pp. 151–157.

Tsatsaronis, G. (1987). A review of exergoeconomic methodologies. In: M. Moran and E. Sciubba (Eds.), *Second Law Analysis of Thermal System*. ASME, New York, pp. 81–97.

Tsatsaronis, G. (1993). Thermodynamic analysis and optimization of energy systems. *Prog. Energy Combus. Sci.*, 19, 227–257.

Tsatsaronis, G. (1999). Strengths and limitations of exergy analysis. In: A. Bejan and E. Mamut (Eds.), *Thermodynamic Optimization of Complex Energy Systems*. Kluwer Academic Publishers, Dordrecht, pp. 93–100.

Tsatsaronis, G. (2007). Comments on the paper 'A brief commented history of exergy from the beginnings to 2004' by E. Sciubba and G. Wall. *Int. J. Thermodyn.*, 10, 187–190.

Tsatsaronis, G. and Winhold, M. (1984). Thermoeconomic Analysis of Power Plants, EPRI AP-3651, RP 2029-8. *Final Report, Electric Power Research Institute*, Palo Alto, CA, USA, August.

Tsatsaronis, G. and Winhold, M. (1985a). Exergoeconomic analysis and evaluation of energy conversion plants. Part I. Anew general methodology. *Energy Int. J.*, 10, 69–80.

Tsatsaronis, G. and Winhold, M. (1985b). Exergoeconomic analysis and evaluation of energy conversion plants. Part II. Analysis of a coal-fired steam power plant. *Energy Int. J.*, 10, 81–94.

Tsatsaronis, G.,Winhold, M. and Stojanoff, C.G. (1986). Thermoeconomic Analysis of a Gasification-Combined-Cycle Power Plant. EPRI AP-4734, RP 2029-8, *Final Report*. Electric Power Research Institute, Palo Alto, CA, USA, August.

Tsatsaronis, G., Lin, L., Pisa, J. and Tawfik, T. (1991). Thermoeconomic Design Optimization of a KRW-Based IGCC Power Plant, *Final Report submitted to Southern Company Services and the U.S. Department of Energy*, DE-FC21-89MC26019. Center for Electric Power, Tennessee Technological University, November.

Tsatsaronis, G., Pisa, J., Lin, L., Tawfik, T., Sears, R.E. and Gallaspy, D.T. (1992). Thermoeconomics in search of cost-effective solutions in IGCC Power Plants (invited paper), *Coal Energy and the Environment, Proceedings of the Ninth Annual International Pittsburgh Coal Conference*, Pittsburgh, Pennsylvania, October 12-16, pp. 369–377.

Tsatsaronis, G., Lin, L. and Pisa, J. (1993). Exergy costing in exergoeconomics. *J. Energy Resourc. Technol.*, 115, 9–16.

Tsatsaronis, G., Tawfik, T. and Lin, L. (1994). Exergetic comparison of two KRW-Based IGCC Power Plants. *J. Eng. Gas Turbines Power*, 116, 291–299.

Tsirlin, A.M. (1974). Averaged optimization and sliding regimes in optimal control problems. *Izv. AN SSSR, Ser Tech. Cybern.*, 2, 143–151.

Tsirlin, A.M. (1997a). *Optimal Cycles and Cyclic Regimes (in Russian)*. Energoatomizdat, Moscow.

Tsirlin, A.M.b (1997). *Methods of Averaged Optimization and Their Applications (in Russian)*. Fizmatlit, Moscow.

Tsirlin, A.M. and Kazakov, V. (2002a). Finite-time thermodynamics: limiting possibilities of irreversible separation processes. *J. Phys. Chem.*, 106(R.S. Bery's issue), 10926–10936.

Tsirlin, A.M. and Kazakov, V. (2002b). Realizability areas for thermodynamic systems with given productivity. *J. Non-Equilib. Thermodyn.*, 27, 91–103.

Tsirlin, A.M. and Leskov, E.E. (2007). Optimization of diffusion systems. *Theor. Found. Chem. Eng.*, 41, 405–413.

Tsirlin, A.M., Mironova, W.A., Amelkin, S.A. and Kazakov, V.A. (1998). Finite-time thermodynamics: conditions of minimal dissipation for thermodynamic processes with given rate. *Phys. Rev. E*, 58, 215–223.

Tsitsiklis, J. (1995). Efficient algorithms for globally optimal trajectories. *IEEE Trans. Automatic Control*, 40(9), 1528–1538.

Tu, M. (1987). Thermodynamic and economic evaluation of a solar unised heat pump. *Int. J. Energy Res.*, 2, 559.

Ulanowicz, R.E. (1997). *Ecology, the Ascendent Perspective*. Columbia University Press, New York.

Ulas, S. and Diwekar, U.M. (2006). Integrating product and process design with optimal control: a case study of solvent recycling. *Chem. Eng. Sci.*, 61, 201–209.

Ullmer, N., Kunde, N., Lassahn, A., Gruhn, G. and Schultz, K. (2005). WADOTM: water design optimization—methodology and software for the synthesis of process water systems. *J. Cleaner Prod.*, 13, 485–494.

Umeda, T., Nida, K. and Shiroko, K. (1979). A thermodynamic approach to heat integration of distillation systems. *AIChE J.*, 25, 423.

Upreti, S.R. and Deb, K. (1997). Optimal design of an ammonia synthesis reactor using genetic algorithms. *Comput. Chem. Eng.*, 1(21), 87–92.

Urbaniec, K., Zalewski, P. and Zhu, X.X. (2000). A decomposition approach for retrofit design of energy systems in the sugar industry. *Appl. Therm. Eng.*, 20, 1431–1442.

Vainberg, M.M. (1964). *Variational Methods for the Study of Nonlinear Operators*. Holden-Day, San Francisco.

Valero, A. (1998). Thermoeconomics as a conceptual basis for energy ecological analysis, *Proc. 1st Intl. Workshop on Advanced Energy Studies*. P. Venere, Italy.

Valero, A. (1999). Qualifying irreversibilities through second law: exergy accounting, *Gordon Research Conference: Modern Developments in Thermodynamics 18-23*, April. Il Ciocco, Tuscany, Italy.

Valero, A. (2003). Thermodynamic process of cost formation. *Enc. of Life-Support Sciences* EOLSS, London.

Valero, A., Lozano, M.A. and Munoz, M. (1986a). A general theory of exergy saving: I. On the exergetic cost. In: Gaggioli, R.A., (Ed.), *Computer-Aided Engineering and Energy Systems Vol. 3: Second Law Analysis and Modelling*. AES Vol. 2-3: ASME Book no. H0341C. ASME, New York, pp. 1–8.

Valero, A., Munoz, M. and Lozano, M.A. (1986b). A general theory of exergy saving: II. On the thermoeconomic cost. In: Gaggioli, R.A. (Ed.), *Computer-Aided Engineering and Energy Systems Vol. 3: Second Law Analysis and Modelling*. AES Vol. 2-3: ASME Book no. H0341C. ASME, New York, pp. 9–15.

Valero, A., Munoz, M. and Lozano, M.A. (1986c). A general theory of exergy saving: III. Energy saving and thermoeconomics. In: Gaggioli, R.A., (Ed.), *Computer-Aided Engineering and Energy Systems Vol. 3: Second Law Analysis and Modelling*. AES Vol. 2-3: ASME Book no H0341C. ASME, New York, pp. 17–21.

Valero, A., Lozano, M.A., Alconchel, J.A., Munoz, M. and Torres, C. (1986d). GAUDEAMO: A system for energetic/exergetic optimization of coal power plants. In: Gaggioli, R.A., (Ed.), *Computer-Aided Engineering and Energy Systems Vol. 3: Second Law Analysis and Modelling. AES* Vol. 2-3: ASME Book no H0341A. ASME, New York, pp. 43–49.

Valero, A., Torres, C. and Lozano, M.A. (1989). On the unification of thermoeconomic theories. In: Boehm, R.F., El-Sayed, Y.M., (Eds.), *Simulation of Thermal Energy Systems, HTD* Vol. 124. ASME Book no H00527. ASME, New York, pp. 63–74.

Valero, A., Torres, C. and Serra, L. (1992). A general theory of thermoeconomics: I. Structural analysis; II. The relative free energy function. In: A. Valero and G. Tsatsaronis (Eds.), *Proceedings of the International Symposium on Efficiency, Costs, Optimization and Simulation of Energy Systems ECOS' 92.* Zaragoza, Spain June 15-18, pp. 137–154.

Valero, A., Serra, L. and Lozano, M.A. (1993). Structural theory of thermoeconomics. In: Richter, R. (Ed.), *Proceedings of the Symposium on Thermodynamics and the Design, Analysis and Improvement of Energy Systems.* November 28–December 3, New Orleans, USA. ASME Book no H00527. ASME, New York, pp. 1–10.

Valero, A., Lozano, M.A., Serra, L., Tsatsaronis, G., Pisa, J., Frangopoulos, Ch. and Von Spakovsky, M.R. (1994a). CGAM problem: definition and conventional solution. *Energy,* 19, 279–286.

Valero, A., Lozano, M.A. and Torres, L.C. (1994b). Application of the exergetic cost theory to the CGAM Problem. *Energy,* 19, 365–381.

Valero, A., Correas, L., Lazaretto, A., Rangel, V., Reini, M., Taccani, R., Toffolo, A., Verda, V. and Zaleta, A. (2004). Thermoeconomic philosophy applied to the operating analysis and diagnosis of energy utility systems. *Int. J. Thermodyn.,* 7, 33–39 (the special issue on thermoeconomic diagnosis).

Valero, A., Serra, L. and Uche, J. (2006). Fundamentals of exergy cost accounting and thermoeconomics. Part i: theory. *J. Energy Resour. Technol.,* 128, 1–8.

van Reisen, J.L.B., Grievink, J., Polley, G.T. and Verheijen, P.J.T. (1995). The placement of two-stream heat exchangers in an existing network through path analysis. *Comput. Chem. Eng.,* 19(Suppl.), S143–S148.

van Reisen, J.L.B., Polley, G.T. and Verheijen, P.J.T. (1998). Structural targeting for heat integration retrofit. *Appl. Therm. Eng.,* 18, 283–294.

Vanderbilt, D. and Louie, S.G. (1984). A Monte Carlo simulated annealing approach to optimization over continuous variables. *J. Comput. Phys.,* 56, 259–271.

Varaiya, P.P. (1972). *Optimization.* Van Nostrand Rheinhold, New York.

Varbanov, P.S. and Klemes, J. (1999). Rules for paths construction for HENs debottlenecking, *Proceedings of PRES'99.* Budapest, Hungary.

Varbanov, P.S., Klemes, J., Boyadijev, Ch., Ivanov, B. and Vaklieva-Bacheva, N. (2000). A new strategy for optimal heat exchanger network retrofit using heuristic paths and superstructures, *Congress CHISA 2000,* Prague, Czech Republic P7.76.

Varbanov, P.S., Doyle, S. and Smith, R. (2004a). Modeling and optimization of utility systems. *Trans IChemE, Part A., Chem. Eng. Res. Des.,* 82(A5), 561–578.

Varbanov, P., Perry, S., Makwana, Y., Zhu, X.X. and Smith, R. (2004b). Top-level analysis of site utility systems. *Trans. IChemE, Part A, Chem. Eng. Res. Des.,* 82(A6), 784–795.

Vargas, J.V.C. and Bejan, A. (2004). Thermodynamic optimization of internal structure in a fuel cell. *Int. J. Energy Res.,* 28, 319–339.

Vargas, J.V.C., Ordonez, J.C. and Bejan, A. (2005). Constructal PEM fuel cell stack design. *Int. J. Heat Mass Transfer,* 48, 4410–4427.

Vasilescu, E.E., Bousseehain, R. and Feidt, M. (2007). Exergy analysis of an adsorption refrigeration machine. *Int J. of Exergy*, 4(2), 197–215.

Venkatasubramanian, V., Chan, K. and Caruthers, J.M. (1994). Computer-aided molecular design using genetic algorithms. *Comput. Chem. Eng.*, 9(18), 833–844.

Verda, V. (2004). Thermoeconomic analysis and diagnosis of energy utility systems from diagnosis to prognosis. *Int. J. Thermodyn*, 7, 73–83 (the special issue on thermoeconomic diagnosis).

Verheyen, W. and Zhang, N. (2006). Design of flexible heat exchanger network for multiperiod operation. *Chem. Eng. Sci.*, 61, 7730–7753.

Vinter, R. (2000). *Optimal Vontrol*. Birkhauser, Basel.

Viswanathan, M. and Evans, L.B. (1987). Studies in heat integration of chemical process plants. *AIChE J.*, 33, 1781–1990.

Visweswaran, V. and Floudas, C.A. (1990). A global optimization algorithm (GOP) for certain cases of nonconvex NLPs – II. Application of theory and test problems. *Comput. Chem. Eng.*, 14(12), 1419–1434.

Von Spakovsky, M.R. (1993). Aspects of the thermoeconomic modelling of energy systems with cogeneration. *Rev. Gen. Therm. Fr.*, 383, 594–605.

Von Spakovsky, M.R. (1994). Application of engineering functional analysis to the analysis and optimization of the CGAM problem. *Energy*, 19, 343–364.

Von Spakovsky, M.R. and Evans, R.B. (1990). The design and performance optimization of thermal systems. *J. Energy Resourc Tech. Trans. ASME*, 112.

Von Spakovsky, M.R. and Frangopoulos, Ch.A. (1993). A global environomic approach for energy systems analysis and optimization – II, *Proceedings of the International Conference Energy Systems and Ecology ENSEC' 93*, Cracow, Poland, July 5–9, pp. 133–144.

Von Spakovsky, M.R. and Olsommer, B. (2002). Fuel cell systems and system modeling and analysis perspectives for fuel cell development. *Energy Convers. Manage*, 43, 1249–1257.

Vujanovic, B. and Jones, S.E. (1988). *Variational Methods in Nonconservative Phenomena*. Academic Press, New York.

Vukalovich, M.P. and Novikov, I.I. (1972). *Thermodynamics*. Mashinostroenije, Moscow.

Wagialla, K.M., Fakeeha, A.H., Elnashaie, S.S.E.H. and Almaktary, A.Y. (1991). Modelling and simulation of energy storage in fluidized beds using the two-phase model. *Energy Sources*, 13, 189.

Wałczyk, K., Jeżowski, J., Bochenek, R. and Jeżowska, A. (2004). Modele optymalizacyjne sieci wymienników ciepła z uwzględnieniem segmentacji strumieni. *Inż. Chem. Proc.*, 25(3–3), 1753–1758.

Wałczyk, K., Bochenek, R., Jeżowski, J. and Jeżowska, A. (2005). Metoda i wspomaganie komputerowe do optymalnej modyfikacji sieci wymienników ciepła. II. Opis programu komputerowego i przykłady zastosowania. *Inż. Chem. Proc.*, 26, 761.

Wall, G. (1986). Thermoeconomic optimization of heat pump system. *Energy*, 11, 957–967.

Wang, Y. and Achenie, L.E.K. (2002). A hybrid global optimization approach for solvent design. *Comput. Chem. Eng.*, 26, 1415–1425.

Wang, B.-C. and Luus, R. (1978). Reliability of optimization procedures for obtaining global optimum. *AIChE J.*, 24(4), 619–626.

Wang, B.-C. and Luus, R. (1997). Optimization of non-unimodal systems. *Int. J. Numer.Met. Eng.*, 11, 1235–1250.

Wang, Y.P. and Smith, R. (1994a). Wastewater minimization. *Chem. Eng. Sci.*, 49(7), 981–1006.

Wang, Y.P. and Smith, R. (1994b). Design distributed effluent treatment systems. *Chem. Eng. Sci.* (49), 3127–3145.

Wang, Y.P. and Smith, R. (1995). Wastewater minimization with flowrate constraints. *Trans. IChemE*, 73(Part A), 889–904.

Wang, C., Quan, H. and Xu, X. (1996). Optimal design of multiproduct batch chemical process using genetic algorithms. *Ind. Eng. Chem. Res.*, 35, 3560–3566.

Wang, K., Löhl, T., Stobbe, M. and Engell, S. (2000). A genetic algorithm for online scheduling of a multiproduct polymer batch plant. *Comput. Chem. Eng.*, 24, 393–400.

Wang, B., Feng, X. and Zhang, Z. (2003). A design methodology for multiplecontaminant water networks with single internal water main. *Comput. Chem. Eng.*, 27, 903–911.

Wang, K., Salhi, A. and Fraga, E.S. (2004). Process design optimisation using embedded hybrid visualisation and data analysis techniques within genetic algorithm optimization framework. *Chem. Eng. Process*, 43, 663–675.

Warner, J.W. and Berry, R.S. (1985). Injection quenching and the high temperature water-splitting reactor. *Solar Energy*, 35, 535–537.

Watowich, S.J. and Berry, R.S. (1986). Optimal current paths for model electrochemical systems. *J. Phys. Chem*, 90, 3631–4624.

Watowich, S.J., Krause, J.L. and Berry, R.S. (1986). Stability analysis of an optimally controlled light-driven engine. *J. Symbolic Computation*, 2, 103.

Wechsatol, W., Ordonez, J.C. and Kosaraju, S. (2006). Constructal dendritic geometry and the existence of asymmetric bifurcation. *J. Appl. Phys.*, 100, 113514.

Wen, Y. and Shonnard, D.R. (2003). Environmental and economic assessments of heat exchanger networks for optimum minimum approach temperature. *Comput. Chem. Eng.*, 27, 1577–1590.

Wenzel, H., Dunn, R.F., Gottrump, L. and Kringelum, J. (2002). Process integration design methods for water conservation and wastewater reduction in industry. Part 3: Experience of industrial application. *J. Clean Technol. Environ. Policy*, 4(1), 16–25.

Westerberg, A.W. (2004). A retrospective on design and process synthesis. *Comput. Chem. Eng.*, 28, 447–458.

Westerberg, A.W. and Subrahmanian, E. (2000). Product design. *Comput. Chem. Eng.*, 24, 959–966.

White, W.B., Johnson, S.M. and Dantzig, G.B. (1958). Chemical equilibrium in complex mixtures. *J. Chem. Phys.*, 5(28), 751–755.

Wibowo, C. and Ng, K.M. (2001). Product-oriented process synthesis and development: creams and pastes. *AIChE J.*, 47(12), 2746–2767.

Wickey, J.S. (1979). The generation of complexity in evolution: a thermodynamic and information—theoretical discussion. *J. Theor. Biol*, 77, 349–365.

Wickey, J.S. (1980). A thermodynamic theory of evolution. *J. Theor. Biol*, 87, 9–23.

Wierzbicki, M. (2009). *Optimization of SOFC based energy system using Aspen Plus^{TM}*, MsD thesis, Warsaw University of Technology, Faculty of Chemical and Process Engineering. Supervised by S. Sieniutycz (Faculty of Chemical and Process Engineering, Warsaw TU) and J. Jewulski (Laboratory of Fuel Cells, Institute of Energetics, Warsaw).

Wilde, D.J. (1964). *Optimum Seeking Methods*. Prentice Hall, Englewood Cliffs.

Wilde, D.J. (1962). *Optimum Seeking Methods*. Prentice Hall, Englewood Cliffs.

Wilde, D.J. and Beightler, Ch.S. (1967). *Foundations of Optimization*. Prentice Hall, Englewood Cliffs.

Williamson, A.G. (1993). The second law of thermodynamics and the economic process. *Ecol. Econ.*, 7, 69–71.

Wintermantel, K. (1999). Process and product engineering—achievements, present and future challenges. *Chem. Eng. Sci.*, 54, 1601–1620.

Witt, U. (1999). Bioeconomics as economics from a Darwinian perspective. *J. Bioecon.*, 1, 19–34.

Wood, R., Wilcox, R.J. and Grossmann, I.E. (1986). A note on the minimum number of units for heat exchanger network synthesis. *Chem. Eng. Commun.*, 14, 105–124.

Wright, S.E. (2004). Comparison of the theoretical performance potential of fuel cells and heat engines. *Renewable Energy*, 29, 179–195.

Wright, S. (2007). Comparative analysis of the entropy of radiative heat transfer and heat conduction. *Int. J. Thermodyn*, 10, 27–35.

Wright, S.E., Scott, D.S., Haddow, J.B. and Rosen, M.B. (2002). The exergy flux of radiative heat transfer for the special case of blackbody radiation. *Exergy Int. J.*, 2, 24–33.

Wu, C. (1989). Power optimization of a finite-time solar radiation heat engine. *Int. J. Ambient Energy*, 10, 145–150.

Wu, C. (1992). Optimal power from a radiating solar powered thermionic engine. *Energy Convers. Manage.*, 33, 59–67.

Wu, X.-J., Zhu, X.-J., Cao, G.-Y. and Tu, H.-Y. (2008a). Dynamic modeling of SOFC based on a T–S fuzzy model. *Simulation Modelling Practice and Theory*, 16, 494–504.

Wu, X.-J., Zhu, X.-J., Cao, G.-Y. and Tu, H.-Y. (2008b). Predictive control of SOFC based on a GA-RBF neural network model. *J. of Power Sources*, 179, 232–239.

Xia, S., Chen, L. and Sun, F. (2009a). Optimal configuration of a finite mass reservoir isothermal chemical engine for maximum work output with linear mass transfer law. *Rev. Mex. Fis.*, 55(5), 399–408.

Xia, S., Chen, L. and Sun, F. (2009b). Maximum power configuration for multi-reservoir chemical engines. *J. Appl. Phys.*, 105(12): 124905.

Xia, S., Chen, L. and Sun, F. (2010a). Finite-time exergy with a finite heat reservoir and generalized radiative heat transfer law. *Revista Mexicana de Fisica*, 56, 287–296.

Xia, S., Chen, L. and Sun, F. (2010b). Optimal paths for minimizing entrance dissipation during heat transfer processes with generalized radiative heat transfer law. *Appl. Math. Modell.*, 34(8), 2242–2255.

Xia, S., Chen, L., Sun, F. (2011c). Endoreversible modeling and optimization of a multistage heat engine system with a generalized heat transfer law via Hamilton-Jacobi-Bellman equations and dynamic programming. Acta Physica Polonica A, 2011, 119(6): 747–760.

Xiao, M.Q. and Basar, T. (1997). Viscosity solutions of a class of Hamilton–Jacobi–Isaacs equations related to nonlinear $H\infty$ control problems, *Proceedings of the 36th IEEE Conference on Decision and Control 2*, pp. 1761–1766.

Xiao, M.Q. and Basar, T. (1999). Optimal control of piecewise deterministic nonlinear system with controlled transitions. Viscosity solutions, their existence and uniqueness, *Proceedings of the 38th Conference on Decision and Control*, Phoenix, Arizona, December.

Xu, W. and Diwekar, U.M. (2005). Improved genetic algorithms for deterministic optimization under uncertainty. Part II. Sovent selection under uncertainty. *Ind. Eng.Chem. Res.*, 44, 7138–7146.

Xue, D. and Dong, Z. (1998). Optimal fuel cell system design considering functional performance and production costs. *J. of Power Sources*, 76, 69–80.

Xue, D., Li, S., Yuan, Y. and Yao, P. (2000). Synthesis of waste interception and allocation networks using genetic-alopex algorithm. *Comput. Chem. Eng.*, 24, 1455–1460.

Xue, X., Tang, J., Sammes, N. and Du, Y. (2005). Dynamic modeling of single tubular SOFC combining heat/mass transfer and electrochemical reaction effects. *Journal of Power Sources*, 142, 211–222.

Ya, W. and Shonnard, D.R. (2000). Environmental and economic assessments of heat exchanger networks for optimum minimum approach temperature. *Comput. Chem. Eng.*, 27, 1577–1590.

Yan, Z. (1993). Comment on "Ecological optimization criterion for finite-time heat engines". *J. Appl. Phys.*, 73, 3583.

Yan, Z. and Chen, J. (1990). Optimal performance of a generalized carnot cycle for another linear heat transfer law. *J. Chem. Phys.*, 92, 1994–1998.

Yan, Z. and Chen, L. (1999). Optimal performance of an irreversible Carnot engine for another linear heat transfer law. In: Ch. Wu, L. Chen and J. Chen (Eds.), *Advance in Recent Finite Time Thermodynamics*. Nova Science, New York, pp. 307–315.

Yan, L. and Ma, D. (2001). Global optimisation of non-convex nonlinear programs using Line-up competition algorithm. *Comput. Chem. Eng.*, 25, 1601–1610.

Yan, X. and Zhao, W. (2006). A novel select-best and prepotency evolution algorithm and its application to develop industrial oxidation reaction macrokinetic model. *Comput. Chem. Eng.*, 30, 807–815.

Yang, Y.H., Lou, H.H. and Huang, Y.L. (2000). Synthesis of an optimal wastewater reuse network. *Waste Manage.*, 20, 311–319.

Yantovski, Y. (2007a). Comments on the paper 'A Brief Commented History of Exergy from the Beginnings to 2004' by E. Sciubba and G. Wall. *Int. J. Thermodyn.*, 10, 193–194.

Yantovski, Y. (2007b). Review of the book: J. Szargut's book: Exergy Method: Technical and Ecological Applications, WIT Press, Southampton, Boston, 2005. *Int. J.Thermodyn.*, 10, 93–95.

Yee, T.F. and Grossmann, I.E. (1987). Optimization model for structural modifications in the retrofit of heat exchanger networks. *Foundations of Computer-Aided Process Operations*. Park City, Utah, pp. 653–662.

Yee, T.F. and Grossmann, I.E. (1988). A screening and optimization approach for the retrofit of heat exchanger networks. *Ann. AIChE Manage.*, Washington DC, paper 81d.

Yee, T.F. and Grossmann, I.E. (1990). Simultaneous optimization models for heat integration II. Heat exchanger network synthesis. *Comput. Chem. Eng.*, 14(10), 1165–1184.

Yee, T.F. and Grossmann, I.E. (1991). A screening and optimization approach for the retrofit of heat-exchanger networks. *Ind. Eng. Chem. Res.*, 30, 146–162.

Yee, T.F., Grossmann, I.E. and Kravanja, Z. (1990a). Simultaneous optimization models for heat integration I. Area and energy targeting and modelling of multi-stream exchangers. *Comput. Chem. Eng.*, 14(11), 1185–1200.

Yee, T.F., Grossmann, I.E. and Kravanja, Z. (1990b). Simultaneous optimization models for heat integration—III. Process and heat exchanger network optimization. *Comput. Chem. Eng.*, 14(10), 1151–1164.

Yong, J. and Zhou, X. (1999). *Stochastic Controls. Hamiltonian Systems and HJB Equations*. Springer-Verlag, New York.

Yoon, S.-G., Lee, J. and Park, S. (2007). Heat integration analysis for an industrial ethylbenzene plant using pinch analysis. *Appl. Therm. Eng.*, 27, 886–893.

Young, J.T. (1991). Is the entropy law relevant to the economics of natural resource scarcity?. *J. Environ. Econ. Manage.*, 21, 169–179.

Yourgrau, Y. and Mandelstam, S. (1968). *Variational Principles in Dynamics and Quantum Mechanics* (3rd edn.). W.B. Saunders, Philadelphia.

Yu, H., Fang, H., Yao, P. and Yuan, Y. (2000). A combined genetic/simulated annealing algorithm for large scale system energy integration. *Comput. Chem Eng.*, 24, 2023–2035.

Zabinsky, Z.B. (1998). Stochastic methods for practical global optimization. *J. Global Optim.*, 13, 433–444.

Zaleta-Aguilar, A., Gallegos-Munoz, A., Rangel-Hernandez, V. and Capilla, A.V. (2004). A reconciliation method based on a module simulator an approach to the diagnosis of energy system malfunctions. *Int. J. Thermodyn.*, 7, 51–60 (the special issue on thermoeconomic diagnosis).

Zalewski, M. (2009). Wpływ dezaktywacji katalizatora na średnie stężenie produktu dla chaotycznej reakcji chemicznej. *Inżynieria i Aparatura Chemiczna*, 6, 196–197.

Zamorra, J.M. and Grossmann, I.E. (1998a). A global MINLP optimization algorithm for the synthesis of heat exchanger networks with no stream splits. *Comput. Chem. Eng.*, 3(22), 367–384.

Zamorra, J.M. and Grossmann, I.E. (1998b). Continuous global optimization of structured process systems models. *Comput. Chem. Eng.*, 12(22), 1749–1770.

Zangwill, W.J. (1967). Minimizing a function without calculating derivatives. *Comput. J.*, 10, 293.

Zangwill, W.J. (1974). *Nonlinear Programming*. Państwowe Wydawnictwa Naukowe, Warsaw.

Zawlocki, I. (1987). Exergy analysis of drying units. *Scient. Bull. Lodz Tech. Univ.*, 14, 170–178.

Zhang, J. and Zhu, X.X. (2000). Simultaneous optimization approach for heat exchanger network retrofit with process changes. *Ind. Eng. Chem. Res.*, 39, 4963–4973.

Zhang, L. and Linninger, A.A. (2006). Towards computer-aided separation synthesis. *AIChE J.*, 52(4), 1392–1409.

Zhang, N. and Xhu, X.X. (2006). Novel modelling and decomposition strategy for total site optimisation. *Comput. Chem. Eng.*, 30, 765–777.

Zhang, W., Croiset, E., Douglas, P.L., Fowler, M.W. and Entchev, E. (2005). Simulation of a tubular solid oxide fuel cell stack using AspenPlus™ unit operation models. *Energy Convers. Manage*, 46, 181–196.

Zhao, Y. and Chen, J. (2009). Modeling and optimization of a typical fuel cell-heat engine hybrid system and its parametric design criteria. *Journal of Power Sources*, 186, 96–103.

Zhao, Y., Ou, C. and Chen, J. (2008). A new analytical approach to model and evaluate the performance of a class of irreversible fuel cells. *Intern. J. of Hydrogen Energy*, 33, 4161–4170.

Zhao, W., Chen, D. and Hu, S. (2000). Optimizing operating conditions based on ANN and modified GAs. *Comput. Chem. Eng.*, 24, 61–65.

Zhao, X.G., O'Neill, B.K., Roach, J.R. and Wood, R.M. (1998a). Heat integration for batch processes. Part 1: Process scheduling based on cascade analysis. *Chem. Eng. Res. Des.*, 76(A), 685–699.

Zhao, X.G., O'Neill, B.K., Roach, J.R. and Wood, R.M. (1998b). Heat integration for batch processes. Part 2: Heat exchanger network design. *Chem. Eng. Res. Des.*, 76(A), 700–710.

Zhelev, T. (2005). Water conservation through energy management. *J. Cleaner Prod.*, 18, 1461–1470.

Zhou, X.Y. (1990). Maximum principle, dynamic programming, and their connection in deterministic control. *J. Optim. Theory Appl.*, 65, 363–373.

Zhou, X.Y. (1991). A unified treatement of maximum principle and dynamic programming in stochastic controls. *Stochastics Stochastics Rep.*, 36, 137–161.

Zhu, H. and Kee, R.J. (2006). Thermodynamics of SOFC efficiency and fuel utilization as functions of fuel mixtures and operating conditions. *Journal of Power Sources*, 161, 957–964.

Zhu, J.Y., Rao, R. and Chuang, K.T. (1999). A new method to determine the best units for breaking heat load loops of heat exchanger networks. *Ind. Eng. Chem. Res.*, 38, 1496–1505.

Zhu, X.X. (1997). Automated design method for heat exchanger network using block decomposition and heuristic rules. *Comput. Chem. Eng.*, 21(10), 1095–1104.

Zhu, X.X. and Asante, D.K. (1999). Diagnosis and optimisation approach for heat exchanger network retrofit. *AIChE J.*, 45(7), 1488–1503.

Zhu, X.X., O'Neill, B.K., Roach, J.R. and Wood, R.M. (1993). Kirchhoff's Law and loop-breaking for the design of heat exchanger networks. *Chem. Eng. Commun.*, 126, 141–152.

Zhu, X.X., O'Neill, B.K., Roach, J.R. and Wood, R.M. (1995a). Area targetin methods for the direct synthesis of heat exchanger networks with unequal film coefficients. *Comput. Chem. Eng.*, 19(2), 223–239.

Zhu, X.X., O'Neill, B.K., Roach, J.R. and Wood, R.M. (1995b). A new method for heat exchanger network synthesis using area targeting procedures. *Comput. Chem. Eng.*, 19(2), 197–222.

Zhu, Y., Wen, H. and Xu, Z. (2000). Global stability analysis and phase equilibrium calculations at high pressures using the enhanced simulated annealing algorithm. *Chem.Eng. Sci.*, 55, 3451–3459.

Ziębik, A. (1986). Mathematical model of an energy balance for the foredesign of the energy management of industrial plants (in Polish). *Problemy Projektowe*: No 3.

Ziębik, A. (1990a). *Mathematical Modelling of Energy Management Systems in Industrial Plants*. Ossolineum, Wrocław.

Ziębik, A. (1990b). *Energy Systems (in Polish)*. Technical University of Silesia, Gliwice.

Ziębik, A. (1991a). Mathematical model of industrial energy systems and its applications, *Proceedings of the International Conference Analysis of Thermal and Energy Systems*, Athens, pp. 745–756.

Ziębik, A. (1991b). *Examples of Calculation Concerning Energy Systems (in Polish)*. Technical University of Silesia, Gliwice.

Ziębik, A. (1995). Applications of second law analysis in industrial energy and technological systems, *Proceedings of the Conference on Second Law Analysis of Energy Systems: Towards the 21 st Century*, Rome, pp. 349–359.

Ziębik, A. (1996). System analysis in thermal engineering. *Arch. Thermodyn.*, 17, 81–97.

Ziębik, A. and Presz, K. (1993). System method of the choice of the energy management structure of an industrial plant (in Polish). *Archiwum Energetyki*, No. 172.

Ziębik, A., Gwóźdź and Presz, K. (1994). Matrix method of calculating the unit costs of energy carriers as a coordination procedure in the optimization of industrial energy systems, *Proceedings of the Second Biennial European Joint Conference on Engineering Systems Design and Analysis ASMB*, London, pp. 19–26.

Ziomek, G., Kaspereit, M., Jeżowski, J., Seidel-Morgenstern, A. and Antos, D. (2005). Effect of mobile phase composition on the SMB processes efficiency. Stochastic optimization of isocratic and gradient operation. *J. Chromatogr. A*, 1070, 11–24.

Zotin, A.A. and Zotin, A.I. (1996). Thermodynamic basis of developmental processes. *J. Non-Equilib. Thermodyn.*, 21, 307–320.

Zoulias, E.I. and Lymberopoulos, N. (2007). *Hydrogen-based Autonomous Power Systems*. Springer Verlag London Limited, London.

Żbikowski, M. (2004). *Thermodynamic Modeling of Electrochemical Processes in Proton-Exchange Fuel Cell*, MsD Thesis, Faculty of Chemical and Process Engineering, Warsaw University of Technology, Supervised by S. Sieniutycz (FChPE, Warsaw University of Technology) and H. Kronberger (Technical University Viena).

Żyłła, R. and Strumiłło, Cz. (1981). Application of heat pumps for the purpose of decreasing the energy use in drying. In: T. Stopińki, (Ed.), *Materials of IV Symposium on Drying*, Instutut Chemii Przemysłowej, Warsaw.

Glossary

Glossary of principal symbols

A, A^{class}	generalized and classical exergy (J, $J\,m^{-3}$, $kJ\,kg^{-1}$)
A_j	chemical or electrochemical affinity of jth reaction ($J\,mol^{-1}$)
A_j^s	affinity of jth reaction in the entropy representation ($J\,mol^{-1}\,K^{-1}$)
$A_C =$	$-\Delta G$ chemical affinity at Carnot point ($J\,mol^{-1}$)
A^k	total exchange area at stage k (m^2)
$A^n(\mathbf{x}^n, \mathbf{u}^n, t^n, \theta^n)$	criterion of stage optimality
a_i	activity of ith species ($mol\,m^{-3}$)
a_v	specific area per unit volume (m^{-1})
b_g, b_s	specific exergy of gas and gas in equilibrium with solid ($kJ\,kg^{-1}$)
b_v	exergy per unit volume ($kJ\,m^{-3}$)
C, c	unit prices of controlled and controlling phase, repectively ($\$\,kg^{-1}$)
c_v	specific heat per unit volume ($J\,m^{-3}\,K^{-1}$)
c_m	mass capacity of active component of fuel (J^{-1})
c_0	propagation speed of thermal waves ($m\,s^{-1}$)
D	diameter (m^{-1}), diffusion coefficient ($m^2\,s^{-1}$), profit ($\$$), disorder ($-$)
$\mathbf{E}$	electric field vector ($kg\,m\,A^{-1}\,s^{-3}$)
E	total energy or energy type function (J)
E^0, E_0	Nernst ideal voltage and idle run voltage, respectively (V)
e	unit energy ($J\,kg^{-1}$), economic value of unit exergy ($\$\,J^{-1}$)
$\mathbf{F}$	force (N), specific external force ($N\,kg^{-1}$), vector of potentials
$F, \mathbf{F}$	free energy and extended free energy, respectively (J)
F	area of surface perpendicular to flux direction (m^2)
F_k	potentials or components of vector $\mathbf{F} = (T^{-1}, -\mu_1 T^{-1}, ..., -\mu_n T^{-1})$
$F^n(\mathbf{x}^n, \lambda)$	optimal drying function at stage n ($kJ\,kg^{-1}$)
$f^n(\mathbf{x}^n)$	optimal performance function of a dynamic programming algorithm
f_0, f_i	intensity of generalized profit and process rates
$\mathbf{G}, G^{jk}$	energy-momentum tensor

G	mass flux ($\mathrm{kg\,s^{-1}}$), molar flux ($\mathrm{mol\,s^{-1}}$), Gibbs free energy (J), gauge, quadratic form
G	Gibbs energy flux driving chemical engine ($\mathrm{J\,s^{-1}}$)
$\dot{G}$	electrochemical resource flux, Chap. 10 ($\mathrm{gs^{-1}}$, $\mathrm{mols^{-1}}$)
g	gradient vector, constraining vector function
g_k, g	partial and overall conductance, respectively ($\mathrm{kJ\,s^{-1}\,K^{-1}}$) ($\mathrm{kg\,s^{-1}}$), ($\mathrm{kmol\,s^{-1}}$)
$g_k, \mathbf{g}_{jk}$	kth constraint and metric tensor, respectively
$\mathbf{H}$	magnetic field, flux density of radiative entropy ($\mathrm{J\,m^{-2}\,K^{-1}\,s^{-1}}$)
$H = (h_{jk})$	Hessian matrix of second derivatives
H	thermodynamic enthalpy of the system (J, kJ)
$H(\mathbf{x}, \mathbf{u}, \mathbf{p}, t)$	Hamiltonian function of a continuous process
$H^{n-1}(\mathbf{x}^n, \mathbf{u}^n, \mathbf{p}^{n-1}, t^n)$	Hamiltonian function of a discrete process at stage n
H_{TU}	height of transfer unit (m)
h	specific enthalpy ($\mathrm{kJ\,kg^{-1}}$), Planck constant, numerical Hamiltonian
h_{v}	enthalpy of unit volume ($\mathrm{Jm^{-3}}$)
h_{σ}	Hamiltonian density in entropy units ($\mathrm{J\,m^{-3}\,K^{-1}}$)
I	molar flux of inert component ($\mathrm{mol\,s^{-1}}$)
$\mathbf{I}_k$	unidirectional flux of kth component ($\mathrm{kg\,s^{-1}}$)
I^n	solid enthalpy at stage n ($\mathrm{kJ\,kg^{-1}}$)
I^n	solid enthalpy at stage n ($\mathrm{kJ\,kg^{-1}}$)
I_{s}	specific enthalpy of solid phase ($\mathrm{kJ\,kg^{-1}}$)
i	electric current density ($\mathrm{Am^{-2}}$)
$i_{\mathrm{g}}, i_{\mathrm{s}}$	specific enthalpy of gas and gas at equilibrium with solid ($\mathrm{kJ\,kg^{-1}}$)
$\mathbf{J}$	mass flux density, conserved mass current ($\mathrm{kg\,m^{-2}\,s^{-1}}$)
$\mathbf{J}_k$	mass flux density of kth component ($\mathrm{kg\,m^{-2}\,s^{-1}}$)
$\mathbf{J}_{\mathrm{e}}, \mathbf{J}_{\mathrm{q}}$	energy flux density and heat flux density, respectively ($\mathrm{J\,m^{-2}\,s^{-1}}$)
j	conductance ratio, g_1/g_2 (−)
K	chemical equilibrium constant, kinetic energy, cost, drying coefficient
$\mathbf{k}^i$	direction vector in ith iteration
k	thermal conductivity ($\mathrm{W\,K^{-1}\,m^{-1}}$), chemical rate constant
k_{B}	Boltzmann constant ($1.380\,658 \times 10^{-23}\,\mathrm{N\,m\,K^{-1}}$)
$\mathbf{L} = L_{ik}$	Onsager's matrix of kinetic coefficients

L	length (m), Lagrangian (J), latent heat $(kJ\,kg^{-1})$
l	transfer area coordinate (m), length coordinate (m), Lagrangian cost
M_i	molar mass of ith species $(kg\,kmol^{-1})$
m	mass (kg)
N	number of moles, number of particles $(-)$
N	total number of stages, number of transfer units $(-)$
N^k	cumulative flux of mole number for stages 1, 2, $\ldots, k$ $(mol\,s^{-1})$
n	current stage number $(-)$, number density (m^{-3}), molar flux $(mol\,s^{-1})$ (Chapter 9)
n	flux of mole number, flux of active component of the fuel mixture $(mol\,s^{-1})$
$\mathbf{P}$	pressure tensor (Pa)
$P(\mathbf{X}, t)$	probability distribution function (probability density p)
P	thermodynamic pressure (Pa), momentum variable, gauged optimal profit (Chapter 2)
P, p	cumulative and local power output, respectively $(J\,s^{-1})$
P^n, p^n	cumulative and local power output for nth stage subprocess $(J\,s^{-1})$
P^n	optimal gauged profit at nth stage in terms of the state $\mathbf{x}^n$ and time t^n (Chapter 2)
$\mathbf{p}$	vector of co-state variables in Maximum Principle
p	photon flux constant $(photons\,cm^{-2}\,K^{-3}\,s^{-1})$
p_i	adjoint (costate) variables, phase space variables, probabilities
Q	electric charge (C)
Q	generalized (total) heat flux including effect of mass transfer $(J\,s^{-1})$
Q^f	optimal performance function in terms of the complete final state $(x_0^f, \mathbf{x}^f, t^f)$
$\mathbf{q}$	heat flux density vector $(J\,s^{-1}\,m^{-2})$
q	sensible heat flux between a stream and power generator $(J\,s^{-1})$
q	sensible heat flux $(J\,s^{-1})$
q_1	driving heat in the engine mode of the stage $(J\,s^{-1})$
$\mathbf{R}, \tilde{\mathbf{R}}$	matrices of chemical and electrochemical resistances
R	universal gas constant $(J\,K^{-1}\,mol^{-1})$

$R^n(\mathbf{x}^n, t^n)$	optimal function of cost type in terms of state, time and number of stages
$R(\mathbf{X}, t)$	optimal work function of cost type in a continuous process $(\mathrm{J\,kg^{-1}})$
$\mathbf{r}$	radius vector
$r_j, \tilde{r}_j$	chemical and electrochemical rates of jth reaction, respectively $(\mathrm{mol\,m^{-3}\,s^{-1}})$
S	performance criterion, entropy $(\mathrm{J\,K^{-1}})$, specific entropy $(\mathrm{J\,kg^{-1}\,K^{-1}})$
S	solid entropy, entropy of controlled phase (solid) $(\mathrm{J\,K^{-1}})$
S_σ	specific entropy production per unit flow $(\mathrm{J\,K^{-1}\,kg^{-1}})\ (\mathrm{J^{-1}\,K^{-1}\,mol^{-1}})$
S_σ	specific entropy production rate $(\mathrm{J\,K^{-1}\,kg^{-1}\,s^{-1}})$
s, s_v	specific and volumetric entropy, respectively $(\mathrm{J\,K^{-1}\,kg^{-1}})$
$\mathbf{T}\ (T, \tau)$	vector composed of the temperature and the number of heat transfer units
$\mathbf{T}^n\ (\mathbf{x}^n, \mathbf{u}^n)$	vector state transformation at nth stage
T	temperature of controlled phase (e.g. solid, resource or fuel stream) (K)
T^n	temperature of phase leaving stage n (K)
T_1, T_2	bulk temperatures of reservoirs 1 and 2 (K)
$T_{1'}, T_{2'}$	temperatures of fluid circulating in engine, active temperatures (K)
T^e	constant equilibrium temperature of environment (K)
T'	Carnot temperature, effective temperature of controlling phase (K)
$\dot{T} = u$	rate of temperature change in nondimensional time (K)
t	physical time, contact time (s)
u	hydrodynamic (barycentric) velocity $(\mathrm{m\,s^{-1}})$
$\mathbf{u} = (u_1, u_2, \ldots, u_r)^T$	control vector
U, u	internal energy (J) and specific internal energy $(\mathrm{J\,g^{-1}})$, respectively
u	value function in HJB equation (Chapter 2), variable controlling fuel consumption
$u^n = \Delta T^n/\Delta \tau^n$	rate of temperature change in nondimensional time (K)
V	volume $(\mathrm{m^3})$, operational voltage (V), scalar potential, optimal generalized profit
$V^n(\mathbf{x}^n, t^n) = \max S$	optimal function of profit type in terms of state, time and number of stages

$\mathbf{v}$	diffusion velocity, relative velocity $(\mathrm{m\,s^{-1}})$
υ	value function (Chapter 2), scalar velocity $(\mathrm{m\,s^{-1}})$
$\upsilon = \rho^{-1}$	specific volume $(\mathrm{m^3\,kg^{-1}})$
$W, \mathcal{W}$	work and cumulative work, respectively (J)
$W = P/G$	total specific work or total power per unit mass flux $(\mathrm{J\,kg^{-1}})$
W	molar work at flow, total power per unit mass flux of inert $(\mathrm{J\,mol^{-1}})$
w	specific or normalized work $(\mathrm{J\,kg^{-1}})$
X	concentration of active component in fuel, moles per mole of inert $(\mathrm{mol\,mol^{-1}})$
X_s	moisture content in solid $(\mathrm{kg\,kg^{-1}})$
X_α	thermodynamic force
$\dot{X} = dX/d\tau_1$	rate change of fuel concentration in time τ_1
$\mathbf{x}, \tilde{\mathbf{x}}$	radius vector and enlarged radius vector, respectively
$\mathbf{x} = (x_1, x_2, \ldots, x_i, \ldots, x_s)$	state vector of a control process
$\tilde{\mathbf{x}} = (x_1, x_2, \ldots, x_i, \ldots, x_s, t)$	enlarged state vector of a control process
x	mass fraction $(\mathrm{kg\,kg^{-1}})$, reaction extent, transfer coordinate
x	molar fraction of active component in the fuel $(\mathrm{mol\,mol^{-1}})$
x_i	concentration of ith species $(\mathrm{mol\,kg^{-1}})$
$x_{1'}$	molar fraction of reactant in chemically active part of the system $(\mathrm{mol\,mol^{-1}})$
$x_{2'}$	molar fraction of product in chemically active part of the system $(\mathrm{mol\,mol^{-1}})$
x_0	performance coordinate
Y_b	absolute humidity in bubble phase $(\mathrm{kg\,kg^{-1}})$
Y_g	absolute humidity of controlling phase (gas) $(\mathrm{kg\,kg^{-1}})$
Y_s	humidity of gas at equilibrium with solid $(\mathrm{kg\,kg^{-1}})$
$y_i = g_i/g_0$	stream flux fraction of ith stream (Chapter 8) $(\mathrm{mol\,mol^{-1}})$
z^i	adjoint variables, variables of adjoint space

Greek symbols

α	asymmetry parameter of energy barrier $(-)$
α'	overall heat transfer coefficient $(\mathrm{J\,m^{-2}\,s^{-1}\,K^{-1}})$
β'	overall mass transfer coefficient $(\mathrm{mol\,m^{-2}\,s^{-1}})$
β_i	multipliers, parameters
χ	time constant (s)

Γ	complexity (−)
γ	cumulative conductance of the system ($J\,s^{-1}\,K^{-a}$)
Δ	nonequilibrium correction, increment
δ	variation, thickness, effective diameter, size measure
ε	total energy flux ($J\,s^{-1}$), Legendre transform of a Lagrangian
ε_i	mechanical displacements, parameters
Φ	factor of internal irreversibility (−), dissipation function ($J\,K^{-1}\,s^{-1}$)
ϕ	relative concentration (−)
$\eta = p/q_1$	first-law thermal efficiency (−)
θ	free interval of an independent variable (s, −)
θ^n	time interval at stage n (s, −)
λ	Lagrangian multiplier, characteristic length
μ_k	chemical potential of kth component ($J\,kg^{-1}$)
μ_1	molar chemical potential of active component of fuel ($J\,mol^{-1}$)
μ'	Carnot chemical potential for active component of fuel ($J\,mol^{-1}$)
μ'	coefficient of outlet gas utilization (−)
ν	stoichiometric coefficient (−), frequency constant, viscosity (Pa s)
$\nu_{ij} = \nu_{ij}^{b} - \nu_{ij}^{f}$	resulting stoichiometric coefficient of jth reaction (−)
$\Pi,\ \Pi_j$	reaction potential vector and potential of jth reaction ($J\,mol^{-1}$, −)
Π_j	unidirectional part of chemical affinity for jth reaction ($J\,mol^{-1}$, −)
ρ	mass density ($kg\,m^{-3}$)
ρ_e	energy density ($kJ\,m^{-3}$)
ρ_s	entropy density ($kJ\,K^{-1}\,m^{-3}$)
σ	Stefan–Boltzmann constant, fraction
σ_s	intensity of entropy production ($J\,K^{-1}\,m^{-3}\,s^{-1}$)
τ	characteristic time, relaxation time (s)
τ	nondimensional time, number of transfer units (x/H_{TU}) (−)
τ^{ik}	viscous stress (Pa)
Ξ	imperfection fraction (−)
ξ	process intensity factor (−)
$\zeta = \mu_{1'} - \mu_{2'}$	efficiency of chemical engine as active part of chemical affinity ($J\,mol^{-1}$)
ζ_{mp}	efficiency of chemical engine at maximum power point ($J\,mol^{-1}$)

ξ_i	reaction progress variable (mol), internal variable
Ω	order variable $(-)$

Subscripts

b	bubbles, backward
C	Carnot state (open circuit)
e	energy
g	gas
i	ith state variable
j	reaction number
m	molar quantity
s	entropy, solid
v	per unit volume
σ	dissipative quantity
0	idle run, profit variable, reference state
1, 2	first and second fluid

Superscripts

a	power exponent (Chapters 6 and 7)
e	environment, equilibrium
f	final state and time, forward
i	initial state and time
int	internal entropy production
k or n	number of kth or nth stage
m	power exponent (Chapter 5)
T	transpose matrix, transform
$'$	Carnot variable, modified quantity
0	ideal (equilibrium) voltage

Abbreviations

BFC	biological fuel cell
CIT	classical irreversible thermodynamics
DMFC	direct methanol fuel cell
DP	dynamic programming
EFC	enzymatic fuel cell
EGM	entropy generation minimization
EIT	extended irreversible thermodynamics
FC	fuel cell
FTT	finite time thermodynamics
GT	gas turbine
H(SOFC + GT)	hybrid of SOFC and GT
MFC	microbial fuel cell

OCT	optimal control theory
PEMFC	polymer electrolyte membrane (PEM) fuel cell
SOFC	solid oxide fuel cell

Symbols in thermal integration algorithms (Chapters 12–20)

A	heat transfer surface area
AT	total heat transfer surface area of HEN
c	cold process stream
C	index set of cold process streams or concentration in water network problem
CAP	investment cost
cu	cooling utility
CU	index set of cooling utilities
cp	heat capacity
CP	heat capacity flow rate
D	index set of sinks (in water networks)
EMAT	minimum temperature approach in heat exchanger (match)
FC	goal function in optimization problem
FP	fitness function in GA optimization
Ft	correction factor for mean temperature difference (MTD) in heat exchanger
F	flow rate
h	hot process stream
H	index set of hot process streams
hu	heating utility
HRAT	minimum temperature approach in heat recovery calculations
HU	index set of heating utilities
I	index set of contaminants (in water network problem)
k	index or number of temperature interval
L	mass load of contaminant
Lp	heat of vaporization/condensation
LMTD	logarithmic mean temperature difference
MTD	mean temperature difference
N	total number of streams exchanging heat in HEN
NC	total number of cold process streams/cold streams
NCU	total number of cooling utilities
NH	total number of hot process streams/hot streams
NHU	total number of heating utilities
Nt	number of tube passes in heat exchanger

Ns	number of shells
NU	number of heat exchangers/matches in HEN
P	index set of water using processes
pcu	unit cost of cooling utility
phu	unit cost of heating utility
Qhu	heat load of heating utility/utilities
Qcu	heat load of cooling utility/utilities
q/Q	heat load
$Qrec$	recovered heat
R	residual flow
S	index set of water using processes
T	temperature or index set of treatment/regeneration operations in water network
x/X	continuous variable/vector of continuous variables
y/Y	discrete variable/vector of discrete variables
UC	utility cost

Greek letters

α	coefficient in formula for investment cost of heat exchanger
β	coefficient in formula for investment cost of heat exchanger
γ	exponent in formula for investment cost of heat exchanger
ΔH	enthalpy change
ΔT	temperature approach in heat exchange/temperature difference in heat exchanger

Subscripts

i	contaminant (in water network problem)
in	inlet
min	minimum value
max	maximum value
out	outlet
*	best (optimum) value

Superscripts

d	sink
fw	freshwater
i	hot process stream/hot stream
j	cold process stream/cold stream

m	heating utility
n	cooling utility
p	water using process
s	source
t	treatment/regeneration process

Abbreviations and acronyms

ARS	adaptive random search (optimization)
BPS	basic process subsystem
CC (plot)	composite curves (plot)
CCC	cold composite curve
DTA	double temperature approach
GA	genetic algorithms (optimization)
GCC (plot)	grand composite curves (plot)
HCC	hot composite curve
HEN	heat exchanger network
LCC	limiting composite curve
LP	linear problem/programming
MEN	mass exchanger network
MER	maximum (heat) energy recovery
MIP	mixed-integer problem/programming
MILP	mixed-integer linear problem/programming
MINLP	mixed-integer nonlinear problem/programming
MNM	minimum number of matches in HEN
MNS	minimum number of shells in HEN
MNU	minimum number of units in HEN
MTA	minimum total area of HEN
MTA-m	minimum total area of HEN for matches
MTA-s	minimum total area of HEN for shells
MUC	minimum utility cost
NLP	nonlinear problem/programming
PT	pinch technology
Pr-T	problem table (method)
SA	simulated annealing (optimization)
TAC	total annual cost
TWN	total water network
US	utility subsystem
WCC	water composite curves (plot)
WN	water network
WNRR	water network with reuse and regeneration
WRN	water reuse network
WSL	water supply line
WTN	wastewater treatment network
WUTN	water usage and treatment network

Index

CPSIA information can be obtained at www.ICGtesting.com
Printed in the USA
LVOW10*1459040913

350690LV00004B/10/P

9 780080 982212